Lecture Notes in Mathematics

Edited by A. Dold and B. Eckmann

1017

Equadiff 82

Proceedings of the international conference
held in Würzburg, FRG, August 23–28, 1982

Edited by H. W. Knobloch and K. Schmitt

Springer-Verlag
Berlin Heidelberg New York Tokyo 1983

Editors

H. W. Knobloch
Mathematisches Institut, Universität Würzburg
Am Hubland, 8700 Würzburg, Federal Republic of Germany

Klaus Schmitt
Department of Mathematics, The University of Utah
Salt Lake City, Utah 84112, USA

AMS Subject Classifications (1980): Main: 34 A-K
Related: 35 B, 70 A-L, 76 A-R, 92 A10, 92 A15

ISBN 3-540-12686-4 Springer-Verlag Berlin Heidelberg New York Tokyo
ISBN 0-387-12686-4 Springer-Verlag New York Heidelberg Berlin Tokyo

Library of Congress Cataloging in Publication Data. Equadiff 82 (1982: Würzburg, Germany)
Equadiff 82. (Lecture notes in mathematics; 1017) 1. Equations, Differential–Congresses.
2. Difference equations–Congresses. I. Knobloch, H. W. (Hans Wilhelm), 1927-. II. Schmitt, Klaus.
III. Title. IV. Series: Lecture notes in mathematics (Springer-Verlag); 1017. QA3.L28 no. 1017
515.3'5 83-16854 [QA370]
ISBN 0-387-12684-4 (U.S.)

Printing and binding: Beltz Offsetdruck, Hemsbach/Bergstr.
2146/3140-543210

Preface

The international conference EQUADIFF 82 was held at the University of Würzburg during the week August 23 to August 28, 1982. It was the fourth in a sequence of international conferences, with focus on the subject of differential equations, which were started in 1970 in Marseille by the late Professor Vogel and then continued in Brussels in 1973 and Florence in 1978.

The program of the conference was prepared by the Scientific committee consisting of R.Conti, W.N.Everitt, J.K.Hale, W.Jäger, J.Kurzweil, J.Mawhin, J.Moser, M.Roseau, K.Schmitt and T.Yoshizawa. All other organizational responsibilities were shared by P.Hagedorn, H.W.Knobloch, H.Kielhöfer, R.Reissig and W.Werner. The official sponsors of the conference were:

Deutsche Forschungsgemeinschaft

and

Bayerisches Staatsministerium

für Unterricht und Kultus.

In keeping with the tradition set by the earlier conferences the scientific committee decided to emphasize the following subjects as conference topics:
Ordinary differential equations, Functional differential equations, Stochastic differential equations, Partial differential equations of evolution type and Difference equations.

Since interest among nonexperts in a highly specialized conference such as EQUADIFF 82 (a name which in German is anything but self-explanatory) could not a priori be assumed, the organizers gratefully acknowledge that all institutions and persons contacted for assistance were most willing to cooperate. Not only was support received from the official sponsors but also from the administration of the University of Würzburg, whose president, Professor Theodor Berchem, and vice-president, Professor Martin Lindauer, showed great interest in the conference. The organizers are particularly indebted to them.

The addition of the number 82 to the title of the conference is a deviation from earlier custom. This was done to indicate that the conference was part of the official program set up to celebrate the founding of the University of Würzburg in the year 1582. This linkage with the fourth centenary celebration of the host institution added a special accent to the conference, particularly to the social program. For this reason a few remarks about the history of the University of Würzburg and about science and mathematics there seem to be in order in these proceedings.

The University of Würzburg was founded as a typical "confessional" university during the course of the great religious conflicts of the sixteenth century. That it did not disappear again - as did many other universities - at the end of the middle ages is mainly due to the common sense and realistic assessment of the value of higher education which the founder of the university and some of his successors had. In the opening decree the rôle of the university was viewed in a remarkably modern way: Above all it should provide the opportunity to acquire knowledge without the necessity to move to distant places or of paying high costs.

Medicine was the first discipline which benefited from the ideas of the founder and first rector of the university. Later - during the age of Enlightenment - his successor freed the faculty of Philosophy from the influence of the Jesuit order. This was the first step in developing a modern faculty of liberal arts and science in Würzburg. The most outstanding and best-known of the many scientific achievements attributed to its members perhaps is the discovery of x-rays by W.C.Röntgen in 1895.

During the 18th and 19th century mathematics, astronomy and theoretical physics were represented by the same chair in Würzburg as was the case in many smaller European universities. As a result mathematical activities were naturally somewhat limited until the middle of the nineteenth century. The rise of modern physics and chemistry then gave new impetus also to mathematics. A second chair was created and given to F.Prym, a student of B.Riemann. His successful efforts to spread and interpret the ideas of his great teacher have influenced the development of modern mathematical thinking in Germany.

The historical alliance between mathematics, astronomy and mechanics during the early history of European science has contributed much to our present knowledge about structures and methods in the theory of differential equations. These roots are still visible in the theory of dynamical systems, a mathematical discipline which encompasses the various topics covered by EQUADIFF 82. The present day notion of a dynamical system is in a very general sense a model of some real system which may be distinguished from its environment and which evolves in time according to some inherent dynamics. The rôle of dynamical systems is hence very similar to the one played by ordinary differential equations in classical mechanics. If one wishes to describe it in simple terms, one is tempted to use almost the same phrases which once were used to describe the merits of university education: By studying the structure of dynamical systems one can gain insight into the real world which otherwise could be obtained only at high cost and at "distant times", that is, through expensive and time consuming experiments. Progress in computer science gives more and more credence to this statement, since it opens new possibilities of exploiting highly complex theoretical results for practical purposes. For this reason the tradition of previous EQUADIFF conferences was carried further and also numerical methods and applications in engineering, physics and biology were represented by special sections in the program.

The conference, which was attended by over 200 scientists from 27 different countries, demonstrated clearly the ever increasing international cooperation in the area of dynamical systems. Cooperation between various institutions was also necessary in order to establish the framework for the scientific program. We here record with gratitude the assistance which we received from the following institutions:

Mathematisches Institut der Universität Würzburg,
Sonderforschungsbereich "Stochastische Mathematische Modelle", Heidelberg,
Sonderforschungsbereich "Approximation und Mathematische Optimierung", Bonn,
Department of Mathematics, University of Utah.

The first three institutions shared the organizational aspects of the conference and the last took on the preparation of these proceedings.

Special thanks are also due to the secretaries
Mrs. Ingrid Böhm and Mrs. Isolde Brugger, Würzburg and
Mrs. Jackie S.Hadley, Salt Lake City
for their invaluable contributions.

Würzburg and Salt Lake City, March 1983

The Editors.

Address list of participants

Allegretto, W.	Dept. of Mathematics, University of Alberta, Edmonton,Alberta T6G 2G1,Canada
Alt, W.	Sonderforschungsbereich 123, Universität Im Neuenheimer Feld 293, D-69oo Heidelberg Germany
Ambrosetti, A.	International School for Advanced Studies (S.I.S.S.A.) Strada Costiera,11 I-34014 Trieste, Italy
Appel, J.	Istituto Matematico "Ulisse Dini", Universita di Firenze, Viale Morgagni, I-50134 Firenze, Italy
Arino, O.	Dept. de Mathēmatiques, Universitē de Pau, Avenue de L. Sallenave, 64ooo Pau, France
Arndt, H.	Institut für Angewandte Mathematik, Universität, Wegelerstr.6, D-5300 Bonn Germany
Arnold, L.	Fachbereich Mathematik, Universität Postfach, D-2800 Bremen 33, Germany
Aronsson, G.	Dept. of Mathematics, University of Lulea S-95187 Lulea, Sweden
Aulbach, B.	Mathematisches Institut, Universität Am Hubland, D-87oo Würzburg, Germany
Balser, W.	Abt. Mathematik V, Universität Ulm (MNH) D-7900 Ulm, Germany
Barbanti, L.	MAT-IME-USP, Cidade Universitaria, Cx. Postal 2o.57o, o55o8 Sao Paulo (SP) Brazil
Bardi, M.	Istituto di Analisi e Meccanica, Univ. di Padova, Via Belzoni 7, I-351oo Padova Italy
Bellen, A.	Istituto di Matematica, Universita di Trieste, I-341oo Trieste, Italy
Besjes, J.G.	Dept. of Mathematics, Technological University Delft, Julianalaan 132 NL-2728 BL Delft, Netherlands
Beyn, W.-J.	Fakultät für Mathematik,Universität Jacob-Burckhardt-Str.17, D-775o Konstanz, Germany
Binding, P.A.	Dept. of Mathem., University of Calgary Calgary, Alberta T2N 1N4, Canada

Bobrowski, D.	Mathematical Institute, Politechnika Poznanska, Sw. Rocha 6-A-7, Pl-61-142- Poznan, Poland
Boudourides, M.	Dept. of Mathematics,Democritus University of Thrace, Xanthi, Greece
Braaksma, B.L.J.	Mathematisch Instituut, Rijksuniversiteit Groningen, Postbus 8oo, NL-97oo AV Groningen, Netherlands
Braess, D.	Mathem.Institut, Ruhr-Universität, Universitätsstr. 15o, D-463o Bochum 1, Germany
Brunovsky, P.	Institute of Applied Mathematics, Comenius University, Mlynska dolina 842 15 Bratislava, Czechoslovakia
van der Burgh, A.H.P.	Dept. of Mathematics, Technological University Delft, Julianalaan 132 NL-2628 BL Delft, Netherlands
Cañada, V.	Dept. de Ecuaciones Funcionales Facultad de Ciencias, Granada, Spain
Capozzi, A.	Istituto di Matematica Applicata Universita degli Studi di Bari, Via Re David, 2oo/A, I-7o125 Bari, Italy
Caristi, G.	Istituto di Matematica, Universita di Trieste, Piazzale Europa 1, I-341oo Trieste Italy
de Castro, A.	Facultad de Matematicas, Universidad de Sevilla, Sevilla-12, Spain
Cerami, G.	Istituto di Matematica, Universita di Palermo, Via Archirafi 34, I-9o123 Palermo, Italy
Cesari, L.	Dept. of Mathematics, University of Michigan, Ann Arbor, Mich. 481o4, USA
Conley, Ch.	Dept. of Mathematics, University of Wisconsin, Van Vleck Hall, 48o Lincoln Drive, Madison, Wisc. 537o6, USA
Conti, R.	Istituto Matematico "Ulisse Dini" University degli studi, Viale Morgagni 67/A, I-5o134 Firenz, Italy
Couot, J.	UER Math., Universite Paul Sabatier 118 Route de Narbonne, F-31o77 Toulouse Cedex, France
Cree, G.C.	Dept. of Mathematics, The University of Alberta, Edmonton, Alberta T6G 2G1 Canada
Crespin, D.	Dept. of Matematicas, P.O.Box 615, Caracas 1o1-A, Venezuela

Dafermos, C.M.	Lefschetz Center for Dyn. Systems, Brown University, Box F, Providence, R.I. o2912, USA
de Pascale, E.	Dipartimento di Matematica, Universita della Calabria, I-87o36 Arcavacata di Rende Italy
Dijksma, A.	Mathematisch Instituut, Rijksuniversiteit Groningen, Postbus 8oo, NL-97oo AV Groningen Netherlands
Djaja, C.	Zahumska 42, YU-11o5o Beograd, Yugoslavia
Doden, Kl.	Mathematisches Seminar, Christian-Albrecht-Universität Kiel, Olshausenstraße 4o-6o, Haus S 12a, D-23oo Kiel 1, Germany
Dominguez, B.	Facultad de Matematicas, Universidad de Sevilla, c/Tarifia s/n, Sevilla, Spain
Eberhard, W.	Universität Duisburg, Mergelskull 25 D-415o Krefeld, Germany
Edelson, A.	Dept. of Mathematics, Univers. of California, Davis, Cal. 95616, USA
Engler, H.	Inst. f. Angew. Mathem., Universität Heidelberg, Im Neuenheimer Feld 293 D-69oo Heidelberg, Germany
Erbe, L.	Dept. of Mathem., Univers. of Alberta, Edmonton, Alberta T6G 2G1, Canada
Farkas, M.	Dept. of Mathem., Budapest Univers. of Technology, H-1521 Budapest, Hungary
Farwig, R.	Inst. für Angew. Mathem., Universität Wegelerstr. 6, D-53oo Bonn 1, Germany
Fenyö, I.	Dept. of Mathem., Technical University Stoczek u. 2, H-1111 Budapest, Hungary
Fiedler, B.	Sonderforschungsbereich 123, Universität Im Neuenheimer Feld 293, D-69oo Heidelberg Germany
Flockerzi, D.	Mathem.Institut der Universität Würzburg Am Hubland, D-87oo Würzburg, Germany
Freedman, H.I.	Dept. of Mathem., University of Alberta Edmonton, Alberta T6G 2G1, Canada
Freiling, G.	FB 11- Mathematik, Universität-Gesamthochschule, Lotharstraße 65, D-41oo Duisburg 1, Germany
Fryszkowsik, A.	Institute of Mathematics, Technical Univ. Pl. J. Robotniczej 1, Pl-oo-661 Warszawa Poland

Furumochi, T.	Dept. of Mathem., Iwate University, Ueda 3-18-33, Morioka City o2o, Japan
Gambaudo, J.M.	Mathem.Dept., I.M.S.P., Universite de Nice, Parc Valrose, F-o6o34 Nice, France
Georgiou, D.	Dept. of Mathem., Democritus University of Thrace, Xanthi, Greece
van Gils, St.	Stichting Mathematisch Centrum, Kruislaan 413, Nl-1o98 SJ Amsterdam, Netherlands
Graef, J.R.	Dept. of Mathem., Mississippi State University, P.O. Drawer MA Mississippi State, MS 39762, USA
Grasman, J.	Dept. of Applied Mathematics, Mathematical Centre, Jruislaan 413, Nl-1o98 SJ Amsterdam, Netherlands
Gripenberg, G.	Institute of Mathematics, Helsinki University of Technology, SF-o215o Espoo 15, Finland
Großmann, U.	Max-Planck-Institut für Systemphysiologie Rheinlanddamm 2o1, D-46oo Dortmund 1, Germany
Gumowski, I.	UER-Math., Universite Paul Sabatier, 118, Route de Narbonne, F-31o62 Toulouse France
Habets, P.	Universite Catholique de Louvain 2, Chemin du Cyclotron, B-1348 Louvain-la-Neuve, Beglium
Haddock, J.R.	Dept. of Mathem. Sciences, Memphis State University, Memphis, Tn 38152, USA
Hagedorn, P.	Institut für Mechanik, Technische Hochschule, Hochschulstraße 1, D-61oo Darmstadt, Germany
Hainzl, J.	Gesamthochschule/Universität Kassel Kaulenbergstr. 8, D-35oo Kassel, Germany
Halbach, U.	Zoologisches Institut der Universität Siesmayerstraße 7o, D-6ooo Frankfurt/M. Germany
an der Heiden, U.	Universität NW 2, D-28oo Bremen 33, Germany
Heil, E.	FB Mathem., Techn.Hochschule, Schloßgartenstr.7, D-61oo Darmstadt, Germany

Herold, H.	FB Mathematik, Universität, Lahnberge D-355o Marburg, Germany
Hino, Y.	Dept. of Mathematics, Chiba University, 1, Yayoicho, Chiba 26o, Japan
Hoppensteadt, F.C.	Dept. of Mathematics, The University of Utah, Salt Lake City, Utah 84112, USA
Hornung, U.	FB 15, Mathematik, Universität, Einsteinstraße 62, D-44oo Münster, Germany
Howes, F.A.	Dept. of Mathem., Univers. of California Davis, CA 95616, USA
Hueber, H.	Fakultät für Mathematik, Universität Bielefeld, Postfach 864o, D-48oo Bielefeld, 1, Germany
Iannacci. R.	Dipartimento di Matematica, Universita della Calabria, I-87o36 Arcavacata di Rende, Italy
Invernizzi, S.	Istituto di Matematica, Universita di Trieste, I-341oo Trieste, Italy
Ize, A.F.	Instituto de Ciencias Matematicas, de Sao Carlos, Caixa Postal 668, 1356o Sao Carlos (Sao Paulo),Brazil
Ize, J.	IIMAS, U.N.A.M, Apartado Postal 2o-726 o1ooo Mexico 2o D.F., Mexico
Jäger, W.	Sonderforschungsbereich 123, Universität, Im Neuenheimer Feld 293, D-69oo Heidelberg, Germany
Jeggle, H.	Fachbereich 3/Mathematik/MA6-3, Technische Universität, Straße des 17. Juni 135, D-1ooo Berlin 12, Germany
Joseph, D.D.	Dept. of Aerospace Engineering and Mechanics, 11o Union Street S.E. Minneapolis, Minnesota 55455, USA
Kaminogo, T.	Mathematical Institute, Tohoku Univers. Sendai 98o, Japan
Kappel, F.	Institut für Mathematik, Karl-Franzens-Universität, Elisabethstraße 11, A-8o1o Graz, Austria
Karim, R.I.I.A.	Mathematics Dept., King Saud University P.O. Bos 2455, Riyad, Saudi Arabia
Kato, J.	Mathem. Institute, Tohoku University Sendai 98o, Japan
Kawohl, B.	Inst. für Angew. Mathem., Universität Martenstr. 3, D-852o Erlangen, Germany
Keller, B.	Koenig & Bauer, Friedrich-Koenig-Str.4 D-87oo Würzburg, Germany

Kielhöfer, H.	Institut für Angew. Mathem. u. Statistik Universität Würzburg, Am Hubland, D-87oo Würzburg, Germany
Kirchgässner, Kl.	Mathematisches Institut A, Universität Pfaffenwaldring 57, D-7ooo Stuttgart, Germany
Kirchgraber, U.	Seminar für Angew. Mathematik der ETH ETH-Zentrum, CH-8o92 Zürich, Switzerland
Kliemann, W.	Forschungsschwerpunkt "Dyn.Systeme" FB Mathematik, Universität, Postf.33o 44o D-28oo Bremen 33, Germany
Knobloch, H.W.	Mathematisches Institut der Universität Am Hubland, D-87oo Würzburg, Germany
Kurzweil, J.	Mathem. Institute of the Czechoslovak Academy of Sciences, Zitna 25, 115 67 Praha 1, Czechoslovakia
Kwapisz, M.	Institute of Mathematics, University of Gdansk, ul. Wita Stwosza 57, Pl-8o-952 Gdansk, Poland
Labonte, G.	Dept. of Mathem. and Computer Sciences Royal Military College of Canada Kingston, Ontario K7L 2W3,Canada
Lange, H.	Mathematisches Institut, Universität Weyertal 86-9o, D-5ooo Köln 41, Germany
Lar'kin, N.	Institute of theoretical and applied mechanics, 63oo9o Novosibirsk 9o, UdSSR
Lasota, A.	Institute of Mathematics, Silesian University, Bankowa 14, Pl-4o-oo7 Katowice, Poland
Liniger, W.	Math. Sciences Dept., IBM-Th.J.Watson Research Center, P.O.Box 218, Yorktown Heights, N.Y. 1o598, USA
Lloyd, N.G.	Dept. of pure Mathem., University College of Wales, Penglais, Aberystwyth, Dyfed Great Britain
Lombet-Goffar, J.	Institut de Mathematique, Universite de Liege, Avenue du Luxembourg 29, B-4o2o Liege, Belgium
Lorenz, J.	Fakultät für Mathematik, Universität Postfach 556o, D-775o Konstanz,Germany
Louis, J.Cl.	Dept. of Mathematics, Facultes Universitäires de Namur, Rempart de la Vierge,8 B-5ooo Namur, Belgium

Maniscalco, C.	Istituto di Matematica, Via Archirafi 34 I-9o123 Palermo, Italy
Marcati, P.	Dipartimento di Matematica, Universita degli Studi di Trento, I-38o5o Povo(Trento) Italy
Marchi, M.V.	Istituto di Matematica, Universita degli Studi, Via Mantica 1, I-331oo Udine,Italy
Martin, R.	Dept. of Mathematics, North Carolina State University, Box 5548, Raleigh, N.C. 2765o, USA
Martinez-Amores, P.	Dept. de Ecuaciones Funcionales Facultad de Cienoias-Matematicas, Granada, Spain
Massabo, I.	Dipartimento di Matematica, Universita della Calabria, I-87o36 Arcavacata di Rende, Italy
Matia, J.	Mathematisches Institut, Ruhr-Universität NA 1/35, D-463o Bochum 1, Germany
Mawhin, J.	Institut de Math. pure et appl., Universite Catholique de Loivain 2, Chemin du Cyclotron, B-1348 Louvain-la-Neuve, Belgium
Medved, M.	Mathematical Institue of the Slovak Academy of Sciences, Obrancov mieru 49 B14 73 Bratislava, Czechoslovakia
Meier, H.D.	Fachbereich Elektrotechnik, Hochschule der Bundeswehr, Postfach 7o o8 22, D-2ooo Hamburg, Germany
van Moerbeke, P.	Institut de Math. pure et appl, Universite Catholique de Louvain, 2,Chemin du Cyclotron, B-1348 Louvain-la-Neuve, Belgium
Morales, P.	Dept. de mathematiques, Universite de Sherbrooke, Sherbrooke, Quebec J1K 2R1, Canada
Moser, J.	Mathematik, ETH-Zentrum, CH-8o92 Zürich, Switzerland
Mubenga-Ngandu, N.	Institut de Math. pure et appl, Universite Catholique de Louvain, 2, Chemin du Cyclotron, B-1348 Louvain-la-Neuve, Belgium
Naito, T.	The University of Electro-Communications 1-5-1, Chofugaoka, Chofu, Tokyo 182,Japan
Nixdorff, K.	Fachbereich Maschinenbau, Hochschule der Bundeswehr, Holstenhofweg 85, D-2ooo Hamburg 7o, Germany

Nocilla, S.
Istituto di Meccanica Razionale Politechnice di Torino, Corso Duca degli Abruzzi, I-1o129 Torino, Italy

Olech, C.
Institute of Mathematics, Polish Academy of Sciences, Sniadeckich 8, P.O.Box 137, oo-95o Warszawa, Poland

Papanicolaou, G.C.
Courrant Institute, New York University 251, Mercer Street, New York, N.Y. 1oo12, USA

Pascali, D.
Fachbereich Mathematik, Ag. 6, Technische Hochschule, Schloßgartenstraße 7, D-61oo Darmstadt, Germany

Paulus, G.
DFF St. Louis, Weil a. Rhein, Germany

Pazy, A.
Institut of Mathem. and Computer Science The Hebrew University of Jerusalem Givat Ram, 919o4 Jerusalem, Israel

Pecher, H.
Fachbereich Mathematik, Gesamthochschule Wuppertal, Postfach 1oo127, D-56oo Wuppertal 1, Germany

Pehkonen, E.
Dept. of Mathematics, University of Helsinki, Hallituskatu 15, SF-oo1oo Helsinki 1o, Finland

Pianigiani, G.
Istituto di Matematica, Universita degli Studi, Via Mantica 1, I-331oo Udine, Italy

Pöppe, Ch.
Sonderforschungsbereich 123, Universität Im Neuenheimer Feld 293, D-69oo Heidelberg Germany

Pöschel, J.
Mathematik, ETH-Zentrum, CH-8o92 Zürich, Schweiz

Pozio, M.A.
Dipartimento di Matematica, Libera Universita degli Studi, I-38o5o Povo(Trento) Italy

Reißig, G.
Institut für Mathematik, Ruhr-Universität Universitätsstr.15o, D-463o Bochum 1, Germany

Reißig, R.
Abteilung für Mathematik, Ruhr-Universität, GEbäude NA, Universitätsstr.15o D-463o Bochum, Germany

Reyn, J.W.
Dept. of Mathematics, Technological University Delft, Julianalaan 132, NL-2628 BL Delft, Netherlands

Riganti, R.
Istituto di Meccanica Razionale Politechnico di Torino, Corso Duca degli Abruzzi, 24, I-1o129 Torino, Italy

Risito, C.	Universita di Parma, Via Bezzecca 12, I-43oo Parma, Italy
Rosinger, E.	National Research Institute for Mathematical Sciences, P.O.Box 395, Pretoria ooo1, South Africa
Rossi, E.	Zentralblatt für Mathematik, Fachinf.-Zentrum Energie, Physik, Mathematik, Hardenbergstr. 29 c, D-1ooo Berlin 12, Germany
Rothe, F.	Institut für Biologie II, Lehrstuhl für Biomathematik, Universität Tübingen, Auf der Morgenstelle 28, D-74oo Tübingen 1, Germany
Salomon, D.	Forschungsschwerpunkt "Dynamische Systeme" FB Mathematik, Universität, Postf.33o 44o D-28oo Bremen 33, Germany
Sanders, J.A.	Wiskundig Seminarium, Vrije Universität P.O. Box 7161, NL-1oo7 MC Amsterdam, Netherlands
Sartori, C.	Istituto di Matematica Applicata, Universita degli studi di Padova. Via Belzoni 7 I-351oo Padova, Italy
Saupe, D.	Forschungsschwerpunkt "Dynamische Systeme" FB Mathematik, Universität, D-28oo Bremen 33 Germany
Schaaf, Kl.	Sonderforschungsbereich 123, Universität Im Neuenheimer Feld 293, D-69oo Heidelberg, Germany
Schaaf, R.	Sonderforschungsbereich 123, Universität Im Neuenheimer Feld 293, D-69oo Heidelberg, Germany
Schäfke, R.	Fachbereich 6 - Mathematik, Universität Essen-GHS, Postfach, D-43oo Essen 1, Germany
Schinas, J.	Dept. of Mathematics, Democritus University of Thrace, Xanthi, Greece
Schmidt, D.	FB Mathematik, Universität, Universitäts-Straße 3, D-43oo Essen, Germany
Schmitt, Kl.	Dept. of Mathematics, The University of Utah, Salt Lake City, Utah 84112, USA
Schneider, A.	Mathem.Institut, Universität, Postfach D-46oo Dortmund, Germany
Schneider, G.	Fachbereich Mathematik, Techn.Hochschule, Schloßgartenstr.7, D-61oo Darmstadt, Germany

Schulz, F.	Mathem.Institut, Universität, Bunsenstr. D-34oo Göttingen, Germany
Schumacher, K.	Sonderforschungsbereich 123, Universität Im Neuenheimer Feld 293, D-69oo Heidelberg, Germany
Seifert, G.	Dept. of Mathem., Iowa State University Ames, Iowa 5oo11, USA
Sell, G.R.	School of Mathematics, Univ. of Minnesota Minneapolis, Minn. 55455, USA
Sergysels-Lamy, A.	Fac. des Sciences Appliquees, Universite Libre de Bruxelles, Avenue F.-D.Roosevelt B-1o5o Bruxelles, Belgium
Shahin, Mazen	Dept. of Mathematics, Institute of Education for Girls, Al-Shamia, Kuweit
Söderbackä, G.	Söderlangvik, SF-2587o Dragsfjärd,Finland
Sree Hari Rao, V.	Dept. of Mathematics, University of Alberta, Edmonton, Alberta T6G 2G1, Canada
Staude, U.	Mathem.Institut, Johannes-Guttenberg-Universität, Saarstraße 21, D-65oo Mainz Germany
Strampp, W.	Fachbereich 17, Mathematik, Gesamthochschule, Wilhelmshöher Allee 73, D-35oo Kassel, Germany.
Stuart, Ch.	Dept. de mathematiques, Ecole Polytechn. 61, av. de Cour, CH-1oo7 Lausanne, Switzerland
Sussmann, H.J.	Dept. of Mathematics, Rutgers University New Brunswick, New Jersey o89o3, USA
Svec, M.	Katedra matematickej analyzy MFF UK Mlynska dolina, 842 15 Bratislava, CSSR
Swinnerton-Dyer, P.	University of Cambridge, St.Catharine's College, Cambridge CB2 1RL, Great Britain
Szulkin. A.	Dept. of Mathematics, University of Stockholm, Box 67o1, S-11385 Stockholm, Sweden
Thieme, H.	Sonderforschungsbereich 123, Universität Im Neuenheimer Feld 293, D-69oo Heidelberg, Germany
Torelli, G.	Istituto di Matematica, Universita di Trieste, I-341oo Trieste, Italy
Troch, I.	Inst. f. Techn. Mathem., Techn. Univers. Gußhausstr.27-29, A-1o4o Wien, Austria

Tubaro, L.	Dipartimento di Matematica, Libera Universita degli Studi, I-38050 Povo, Italy
Urmann, D.	Fachbereich Maschinenbau, Hochschule der Bundeswehr, Holstenhofweg 85, D-2000 Hamburg 70, Germany
Vanderbauwhede, A.	Inst. voor Theoretische Mechanica, Rijksuniversiteit Gent, Krijgslaan 271 B-9000 Gent, Belgium
Vegas, J.M.	Facultad de Matematicas, Universidad Complutense, Madrid-3, Spain
Vogt, Ch.	Sonderforschungsbereich 123, Universität Im Neuenheimer Feld 293, D-6900 Heidelberg, Germany
Volkmann, P.	Mathematisches Institut I, Universität Postfach 6380, D-7500 Karlsruhe 1, Germany
Volkmer, H.	Fachbereich 6, Mathematik, Universität Essen-Gesamthochschule, Postf. 103764 D-4300 Essen 1, Germany
Vosmansky, J.	Dept. of Mathem.Analysis, J.E.Purkyne University, Janackova nam. 2a, 66395 Brno, CSSR
Walther, H.O.	Mathem. Institut, Universität, Theresienstraße, D-8000 München, Germany
Waltman, P.	Dept. of Mathematics, Univ. of Iowa Iowa City, Iowa, USA
Weber, H.	Rechenzentrum der Johannes-Gutenberg-Universität, Bentzelweg 12, D-6500 Mainz1 Germany
Wedig, W.	Institut für Techn. Mechanik, Universtät Karlsruhe, Kaiserstraße 12, D-7500 Karlsruhe, Germany
Welk, R.	Zentralblatt für Mathematik, Fachinf.-Zentrum Energie Physik Math., Hardenbergstraße 29c, D-1000 Berlin 12, Germany
Wendland, W.	Fachbereich Mathematik, Techn.Hochschule Schloßgartenstraße 7, D-6100 Darmstadt, Germany
Werner, H.	Inst. f. Angew. Mathematik, Universität Wegelerstraße 6, D-5300 Bonn, Germany
Wilcox, C.	Dept. of Mathem., University of Utah Salt Lake City, Utah 84112, USA
Wildenauer, P.	FB 17, Gesamthochschule Kassel, Postfach 101380, D-3500 Kassel,Germany

Wihstutz, Volker	Studienbereich 4, Mathematik, Kufsteinerstraße, 28oo Bremen 33
Willem, M.	Institut Mathematique, Universite Catholique de Louvain, 2, Chemin du Cyclotron B-1348 Louvain-la-Neuve, Belgium
Yoshizawa, T.	Mathematical Institute, Tohoku University, Sendai 98o, Japan
Zanolin, F.	Istituto Matematica, Universita degli Studi, P.Le Europa 1, I-341oo Trieste, Italy
Zennaro, M.	Istituto di Matematica, Universita di Trieste, I-341oo Trieste, Italy
Zhang, K.P.	Institut für Mathematik, Karl-Franzens-Universität, Elisabethenstr. 11, A-8olo Graz, Austria
Zwiesler, H.J.	Mörikestr. 15, D-792o Heidenheim 1, Germany

C O N T E N T S

A CONTRACTION-DISASSEMBLY MODEL FOR INTRACELLULAR ACTIN GELS

Wolfgang Alt
Inst. Appl. Math.
Heidelberg (FRG)

Micah Dembo*
Los Alamos Sc. Lab.
New Mexico (USA)

Cell motility apparently depends on the process of changing shape and formation of the cell plasma membrane. There is evidence that in most motile cells these deformations are primarily caused by filamenteous contractile systems being concentrated in regions near the plasma membrane, possibly interacting with it and with the intracellular cytoskeleton, cp. [1] and Fig. 1. The formation of membrane protrusions, for example, can be explained by local contraction of an actin-myosin-system at the site of protrusion, thereby leading to a flow of material towards the center of contraction. It may then be postulated, as a part of the theoretical model, that high densities of actin filaments cause their depolymerization and rapid redistribution of actin monomers within the cell plasma. See [1] for more biophysical and chemical details.

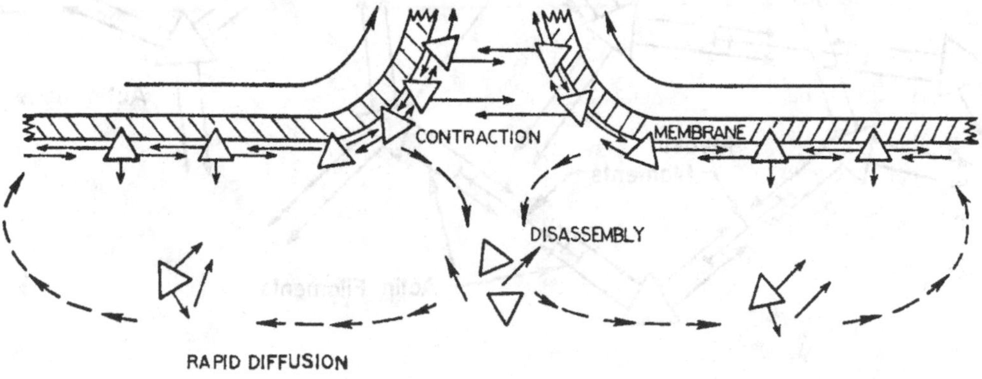

Figure 1. Contraction-disassembly cycle of actin filaments possibly inducing a membrane protrusion, from [1], fig. 6.

A mathematical description of this basic contraction-disassembly cycle should be able to simulate the following (also experimentally) observed phenomena of contractile actin gels: (1) spontaneous or induced formation of contraction centers, (2) non-excitability of regions near a contraction center, and under certain conditions (3) the stability of steady or oscillatory contraction patterns.

Similar to the fiber-fluid model presented by Odell [2] we will start with a "microscopic" model describing the different interactions between actin polymers. From the constitutive "mean field" equations for the probability distributions we will derive a hyperbolic-elliptic system of differential equations for the mean density and mean velocity of the "macroscopic" actin gel, provided the gel is of high viscosity.

Constitutive "microscopic" equations.

We base our mathematical model on the hypothesis that actin filaments are created (nucleated) at certain actin-binding proteins and grow from these in certain directions, thus forming polymer "bundles", which may be called actin nodes (see figure 2 below). Let $\rho(t,x,v)$ measure the probability density of such actin nodes to move

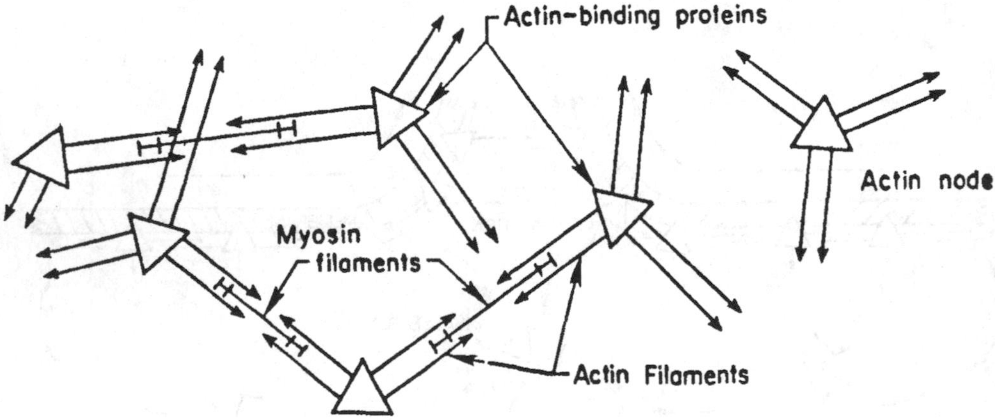

Figure 2. Hypothetical structure of a contractile actin-myosin-system, [1].
(1) Nucleation of actin nodes
(2) Polymerization of actin filaments controlled by steric inhibition
(3) Mutual attraction of actin nodes via binding to myosin filaments
(4) Mutual friction of actin nodes by shearing forces

with velocity $v \in \mathbb{R}^n$ at time t and location $x \in \mathbb{R}^n$, $n \geq 1$ being the dimension of the polymer bundles and of the resulting movement. Denote by

$$\overline{\rho}(t,x) = \int \rho(t,x,v)dv \qquad \text{the } \underline{\text{mean density}}$$

and by

$$\overline{v}(t,x) = \int \rho(t,x,v)v dv \Big/ \overline{\rho}(t,x) \qquad \text{the } \underline{\text{mean velocity}}$$

of actin nodes. We assume that both

(1) nucleation rate $N_+(\overline{\rho})$ and dissociation rate $N_-(\overline{\rho})$ of an actin node essentially depend on the mean density $\overline{\rho}$ only. The same "mean field" assumption is made for the acceleration $b = b(t,x,v)$ of an actin node. Then the constitutive equation (forward Kolmogorov equation) for ρ is

$$(*) \qquad \partial_t \rho + v \cdot \nabla_x \rho + \nabla_v (b\rho) = N_+(\overline{\rho}) - \rho\, N_-(\overline{\rho})$$

Furthermore we suppose that

(2) due to a high polymerization rate R_+ of actin filaments the mean length z of filaments in an actin node rapidly adjusts to a pseudo steady state $z = z(t,x)$ by balancing R_+ with a depolymerization rate R_- : Depolymerization is caused by "steric inhibition" when an actin node overlaps with neighboring ones. Using the simple kernel

$$k_z(x,y) = [z(t,x) + z(t,y) - |x-y|]_+$$

to measure the degree of overlapping of two actin nodes located at x and y, we may model

$$R_- = r_- \iint k_{z(t,\cdot)}(x,y)\rho(t,y,v')dv'\,dy$$

with r_- being the depolymerization rate of one filament in a bundle per unit length of overlapping with another filament.

Since the length scale is macroscopic, r_- usually is large compared to R_+ :

$$r_- = \frac{\alpha}{\varepsilon^{n+1}} R_+ \qquad \text{with some small } \varepsilon > 0 .$$

Then by scaling

$$z(t,x) = \varepsilon\, \zeta(t,x)$$

the pseudo steady state equation $R_+ = R_-$ results in

$$1 = \bar{\rho}(t,x) \cdot \alpha \int [2\zeta(t,x) - r]_+ r^{n-1} dr + O(\varepsilon)$$

$$= \bar{\rho}(t,x) \frac{2^{n+1}\alpha}{n(n+1)} \cdot \zeta(t,x)^{n+1} + O(\varepsilon) ,$$

thus the scaled mean length of actin filaments is approximately

$$(2)_0 \qquad \zeta(t,x) = \left(\frac{K}{\bar{\rho}(t,x)}\right)^{\frac{1}{n+1}} + O(\varepsilon^{\frac{1}{n+1}}) .$$

Apart from this purely chemical interaction of polymer bundles there is a physico-chemical interaction, namely the mutual attraction of overlapping filaments via temporary binding with myosin molecules, and a purely physical interaction, namely the shearing force between two bundles moving with different velocities. Therefore in equation (*) the acceleration of an actin node due to attractive and shearing forces by surrounding nodes can be written as $b = b_{attr} + b_{shear}$, where the two contributions may be modeled using the same kernel k_z as above:

$$(3) \qquad b_{attr}(t,x) = \frac{\beta}{\varepsilon^{n+3}} \iint k_{z(t,\cdot)}(x,y) \frac{y-x}{|y-x|} \rho(t,y,v')dv'\,dy$$

and

$$(4) \qquad b_{shear}(t,x,v) = \frac{\gamma}{\varepsilon^{n+4}} \iint k_{z(t,\cdot)}(x,y)(v' - v)\rho(t,y,v')dv'\,dy$$

For small $\varepsilon > 0$ the support of the kernel $\frac{1}{\varepsilon^{n+1}} k_{\varepsilon\cdot\zeta}$ shrinks to a neighborhood of the diagonal $x = y$ and the acceleration b turns out to be of size $\frac{1}{\varepsilon}$, that means the viscosity of the actin gel dominates the inertial effects during its motion This results in the following

Derivation of an approximative differential equation system

Integration of (*) with respect to v gives the

(M) mass balance equation: $\partial_t\bar{\rho} + \nabla_x \cdot (\overline{\rho v}) = f(\bar{\rho})$

where $f(u) = N_+(u) - u\,N_-(u)$. Multiplication of (*) by v and integration with respect to v gives, for small $\varepsilon > 0$, the

(F) force balance equation

$$0(\varepsilon) = \varepsilon\int b(t,x,v)\rho(t,x,v)dv$$

$$= \frac{1}{\varepsilon^{n+2}}\int k_{z(t,\cdot)}(x,y)\left[\beta\frac{y-x}{|y-x|} + \gamma\frac{\bar{v}(t,y)-\bar{v}(t,x)}{\varepsilon}\right]\bar{\rho}(t,y)\cdot\bar{\rho}(t,x)dy$$

$$= \sum_{i=1}^{n}\partial_{x_i}\left\{\bar{\rho}^2(t,x)\iint_{r,s\geq 0}[2\zeta(t,x)-h]_+ s^2 r^{n-2}\left(\frac{\beta}{h}e_i + \gamma\partial_{x_i}\bar{v}(t,x)\right)drds + 0(\varepsilon)\right\}$$

where $h = \sqrt{r^2+s^2}$. For the derivation of the approximative divergence formula we used the skew symmetry of the integrand with respect to x and y . Evaluation of the integral yields the 0-th order

$(F)_0$ approximate force balance equation

$$\nabla_x\cdot(\zeta^{n+3}\bar{\rho}^2\nabla_x\bar{v}) + \tilde{\beta}\nabla_x(\zeta^{n+2}\bar{\rho}^2) = 0$$

with $\tilde{\beta} = k\frac{\beta}{\gamma}$, k being a numerical constant only depending on n .

Inserting the expression $(2)_0$ into this differential equation we finally get an approximative

Hyperbolic-elliptic system for $\bar{\rho}$ and $\bar{v}$:

$$\partial_t\bar{\rho} + \nabla_x\cdot(\overline{\rho v}) = f(\bar{\rho})$$

(5)

$$\nabla_x\cdot(\bar{\rho}^{\frac{n-1}{n+1}}\nabla_x\bar{v}) + \psi\nabla_x(\bar{\rho}^{\frac{n}{n+1}}) = 0$$

where $\psi = \tilde{\beta}/K^{\frac{1}{n+1}}$. If these differential equations are valid in an open bounded domain $\Omega\subset\mathbb{R}^n$, we impose boundary conditions, for example

$$\bar{v} = 0 \text{ on } \partial\Omega ,$$

to model no-slip flow at the boundary.

In this paper we want to investigate only the situation where the actin filaments are extended in one dimension in a parallel alignment, that is $n = 1$. Considering a homogeneous transversal distribution in a tube for example, system (5) becomes one-dimensional and the two quantities $u: = \bar{\rho}$ and $v: = \bar{v}$ satisfy the following equations on the interval $[0,1]$:

(6)
$$\partial_t u + \partial_x (uv) = f(u)$$
$$\partial_x v = \overline{g(u)} - g(u) \ , \ v(\cdot,0) = 0$$

where $g(u) = 2\psi u^{\frac{1}{2}}$ and the bar means integration over $x \in [0,1]$. The expression for $\partial_x v$ just generates the second boundary condition $v(\cdot,1) = 0$. $u(t,\cdot) \in L^\infty([0,1])$ can be defined as a weak solution of (6), and even $u(t,\cdot) \in L^1([0,1])$ provided f and g are sublinear functions.

Introducing characteristics $\theta(t,y)$ satisfying

$$\left.\begin{aligned} \partial_t \theta(t,y) &= v(t,\theta(t,y)) \\ \theta(0,y) &= y \end{aligned}\right\} y \in [0,1]$$

we see that system (6) is equivalent to the "<u>characteristic</u>" <u>differential integral system</u> for

$$\left.\begin{aligned} U(t,y) &= u(t,\theta(t,y)) \\ \Phi(t,y) &= \partial_y \theta(t,y) \end{aligned}\right\} y \in [0,1] \ ,$$

namely

(7)
$$\partial_t U = (g(U) - \overline{\Phi g(U)})U + f(U)$$
$$\partial_t \Phi = -(g(U) - \overline{\Phi g(U)})\Phi$$

with initial data

$$U(0,y) = u_0(y) \ , \ \Phi(0,y) = 1 \ .$$

The equation for Φ in (7) implies $\overline{\Phi} \equiv 1$, $\Phi > 0$, meaning that the characteristics $\theta(t,y) = \int_0^y \Phi(t,\cdot)$ do not cross the boundary of $[0,1]$. System (7) can be regarded as an ordinary differential equation with values in the Banach space $F \times F$, where the function space F can be any Banach algebra contained in $L^\infty([0,1])$ such that $U \in F$ implies $f(U), g(U) \in F$. Here we suppose that f and g are smooth functions.

Depending on the initial function $u_0 \in F$ there exists a maximal time $T \leqslant \infty$

so that the solution is uniquely determined on $[0,T)$. For each $y \in [0,1]$ positivity and regularity properties of $U(t,y)$ and $\Phi(t,y)$ are preserved for all $t < T$. Moreover, in the case $F = C^1([0,1])$ the resulting differential equations for $\partial_y U$ and $\partial_y \Phi$ show that also the monotonicity of U is preserved, and that the Jacobian $1/\Phi$ is isotone to U for $0 < t < T$. This means that the characteristics condense toward maxima of the density function $u(t,\cdot)$, therefore resulting in a flow toward the "contraction centers". Simple comparison arguments prove the following

Proposition 1 (Boundedness of solutions)

Assume $f(0) > 0$, $g \geqslant 0$ and the existence of a minimal $u^* > 1$ fulfilling

$$g(u^*)u^* + f(u^*) = 0 .$$

Then all solutions of (6) or (7) with initial values $0 < u_0 \leqslant u^*$ satisfy $u_* \leqslant u \leqslant u^*$ for all times $t > 0$, with some positive u_*.

Linearization of system (5) yields the

Proposition 2 (Local stability of the constant steady state)

Suppose that f is monotone decreasing with $f(1) = 0$ and $f'(1) = -\eta < 0$. Then the unique constant steady state $u \equiv 1$ is locally asymptotically stable iff $g'(1) < \eta$.

According to propositions 1 and 2 the question arises how, in the unstable case $g'(1) > \eta$, (small) perturbations of the homogeneous state $u \equiv 1$ evolve while staying bounded for all time. Numerical calculations suggest stable steady contraction centers at one side as well as competing oscillatory contractions towards both ends of the interval, see [1].

In order to study the qualitative behavior of solutions by analytic means we consider, for simplicity, a linear dependence

$$g(u) = \psi u$$

and simple step functions

$$u(t,x) = \begin{cases} u_1(t): & 0 \leqslant x < s(t) \\ u_2(t): & s(t) < x \leqslant 1 . \end{cases}$$

Supposing u_1, u_2 and s are smooth functions, then the density u is a (weak) solution of (6) iff u_1, u_2 and s satisfy the following

<u>ODE-system</u>

$$\dot{u}_1 = \psi(1-s)(u_1-u_2)u_1 + f(u_1)$$

(8)
$$\dot{u}_2 = \psi s(u_2-u_1)u_2 + f(u_2)$$

$$\dot{s} = \psi s(1-s)(u_2-u_1)$$

$\dot{s}$ is precisely the extremum of the piecewise linear flow $v(t,x)$ at $x = s(t)$. The constant steady state $u_1 = u_2 = 1$ and $0 < s < 1$ is unstable iff $\psi > \eta$.

Analysis of system (8) in this situation shows that for initial data $0 < u_1 < u_2 < u^*$ the solutions converge (as $t \to \infty$) to the steady state

$$u_1 = 1 , \; u_2 = \hat{u} , \; s = 1$$

provided $u^* > 1$ is as in Prop. 1 and there exists a unique $1 < \hat{u} < u^*$ with

$$\psi(\hat{u}-1)\hat{u} + f(\hat{u}) = 0 .$$

Since the whole mass with density level u_2 is $m_2 = (1-s)u_2$ and thus $m_2 \to 0$, the step solution $u(t,\cdot)$ converges in the function space $L^1([0,1])$ towards the <u>unstable</u> steady state $u \equiv 1$. Thus we have constructed a three-dimensional <u>family of homoclinic orbits</u> in L^1 as solutions of (7), modeling the <u>spontaneous formation of transient contractions</u> at one end of the interval. Small perturbations of this solution at the other side of the interval first die out when applied in an early stage, whereas they will grow to an own transient contraction center when the first contraction mass m_2 is already small enough. This can be seen by investigating an enlarged ODE-system for threefold step functions.

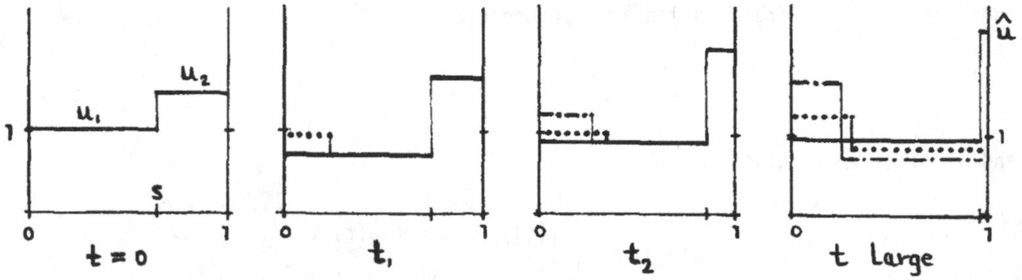

Figure 3. Formation of transient contraction centers by small excitations. Perturbation (•••) at early times, (-•-•) for larger times.

Induced formation of steady contraction patterns

In contrast to the previous situation let us assume the stability condition $\psi < \eta$ and the asymptotic linearity estimate

$$f(u) \geqslant -\eta_{\infty} u \quad \text{for} \quad u \to \infty$$

with some positive η_{∞}. Then small perturbations of the locally stable homogeneous state $u \equiv 1$ will be damped out, whereas for sufficiently large initial values $u_2 > u_1$ we can prove the existence of positive constants c, U and M such that

$$\left.\begin{aligned} u_2(t) &\sim (c - \psi t)^{-1} \\ m_2(t) &\to M \\ u_1(t) &\to U \\ s(t) &\to 1 \end{aligned}\right\} \quad \text{as} \quad t \to T := \frac{c}{\psi}$$

Thus, after finite time, the solution approaches, in a distributional sense, the measure $U + M \cdot \delta_1$, where δ_1 represents the unit Dirac measure concentrated at $x = 1$.

This models the formation of final contraction patterns with point mass at $x = 1$, which can be induced by super-threshold initial excitation. Although these results cannot be carried over to the behavior of smooth solutions of system (6) they partially show how the simple hyperbolic-elliptic system can simulate spontaneous or induced formation of contraction patterns. For more complicated systems of differential equations modeling similar contraction phenomena we refer to recent work by Odell and Oster, see [3].

REFERENCES

1. Dembo, M., Harlow, F. and Alt, W.: The biophysics of cell surface motility. In: "Cell surface phenomena". Eds. DeLisi, Wiegel, Perelson. Marcel Dekker, New York 1982.

2. Odell, G.: Amoeboid motions. In: "Modern modeling of continuum phenomena". Lect. Appl. Math. Vol. 16. AMS, Providence 1977.

3. Odell, G. and Oster, G.: The mechanical basis of morphogenesis III: A continuum model of epithelial cell sheets (to appear).

DIFFERENTIAL EQUATIONS WITH MULTIPLE SOLUTIONS AND NONLINEAR FUNCTIONAL ANALYSIS

Antonio Ambrosetti

Scuola Internazionale Superiore
di Studi Avanzati
Trieste, I-34100 Italy

§1. INTRODUCTION

An interesting feature of several nonlinear problems is to posses multiple solutions and in the last years a remarkable progress has been made in the study of methods of Nonlinear Functional Analysis in order to obtain those non-uniqueness results.

Our purpose in the present paper is to discuss some of those topics, mainly in connections with the applications to Differential Equations.

The work is divided into two parts: Part I, containing sections 2, 3 and 4, is devoted to the abstract setting, while Part II (sections 5, 6, 7, 8 and 9) is concerned with the applications.

What happens, in practice, is that for each problem there is a particular tool which is more appropriate and this is the case in our applications, too. Motivated by what is needed in Part II, we bring out briefly some aspects of the following topics: topological degree and global branching of solutions, Morse theory and min-max principles in critical points theory.

As for the applications, to limit the survey to a reasonable length, we consider here Dirichlet Boundary Value Problems like

$$(BVP) \quad \begin{aligned} Lu &= f(x,u) & x &\in \Omega \\ u &= 0 & x &\in \partial\Omega \end{aligned}$$

where Ω is a smooth, bounded domain in R^N and L is an elliptic differential operator.

The results on (BVP) depend mainly on the behaviour of $f(x,u)$ as $|u| \to \infty$, and our discussion is divided accordingly. Other questions related to some nonlinear eigenvalue problems are also investigated.

Of course, we do not cover many other topics concerning multiplicity results, as: Hamiltonian Systems, wave equations, bifurcation problems, etc. Some of them are discussed in a recent, nice survey by L.Nirenberg [1].

As for the bibliografy, we have listed only those works strictly related to the matter exposed here. The interested reader is referred there to have a more exhaustive list of papers.

PART I : ABSTRACT SETTING

Usually, to solve (BVP) one transforms the problem into an abstract equation $\Phi(u) = 0$, for u in a suitable Banach space. In section 2 we use the Leray-Schauder topological degree to study such Φ, assuming Φ = Identity - Kompact. One of the typical results in this field, namely the existence of global branches of solutions, is discussed, in connection with the applications to nonlinear eigenvalue problems (§7).

Sections 3 and 4 deal with *Potential Operators*, namely with Φ which are gradients of suitable functionals. In view of applications to *Variational BVP's*, we sketch some results of Morse Theory (§3), while in §4 we outline some Min-max theorems closer in spirit to the Lusternik-Schnirelman theory.

§2. TOPOLOGICAL DEGREE AND GLOBAL BRANCHES OF SOLUTIONS.

Let X be a Banach space, U an open subset of X and $\Phi \in C(\bar{U},X)$. Suppose there exists a compact mapping $\Psi : \bar{U} \to X$ such that $\Phi(u) = u - \Psi(u)$ and that $\Phi(u) \neq 0$ for all $u \in \partial U$. Then it is possible to define a non-negative integer, $\deg(\Phi,U,0)$, the Leray-Schauder topological degree of Φ with respect to U and 0, with the property that the equation $\Phi(u) = 0$ has a solution in U provided $\deg(\Phi,U,0) \neq 0$.

An important property is the "invariance by homotopy". Let $\Phi(t,u)$ be a family of continuous mappings from $[0,1]\times\bar{U}$ to X such that, for each $t \in [0,1]$, $\Phi(t,u) = u - \Psi(t,u)$, xith $\Psi(t,.)$ compact. If

$$(2.1) \qquad \Phi(t,u) \neq 0 \qquad \forall (t,u) \in [0,1]\times\partial U,$$

then

$$(2.2) \qquad \deg(\Phi(0,.),U,0) = \deg(\Phi(1,.),U,0)$$

The usual way to apply (2.2) is to take a homotopy $\Phi(t,u)$ such that $\Phi(0,.) = \Phi$ and $\Phi(1,.)$ has a known degree. Obviously, the "a priori" bound (2.1) has to be verified.

If $v \in U$ is an isolated solution of $\Phi(u) = 0$, then the

$$\lim_{\varepsilon\to 0} \deg(\Phi,B_\varepsilon(v),0)$$ [1]

exists and is denoted by $\mathrm{ind}(\Phi,v)$ (the Leray-Schauder index of Φ with respect to v). If $\Phi \in C^1(U,X)$ and the linear mapping $\Phi'(v)$ is invertible then v is an isolated solution of $\Phi = 0$ and the $\mathrm{ind}(\Phi,v)$ can be evaluated by "linearization":

$$(2.3) \qquad \mathrm{ind}(\Phi,v) = (-1)^h$$

where h is the sum of the multiplicities of the characteristic values μ of $\Psi'(v)$, $0<\mu<1$. Above $\Phi'(v)$, $\Psi'(v)$ denote the Frechet-derivative of Φ, Ψ, evaluated at v.

General references for the topological degree can be, e.g., [2,3,4].

---

[1] Here and in the sequel we set $B_\varepsilon(v)=\{u\in X: \|u-v\| <\varepsilon\}$ and $B_\varepsilon=B_\varepsilon(0)$.

A typical result which can be obtained by the topological degree is the existence of global bifurcations. Roughly, this consists in what follows. Suppose $\Phi \in C(R\times X,X)$ is such that $\Phi(\lambda,0) = 0$ for all λ and set:

$$\Sigma = \text{closure of } \{(\lambda,u) \in R\times X : \Phi(\lambda,u) = 0,\ u \neq 0\}.$$

A bifurcation point (for Φ from the trivial soultion u=0) is defined as a λ_o such that $(\lambda_o,0) \in \Sigma$.

If $\Phi(\lambda,.)$ is of the form Identity-Kompact, is C^1 and $\Phi'_u(\lambda,0) = I - K$, I identity and K compact, then, by a classical result of Krasnoselski [5] each *odd* characteristic value of K is a bifurcation point. This depends upon the fact that, if λ_o is such a value, then, using (2.3), the $\text{ind}(\Phi(\lambda,.),0)$ for $\lambda_o-\varepsilon<\lambda<\lambda_o$ ($\varepsilon>0$ small enough) differs for the factor -1 from the $\text{ind}(\Phi(\lambda,.),0)$ when $\lambda_o<\lambda<\lambda_o+\varepsilon$. The global nature of this result has been remarked by Rabinowitz [6] who proved that from a odd characteristic value of K bifurcates a closed, connected component $\Sigma_o \subset \Sigma$ such that: either Σ_o is unbounded or $(\tilde{\lambda},0) \in \Sigma_o$ for a characteristic value $\tilde{\lambda} \neq \lambda_o$ of K.

In our application, however, we will deal with mappings which are possibly not differentiable at u=0. Below we discuss the results we need later.

Let $\Phi \in C(R^+\times X,X)$, $R^+ = [0,\infty)$, be such that:

(H1) $\quad \Phi(\lambda,u) = u - \lambda\Psi(u)$, Ψ compact.

Moreover, let us suppose there exists $\lambda_o \geq 0$ such that:

(H2) $\quad$ for every compact interval $\Lambda \subset R^+$, $\lambda_o \notin \Lambda$, there is $\varepsilon>0$ such that $\Phi(\lambda,u) \neq 0$ for all $\lambda \in \Lambda$ and all $0 < \|u\| \leq \varepsilon$.

From (H2) it follows that for each $\lambda \geq 0, \lambda \neq \lambda_o$, the $\text{ind}(\Phi(\lambda,.),0)$ is defined. Further we assume:

(H3) $\quad \text{ind}(\Phi(\lambda,.),0) \neq 1 \qquad \forall\lambda > \lambda_o$

THEOREM 2.1. Suppose Φ satisfies (H1) and let $\lambda_o > 0$ be such that (H2-3) hold. Then λ_o is the only bifurcation from $u=0$ and from $(\lambda_o,0)$ bifurcates a closed, connected, unbounded component $\Sigma_o \subset \Sigma$.

Proof. First of all, (H2) implies λ_o is the only possible bifurcation point for Φ. Moreover, using again (H2), one has:

$$\text{(2.4)} \qquad \operatorname{ind}(\Phi(\lambda,.),0) = 1 \qquad \text{for all} \quad 0 \leq \lambda < \lambda_o$$

In fact, let $0 < \lambda < \lambda_o$ (if $\lambda=0$ $\Phi(0,u)=u$ and (2.4) holds directly). Taking $\Lambda = [0,\lambda]$, (H2) implies:

$$u - t\lambda\, \Psi(u) \neq 0 \qquad \text{for all } 0 \leq t \leq 1\ ,\ 0 < \|u\| \leq \varepsilon$$

By homotopy, this implies: $\deg(\Phi(\lambda,.),B_\varepsilon,0) = \deg(I,B_\varepsilon,0) = 1$, as required. By (2.4) and (H3) it follows that the $\operatorname{ind}(\Phi(\lambda,.),0)$ has a gap as λ passes through λ_o. The argument is now as in [6]. ■

We say that $\lambda_\infty > 0$ is a *bifurcation from* ∞ (for Φ) if there is a sequence $(\lambda_n,u_n)\to\infty$ with $\lambda_n \to \lambda_\infty$ and $\|u_n\| \to \infty$, such that $\Phi(\lambda_n,u_n) = 0$.

To find the possible bifurcation from ∞, we can perform the change of variable

$$\text{(2.5)} \qquad z = \frac{u}{\|u\|^2}\ , \quad u \neq 0.$$

By (2.5) Φ is transformed into

$$\tilde{\Phi}(\lambda,z) = z - \lambda\, \|z\|^2\, \Psi\!\left(\frac{z}{\|z\|^2}\right)$$

and setting $\tilde{\Sigma} = \{(\lambda,z) \in R^+\times X : \tilde{\Phi}(\lambda,z) = 0\}$, one has that $(\lambda,u)\in\Sigma$, $u\neq 0$ if and only if $(\lambda,z) \in \tilde{\Sigma}$, with z given by (2.5). Hence: λ_∞ is a bifurcation from ∞ if and only if λ_∞ is a bifurcation from $z=0$ for $\tilde{\Phi}$; corresponding to Theorem 2.1, the following holds:

THEOREM 2.2. Suppose Φ satisfies (H1) and

(H2') there exists $\lambda_\infty > 0$ such that for every compact interval $\Lambda \subset R^+$, $\lambda_\infty \notin \Lambda$, there exists $r>0$ such that $\Phi(\lambda,u) \neq 0$ for all $\lambda \in \Lambda$ $\|u\| \geq r$;

(H3') for all $\lambda > \lambda_\infty$, $\deg(\Phi(\lambda,.),B_r,0) \neq 1$.

Then λ_∞ is the only bifurcation from ∞. Precisely from λ_∞ bifurcates a closed, connected component $\Sigma_\infty \subset \Sigma$ such that $\tilde{\Sigma}_\infty$, obtained transforming Σ_∞ through (2.5), is unbounded.

§3. MORSE THEORY.

Here E is a Hilbert space with scalar product (.,.) and $\Phi: E \to E$ is variational, namely there exists a functional $J \in C^1(E,R)$ such that $J'(u) := \text{grad } J(u) = \Phi(u)$. The solutions of $\Phi(u) = 0$ are hence the critical points of J, i.e. the $u \in E$ such that $J'(u)=0$.

Some topics of the Morse Theory which will be used in sections 6 and 8 are discussed below. For a more complete setting, including the case of critical points constrained on Manifolds, see, e.g., [2,7,8].

Let $J \in C^2(E,R)$ and set $E_a = \{u \in E : J(u) \leq a\}$, $Z = \{u \in E : J'(u) = 0\}$, $Z_c = \{u \in Z : J(u) = c\}$.

Let $u \in Z$ and denote by $A(u)$ the linear operator induced in E by setting $J''(u)[h,k] = (A(u)h,k)$ for all $h,k \in E$. The critical point u is said *non-degenerate* if $A(u)$ is invertible; the Morse index of u, M-ind(u), is defined as the dimension of the linear manifold where $A(u)$ is negative-defined. We denote by $C_q(E_a)$ the number of those $u \in Z \cap E_a$ which are non-degenerate with M-ind(u)=q, and set $C_q = C_q(E)$.

The following compactness condition will be imposed on J:

(PS) let $u_n \in E$ be such that $J(u_n)$ is bounded and $J'(u_n) \to 0$; then u_n has a converging subsequence.

In some cases weakenings of (PS) have shown to be useful (e.g. [8, pag.5]); however the above form is enough for our purposes.

It is easy to see that if J satisfies (PS) and is bounded from below, then the $\min\{J(u) : u \in E\}$ is attained at some $\bar{u}$, which is a critical point of J.

It is remarkable that (see §9) a functional J could satisfy (PS) even if the corresponding problem $J'(u)=0$ has no "a priori" bounds for the solutions.

From the connections between $C_q(E_a)$ and the Homology groups $H_q(E_a)$ (the usual notations for those groups are emploied, see, e.g. [9]) the following *Morse Inequalities* can be deduced. Besides of (PS) we suppose:

(J1) J has only non-degenerate critical points;

(J2) J is bounded from below.

Then, for all integer $q \geq 0$ and all $a > \inf\{J(u) : u \in E\}$, it results:

(3.1) $$\operatorname{rank} H_q(E_a) - \operatorname{rank} H_{q-1}(E_a) + \ldots + (-1)^q \operatorname{rank} H_o(E_a) \leq$$
$$\leq C_q(E_a) - C_{q-1}(E_a) + \ldots + (-1)^q C_o(E_a).$$

As consequence we get:

THEOREM 3.1. Let $J \in C^2(E,R)$ satisfy (PS) and (J1-2). Moreover, suppose there is $a \in R$ such that $Z \subset E_a$. Then, for all integer $q \geq 0$ it results:

(3.2) $$(-1)^q \leq C_q - C_{q-1} + \ldots + (-1)^q C_o.$$

Proof. If $Z \subset E_a$ then $C_q(E_a) = C_q$. Moreover E_a is a deformation retract of E [8, lemma 3.3-a]. Hence: $\operatorname{rank} H_q(E_a) = \operatorname{rank} H_q(E)$ for all $q \geq 0$. This latter is 1 for $q=0$ and 0 otherwise. Substituting in (3.1), one obtains (3.2). ■

Remark 3.2. In the preceding theorem one can allow J to possess a finite number of possibly degenerate, isolated minima. In such a case, those minima have to be counted in C_o; see [10, §2].

An appropriate use of the Sard Lemma permits to handle some cases of degeneracy. Following [8, §2] we sketch below a result of this sort.

PROPOSITION 3.3. Suppose $J \in C^2(E,R)$ satisfies (PS) and has at u_o an isolated critical point such that the Frechet derivative $dJ(u_o)$ is Fredholm of index 0. Then there exists $\delta > 0$ such that for all $0 < \varepsilon < \delta$ there is $J_\varepsilon \in C^2(E,R)$ such that:

i) J_ε satisfies (PS);

ii) $J_\varepsilon(u) = J(u)$ for all $\|u\| \geq \varepsilon$;

iii) J_ε has a finite number of non-degenerate critical points in $B_\varepsilon(u_o)$.

Proof. Let $\omega_\varepsilon \in C^\infty(R^+)$ be such that $\omega_\varepsilon(t) = 1$ for all $0 \leq t \leq \frac{\varepsilon}{2}$ and $\omega_\varepsilon(t) = 0$ for $t \geq \varepsilon$. Set:

$$J_\varepsilon(u) = J(u) - \omega_\varepsilon(\|u\|)\,(u,y)$$

for some $y \in E$ specified later. For all $\|u\| \geq \varepsilon$, $J_\varepsilon = J$ and hence for ε small enough, J_ε has the some critical points $u \neq u_o$ than J. Moreover, it is easy to see that there is $\delta' > 0$ such that $J'_\varepsilon(u) \neq 0$ for all $\frac{\varepsilon}{2} \leq \|u\| \leq \varepsilon$, provided $\|y\| < \delta'$. In $B_{\varepsilon/2}(u_o)$ one has:

$$J'_\varepsilon(u) = J'(u) - y. \tag{3.3}$$

By the Sard lemma it is possible to choose $y \in E$, $\|y\| < \delta'$, in such a way that J'_ε has only a finite number of non-degenerate critical points. Lastly, using the fact that $dJ(u_o)$ is Fredholm of index 0, one shows that (taking ε possibly smaller) J_ε satisfies (PS). ■

§4. MIN-MAX PRINCIPLES

Another method developed to study critical points is the Lusternik-Schnirelman critical Points Theory [11]. With respect to the Morse theory, it does not require the non-degeneracy of the critical points even if it gives rise to less precise results. The idea is to find critical levels, namely levels c such that $Z_c \neq \emptyset$, using min-max procedures on suitable classes of sets defined by means of a topological tool: the

Lusternik-Schnirelman category; see [12,13,14]. In [15] a critical point theory has been developed, under a similar spirit, in order to study unbounded (both from above and from below) functionals.

To illustrate, roughly, the idea of the simplest result of this sort, let us take $J \in C^2(E,R)$, satisfying (PS) and bounded from below. If, besides of the global minimum, J has another (isolated) local minimum, then (3.2) with q=1 and Remark 3.2 imply the existence of another critical point: for, otherwise, $C_o=2$ and $C_1=0$ are in contradiction with: $-1 \leq C_1 - C_o$.

This result actually holds in much greater generality [15]:

<u>THEOREM 4.1</u>. Let $J \in C^1(E,R)$ satisfy (PS). Suppose:

(J3) there are $v_1, v_2 \in E$, $\varepsilon>0$ and $\kappa \in R$ such that $v_2 \in B_\varepsilon(v_1)$ and $\inf\{J(u) : \|u-v\| = \varepsilon\} \geq \kappa > \max(J(v_1), J(v_2))$

Then, setting

(4.1) $$b = \inf_{p \in P} \max\{J(p(t)) : 0 \leq t \leq 1\}$$

where P denotes the class of paths from v_1 to v_2, one has: $b \geq \kappa$ and $Z_b \neq \emptyset$.

In spite of its simplicity, Theorem 4.1 has found applications in several interesting cases: besides of (BVP), see §9, we recall Hamiltonian systems [16], wave equations [17], to cite only few of them.

Improvements of Theorem 4.1 can be found in [18, 19]. The sharpening we report is related with the Morse Theory and the remarks before theorem 4.1. The proof [20] makes use of the Morse lemma jointly with the usual deformation arguments for min-max principles, and is briefly sketched below.

<u>THEOREM 4.2</u>. Let $J \in C^2(E,R)$ satisfy (PS) and (J3). Moreover,

for b as in (4.1), let $Z_b = \{\hat{u}\}$ and let $\hat{u}$ be non degenerate. Then M-index$(\hat{u}) = 1$.

Proof. For simplicity of notations we set $\hat{u}=0$ and $b=J(\hat{u})=0$. By contradiction, let M-ind$(0)=q \geq 2$. Using the Morse lemma, one finds linear subspaces N and M, dim N = q, a neighborhood $U := \{v+w,\ v \in N,\ w \in M: \|v\| < \alpha,\ \|w\| < \beta\}$ and $\nu > 0$, such that:

(i) $J(u) \geq \nu$ for all $u \in \bar{U}$: $\|w\| = \beta$;

(ii) the mapping $s \to J(v+(1-s)w)$ with $\|v\| = \alpha$, $\|w\| < \beta$ is non-decreasing.

Taken $\varepsilon < \nu$, let $p \in P$ be such that

$$\text{(4.2)} \qquad \max \{J(p(t)) : 0 \leq t \leq 1\} < \varepsilon$$

Letting $|p| = \{p(t) : 0 \leq t \leq 1\}$, suppose $|p| \cap U \neq \emptyset$ (otherwise the proof follows more directly) and let t_1 (resp. t_2) the first (last) value of t such that $p(t) \in \bar{U}$. For j=1,2, set $z_j = p(t_j)$, $z_j = v_j + w_j$. By (i) and (4.2) it follows that $\|w_j\| < \beta$ and $\|v_j\| = \alpha$. We set:

$$\sigma_j = \{u = v_j + (1-s) w_j : 0 \leq s \leq 1\}.$$

Since dim $N \geq 2$, v_1 and v_2 can be connected by a path π on $\partial U \cap N$. Lastly, we substitute p with $\hat{p}$, defined by $\hat{p}(t) = p(t)$ for $t < t_1$ and $t > t_2$ and $\hat{p} = \sigma_1 \cup \sigma_2 \cup \pi$ outside. Using (ii) and taking α, β small enough it is possible to show that:

$$\max \{J(\hat{p}(t)) : 0 \leq t \leq 1\} < \varepsilon.$$

Since now $|\hat{p}| \cap U = \emptyset$, one can deforme (see the "Deformation Lemma" 1.3 in [15]) $\hat{p}$ in $\tilde{p} \in P$ such that

$$\max \{J(\tilde{p}(t)) : 0 \leq t \leq 1\} < -\varepsilon$$

in contradiction with the definition (4.1) of b.∎

We end this section, stating a multiplicity result in the case in which J is even (see [15]).

THEOREM 4.3. Let $J \in C^2(E,R)$ satisfy (PS) and:

(J4) there exist $\rho>0$ and $\kappa>0$ such that $J>0$ in $B_\rho - \{0\}$ and $J \geqq \kappa > 0$ on ∂B_ρ;

(J5) $J(-u) = J(u)$;

(J6) for any finite dimensional subspace E' of E the set $E' \cap \{u \in E : J(u) \geqq 0\}$ is bounded.

Then J has an increasing sequence of critical values $b_m \to \infty$.

PART II : APPLICATIONS

This part is organized as follows: section 5 contains some preliminary results, section 7 is devoted to the existence of positive solutions of a class of nonlinear eigenvalue problems; the remaining sections deal with (BVP); precisely: section 6: sublinear bvp, section 8: jumping nonlinearities, section 9: superlinear bvp.

§5. PRELIMINARIES

In the sequel Ω will denote a bounded domain in R^N with smooth boundary $\partial\Omega$. The second order deifferential operator L, either in the form

(L1) $$L = \sum D_i(a_{ij}D_j),$$

or

(L2) $$L = \sum a_{ij}D_iD_j + \sum a_jD_j + a_o$$

will be assumed uniformly elliptic in Ω, with smooth coefficients $a_{ij} = a_{ji}$ and in (L2) $a_o \geqq 0$.

Suppose first (L1). Let $m \in L^{\infty}(\Omega)$; the linear eigenvalue problem

$$(5.1) \qquad \begin{array}{ll} Lu = \lambda mu & x \in \Omega \\ u = 0 & x \in \partial\Omega \end{array}$$

has a sequence of eigenvalues $\mu_k(m)$, k=1,2,... Setting $\lambda_k = \mu_k(1)$, with $0 < \lambda_1 < \lambda_2 \leqq \lambda_3 \leqq \ldots$, we indicate with ϕ_k an eigenfunction associated to λ_k, taken such that $(\phi_i,\phi_j)_{L^2} = 0$, $\|\phi_k\|_{L^2} = 1$. Moreover ϕ_1 does not change sign in Ω (and we will fix $\phi_1 > 0$) and λ_1 is the only eigenvalue with this property. It is known that $\mu_k(m)$ has a variational characterization (see [21,pag.398 and foll.] and [22]):

$$(5.2) \qquad (\mu_k(m))^{-1} = \max\left\{ \frac{\int mu^2}{\int \sum a_{ij} D_i u D_j u} : u \in W_0^{1,2}(\Omega), \quad \int u\, \phi_h = 0, \text{ for } h = 1,2,\ldots,k-1 \right\}.$$

Moreover, if $m \geq p$ in Ω and $m(x) > p(x)$ for $x \in \Omega' \subset \Omega$, meas$\Omega'>0$, then $\mu_k(m) < \mu_k(p)$ for all k=1,2,...(*Comparison Property*).

Next, let $m_\varepsilon \to m$ in $L^{N/2}(\Omega)$. Since $W_0^{1,2}(\Omega) \subset L^{2N/N-2}(\Omega)$ (for N>2, otherwise the argument is the same, using the stronger form of the Sobolev Imbedding Theorems) and using the Hölder inequality, one has:

$$(5.3) \qquad \int |m_\varepsilon - m|\, u^2 \leqq \|m_\varepsilon - m\|_{L^{N/2}} \|u^2\|_{L^{N/N-2}}$$

By (5.3) we deduce readily the *Continuity Property*:

$$(5.4) \qquad m_\varepsilon \to m \text{ in } L^{N/2} \text{ implies } \mu_k(m_\varepsilon) \to \mu_k(m), \quad k=1,2,\ldots$$

We will consider the following

$$(\mathrm{BVP}) \qquad \begin{array}{ll} Lu = f(x,u) & x \in \Omega, \\ u = 0 & x \in \partial\Omega \end{array}$$

where $f\colon \bar\Omega \times R \to R$.

If (L1) holds, then, letting $E = W_0^{1,2}(\Omega)$ and $F(x,z) = \int_0^z f(x,t)\,dt$

each critical point (formally) on E of

$$(5.5)\qquad J(u) := \frac{1}{2}\int\sum a_{ij} D_i u\, D_j u \;-\; \int F(x,u)$$

is an $u \in E$ such that

$$\int\sum a_{ij} D_i u\, D_j v \;-\; \int f(x,u)v = 0$$

for all $v \in E$. In other words, such u is a weak solution of (BVP). Assuming f is Hölder-continuous a "bootstrap" argument lieds to show that u is a classical solution. Of course, suitable restrictions have to be imposed on f in order J to be well defined for $u \in E$. Assuming for the moment this is the case, we remark here that, if $f \in C^1$ in u, then $u \in Z$ is non-degenerate and M-ind(u)=q provided for the linearized problem

$$(5.6)\qquad \begin{aligned} Lv &= \lambda f'_u(x,u)v \qquad & x \in \Omega \\ v &= 0 & x \in \partial\Omega \end{aligned}$$

it results:

$$(5.7)\qquad \mu_q(f'_u(x,u)) < 1 < \mu_{q+1}(f'_u(x,u))$$

In (5.7) we mean $1 < \mu_1(f'_u(x,u))$ if q=0. Moreover, let us point out that (5.7) shows we can distinguish by the Morse index two solutions of (BVP) whose Leray-Schauder index is possibly equal: compare (2.3) with the preceding remark. This greater precision permits to obtain much stronger results in the variational case.

Lastly, suppose (L2) holds. In this case we indicate by λ_1 the *principal eigenvalue* of

$$(5.8)\qquad \begin{aligned} Lu &= \lambda u \qquad & x \in \Omega \\ u &= 0 & x \in \partial\Omega. \end{aligned}$$

The Krein-Rutman theorem [23] ensures each eigenvalue $\lambda \neq \lambda_1$ of (5.8) is such that $\mathrm{Re}(\lambda) > \lambda_1$; moreover λ_1 is simple with eigenfunction ϕ_1, which can be taken positive in Ω, and it is the only eigenvalue with this property.

L induces on $X:=C(\bar{\Omega})$ a closed operator with domain $W_0^{1,2}(\Omega)\cap W^{2,2}(\Omega)$ and compact inverse K. Letting F the Nemitski operator induced by f on X, (BVP) turns out to be equivalent to

$$(5.9) \qquad \Phi(u) := u - KF(u) = 0.$$

§6. SUBLINEAR PROBLEMS

We suppose (L1) and

$$(f1) \qquad \lim_{|u|\to\infty} \frac{f(x,u)}{u} < \lambda_1$$

uniformly in $x \in \Omega$. As remarked in section 5, since no restrictions are imposed on the growth of f in the case $f(x,u)\to -\infty$ (resp. ∞) as $u\to\infty$ (resp. $-\infty$), the $\int F(x,u)$ in (5.5) could make no sense for $u \in E$. In such a case a truncation is needed. Precisely, let $u'>0$ be such that $f(x,u') \leq 0$. We substitute f with $\hat{f}(x,u)$ which has the same regularity of f, $\hat{f}(x,u)= f(x,u)$ for all $u \leq u'$ and $0 \geq \hat{f}(x,u) \geq c$ for all $u \geq u'$. Here and throughout in the sequel $c, c_1, c_2, \ldots$ denote (possibly different) constants. By the maximum principle, the solutions u of

$$\begin{aligned} Lu &= \hat{f}(x,u) & x \in \Omega \\ u &= 0 & x \in \partial\Omega \end{aligned}$$

satisfy $u \leq u'$ and hence solve (BVP). Similar is the case in which there is $u''<0$ such that $f(x,u'') \geq 0$, or both. In the following we always understand that such a truncation has been done, if necessary, even if we will continue to use, for simplicity, the same symbol f.

Coming back to J given by (5.5), we remark that $J \in C^k(E,R)$ provided $f \in C^{k-1}(\Omega\times R)$ (in fact, the measurability with respect to x would suffice). Moreover, from (f1) it follows:

$$(6.1) \qquad J(u) \geq \frac{1}{2} \| u\|_E^2 - \frac{1}{2} \lambda_1 \|u\|_{L^2}^2 - c.$$

By the Poincaré inequality: $\lambda_1 \|u\|^2_{L^2} \leq \|u\|^2_E$, and from (6.1) one deduces J is bounded from below on E. Lastly it is readily verified that J satisfies (PS). Then J atteins the minimum on E and hence:

THEOREM 6.1. Suppose $f:\bar{\Omega}\times R \to R$ is Hölder-continuous and satisfies (f1). Then (BVP) has (at least) a solution.

REMARK 6.2. If L is in the form (L2) the same existence result holds. In this case (f1) provides an "a priori" bound for Φ (see (5.9)) and by homotopy with the Identity, one shows $\deg(\Phi, B_r, 0) = 1$ for r large enough.

To discuss the existence of multiple solutions, we will assume $f(x,0)=0$ and look for nontrivial (namely $\not\equiv 0$) solutions. For the sake of simplicity, we take f independent on x; it is easy to extend the results to the general case.

THEOREM 6.3. Suppose $f \in C^1(R)$ and let (f1) hold. Moreover we assume $f(0) = 0$ and let $f'(0) = \lambda$. Then:

(i) for all $\lambda > \lambda_1$ (BVP) has a positive (resp. negative) solution u_1 (u_2).

Further, suppose:

$$\frac{f(u)}{u} > f'(u) \quad \text{for all} \quad u \neq 0. \tag{6.2}$$

Then:

(ii) for all $\lambda>\lambda_2$ (BVP) has another nontrivial solution $u_3 \neq u_1, u_2$.

Proof. From $f(0)=0$, $f'(0)=\lambda$ we get:

$$J(t\phi_1) = \frac{1}{2} t^2 \int\sum a_{ij} D_i\phi_1 \, D_j\phi_1 - \frac{1}{2}\lambda t^2 + o(t^2) =$$
$$= \frac{1}{2}\lambda_1 t^2 - \frac{1}{2}\lambda t^2 + o(t^2) \qquad \text{as } t \to 0$$

Hence, if $\lambda>\lambda_1$, min $J(u) < 0$ on E and (BVP) has a nontrivial solution. To get $u_1>0$ (resp. $u_2<0$) one substitutes f with its positive (negative) part and uses the maximum principle to show that the corresponding minimum is positive (negative) in Ω (and therefore solves (BVP)).

To prove (ii) we first deduce from (6.2) that u_1 (same argument for u_2) is a local minimum for J (up to now one only knows that $J(u_1) \leq J(u)$ for all $u>0$ in Ω). From $Lu_1 = f(u_1)$ and $u_1 > 0$, it follows that

$$\mu_1(\frac{f'(u_1)}{u_1}) = 1.$$

Using (6.2) and the Comparison property of μ_k, one gets:

$$1 = \mu_1(\frac{f'(u_1)}{u_1}) < \mu_1(f'(u_1)).$$

According to (5.7) this proves $M\text{-ind}(u_1) = 0$. Now, suppose $Z = \{0,u_1,u_2\}$ and let us fix, first, $\lambda \neq \lambda_k$. In this case $u=0$ is non-degenerate and $M\text{-ind}(0) \geq 2$, because $\lambda>\lambda_2$. Applying Theorem 3.1 and (3.2) with $q=1$, we get a contradiction, because here $C_o=2$ and $C_1=0$. If $\lambda = \lambda_k$ for some $\lambda_k > \lambda_2$, we use proposition 3.3 with $u_o=0$. Remark that J' is of the type Identity - Kompact and hence those assumptions are verified. Besides of u_1 and u_2 the perturbed functional J_ε has in B_ε a finite number of non-degenerate critical points. Taking into account that J'_ε has the form (3.3), we can use (5.7) to evaluate the M-index of those critical points. Now, if u_ε is any of such points, then $u_\varepsilon \to 0$ in E as $\varepsilon \to 0$; thus, by the continuity property (5.4) we get:

$$\mu_2(f'(u_\varepsilon)) \to \mu_2(f'(0)) = \frac{\lambda_2}{\lambda_k} < 1.$$

Therefore $M\text{-ind}(u_\varepsilon) \geq 2$ for ε small enough, and again for J_ε we have a contradiction, because $C_o=2$ and $C=0$. ∎

Theorem 6.3 has been first proved by Struwe [24] by means of different arguments. The proof above follows [10] where (6.2) is eliminated using the procedure sketched before jointly with a Liapunov-Schmidt reduction.

<u>REMARK 6.4.</u> If (6.2) holds then it can be shown (BVP) has precisely 3 solutions $(0,u_1,u_2)$ for $\lambda_1 < \lambda \leq \lambda_2$ [25].

REMARK 6.5. If one knows that u_3 is non-degenerate, then it could be shown that (BVP) has another solution u_4 for all $\lambda > \lambda_2$. In fact, if $\lambda_2<\lambda<\lambda_3$ and $Z=\{0,u_1,u_2,u_3\}$ then $M\text{-ind}(u_{1,2})=0$, $M\text{-ind}(0)=2$ and (3.2) with q=1 imply $C_1\geq 1$, i.e. $M\text{-ind}(u_3)=1$. Since $C_3=0$, (3.2) with q=2,3 gets: $1=C_2-C_1+C_0=2$, a contradiction. If $\lambda_k<\lambda<\lambda_{k+1}$ (if $\lambda=\lambda_k$ one uses Proposition 3.3), k>2, one has $M\text{-ind}(0)=k$ and again $C_0=2$, $C_1=1$. Since $C_{k-1}=C_{k+1}=0$, (3.2) with q=k,k+1, gets:

$$(-1)^k = C_k + \hat{C},$$

where $\hat{C}=C_{k-2}-C_{k-3}+\ldots+(-1)^k C_0$.
Using (3.2) with q=k+1 and k-2, it follows $\hat{C} = (-1)^k$ and $C_k=0$, a contradiction.

REMARK 6.6. If $f(-u) = -f(u)$, then (BVP) has k pairs of nontrivial solutions, provided $\lambda > \lambda_k$ [26,27,28]. It is an open problem to get some kind of perturbation result, or also to investigate how the number of solutions increase as $\lambda \to \infty$.

§7. ASYMPTOTICALLY LINEAR PROBLEMS

In the present section we assume (L2). On f we suppose $f(x,u) = au + g(x,u)$ with g bounded. If L-a is invertible, it is trivial to see, by a direct homotopy, that (BVP) has always a solution. On the contrary, if $\text{Ker}(L-a) \neq \{0\}$, additional restrictions have to be imposed in order (BVP) to be solvable. Beginning with the well known paper by Landesman and Lazer [29], a great amount of work has been done on those *Problems at resonance*. The interested reader is referred, e.g., to the book by Fucik [30] and references therein.

We point out that the global feature of the topological degree has also shown to be useful, see, e.g. [31,32].

Here we discuss, following [33], a specific case concerning the existence of positive solutions of

$$(7.1) \qquad \begin{aligned} Lu &= \lambda f(u) \qquad && x \in \Omega \\ u &= 0 && x \in \partial\Omega \end{aligned}$$

where $\lambda \in R^+$ and $f: R^+ \to R$ is Hölder-continuous and:

(f2) $f(0)=0$, $f'_+(0)=m_o > 0$ and $f(u)=m_\infty u + g(u)$ with $m_\infty > 0$ and g bounded.

By a *positive solution* of (7.1) we mean a pair (λ,u) with $\lambda > 0$, $u \geq 0$ in Ω, $u \in C^2(\Omega)$ and such that (7.1) holds pointwise.

First of all, we extend f to $\bar{f}: R \to R$ setting $\bar{f}(u) = f(u)$ for all $u \geq 0$ and $\bar{f}(u) = 0$ for all $u < 0$. If

$$(7.2) \qquad \begin{aligned} Lu &= \lambda \bar{f}(u) \qquad && x \in \Omega \\ u &= 0 && x \in \partial\Omega \end{aligned}$$

for some $\lambda>0$, then the maximum principle implies $u > 0$ in Ω and (λ,u) is a positive solution of (7.1).

We will apply to (7.2) the results of section 2, whose notations, as well as those of section 5 will be emploied. In particular, we take $X = C(\bar{\Omega})$ with the "sup norm" denoted by $|u|_\infty$, and

$$\Phi(\lambda,u) := u - \lambda\Psi(u), \qquad \Psi(u) := K\,\bar{F}(u)$$

according to (5.9). Remark that the compactness of K and the continuity of $\bar{F}$ on X imply (H1) holds.

<u>LEMMA 7.1</u>. If (f2) holds then, letting

$$(7.3) \qquad \lambda_o = \frac{\lambda_1}{m_o} \qquad \text{and } \lambda_\infty = \frac{\lambda_1}{m_\infty}$$

Φ (resp. $\tilde{\Phi}$) verifies (H2) (resp. (H2')).

To show that (H3), resp. (H3') holds, we need:

<u>LEMMA 7.2</u>. Suppose (f2) and let $\lambda > \lambda_o$ (resp. $\lambda > \lambda_\infty$). then there exists $\varepsilon>0$ ($r > 0$) such that for all $t \geq 0$ and all $0 < |u|_\infty \leq \varepsilon$ ($|u|_\infty \geq r$) one has:

$\Phi(\lambda,u) \neq t\phi_1$.

For the proof, see [33].

From Lemma 7.2, taking $t = \tau|u|_\infty^2$, $0 \leq \tau \leq 1$, we get, for all $|u|_\infty \geq r$:

(7.4) $\qquad \Phi(\lambda,u) \neq \tau|u|_\infty^2 \phi_1$

Hence $\tilde{\Phi}(\lambda,z) \neq \tau\phi_1$ for all $0 < |z|_\infty \leq \frac{1}{r}$; by homotopy one has:

(7.5) $\qquad \deg(\tilde{\Phi}(\lambda,.),B_{1/r},0) = \deg(\tilde{\Phi}-\phi_1,B_{1/r},0)$

The right hand side in (7.5) is =0, because (7.4) with τ=1 implies $\tilde{\Phi}(\lambda,z) \neq \phi_1$ in $B_{1/r}$. This shows (H3') holds. Similar argument for (H3).

We are now in position to apply Theorems 2.1 and 2.2. Recall that, as pointed out before, Σ consists actually of positive solutions of (7.1)

THEOREM 7.3. Suppose $f:R^+ \to R$ is Hölder-continuous and satisfies (f2). Let λ_o and λ_∞ be given by (7.3). Then the statements of Theorems 2.1 and 2.2 are true. Moreover, if $f(u) > 0$ for all $u > 0$, then there exists $\lambda^* > 0$ such that $\Sigma \subset [0,\lambda^*]\times X$ and Σ_o can be taken as Σ_∞. If $f(u) \leq 0$ for $s' \leq u \leq s''$, then $\Phi(\lambda,u) \neq 0$ for all $s' \leq |u|_\infty \leq s''$. Hence in this case $\Sigma_o \cap \Sigma_\infty = \emptyset$ and, in particular, (7.1) has at least 2 positive solutions for $\lambda > \max(\lambda_o,\lambda_\infty)$.

REMARK 7.4. It is possible to study the case in which $f(0) > 0$ or $f'_+(0) \leq 0$, see [33]. Moreover, the behaviour of Σ_∞ near (λ_∞,∞) can be greatly precised: Σ_∞ bifurcates to the right (left) provided $\lim_{u\to\infty} g(u) < 0$ (> 0).

REMARK 7.5. It is easy to extend the above results to f depending on x, and such that $m_\infty(x) > 0$, $f'_+(x,0) > 0$. When $m_\infty(x)$ and/or $f'_+(x,0)$ change sign in Ω, the problem requires a new version of the Krein-Rutman Theorem [34,35].

REMARK 7.6. Besides of improvements, a remarkable study of problems like (7.1) from numerical point of view has been done in [36,37]

§8. JUMPING NONLINEARITIES

We begin assuming (L2). By a "jumping nonlinearity" we mean a $f:\bar{\Omega}\times\times R \to R$ such that:

$$\limsup_{u \to -\infty} \frac{f(x,u)}{u} \neq \liminf_{u \to \infty} \frac{f(x,u)}{u}$$

More specifically, we suppose $f: \bar{\Omega}\times R \to R$ is Hölder continuous and there exist $\alpha,\beta \in R$,

$$(8.1) \qquad \alpha < \lambda_1 < \beta$$

and g bounded such that:

$$(f3) \qquad f(x,u) = \beta u^+ - \alpha u^- + g(x,u);$$

above $u^+ = \max(u,0)$ and $u^- = u^+ - u$.

Introducing a parameter $t \in R$, we consider:

$$(BVP)_t \qquad \begin{aligned} Lu &= f(x,u) + t\phi_1 & x \in \Omega \\ u &= 0 & x \in \partial\Omega \end{aligned}$$

The feature of this kind of bvp's is described in the following:

THEOREM 8.1. Suppose $f:\bar{\Omega}\times R \to R$ is Hölder-continuous and (f3) and (8.1) hold. Then there exists $T \in R$ such that:

(i) for all $t < T$ $(BVP)_t$ has at least 2 solutions;

(ii) for all $t > T$ $(BVP)_t$ has no solutions;

(iii) for $t = T$ $(BVP)_T$ has at least one solution.

Theorem 8.1 has been proved by Amann and Hess [38], improving a preceding result [40]. Topological degree jointly with lower and upper solutions is emploied. The first paper on (BVP) with (convex) jumping nonlinearties satisfying (8.1) has been [40]: it is shown that the statements (i)-(iii) above are precise, provided (L1) holds and $\beta < \lambda_2$. In fact, that result is obtained as consequence of more general abstract theorems concerning the invertibility in the large of mappings when singularities arise. From this point of view, one expects that, when β passes through λ_2, λ_3, etc., new singularities appear and, correspondingly, new pairs of solutions arise. But this is far to be proved. Next, we discuss a result in this direction.

Precisely, we assume (L1) and take $\beta > \lambda_2$. Improving [41], see also [42], we show:

THEOREM 8.2. Assume $f \in C^1(\Omega \times R)$ verifies (f3) and (8.1). Moreover suppose (L1) and let $\beta > \lambda_2$, $\beta \neq \lambda_k$. Then there exists T' such that for all $t < T'$ $(BVP)_t$ has a positive solution $\bar{u}_t$, a negative solution $\hat{u}_t$ and a third solution $\tilde{u}_t \neq \bar{u}_t, \hat{u}_t$.

Proof. Following [20], we will apply Theorem 4.2. First, since $\beta \neq \lambda_k$ and g is bounded, it follows that

$$(8.2)_t \qquad \begin{aligned} Lu &= \beta u + g(x,u) + t\,\phi_1 && x \in \Omega \\ u &= 0 && x \in \partial\Omega \end{aligned}$$

has a solution, say $\bar{u}_t$, for all $t \in R$. By a direct calculation, letting $\bar{v}_t = \bar{u}_t - t(\lambda_1 - \beta)^{-1}\phi_1$, one has:

$$\begin{aligned} L\bar{v}_t &= \bar{v}_t + g(x,\bar{v}_t) && x \in \Omega \\ \bar{v}_t &= 0 && x \in \partial\Omega \end{aligned}$$

Moreover, by a "boot-strap" argument using regularity theory of elliptic equations, it results: $\|\bar{v}_t\|_{C^1} \leq c$. Hence, being $\beta > \lambda_1$, there exists $\tau < 0$ such that for all $t < \tau$ it results $\bar{u}_t > 0$, so that $\bar{u}_t$ solves not only $(B.2)_t$ but also $(BVP)_t$. Same arguments, replacing β with α, permit to find a solution $\hat{u}_t < 0$.

From $\bar{u}_t = \bar{v}_t + t(\lambda_1 - \beta)^{-1}\phi_1$ with $\|\bar{v}_t\|_{C^1} \leq c$, it follows that $f'_u(x,\bar{u}_t) \to \beta$ as $t \to -\infty$ in L^p for all $p > 1$. Using (5.4), one has, let ting $\lambda_k < \beta < \lambda_{k+1}$:

$$\mu_k(f'_u(x,\bar{u}_t)) \to \mu_k(\beta) = \lambda_k \beta^{-1} < 1,$$

and

$$\mu_{k+1}(f'_u(x,u_t)) \to \mu_{k+1}(\beta) = \lambda_{k+1}\, \beta^{-1} > 1.$$

Similarly, from $f'_u(x,\hat{u}_t) \to \alpha$ as $t \to -\infty$, one gets:

$$\mu_1(f'_u(x,\hat{u}_t)) \to \alpha\, \lambda_1^{-1} < 1.$$

This shows there exists $T'' \leqq \tau$ such that for all $t \leqq T''$:

(8.3) $\quad \text{M-ind}(\bar{u}_t) = k \geqq 2$

(8.4) $\quad \text{M-ind}(\hat{u}_t) = 0.$

In order to apply Theorem 4.2, we let $E := W_o^{1,2}(\Omega)$ and

$$J_t(u) = \frac{1}{2} \int \sum a_{ij} D_i u\, D_j u \;-\; \int F(x,u) \;-\; t \int u\phi_1$$

It is easy to see that J_t is C^2 and (PS) holds. Moreover, (f3) implies for all $\rho > 0$:

$$J_t(\rho\phi_1) \leqq \frac{1}{2}\lambda_1\rho^2 - \frac{1}{2}\beta\rho^2 - t\rho + c.$$

Thus, from $\beta > \lambda_1$, one gets $J_t(\rho\phi_1) \to -\infty$ as $\rho \to \infty$, and, for $t \leqq T''$, (J3) holds with $v_1 = \hat{u}_t$ (according to (8.4)) and $v_2 = \rho\phi_1$, ρ large enough. If J_t has only $\hat{u}_t$ and $\bar{u}_t$ as critical points, (8.3) contradicts Theorem 4.2, proving the existence of a third solution for those t.■

REMARK 8.3. In the preceding proof we have not used the assumption $\beta > \lambda_2$ to find $\bar{u}_t$ and $\hat{u}_t$. Therefore the following sharpening of Theorem 8.1 -(i) holds: if $\beta > \lambda_1$, $\beta \neq \lambda_k$, then there exists $\tau < 0$ such that

The proof of theorem 8.2 is sketched under the assumption that $g'_u(u) \to 0$ as $|u| \to \infty$.

for all $t < \tau$ $(BVP)_t$ has a positive and a negative solution.

REMARK 8.4. Hofer [43] has proved by different methods the existence of 4 solutions under the assumptions of Theorem 8.2. It would be interesting to study how the number of solutions of $(BVP)_t$ increases as $\beta \to \infty$.

Only few words concerning the general case, namely when (f3) holds but α, β do not satisfy (8.1). It is trivial to see, using a direct homotopy argument, that : if the interval (α, β) does not contain any λ_k, then

$$\begin{aligned} Lu &= f(x,u) \qquad && x \in \Omega \\ u &= 0 && x \in \partial\Omega \end{aligned}$$

has at least one solution.

When $\lambda_{k-1} < \alpha < \lambda_k < \beta < \lambda_{k+1}$ (in particular λ_k is simple), some result has been obtained (e.g. [44,45]); however, very few is known when $\lambda_k \in (\alpha, \beta)$ is not simple or when many λ_k belong to (α, β).

§9. SUPERLINEAR PROBLEMS

In this last section we assume (L1) and begin dealing with $f: \bar{\Omega} \times R^+ \to R$ such that, letting $F(x,u) = \int_0^u f(x,t)\,dt$:

(f4) there exist $a > 0$ and $\theta > 2$ such that $uf(x,u) \geq \theta F(x,u)$ for all $u \geq a$ and all $x \in \Omega$.

Let us remark that from (f4) it follows : $F(x,u) \geq c\,u^\theta$ for $u \geq a$ and hence, being $\theta > 2$, f is "superlinear".

At u=0 we suppose:

(9.1) $f(x,0) = 0$.

Thus (BVP) has $u \equiv 0$ as solution and we look for *positive solutions*. As in section 7 we extend f setting $f(x,u) \equiv 0$ for $u < 0$, and this will be understood in what follows.

Critical point theory, in particular the results of section 4, will be emploied here to study (BVP). Set $E:=W_0^{1,2}(\Omega)$ and let J be given by (5.5). Since now $f(x,u) \to \infty$ as $u \to \infty$, some growth restrictions have to be imposed in order $\int F(x,u)$ make sense. Recall that by the Sobolev imbedding theorems, $E \subset L^{2^*}$, $2^* = 2N(N-2)^{-1}$ and hence $F(x,u) \in L^1$ provided $|F(x,u)| \leq c_1 + c_2|u|^{2^*}$. More precisely, we will suppose:

$$(9.2) \qquad |f(x,u)| \leq c_3 + c_4|u|^p \qquad 1 < p < \frac{N+2}{N-2}$$

Above we have taken N>2, and this will be the case throughout in the following. If N=2, p can be taken arbitrary.

It is possible to show that (9.2) implies J is well defined on E and of class C^1, provided f is continuous. Moreover, since $p < \frac{N+2}{N-2}$, the compactness of the Sobolev imbedding jointly with (f4) can be used to prove (PS).

Next, streghtening (9.1), we assume:

$$(9.3) \qquad f(x,u) = o(u) \qquad \text{as } u \to 0^+.$$

It is immediate to verify that (9.3) implies $u=0$ is a local strict minimum, in the sense that there are $a,b > 0$ such that $J > 0$ on $B_a - \{0\}$ and $J \geq b$ for $\|u\| = a$. Lasltly, (f4) implies $J(t\phi_1) \to -\infty$ as $t \to \infty$ and therefore (J3) holds with $v_1 = 0$ and $v_2 = t\phi_1$, t large enough. Then applying Theorem 4.1 one gets:

THEOREM 9.1. Suppose $f: \bar{\Omega} \times R^+ \to R$ is Hölder-continuous, satisfies (f4), (9.2) and (9.3). Then (BVP) has a positive solution.

REMARK 9.2. Condition (9.2) is in some sense necessary for (BVP) to have a (nontrivial) solution. In fact, as consequence of an integral identity, Pohozaev [46] has shown that the bvp

$$\begin{aligned} -\Delta u &= \lambda|u|^p & x &\in \Omega \\ u &= 0 & x &\in \partial\Omega \end{aligned}$$

$\lambda>0$ has the trivial solution only when $p \geq \frac{N+2}{N-2}$ and Ω is star-shaped with respect to the origin in R^N. For a study of (BVP) when f behaves as $|u|^{\frac{N+2}{N-2}}$ at ∞, see [47].

If f is odd in u, Theorem 4.3 lieds to find infinitely many solutions:

THEOREM 9.3. Suppose $f:\bar{\Omega}\times R \to R$ is Hölder-continuous, $f(x,-u) = -f(x,u)$ and satisfies (f4) and (9.2). Then (BVP) has infinitely many solutions u_m.

It is possible to show that actually $\|u_m\| \to \infty$. Therefore in the present case there are no "a priori" bounds for the solutions of (BVP). Compare with the remark in section 3.

It is natural to try to eliminate the oddness of f in Theorem 9.3. A first result in this direction [48] can be stated, roughly, as follows:

Consider

$$(9.4) \qquad \begin{aligned} Lu &= u|u|^{p-1} + \varepsilon\psi(x,u) & x &\in \Omega \\ u &= 0 & x &\in \partial\Omega \end{aligned}$$

where ψ is bounded and $1 < p < \frac{N+2}{N-2}$; Then, for any integer $n > 0$ there exists $\varepsilon_n > 0$ such that for all $0 \leq \varepsilon \leq \varepsilon_n$, (9.4) has at least n solutions.

Such result has been improved in [49,50,51]. The following is a kind of result proved:

THEOREM 9.4. Let σ be the greatest root of $(2N-2)s^2 - (N+2)s - N = 0$. Then, if $1 < p < \sigma$ and ψ is bounded, the bvp

$$\begin{aligned} Lu &= u|u|^{p-1} + \psi(x,u) & x &\in \Omega \\ u &= 0 & x &\in \partial\Omega \end{aligned}$$

has infinitely many solutions.

REMARK 9.5. Let us point out that $\sigma < \frac{N+2}{N-2}$, while the range of admissible p is all $(1,\frac{N+2}{N-2})$ in the result preceding Theorem 9.4. It would be interesting to see wether or not such teorem holds for $1 < p < \frac{N+2}{N-2}$. In this direction, we want to recall that by a "generic" result of Bahri [52], the bvp

$$\begin{aligned} Lu &= u|u|^{p-1} + h(x) \qquad x \in \Omega \\ u &= 0 \qquad x \in \partial\Omega \end{aligned}$$

is solvable for $1 < p < \frac{N+2}{N-2}$ and h in a dense subset of $L^2(\Omega)$.

REFERENCES

1. L.NIRENBERG, Variational and topological methods in nonlinear problems. Bull.A.M.S., 4-3 (1981), 267-302.
2. J.LERAY and J.SCHAUDER, Topologie et èquations fonctionelle. Ann. Sci. Ecole Norm. Sup. , 51 (1934), 45-78.
3. H.AMANN, Lectures on some fixed point theorems. Monog. de Matem., IMPA, Rio de Janeiro.
4. J.T.SCHWARTZ, Nonlinear Functional Analysis, Gordon&Breach, New York, 1969.
5. M.A.KRASNOSELSKII, Topological Methods in the theory of nonlinear integral equations. McMillan, New York, 1964.
6. P.H.RABINOWITZ, Some global results for nonlinear eigenvalue problems. J. Func. Anal., 7 (1971), 487-513.
7. R.PALAIS, Morse theory on Hilbert manifolds, Topology 2 (1963), 299-340.
8. A.MARINO and G.PRODI, Metodi perturbativi nella teoria di Morse, Boll.U.M.I., 11 (1975), 1-32.
9. E.H.SPANIER, Algebraic Topology, McGraw&Hill Co., 1966.
10. A.AMBROSETTI and D.LUPO, On a class of nonlinear Dirichlet problems with multiple solutions, Nonlin. Anal. TMA, to appear.
11. L.A.LUSTERNIK and L.G.SCHNIRELMAN, Topological methods in variational problems. Trudy Inst. Math.Mech., Moscow State Univ., (1930), 1-68.
12. F.E.BROWDER, Infinite dimensional manifolds and nonlinear ellptic eigenvalue problems. Ann. of Math., 82 (1965), 459-477.
13. R.S.PALAIS, Lusternik-Schnirelman theory on Banach manifolds, Topology, 5 (1966), 115-132.
14. J.T.SCHWARTZ, Generalizing the Lusternik-Schnirelman theory of critical points. Comm. Pure Appl. Math., 82 (1964), 307-315.
15. A.AMBROSETTI and P.H.RABINOWITZ, Dual variational methods in criti-

cal point theory and applications, J. Funct. Anal., 14 (1973) 349-381.
16. I.EKELAND, Periodic solutions of Hamiltonian equations and a theorem of P.Rabinowitz, J. Diff. Eq., 34 (1979), 523-534.
17. H.BREZIS, J.M.CORON and L.NIRENBERG, Free vibrations for a nonlinear wave equations and a theorem of P.Rabinowitz. Comm. Pure Appl. Math., 33 (1980), 667-684.
18. W.M.NI, Some minimax principles and their applications in nonlinear elliptic equations. J. d'Analyse Math., 37 (1980), 248-275.
19. V.BENCI and P.H.RABINOWITZ, Critical point theorems for indefinite functionals. Invent.Math., 52 (1979), 241-273.
20. A.AMBROSETTI, Elliptic equations with jumping nonlinearities, J. Math. Phys. Sci., to appear.
21. S.COURANT and D.HILBERT, Methods of Mathematical Physics. Interscience, New York, 1965.
22. A.MANES and AM.MICHELETTI, Un'estensione della teoria variationale classica degli autovalori per operatori ellittici del secondo ordine. Boll. U.M.I. 7 (1973), 285-301.
23. MG.KREIN and M.A.RUTMAN, Linear operators leaving invariant a cone in a Banach space. Am.Math.Soc.Transl. 1-10 (1950), 199-325.
24. M.STRUWE, A note on a result of Ambrosetti and Mancini, Ann. Mat. Pura Appl., to appear.
25. A.AMBROSETTI and G.MANCINI, Sharp nonuniqueness results for some nonlinear problems. Nonlin. Anal.TMA 3-5 (1979),635-645.
26. A.AMBROSETTI, On the existence of multiple solutions for a class of nonlinear boundary value problems. Rend.Sem.Mat.Univ.Padova, 49 (1973), 195-204.
27. P.H.RABINOWITZ, Variational methods for nonlinear eigenvalue problems. Ind.Univ.Math.J. 23 (1974), 729-754.
28. D.C.CLARK, A variant of the Lusternik-Schnirelman theory. Ind.Univ. Math.J. 22 (1972), 65-74.
29. E.M.LANDESMAN and A.C.LAZER, Nonlinear perturbations of linear elliptic problems at resonance. J.Math.Mech. 19 (1970), 609--623.
30. S.FUCIK, Solvability of nonlinear equations and boundary value problems. D.Reidel Publ. Co., Dordrecht, 1980.
31. H.AMANN,A.AMBROSETTI and G.MANCINI, Elliptic equations with noninvertible Fredholm linear part and bounded nonlinearities. Math.Zeit. 158 (1978), 179-194.
32. M.FITZPATRICK, Existence results for equations involving noncompact perturbations of Fredholm mappings with applications to differential equations. J.Math.Anal.Appl. 66 (1978), 151-177.
33. A.AMBROSETTI and P.HESS, Positive solutions of asymptotically linear ellptic eigenvalue problems. J.Math.Anal.Appl. 73 (1980), 411-422.
34. T.KATO and P.HESS, On some linear and nonlinear eigenvalue problems with an indefinite weight function. Comm.P.D.E. to appear.
35. P.HESS, On bifurcation from infinity for positive solutions of second order elliptic eigenvalue problems. To appear.
36. H.O.PEITGEN and K.SCHMITT, Global topological perturbation of nonlinear elliptic eigenvalue problems. To appear.
37. H.O.PEITGEN, J.SAUPE and K.SCHMITT, Nonlinear elliptic boundary va-

lue problems versus their finite difference approximations: Numerical irrilevant solutions. J.Reine Ang.Math.,to appear
38. H. AMANN and P.HESS, A multiplicity result for a class of elliptic boundary value problems. Proc.Royal Soc. Ed. 84-A (1979), 145-151.
39. J.L.KASDAN and F.W.WARNER, Remarks on some quasilinear elliptic equations. Comm.Pure Appl.Math. 28 (1975), 567-597.
40. A. AMBROSETTI and G.PRODI, On the inversion of some differentiable mappings with singularities between Banach spaces. Ann.Mat. Pura Appl. 93 (1973), 231-247.
41. A.C.LAZER and P.J.MCKENNA, On the number of solutions of a nonlinear Dirichlet problem. J.Math.Anal.Appl. 84 (1981), 282-294.
42. S.SOLIMINI, Existence of a third solution for a class of bvp with jumping nonlinearities, to appear.
43. H.HOFER, Variational and topological methods in partially ordered Hilbert spaces. Math. Annalen, to appear.
44. E.N.DANCER, On the Dirichlet problem for weakly nonlinear elliptic partial differential equations. Proc. Royal Soc.Ed. 76-A (1977), 283-300.
45. B.RUF, On nonlinear elliptic problems with jumping nonlinearities. Ann.Mat. Pura Appl., to appear.
46. S.I.POHOZAEV, Eigenfunctions of the equation $-\Delta u+\lambda f(u)=0$. Soviet Math. 5 (1965), 1408-1411.
47. H.BREZIS and L.NIRENBERG, to appear.
48. A.AMBROSETTI, A perturbation theorem for superlinear boundary value problems. M.R.C. Univ. of Wisconsin - Madison, Tech.Summ. Report N.1446, 1974.
49. A.BAHRI and H.BEERSTYCKI, A perturbation method in critical point theory and applications. Trans.A.M.S. 267-1 (1981), 1-32.
50. M.STRUWE, Infinitely many critical points for functionals which are not even and applications to superlinear boundary value problems. Manus. Math. 32 (1980), 335-364.
51. P.H.RABINOWITZ, Multiple critical points of perturbed symmetric functionals, to appear.
52. A.BAHRI, Topological results on a certain class of functionals and applications. J.Funct.Anal. 41 (1981), 397-427.

"GENAUE" FIXPUNKTSÄTZE UND NICHTLINEARE STURM - LIOUVILLE - PROBLEME

Jürgen Appell
Istituto Matematico "U. Dini"
Università di Firenze
Viale Morgagni 67/A
I-50134 Firenze, Italia

Ziel des Vortrags ist es, die nichtlineare Operatorgleichung

(1) $$Lu = F(u)$$

mit Hilfe "genauer" Fixpunktsätze zu lösen, wo L ein linearer Sturm-Liouville-Operator und F ein nichtlinearer Nemytskij-Operator zwischen zwei Funktionenräumen X und Y über einem unbeschränkten Gebiet Ω ist. Gelingt es die Räume X und Y so zu wählen, dass der Operator L invertierbar ist, so ist (1) zum Fixpunktproblem

(2) $$u = Au$$

($A := L^{-1}F$) im Raum X äquivalent. Wegen der Unbeschränktheit des zugrundeliegenden Gebietes Ω ist es dabei oft notwendig, statt des klassischen Schauderschen Fixpunktprinzips eine seiner Verallgemeinerungen für nichtkompakte Operatoren heranzuziehen. Das einfachste solche Fixpunktprinzip liefert der folgende

<u>*SATZ:*</u> *([1],[2]) Sei X ein Banachraum, B_r eine abgeschlossene Kugel mit Radius $r>0$ in X, und A: $B_r \longrightarrow B_r$ stetig und kondensierend. Dann hat A einen Fixpunkt in B_r.*

Hierbei bedeutet "A kondensierend", dass A das Nichtkompaktheitsmass $\chi(M) = \inf\{\varepsilon: \varepsilon>0$, M besitzt ein endliches ε-Netz$\}$ verkleinert, d. h. es gilt ein $k<1$ derart, dass $\chi(AM) \leq k\chi(M)$ ist für alle $M \subseteq B_r$.

Eine genauere Betrachtung dieses Satzes zeigt, dass er - wie die meisten Fixpunktprinzipien - aus zwei Teilen besteht, nämlich einem "geometrischen" (Existenz einer invarianten Kugel für A) und einem "topologischen" (Verkleinerung des Nichtkompaktheitsmasses durch A). Um diese beiden Eigenschaften von A messbar zu machen, setzen wir

(3) $\mu(A,r) := \sup\{\|Au\| : u \in B_r\}$

und

(4) $\chi(A) := \sup\left\{\frac{\chi(AM)}{\chi(M)} : M \text{ nicht relativkompakt}\right\}.$

Unter "genauen" Fixpunktsätzen verstehen wir nun Formeln, die die Berechnung der Funktionen $\mu(F,r)$, $\chi(L^{-1})$ und $\chi(F)$ erlauben; die Bedingung $\|L^{-1}\|\mu(F,r) \leq r$ bedeutet dann gerade die Invarianz von B_r unter A und liefert zusammen mit der Bedingung $\chi(L^{-1})\chi(F) < 1$ die Existenz eines Fixpunkts.

Der Rest des Vortrages ist drei Beispielen in aufsteigender Allgemeinheit gewidmet. Hierbei erhalten wir für den nichtlinearen Operator $F: X \dashrightarrow Y$ folgende Ergebnisse:

I) Gleichheiten für $\mu(F,r)$ *und* $\chi(F)$ *im Fall* $X = Y = L_p$,

II) eine Gleichheit für $\mu(F,r)$ *und eine Abschätzung für* $\chi(F)$ *im Fall* $X = L_p$, $Y = L_q$ $(p \neq q)$,

III) Abschätzungen für $\mu(F,r)$ *und* $\chi(F)$ *im Fall* $X = L_M$, $Y = L_N$ (Orlicz-Räume).

BEISPIEL I: Wir betrachten die Differentialgleichung

(5) $$-\frac{d}{dx}[p(x)\frac{d}{dx}u(x)] + q(x)u(x) = f(x,u(x)) \qquad (0 \leq x < \infty)$$

als Operatorgleichung $Lu = F(u)$ in den durch das Diagramm

$$L_2 \overset{F}{\dashrightarrow} L_2 \overset{L^{-1}}{\dashrightarrow} L_2$$

angedeuteten Lebesgue-Räumen. Hier ergibt sich:

(6) $\mu(F,r) = \inf\{\|a\|_{L_2} + br : |f(x,u)| \leq a(x) + b|u|\}$,

(7) $\chi(F) = \inf\{\|k\|_{L_\infty} : |f(x,u)-f(x,v)| \leq k(x)|u-v|\}$

und

(8) $\chi(L^{-1}) = [\inf \sigma_e(L)]^{-1}$,

wo $\sigma_e(L)$ das wesentliche Spektrum von L im Raum L_2 bezeichnet; zur Berechnung des inneren Spektralradius' kann die Formel

(9) $$\inf \sigma_e(L) = \varliminf_{x\to\infty} \{q(x) + [4p(x)\{\int_0^x p(t)^{-1}dt\}^2]^{-1}\}$$

herangezogen werden (siehe z. B. [3, S. 1501]), zur Berechnung der rechten Seite in (6) gibt es explizite Verfahren (für den Fall X = Y = = L_1 siehe [4]).

BEISPIEL II: Wir betrachten die Differentialgleichung

(10) $-\Delta u(x) + q(x)u(x) = f(x,u(x),\nabla u(x))$ $(x \in \mathbb{R}^N,\ N\geq 3)$

als Operatorgleichung $Lu = F(J_p u)$ in den durch das Diagramm

$$H_2 \xrightarrow{J_p} (L_p)^{N+1} \xrightarrow{F} L_2 \xrightarrow{L^{-1}} H_2$$

angedeuteten Sobolev-Räumen; der durch $J_p u := (u,\nabla u)$ definierte Operator ist stetig vom Sobolev-Raum H_2 in das Produkt $(L_p)^{N+1}$, falls $2<p<2N/(N-2)$ ist. Hier ergibt sich:

(11) $$\mu(F,r) = \inf\{\|a\|_{L_2} + br^{p/2} : |f(x,u,\xi)| \leq a(x) + \ldots$$
$$\ldots + b[|u| + \sum_{i=1}^{N}|\xi_i|]^{p/2}\},$$

(12) $$\chi(F) \leq \inf\{\|k\|_{L_{2p/(p-2)}} : |f(x,u,\xi) - f(x,v,\eta)| \leq$$
$$\leq k(x)[|u-v| + \sum_{i=1}^{N}|\xi_i - \eta_i|]\}$$

und

(13) $$\chi(L^{-1}) = [\inf \sigma_e(L)]^{-1};$$

zur Berechnung des inneren Spektralradius' kann hier die Formel

(14) $$\inf \sigma_e(L) = \sup_{K\subset\subset\mathbb{R}^N} \inf \left\{ \frac{(L\phi,\phi)}{\|\phi\|^2_{L_2}} : \phi \in C_0^\infty(\mathbb{R}^N \setminus K) \right\}$$

herangezogen werden, die an die klassische Formel

(15) $$\inf \sigma(L) = \inf \left\{ \frac{(L\phi,\phi)}{\|\phi\|^2_{L_2}} : \phi \in C_0^\infty(\mathbb{R}^N) \right\}$$

für das gesamte Spektrum $\sigma(L)$ erinnert. Im Unterschied zu (7) steht in (12) nur eine Ungleichheit; in der Tat, es existieren sogar Beispiele von Nichtlinearitäten f, bei denen die rechte Seite von (12) unendlich ist, $\chi(F)$ jedoch endlich (siehe [5]).

BEISPIEL III: Wir betrachten die Differentialgleichung

(16) $-u''(x) + q(x)u(x) = f(x,u(x))$ $(-\infty<x<\infty)$

als Operatorgleichung $Lu = F(u)$ in den durch das Diagramm

$$L_M \xrightarrow{F} L_N \xrightarrow{L^{-1}} L_M$$

angedeuteten Orlicz-Räumen. Die Youngfunktionen M und N werden hierbei entsprechend dem Wachstum der Nichtlinearität f gewählt. Dieses Beispiel kann durch eine einfache Transformation auf den Fall $X = Y = L_1$ zurückgeführt werden: In der Tat, setzen wir

$$(17) \qquad \phi(x,u) := RN[\tfrac{1}{R}f(x,rM^{-1}(u/r))] \qquad (r,R>0),$$

so bildet der Nemytskij-Operator F die Kugel $B_r \subset L_M$ in die Kugel $B_R \subset L_N$ genau dann ab, wenn der Nemytskij-Operator Φ (erzeugt durch ϕ) dieselben Kugeln - jedoch im Raum L_1 - ineinander abbildet. Unter Benutzung dieses Ergebnisses ergibt sich für den Operator F:

$$(18) \qquad \mu(F,r) \le C_1 \inf \{\|a\|_{L_N} + bN^{-1}[M(r)]: |f(x,u)| \le a(x)+bN^{-1}[M(u)]\}$$

und

$$(19) \qquad \chi(F) \le C_2 \inf \{\|k\|_{L_P} : |f(x,u)-f(x,v)| \le k(x)|u-v|\};$$

hierbei liegt die Funktion k in (19) in einem Orlicz-Raum L_P, der "Multiplikator" zu L_M bezüglich L_N ist (d. h. es existieren zueinander konjugierte Youngfunktionen Q und R sowie Konstanten $\alpha,\beta,\omega>0$ derart, dass

$$Q(u) < N^{-1}[P(\alpha u)], \qquad R(u) < N^{-1}[M(\beta u)]$$

für $u \ge \omega$ ist, vgl. z. B. [6, S. 120]). Wir bemerken noch, dass die Konstante C_1 in (18) von M und N abhängt, die Konstante C_2 in (19) dagegen von N, Q, R, α, β und ω.

LITERATUR

[1] Darbo, G.: *Punti uniti in trasformazioni a codominio non compatto,* Rend. Sem. Mat. Univ. Padova 24, 84-92 (1955)

[2] Sadovskij, B. N.: *Nichtkompaktheitsmasse und kondensierende Operatoren (Russisch)*, Probl. Mat. Anal. Slozhn. Sistem, Voronezh, 89-119 (1968)

[3] Dunford, N. / Schwartz, J.: *Linear operators II,* Int. Publ. 1963

[4] Appell, J. / Zabrejko, P. P.: *On a theorem of M. A. Krasnosel'skij,* erscheint in Nonlin. Anal. TMA

[5] Appell, J.: *On the solvability of nonlinear noncompact problems in function spaces with applications to integral and differential equations,* erscheint in Boll. Unione Mat. Ital.

[6] Krasnosel'skij, M. A. / Rutitskij, Ja. B.: *Convex functions and Orlicz spaces*, Noordhoff Groningen 1961

"SHARP" FIXED POINT THEOREMS AND NONLINEAR STURM - LIOUVILLE PROBLEMS

Summary: Given a linear Sturm-Liouville operator Lu = (pu')' - qu and a nonlinear Nemytskij operator F(u) = f(,u,u'), we prove existence theorems for the equation Lu = F(u) by means of sharp fixed point theorems for the (usually noncompact) operator $A = L^{-1}F$. By a "sharp fixed point theorem" we mean the problem of finding or estimating the highest admissible growth and non-compactness of A still guaranteeing the existence of fixed points.

Key words: Fixed point theorems, Nemytskij operator, Sturm-Liouville problem

AMS Classification: 47H10 , 47H09 , 47H15 , 34A34 , 35J60

ASYMPTOTIC INTEGRATION OF FUNCTIONAL DIFFERENTIAL SYSTEMS WHICH ARE ASYMPTOTICALLY AUTONOMOUS

O. ARINO
Université of PAU
FRANCE

I. GYÖRI
Medical University of Szeged
HUNGARY

INTRODUCTION

In this paper we study the asymptotic behaviour of the solutions of non-autonomous retarded linear or lipschitzean differential systems.

The following equations are typical of the framework and the aims of this work :

(1) $\dot{x} = p(t).(x(t)-x(t-1))$, with $p(\cdot)$ in L^2 or $\limsup_{s \to +\infty} |p(s)| < 1$;

(2) $\dot{x} = ax(t-r(t))$, with $r(\cdot)$ in L^p , and $r(t) \to 0$, $t \to +\infty$;

(3) $\dot{x} = (\lambda + a(t))\,x(t) + b(t)x(t-r)$, with $a(\cdot)$ and $b(\cdot)$ in L^2 .

The asymptotic behaviour of the solutions of these equations is well-known (at least, in the scalar case) :
The equation (2) has been studied by K.L. COOKE ([3] , [4]) in the case $1 \le p < 2$, also by I. GYÖRI ([8] , [9]) and recently by O. ARINO and P. SEGUIER ([1]).

The equation (1) has what PANKOV ([17]) calls a balanced term : a term which is zero along the constants. This equation, under many forms and more or less specific assumptions, has received a lot of attention : it appears in compartmental systems and -in particular- in epidemic models : [5] ; many systems can be reduced to this form as is done in [1] [20] . With the assumption given above it has been studied by HADDOCK and SACKER in the case L^2 ([11]), and by ATKINSON and HADDOCK in the case $\limsup |p| < 1$ ([2]).

The equation (3) -in the case $b = 0$- has been studied by many authors [13],[14]. A first approach -in the scalar case- with $b \neq 0$-has been made by HADDOCK and SACKER ([11]).

All these equations are, in the first place, perturbations of ordinary autonomous ones :

$\dot{x} = 0$ for (1) ; $\dot{x} = ax(t)$ for (2) ,

$\dot{x} = \lambda x(t)$ for (3) ; and, we would expect results of the same nature as for the non-perturbed forms.

But, in reality, it is not so easy. If equations (1) and (2) (with L^1) have the same behaviour as the ordinary equations which they resemble, this is not true for equation (3).

The question depends heavily on the class of the perturbation, also on the way it acts in the system (i.e : if it affects the amplitude $(p(t))$ or the argument $(r(t))$, or both).

Nevertheless the results obtained up to now suggest that after convenient transformations such diverse perturbations can be reduced to a unique larger class.

This already led some authors to study certain equations by transforming them back into other ones : for instance, J. KATO in [15] used results for perturbation of equations (1) to deduce special solutions which he calls "O. curves" of equations (2) ; HADDOCK and SACKER in [11] proved an asymptotic formula for a scalar equation (3) by transforming it into an equation (1).

It is precisely this same idea which motivated our work : i.e, to define a class of systems general enough to take into account equations (1), (2) and (3). The objective is partly achieved in the sense that the results on equations (1), (2) and some cases of (3) are obtained as applications of a single fundamental theorem of perturbation (theorem 1).

This theorem applies even for vectorial systems on condition that -roughly- it is possible to order the components of solutions with respect to their behaviour $\to t \to +\infty$, the simplest example being the scalar case or a purely diagonal system.

To treat a more general case, we established another fundamental theorem (theorem 2) which, in the class of L^p-perturbations- extends theorem 1.

In the next section we will state our theorems of perturbation and then, in the last section, we will indicate how they apply to equations of types (1), (2) and (3).

SECTION 1. TWO PERTURBATION THEOREMS.

Before stating the theorems, let us say that we consider as classical notations, definitions and results in the theory of functional differential equations those used in J.K. HALE'S book [12].

In particular, we will use the following :

$C = C([-r,0],\mathbb{R})$;

$C_n = C([-r,0],\mathbb{R}^n)$, $n \in \mathbb{N}$;

$x_t = \{\theta \in [-r,0] \to x_t(\theta) = x(t+\theta)\}$.

THEOREM 1 :

Consider the equation : (F) $\dot{x}(t) = F(t,x_t)$, in which $F : [t_o,+\infty)\times C_n \to \mathbb{R}^n$, for some t_o , can be written as :

$F(t,\varphi) = B(t,\varphi) + P(t,\varphi)$, and :

a) B is a "balanced" term :

i.e : $B(t,c) = 0$, $c \in \mathbb{R}^n$

moreover : $|B(t,\varphi) - B(t,\psi)| \leq \int_{-r}^{0} b(t,s)|\dot{\varphi}(s)-\dot{\psi}(s)|ds$,

for φ , ψ , absolutely continuous ,

where $\limsup_{t\to\infty} \int_{-r}^{0} b(t-s,s)ds < 1$;

b) $|P(t,\varphi) - P(t,\psi)| \leq p(t)|\varphi-\psi|_{C_n}$,

with : $p(\cdot)$ in L^1 ; $P(t,0)$ in $L^1(t_o,+\infty)$.

Then : each solution x tends to a limit, at $+\infty$; moreover for each c in $\mathbb{R}^n$, t_1 large enough there exists a solution defined on $[t_1-r,+\infty)$, which tends to c at $+\infty$.

Sketch of proof : The convergence of $x(t)$ is proved by showing that $|\frac{dx}{dt}|$ is in L^1.

To do that, we first reduce the equation to the following inequation :

$$(*) \quad |\dot{x}| \leq b(t)\int_{t-r}^{t} |\dot{x}|\, ds + p(t)|x_t| + p(t)$$

where $p(t) = |P(t,0)|$.

Then, we use a lemma which slightly refines a recent result by ATKINSON and HADDOCK [2] :

LEMMA : Suppose there exists a positive non-decreasing function $q(\cdot)$ such that :

$$\int_t^{t+r} b(s)ds \le \frac{q(t)-1}{q(t+r)}$$

and : $q(t).b(t)(\int_{t-r}^{t} (p(s)+p(s))ds$ is in L^1.

Then : any solution of $(*)$ has its derivative in L^1.

To prove the second part of theorem 1, we introduce an operator, parametrised by c, the fixed points of which being, for given c, solutions tending to c. Precisely, put $X = \{y , y \in L^1(t_1-r,+\infty) , y/(t_1-r,t_1) = 0\}$ with, as a norm, $|y|_X = \int_{t_1}^{+\infty} |y(s)|ds$; consider the operator U on X défined by :

$$(Uy)(t) = F(t,-\int_{t+.}^{+\infty} y(\tau)d\tau + c), \quad t \ge t_1$$

$$(Uy)(t) = 0 , \quad t_1-r < t < t_1 .$$

This operator simply comes from the integration on $(t,+\infty)$ of the derivative of a solution of (F) having c as a limit at $+\infty$ with its derivative in L^1. We can see easily that :

$$U(0)(t) = P(t,c) , \; t \ge t_1 , \text{ so : } U(0) \in X ;$$

$$|U(y_1)-U(y_2)|_X \le \left[\int_{t_1}^{+\infty} p(s)ds + \sup_{t\ge t_1} \int_{-r}^{0} b(t-s,s)ds\right] . |y_1-y_2|_X .$$

So : U is well-defined as an operator from X into X, and for t_1 large enough U is a strict contraction, therefore has a unique fixed point, y.

Now, the function X defined by :

$$x(t) = 0 , \quad t_1-r \le t < t_1$$

$$x(t) = -\int_t^{+\infty} y(\tau)d\tau + c , \quad t \ge t_1$$

is a solution of (F) on $(t_1,+\infty)$ and has c as a limit at $+\infty$.

Remarks : 1) The solutions obtained in part 2 of the proof have a jump at the initial point. But, it would have been possible to show the existence of a continuous data, or an even more regular solution, depending in fact on the regularity of F . We can find, for example, for each φ in C_n , a data of the form $\varphi + d$, d in $\mathbb{R}^n$.

For that, we replace $\int_t^{+\infty} y(\tau)d\tau$ in the definition of U by

$$(Jy)(t) = Y(t_1-t)(\varphi(t_1)-\varphi(t)) + \int_{\max(t_1,t)}^{+\infty} y(\tau)d\tau$$

in which $Y(u) = 1,\ u > 0$

$= 0,\ u \leq 0$.

2) The property for systems of verifying the second part of theorem 1 is called completness by POPOV [18] and ZVERKIN [23]. When it is possible to go to any point of $\mathbb{R}^n$ starting from a data of the form $\varphi + d$, for any φ in C_n , and some d in a prescribed finite dimensional subspace of data, the system is called ϕ-complete (cf. I. GYÖRI [10]). These notions are interesting in the control theory.

3) In general it is impossible to make the statement of completeness independent of t_1. Suppose, for example, that on a large interval the system coincides with a degenerate one, such as the system given by POPOV [18] ; then the solutions starting from a preceding time will be eventually on a finite dimensional space, with dimension less than n , and so cannot cover $\mathbb{R}^n$ at $+\infty$.

THEOREM 2 : Consider the linear system (in $\mathbb{R}^n \times \mathbb{R}^m$)

$$(E)\quad \begin{cases} \dot{x}(t) = L(t,x_t) + M(t,y_t) \\ \dot{y}(t) = P(t,x_t) + Q(t,y_t) \end{cases}$$

in which the system (L) : $\dot{x}(t) = L(t,x_t)$ is stable ;

the system (Q) : $\dot{y}(t) = Q(t,y_t)$ is exponentially stable ;

and M and P are L^2-perturbations (that is : $|M(t,\cdot)|_{C_m}$, $|P(t,\cdot)|_{C_n}$ are in $L^2(t_0,\infty)$, for some t_0 .

If (x,y) is a solution of (E), then : x is bounded and $y(t)$ tends to zero at $+\infty$.

Moreover, if the solutions of (L) converge, the same holds for the solutions of (E).

<u>Sketch of proof</u> : We introduce the resolvents (or : kernel-functions) $U(t,s)$, $V(t,s)$, $t \geq s$, associated to (L), (Q).

We have : $|U(t,s)| \leq K$,

$|V(t,s)| \leq K.\exp{-\alpha(t-s)}$, $t \geq s$

for some K, $\alpha > 0$ (cf. for example [12]).

Set : $u(t) = |x_t|$, $v(t) = |y_t|$,

$m(t) = |M(t,\cdot)|_{C_m}$, $p(t) = |P(t,\cdot)|_{C_n}$.

From (E), we deduce the following system of inequalities :

$$(*)_1 : u(t) \leq K\, u(t_o) + K \int_{t_o}^{t} m(s)v(s)ds$$

$$(*)_2 : v(t) \leq Ke^{-\alpha(t-t_o)}.v(t_o) + K\int_{t_o}^{t} e^{-\alpha(t-s)}p(s)u(s)ds .$$

Replacing v in $(*)_1$ by the right hand-side of $(*)_2$ we can show that U is bounded.

Coming back to $(*)_2$, it gives us :

$$v(t) \leq Ke^{-\alpha(t-t_o)}v(t_o) + \frac{K}{(2\alpha)^{1/2}} \left(\int_{t_o}^{+\infty} (p(s))^2 ds \right) . \sup_{s \geq t_o} u(s) ,$$

from which it follows that $v(t)$ tends to zero at $+\infty$.

<u>COROLLARY 1</u> : Under the assumptions of theorem 2 , on M,P,Q and, in addition, $|L(t,\cdot)|$ being in L^1 , let (x,y) be a solution of (E). Then : $\lim_{t\to\infty} x(t)$ exists, $\lim_{t\to\infty} y(t) = 0$. Moreover, for each c in $\mathbb{R}^n$, and t_1 large enough, there exists a solution defined for $t \geq t_1$ such that : $\lim_{t\to\infty} x(t) = c$.

SECTION 2. APPLICATIONS.

THEOREM 1 applies directly to equations (1).

Two classes of systems which contain equations (1) and easily verify the assumptions of theorem 1 are :

COROLLARY 2 : Consider the equation $\dot{x}(t) = \sum_{i=1}^{k} P_i(t,x_t)$, in which each P_i verifies such an estimate :

$$|P_i(t,\varphi) - P_i(t,\psi)| \leq p_i(t).|\varphi-\psi|_{C_n} ,$$

with p_i in L^{q_i} , $1 \leq q_i \leq 2$, and P_i is balanced if $q_i > 1$,

and $\sum_{i=1}^{k} P_i(t,0) \in L^1$.

Then the conclusions of theorem 1 hold.

COROLLARY 3 : Consider equation (F) , in which we suppose that :

$$B(t,\varphi+c) = B(t,\varphi) , \quad \varphi \in C_n , \quad c \in \mathbb{R}^n ,$$

$$B(t,0) = 0$$

$$|B(t,\varphi)-B(t,\psi)| \leq k(t).|\varphi-\psi|_{C_n} ,$$

with : $\limsup_{t \to \infty} k(t) < \frac{1}{r}$;

P verifies assumptions b) of theorem 1.

Then : the conclusions of theorem 1 hold.

To apply theorem 1 to equations (2) we have to transform these equations. The general idea is to find a family of isomorphisms $(J(t))_{t \geq t_0}$ on $\mathbb{R}^n$ such that the new functions y , defined by : $x(t) = J(t).y(t)$ satisfy an equation verifying the assumptions of theorem 1 .

When it exsts, J(t) can be written as :

$$J(t) = \exp \Lambda(t) = \exp \int^t \lambda(s)ds ,$$

in which $\lambda(\cdot)$ is a continuous, complex-valued matrix function.

DEFINITION : In analogy to the completeness we will say that the equation in x is exponentially complete if there exists such a function $\lambda(\cdot)$ such that the equation in y is complete. In applications to equations (2) we only examined cases in which $\lambda(\cdot)$ can be taken as $\lambda(t)I$, with $\lambda(\cdot)$ a real function.

We look now at this case, with a linear equation that we write in a general form :

$$(L) : \dot{x}(t) = L(t,x_t) .$$

After changing x into y, we obtain the following :

$$(\tilde{L}) : \dot{y}(t) = \left[-\lambda(t)+L(t,\exp-\int_{t+.}^{t} \lambda(s)ds)\right]\cdot y(t) + L(t,\exp(-\int_{t+.}^{t} \lambda(s)ds).(y_t-y(t))),$$

In such a form, the second term is of a balanced type. Using theorem 1 readily gives the following condition :

PROPOSITION 1 : A sufficient condition for the equation (L) to be exponentially complete is that there exists a locally integrable function $\lambda(\cdot)$ such that :

i) $-\lambda(t)I + L(t,\exp(-\int_{t+\cdot}^{t} \lambda(s)ds).I)$ is in L^1 ,

ii) $\limsup_{t \to \infty} \left[\ell(t) \sup_{-r\le s\le 0} \exp(-\int_{t+s}^{t} \lambda(u)du)\right] < \frac{1}{r}$ where $\ell(t) = |L(t,\cdot)|$.

DEFINITION : In analogy to the linear autonomous case, we will say that $\lambda(\cdot)$ is an exponent.

Such functions $\lambda(\cdot)$ are not uniquely determined. In particular, we can add to one any L^1 function- In fact, we can see that if λ_1 and λ_2 are two exponents of a same system, then :

$$\lim_{t \to \infty} \int_{t_o}^{t} (\lambda_1(s)-\lambda_2(s))\, ds \quad \text{exists.}$$

As an example consider the scalar equation :

$$(L_2) : \quad \dot{x}(t) = a(t)\ x(t-r(t)) \quad ;$$

Conditions i) and ii) of proposition 1 can be expressed as :

(i) : $-\lambda(t) + a(t)\exp - \int_{t-r(t)}^{t} \lambda(s)ds$ is in L^1

(ii): $\limsup_{t\to\infty} \left[r(t)\ |a(t)|\cdot \sup_{-r(t)\le s\le 0} \exp-(\int_{t+s}^{t} \lambda(u)du)\right] < 1$.

The next proposition is now straightforward :

PROPOSITION 2 : Suppose that r is in L^1 and $\lim_{t\to\infty} r(t) = 0$, and a is bounded. Then, the equation (L_2) is exponentially complete, with $a(\cdot)$ is an exponent.

Remarks : 1) The result of prop.2 is the same as the first one given in 1966-67 by K.L. COOKE. We can also obtain the statement corresponding to r in L^p , $1 < p \leq 2$.

2) Concerning the extension of exponential solutions and characteristic exponents to non-autonomous systems ; it has already been considered by RYABOV [19] and UVAROV [22] who introduced the notion of special solutions with exponentially bounded growth. This notion has been shown by R. DRIVER [6] to be useful in the study of asymptotic behaviours.

Also more recently PANKOV ([17]) extended the notion of characteristic exponents to a non-autonomous, and even nonlinear, situation.

We now pass to equations (3). In search of an extension of results for ordinary systems by P. HARTMAN and A. WINTNER [14], W.A. HARRIS and D.A. LUTZ [13], HADDOCK and SACKER in [11] conjectured an asymptotic formula for the solutions of the following equation :

$$\dot{x}(t) = (\Lambda + A(t)x(t) + B(t)x(t-r) ,$$

in which Λ is a diagonal matrix, $\Lambda = (\lambda_i)_{1\leq i\leq n}$, $\lambda_i \neq \lambda_j$ $i \neq j$, and we will assume that $A(\cdot)$ and $B(\cdot)$ are L^2 perturbations .

They stated that there exists a matrix function $F(\cdot)$, $F(t) \to 0$, $t \to \infty$ such that : for any x solution of the equation there exists a constant c , such that $x(t)$ can be written as :

$$x(t) = (I+F(t)) \exp \Lambda(t) . [c + O(1)]$$

where $\Lambda(t)$ is the diagonal matrix with

$$\Lambda_i(t) = \int^t \lambda_i(s)ds \quad ,$$

$$\lambda_i(t) + \lambda_i + a_{ii}(t) + b_{ii}(t)e^{-\lambda_i r} \quad .$$

We consider a more general system :

$$(L_3) : \dot{x}(t) = \Lambda x(t) + L(t,x_t) ,$$

with the same assumptions as above on Λ ; also, we assume that the λ_i's are ordered : $\lambda_i > \lambda_j$ if $i < j$; $(t,\varphi) \to L(t,\varphi)$ is continuous on $(t_o,+\infty) \times C_n$, linear in φ , and : $|L(t,\cdot)|$ is in $L^2(t_o,+\infty)$.

Notation : for a matrix A , we denote by $\mathrm{diag}\{A\}$ the diagonal matrix which has the same diagonal as A .

We distinguish two cases :

- the "quasi-triangular" case. It is an extension of the situation in which $L(t,.)$ is triangular (independently of t) :

PROPOSITION 3 : Suppose that for some $\varepsilon > 0$ and each $i,j,\ i>j$, we have :

$$|L_{ij}(t,\cdot)| = O(\exp(\lambda_i-\lambda_j-\varepsilon)t) .$$

Then : the system (L_3) is exponentially complete, with $\lambda(t)=\Lambda+\mathrm{diag}\{L(t,\exp(\Lambda\cdot))\}$ as a matrix exponent.

The proof is done by showing that after changing the variable x into y , $x(t) = \exp\int^t \lambda(s)ds.\ y(t)$, the system in y satisfies the assumptions of theorem 1.

Remarks : 1) Obviously, prop.3 applies to triangular systems and, in particular, t_o the scalar equations, which is the result given by HADDOCK and SACKER.

2) The statement of prop.3 is stronger than the conjecture since here $F(t) = 0$.

- the general case. We have the following result :

PROPOSITION 4 : Consider the equation (L_3), in which the λ_i's are all distinct (and ordered as before), $|L(t,\cdot)|$ is in L^2. Then : there exists a $n\times n$ matrix $F(s,t)$ of functionals on $C_n([s-r,t])$, $s \le t$, $|F(s,t)|_{C_n([s-r,t])} \to 0$ as $t\to+\infty$ uniformly in s , $s \le t$, such that : each solution x of (L_3) starting from to can be written as :

$$x(t) = (I+F(s,t))\exp\int^t \lambda(\sigma)d\sigma.\ (c+\psi(\cdot)) , \quad \text{for } t_o \le s \le t ,$$

with $c = c(x)$ a constant, $\psi(t) \to 0$, $t \to \infty$ and $\lambda(\cdot)$ as in prop.3.

Moreover, for each c in $\mathbb{R}^n$, there exists a solution x of (L_3) such that : $c = c(x)$.

The proof is done by induction. Set : $x = (x_1, z)$, $x_1 \in \mathbb{R}$, $z \in \mathbb{R}^{n-1}$,

and $x(t) = \exp(\int^t \lambda_1(s)ds).u(t)$, $z(t) = \exp(\int^t \lambda_1(s)ds).y(t)$

where $\lambda_1(t) = \lambda_1 + L_{11}(t, e^{\lambda_1 \cdot})$.

The system verified by (u,y) takes the form :

$$\dot{u} = \alpha(t,u_t) \neq \beta(t,y_t) \quad , \quad \dot{y} = \delta(t,u_t) + \gamma(t,y_t)$$

where : $|\alpha(t,\cdot)|$ is in L^1 ; $|\beta(t,\cdot)|$, $|\delta(t,\cdot)|$ are in L^2 ; $\gamma(t,.) = (\Lambda' - \lambda_1(t)I)\delta_o + \gamma_1(t,.)$, where δ_o is the Dirac distribution , Λ' is the $n-1 \times n-1$ diagonal matrix with λ_i , $2 \le i \le n$ as eigenvalues, and $|\gamma_1(t,\cdot)|$ is in L^2. Hence the system $\dot{u} = \alpha(t,u_t)$ is stable, the system $\dot{y} = \gamma(t,y_t)$ is exponentially stable ; all the conditions of theorem 2 are satisfied (as well as corollary 1), so that we have : $\lim_{t\to\infty} y(t) = 0$, $\lim_{t\to\infty} u(t) = c_1$.

This gives :

$$x_1(t) = \exp\int^t \lambda_1(s)ds \ . \quad [c_1 + O(1)] \ ,$$

so that : $F_{1j} = 0$, $1 \le j \le n$. z can be written as a sum : $z = z^{(1)} + z^{(2)}$.

$$z^{(2)}(t) = \exp(\int^t \lambda_1(\sigma)d\sigma). \int_{t_o}^t (V(t,s)\ Y_o)(0)\delta(s,(c_1+O(1)))\,ds$$

$$= \exp\int^t \lambda_1(\sigma)d\sigma \ . \quad \varepsilon_1(t,\cdot)\ [c_1+O(1)] \ ,$$

which $\varepsilon_1(t,\cdot)$ is defined on $C([t_o - r, t])$ by :

$$\varepsilon_1(t,u) = \int_{t_o}^t (V(t,s).Y_o)(0)\delta(s,u_s)ds \ ,$$

V being the resolvent of the equation $\dot{y} = \gamma(t,y_t)$.

We have : $|\varepsilon_1(t,\cdot)|_{C([t_o-r,t])} \to 0$, $t \to \infty$ (and is in fact in L^2).

So, $z^{(2)}(t) = \varepsilon_1(t,\cdot)\cdot \exp\int^t \lambda_1(\sigma)d\sigma\cdot\ (c_1+O(1))$, which gives :

$$(F_{j,1}(t))_{2\le j\le n} = \varepsilon_1(t,.) \ .$$

Now, $z^{(1)}$ is a solution of the equation :

$$\dot{z}^{(1)}(t) = \Lambda'\ z^{(1)}(t) + L'(t,z_t^{(1)}) \ , \ z_{t_o}^{(1)} = \exp(\int^{t_o+\cdot} \lambda_1(s)ds).y_{t_o} \ ,$$

in which $L'(t,\cdot)$ is the functional defined on C_{n-1} by $L'_{ij}(t)\varphi_2,\dots,\varphi_n) = L_{ij}(t,0,\varphi_2,\dots,\varphi_n)$. If we make the assumption that the formula holds for $n-1$, then we can express $Z^{(1)}$ in the form :

$$Z^{(1)}(t) = (I' + P'(s,t))\exp(\int^t \lambda'(\sigma)d\sigma).(c'+\psi'(.))$$

and putting the expressions of $X_1, Z^{(1)}, Z^{(2)}$ together we obtain the formula of prop.4 for n.

CONCLUSION

Our purpose in this work was to unify a number of results on asymptotic behaviour by using perturbation technics . The most interesting tool when handling these perturbations is theorem 1 : it is simple and efficient. This theorem is in fact a compromise between some earlier statements such as in [10], [21] which it extends, and some finer results such as in [2], the conditions of which cannot be easily computed. Theorem 2 is also a fundamental result. Amongst results that we can cover with these theorems we could also quote these given by SVEC [21] , R.B. EVANS [7], LADDE [16] partly. As a final word, we think that some extensions in this direction are of interest ; for example, the notion of "balanced terms". Also some other results could certainly be viewed as perturbation results : as for example the results given by R.D. DRIVER [6] using "special solutions".

REFERENCE LIST

[1] O. ARINO : Contribution à l'étude des comportements des solutions d'équations différentielles à retard par des méthodes de monotonie et de bifurcation. Thèse d'Etat ; Bordeaux I, Octobre 1980.

[2] ATKINSON - HADDOCK : Conditions for asymptotic convergence of solutions of functional differential equations (preprint 1981).

[3] K.L. COOKE : Functional differential equations close to differential equations. Bull. Amer. Math. Soc. 72 (1966). 285.

[4] K.L. COOKE : Asymptotic theory for the delay differential equation : $\frac{du}{dt} = -au(t-r(t))$. Journ. of Math. Anal. and Appl.19 (1967). 160-175.

[5] K.L. COOKE and J.A. YORKE : Some equations modelling growth processes and gonorrhea epidemics. Math. Biosci (1973).

[6] D.R. DRIVER : Linear differential systems with small delays. J.D.E. 21 (1976) 149-167.

[7] R.B. EVANS : Asymptotic equivalence of linear functional differential equations. Journ. of M. An. Appl.51 (1975) 223-228.

[8] I. GYÖRI : Asymptotic behaviour of solutions of unstable-type first order differential equations with delay. Stud. Sci. Math. Hungarica. 8 (1973) 125-132.

[9] I. GYÖRI : Asymptotic behaviour of solutions of functional differential equations Candidate thesis (in Hugarian) Szeged (1974).

[10] I. GYÖRI : On existence of the limits of solutions of functional differential equations. Coll. Mat. Soc. J. Bolyai 30. Qual. Theory of Diff. Equat. (1979).

[11] J.R. HADDOCK and R. SACKER : Stability and asymptotic integration for certain linear systems of functional differential equations. Journ. of Math. An. and Appl. 76. 328-338 (1976).

[12] J.K. HALE : Theory of functional differential equations. Applied Mathematical Sciences 3. Springer-Verlag (1977).

[13] W.A. HARRIS and D.A. LUTZ : A unified theory of asymptotic integration. J.M. A.A. 57 (1977) 571-586.

[14] P. HARTMAN and A. WINTNER : Asymptotic integration of linear differential equations. Amer. J. Math. 77 (1955). 45-86.

[15] J. KATO : On the existence of O-curves II. Tohoku Math. J. 19 (1967).126-140.

[16] G.S. LADDE : Class of functional equations with applications. Non lin. Anal. Theor. Meth. and Appli. vol.2. N°2 (1978). 259-261.

[17] P.S. PANKOV : Diff. Uravnenia. XIII, 8 (1977). 455-462.

[18] V.M. POPOV : Pointwise degeneracy of linear time-invariant delay differential equations. J.D.E. 11 (1972). 541-561.

[19] RYABOV : Certain asymptotic properties of linear systems with small time lag (in Russian). Trudy Sem. Teor. Diff. Urav.s Otklon. Argumentom. Univ. Druzby Norodov Patrisa Lumumby.3. (1965).153-165.

[20] G.L. SLATER : The differential-difference equation : $\frac{dw}{ds} = g(s)[w(s-1)-w(s)]$. Proc. of Royal Soc. of Edinburgh. 78A. (1977). 41-55.

[21] M. ŠVEC : Some properties of functional differential equations. Boll. U.M.I. (4). 11. Suppl. Fasc. 3. (1975).467-477.

[22] V.B. UVAROV : Asymptotic properties of the solutions of linear differential equations with retarded arguments (in Russian). Diff. Uravn.4 (1968). 659-663.

[23] A.M. ZVERKIN : Pointwise completeness of systems with delay (in Russian) Diff. Uravn.9 (1973). 430-436.

Approach to Hyperbolic Manifolds of Stationary Solutions

Bernd Aulbach

Mathematisches Institut
der Universität
D-8700 Würzburg
Federal Republic of Germany

1. Introduction

In this paper we discuss a problem that arises when a solution of an autonomous system of ordinary differential equations approaches a manifold of stationary solutions, a situation which occurs in a number of biological systems, particularly in population genetics (see e.g. [5]). The problem is whether a solution approaching the manifold M of equilibria converges to some equilibrium on M or whether it ignores the stationary character of the flow on M by moving along near M for all future. In this paper we discuss questions of the following kind. What conditions imply convergence to a stationary solution? Are those conditions generic? Are they necessary? Are they all needed?

To begin with we choose a simple version of the theorem that will be proved later on in this paper. Consider an autonomous system

$$\dot{x} = f(x) \tag{1.1}$$

where $f \in C^3(\mathbb{R}^n, \mathbb{R}^n)$ and suppose it admits an m-dimensional $(0 \leq m \leq n)$ differentiable manifold M of stationary solutions.

THEOREM 1: Let $x(t)$ be a solution of (1.1) and denote its ω-limit set by Ω. Suppose

(1) $\Omega \neq \emptyset$,

(2) $\mathrm{dist}(x(t),M) \to 0$ as $t \to \infty$,

(3) for each $x_0 \in M$ n-m eigenvalues of the Jacobian $f_x(x_0)$ of f at x_0 have real parts different from 0.

Then $x(t)$ converges as $t \to \infty$ to a stationary solution on M.

Results closely related to this theorem can be found in [2],[3] and [4]. In [2] M is a manifold generated by a family of periodic solutions with amplitude-independent period whereas in [3, Theorem 4] the period is allowed to depend continuously on the amplitude. The most general case is treated in [4] where M is a compact manifold carrying a flow which is parallel in a certain sense. Since in [2],[3],[4] the intrinsic features of the proof of Theorem 1 are obscured by technical difficulties arising from the nonstationary flow on M it seems worthwhile to give a separate proof for the simple situation described in Theorem 1. In particular it shows that this proof is general enough for Theorem 2 below, a generalization of Theorem 1.

A further stimulation for this paper was given by Hale and Massatt [7] who considered the above described situation in case M is one-dimensional. Our approach is different from theirs, in particular we give a rigorous proof of the intuitively obvious idea, that a solution approaching M eventually enters one of the stable manifolds associated with the family of stationary solutions on M. We return to this point in the remarks succeeding Theorem 2.

2. Discussion of Theorem 1

The fact that M is an m-dimensional manifold of stationary solutions of (1.1) implies that for each $x_0 \in M$ the Jacobian $f_x(x_0)$ has 0 as an eigenvalue of multiplicity m. Thus condition (3) (the normal hyperbolicity assumption) says that all eigenvalues lie off the imaginary axis apart from those which are necessarily bound to this axis by the manifold property of the set M of stationary solutions. In this sense the approach to M "with asymptotic phase" as asserted in the theorem is generic.

The sufficiency of the set of conditions (1),(2),(3) is stated in Theorem 1 and will be proved later in this paper. How about necessity? Obviously (1) and (2) are necessary while (3) is not. To see the latter consider the simple system $\dot{x} = 0$, $\dot{y} = -y^3$ where the x-axis is a stationary manifold with asymptotic phase although it is not normally hyperbolic.

The next question is whether one or the other of the three assumptions of Theorem 1 is redundant. By means of three examples we demonstrate that neither of these hypotheses may be dropped.

EXAMPLE 1: Consider the two-dimensional system

$$\dot{x} = x\,y^2$$
$$\dot{y} = -\frac{y}{x}$$

in the half-plane $x > 0$. It has the positive x-axis as one-dimensional manifold M of stationary solutions and each point $(x_o,0)$ on it has the eigenvalues 0 and $-x_o^{-1}$. As $x(t)$ we choose any solution whose trajectory $y = (2/x)^{1/2}$ solves the corresponding scalar equation $dy/dx = -x^{-2}y^{-1}$. Only condition (1) is violated, $x(t)$ does not converge to a point on M (see fig.1).

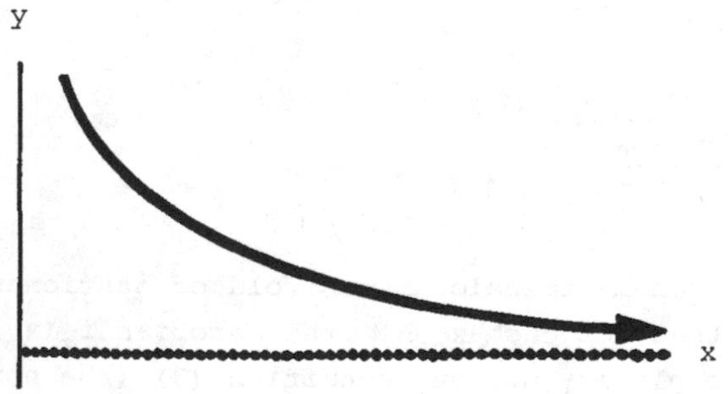

Fig.1

EXAMPLE 2: (see Andronov et al. [1], p.37) The parameter depending system

$$\dot{x} = 2y - \mu(y^2 - 2x^2 + x^4)(4x - 4x^3)$$
$$\dot{y} = 4x - 4x^3 + \mu(y^2 - 2x^2 + x^4)$$

has $V(x,y) = y^2 - 2x^2 + x^4$ as Ljapunov function by means of which the part of the phase portrait exhibited in fig.2 can be proved (for $\mu < 0$).

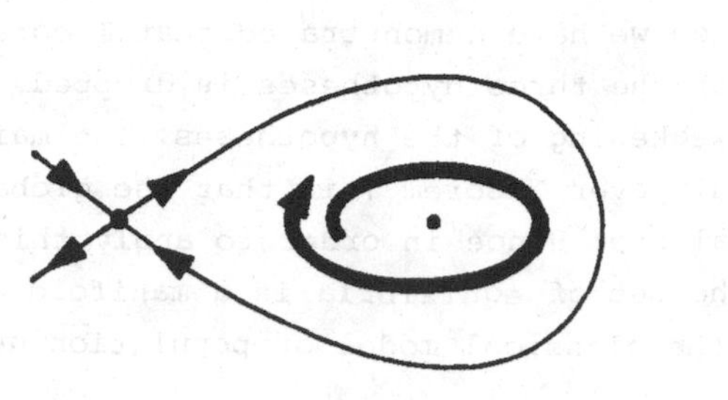

Fig.2

The coordinate origin (chosen as M, m = 0) is hyperbolic and ω-limit point of any nonconstant solution x(t) inside the homoclinic orbit. Nevertheless x(t) does not converge to the origin.

EXAMPLE 3: The plane polar coordinate system

$$\dot{r} = (1 - r)^3$$
$$\dot{\varphi} = (1 - r)^2$$

has the unit circle $r = 1$ as one-dimensional manifold of stationary so-

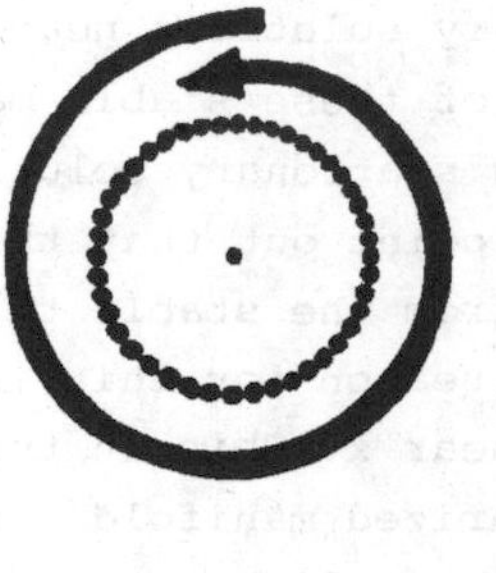

Fig.3

lutions. The nonconstant trajectories $1+(r_o-1)e^{-\varphi}$ spiral onto the unit circle (thus satisfying (1) and (2)) without converging to a point on it (see fig.3).

3. Main theorem

In the previous section we have demonstrated that Theorem 1 becomes false if either one of the three hypotheses is dropped. Nevertheless this theorem allows weakening of the hypotheses. The main advantage of this generalized result over Theorem 1 is that the global assumption (2) is replaced by a local one. Hence in order to apply this theorem it is not necessary that the set of equilibria is a manifold as a whole. This is the case e.g. in the classical model of population genetics (see [5]).

THEOREM 2: Let $x(t)$ be a solution of (1.1) and denote its ω-limit set by Ω. Assume that

(1) there exists a point $x^* \in \Omega$,

(2) there exists a neighborhood U of x^* such that $\Omega \cap U \subset M$,

(3) $n-m$ eigenvalues of $f_x(x^*)$ have real parts different from 0.

Then $\lim_{t\to\infty} x(t) = x^*$.

REMARKS: 1) If for some $\bar{x} \in M$ $n-m$ eigenvalues of $f_x(\bar{x})$ have negative real parts then, as is well known for a long time (see Malkin [10]), any solution starting sufficiently close to $\bar{x}$ tends to a stationary solution near $\bar{x}$ on M. In modern terminology the proof of this statement can be sketched in one sentence: There exists a neighborhood of $\bar{x}$ which is positively invariantly fibered by a family of stable manifolds (corresponding to the family of stationary solutions near $\bar{x}$ on M) and thus each solution near $\bar{x}$ decays on one of these stable manifolds at an exponential rate to the corresponding stationary solution near $\bar{x}$ on M. It is the purpose of this remark to point out that this simple geometric idea of proof does not carry over from the stable to the hyperbolic case as considered in this paper. The reason for this is the lacking invariant fibration of the phase space near $\bar{x}$. This in turn is based on the fact that the invariance of a linearized manifold carrying a hyperbolic flow does not "survive" in general a nonlinear perturbation. An example in

[9] shows the possibility of this phenomenon. Hence in view of this remark it is intelligible that the proof we give for Theorem 2 is not direct and geometric but indirect and mainly analytic.

2) If, under the assumptions of Theorem 2, k ($\leq n-m$) eigenvalues of $f_x(x^*)$ have negative real parts then each stationary solution x_0 in a (relative) neighborhood N of x^* on M is associated with a k-dimensional stable manifold S_{x_0}. All solutions on S_{x_0} near x_0 decay to x_0 as $t \to \infty$ at an exponential rate. In the hyperbolic case $0 < k < n-m$ there is "a lot of space" between the members of the family S_{x_0}, $x_0 \in N$, of stable manifolds and the question arises whether there is another possibility of approaching x^* rather than through the stable manifold S_{x^*}. That this is not the case is a consequence of Theorem 2 (and not the reason for its validity). The proving argument is as follows: Since by Theorem 2 $x(t) \to x^*$ as $t \to \infty$ the solution $x(t)$ remains for all future near x^* and thus it lies on a center-stable manifold of x^* which under our assumptions is made up of members of the family S_{x_0}, $x_0 \in N$, of stable manifolds. If $x(t)$ were not on S_{x^*} it would be on some S_{x_0}, $x_0 \neq x^*$ and consequently it would converge to x_0, contradiction. The résumé of this remark is that under the assumptions of Theorem 2 only the solutions on the stable manifold of x^* converge to x^* as $t \to \infty$.

4. Proof of Theorem 2

Throughout this section let N,P denote real square matrices whose eigenvalues have negative, positive real parts, respectively. Let α,γ be corresponding positive constants such that

$$\| e^{N(t-s)} \| \leq \gamma e^{-\alpha(t-s)} \quad \text{for all } t \geq s,$$
$$\| e^{P(t-s)} \| \leq \gamma e^{\alpha(t-s)} \quad \text{for all } t \leq s.$$

The proof of our main result needs a bit of preparation. In order to exhibit the explicit dependence of the parameters involved we first state a well known result on linearly perturbed linear systems. The proof is based on a straightforward application of Gronwall's inequality (see Coppel [6], Prop.1.1) and is therefore omitted.

LEMMA 1: Let A(t) be a real square matrix whose elements are continuous

functions on an interval J such that $\|A(t)\| \le \delta$ on J for some $\delta > 0$. Then
(a) the principal fundamental matrix $\Phi_N(t,s)$ of $\dot{x} = [N + A(t)]x$ satisfies

$$\|\Phi_N(t,s)\| \le \gamma e^{(-\alpha+\delta\gamma)(t-s)} \quad \text{for all } t,s \in J \text{ with } t \ge s,$$

(b) the principal fundamental matrix $\Phi_P(t,s)$ of $\dot{x} = [P + A(t)]x$ satisfies

$$\|\Phi_P(t,s)\| \le \gamma e^{(\alpha-\delta\gamma)(t-s)} \quad \text{for all } t,s \in J \text{ with } t \le s.$$

The next lemma gives an inequality relating the boundary value components for solutions of a certain class of linear differential systems. This lemma will play an essential role in the proof of our main theorem.

<u>LEMMA 2</u>: Consider the class of linear differential systems

$$\begin{aligned} \dot{u} &= [N + A_1(t)]u + A_2(t)v \\ \dot{v} &= [P + A_3(t)]v \\ \dot{w} &= A_4(t)u + A_5(t)v \end{aligned} \tag{4.1}$$

where the matrices $A_i(t)$, $i=1,\dots,5$ are continuous and bounded (in norm) above by $\frac{\alpha}{2\gamma}$ on a compact interval $I := [t_0, T_0]$.
Then any solution $(u(t), v(t), w(t))$ of (4.1) satisfies the inequality

$$\|w(T_0)\| \le \|w(t_0)\| + \|u(t_0)\| + \tfrac{3}{2}\|v(T_0)\| .$$

<u>REMARK</u>: The crucial point of this lemma is that the asserted inequality holds uniformly for any solution of any system of the form (4.1) as long as the matrices $A_i(t)$, $i=1,\dots,5$ are bounded by $\frac{\alpha}{2\gamma}$, a number only depending upon N and P.

<u>PROOF</u>: Later in this proof we need the following estimates which are easily verified:

$$\int_{t_0}^{T_0} e^{-\frac{\alpha}{2}(\tau - t_0)}\, d\tau < \frac{2}{\alpha}, \qquad \int_{t_0}^{T_0} e^{\frac{\alpha}{2}(\tau - T_0)}\, d\tau < \frac{2}{\alpha},$$

$$\int_{t_o}^{T_o}\int_{t_o}^{\tau} e^{-\frac{\alpha}{2}(\tau-\sigma)}\, e^{\frac{\alpha}{2}(\sigma-T_o)}\, d\sigma d\tau < \frac{2}{\alpha^2}.$$

By Lemma 1 we get for the principal fundamental matrices $\Phi(t,s)$, $\Psi(t,s)$ of $\dot{u} = [N + A_1(t)]u$, $\dot{v} = [P + A_3(t)]v$, respectively, the estimates

$$\|\Phi(t,s)\| \leq \gamma e^{-\frac{\alpha}{2}(t-s)} \qquad \text{for } t,s \in I,\ t \geq s,$$

$$\|\Psi(t,s)\| \leq \gamma e^{\frac{\alpha}{2}(t-s)} \qquad \text{for } t,s \in I,\ t \leq s.$$

For the components of any solution $(u(t),v(t),w(t))$ of (4.1) we get for each $t \in I$ the relations

$$u(t) = \Phi(t,t_o)u(t_o) + \int_{t_o}^{t} \Phi(t,\sigma)A_2(\sigma)v(\sigma)d\sigma,$$

$$v(t) = \Psi(t,T_o)v(T_o),$$

$$w(t) = w(t_o) + \int_{t_o}^{t} [A_4(\tau)u(\tau) + A_5(\tau)v(\tau)]d\tau,$$

whose combination leads to

$$w(T_o) = w(t_o) + \int_{t_o}^{T_o} A_4(\tau)\Phi(\tau,t_o)u(t_o)d\tau +$$

$$\int_{t_o}^{T_o}\int_{t_o}^{\tau} A_4(\tau)\Phi(\tau,\sigma)A_2(\sigma)\Psi(\sigma,T_o)v(T_o)d\sigma d\tau + \int_{t_o}^{T_o} A_5(\tau)\Psi(\tau,T_o)v(T_o)d\tau.$$

This in turn yields the estimate

$$\|w(T_o)\| \leq \|w(t_o)\| + \frac{\alpha}{2}\|u(t_o)\| \int_{t_o}^{T_o} e^{\frac{\alpha}{2}(\tau-t_o)} d\tau +$$

$$\frac{\alpha^2}{4}\|v(T_o)\| \int_{t_o}^{T_o}\int_{t_o}^{\tau} e^{-\frac{\alpha}{2}(\tau-\sigma)}\, e^{\frac{\alpha}{2}(\sigma-T_o)} d\sigma d\tau + \frac{\alpha}{2}\|v(T_o)\| \int_{t_o}^{T_o} e^{\frac{\alpha}{2}(\tau-T_o)} d\tau.$$

Application of the above integral inequalities proves Lemma 2. ■

Now we are ready for the PROOF OF THEOREM 2. First we shift the ω-limit point x^* of $x(t)$ into the coordinate origin and introduce curvilinear local (u,v,w)-coordinates such that a neighborhood of x^* on the manifold M corresponds to a (relative) neighborhood of the $\mathbb{R}^n$-origin in the linear subspace with vanishing u- and v-coordinates. Hence we may consider instead of (1.1) near $x^* \in M$ a differential system of the form

$$\begin{aligned} \dot{u} &= Nu + r_1(u,v,w) \\ \dot{v} &= Pv + r_2(u,v,w) \\ \dot{w} &= \quad\;\; r_3(u,v,w) \end{aligned} \tag{4.2}$$

near (0,0,0) where the C^2-functions r_i satisfy $r_i(0,0,w) \equiv 0$, $i=1,2,3$ for small $\|w\|$. The spectral assumption (3) justifies that we denote the matrices of the linear part by the letters N and P with the fixed meaning of this section. System (4.2) admits a local center-stable manifold with C^2-representation $v=s(u,w)$. The transformation $(u,v,w) \to (u,v-s(u,w),w)$ preserves all essential properties of system (4.2) and moreover normalizes the equation for the center-stable manifold to $v=0$ which means that $r_2(u,0,w) \equiv 0$ for small $\|u\|$ and $\|w\|$ (for details see [3], p.362). The particular properties of the nonlinearities r_i allow (see Hartman [8], Ch.V, Lemma 3.1) to write system (4.2) in the quasilinear form

$$\begin{aligned} \dot{u} &= [N + B_1(u,v,w)]u + B_2(u,v,w)v \\ \dot{v} &= [P + B_3(u,v,w)]v \\ \dot{w} &= B_4(u,v,w)u + B_5(u,v,w)v \end{aligned} \tag{4.3}$$

where $B_i(u,v,w)$, $i=1,\dots,5$ are continuous matrices vanishing as (u,v,w) tends to (0,0,0).

Before completing the proof we restate assumptions (1) and (2) of the theorem in terms of the (u,v,w)-coordinate system: There exists a solution $(u(t),v(t),w(t))$ of (4.3) and a sequence $t_\nu \to \infty$ as $\nu \to \infty$, $\nu \in \mathbb{N}$, such that

$$\lim_{\nu\to\infty} (u(t_\nu),v(t_\nu),w(t_\nu)) = (0,0,0). \tag{4.4}$$

Furthermore there exists a ρ-ball B_ρ around (0,0,0) with the property that all ω-limit points of $(u(t),v(t),w(t))$ in this ball have vanishing u- and v-components.

The final goal of the proof is to show that the ω-limit point (0,0,0) is the limit of $(u(t),v(t),w(t))$ as $t\to\infty$. To this end we suppose the contrary. This means we may assume that there exists a sequence of intervals $I_\nu := [t_\nu, T_\nu]$, $t_\nu < T_\nu$, and a positive constant $\sigma < \rho$ such that

$$\|(u(t),v(t),w(t))\| < \sigma \text{ for all } t \in [t_\nu, T_\nu),$$

$$\|(u(T_\nu),v(T_\nu),w(T_\nu))\| = \sigma \text{ for all } \nu.$$

Without loss of generality we may take σ so small that

$$\|B_i(u(t),v(t),w(t))\| \leq \frac{\alpha}{2\gamma} \text{ on each } I_\nu,\ i=1,..,5,\ \nu\in\mathbb{N}.$$

Thus, for each ν, $(u(t),v(t),w(t))$ is a solution of the linear system

$$\begin{aligned} \dot{u} &= [N + C_1(t)]u + C_2(t)v \\ \dot{v} &= [P + C_3(t)]v \\ \dot{w} &= C_4(t)u + C_5(t)v \end{aligned}$$

on I_ν where $C_i(t) := B_i(u(t),v(t),w(t))$ and we may apply Lemma 2. This provides the estimates

$$\|w(T_\nu)\| \leq \|w(t_\nu)\| + \|u(t_\nu)\| + \frac{3}{2}\|v(T_\nu)\| \quad \text{for all } \nu. \tag{4.5}$$

Since $(u(T_\nu),v(T_\nu),w(T_\nu))$ is a bounded sequence there exists a convergent subsequence $(u(T_{\nu_\mu}),v(T_{\nu_\mu}),w(T_{\nu_\mu}))$ with limit $(u_\infty,v_\infty,w_\infty)$, say. This point, on the other hand, is an ω-limit point of $(u(t),v(t),w(t))$ in the ρ-ball B_ρ and thus u_∞ and v_∞ are zero. This implies

$$\lim_{\mu\to\infty} v(T_{\nu_\mu}) = 0, \tag{4.6}$$

$$\|\lim_{\mu\to\infty} w(T_{\nu_\mu})\| = \sigma > 0, \tag{4.7}$$

and finally, combining (4.4),(4.5) and (4.6), we get

$$\lim_{\mu\to\infty} w(T_{\nu_\mu}) = 0$$

which contradicts (4.7) and completes the proof of Theorem 2. ■

References

[1] A.A.Andronov, E.A.Leontovich, I.I.Gordon and A.G.Maier, Qualitative theory of second-order dynamic systems. Wiley, New York 1973.

[2] B.Aulbach, Asymptotic amplitude and phase for isochronic families of periodic solutions, in "Analytical and Numerical Approaches to Asymptotic Problems in Analysis", 265 - 271, North Holland, Amsterdam 1981.

[3] B.Aulbach, Behavior of solutions near manifolds of periodic solutions. J.Differential Equations 39 (1981), 345 - 377.

[4] B.Aulbach, Invariant manifolds with asymptotic phase. J.Nonlinear Analysis 6 (1982), 817 - 827.

[5] B.Aulbach and K.P.Hadeler, Convergence to equilibrium in the classical model of population genetics. Preprint No.83, Math. Inst. Univ. Würzburg 1982.

[6] W.A.Coppel, Dichotomies in stability theory. Lecture Notes in Mathematics No.629, Springer, Berlin 1978.

[7] J.K.Hale and P.Massatt, Asymptotic behavior of gradient-like systems, in "Univ. Florida Symp. Dyn. Syst. II", Academic Press, New York 1982.

[8] P.Hartman, Ordinary differential equations. Wiley, New York 1964.

[9] H.W.Knobloch and B.Aulbach, The role of center manifolds in ordinary differential equations, to appear.

[10] I.G.Malkin, Theory of stability of motion (Russian), Moscow 1952.

LINEAR VOLTERRA-STIELTJES INTEGRAL EQUATIONS AND CONTROL

L.Barbanti
Instituto de Matemática e Estatística
Universidade de São Paulo
05508-São Paulo(SP) - Brasil

1.Introduction

The purpose of this paper is to study some aspects on controllability concerning the linear Volterra-Stieltjes integral equations.

Volterra-Stieltjes integral equations are considered in many works:see [1] to [16] .

Here we work in the context by Hönig.

The development of the control theory for this type of equation has an intrinsic interest since it encloses very general classes of evolutive systems.It comprises,for instance,the linear Stieltjes integral equations,

$$\text{(L)} \qquad y(t) - x + \int_a^t dA(s).y(s) = g(t) - g(a), \qquad (a \leqslant t \leqslant b),$$

the linear delay differential equations and Volterra integral equations (see [3] ,pp.81-94,and [7]).Moreover,it is easy to construct very simple models of a perturbated control system,where the process that describes the transfering of the optimal instantaneous controls (with respect to the perturbation) is a linear Volterra-Stieltjes type(see for this direction [17] ,p.67).

2.Linear Volterra-Stieltjes integral equations

Given $[a,b] \subset \mathbb{R}$,and X a Banach space,we define the semi-variation of $g:[a,b] \longrightarrow L(X)$ as

$$SV[g] = \sup_{d \in D} \sup\{\sum_{1}^{|d|}{}_i \|(g(t_i)-g(t_{i-1})).x_i\| \;; x_i \in X, \|x_i\| \leqslant 1\} ,$$

where D is the set of all partitions

$$d = \{a = t_0 < t_1 < \dots < t_{|d|} = b\} ,$$

of the interval $[a,b]$.

If $SV[g] < \infty$ we say that g is of bounded semi-variation, and we write $g \in SV([a,b],L(X))$. Note that SV is a seminorm.

We say that $f: [a,b] \longrightarrow X$ is regulated and write $f \in G([a,b],X)$ if f has only discontinuities of first kind.

For $g \in SV([a,b],L(X))$ and $f \in G([a,b],X)$ there exists the interior (or Dushnik type) integral

$$F_g(f) = \int_a^b \cdot dg(t) \cdot f(t) = \lim_{d \in D} \sum_{1}^{|d|} {}_i (g(t)-g(t_{i-1})) f(\mathring{s}_i) \in X ,$$

where $\mathring{s}_i \in (t_{i-1}, t_i)$ (see [5] ,Th.1.11).

Given

$$Q = \{ (t,s) \in [a,b] \times [a,b] \; ; a \leq s \leq t \leq b \} \subset \mathbb{R}^2 ,$$

and a mapping $T: Q \longrightarrow L(X)$, and putting $T^t(s) = T_s(t) = T(t,s)$, we write $T \in G_0 \cdot SV^u(Q,L(X))$ if T satisfies:

(D^0) $T(t,t) = 0$,

(G) for every $s \in [a,b]$,and all $x \in X$, we have $T_s \cdot x \in G([a,b],X)$, (where we define $T_s \cdot x(t) = T(t,s)x$), and

(SV^u) $SV^u[T] = \sup_{a \leq t \leq b} [K^t] < \infty$.

If instead of (D^0), T satisfies:

(D^I) $T(t,t) = I_X$,

we write $T \in G_I \cdot SV^u(Q,L(X))$.

We can state now, the

<u>Prop.1</u> (Th.3.4. in [5]) Suppose that the kernel of the linear Volterra-Stieltjes integral equation

$$(K) \qquad x(t) + \int_a^t \cdot d_s K(t,s) x(s) = u(t) \qquad (a \leq t \leq b),$$

is such that $K \in G_0 \cdot SV^u(Q,L(X))$, then the following properties are equivalent:

(1) for every $u \in G([a,b],X)$, (K) has one and only one solution x_u, and the operator $u \longmapsto x_u$ is causal,

(2) there exists one and only one mapping $R \in G_I.SV^u(Q,L(X))$ such that all solutions x_u can be represented as

$$x_u(t) = u(t) - \int_a^t .d_s R(t,s)u(s) \qquad (a \leqslant t \leqslant b),$$

(3) there exists an $R \in G_I.SV^u(Q,L(X))$ that satisfies

$$(R_*) \qquad R(t,s)x - K(t,s)x - x + \int_s^t .d_r R(t,r).K(r,s)x = 0$$

for all $x \in X$, and $a \leqslant s \leqslant t \leqslant b$.

Notice that in [7] (Th.3.1. in [6]), Hönig gave a sufficient condition for the existence of such an R in the proposition above, that encompasses all the known existence theorems for the resolvent of the integral equation (K).

In the following, we assume that (K) satisfies the properties (1)-(3) above.

Finally, for an axhaustive list of historical remarks for Volterra-Stieltjes integral equations, see the introduction of [5] .

3. Control process

In this part, we will show some relations between concepts of controllability for the process (K).

Fixed an $c \in (a,b]$, we have:

Def.1 The process (K) is (exact) controllable in time c (and we denote this property by (C)) if for every $x,y \in X$, there exists an element $u \in G([a,b],X)$, such that the solution x_u satisfies $x_u(a) = x$ and $x_u(c) = y$.

Def.2 The process (K) is controllable into zero (from zero) in time c if for every $x \in X$, there exists an $u \in G([a,b],X)$ (an $v \in G([a,b],X)$) such that $x_u(a) = x$ and $x_u(c) = 0$ ($x_u(a) = 0$ and $x_u(c) = x$)

We denote by $(C)_0$ and $(C)^0$ the property of (K) to be controllable into zero or controllable from zero, in time c, respectively.

By comparing the properties (C), $(C)_0$ and $(C)^0$ we have immediately

<u>Prop.2</u> Given the process (K), the following properties are equivalent: (i)-(C) ,(ii)-$(C)_o$ + $(C)^o$.

<u>Proof:</u> From (ii), we have that for all $x,y \in X$, there are u and v such that $x_u(a) = x$, $x_v(a) = x_u(c) = 0$ and $x_v(c) = y$. So, u + v tranfers x to y.

In the finite dimensional context, we have for O.D.E. that (C), $(C)_o$ and $(C)^o$ are equivalent properties. But in the linear Volterra-Stieltjes integral equations context, is not always possible to say if the operator R(t,s) is invertible nor if invertible, $R(t,s)^{-1}=R(s,t)$ is true, (see example in [3] ,p.99).

Let us define,

<u>Def.3</u> Given the process (K), $x \in X$ is a <u>returning point</u> at time c (for (K)), if there exists an $u \in G([a,b],X)$ such that $x_u(a) = x_u(c) = x$.

If all $x \in X$ are returning points at time c for (K), we denote this property by (R.P).

Then it is possible to state,

<u>Prop.3</u> Given the process (K), the following properties are equivalent: (i)-(C) , (ii)-$(C)_o$ + (R.P) and (iii)-$(C)^o$ + (R.P) .

<u>Proof:</u> We prove only -(ii) implies (i)-. The proof of -(iii) implies (i)-is analogous.

For all $x,y \in X$, we have from $(C)_o$ that there exists an element $u \in G([a,b],X)$, such that

(I) $\qquad x_u(a) = x-y$ and $x_u(c) = 0$,

and on other hand, from (R.P), that there exists an v such that

(II) $\qquad x_v(a) = x_v(c) = y$.

Finally, by (I) and (II) we have, $x_{u+v}(a) = x$ and $x_{u+v}(c) = y$.

It is possible to give conditions for the function R, in order to have (R.P).

<u>Prop.4</u> Given the process (K), if

(*) $$R(t,s)oR(s,r) = R(t,r) \qquad t,s,r \in [a,c],$$

then (K) satisfies (R.P).

Proof: For all $x\in X$, define $u\in G([a,b],X)$,

$$u_x(t) = (R(t,a) - K(t,a))x .$$

By the property (R_*) we have

$$u_x(c) - \int_a^c .d_s R(c,s).K(s,a)x = x .$$

According with (*), we have $R(c,s)oR(s,a) = R(c,a)$. Then,

$$u_x(c) - \int_a^c .d_s R(c,s)u_x(s) = x .$$

As an immediate corollary, we have,

Prop.5 For a linear Stieltjes integral control process (L), the properties (C), $(C)_o$ and $(C)^o$ are eqivalent.

Proof: The mapping R generated by A in equation (L), satisfies the identity (*) above, (see [3] ,p.119).

We point out that the notion of returning point was exploited in the context of O.D.E. in [18] ,and in another stronger sense in [19] .

Finally we state another proposition according the spirit of the previous results.

Let us define the subset of X,

$$W(c) = \{x_u(c); u\in G([a,b],X): u(a) = 0: \sup_{a\leq t\leq b} \|u(t)\| < 1\}.$$

Then we have,

Prop.6 Given (K), the following properties are equivalent: (i)-$(C)^o$ and (ii)- $0\in \text{int } W(c)$.

Proof: The statment is an immediate consequence of the Open Mapping Theorem.

Acknowledgment

Work supported by Fundação de Amparo à Pesquisa do Estado de São Paulo (FAPESP)-Proc.82/1181-9

References

[1] -D.B.Hinton, A Stieltjes-Volterra integral equations theory,Cana-J.Math.,18(1966),314-331.

[2] -C.W.Bitzer, Stieltjes-Volterra integral equations, Illinois J. of Math.,14(1970),434-451.

[3] -C.S.Hönig, Volterra Stieltjes-integral equations, Mathematics Studies,16,North-Holland Publishing Comp.,Amsterdam,1975.

[4] -C.S.Hönig, Volterra-Stieltjes integral equations with linear constraints and discontinuous solutions,Bull.Amer.Math.Soc., 81(1975), 593-598.

[5] -C.S.Hönig, Volterra-Stieltjes-Integral equations,Springer Lecture Notes in Mathematics,vol.799(1980),173-216.

[6] -C.S.Hönig, The resolvent of linear Volterra Stieltjes-integral equations,preprint,1981.

[7] -C.S.Hönig, Equations intégrales généralisées et applications, Publications Mathématiques d'Orsay,(1981-82).To appear.

[8] -S.E.Arbex, Equações integrais de Volterra-Stieltjes com núcleos descontínuos,Doctor Thesis,Instituto de Matemática e Estatística -Univ. de São Paulo,1976.

[9] -J.B.F.Gomes, O índice de equações integrais lineares de Volterra Stieltjes,Doctor Thesis,Institutode Matemática e Estatística -Univ.de São Paulo,1980.

[10] -M.I.de Souza, Equações diferencio-integrais do tipo Riemann - Stieltjes em espaços de Banach com soluções descontínuas, Master Thesis,Instituto de Matemática e Estatística-Univ. de São Paulo,1974.

[11] -C.S.Cardassi,Dependência diferenciável das soluções de equações integro-diferenciais em espaços de Banach,Master Thesis,Instituto de Matematica e Estatistica-Univ. de São Paulo,1975.

[12] -L.Fischman, Equações lineares de Volterra-Stieltjes,com condições de contorno (adjunta),Master Thesis,Instituto de Matemática e Estatística-Univ.de São Paulo.In preparation.

[13] -St.Schwabik, On Volterra-Stieltjes integral equations,Čas.Pěst. Mat.,99(1974),225-278.

[14] -St.Schwabik, Note on Volterra-Stieltjes integral equations,Čas. Pěst.Mat.,102(1977),275-279.

[15] -St.Schwabik,On an integral operator in the space of functions with bounded variation, Čas.Pěst.Mat.,97(1972),297-330.

[16] -St.Schwabik,M.Tvrdy and O.Vejdova, Differential and integral equations.Academia Praha,Praha,1979.

[17] -R.K.Miller, Nonlinear Volterra integral equations,Mathematics Lecture Note Series,W.A.Benjamin,Inc.,1971.

[18] -R.Conti, Return sets of a linear control process,JOTA.To appear.

[19] -A.Bacciotti, Autoacessibilité par familles symétriques de champs de vecteurs,Ricerche di Automatica,7(1976),189-197.

EXCHANGE OF STABILITY ALONG A BRANCH OF PERIODIC SOLUTIONS OF A SINGLE SPECIE MODEL

Martino Bardi

(Istituto di Analisi e Meccanica, Università di Padova)

1. Introduction

In this paper I study some mathematical problems which arise when a well-known biological phenomenon, the Allee-Robertson effect, is taken into account in the differential equations describing the growth of single populations. These problems are the existence of turning points in the branches of steady or periodic solutions and the exchange of stability of solutions at such points; to my knowledge they have not been studied until now in the context of biomathematics, but are well known in other fields as fluid dynamics.

A brief survey of the biological literature about the Allee-Robertson phenomenon, some recent experimental data and models are given in [2]. Clark [4] describes this phenomenon, naming it depensation, and points out its important consequences in fishing problems.

I will say that a single specie model exhibits depensation or Allee effect if the relative rate of increase of the population (i.e. N'/N, where $N(t)$ is the density of the population at time t) is a decreasing function of the density when the density is sufficiently high, but is increasing when the density is low. On the contrary in logistic-type models the relative rate of increase is always a decreasing function of the density.

The new features of depensation models can be pointed out by considering the simple ordinary autonomous equation

$$N' = N(r-g(N)) \quad , \tag{1.1}$$

where g is the crowding function which is supposed to have a minimum $M>0$, which is the optimal density, and to be increasing for $N>M$. Since

I will discuss the asymptotic behaviour of solutions using r as a parameter, I remark that this is equivalent to discussing the equilibria of an exploited population which obeys the equation

$$N' = N(r_0 - g(N)) - EN \quad , \tag{1.2}$$

upon varying the constant harvesting effort E. It is easily seen that the equilibrium solutions of (1.1) and (1.2) as a function of $r=r_0-E$ are given by the bifurcation diagram in fig.1 which is obtained by simply rotating the diagram of g (dashed and continuous lines represent respectively unstable and asymptotically stable equilibria). An easy

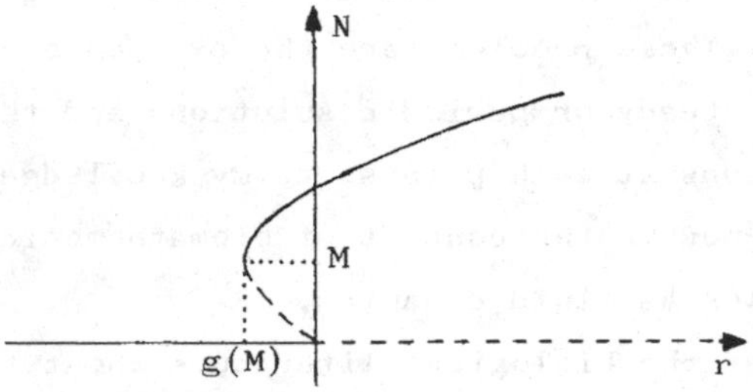

Figure 1

calculation shows that the largest equilibrium is asymptotically stable and that an exchange of stability occurs along the curve of equilibria at the turning point (g(M),M). If the harvesting effort satisfies $E < r_0 - g(M)$ and the initial population density is not too small, then the population tends to a positive equilibrium, when E is continuously increased towards the critical value $r_0-g(M)$, this equilibrium continuously tends to the optimal density M, but when E is increased beyond $r_0-g(M)$ the continuity is broken, there is not any positive equilibrium and the population becomes extinct. Summarizing: (A) equation (1.1) has essentially the same behaviour of a logistic-type equation for large r and large N; when r is small it exhibits the following new features: (B) there exist multiple equilibrium solutions, (C) there is a "structural instability" at a new critical value of the parameter r.

In two recent papers [2,3] I studied the extension of these facts to more general single specie models including a time-periodic fluctuation of the environment and, in [3], the presence of delayed effects. In such models equilibria are replaced by periodic solutions and the bifurcation parameter is the average of r(t) or the average of the periodic harvesting effort. The main results of the cited papers are reported in §2. They concern mostly points (A) and (B), but the existence of a new critical point (in a suitable sense) for a quite general periodic functional differential equation with Allee effect is established too. In §3 I investigate in more detail the structure of the set of periodic solutions around such a critical point, giving conditions under which this set is a curve, the critical point is a turning point (according to an appropriate definition) and an exchange of stability occurs. For simplicity the results are proved in the ordinary case, but the method works for functional d.e. with bounded delay as well. In the delay case however, one has to assume several spectral properties which are automatically satisfied in the ordinary case.

The theory of stability along curves of equilibrium solutions for autonomous ordinary equations was developed by Rosenblat [14] and Joseph [11,9]. The extension to periodic evolution equations was treated by Joseph [10]. Results on the stability at turning points for evolution equations were obtained by Sattinger [15], Crandall and Rabinowitz [5], Joseph and Nield [12] and Weinberger [16].

2. Periodic solutions

In order to take into account periodic fluctuations of the environment, I consider the following equation which generalizes (1.1):

(2.1) $$N' = Nf(t,N) \quad , \quad f(t,N) = r(t) - g(t,N) \quad ,$$

where r is continuous and periodic of period $\omega > 0$, $g: \mathbb{R} \times \mathbb{R}_+ \to \mathbb{R}$ is continuous , ω-periodic in t and locally lipschitzian in N, and satisfies $g(t,0) \equiv 0$. I assume the following hypotheses:

(o1) there exists an optimal density, i.e. a continuous and ω-periodic function $M: \mathbb{R} \to \mathbb{R}_+$ such that $r(t) \le f(t,N) \le f(t,M(t))$ for each t if $0 \le N \le M(t)$, and f(t,N) is strictly decreasing in N for all t if $N > M(t)$;

(o2) for every t such that $f(t,M(t)) \ge 0$ there exists $K(t) \ge M(t)$ such that $f(t,K(t)) = 0$.

In [2] this model is studied using the Poincarè map. This approach needs a further assumption which provides a flow-invariant set. One of the two following hypotheses is assumed:

(H) $f(t,M(t))>0$ for all t and $M_{max}:=\max M(t)<\min K(t)$;

(I) M is differentiable and $M'(t)<M(t)f(t,M(t))$ for all t .

All the four hypotheses above have a biological meaning which is discussed in [2,§2].

Denote by $<\varphi>=<\varphi(t)>=\omega^{-1}\int_0^\omega \varphi(t)dt$ the average of the continuous function φ and define by S the set of positive solutions of (2.1), by S_H the set of the solutions such that $N(s)\geq M_{max}$ for some s, and by S_I the set of the solutions such that $N(s)\geq M(s)$ for some s.

THEOREM 2.2 [2]. If (o1-2) and (H) (resp. (I)) hold, then:

(a) there exists an ω-periodic solution N^* of (2.1) which is globally asymptotically stable in S_H (resp. S_I); if $r(t)>0$ for all t, then N^* is globally asymptotically stable in S ;

(b) if $M\not\equiv 0$ (resp. $M(t)>0$ for all t) and $<r><0$, then there exist in $S\setminus S_H$ (resp. $S\setminus S_I$) an unstable ω-periodic solution N_* of (2.1) .

For a comparison with logistic-type models see [7]. A question left open is the existence of "unexpected"periodic solutions, i.e. solutions which have not a corresponding equilibrium solution in the autonomous time-averaged equation, as was observed in the Volterra competition system [8]. Such solutions do not exist in logistic-type models, but here they might appear in $S\setminus S_H$ (resp. $S\setminus S_I$).

A more general single specie model which takes into account both environmental oscillations and delayed effects is the following functional differential equation with unbounded delay:

$$N'(t)=N(t)(r(t)-a(t)N(t)-H(t,N_t)) \quad , \tag{2.3}$$

where $N_t(s)=N(t+s)$ for $s\in(-\infty,0]$, $r,a\in B$, B is the Banach space of continuous ω-periodic functions endowed with the sup-norm, $a>0$, H: $\mathbb{R}\times E\rightarrow\mathbb{R}$ satisfies $H(t,\varphi)=H(t+\omega,\varphi)$ for every t and φ , E is the Banach space of continuous and bounded functions $(-\infty,0]\rightarrow\mathbb{R}$ with the sup-norm. I assume that: (f1) for any function $x:(-\infty,T]\rightarrow\mathbb{R}$ continuous and bounded $t\mapsto H(t,x_t)$ is continuous, (f2) H is locally lipschitzian in the second argument, (f3) there exists $b:\mathbb{R}_+\rightarrow\mathbb{R}_+$ continuous and nondecreasing such that $b(0)=0$ and $|H(t,\varphi)|\leq b(\|\varphi\|)$ for all t and φ , (f4) there exist $m,p,q \in B$, $m,p,q \geq 0$, $p < a$ for all t, such that $H(t,\varphi) \geq -p(t)\varphi(0)-q(t)$ if $\varphi(0) \geq m(t)$. Then there exists a unique global solution of the Cauchy problem for (2.3) which is positive if the initial datum $\varphi \in E$ satisfies $\varphi(0) > 0$ [3,§2]. The two following very weak hypotheses define the presence of Allee effect in (2.3) (compare with (o1-2)) : (f5) the derivative at 0 of the map $\varphi \mapsto H(t,\varphi)$, denoted by $DH(t,0)$, satisfies the inequality:

$$< aN + DH(t,0)N_t > < 0 \qquad \text{for } N \in B \text{ , } N > 0 \text{ ;}$$

(f6) there exists $c \in \mathbb{R}$ such that

$$< aN + H(t,N_t) > \geq c \qquad \text{for } N \in B \text{ , } N \geq 0 \text{ .}$$

THEOREM 2.4 [3]. Assume (f1-6) and suppose H twice differentiable in the second argument around 0 and the second partial derivative bounded uniformly in t and uniformly continuous. Then:

(a) there exists $\lambda_0 \geq c$ such that, if $<r> \geq \lambda_0$, then there is a positive ω-periodic solution of (2.3) ;

(b) there exists $r_0 < 0$ such that, if $r_0 < <r> < 0$, then there are two distinct positive ω-periodic solutions of (2.3) ;

(c) there exist $N^{\circ} \in B$, which is a positive solution of (2.3) when $<r> = \lambda_0$, and a neighbourhood V of N° in B , such that if $<r>$ is sufficiently close to λ_0 and $<r> < \lambda_0$, then there are no solutions of (2.3) in V, but in every neighbourhood U of N° there is a solution of (2.3) when $<r> > \lambda_0$ and $<r>$ is sufficiently close to λ_0 .

For a comparison with logistic-type models see [6,1]. Point (c) is a structural instability result and says that (λ_0, N°) is a bifurcation point in a generalized sense. It completes also the study of equation (2.1) which satisfies the hypotheses of thm. 2.4 if (o1-2) hold and $g(t,N)=a(t)N+h(t,N)$ with h twice continuously differentiable with respect to N around 0 . I can choose the constant c in (f6) as $c= <g(t,M(t))>$. Notice that in the autonomous case (1.1) I have $c=\lambda_0$, but it can be easily proved [3,§4] that in the periodic case $\lambda_0 > c$ except in the unrealistic case that the optimal density M(t) is a solution of (2.1) for $<r> = c$. Hence the critical value λ_0 shifts to the right as the parameters of (1.1) are periodically perturbed. This phenomenon looks quite interesting for the harvesting problem described in the introduction.

I explain briefly the idea of the proof of thm. 2.4 because I will follow the same approach in §3. Set $r(t) = \mu + \theta(t)$, $\mu = \langle r \rangle + 1$, $\langle \theta \rangle = -1$. For each $b \in B$ there exists a unique periodic solution of the linear equation $x' = \theta x + b$, this solution is

$$Lb(t) := \int_0^{\omega} k(t,s) b(s) ds \quad , \tag{2.5}$$

where

$$k(t,s) = \begin{cases} \exp(\int_s^t \theta(\tau) d\tau) \cdot e/(e-1) & \text{for } s \leq t \\ \exp(\int_s^{t+\omega} \theta(\tau) d\tau) \cdot e/(e-1) & \text{for } t < s \; . \end{cases} \tag{2.6}$$

$N(t)$ is an ω-periodic solution of (2.3) if and only if it satisfies the equation in the space B :

$$N = \mu LN + LG(N) \quad , \tag{2.7}$$

where $G(N)(t) = -a(t)N^2(t) - N(t)H(t,N_t)$. (2.7) is a nonlinear eigenvalue problem and, since L is compact, I have available both local and global results on the existence of solutions. For the complete proof, see [3].

3. The turning point

Define $\mathcal{F}: \mathbb{R} \times B \to B$ by $\mathcal{F}(\mu,N) = \mu LN - N + LG(N)$. (2.7) is equivalent to

$$\mathcal{F}(\mu,N) = 0 \quad . \tag{3.1}$$

I have to solve this equation around (μ_0, N°) , $\mu_0 = \lambda_0 + 1$, a point where $D_2\mathcal{F}(\mu_0, N^\circ)$ has not a bounded inverse ($D_i\mathcal{F}(\mu, N)$ indicates the partial derivative with respect to the i-th variable). By a theorem of Crandall and Rabinowitz [5,thm.3.2] I have that if (i) $\mathcal{F}$ is continuously differentiable, (ii) $\mathrm{Ker}(D_2\mathcal{F}(\mu_0,N^\circ)) = \mathrm{span}\{P\}$ is one-dimensional, (iii) $\mathrm{codim}\, R(D_2\mathcal{F}(\mu_0,N^\circ)) = 1$ and (iv) $D_1\mathcal{F}(\mu_0,N^\circ) \notin R(D_2\mathcal{F}(\mu_0,N^\circ))$, (Ker and R denote the kernel and the range respectively), then there exist two continuously differentiable functions $\tau : [-1,1] \to \mathbb{R}$, $z : [-1,1] \to Z$, where Z is a complement of $\mathrm{span}\{P\}$ in B, such that $\tau(0) = \tau'(0) = 0$, $z(0) = z'(0) = 0$ and the curve $(\mu_0 + \tau(\xi), N^\circ + \xi P + z(\xi))$ is the set of the solutions of (3.1) in a neighbourhood of (μ_0, N°). In fact the operator $\mathcal{G}: \mathbb{R} \times \mathbb{R} \times Z \to B$, $\mathcal{G}(\xi,\tau,z) = \mathcal{F}(\mu_0+\tau, N^\circ + \xi P + z)$ satisfies the conditions of the implicit function theorem at (0,0,0), because the derivative at (0,0,0) of $(\tau,z) \mapsto \mathcal{G}(0,\tau,z)$ is the operator

$$\Phi : (\tau,z) \mapsto \tau D_1\mathcal{F}(\mu_0,N^\circ) + D_2\mathcal{F}(\mu_0,N^\circ)z \quad ,$$

which is a linear homeomorphism. If $\mathcal{F}$ is twice continuously differenti able, the following formula for the second derivatives of $\tau(\xi)$ and $z(\xi)$ can be computed by twice differentiating at 0 the equality $\mathcal{G}(\xi,\tau(\xi),z(\xi))=0$:

$$D_2D_2\mathcal{F}(\mu_o,N^\circ)(P,P) = -\Phi(\tau''(0),z''(o)) \quad . \tag{3.2}$$

The derivatives of $\mathcal{F}$ are easily computed:

$$\begin{aligned} D_1\mathcal{F}(\mu_o,N^\circ) &= LN^\circ \quad , \quad D_2\mathcal{F}(\mu_o,N^\circ)= \mu_o L - I + LDG(N^\circ) \quad , \\ D_2D_2\mathcal{F}(\mu_o,N^\circ) &= LD^2G(N^\circ) \quad . \end{aligned} \tag{3.3}$$

Since $\mu_o L + LDG(N^\circ)$ is a compact operator I have that dim $\mathrm{Ker}(D_2\mathcal{F}(\mu_o,N^\circ))$ and codim $R(D_2\mathcal{F}(\mu_o,N^\circ))$ are finite. Moreover if I establish condition (ii) and prove that $(D_2\mathcal{F}(\mu_o,N^\circ))^{-1}(P) \in \mathrm{span}\{P\}$, then also (iii) is verified and I can take $Z= R(D_2\mathcal{F}(\mu_o,N^\circ))$.

If (2.3) is a functional d.e. with bounded delay, one can give now a set of assumptions which ensure (i)-(iv). For simplicity I limit myself to the ordinary case.

<u>THEOREM 3.4</u> . <u>Assume that equation</u> (2.3) <u>is ordinary, i.e.</u> $H(t,N_t)= h(t,N(t))$, <u>with</u> h <u>continuously differentiable in</u> N. <u>Then the solutions of</u> (3.1) <u>in a neighbourhood of</u> (μ_o,N°) <u>in</u> $\mathbb{R}\times B$ <u>form a differentiable curve.</u>

<u>Proof</u>. Define

$$g(t,N(t))= -a(t)N^2(t) - N(t)h(t,N(t)) \tag{3.5}$$

and $j(t)= g'_N(t,N^\circ(t))$. (i) is satisfied and $DG(N^\circ)y= jy$. Since $\mu_o L-I+DG(N^\circ)$ has not a bounded inverse, the linearization of (2.3) around N°, that is

$$y' = (\mu_o+\theta+ j)y \quad , \tag{3.6}$$

has a one-dimensional vector space of periodic solutions, spanned by $P(t)= \exp(\int_0^t \psi\, ds)$, where $\psi(s)= \mu_o+ \theta(s) + j(s)$. (ii) is proved. Notice that $\langle\psi\rangle = 0$. Let $z \in R(D_2\mathcal{F}(\mu_o,N^\circ))$, i.e. $z= \mu_o Lx-x+LDG(N^\circ)x$, and $y= z+x$. By the definition of L, y satisfies the equation $y'= \psi y-(\mu_o-j)z$ which has a periodic solution if and only if

$$u(z):=\langle z(t)(\mu_o+ j(t))\exp(-\int_0^t \psi\, ds)\rangle = 0 \quad .$$

Hence $R(D_2\mathcal{F}(\mu_o,N^\circ))$ is the kernel of the functional u, which proves (iii). Now I have to prove that $u(LN^\circ) \neq 0$. By (2.5) and Fubini's Theorem I have that $u(LN^\circ)= \int_0^\omega q(t)N^\circ(t)dt$, where

$$q(t)= \int_0^\omega k(v,t)(\mu_o+ j(v))\exp(-\int_0^v \psi\, ds)dv \quad .$$

By (2.6)

$$q(t)= (e-1)^{-1}\exp(\int_0^t -\theta ds) \int_0^t \exp(-\int_0^v (\psi-\theta)ds)(\mu_o+j(v))dv +$$

$$e\cdot(e-1)^{-1}\exp(\int_0^t -\theta ds) \int_t^\omega \exp(-\int_0^v (\psi-\theta)ds)(\mu_o+j(v))dv \quad ,$$

and, since $\mu_o+j=\psi-\theta$, integrating one finds

$$q(t)=(e-1)^{-1}\exp(\int_0^t -\theta ds)(1 - \exp(-\int_0^t (\mu_o+j)ds)) +$$

$$e\cdot(e-1)^{-1}\exp(\int_0^t -\theta ds)(\exp(-\int_0^t (\mu_o+j)ds) - e^{-1})= \exp(\int_0^t -\psi ds) \quad ,$$

and hence $u(LN^\circ)>0$, which proves (iv). Q.E.D.

<u>Remark 3.7</u>. The proof of thm. 3.4 implies that in the ordinary case the set of positive solutions of (3.1) which bifurcates from (1,0) is a one-dimensional $\mathscr{C}^1$ manifold in $\mathbb{R}\times B$ and it has not secundary bifurcations.

Let $\mathscr{E}(\mu,x)=0$ be an equation in a Banach space which gives the periodic solutions of a periodic differential equation depending on a parameter μ, and suppose that it has a curve Γ of solutions with a continuously differentiable parametrization $\xi\mapsto(\overline{\mu}(\xi),\overline{x}(\xi))$ with non-vanishing derivative. A <u>turning point</u> of Γ is a point at which $\overline{\mu}'(\xi)$ changes sign, it is a <u>simple turning point</u> if $\overline{\mu}''(\xi)\neq 0$, it is a <u>regular turning point</u> if $D_1\mathscr{E}(\overline{\mu}(\xi),\overline{x}(\xi))\neq 0$. These definitions generalize those given in [9,11,14] for branches of equilibria of autonomous ordinary differential equations, where $\mathscr{E}$ is simply the vector field of the equation. In the periodic case there are several possible choices of the Banach space and of the operator $\mathscr{E}$, for a survey see [13]. Notice that for $\mathscr{E}=\mathscr{F}$ a turning point with $N^\circ>0$ is automatically regular.

<u>THEOREM 3.8</u>. <u>Assume that eq.</u> (2.3) <u>is ordinary,</u> h <u>is twice continuously differentiable with respect to</u> N, <u>and</u>

(3.9) $\quad N^\circ(t)h''_{NN}(t,N^\circ(t))+ 2a(t)+ 2h'_N(t,N^\circ(t))\geq 0$ <u>and</u> $\not\equiv 0$.

<u>Then</u> (μ_o,N°) <u>is a simple turning point of the curve of solutions of</u> (3.1) <u>found in thm.</u> 3.4.

<u>Remark 3.10</u>. (3.9) is a concavity condition on the right hand side of (2.3). In the autonomous case, i.e. $h(t,N)=h(N)$, (3.9) reduces to $h''(N^\circ)>0$, which is exactly the condition for (μ_o,N°) to be a simple turning point according to Rosenblat's definition [14].

<u>Proof of (3.8)</u>. (3.2) and (3.3) imply that $LD^2G(N^\circ)(P,P)+\tau''(0)LN^\circ \in Z$, and thus, since I have proved in thm. 3.4 that $LN^\circ\notin Z$, $\tau''(0)\neq 0$ if and only if $LD^2G(N^\circ)(P,P)\notin Z$. It is easy to prove that

$D^2G(N^\circ)(P,P)(t) = g''_{NN}(t,N^\circ(t))P^2(t)$, and $g''_{NN}(t,N^\circ(t)) = -2a(t)-$ $2h'_N(t,N^\circ(t)) - N^\circ(t)h''_{NN}(t,N^\circ(t))$. (3.9) implies that $u(LD^2GN^\circ(P,P)) < 0$, where the functional u defined in thm. 3.4 is such that $Z = \mathrm{Ker}(u)$. This completes the proof. Q.E.D.

The last theorem proves the exchange of stability at a turning point:

THEOREM 3.11 . Let (2.3) be ordinary and $(\bar{\mu}(0),\bar{N}(0))$, $\bar{N}(0) > 0$, be a turning point of a curve $(\bar{\mu}(\xi),\bar{N}(\xi))$ of solutions of (3.1). Then the characteristic multiplier $\zeta(\xi)$ of the linearization around $\bar{N}(\xi)$ of (2.3) is such that $\zeta(0) = 1$ and $\zeta(\xi)-1$ changes sign at 0 . Under the hypotheses of thm. 3.8 there exist $\mu_+ > \mu_0$ and a neighbourhood U of N° such that, if $<r> \in (\mu_0,\mu_+)$, then there are exactly two periodic solutions of (2.3) in U , N^* and N_* , with $N^*(t) > N_*(t)$ for all t, N^* is asymptotically stable and N_* is unstable.

Proof. Let $\bar{P}(\xi)$ denote the derivative of $\xi \mapsto \bar{N}(\xi)$ and set $j_\xi(t) = g'_N(t,\bar{N}(\xi)(t))$, where g is defined by (3.5). Notice that $\bar{P}(\xi) \neq 0$ for every t for ξ close to 0 . Differentiating $\mathcal{F}(\bar{\mu}(\xi),\bar{N}(\xi)) = 0$ one obtains

$$\bar{\mu}'(\xi)L\bar{N}(\xi) + \bar{\mu}(\xi)L\bar{P}(\xi) - \bar{P}(\xi) + LDG(\bar{N}(\xi))\bar{P}(\xi) = 0 \quad ,$$

which means that $\bar{P}(\xi)$ is an ω-periodic solution of

$$(3.12) \qquad x' = (\theta + \bar{\mu}(\xi) + j_\xi)x + \bar{\mu}'(\xi)\bar{N}(\xi) \quad .$$

Notice that the homogeneus equation associated to (3.12) is the linearization of (2.3) around $\bar{N}(\xi)$ and hence $\zeta(\xi) = \exp(<\psi_\xi>)$, where $\psi_\xi(t) = \theta(t) + \bar{\mu}(\xi) + j_\xi(t)$. The variation of constants formula for (3.12) gives:

$$x(t) = x(0)\exp\left(\int_0^t \psi_\xi ds\right) + \int_0^t \bar{\mu}'(\xi)\bar{N}(\xi)(v)\exp\left(\int_v^t \psi_\xi ds\right)dv \quad .$$

Thus, taking for instance $\bar{P}(\xi)$ positive, the periodicity of $\bar{P}(\xi)$ implies that $\int_0^\omega \bar{\mu}'(\xi)\bar{N}(\xi)(v)\exp\left(\int_v^\omega \psi_\xi ds\right)dv < 0$ if and only if $\exp(<\psi_\xi>) > 1$ and hence $<\psi_\xi>$ changes sign as $\bar{\mu}'(\xi)$ changes sign. In the hypotheses of thm. 3.8 $\bar{\mu}''(0) = \tau''(0) > 0$ and $\bar{N}(\xi) = N^\circ + \xi P + z(\xi)$ where $P = \bar{P}(0)$. Since P is positive, for small $\xi > 0$ I have $\bar{\mu}'(\xi) > 0$, $\zeta(\xi) < 1$ and $\bar{N}(\xi) > N^\circ$, and for $\xi < 0$ all the inequalities are reversed. The conclusion is achieved because there exists ξ_+ such that $\tau'(\xi) \neq 0$ in both $(0,\xi_+)$ and $(-\xi_+,0)$ and hence $\tau(\xi)$ is invertible in these intervals. Q.E.D.

References

1. Badii,M., Schiaffino,A.: Asymptotic behaviour of positive solutions of periodic delay logistic equations. J.Math.Biol. 14, 95-100 (1982)
2. Bardi,M.: An equation of growth of a single specie with realistic dependence on crowding and seasonal factors. Preprint
3. Bardi,M.: A nonautonomous nonlinear functional differential equation arising in the theory of population dynamics. Preprint
4. Clark,C.W.: Mathematical bioeconomics: the optimal management of renewable resources. New York: Wiley 1976
5. Crandall,M.G., Rabinowitz,P.H.: Bifurcation, perturbation of simple eigenvalues, and linearized stability. Arch.Rat.Mech.Anal. 52, 161-180 (1973)
6. Cushing,J.M.: Stable positive periodic solutions of the time-dependent logistic equation under possible hereditary influences. J.Math. Anal.Appl. 60, 747-754 (1977)
7. de Mottoni,P., Schiaffino,A.: Bifurcation of periodic solutions for some systems with periodic coefficients. In: Nonlinear differential equations (P.de Mottoni and L.Salvadori, eds.), pp.327-338. New York: Academic Press 1981
8. de Mottoni,P., Schiaffino,A.: Competition systems with periodic coefficients: a geometric approach. J.Math.Biol. 11, 319-335 (1981)
9. Iooss,G., Joseph,D.D.: Elementary stability and bifurcation theory. New York: Springer Verlag 1980
10. Joseph,D.D.: Factorization theorems, stability, and repeated bifurcation. Arch.Rat.Mech.Anal. 66, 99-118 (1977)
11. Joseph,D.D.: Factorization theorems and repeated branching of solutions at a simple eigenvalue. Ann.New York Acad.Sci. 316, 150-167 (1979)
12. Joseph,D.D., Nield,D.A.: Stability of bifurcating time-periodic and steady solutions of arbitrary amplitude. Arch.Rat.Mech.Anal. 58, 369-380 (1975)
13. Krasnosel'skii,M.A.: The theory of periodic solutions of non-autonomous differential equations. Russian Math.Surveys 21, 53-74 (1966)
14. Rosenblat,S.: Global aspects of bifurcation and stability. Arch. Rat.Mech.Anal. 66, 119-134 (1977)
15. Sattinger,D.H.: Stability of solutions of nonlinear equations. J. Math.Anal.Appl. 39, 1-12 (1972)
16. Weinberger,H.F.: The stability of solutions bifurcating from steady or periodic solutions. Univ.Florida Internat.Symp.Dynamical Systems. New York: Academic Press 1977

ON ASYMPTOTICALLY QUADRATIC HAMILTONIAN SYSTEMS.

V. Benci A. Capozzi D. Fortunato

Let $H\in C^1(\mathbb{R}^{2n},\mathbb{R})$ and consider the Hamiltonian system of 2n ordinary differential equations

$$(1) \qquad \dot{p} = -H_q(p,q) \ , \ \dot{q} = H_p(p,q) \ ,$$

where p and q are n-tuples, $\cdot$ denotes $\frac{d}{dt}$ and $H_q = \text{grad}_q H$, $H_p = \text{grad}_p H$. This system can be represented more concisely as

$$(2) \qquad -J\dot{z} = H_z(z)$$

where $z = (p,q)$ and $J = \begin{pmatrix} 0 & -I \\ I & 0 \end{pmatrix}$, I being the identity matrix in $\mathbb{R}^n$.

There are many types of questions, both local and global, in the study of periodic solution of such systems (cf. [5] and its references).

Here we are concerned about the existence of periodic solutions of (2) when the period $T = 2\pi\omega$ is prescribed. Making the change of variable $t \longrightarrow \frac{1}{\omega} t$, (2) becomes

$$(3) \qquad -J\dot{z} = \omega H_z(z)$$

and we seek 2π-periodic solutions of (3), which, of course, correspond to the $2\pi\omega$-periodic solutions of (2). These solutions are the critical points of the functional of the action

(4) $$f(z) = \int_0^{2\pi} (\frac{1}{2}(-J\,\dot{z},z)_{\mathbb{R}^{2n}} - \omega H(z))\,dt .$$

If there exist positive constants k_1, k_2, α such that

(5) $$|H_z(z)| \leq k_1 + k_2|z|^{\alpha}$$

it is easy to see that f is continuously Fréchet-differentiable on the space $W^{1/2}(S^1,\mathbb{R}^{2n})$ of 2n-tuples of 2π-periodic functions, which possess square integrable "derivative of order 1/2".(*)
The spectrum of the linear operator $z \longrightarrow -J\,\dot{z}$ (with periodic conditions) consists of infinitely many positive and negative eigenvalues. For this reason the functional (4) is indefinite in a strong sense, i.e. it is not bounded from above or from below, even modulo weakly continuous perturbations (cf. [2], [4]).

In this paper we are concerned with the case in which H(z) is asymptotically quadratic, i.e. there exists a linear operator $H_{zz}(\infty):\mathbb{R}^{2n} \longrightarrow \mathbb{R}^{2n}$ such that

---

(*)
We set $L^t = L^t(S^1,\mathbb{R}^{2n})$, $t \geq 1$, and for every $\alpha \in \mathbb{R}$ we shall set

$$W^s = \{u \in L^2 \mid \sum_{\substack{j \in Z \\ k=1,\dots,2n}} j^{2s}|u_{jk}|^2 < +\infty\},$$

where u_{jk} are the Fourier components with respect to the basis $\psi_{jk} = e^{jtJ}\Phi_k$ ($j \in Z$; Φ_k $(k=1,\dots,2n)$ is the standard basis in $\mathbb{R}^{2n}$). W^s is an Hilbert space with the inner product $(u|v)_{W^s} = \sum_{j,k} (1+|j|)^{2s} u_{jk} v_{jk}$.

(6) $$H_z(z) = H_{zz}(\infty)z + 0(z)$$

where $0(z) \longrightarrow 0$ as $|z| \longrightarrow \infty$. Moreover we suppose that

(7) $H(z)$ is twice differentiable for $z=0$ and $H(0)=0$

The aim of this paper is to give a lower bound to the number of $2\pi\omega$-periodic solutions of (3) by the comparison between the operator $H_{zz}(0)$ and $H_{zz}(\infty)$. We define as in [2] an even integer number $\theta(\omega H_{zz}(0), \omega H_{zz}(\infty))$, which will provide such bound. Given two Hermitian operators $A,B:\mathbb{C}^{2n} \longrightarrow \mathbb{C}^{2n}$, we set

$$N(A) = \{\text{number of negative eigenvalues of } A\}$$

$$\overline{N}(A) = \{\text{number of nonpositive eigenvalues of } A\} .$$

and

$$\theta(A,B) = \sum_{k\in Z} N(ikJ+A)-\overline{N}(ikJ+B) .$$

Observe that $\theta(A,B)$ is a finite number. In fact for k big enough $N(ikJ+A) = \overline{N}(ikJ+B) = n$. Let $\sigma(A)$ denote the spectrum of an Hermitian matrix A. If

(8) $$\sigma(i\omega JH_{zz}(\infty)) \cap Z = \emptyset,$$

(9) $$\sigma(i\omega JH_{zz}(0)) \cap Z = \emptyset,$$

then $\theta(\omega H_{zz}(\infty), \omega H_{zz}(0)) = -\theta(\omega H_{zz}(0), \omega H_{zz}(\infty))$.

The following theorem holds:

THEOREM 1. *Suppose that* H *satisfies* (6), (7), (8) *and*

(10) $H_{zz}(\infty)$ *is positive definite*

(11) $H(z) \geq 0$ *for every* $z \in \mathbb{R}^{2n}$ *s.t.* $H_z(z)=0$,

then (3) *has at least* $\frac{1}{2}\,\theta(\omega H_{zz}(\infty),\ \omega H_{zz}(0))$ *non constant* $2\pi\omega$-*periodic solutions whenever* $\theta(\omega H_{zz}(\infty),\ \omega H_{zz}(0))>0$.

If the assumptions (10) *and* (11) *are replaced by the following ones*

(10)' $H_{zz}(0)$ *is positive definite*

(11)' $H(z) \leq 0$ *for every* $z \in \mathbb{R}^{2n}$ *s.t.* $H_z(z)=0$,

then (3) *has at least* $\frac{1}{2}\,\theta(\omega H_{zz}(0),\ \omega H_{zz}(\infty))$ *non constant* $2\pi\omega$-*periodic solutions whenever* $\theta(\omega H_{zz}(0),\ \omega H_{zz}(\infty))>0$.

Remark 1. The second part of theorem (1) completes Theorem 5.1 in [2]. In spite of (10)' and (11)' are the dual of (10) and (11), the proof of this part needs more technicality.

Remark 2. The assumption (8) is a non-resonance condition. If (8) does not hold the same conclusion of theorem (1) holds if we replace (8) by the following assumptions

(12) $H(z) - \frac{1}{2}(H_z(z)\,|\,z)_{R^{2n}} \geq c_1|z|^{\alpha} - c_2$

(13) $|H_z(z)| \leq c_3 + c_4|z|^{\beta}$

where $\alpha>\beta>0$.

From theorem (1) the following corollary easily follows:

COROLLARY 1. If $H(z)$ satisfies (6), (7), (8), (9), (10), (10)' and

(14) $$H_z(z)\neq 0 \quad \text{for every} \quad z\in\mathbb{R}^{2n}, \quad z\neq 0$$

then the system (3) has at least

$$\frac{1}{2}|\theta(\omega H_{zz}(\infty),\ \omega H_{zz}(0))|$$

$2\pi\omega$-periodic solutions.

Amman and Zehnder in [1] have obtained a similar result using, instead of (10) and (10)', the stronger assumption of uniform convexity of $H(z)$.

In order to prove the Theorem 1 we need the following abstract critical point theorem (cf. [3]):

THEOREM 2. Let X be a real Hilbert space, on which a unitary representation T_g of the group S^1 acts. Let $f\in C^1(X,\mathbb{R})$ be a functional on X satisfying the following assumptions:

f_1) $f(u)=\frac{1}{2}(Lu|u)_X-\psi(u)$, where $(\cdot|\cdot)_X$ is the inner product in X, L is a bounded selfadjoint operator and $\psi\in C^1(X,\mathbb{R})$, $\psi(0)=0$, is a functional whose Fréchet derivative is compact. We suppose that both L and ψ' are equivariant with respect to the action of the group S^1.

f_2) 0 does not belong to the essential spectrum of L.

f_3) Every sequence $\{u_n\}\subset X$, for which $f(u_n)\to c\in]0,+\infty[$ and $\|f'(u_n)\|\cdot\|u_n\|\to 0$, possesses a bounded subsequence.

f_4) There are two closed subspaces S^1-invariant $V,W\subset X$ and $R,\delta>0$ s.t.

a) W is L-invariant, i.e. $LW=W$

b) Fix $(S^1)^* \subset V$ or Fix $(S^1) \subset W$

c) $f(u)<\delta$ for $u \in$ Fix (S^1)

d) f is bounded from above on W

e) $f(u) \geq \delta$ for $u \in V$ s.t. $\|u\|=R$

f) codim $(V+W)<+\infty$, dim $(V\cap W)<+\infty$

Under the above assumptions there exists at least

$$\frac{1}{2}\ (\dim(V\cap W)-\operatorname{codim}(V+W))$$

orbits** of critical points, with critical values greater or equal than δ.

Proof of Theorem 1.

We denote by L_∞ and L_o the self-adjoint operators $W^{1/2}$ defined as follows

$$L_\infty z = -J\dot{z}-\omega H_{zz}(\infty)z \qquad z\in W^{1/2}$$
$$L_o z = -J\dot{z}-\omega H_{zz}(0)z$$

It is easy to see that the spectrum of L_∞ and L_o consists of eigenvalues of finite multiplicity and $\sigma_e(L_\infty) = \sigma_e(L_o) = \{-1,+1\}$.
Let M_μ^o (resp. M_μ^∞) denote the eigenspace of L_o (resp. L_∞) corresponding to the eigenvalue μ. We set

---

*$\text{Fix}(S^1)=\{u\in X|T_g u=u$ for every $g\in S^1\}$

**If $u\in X$ the "orbit" of u is the set $\{T_g u: g\in S^1\}$.

$$W_o^+ = \overline{\bigoplus_{\mu>0} M_\mu^o}\ ,\quad W_o^- = \overline{\bigoplus_{\mu<0} M_\mu^o}\ ,\quad W_\infty^+ = \overline{\bigoplus_{\mu>0} M_\mu^\infty}\ ,\quad W_\infty^- = \overline{\bigoplus_{\mu<0} M_\mu^\infty}$$

where the closures are taken in $W^{1/2}$.

We initially suppose that, besides (6), (7), (8), the Hamiltonian H satisfies (10) and (11).

We can write the action functional as follows:

$$f(z) = +\frac{1}{2}(L_\infty z|z) - \omega\int_0^{2\pi}(H(z) - \frac{1}{2}(H_{zz}(\infty)z|z)_{\mathbb{R}^{2n}})dt$$

We shall show that f satisfies the assumptions $f_1), \ldots, f_4)$ of Theorem 2. with:

$$L = L_\infty\ ,\quad \psi(z) = \omega\int_0^{2\pi}(H(z) - \frac{1}{2}(H_{zz}(\infty)z|z)_{\mathbb{R}^{2n}})dt$$

$$V = W_o^+ \quad \text{and} \quad W = W_\infty^-$$

Then the conclusion of Theorem 1. will easily follow from Theorem 1. and from Lemma 5.6 of [2].

It is easy to see that $f_1)$, $f_2)$ are satisfied. Moreover, by virtue of the non resonance assumption (8), it can be shown that also $f_3)$ is satisfied (cf. the proof of Theorem 5.1 and remark 4.10 in [2]).

Let us now prove that also $f_4)$ is satisfied.

$(f_4.a)$ is obviously satisfied, moreover, because $L_\infty - L_o$ is compact, also $(f_4.f)$ holds.

Because $H_{zz}(\infty)$ is positive definite, we have

$$\mathrm{Fix}\,(S^1) = \mathbb{R}^{2n} \subset W_\infty^- \quad .$$

Then also $(f_4.b)$ is satisfied.

Let $z \in W_o^+$ then,

$$f(z) = f(o)+<f'(o),z>+ \frac{f''(o)}{2}[z,z]+0(\|z\|^2)=\frac{1}{2}(L_o z|z)+0(\|z\|^2)$$

$$\geq \frac{\mu_o}{2}\|z\|^2+0(\|z\|^2) \quad \text{as} \quad \|z\|\longrightarrow o$$

where $\mu_o=\min\{\mu\in\sigma(L_o)|\mu>0\}$.

So also assumption $(f_4.e)$ holds. Moreover, by (11), assumption $(f_4.c)$ holds.

Let us finally verify that $(f_4.d)$ is satisfied.

Let $z\in W_\infty^-$ then

(15) $$f(z)\leq\mu_1\|z\|^2-\omega\int_o^{2\pi}(H(z)-\frac{1}{2}(H_{zz}(\infty)z|z)_{R^{2n}})dt$$

where $\mu_1=\max\{\mu\in\sigma(L_\infty)|\mu<0\}$.

Now $\forall z\in\mathbb{R}^{2n}$ $H_z(z)-H_{zz}(\infty)z=g(z)$ where

(16) $$|g(z)|\longrightarrow o \text{ as } \|z\|\longrightarrow\infty .$$

Then

$$\int_o^1(H_z(sz)-H_{zz}(\infty)(sz)|z)_{R^{2n}}ds=\int_o^1(g(sz)|z)_{\mathbb{R}^{2n}}ds$$

So

$$H(z)-\frac{1}{2}(H_{zz}(\infty)z|z)_{R^{2n}}=\int_o^1(g(sz)|z)_{R^{2n}}ds$$

From which we deduce that

(17) $$\forall z\in\mathbb{R}^{2n}, |H(z)-\frac{1}{2}(H_{zz}(\infty)z|z)_{R^{2n}}|\leq|z|\int_o^1|g(sz)|ds .$$

If $\varepsilon>0$ by (16) there exists $M>0$ s.t.

(18) $$|g(z)|\leq\varepsilon|z| \quad \text{for} \quad |z|\geq M$$

Let $|z|\geq M$ and set

$$A_1(z) = \{t\in[0,1] \,|\, |tz|<M\}$$
$$A_2(z) = \{t\in[0,1] \,|\, |tz|\geq M\}$$

Then, by (18), we have

$$\int_0^1 |g(sz)|ds = \int_{A_1(z)} |g(sz)|ds+\int_{A_2(z)} |g(sz)|ds\leq c_1+\frac{\varepsilon}{2}|z| \tag{19}$$

where $c_1 = \sup\{|g(z)| \,|\, |z|<M\}$.

Then, by (17) and (19) we have

$$\forall z\in\mathbb{R}^{2n}, |z|\geq M, |H(z)-\frac{1}{2}(H_{zz}(\infty)z|z)_{R^{2n}}|\leq c_1|z|+\frac{\varepsilon}{2}|z|^2 \tag{20}$$

Then, by (15) and (20) we easily deduce that

$$\forall z\in W_\infty^-, f(z)\leq \mu_1\|z\|^2+\omega(\|z\|_{L^1}+\frac{\varepsilon}{2}\|z\|^2_{L^2})+c_2 \tag{21}$$

where c_2 is a positive number depending on ε.

So if we choose ε sufficiently small, we deduce by (21) that f is bounded from above on W_∞^- .

Let us now suppose that the Hamiltonian H satisfies assumptions (10)', (11)' instead of (10) and (11).

In this case

$$\mathbb{R}^{2n} = \text{Fix }(S^1)\subset W_0^-$$

Moreover, by following analogous arguments as before, it can be proved that the opposite of the action, i.e. the functional

$$\phi(z) = -f(z) = \frac{1}{2}\int_0^{2\pi} [(J\dot{z}|z)_{R^{2n}}+\omega H(z)]dt$$

satisfies the assumptions of Theorem 2., with

$$L = -L_\infty \quad , \quad \psi(z) = \omega\int_0^{2\pi} (-H(z)+\frac{1}{2}(H_{zz}(\infty)z|z)_{R^{2n}})dt$$

$$V = W_0^- \quad \text{and} \quad W = W_\infty^+ \quad .$$

Then the conclusion easily follows from Theorem 2. and Lemma 5.6 of [2].

Q.E.D.

REFERENCES

[1] Amann, H., E. Zehnder, Multiple periodic solutions of asymptotically linear Hamiltonian equations, preprint.

[2] Benci, V., On the critical point theory for indefinite functionals in the presence of symmetry, to appear on Trans. Amer. Math. Soc.

[3] Benci, V., A. Capozzi, D. Fortunato, Periodic solutions of Hamiltonian systems with a prescribed period, Preprint.

[4] Benci, V., D. Fortunato, "Soluzioni periodiche multiple per equazioni differenziali non lineari relative a sistemi conservativi", Proceedings of the Symposium "Metodi asintotici e topologici in problemi diff. non lineari", l'Aquila (1981).

[5] Rabinowitz, P.H., Periodic solutions of Hamiltonian systems: a survey, Math. Research Center Technical Summary Report, University of Wisconsin-Madison.

Author's address: Istituto di Matematica Applicata - Facoltà di Ingegneria - Via Re David, 200 - 70125 Bari - ITALY.
Istituto di Matematica Applicata - Facoltà di Ingegneria - Via Re David, 200 - 70125 Bari - ITALY.
Istituto di Analisi Matematica - Facoltà di Scienze Palazzo Ateneo - Via Nicolai, 2 - 70122 Bari-ITALY.

NUMERICAL ANALYSIS OF SINGULARITIES IN A DIFFUSION REACTION MODEL

W.-J. Beyn
Fakultät für Mathematik der
Universität Konstanz
D-7750 Konstanz, F.R. of Germany

Abstract: In a recent paper, Bigge and Bohl [2] found some interesting bifurcation diagrams for a discrete diffusion reaction model. We give an interpretation of their results from the view of singularity theory and we will also indicate how this theory may be used to set up numerical methods for singular solutions such as bifurcation points or isolated points.

1. Introduction. This paper is intended to show how singularity theory (e.g. [4, 5, 6]) may help to understand bifurcation diagrams that have been obtained numerically for a finite dimensional system of N equations in $N+1$ variables

(1) $T(z)=0,\ T:\mathbb{R}^{N+1}\to\mathbb{R}^{N},\ z=$ state variable.

In contrast to the standard situation in the theory we assume that we do neither know the singularity of (1) (i.e. a point $z_0\in\mathbb{R}^{N+1}$ such that $T(z_0)=0$, rank $(T'(z_0))<N$) nor do we know its type. Moreover, we are usually not even given a system (1) which has a singularity, but rather a parametrized system of equations

(2) $T(z,c)=0,\ T:\mathbb{R}^{N+1}\times\mathbb{R}^{P}\to\mathbb{R}^{N}$, $c=$ control variable

for which solution diagrams have been computed for various values of c. The problem then is to guess from these data the type of a singularity z_0 of $T(z,c_0)=0$, where c_0 is close to the numerical c-values. The choice of a correct singularity depends on its universal unfolding (cf. [5,6]) which should generate in a qualitative way the bifurcation pictures which have been observed numerically. Once a proper singularity is detected we can use the theory of unfolding to predict further types of solution curves of the system (2) and try to find them numerically by varying c.

We will illustrate this process for a specific example recently discussed in [2]

$$(3)\quad \begin{aligned}&x_1=x_N=0 \text{ and for } i=2,\dots,N-1\\ &h^{-2}(-x_{i-1}+2x_i-x_{i+1})+\nu h^{-1}(x_i-x_{i-1})=10^{\mu}(\beta-x_i)\exp\left(-\frac{\lambda}{1+x_i}\right)\end{aligned}$$

where $h=(N-1)^{-1}$. We put (3) into the form (2) by setting

(4) $z=(x_1,x_2,\dots,x_N,\lambda),\ c=(\nu,\mu,\beta)$.

The system (3) may either be viewed as a discrete cell model with diffusion and an exothermic reaction or as a discretization of a corresponding boundary value problem ([2, 3]). Our special singularity proposed here is based on the results of [2] and should also appear in the more general equations described therein.
Finally, we show how the above interplay of singularity theory and numerical computations can be made more rigorous. The crucial point here is to actually compute a singularity as a solution of a so called defining equation (cf. the inflated systems for bifurcation points in [7, 10]). In the case of bifurcation points and isolated points we will present some defining equations and show their relation to singularity theory.

Acknowledgement: I am particularly indebted to Dipl. Math. J. Bigge for providing me with several solution branches of the example (3) which are not contained in [2].

2. Finding and analysing the singularity in the discrete model

Let us first recall some solution branches of (3) from [2] as given in figure 1 for the values $N = 11$, $\nu = 0$, $\mu = 7$ and $\mu = 12$, $\beta = 1$.

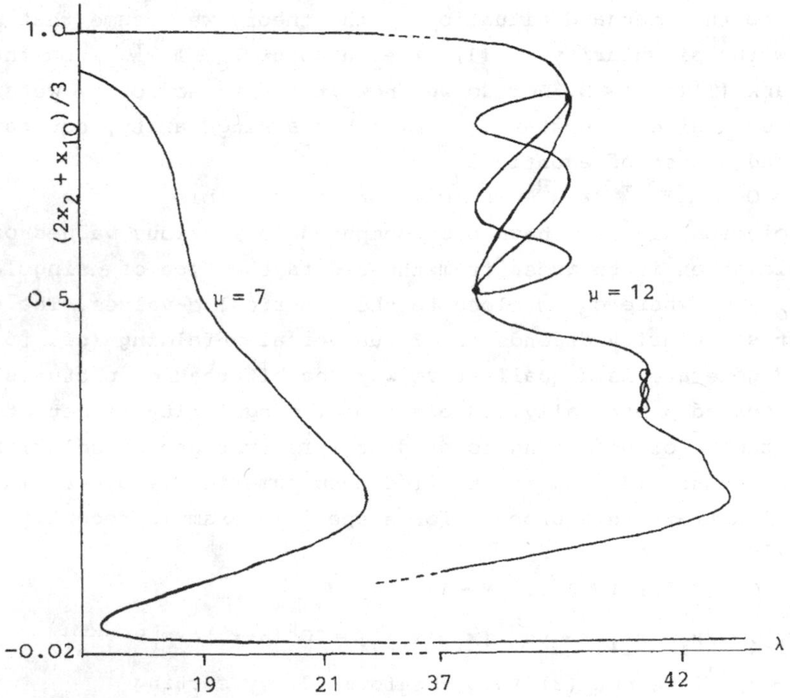

Fig. 1: solutions of (3) for $\mu = 7,12$

We are interestd in the upper configuration for $\mu = 12$ which was called a double-figure-eight in [2]. Here the main branch of symmetric solutions (i.e. $x_i = x_{N+1-i}\ \forall i$) is intersected by a closed loop of unsymmetric solutions at two bifurcation points (marked by a dot). The topological type of these branches is more clearly seen when projecting them onto a (x_{10}, x_2)-plane as is done in figure 2.

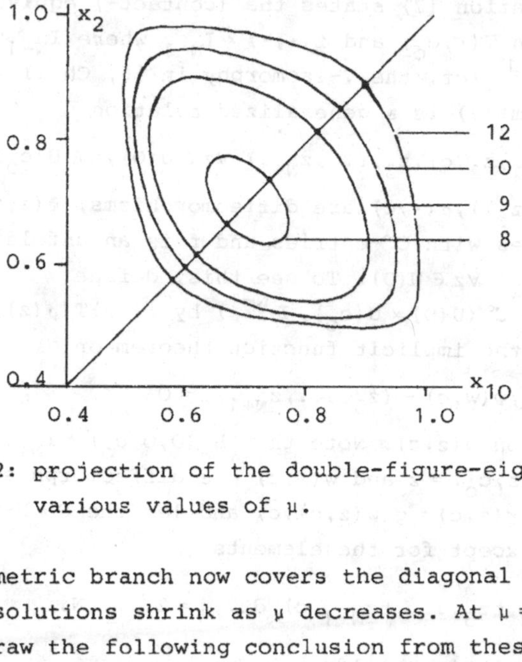

Figure 2: projection of the double-figure-eight onto (x_{10}, x_2) for various values of μ.

The symmetric branch now covers the diagonal and the loops of unsymmetric solutions shrink as μ decreases. At $\mu = 7$ they have vanished.

We draw the following conclusion from these pictures. There should exist a set of parameters

$c_o = (0,\ \mu_o,\ 1)$ where $7 < \mu_0 < 8$

and a singularity z_0 of (3) at $c = c_0$ such that an upward perturbation of μ instantaneously creates the bifurcation diagram of fig. 2.

In order to find the type of the singularity we use the model equations

(5) $\lambda = x^3 - ax + b,\ \lambda = y^3 - cy + d,\ a,b,c,d > 0$

which were derived in [2] from the numerical values and shown to create the double-figure-eight as well as some of its perturbations. Upon eliminating λ from (5) and setting $a = b = c = d = 0$ we end up with

(6) $f(x,y) := x^3 - y^3 = 0,$

which has $(x,y) = (0,0)$ as a singularity.

Our hypothesis is that (6) gives the correct type of the singularity z_0 for the equation $T(z, c_0) = 0$. More precisely, we assume that there exists a relation

(7) $\tau(z)T(\rho(z),c_o) = (f(z_1,z_2),z_3,\ldots,z_{N+1}) \quad \forall z \in U(0) \subset \mathbb{R}^{N+1}$

where $\rho \in C^\infty(U(0), U(z_o))$ is diffeomorphic, $\rho(0) = z_o$ and $\tau(z)$ are nonsingular $N \times N$-matrices infinitely differentiable with respect to z. Here and in what follows $U(0)$, $U(z_o)$ and $U(c_o)$ denote suitable neighbourhoods not always the same at different occurrences. In terms of singularity theory, the relation (7) states the (contact-) equivalence of the germs associated with $T(\cdot,c_o)$ and $f(\cdot,\cdot) \times I_{N-1}$ where I_{N-1} denotes the identity in $\mathbb{R}^{N-1}$ (cf. the V-isomorphy in [6, Ch.II]).

Our first conclusion from (7) is a generalized relation

(8) $\tilde\tau(z,c)T(\tilde\rho(z,c),c) = (\tilde f(z_1,z_2,c),z_3,\ldots,z_{N+1}) \quad \forall z \in U(0), c \in U(c_o)$

where $\tilde\rho \in C^\infty(U(0) \times U(c_o), U(z_o))$, $\rho(\cdot,c)$ are diffeomorphisms, $\tilde\tau(z,c)$ are nonsingular $N \times N$-matrices with C^∞-entries and $\tilde f$ is an unfolding of f, i.e. $\tilde f(z_1,z_2,c_o) = f(z_1,z_2)\ \forall z \in U(0)$. To see this, define $g \in C^\infty(U(0) \times U(c_o),\mathbb{R})$ and $\Phi \in C^\infty(U(0) \times U(c_o), \mathbb{R}^{N-1})$ by $\tau(z)T(\rho(z),c) = (g(z,c), \Phi(z,c))$ and use the implicit function theorem on

$$H(w,z,c) := (w_1-z_1,w_2-z_2,\Phi(w,c) - (z_3,\ldots,z_{N+1})) = 0$$

in order to obtain a function $w(z,c)$. Note that $H_w(0,0,c_o) = I_{N+1}$ and $H(z,z,c_o) = 0$ hold, hence $w(z,c_o) = z$ and $w(\cdot,c)$ are diffeomorphisms for $c \in U(c_o)$. Finally, let $\gamma(z,c) = g(w(z,c),c)$ and let the $N \times N$-matices $\hat\tau(z,c)$ be identical to I_N except for the elements

$$\hat\tau_{1j}(z,c) = -\int_0^1 \frac{\partial\gamma}{\partial z_{j+1}}(z_1,z_2,tz_3,\ldots,tz_{N+1},c)\,dt, \quad j = 2,\ldots,N.$$

Then a straightforward calculation yields

$\hat\tau(z,c)\tau(w(z,c))T(\rho(w(z,c)),c) = (g(z_1,z_2,0,\ldots,0,c), z_3,\ldots,z_{N+1})$

and hence (8).

From the relation (8) we obtain a local correspondence $z = \tilde\rho(x,y,0,\ldots,0,c)$ between the solutions z of (2) and (x,y) of

(9) $\tilde f(x,y,c) = 0$.

For qualitative purposes it is therefore sufficient to consider (9) instead of (2). In addition, if $f(x,y,\alpha)$ is a universal unfolding of $f(x,y)$ then to each c close to c_o there exists an α close to 0 such that the solution curves of (9) and of

(10) $f(x,y,\alpha) = 0$

are diffeomorphic (cf.[6, Ch.II]). For the particular case (6), a universal unfolding needs at least 4 parameters and one such is

(11) $f(x,y,\alpha) = x^3 - y^3 + \alpha_4 xy + \alpha_3 y + \alpha_2 y + \alpha_1, \quad \alpha = (\alpha_1,\alpha_2,\alpha_3,\alpha_4)$.

Note that (11) with $\alpha_1 = 0$ is the hyperbolic umbilic in catastrophe theory [8]. Figures 3 and 4 show two three dimensional projections of the bifurcation set

$$B = \{\alpha \in \mathbb{R}^4 : \exists x,y \in \mathbb{R} \text{ such that } f(x,y,\alpha) = f_x(x,y,\alpha) = f_y(x,y,\alpha) = 0\}$$

along with some (x,y)-solution curves of (10) associated with special values $\alpha \in B$ (indicated by arrows) and $\alpha \in \mathbb{R}^4 \setminus B$.

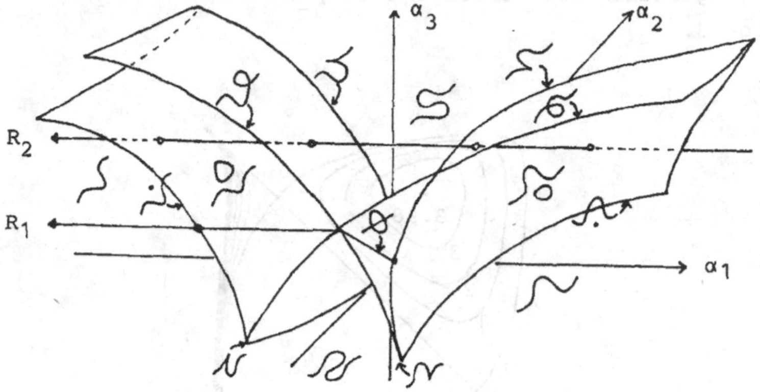

Fig. 3: Projection of B onto $\alpha_4 = 0$

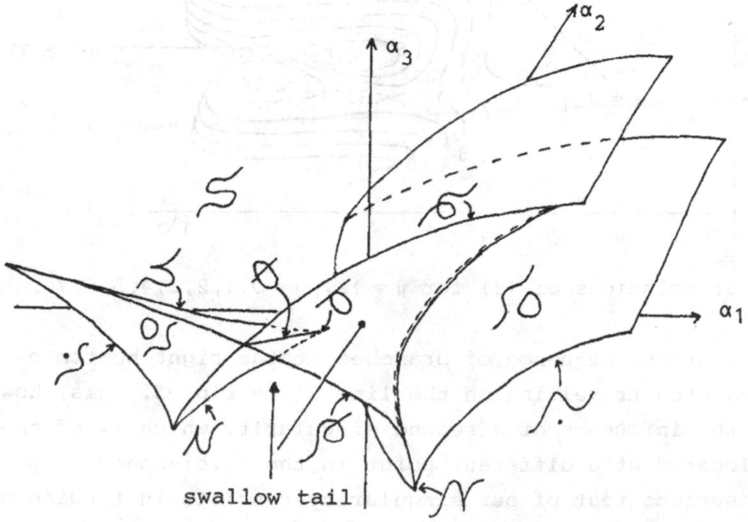

Fig. 4: Projection of B onto $\alpha_4 > 0$ fixed.

3. Testing the singularity

Let us first note that we can recover the curves of fig. 2 from fig. 3 if we let α move towards the origin on the line $\alpha_4 = 0$, $\alpha_3 = -\alpha_2 > 0$, $\alpha_1 = 0$.

Here we have $f(x,y,0,-\alpha_2,\alpha_2,0) = (x-y)(x^2+xy+y^2+\alpha_2)$ so that (10) describes an ellipse cut by a straight line. Our difference equations (3) correspond to this 'nongeneric' set of parameter values because of the inherent symmetry in the case $\nu = 0$ ((3) is invariant under the transformation $x_i \to x_{N+1-i}$).
If we fix $\mu = 12$ and let ν increase then this symmetry is destroyed and the curves of fig. 5 show up numerically. The perturbations of the upper configuration now correspond to parameters α moving on the ray R_1 in fig. 3.

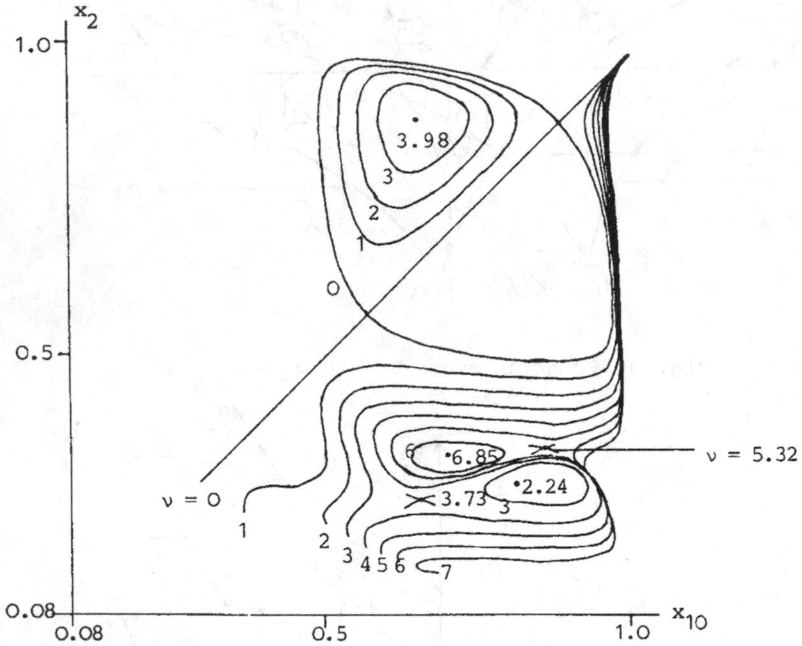

Fig. 5: solutions of (3) for $\mu = 12$, $\nu = 0,1,2,3,4,5,6,7$

Note also that the sequence of branches at the right bottom of fig. 5 is obtained when travelling on the line R_2 in fig. 3. This, however, should be the influence of a second singularity which is of the same type but located at a different point in the (z,c)-space.

A more serious test of our singularity consists in finding numerically the singular solution branches in the ν-sequence, in particular those singularities predicted by the intersections of the lines R_1 and R_2 with the bifurcation set B. The resulting isolated points and bifurcation points are shown in fig. 5. These were computed by tracing one of the curves in fig. 5 and switching to a defining equation at various points. Defining equations for bifurcation points have been set up in [7, 10] by inflating the system (2). The approach taken here rather uses a deflation of (2), which seems to be simpler conceptionally although not computationally. The numerical details and a proof of the theorem below will be contained in a forthcoming paper.

We consider an equation (2) with one parameter $c(p=1)$ and let $T \in C^\infty(\mathbb{R}^{N+2},\mathbb{R}^N)$ for simplicity. Let us decompose $\mathbb{R}^{N+1} = \Phi \oplus V$, $\mathbb{R}^N = \Psi \oplus W$ into subspaces such that $\dim\Phi = 2$, $\dim\Psi = 1$ and let $P : \mathbb{R}^N \to W$ be the projector along Ψ. Further assume that there exists a solution $(\bar{\varphi},\bar{v},\bar{c}) \in (\Phi \oplus V) \times \mathbb{R}$ of (2) such that $PT_v(\bar{\varphi},\bar{v},\bar{c}) : V \to W$ is nonsingular. Now we can define an implicit function $v(\varphi,c) \in V$ in some open neighbourhood U of $(\bar{\varphi},\bar{c})$ by $PT(\varphi,v(\varphi,c),c) = 0$. The function $S : \Phi \times \mathbb{R} \to \Psi$, $S(\varphi,c) = (I_N - P)T(\varphi,v(\varphi,c),c)$ may then be considered as a Liapunov-Schmidt type reduction of T [9].

The defining equations for a simple bifurcation point or an isolated point of S are simply the three equations

(12) $S(\varphi,c) = 0$, $S_\varphi(\varphi,c) = 0$,

for which we have the following result:

<u>Theorem:</u> $(\varphi_0,c_0) \in U$ is a regular solution of (12) if and only if $z_0 = (\varphi_0, v(\varphi_0,c_0))$ is either a simple bifurcation point or an isolated point of $T(z,c_0) = 0$ (i.e. $T(z,c_0)$ is equivalent to either $z_1^2 - z_2^2$ or $z_1^2 + z_2^2$ in the sense of (7)) and $T(z,c)$ is a universal unfolding of $T(z,c_0)$.

The regularity of (φ_0,c_0) means that the system (12) has a nonsingular Jacobian at (φ_0,c_0) which implies that Newton's method is locally quadratically convergent (note however that the evaluation of S needs the solution of an implicit equation). The right-hand side of our above equivalence may be more conventionally written down in terms of the null space N_0 of $T_z(z_0,c_0)$, the range W_0 of $T_z(z_0,c_0)$ and the projector Q_0 onto a complementary space Ψ_0 of W_0: $\dim N_0 = 2$, $T_c(z_0,c_0) \notin W_0$ and the quadratic form $Q_0T_{zz}(z_0,c_0) : N_0 \times N_0 \to \Psi_0$ is nondegenerate (i.e. there exist two nonzero eigenvalues either of the same or of the opposite sign [1,10]).

<u>References:</u>

[1] Beyn, W.-J.: On discretizations of bifurcation problems. pp. 46-73 in Bifurcation problems and their numerical solution (Eds.: H.D. Mittelmann, H. Weber), ISNM 54, Birkhäuser Verlag, 1980.
[2] Bigge, J., Bohl, E.: On the steady states of finitely many chemical cells (submitted).
[3] Bohl, E.: Finite Modelle gewöhnlicher Randwertaufgaben. LAMM Bd. 51, Teubner Verlag, Stuttgart, 1981.

[4] Golubitsky, M., Guillemin, V.: Stable mappings and their singularities. Graduate Texts in Mathematics No. 14, Springer Verlag, New York, 1974.
[5] Golubitsky, M., Schaeffer, D.: A theory for imperfect bifurcation via singularity theory. Commun. Pure Appl. Math. 32, 21-98, 1979.
[6] Martinet, J.: Deploiements versels des applications differentiables et classification des applications stables. pp. 1-44, Lecture Notes in Mathematics 535, Springer Verlag, 1976.
[7] Moore, G.: The numerical treatment of non-trivial bifurcation points. Numer. Funct. Anal. Optimiz. 2, 441-472, 1980.
[8] Poston, T., Stewart, I.:Catastrophe theory and its applications. Pitman, London, 1978.
[9] Vainberg, M.M., Trenogin, V.A.: The methods of Lyapunov and Schmidt in the theory of non-linear equations and their further development. Russian Math. Surveys 17, 1-60, 1962.
[10] Weber, H.: On the numerical approximation of secondary bifurcation points. pp. 407-425, Lecture Notes in Mathematics 878, Springer Verlag, 1981.

HYPERBOLIC LIPSCHITZ HOMEOMORPHISMS AND FLOWS

M. A. Boudourides
Department of Mathematics
Democritus University of Thrace
Xanthi, Greece

Let f be a Lipschitz homeomorphism of a Banach space E vanishing at 0. If, moreover, E admits the direct sum decomposition $E = E^s \oplus E^u$ with E^s and E^u invariant under f and the restriction of f on E^s (resp. E^u) a contraction (resp. expansion) on E^s (resp. E^u) with respect to an equivalent norm, then we say that f is a hyperbolic Lipschitz homeomorphism. Apparently, all hyperbolic linear automorphisms are hyperbolic Lipschitz homeomorphisms. Nevertheless, one can easily construct a nonlinear map, which is a hyperbolic Lipschitz homeomorphism, but which is not differentiable at 0 and, therefore, for which the theory of differentiable dynamical systems cannot apply. However, it is our purpose to show that the nonlinear hyperbolic Lipschitz homeomorphisms share many common properties with the hyperbolic linear automorphisms. In addition, we are going to consider a class of maps, which are monotone in a certain sense, and the time t maps of the flows, generated by the differential equations with right-hand sides in this class, are hyperbolic Lipschitz homeomorphisms.

We start by introducing a series of sets of maps between two real Banach spaces X, Y both with norm $|.|$:

$$\mathrm{Lip}(X,Y) = \{f:X\to Y \text{ such that } \|f\| = \sup_{x\neq y}\frac{|f(x)-f(y)|}{|x-y|} < \infty\},$$
$$\mathrm{Lip}(X) = \mathrm{Lip}(X,X),$$
$$\mathrm{GLip}(X) = \{f\in\mathrm{Lip}(X) \text{ such that } f^{-1}\in\mathrm{Lip}(X)\},$$
$$\Lambda(X,Y) = \{f\in\mathrm{Lip}(X,Y) \text{ such that } f(0)=0\},$$
$$\Lambda(X) = \Lambda(X,X),$$
$$G\Lambda(X) = \{f\in\Lambda(X) \text{ such that } f^{-1}\in\Lambda(X)\},$$
$$B(X,Y) = \{f:X\to Y \text{ bounded, i.e. } |||f||| = \sup_{x\in X}|f(x)| < \infty\},$$
$$B(X) = B(X,X),$$
$$\mathrm{BLip}(X,Y) = \{f\in\mathrm{Lip}(X,Y) \text{ such that } f\in B(X,Y)\},$$
$$\mathrm{BLip}(X) = \mathrm{BLip}(X,X),$$
$$L(X,Y) = \{f\in B(X,Y) \text{ linear}\},$$
$$L(X) = L(X,X).$$

We note that Lip(X,Y) is a seminormed linear space with seminorm $\|.\|$

and that $\Lambda(X,Y)$, $B(X,Y)$ are Banach spaces with norms $\|.\|$, $|||.|||$, respectively.

We denote by $S(X)$ the unit sphere in X. If $f \in G\Lambda(X)$, $\|f\| < 1$, we denote by $D(f)$ (called the fundamental domain of f) the closed annulus between $S(X)$ and $f(S(X))$. Clearly, the images of $D(f)$ under positive and negative powers of f cover $X\setminus\{0\}$.

Let us also mention here that, in what follows, i is going to denote the identity map and juxtaposition of maps is going to stand for composition of maps.

From now on, let E be a real Banach space with norm $|.|$. Whenever E admits a direct sum decomposition $E = E^s \oplus E^u$, we write $x = x_s + x_u$ and, given $f: E \to E$, we denote by $f_s: E \to E^s$, $f_u: E \to E^u$ the functions defined by $f_s(x) = (f(x))_s$, $f_u(x) = (f(x))_u$, respectively. Moreover, in this case, we are going to employ the product norm $|x| = \max\{|x_s|, |x_u|\}$.

Among the maps in the set $G\Lambda(X)$ of Lipschitz homeomorphisms of E vanishing at 0, we define hyperbolicity as it follows.

Definition 1. *A map* $f \in G\Lambda(X)$ *is called hyperbolic if there exists an equivalent norm* $|.|_*$ *on* E *and a direct sum decomposition* $E = E^s \oplus E^u$ *such that* (i) E^s *and* E^u *are each invariant under* f *and* (ii) $\alpha = \max\{\|f_s\|_*, \|f_u^{-1}\|_*\} < 1$ *(where* $\|.\|_*$ *denotes the Lipschitz norm corresponding to* $|.|_*$*).*

In the sequel, we are going to drop the asterisk from the equivalent norm defining hyperbolicity. We remark that, if f is hyperbolic, then f_s (resp. f_u) is identified (because of (i)) with the restriction of f on E^s (resp. E^u) and (because of (ii)) it is a contraction (resp. expansion) on E^s (resp. E^u) with respect to the equivalent norm.

An immediate consequence of Definition 1 is the following lemma.

Lemma 1. *If* $f \in G\Lambda(E)$ *is hyperbolic, then* $i - f \in G\Lambda(E)$ *with* $\|(i-f)^{-1}\| \leq (1-\alpha)^{-1}$.

Proof. By the Lipschitz inverse function theorem (cf. [3], p.67), $i_s - f_s \in G\Lambda(E^s)$ with $\|(i_s-f_s)^{-1}\| \leq (1-\alpha)^{-1}$ and $i_u-f_u \in G\Lambda(E^u)$ with $\|(i_u-f_u)^{-1}\| \leq (1-\alpha)^{-1}$. Therefore, the map $(i-f)^{-1}$ defined by $(i-f)^{-1}(x) = (i_s-f_s)^{-1}(x_s) + (i_u-f_u)^{-1}(x_u)$ is the inverse of $i-f$ in $G\Lambda(E)$ and $\|(i-f)^{-1}\| = \max\{\|(i_s-f_s)^{-1}\|, \|(i_u-f_u)^{-1}\|\} \leq (1-\alpha)^{-1}$.

We denote by $H\Lambda(E)$ the set of all hyperbolic $f \in G\Lambda(E)$. It is well known (cf. [2] and [5]) that the set $HL(E)$ of all hyperbolic linear automorphisms of E is in $H\Lambda(E)$ (a linear automorphism is called hyperbolic if its spectrum does not intersect the unit circle in $\mathbb{C}$).

An example of a (nonlinear) $f \in H\Lambda(E)$ not in $HL(E)$ is the following. For $E = \mathbb{R}^2$, we take $f(x_1,x_2) = (\frac{1}{3}|x_1| + \frac{1}{2}x_1, |x_2| + 3x_2)$. Clearly, such a map is hyperbolic according to Definition 1, but it is neither linear nor linearizable at 0, since it is not differentiable at 0.

By Banach's fixed point theorem, 0 is the unique fixed point of any map in $H\Lambda(E)$. However, such a property is stable under suitable Lipschitz perturbations of maps in $H\Lambda(E)$, as it is shown by the following result.

Lemma 2. *Let* $f \in H\Lambda(E)$, $g \in \mathrm{Lip}(E)$ *with* $\|g\| < 1-\alpha$. *Then* $f+g$ *has a unique fixed point.*

Proof. Indeed, any fixed point of $f+g$ is a fixed point $(i-f)^{-1}g$, which is a contraction (by Lemma 1).

For reasons of classification, we need to define a notion of "equivalence". We say that two maps $f:X \to X$ and $g:Y \to Y$ (X,Y Banach spaces) are *topologically equivalent* if there is a homeomorphism $h:X \to Y$ such that $fh = hg$.

First, we are trying to characterize the topologically equivalent maps in $H\Lambda(E)$. Whenever $f,g \in H\Lambda(E)$, a subscript (or a superscript) f or g will refer to the decomposition of E for f or g.

Proposition 1. *Let* $f,g \in H\Lambda(E)$. *Then* f *and* g *are topologically equivalent if and only if there exist homeomorphisms* $h_s:D(f_s) \to D(g_s)$ *and* $h_u: D(f_u^{-1}) \to D(g_u^{-1})$ *such that* $h_s f_s(x) = g_s h_s(x)$, *for all* $x \in S(E_f^s)$, *and* $h_u f_u(x) = g_u h_u(x)$, *for all* $x \in S(E_f^u)$.

Proof. By Definition1, it suffices to prove that f_s and g_s are topologically equivalent if and only if there exists a homeomorphism $h_s:D(f_s) \to D(g_s)$ such that $h_s f_s = g_s h_s$ on $S(E_f^s)$.

To prove the necessity, let $h: E_f^s \to E_g^s$ be a homeomorphism such that $hf_s = g_s h$ on E_f^s. Since for each $x \in D(f_s)$ there exist $y \in E_f^s$ and an integer n such that $x = f_s^n(y)$ and $g_s^n h(y) \in D(g_s)$, the map $h_s(x) = hf_s^n(y)$ defines a homeomorphism $h_s:D(f_s) \to D(g_s)$ such that $h_s f_s = g_s h_s$ on $S(E_f^s)$ (because, for any integer m, $hf_s^m = g_s^m h$).

For the suffeciency, suppose that $h_s:D(f_s) \to D(g_s)$ is a homeomorphism such that $h_s f_s = g_s h_s$ on $S(E_f^s)$. We define $h: E_f^s \to E_g^s$ by $h(0)= 0$ and, for $x \in E_f^s\setminus\{0\}$, $h(x) = g_s^{-n} h_s f_s^n(x)$, where $f_s^n(x) \in D(f_s)$. Clearly, $hf_s = g_s h$ on E_f^s. Since h maps the set $\{f_s^n(x): x \in E_g^s, |x| \leq 1, n \text{ integer}\}$ onto the set $\{g_s^n(x): x \in E_g^s, |x| \leq 1, n \text{ integer}\}$, i.e. a basis of neighborhoods of 0 in E_f^s onto a similar basis in E_g^s, it is implied that h is continuous at 0. Similarly, we construct the inverse $h^{-1}:E_g^s \to E_f^s$ from $h_s^{-1}:D(g_s) \to D(f_s)$ and the proof is completed.

The above result indicates that maps in $H\Lambda(E)$ behave like the ones in $HL(E)$. Now, it will be seen that any appropriate perturbation of a map in $H\Lambda(E)$ behaves like the map itself too. Actually, the following more general result is proved (which generalizes what in the linear case is known as Hartman's linearization theorem for homeomorphisms, cf. [2], [5] and [6]).

Proposition 2. *Let* $f \in H\Lambda(E)$, $g,k \in BLip(E)$ *with* $\max\{\|g\|,\|k\|\} < \min\{1-\alpha, \|f_s^{-1}\|^{-1}\}$. *Then* $f+g$ *and* $f+k$ *are topologically equivalent; actually, there exists a unique homeomorphism* $h:E \to E$ *such that* $h, h^{-1} \in BLip(E)$ *and* $(f+g)h = h(f+k)$.

Proof. First, we remark that, since $\|k\| \leq \|f_s^{-1}\|^{-1} = \|f^{-1}\|^{-1}$, the Lipschitz inverse function theorem implies that $f+k$ is invertible. We consider the map $T:BLip(E) \to BLip(E) = BLip(E,E^s) \times BLip(E,E^u)$ defined by $Th = (T_s h, T_u h)$, where

$$T_s h = (f_s + g_s)h(f+k)^{-1}$$
$$T_u h = f_u^{-1}\{h_u(f+k) - g_u h\} .$$

For any $h, \tilde{h} \in BLip(E)$, we obtain

$$|||Th - T\tilde{h}||| \leq \max\{\|f_s\| + \|g\|, \|f_u^{-1}\|(1+\|g\|)\} |||h - \tilde{h}||| .$$

Consequently, $T \in Lip(BLip(E))$ and since $\|f_s\| + \|g\| < \alpha + 1 - \alpha = 1$ and $\|f_u^{-1}\|(1+\|g\|) < \alpha^{-1}(1+1-\alpha) < \alpha^{-1}(1+\alpha^{-1}-1) = 1$, we get $\|T\| < 1$, i.e. T is a contraction in the Banach space $BLip(E)$. Hence, by Banach's fixed point theorem, there exists a unique fixed point $h \in BLip(E)$ of T, which is easily seen to satisfy the relation $(f+g)h = h(f+k)$. In the same way, interchanging g and k, we obtain that there exists a unique $\bar{h} \in BLip(E)$ such that $(f+g)h\bar{h} = h(f+k)\bar{h} = h\bar{h}(f+k)$. Since $(f+g)i = i(f+g)$, the uniqueness of the above fixed point (when $k=g$) implies $h\bar{h} = i$. Similarly, one gets $\bar{h}h = i$, i.e. h is the required homeomorphism.

Finally, we are going to construct a class of nonlinear maps such that the differential equations with right-hand sides in this class generate hyperbolic (according to Definition 1) flows.

For this purpose, we need to introduce the two-sided Gateaux variation of the norm $|.|$ of an arbitrary real Banach space X :

$$m_\pm[x,y] = \lim_{h \to 0\pm} \frac{1}{h}(|x+hy| - |x|) .$$

We refer to [3] for properties of $m_\pm$.

We denote by $M(X)$ the set of continuous maps $F:X \to X$ such that there exist real numbers $\mu^-[F]$ and $\mu^+[F]$ such that

$$m_-[x-y, F(x)-F(y)] \leq \mu^-[F]|x-y|$$
$$m_+[x-y, F(x)-F(y)] \geq \mu^+[F]|x-y| .$$

Clearly, any $A \in L(X)$ belongs to $M(X)$ with $\mu^-[A] = \lambda[A]$ and $\mu^+[A] = -\lambda[-A]$, where $\lambda[A] = \lim_{h \to 0} h^{-1}(\|i+hA\|-1)$ is the logarithmic norm of A (cf. [1]).

Also, any $F \in Lip(X)$ belongs to $M(X)$ with $\mu^-[F] = \|F\|$ and $\mu^+[F] = -\|F\|$. One can find in [4] an example of a nonlinear map, which belongs to $M(X)$ but which is not in $Lip(X)$.

According to a result of Martin [3], if $F \in M(X)$, then, for each $x \in X$, there is a unique solution $s(t,x)$ to the differential equation $x' = F(x)$ such that $s(t,x)$ is defined for all $t \in \mathbb{R}$, $s(0,x) = x$, and, for all $x,y \in X$, $t \in \mathbb{R}$,

$$e^{\mu^+[F]t}|x-y| \leq |s(t,x)-s(t,y)| \leq e^{\mu^-[F]t}|x-y|, \text{ for } t \geq 0$$
$$e^{\mu^-[F]t}|x-y| \leq |s(t,x)-s(t,y)| \leq e^{\mu^+[F]t}|x-y|, \text{ for } t \leq 0 .$$

Therefore, for any $F \in M(X)$ one can define the *flow* generated by F as the map $f: \mathbb{R} \times X \to X$ defined by $f(t,x) = s(t,x)$. The map f^t, such that $f^t(x) = f(t,x)$, is called the *time* t *map* of the flow generated by F. Clearly, for any $t \in \mathbb{R}$, f^t is a homeomorphism such that $f^{s+t} = f^s f^t$ and $f^0 = i$. Now, the above inequalities imply some further properties of f^t.

<u>Lemma</u> 3. *If* $F \in M(X)$, *then, for any* $t \in \mathbb{R}$, $f^t \in GLip(X)$ *with* $\|f^t\| \leq e^{\mu^-[F]t}$ *for* $t \geq 0$, $\|f^t\| \leq e^{\mu^+[F]t}$ *for* $t \leq 0$.

Let us return again to the Banach space E. We intend to define hyperbolicity among the nonlinear maps from E to itself vanishing at 0.

<u>Definition</u> 2. *A map* $F: E \to E$, $F(0) = 0$, *is called hyperbolic if there is an equivalent norm* $|.|_*$ *on* E *and a direct sum decomposition* $E = E^s \oplus E^u$ *such that* (i) E^s *and* E^u *are each invariant under* F *and* (ii) $F_s \in M(E^s)$ *with* $\mu^-[F_s] < 0$ *and* $F_u \in M(E^u)$ *with* $\mu^+[F_u] > 0$ *(where the* μ*'s are computed in terms of* $|.|_*$*)*.

We denote by $HM(E)$ the set of all hyperbolic maps from E to itself vanishing at 0. Definition 2 says that if $F \in HM(E)$, then F_s is strictly dissipative on E^s and F_u is strictly accretive on E^u. It is well known (cf. [2]) that any $A \in L(E)$, which is hyperbolic (i.e. the spectrum of A does not intersect the imaginary axis in $\mathbb{C}$), belongs to $HM(E)$. An example of a map $F \in HM(E)$, which is not linear, would be the following. For $E = \mathbb{R}^2$, we take $F(x_1,x_2) = (|x_1| - 2x_1, |x_2| + 2x_2)$ (one easily computes $\mu^-[F_s] = -1$ and $\mu^+[F_u] = 1$).

By a direct verification of the conditions of Definition 1 we get:

<u>Proposition</u> 3. *If* $F \in HM(E)$, *then, for any* $t \in \mathbb{R}\setminus\{0\}$, $f^t \in H\Lambda(E)$.

REFERENCES

[1] - COPPEL, W.A., "Stability and Asymptotic Behavior of Differential Equations", Heath and Co., Boston, 1965.

[2] - IRWIN, M.C., "Smooth Dynamical Systems", Academic Press, 1980.

[3] - MARTIN Jr., R.H., "Nonlinear Operators and Differential Equations in Banach Spaces", John Wiley, New York, 1976.

[4] - MURAKAMI, H., On non-linear ordinary and evolution differential equations, Funkcial. Ekvac., 9 (1966), 151-162.

[5] - PALIS, J., On the local structure of hyperbolic points in Banach spaces, An. Acad. Brasil Ci., 40 (1968), 263-266.

[6] - PUGH, C., On a theorem of P. Hartman, Amer. J. Math., 91 (1969), 363-367.

SOME EVOLUTION EQUATIONS ARISING IN PHYSICS

T.BRUGARINO *, A.CANINO ** and P.PANTANO **

(*) Istituto Matematico, Facoltà di Ingegneria-Università di Palermo, Italy.

(**) Dipartimento di Matematica, Università della Calabria-Cosenza, Italy.

Abstract. In this paper we consider a new series of evolution equations generalizing the Korteweg-deVries (KdV) and Burgers equations, and we report recent advances on these equations together with the physical phenomena where they arise. In particular we consider a generalized Burgers' equation and we sketch a method for solution in series by using the theory of Sobolevskij and Tanabe. Then we study the KdV equation with nonuniformity terms and we describe various physical interpretation of this equation. We consider various particular cases in which varying solitonic solutions exist. Also we sketch a unicity theorem. Finally modified Burgers-KdV equations are considered.

1. Introduction.

In the last years a large class of partial differential equations has been found to describe very important phenomena in physics as the Korteveg-deVries equation, the sine-Gordon, the KdV-Burgers' equation ect. The mathematical tool of the Inverse Scattering Transform was introduced to solve these equations . In this paper we consider a new series of equations generalizing the KdV and Burgers' and we report recent advances on these equations together with physical phenomena where these equations arise. In sect.2 we consider the generalized Burgers' equation and we sketch a method for solution in series by using the theory of Sobolevskij and Tanabe. In sect.3 we treat the KdV equation with nonuniform terms and we describe various physical interpretation of this equation. We consider some particular cases in which solitons with varying characteristics exist. Finally we sketch a unicity theorem . In sect.4 we consider KdV or Burgers' equations in more than one dimension and we report recent advances on these equations. Finally, when a dissipation with a dispersion is considered, we have a new class of equations, and other problems can be considered.

2. Burgers-type equations.

2.1. Where the Burgers-types equations arise.

As is well known, the Burgers' equation is the simplest dissipative and nonlinear partial differential equation; it can be integrated by using the Cole-Hopf transformation to obtain as solutions shocks with structure, N-waves ect, [1] . This equation describes the one-dimensional propagation of waves in many physical systems [2], and in particular of sonic waves in thermoviscous fluids [3] . When we consider the one-dimensional propagation of waves in dissipative and nonhomogeneous media [4], or the three-dimensional propagation in the homogeneous and dissipative media [5]-[6] , the equation ruling the evolution of such waves is a Burgers' equation with a linear term due to the nonhomogeneity or to the geometry of wave surfaces respectively. In the particular case of three-dimensional propagation of acoustic waves in thermoviscous fluids, the equation involved is the following [7]-[8]

$$u_\sigma + auu_\xi + \frac{\theta'}{\theta}u = bu_{\xi\xi} \tag{1}$$

where u_σ is the derivative of u with respect to a time σ contracted along the rays, ξ is the phase, $\frac{\theta'}{\theta}$ is a geometrical term with $\theta' = \frac{du}{d\sigma}$ where θ is the expansion parameter, a and b are two constants. For this equation, up to now, does not exists a linearising transformation similar to that of Cole-Hopf, and the main informations is obtained by numerical integration , similarity solutions or by asymptotic expansion by supposing $\frac{\theta'}{\theta}$ small [5 and quoted literature].

2.2. An existence and unicity theorem.

The equation (1) can be written as

$$v_t + vv_\xi - \tilde{b}(t)v_{\xi\xi} = 0 \tag{2}$$

where $v = \theta u$, $dt = \frac{a}{\theta}d\sigma$ and $\tilde{b}(t) = \frac{a}{b}\theta$.

The boundary conditions are

$$v(\rho_1,t) = v(\rho_2,t) = 0 \tag{3}$$

and the initial value of v is

$$v(\xi,0) = v_0(\xi) \tag{4}$$

If we consider expansive waves, $\tilde{b}(t)$ is monotonically increasing.

The initial value problem (IVP) (2)-(4) can be written, in an abstract form [9] - [10], in the Hilbert space $L^2(\rho_1,\rho_2)$, $-\infty<\rho_1<\rho_2<+\infty$, as

$$(5)\begin{cases}\frac{dv}{dt} - A(t)v = g(v) & t_0 < t < t_1 \\ v(t_0) = v_0\end{cases}$$

where $g(v) = vDv$, with $D = \frac{\partial}{\partial \xi}$, and $A(t)$ is a family of operators with domain on H where

(6) $H = H(A(t)) = \{v\in L^2 | v$ absolutely continous, $Dv\in L^2$, $D^2v\in L^2$; $v(\rho_1) = v(\rho_2) = 0\}$

(7) $A(t)v = -\tilde{b}(t)\frac{\partial^2 v}{\partial \xi^2}$; $\forall v\in H$.

We give now an existence and unicity theorem for bounded domains : If IVP is defined by (5) , then there existes a unique solution.

Nonhomogeneous boundary conditions lead to a similar IVP through a suitable change of the unknown function. The theorem is easy proved because the theory of Sobolevskij [11] - [12] for quasilinear evolution equations is applicable. The solution of the IVP is given by the sum

(8) $\sum_{n=0}^{\infty} v_n(t) = v(t)$

with

(9) $v_0(t) = u(t,t_0)v_0$, $\quad v_{n+1}(t) = \int_0^t u(t,s)g(A_0^{-\alpha}v_n(s))\,ds$

where $v_0\in H(A_0^{-\alpha})$, $A_0^{-\alpha}$ being a suitable fractional power $A_0 = A(t_0)$ and $u(t,s)$ the evolution operator generated by $A(t)$.

3. Korteweg-deVries type equations.

3.1. Generalities

As is well known, the KdV equation appears when a balance between the nonlinearity and dispersion exists [2]- [3].

Solitons were found first as solutions for this equation [13] - [15] . Such a balance occurs in many physical systems as water waves, plasma waves ect. [16] - [18]. If other physical phenomena are considered, we hawe a large class of KdV-type equations, that can be written in a general form [19]

(10) $\alpha(\sigma)v + \beta(\sigma)\xi v_{\xi} + \gamma(\sigma)v_{\xi} + \delta(\sigma)vv_{\xi} + \epsilon(\sigma)v_{\xi\xi\xi} + \eta(\sigma)v_{\sigma} = 0$.

In general there does not exist a solution for the eq. (10), but if an appropriate condition is satisfied between the coefficients, eq. (10) reduces to a KdV eq. with constant coefficients by means of an opportune transformation [19] . This occurs, for example, in the case of the cylindrical ion-acoustic waves [20] , ect. Another interesting KdV of type (10) occurs when we consider the two-dimensional propagation of waves in shallow water of variable depth [21],

(10') $\Pi_{\sigma} + 3/2\Pi\Pi_{\xi} + 1/6H^{\frac{1}{2}}\Pi_{\xi\xi\xi} + \Pi(\ln\theta H)_{\sigma} = 0$

where $H = H(\sigma)$ is the variable water depth. This equation can be reduced to

(10'') $v_{\tau} + vv_{\xi} + v_{\xi\xi\xi} = -\Gamma(\tau)v$

where $\Gamma(\tau) = (\ln\theta H^{\frac{3}{2}})_{\tau}$.

In the particular case in which $\theta = H^{\upsilon}$ and $H = (a - \tau)^{m}$ [22]we have,

i) $\Gamma(\tau') = -1/2\tau'$ for $m = \upsilon = 1$

ii) $\Gamma(\tau') = -1/\tau'$ for $m = 1, \upsilon = \frac{1}{2}$

iii) $\Gamma(\tau') \approx \delta$ for $m = 1$ and $\upsilon \approx 3/2$ but $\upsilon \neq 3/2$, where δ is a small parameter

iv) $\Gamma(\tau') = 0$ for $m = 1$ and $\upsilon = 3/2$

where $\tau' = \tau - a$. In the case iv), the eq. (10'') reduces to a KdV eq. with constant coefficients, and the eq. (10') has the following solutions

$$\Pi = \frac{27}{H(0)}uH^{\frac{1}{2}}(\sigma)\,\mathrm{sech}^{2}\{\tfrac{9}{2}(\frac{u}{H(0)})^{\frac{1}{2}}(\xi - \frac{81}{H(0)}u\int_{0}^{\sigma}\frac{H(\sigma')^{\frac{1}{2}}}{6}d\sigma')\}$$

that represents solitons with varying characteristics. They result as a strain balance between two geometrical terms due to the varying depth and the wave front geometry.

3.2. Water waves and plasma waves

The eq. (10) was found in studying one-dimensional propagation of waves in shallow water of variable depth [21] - [23] where $\Gamma = \frac{9}{4}H'/H$. The same equation was obtained in three-dimensional propagation of ion-acoustic waves in a collisionless plasma [24]; Γ for cylindrical or spherical symmetry reduces to the i) or ii), respectively. Also for this simple equation, in general, there does not exist some solution, and the main information can be obtained by numerical integration, similarity solutions and perturbation methods. In the last case one supposes that $\Gamma = \delta$, i.e. one considers quasi-plane waves . The theory was developed by Kaup-Newell [25] and Karpmann-Maslov [26] ; by working on IST and using the conservation laws, they prove that a solitonic initial data evolves retaining their solitonic main shape with varying characteristics but behind can be noted a shelf. Various questions on this field are opened.

3.3. Well posed Cauchy problem and a unicity theorem.

We now consider the KdV eq. of the type (10") :

(11) $\frac{d}{d\sigma}v + D^3 v + vDv + \Gamma(\sigma)v = 0$ with $\sigma \geq 0$, $\xi \in \Omega$ $D=\frac{d}{d\xi}$

(12) $v(0,\xi) = \Phi(\xi)$

we are interested when the problem (11)-(12) is well posed. Although the well-posedness is not a precisely defined notion [27], we shall consider the problem "globally well posed" in a space Y if $\Phi \in Y$ implies the existence of a unique solution v such that $v \in C([0,+\infty[,Y)$ and if the map $\Phi \to v$ is continous in the associated topologies. The problem is "locally well posed" in Y if the results stated above hold with $[0,+\infty[$ replaced with $[0,T[$, that can depend on Φ . Now, because the eq. (10") occurs in one-dimensional propagation of water waves of variable depth, or in the three-dimensional propagation of ion-acoustic waves in a collisionless plasma, we investigate the problem in these two different physical cases.

3.3.1. Water waves.

In this case $\xi \in R$, and $\Gamma(\sigma)$, representing the variable depth can have an opportune dependence on σ . In general this is a monotonic function. The well posed Cauchy problems can be investigated by substituting Ω with R .

3.3.2. Three-dimensional plasma waves.

In this case we must distinguish between expansive or implosive waves. For expansive waves $\xi \in R^+$, $\Gamma(\sigma)$ is a positive monotonic function, with $\Gamma'(\sigma) < 0$. Asymptotically $\Gamma' \to 0$. A globally well posed Cauchy problem can be considered. For imploding waves $\Gamma(\sigma) < 0$ and $|\Gamma'(\sigma)| > 0$, $\xi \in R^+$, and a global problem cannot be considered because we have a caustic at a finite time .Only local problems can be considered . The search of such spaces where (11)-(12) is well posed is actual.

3.3.3. An unicity theorem.

Now we sketch the demonstration of the following uniqueness theorem : If $T > 0$, $b > 0$ and $e^{b\xi}\Phi \in L^2$, then there is at most one solution of I.V.P. (11)-(12) satisfyng the following conditions:

i) $(1 + e^{b\xi})v \in L^{\infty}((0,T);L^2)$

ii) $e^{b\xi}Dv \in L^2_{loc.}((0,T);\ L^2)$

iii) $e^{b\xi}(v(\sigma) - \Phi) \to 0$ in L^2 as $\sigma \to 0$.

To prove the theorem we suppose that there is another function $\tilde{v}$ with the same properties. Then, $w = v - \tilde{v}$ satisfies

$$\frac{dw}{d\sigma} + D^3 w + vDw + wD\tilde{v} + r(\sigma)w = 0 .$$

Following [27], we have that

(13) $\frac{d}{d\sigma}\|e^{b\xi} w\|^2 \leqslant 2K \|e^{b\xi} w\|^2 \qquad 0 < \sigma < T$

where K is a constant, because holds

$\|r(\sigma) e^{b\xi}\|^2 \leqslant |r(\tilde{\sigma})| \|e^{b\xi} w\|^2$, with $K = |r(\tilde{\sigma})|$, where $\tilde{\sigma} = 0$, if $r(\sigma) > 0$ and $r'(\sigma) > 0$ or $\tilde{\sigma} = T$ if $r(\sigma) < 0$ and $r'(\sigma) < 0$. Since condition iii) implies that $\|e^{b\xi}w\| \to 0$ as $\sigma \to 0$, it follows from (19) that $\|e^{b\xi}w\| = 0$ for $0 < \sigma < T$. This proves that $v = \tilde{v}$ as required.

4. Multi-dimensional equations and the KdV-Burgers' equation.

Finally, in this section we introduce a new class of equations modifying Burgers' or KdV equations on which several studies are current . In sect. 2 and sect. 3, when we have introduced modified KdV or Burgers' equations describing multi-dimensional propagation of waves, we have supposed that all the spatial variables are equivalent. But, if we think that, in two-dimensional propagation, the two spatial variables have different scaling, a new class of equations arise: the two-dimensional KdV equation [28]-[29]

(14) $(u_t + uu_x + u_{xxx})_x + u_{yy} = 0$

and the two-dimensional Burgers' equation

(15) $(u_t + uu_x + u_{xx})_x + u_{yy} = 0$.

The eq. (14), also known as Kadomtsev-Petviashvili equation, arises in in plasma physics or in studying the water waves [30] . The eq. (15) arises in the two-dimensional propagation in dissipative media [31]. The eq. (14) and (15) are the objects of several current searchs, and only for the (14) pioneristic results are found. The eqs. (14)-(15) can be generalized in considering non homogeneous media or propagation in more than two dimensions [32]. In the last case we obtain the following equation

$$(16) \quad (u_t + g(t)u + uu_x + u_{xxx})_x + u_{yy} = 0$$

where we have supposed that the spatial variables have different scaling.

In the nonlinear and dispersive media can exist various dissipative effects which enter in the balance equations. The result is that a large number of new equations arise [33]-[35]. In general the equation ruling the propagation of waves in such media explains a balance between the nonlinearity, wave front geometry, dispersion and various types of dissipation. A general resulting equation that describes a such balance is the following

$$(17) \quad \eta(\sigma)v_\sigma + \alpha(\sigma)v + \beta(\sigma)\xi v_\xi + \gamma(\sigma)vv_\xi + \delta(\sigma)v_{\xi\xi} + \mu(\sigma)v_{\xi\xi\xi} = 0 .$$

When the coefficients are constant and $\alpha=\beta=0$, (17) is the well known KdV-Burgers' equation.

REFERENCES

[1] G.B.WHITHAM: "Linear and nonlinear waves" John Wiley (1974) New York.
[2] T.TANIUTI-C.C.WEI: J. Phys. Soc. Jpn. 24, (1968) 941.
[3] T.TANIUTI: Suppl. Progr. Phys. 55, (1974) 1.
[4] N.ASANO-H.ONO: J.Phys. Soc. Jpn. 31, (1971) 1830.
[5] D.G.CRIGHTON: Ann. Rev. Fluid Mech. 11, (1979) 11.
[6] S.GIAMBO'-A.PALUMPO-P.PANTANO: Ann. di Matem. Pura ed Appl. (1981).
[7] S.GIAMBO'-A.GRECO-P.PANTANO: C.R. Acad. Sc. Paris 298 A,(1979) 553.
[8] S.GIAMBO'-A.GRECO-P.PANTANO: C.R. Acad. Sc. Paris 288 A,(1979) 85.
[9] G.BUSONI-P.PANTANO: Comm. 9° ISNA Leeds (1981).
[10] R.M.MARTIN Jr: "Nonlinear Operators and Differential Equations in a Banach Spaces" John Wiley (1976) New York.
[11] A.FRIEDMANN: "Partial Differential Equations" R.E.Krieger (1976).
[12] P.E.SOBOLEVSKIJ: Am. Math. Soc. Transl. 2, (1965) 1.
[13] N.J.ZABUSKI-M.D.KRUSKAL: Phys. Rev. Lett. 15, (1965). 240.
[14] C.S.GARDNER-J.M.GREEN-M.D.KRUSKAL-R.M.MIURA:Comm. Pure Appl. Math. 21, (1968) 467.
[15] C.S.GARDNER-J.M.GREEN-M.D.KRUSKAL-R.M.MIURA:Phys. Rev. Lett.19,(1974)97.
[16] A.C.SCOTT-F.V.F.CHU-D.W.MCLAUGHLIN: IEEE 61, (1973) 1443.
[17] T.KAKUTANI: Suppl. Progr. Phys. 55, (1974) 92.
[18] H.LONNGREN:"Soliton in action" Acad. Press (1978) London.
[19] T.BRUGARINO-P.PANTANO: Phys.lett. 80 A , (1980) 223.
[20] S.MAXON-J.VIECELLI: Phys. Fluids 17, (1974) 1614.
[21] T.BRUGARINO-P.PANTANO: Phys. Lett. 86 A , (1981) 478.
[22] J.W.BARKER-G.B.WHITHAM: Comm. Pure Appl. Math. XXXIII, (1980) 447.
[23] T.KAKUTANI: J. Phys. Soc. Jpn. 30, (1971) 272.
[24] S.GIAMBO'-P.PANTANO: Lett. Nuovo Cimento 34, (1982) 380 .
[25] D.J.KAUP-A.C.NEWELL: Proc. Roy. Soc. London 361 A , (1978) 413.
[26] V.L.KARPMANN-E.M.MASLOV: Phys. Lett. 60 A , (1977) 307.
[27] T.KATO: Res. Not. in Math. 53, Pitman ed. , (1980) 293.
[28] B.B.KADOMTSEV-V.I.PETVIASHVILI: Sov. Phys. Dokl. 15, (1970) 539.
[29] V.S.DRYUMA: Sov. Phys. JETP Lett. 19, (1974) 382.
[30] R.S.JOHNSON: J. Fluid Mech. 97, (1980) 701.
[31] T.BRUGARINO-P.PANTANO: To be published.
[32] P.PANTANO: In preparation.
[33] E.OTT-R.N.SUDAN: Phys. Fluids 12, (1969) 2388.
[34] E.OTT-R.N.SUDAN: Phys. Fluids 13, (1970) 1432.
[35] F.BAMPI-A.MORRO: Il Nuovo Cimento 46, (1981) 551.

PERIODIC SOLUTIONS OF NEUTRAL FUNCTIONAL DIFFERENTIAL EQUATIONS

by

A. CAÑADA and P. MARTINEZ-AMORES
Departamento de Ecuaciones Funcionales
Universidad de Granada, Granada, Spain.

1. INTRODUCTION.

In this paper we prove the existence of periodic solutions for a class of neutral functional differential equations of the form

$$\frac{d}{dt} D(t,x_t) = g(t,x_t)$$

where D is a uniformly stable operator, linear in the second variable and g is a nonlinear mapping.

Our results are inspired and generalize whose by Hale and Mawhin [6], where they consider quasibounded nonlinearities. Our theorems are based in some continuation theorems in the frame of the coincidence degree developed by Mawhin [8] and in an existence theorem for operator equations proved in [1], where we consider nonlinearities that include quasibounded terms and also, terms of exponential type.

Some related results to this paper are whose by Hetzer [7] and Sadovskii [9] (see also Hale [4]), where the nonlinear mapping g can depend of the derivative $\dot{x}_t$, but our nonlinearities are not included in their results.

Also, in the scalar case our results include the asymptotic conditions of Landesman-Lazer type such as it is done in [2].

2. EXISTENCE THEOREMS FOR NONLINEAR OPERATOR EQUATIONS.

Let X and Z be real normed spaces and $L : \mathrm{dom}\, L \subset X \to Z$ a linear Fredholm mapping of index zero. Then there exist continuous projections $P : X \to X$ and $Q : Z \to Z$ such that $\mathrm{Im}\, P = \ker L$, $\mathrm{Im}\, L = \ker Q$. Moreover the mapping $L_P : \mathrm{dom}\, L \cap \ker P \to \mathrm{Im}\, L$ is invertible; denote its inverse by $K : \mathrm{Im}\, L \to \mathrm{dom}\, L \cap \ker P$.

Let Ω be an open bounded subset of X such that $\mathrm{dom}\, L \cap \Omega \neq \phi$ and $N : \bar{\Omega} \to Z$ a nonlinear mapping. The mapping N is said to be L-compact on $\bar{\Omega}$ if the mappings $QN : \bar{\Omega} \to Z$ and $K(I - Q)N : \bar{\Omega} \to X$ are compact, i.e. continuous and $QN(\bar{\Omega})$, $K(I - Q)N(\bar{\Omega})$ are relatively compact.

Let $C_L(\Omega)$ denote the class of mappings F : dom $L \cap \overline{\Omega} \to Z$ which are of the form $F = L - N$ with $N : \overline{\Omega} \to Z$ L-compact on $\overline{\Omega}$ and which satisfy the condition $0 \notin F(\text{dom } L \cap \partial\Omega)$. Then, if $L - N \in C_L(\Omega)$, by using the Leray-Schauder degree, it is possible to define an integer that we shall denote by $d_M[(L,N), \Omega]$ and called the coincidence degree of L and N in Ω, not identically zero and satisfying the classic conditions of the topological degree (see [8]).

Consider now the equation

$$Lx = Nx \tag{2.1}$$

By using the previous concept and a Leray-Schauder type existence theorem in [8], the following theorems are proved in [1]:

<u>Theorem 2.1</u>. Let X,Z be real normed spaces and denote by $|\cdot|$ the corresponding norms. Let L : dom $L \subset X \to Z$ a linear Fredholm mapping of index zero and $N : X \to Z$ a nonlinear L-compact mapping on bounded subsets of X verifying the following assumptions :

1) K is continuous and there exists a linear functional $\gamma: Z \to R$ with $\text{Im } L \subset \ker\gamma$ and constants $\alpha_1 \geq 0$, $\beta_1 \geq 0$, such that
$$|Nx| \leq \gamma(Nx) + \alpha_1|x| + \beta_1 \tag{2.2}$$
for every $x \in X$.
2) There exists a mapping $\Phi: X \to Z$ which is L-compact on bounded subsets of X and constants $\alpha_2 \geq 0$, $\beta_2 \geq 0$ such that
$$|\Phi x| \leq \gamma(\Phi x) + \alpha_2|x| + \beta_2$$
for every $x \in X$.
3) Every possible solution x of the equation $\lambda QNx + (1-\lambda)Q\Phi x = 0, \lambda \in]0,1[$ satisfies the relation
$$|Px| < \nu|(I - P)x| + R \tag{2.3}$$
for some $R > 0$ and some $\nu \geq 0$.
4) $d_M[(L,\Phi), B_X(s)] \neq 0$, for every $s \geq R$, where $B_X(s)$ is the open ball of center 0 and radius s in X.

Then there is $\alpha_0 > 0$ such that equation (2.1) has at least one solution provided $\alpha_1 + \alpha_2 \leq \alpha_0$.

<u>Theorem 2.2</u>. Let L,N be as in Theorem 2.1 and suppose that the following assumptions hold .

a) Assumption 1) of Theorem 2.1.
b) Every possible solution x of the equation $QNx = 0$ satisfies the relation (2.3).
c) The Brouwer degree $d_B(JQN, B_X(s) \cap \ker L, 0)$ is not zero, for every $s \geq R$.

Then there is $\alpha_0 > 0$ such that equation (2.1) has at least one solution provided $\alpha_1 \leq \alpha_0$.

3. SOME NOTATIONS AND MATHEMATICAL BACKGROUND.

For $h \geq 0$, let $C = C([-h,0],R^n)$ be the space of continuous functions mapping $[-h,0]$ into R^n with the topology of uniform convergence. The norm in C will be designated by $\|\psi\| = \sup_{\theta\in[-h,0]} |\psi(\theta)|$, $\psi\in C$, where $|\cdot|$ is any norm in R^n. For x in $C([-h,b],R^n)$, $b > 0$, let $x_t \in C$, $t\in[0,b]$ be defined by $x_t(\theta) = x(t+\theta)$, $\theta\in[-h,0]$.

Suppose $D : R\times C \to R^n$ is a continuous function, linear in the second variable, such that $D(t,\psi) = \psi(0) - l(t,\psi)$, where

$$l(t,\psi) = \int_{-h}^{0} [d_\theta \mu(t,\theta)]\psi(\theta)$$

satisfies

$$\left| \int_{-s}^{0} [\, d_\theta \; \mu(t,\theta)\,]\psi(\theta) \right| \leq m(s)\|\psi\| \quad , \; s\in[0,h]$$

where μ is a nxn matrix function of bounded variation in θ and m is a continuous nondecreasing function from $[0,h]$ into R with $m(0) = 0$.

If $g : R\times C \to R^n$ is a continuous function, then the relation

$$\frac{d}{dt} D(t,x_t) = g(t,x_t) \tag{3.1}$$

is a neutral functional differential equation.

From now we shall assume that g takes bounded sets into bounded sets and that, for a fixed $T > 0$, D and g are T-periodic with respect to t. We shall concerned with the existence of T-periodic solutions of (3.1).

Let X denote the Banach space of mappings $x : R \to R^n$ which are continuous and T-periodic with the norm

$$\|x\|_X = \sup_{t\in R} |x(t)| = \sup_{t\in[0,T]} |x(t)| \quad .$$

By Z we denote the Banach space of continuous mappings $z : R \to R^n$, with $z(0) = 0$, which are of the form $z(t) = dt + x(t)$ for some $d\in R^n$ and $x\in X$, with the norm $\|z\|_Z = |d| + \|x\|_X$. If we define the operators $L : X \to Z$, $N : X \to Z$ by

$$(Lx)(t) = D(t,x_t) - D(0,x_0), \quad (Nx)(t) = \int_0^t g(s,x_s)ds, \; t\in R$$

then our problem is equivalent to solving the operator equation $Lx = Nx$.

It is proved in [6] (see also [5]) that if the operator D is uniformly stable, i.e. the solutions of the functional equation $D(t,x_t) = 0$ are uniformly asymptotically stable, then, for any $c\in R^n$, there is a unique T-periodic solution Mc of the equation $D(t,x_t) = c$. Furthermore, $M : R^n \to X$ is a continuous linear operator and

$$\ker L = \{x\in X : \text{there exists } c\in R^n \text{ with } x = Mc\} .$$

If we define the operators

$$P : X \to X,\ x \to Px = M(D(0,x_0))$$
$$Q : Z \to Z,\ z \to Qz = \frac{1}{T}\, z(T)t,\ t \in R$$

then P and Q are continuous projectors such that Im P = ker L, ker Q= Im is closed in Z and dim ker L = codim ker Q. So L is a continuous linear Fredholm operator of index zero.

Also, the mapping L_P : ker P → Im L has a continuous linear inverse K : Im L → ker P and the mapping N is compact. Hence N is L-compact on bounded subsets of X.

Now, we are in position to apply to equation (3.1) the results of Section 2.

4. EXISTENCE THEOREMS OF PERIODIC SOLUTIONS.

We shall consider in this section T-periodic equations of the form (3.1), where D and g satisfy the conditions in Section 2.

Theorem 4.1. Assume that the following conditions hold.

a) There exist $a \in R^n$, $\alpha_1 \geq 0$, $\beta_1 \geq 0$, such that

$$|g(t,\psi)| \leq \langle a,g(t,\psi)\rangle + \alpha_1 \|\psi\| + \beta_1$$

for every $(t,\psi) \in R\times C$.

b) There exist a continuous, T-periodic mapping taking bounded sets into bounded sets $f : R\times C \to R^n$, $(t,\psi) \to f(t,\psi)$, and constants $\alpha_2 \geq 0$, $\beta_2 \geq 0$, such that

$$|f(t,\psi)| \leq \langle a,f(t,\psi)\rangle + \alpha_2 \|\psi\| + \beta_2$$

for every $(t,\psi) \in R\times C$.

c) Let $\Phi: X \to Z$ be defined by $(\Phi x)(t) = \int_0^t f(s,x_s)ds$, $t \in R$, and suppose that every solution x of the equation

$$\lambda QNx + (1-\lambda)Q\Phi x = 0, \lambda \in]0,1[$$

satisfies the relation

$$\|Px\|_X < R + \nu\|(I-P)x\|_X$$

for some $R > 0$ and some $\nu \geq 0$.

d) $d_M[(L,\Phi),B_X(s)] \neq 0$, for every $s \geq R$.

Then there is $\alpha_0 > 0$ such that equation (3.1) has at least one T-periodic solution provided $\alpha_1 + \alpha_2 \leq \alpha_0$.

Proof. It consists in showing that conditions of Theorem 2.1 are satisfied. We know that K is continuous and N,Φ are L-compact on bounded sets of X. Since conditions c) and d) are the same as 3) and 4) in Theorem 2.1, it only have to prove that 1) and 2) follow from a) and b).

If we define the linear functional $\gamma: Z \to R$ by

$$\gamma(z) = (1 + 2T) \langle a, \tfrac{1}{T} z(T) \rangle$$

then Im $L \subset \ker \gamma$.

If $x \in X$, then

$$\|Nx\|_Z = \left| \frac{1}{T} \int_0^T g(s,x_s)ds \right| + \sup_{t \in [0,T]} \left| \int_0^t (g(s,x_s) - \frac{1}{T}\int_0^T g(s,x_s)ds)ds \right| =$$

$$\leq (\tfrac{1}{T} + 2) \int_0^T |g(s,x_s)| ds$$

$$\leq (1 + 2T) \langle a, \frac{1}{T} \int_0^T g(s,x_s)ds \rangle + \bar{\alpha}_1 \|x\|_X + \bar{\beta}_1$$

$$= \gamma(Nx) + \bar{\alpha}_1 \|x\|_X + \bar{\beta}_1,$$

where $\bar{\alpha}_1 = \alpha_1(2T + 1)$ and $\bar{\beta}_1 = \beta_1(2T + 1)$, so that inequality (2.2) in Theorem 2.1 is verified. Analogously, condition 2) follows from assumption b), and the proof is complete.

It is possible to substitute condition c) in Theorem 4.1 by an asymptotic condition expresed in terms of nonstrict inequalities on the nonlinearities g and f if the operator M associated with the equation $D(t,x_t) = c$, $c \in R^n$, has the following property.

Definition [6]. The operator M has property μ if there exists $\mu > 0$ such that $|(Mc)(t)| \geq \mu |c|$ for every $t \in R$ and every $c \in R^n$.

Theorem 4.2. Assume that the operator M has property μ and that conditions a), b) and d) in Theorem 4.1 hold. Also, assume that the following condition is satisfied :

c') There exists $r > 0$ such that, for all $x \in X$ with $\min_t |x(t)| \geq r$, one h

$$\langle \int_0^T g(t,x_t)dt, \int_0^T f(t,x_t)dt \rangle \geq 0$$

where both integrals are not simultaneously the null vector.

Then there is $\alpha_0 > 0$ such that equation (3.1) has at least one T-periodic solution provided $\alpha_1 + \alpha_2 \leq \alpha_0$.

Proof. We shall prove that condition c) of Theorem 4.1 follows from condition c'). If $x \in X$, $\lambda \in]0,1[$ are such that

$$\lambda QNx + (1 - \lambda) Q\Phi x = 0$$

then

$$\lambda \frac{1}{T}(Nx)(T)t + (1 - \lambda)\frac{1}{T}(\Phi x)(T)t = 0$$

that is

$$\lambda \int_0^T g(t,x_t)dt + (1 - \lambda) \int_0^T f(t,x_t)dt = 0$$

By c'), there must exist some $t \in [0,T]$ for which $|x(t)| < r$. Hence, if $c \in R^n$ is such that $Px = Mc$, we have

$\mu|c| \leq |(Mc)(t)| \leq |x(t)| + |((I - P)x)(t)|$

which implies

$\|Px\|_X \leq |M||c| \leq |M|\mu^{-1}(r + \|(I - P)x\|_X) = R + \nu\|(I - P)x\|_X$,

where $R = |M|\mu^{-1}r$ and $\nu = |M|\mu^{-1}$.

The problem of computing the coincidence degree in condition d) of Theorem 4.1 can be reduced to the study of Brouwer degree of some well defined finite-dimensional mapping such as is done in [8],[1]. Thus, by using Theorem 2.2 we have the following corollary which generalizes Theorem 5.2. of Hale and Mawhin [6], where they suppose quasibounded nonlinearities.

<u>Corollary 4.1</u>. Assume that the following conditions hold.

1) There exist $a \in R^n$, $\alpha_1 \geq 0$, $\beta_1 \geq 0$, such that

$$|g(t,\psi)| \leq \langle a, g(t,\psi)\rangle + \alpha_1 \|\psi\| + \beta_1$$

for every $(t,\psi) \in R\times C$.

2) The operator M has property μ.

3) There exists $r > 0$ such that, for all $x \in X$ with $\min_t |x(t)| \geq r$, one has

$$\int_0^T g(t,x_t)dt \neq 0$$

4) The Brouwer degree $d_B(\mathcal{H}, B_X(s) \cap \ker L, 0) \neq 0$, for every $s \geq |M|\mu^{-1}r$,

where $\mathcal{H}: R^n \to R^n$, $c \to \frac{1}{T}\int_0^T g(t,(Mc)_t)dt$

Then, there is $\alpha_0 > 0$ such that equation (3.1) has at least one T-periodic solution provided $\alpha_1 \leq \alpha_0$.

We note that if $D(t,\psi)$ is independent of t, then the operator M has always property μ and we can take M = I, the identity operator, and $\mu = 1$. In this case $\ker L = \{$ constant functions in X $\}$.

For the scalar case, n = 1, we can reduce conditions in Corollary 4.1 to conditions of Landesman-Lazer type such as it is done in [1]

Consider the scalar equation

$$\frac{d}{dt}[x(t) - \sum_{i=1}^m a_i x(t - h_i)] = g(t,x_t) \qquad (4.1)$$

where $h_i \in [0,h]$, $\sum_{i=1}^m |a_i| < 1$ and $g : R\times C \to R$ is T-periodic with respect to t, continuous and takes bounded sets into bounded sets. Then $D(\psi) = \psi(0) - \sum_{i=1}^m a_i\psi(-h_i)$ is uniformly stable (see [3]) and M has property μ.

<u>Corollary 4.2</u>. Assume that the following conditions hold .

1) There exist $\alpha_1 \geq 0$, $\beta_1 \geq 0$, such that

$$|g(t,\psi)| \leq g(t,\psi) + \alpha_1 \|\psi\| + \beta_1.$$

for all $(t,\psi) \in R\times C$.

2) There exists $r > 0$ such that, for all $x \in X$ with $\min_t |x(t)| \geq r$, one has

$$\operatorname{sign} x(t) \int_0^T g(t,x_t)dt \geq 0$$

Then there is $\alpha_0 > 0$ such that equation (4.1) has at least one T-periodic solution provided $\alpha_1 \leq \alpha_0$.

In the case that the nonlinear term g be of the particular form $g(t,x(t-h))$, i.e. $g: R\times R \to R$, $(t,y) \to g(t,y)$, it is continuous and T-periodic with respect to t and there are constants $\delta+$ and $\delta-$ with $g(t,y) \geq \delta+$ for $y \geq 0$ and $g(t, \ ) \leq \delta-$ for $y \leq 0$, we can substitute (see[2]) condition 2) in the previous corollary by the condition

2') $\int_0^T G_-(t)dt < 0 < \int_0^T G_+(t)dt$

where $G_\pm : R \to R \cup \{+\infty, -\infty\}$ are defined by

$$G_+(t) = \liminf_{y \to +\infty} g(t,y), \quad G_-(t) = \limsup_{y \to -\infty} g(t,y).$$

For example, if the function $g: R\times C \to R$ in (4.1) is defined by $g(t,\psi) = e^{\psi(-h)} + f(t)$, $f \in X$, and $\int_0^T f(t)dt < 0$, then the equation

$$\frac{d}{dt}\left[x(t) - \sum_{i=1}^m a_i x(t-h_i)\right] = e^{x(t-h)} + f(t)$$

will have a T-periodic solution.

REFERENCES

1. A. CAÑADA and P. MARTINEZ-AMORES. Solvability of some operator equations and periodic solutions of nonlinear functional differential equations. To appear in J. Diff. Eqns.
2. A. CAÑADA and P. MARTINEZ-AMORES. Periodic solutions of nonlinear vector ordinary differential equations of higher order at resonance. To appear in Nonlinear Anal.
3. M.A. CRUZ and J.K. HALE. Stability of functional differential equations of neutral type. J. Diff. Eqns. 7,(1.970),334-355.
4. J.K. HALE. α-contractions and differential equations. Equations differentielles et fonctionelles non linéaires, 15-42. Hermann, París, 1.973.
5. J.K. HALE. Oscillations in neutral functional differential equations. In Nonlinear Mechanics. C.I.M.E., June, 1.972.
6. J.K. HALE and J. MAWHIN. Coincidence degree and periodic solutions of neutral equations. J. Diff. Eqns. 15, (1.974), 295-307.
7. G. HETZER. Some applications of the coincidence degree for k-set contractions to functional differential equations of neutral type. Comment. Math. Univ. Carolinae, 16, (1.975), 121-138.
8. J. MAWHIN. Topological degree methods in nonlinear boundary value problems. C.B.M.S. Reg. Conf. Series in Math. 40, Amer. Math. Soc., 1.978.
9. B.N. SADOVSKII. Limit compact and condensing operators. Russian Math. Surveys, (1.972), 85-146.

ON SUBQUADRATIC NOT-AUTONOMOUS HAMILTONIAN SYSTEMS

A. Capozzi - BARI

0. Introduction.

Consider the Hamiltonian system of 2n ordinary differential equations

(0.1) $\dot{p} = -H_q(t,p,q) \qquad \dot{q} = H_p(t,p,q)$,

where $H \in C^1(\mathbb{R} \times \mathbb{R}^{2n}, \mathbb{R})$, p and q are n-tuples, $t \in \mathbb{R}$, $H_q = \text{grad}_q H$, $H_p = \text{grad}_p H$ and $\cdot$ denotes $\frac{d}{dt}$. This system can be represented more concisely as

(0.2) $-J\dot{z} = H_z(t,z)$,

where $z=(p,q)$, $H_z = \text{grad}_z H$ and $J = \begin{pmatrix} 0 & -I \\ I & 0 \end{pmatrix}$, I being the identity matrix in $\mathbb{R}^n$. In the sequel we suppose that $H(t,z)$ is T-periodic in t.

There are many types of question in the study of periodic solutions of such systems (cf. [15] and its references). We will be concerned with the existence of periodic solutions of (0.2) when the period T is prescribed and H(t,z) is "subquadratic", i.e. grows less than $|z|^2$ as $|z| \to \infty$ uniformly in t. The results known to the author, concerning the "subquadratic" case for general Hamiltonian systems are contained in Amann [1], Amann - Zehnder [2-3], Benci-Rabinowitz [9], Clarke-Ekeland [12] and Coron [13]. In [12] and [13] the Hamiltonian H(t,z) is required to be convex. In [1], [2] and [3] the Hessian matrix $H_{zz}(t,z)$ is bounded. As in [9] we require that H(t,z) is only differentiable and we shall prove the following theorem:

Theorem 0.3. Let $H \in C^1(\mathbb{R} \times \mathbb{R}^{2n}, \mathbb{R})$ and suppose that there exist positive constants M, r_1, r_2, ε such that

H_1) $|H_z(t,z)| \leqslant \varphi_1(|z|)$ for $z \in \mathbb{R}^{2n}$, $|z| > r_1$ and $t \in \mathbb{R}$

H_2) $H(t,z) \geqslant \varphi_2(|z|) - M$ for $z \in \mathbb{R}^{2n}$, and $t \in \mathbb{R}$

H_3) $H(t,z) - \frac{1}{2}(z, H_z(t,z))_{\mathbb{R}^{2n}} \geqslant \varphi_1^{1+\varepsilon}(|z|)$ for $z \in \mathbb{R}^{2n}$, $|z| > r_2$ and $t \in \mathbb{R}$

where φ_1 and φ_2 are two positive continuous functions satisfying the following assumptions

i) $\frac{\varphi_1(|z|)}{|z|} \to 0$ as $|z| \to \infty$ and φ_1 is increasing

ii) $\frac{\varphi_2(|z|)}{\varphi_1^{1+\varepsilon}(|z|)} \to +\infty$ as $|z| \to \infty$

iii) $\varphi_1^{1+\varepsilon}(|z|)$ is Lip. continuous

iv) $\varphi_2(|z|) \to +\infty$ as $|z| \to \infty$

Then (0.2) possesses at least a T-periodic solution.

Remark. Benci-Rabinowitz have obtained the similar results in [9] (theorems 4.1 and 4.11).

Th. 0.3. is a variant of theorem 4.11 and in comparison with theorem 4.1 we don't require the strong assumption of boundness of H_z.

Example: $H(z) = |z| \lg(1+|z|)$ doesn't verify the assumptions of theorems 4.1 and 4.11 but verifies the assumptions of theorem 0.3.

1. Notations and preliminaries.

If E is a real Hilbert space we denote by $(\cdot,\cdot)$ the scalar product in E, by $||\cdot||$ the norm in E, by $C^1(E, \mathbb{R})$ the space of cont. Fréchet differ. maps from E to $\mathbb{R}$ and, if $f \in C^1(E, \mathbb{R})$, by $f'(u)$ its derivative at $u \in E$. We shall identify E with its dual E' and we denote by $\langle\cdot,\cdot\rangle$ the pairing between E' and E. If $R > 0$ we set $B_R = \{u \in E \mid ||u|| < R\}$.

Observe that making the change of variable $t \to \frac{2\pi t}{T} = \frac{t}{\lambda}$, (0.2) becomes

$$(1.1) \qquad -J\dot{z} = \lambda H_z(\lambda t, z)$$

and the 2π-periodic solutions of (1.1) correspond to the T-periodic solutions of (0.2).

We set $\tilde{E} = \{z(t) \in C^\infty(\mathbb{R}, \mathbb{R}^{2n}) \mid z(t) \text{ is } 2\pi\text{-periodic}\}$. If $z(t) \in \tilde{E}$, then z(t) has a Fourier expansion $z = \sum_{-\infty}^{+\infty} a_k e^{ikt}$ with $a_k \in \mathbb{C}^{2n}$ and $a_{-k} = \bar{a}_k$.

In the sequel we denote by E the closure of $\tilde{E}$ under the norm

$$||z|| = \left(\sum_{k\in\mathbb{Z}} (1 + |k|)\ |a_k|^2\right)^{\frac{1}{2}}$$

and we observe that E can be identified with Sobolev space $W^{\frac{1}{2},2}(S^1)$ obtained by interpolating between $L^2(S^1)$ and $W^{1,2}(S^1)$ (cf., e.g., [7] and appendix of [8]).

Suppose now that there exist constants $\bar{c}_1, \bar{c}_2, \mu > 0$ such that

$$|H_z(t,z)| \leqslant \bar{c}_1 + \bar{c}_2\ |z|^\mu \qquad \text{for every } (t,z) \in \mathbb{R} \times \mathbb{R}^{2n} \tag{1.2}$$

and let $L : E \to E$ be the continuous linear operator such that

$$\frac{1}{2}(L\ z,\ z) = \int_0^{2\pi} (p,\ \dot{q})_{\mathbb{R}^n}\ dt\ . \tag{1.3}$$

Standard arguments (cf., e.g., [7], [9]) show that the functional

$$f(z) = \frac{1}{2}(L\ z,\ z) - \psi(z) \qquad z \in E\ , \tag{1.4}$$

where $\psi(z) = \lambda\int_0^{2\pi} H(\lambda t,\ z)\ dt$, satisfies the following assumptions:

f_0) $f \in C^1(E, \mathbb{R})$

f_1) i) L is a continuous self-adjoint operator on E

ii) $\psi \in C^1(E, \mathbb{R})$ and ψ' is a compact operator

f_2) 0 is a finite multiplicity isolated eigenvalue of L.

Moreover the critical points of the functional (1.4) correspond to the 2π-periodic solutions of (1.1).

For $z \in E$ we set

$$A(z) = \frac{1}{2}(L\ z,\ z) \tag{1.5}$$

and if E^-, E°, E^+ are the subspaces of E on which A is negative, null, positive definite, then it can be verified that $E = E^- \oplus E^\circ \oplus E^+$ (cf., e.g., [14]). Moreover if $z = z^- + z^\circ + z^+ \in E$ we can take (cf. [9]) the equivalent norm in E

(1.6) $$||z||^2 = |z^\circ|^2 + A(z^+) - A(z^-)$$

and we observe that if $\beta \in [1, \infty)$ and $z \in L^\beta(S^1)$, then there exists $a_\beta > 0$ such that

(1.7) $$||z||_{L^\beta} \leq a_\beta ||z||$$

with the embedding of E in L^β being compact.

By (f_1, i) we can consider the following decomposition of E

(1.8) $$E = \bigoplus_{j \in \mathbb{Z}} E_j$$

where E_j $(j \in \mathbb{Z})$ is the eigenspace corresponding to eigenvalue λ_j of L.

We set

(1.9) $$\mathscr{U} = \left\{ U : E \to E \mid U \text{ is a bounded homeomorphism of the form } U = e^{\alpha(\cdot) L}[\cdot], \text{ where } \alpha(\cdot) \in C(E, \mathbb{R}) \right\}$$

(1.10) $$\mathscr{B} = \left\{ b : E \to E \mid b \text{ is a bounded continuous operator such that for every } R > 0 \;\; b(B_R) \subset \bigoplus_{i \in I(R)} E_j \text{ for a finite set } I(R) \subset \mathbb{Z} \right\}$$

(1.11) $$\mathscr{H} = \left\{ h : E \to E \mid h \text{ is a bounded homeomorphism such that } h = U + b,\; h^{-1} = U' + b', \text{ where } U, U' \in \mathscr{U} \text{ and } b, b' \in \mathscr{B} \right\}$$

Given a constant β and a functional $f : E \to \mathbb{R}$ we set

(1.12) $$\mathscr{L}(f, \beta) = \{ h \in \mathscr{H} \mid U(u) = u,\; b(u) = 0 \text{ for } u \notin f^{-1}[\beta, +\infty) \}.$$

<u>Definition 1.13.</u> Given two Hilbert manifolds S and Q, we say that S and ∂Q "$\mathscr{L}(f,\beta)$ - link" if

i) $\partial Q \cap S = \emptyset$

ii) $h(Q) \cap S \neq \emptyset$

for every $h \in \mathscr{L}(f,\beta)$ for which $h(u) = u$ on ∂Q.

The following results hold:

Theorem 1.14. Given two constants c_α, c_β *let* $f \in C^1(E, \mathbb{R})$ *be a functional which satisfies* (f_1), (f_2) *and*

f_3) *given* $\bar{c} \in]c_\alpha, c_\beta[$, *every sequence* $\{u_n\}$, *for which* $\{f(u_n)\} \to \bar{c}$ *and* $||f'(u_n)|| \cdot ||u_n|| \to 0$, *possesses a bounded subsequence.*

Moreover, given two Hilbert manifolds S *and* Q *and two constants* α *and* β *with* $\alpha \ngtr \beta$ *we suppose that*

i) $f(u) \geq \alpha$ *on* S

ii) $f(u) \leq \beta$ *on* ∂Q

iii) $\sup_Q f(u) < +\infty$

iv) S *and* ∂Q "$\mathscr{L}(f,\beta)$ - link"

Then f *possesses a critical value* $c \geq \alpha$.

Proof: This theorem can be proved by using some abstract results obtained in [7]. We refer to [10] for a detailed proof.

Lemma 1.15. Let $f : E \to \mathbb{R}$ *be a functional of the form* (1.4) *and let* E_1, E_2 *be two closed L-invariant subspaces such that* $E_1 \oplus E_2 = E$. *Moreover, given two constants* $R > 0$ *and* β, *suppose* $Q = B_R \cap E_1$, $q \in Q$, $S = q + E_2$ *and* $f(u) \leq \beta$ *on* ∂Q. *Then* S *and* ∂Q "$\mathscr{L}(f,\beta)$-link".

The proof of this lemma is in [10].

By theorem 1.14 and lemma 1.15 we get the following theorem:

Theorem 1.16. Let $g \in C^1(E, \mathbb{R})$ *a functional satisfying* (f_1), (f_2) *and* (f_3). Moreover we suppose that there exist constant R, α, β with $R > 0$ and $\alpha > \beta$ *such that*

i) $g(u) \geq \alpha$ *for any* $u \in S$

ii) $g(u) \leq \beta$ *for any* $u \in \partial Q$

iii) $\sup_{Q} g(u) < +\infty$

<u>where</u> S <u>and</u> Q <u>are defined as in lemma 1.15. Then</u> g <u>possesses at least a critical value</u> $c \geqslant \alpha$.

2. <u>Proof of the theorem.</u>

In order to prove theorem 0.3 we set $g(z) = -f(z)$, where $f(z)$ is the functional (1.4). Obviously $g(z) \in C^1(E, \mathbb{R})$ and satisfies (f_1) and (f_2). We will prove that $g(z)$ satisfies (f_3).

Let $\{z_m\} \subset E$ a sequence such that

$$g(z_m) \to c \tag{2.1}$$

$$||g'(z_m)|| \cdot ||z_m|| \to 0. \tag{2.2}$$

By (2.2) there exists a subsequence, which we indicate always with $\{z_m\}$, and two positive constants $\overline{K}$ and υ such that for any $m > \upsilon$

$$|(g'(z_m), z_m^+)| \leqslant \overline{K} \tag{2.3}$$

Observe that

$$< g'(z), z^+ > \; = \lambda \int_0^{2\pi} (H_z(t,z), z^+)_{\mathbb{R}^{2n}} \, dt - (Lz, z^+) \tag{2.4}$$

then, dropping subscripts, by (2.3) and (2.4) we have

$$2A(z^+) - \lambda \int_0^{2\pi} (H_z(t,z), z^+)_{\mathbb{R}^{2n}} \, dt \leqslant \overline{K} \tag{2.5}$$

By (2.5), (H_1) and (1.7) it follows that ultimately

$$\begin{aligned} 2||z^+||^2 &= 2A(z^+) \leqslant \lambda \int_0^{2\pi} (H_z(t,z), z^+)_{\mathbb{R}^{2n}} \, dt + \overline{K} \leqslant \\ &\leqslant \lambda \int_0^{2\pi} \varphi_1(|z|) \cdot |z^+| \, dt + \overline{K} \leqslant \\ &\leqslant \lambda \, c_0 \, ||\varphi_1(|z|)||_{L^{1+\varepsilon}} \cdot ||z^+|| + \overline{K} . \end{aligned} \tag{2.6}$$

Moreover by (2.1) and (2.2) it follows that (cf. [7])

$$\int_0^{2\pi}\left(H(t,z) - \frac{1}{2}\,(z\,,H_z(t,z))_{\mathbb{R}^{2n}}\right)dt \quad \text{is bounded}\,. \tag{2.7}$$

By (H_3) we have that ultimately

$$\int_0^{2\pi}\left(H(t,z) - \frac{1}{2}(z,H_z(t,z))_{\mathbb{R}^{2n}}\right)dt \geq \int_0^{2\pi} \varphi_1^{\,1+\varepsilon}(|z|)\,dt = ||\varphi_1(|z|)||^{1+\varepsilon}_{L^{1+\varepsilon}}\,. \tag{2.8}$$

By (2.7) and (2.8) we get that

$$||\varphi_1(|z|)||_{L^{1+\varepsilon}} \quad \text{is bounded.} \tag{2.9}$$

Finally by (2.9) and (2.6) we conclude that $||z^+||$ is bounded. By the same arguments also $||z^-||$ is bounded.

Observe now that by (2.1) there exists a constant $\overline{M} > 0$ such that

$$g(z) \leq \overline{M} \qquad \text{for any } z\,. \tag{2.10}$$

Moreover

$$\begin{aligned} g(z) &= \lambda\int_0^{2\pi} H(t,z)\,dt - A(z^+) - A(z^-) = \\ &= \lambda\int_0^{2\pi} H(t,z^\circ)\,dt + \lambda\int_0^{2\pi}(H(t,z) - H(t,z^\circ))\,dt - ||z^+||^2 + ||z^-||^2 \end{aligned} \tag{2.11}$$

and by (i)

$$|H(t,z) - H(t,z^\circ)| \leq \varphi_1\,(|\xi|)\cdot|\tilde z| \tag{2.12}$$

where $|\tilde z| = |z - z^\circ|$ and $|\xi(t)| = \sup\{|z(t)|,\ |z^\circ(t)|\}$.

Then by (2.11), (2.12), (H_1), (H_2) and (1.7) we have that ultimately

$$\begin{aligned} g(z) &\geq \lambda\int_0^{2\pi}\varphi_2(|z^\circ|)\,dt - \lambda\int_0^{2\pi}\varphi_1(|\xi|)\cdot|\tilde z|\,dt - c_1 \geq \\ &\geq 2\pi\,\lambda\,\varphi_2(|z^\circ|) - \lambda\,c_2||\varphi_1(|\xi|)||_{L^{1+\varepsilon}}\cdot||\tilde z|| - c_1\,. \end{aligned} \tag{2.13}$$

If $||\varphi_1(|\xi|)||_{L^{1+\varepsilon}}$ is bounded, then by the boundness of $||z^-||$ and $||z^+||$, by (iv) and by (2.10) it follows that $|z^\circ|$ is bounded.

Consider now the case in which $||\varphi_1(|\xi|)||_{L^{1+\varepsilon}}$ is not bounded. By (2.13) we have that

$$g(z) \geqslant 2\pi\lambda\, \varphi_2(z^\circ) - \lambda c_3 \, \|\varphi_1(|\xi|)\|^{1+\varepsilon}_{L^{1+\varepsilon}} - c_1$$

(2.14)

$$= 2\pi\lambda \left(\varphi_2(|z^\circ|) - c_3\, \varphi_1^{1+\varepsilon}(|z^\circ|)\right) + \lambda c_3 \int_0^{2\pi} \left(\varphi_1^{1+\varepsilon}(|z^\circ|) - \varphi_1^{1+\varepsilon}(|\xi|)\right) dt - c_1$$

Since $\left||z^\circ| - |\xi|\right| < |\tilde{z}|$, then by (iii), (1.7) and boundness of $\|z^-\|$ and $\|z^+\|$ we have that

(2.15) $$\int_0^{2\pi} \left(\varphi_1^{1+\varepsilon}(|z^\circ|) - \varphi_1^{1+\varepsilon}(|\xi|)\right) dt \leqslant c_4 \|\tilde{z}\| \leqslant c_5 .$$

Finally by (2.14), (2.15), (ii) and (2.10) it follows that $|z^\circ|$ is bounded.

Now we shall prove that $g(z)$ satisfies the assumptions (i), (ii) and (iii) of theorem 1.16, where $Q = E^+ \cap B_R$ with $R > 0$ suitable and $S = E^- \oplus E^\circ$.

By (H_2) we have that if $z \in E^- \oplus E^\circ$

$$g(z) = \lambda \int_0^{2\pi} H(t,z)\, dt + \|z^-\|^2 \geqslant$$

(2.16)

$$\geqslant \lambda \int_0^{2\pi} \varphi_2(|z|)\, dt - 2\pi\lambda M + \|z^-\|^2 \geqslant \alpha .$$

Moreover by (H_1) we have that

(2.17) $$H(t,z) \leqslant c_6\, |z| + c_7\, \varphi_1(|z|) \cdot |z| .$$

Then by (2.17) we have that if $z \in E^+$

$$g(z) = \lambda \int_0^{2\pi} H(t,z^+)\, dt - \|z^+\|^2 \leqslant$$

(2.18)

$$\leqslant c_8 \|z^+\| + \lambda\, c_7 \int_0^{2\pi} \varphi_1(|z^+|) \cdot |z^+|\, dt - \|z^+\|^2 .$$

Finally by (i) and by (2.18) we get

(2.19) $$\sup_Q g(z) < +\infty$$

(2.20) $g(z) \leqslant \beta$ on ∂Q

where $Q = E^+ \cap B_R$ and R is a suitable positive constant.

Now by theorem 1.6 the conclusion of theorem 0.3 follows.

References

[1] H. AMANN, Saddle points and multiple solutions of differential equations, Math. Z., 169, (1979), 127-166.

[2] H. AMANN, - E. ZEHNDER, Nontrivial solutions for a class of nonresonance problems and applications to nonlinear differential equations, Ann. Sc. Norm. Sup. Pisa, Cl. Sci. IV Ser. 7, 539-603 (1980) .

[3] H. AMANN, - E. ZEHNDER, Multiple periodic solutions of asymptotically linear Hamiltonian equations, preprint.

[4] V. BENCI, Some critical points Theorem ad Applications, Comm. Pure Appl. Math., 33 (1980).

[5] V. BENCI, A geometrical Index for the group S^1 and some applications to the study of periodic solutions of ordinary differential equations, Comm. Pure Appl. Math., 34, (1981), 393-432.

[6] V. BENCI, On the critical point theory for indefinite functionals in the presence of symmetry, to appear on Trans. Amer. Math. Soc.

[7] V. BENCI - A. CAPOZZI - D. FORTUNATO, Periodic solutions of Hamiltonian systems with a prescribed period, preprint.

[8] V. BENCI - D. FORTUNATO, The dual method in critical point theory - Multiplicity results for indefinite functionals -, to appear on Annali Mat. Pura e App.

[9] V. BENCI - P.H. RABINOWITZ, Critical point theorems for indefinite functionals, Inv. Math., 52, (1979), 336-352.

[10] A. CAPOZZI,On subquadratic Hamiltonian systems. preprint.

[11] D.C. CLARK, Periodic solutions of variational systems of ordinary differential equations, J. Diff. Eq., 28, (1978), 354-368.

[12] F.H. CLARKE - I. EKELAND, Hamiltonian trajectories having prescribed minimal period, Comm. Pure Appl. Math., 33, (1980), 103-116.

[13] J. M., CORON, Resolution de l'équation Au + Bu = f où A est linéare auto-adjoint et B deduit d'un potential convex, to appear. Ann. Fac. Sci. Toulouse.

[14] P.H. RABINOWITZ, Periodic solutions of Hamiltonian systems, Comm. Pure Appl. Math., 31, (1978), 157-184.

[15] P.H. RABINOWITZ, Periodic solutions of Hamiltonian systems: a survey, Math. Research Center Technical Summary Report, University of Wisconsin-Madison.

Author's address: Istituto di Matematica Applicata - Facoltà di Ingegneria
Via Re David, 200 - 70125 - BARI, (Italy).

A NOTE ON A CLASS OF AUTONOMOUS HAMILTONIAN SYSTEMS WITH STRONG RESONANCE AT INFINITY

A. CAPOZZI - BARI A. SALVATORE - TRIESTE

0.

In this note we are looking for solutions $x(t) \in C^2(\mathbb{R}, \mathbb{R}^n)$ of the equations

$$(0.1) \qquad \begin{cases} -\ddot{x} = \nabla U(x) \\ x(o) = x(T) \\ \dot{x}(o) = \dot{x}(T) \end{cases}$$

where $T > o$ is a given period and $U(x) \in C^2(\mathbb{R}^n, \mathbb{R})$, in presence of "strong resonance" at infinity (i.e. $U(x)$ satisfies the assumptions $(I_2.ii)$ and (I_3). The problem (0.1) has been studied by many authors in the non-resonance case (cf.[5]) and in the resonance case (cf.[1] and its references). The results known to the authors concerning the strong-resonance case are contained only in [3] and in [6], but in these papers the system (0.1) is not-autonomous and the potential $U(t,x)$ is even in x.

With the change of variable $t \to \frac{1}{\omega} t$, where $\omega = \frac{T}{2\pi}$, the solutions of the problem (0.1) correspond to the solutions of the problem

$$(0.2) \qquad \begin{cases} -\ddot{x} = \omega^2 \nabla U(x) \\ x(o) = x(2\pi) \\ \dot{x}(o) = \dot{x}(2\pi). \end{cases}$$

We denote by U_{xx} the Hessian matrix of $U(x)$ and we assume that

$$U_{xx}(x) \to M \qquad \text{as } |x| \to \infty$$

where M is an $[n \times n]$ symmetric matrix. If we set $\nabla U(x) = Mx - \nabla V(x)$ the problem (0.2) becomes

$$(0.3) \qquad \begin{cases} -\ddot{x} = \omega^2 Mx - \omega^2 \nabla V(x) \\ x(o) = x(2\pi) \\ \dot{x}(o) = \dot{x}(2\pi). \end{cases}$$

We denote by $\mathscr{L}$ the self-adjoint realization in $L^2((0,2\pi), \mathbb{R}^n)$ of the operator $x \to -\ddot{x} - \omega^2 Mx$ with periodic conditions.

We assume that

I_1) i) $V(o) = o$

ii) $\nabla V(o) = o$

I_2) i) $V(x) \to o$ as $|x| \to \infty$

ii) $(\nabla V(x), x) \to o$ as $|x| \to \infty$

I_3) $o \in \sigma(\mathscr{L})$

I_4) $V_{xx}(o)$ is a positive definite matrix

I_5) there exists $\lambda_h \in \sigma(\mathscr{L})$ $\lambda_h \leqslant o$ s.t. $\lambda_h + \omega^2 \mu > o$,

where $\mu = \min \sigma(V_{xx}(o))$

I_6) M is positive semidefinite or $\nu \leqslant -\frac{\lambda_h}{\omega^2}$

where $\nu = \max \sigma(M)$

I_7) $V(x) \leqslant \frac{1}{2}(Mx,x)$ $\forall x \in \mathbb{R}^n$ s.t. $\nabla V(x) = Mx$

Remark 0.1. By (I_3) M possesses at least one eigenvalue greater or equal than zero. (I_6) and (I_7) are technical assumptions.

We consider the operator $x \to -\ddot{x} - \omega^2 \nabla U(x)$ linearized at infinity and at origin and we set

$$L_\infty(x) = -\ddot{x} - \omega^2 Mx \qquad L_o(x) = -\ddot{x} - \omega^2 Mx + \omega^2 V_{xx}(o)\, x.$$

We denote by m_∞ (resp. m_0) the maximal dimension of subspaces where L_∞ (resp. L_0) is negative semidefinite.

The following theorems hold:

Theorem 0.1. If (I_1) - (I_7) hold, then the problem (0.3) possesses at least m distinct orbits of non-constant solutions with

$$m = \frac{m_\infty - m_0}{2} .$$

Theorem 0.2. Under the same assumptions of theorem 0.1, where I_4 - I_7 are replaced by

I_4') $V_{xx}(o)$ is a negative definite matrix

I_5') there exists $\lambda_s \in \sigma(\mathcal{L})$ $\lambda_s \geq o$ s.t. $\lambda_s + \omega^2 \mu < o$, where $\mu = \max \sigma(V_{xx}(o))$

I_6') M is negative semidefinite or $\nu \geq - \frac{\lambda_s}{\omega^2}$, where $\nu = \min \sigma(M)$

I_7') $\frac{1}{2}(Mx,x) \leq V(x)$ $\forall x \in \mathbb{R}^n$ s.t. $\nabla V(x) = Mx$

the problem (0.3) possesses at least m distinct orbits of non-constant solutions with

$$m = \frac{m_0 - m_\infty}{2} .$$

Remark 0.2. If $M = \lambda_k I$, where I is the identity matrix in $\mathbb{R}^n$ and $\lambda_k \in \sigma(\mathcal{M})$, $\mathcal{M}$ being the self-adjoint realization in $L^2((0,2\pi), \mathbb{R}^n)$ of the operator $x \to -\ddot{x}$ with periodic conditions, (I_6) and (I_6') are verified.

In order to prove theorem (0.1) and theorem (0.2) we shall use an abstract theorem proved in [2].

1.

We set $L^2 = L^2((0,2\pi), \mathbb{R}^n)$, $H^1 = H^1((0,2\pi), \mathbb{R}^n)$ and denote by

$$(\cdot,\cdot), \quad (\cdot,\cdot)_{L^2}, \quad (\cdot,\cdot)_{H^1}$$

respectively the scalar products on $\mathbb{R}^n$, L^2, H^1. We set $H = \{u \in H^1 \mid u(o) = u(2\pi)\}$ equipped with the scalar product

$$(u,v)_H = (u,v)_{H^1}.$$

In the sequel we shall use the unique symbol $||\cdot||$ for the norms in H and in its dual H'. We say that $f \in C^1(H, \mathbb{R})$ if f is cont. Fréchet differ. on H and we denote by $f'(u)$ its derivative at $u \in H$. If $R > o$ we set $B_R = \{u \in H \mid ||u|| \leq R\}$ and $S_R = \delta B_R$.

By standard arguments it can be proved that the classical solutions of (0.3) correspond to the critical points of the functional

$$(1.1) \qquad f(u) = \frac{1}{2} \int_0^{2\pi} |\dot{u}|^2 \, dt - \frac{\omega^2}{2} \int_0^{2\pi} (Mu,u) \, dt + \omega^2 \int_0^{2\pi} V(u) \, dt.$$

Obviously $f \in C^2(H, \mathbb{R})$.

We recall the following theorem (cf. [2]):

Theorem 1.1. Let X be a real Hilbert space, on which a unitary representation T_g of the group S^1 acts. Let $f \in C^1(X,\mathbb{R})$ be a functional on X satisfying the following assumptions:

f_1) $f(u) = \frac{1}{2}(Lu\,|u)_X - \psi(u)$, where $(\cdot|\cdot)_X$ is the inner product in X, L is a bounded selfadjoint operator and $\psi \in C^1(X,\mathbb{R})$, $\psi(o) = o$, is a functional whose Fréchet derivative is compact. We suppose that both L and ψ' are equivariant with respect to the action the group S^1.

f_2) 0 does not belong to the essential spectrum of L.

f_3) Every sequence $\{u_n\} \subset X$, for which $f(u_n) \to c \in]o, +\infty[$ and $||f'(u_n)||\cdot||u_n|| \to o$, possesses a bounded subsequence.

f_4) There are two closed subspaces S^1 - invariant $V, W \subset X$ and $R, \delta > o$ s.t.

i) W is L-invariant, i.e. $LW = W$

ii) Fix $(S^1)^{x} \subset V$ <u>or</u> Fix $(S^1) \subset W$

iii) $f(u) < \delta$ <u>for</u> $u \in$ Fix (S^1) s.t. $f'(u) = o$

iv) f <u>is bounded from above on</u> W

v) $f(u) \geq \delta$ <u>for</u> $u \in V$ <u>s.t.</u> $||u|| = R$

vi) codim $(V+W) < +\infty$, dim $(V \cap W) < +\infty$.

<u>Under the above assumptions there exists at least</u>

$$\frac{1}{2}\ (\dim (V \cap W) - \operatorname{codim} (V+W))$$

<u>orbits</u>[xx] <u>of critical points, with critical values greater or equal than</u> δ.

We set $\beta = \nu\,\omega^2$ (cf. (I_6)) and we consider the bilinear form defined by

$$a(u,v) = \int_0^{2\pi} (\dot{u},\dot{v})\,dt + \int_0^{2\pi} (u,v)\,dt - \omega^2 \int_0^{2\pi} (Mu,v)\,dt + \beta \int_0^{2\pi} (u,v)\,dt.$$

By easy computations it can be proved that a is continuous and coercive on H. Then by standard theorems (cf. [4]) there exists a unique bounded linear operator $S : H \to H$ with a bounded linear inverse S^{-1} such that $(Su,v)_H = a(u,v)$ $\forall\, u, v \in H$. We set $\mathcal{D}(\mathcal{S}) = \{u \in H \mid Su \in L^2\}$ and $\mathcal{S} = S_{|\mathcal{D}(\mathcal{S})}$. $\mathcal{S}$ is a linear self-adjoint operator with compact resolvent. Then $\sigma(\mathcal{S})$ consists of a positively divergent sequence of isolated eigenvalues with finite multiplicities (cf. [4]). If we denote by $s_o < s_1 < \ldots < s_j < \ldots$ the eigenvalues of $\mathcal{S}$ and by M_j the corresponding eigenspaces, then $L^2 = \bigoplus_j M_j$. We have that $\mathcal{L} = \mathcal{S} - (\beta+1)\ I$, where $I : L^2 \to L^2$ is the identity map, then $\lambda_j = s_j - (\beta + 1)$. We observe that the eigenspace corresponding to λ_j is M_j for every j.

By (I_3) there exists $\lambda_k \in \sigma(\mathcal{L})$ s.t. $\lambda_k = o$. We set

---

x Fix $(S^1) = \{u \in X \mid T_g\, u = u$ for every $g \in S^1\}$

xx If $u \in X$ the "orbit" of u is the set $\{T_g\, u : g \in S^1\}$

$$(1.2)\quad \begin{cases} W = H^-(k) = \bigoplus_{j\leq k} M_j \\ V = H^+(h) = \overline{\bigoplus_{j\geq h} M_j} \end{cases}$$

where the closure is under the norm of H.

Now we shall prove that the assumptions of theorem 1.1 are verified. If we set

$$(Lu,\, u)_H = \int_0^{2\pi} |\dot{u}|^2\, dt - \omega^2 \int_0^{2\pi} (Mu,\, u)\, dt$$

$$\psi(u) = -\ \omega^2 \int_0^{2\pi} V(u)\, dt$$

by the previous considerations it follows that the functional (1.1) satisfies (f_1) and (f_2).

By the same arguments used in [3] it can be proved that the functional (1.1) satisfies (f_3) and that the subspaces V and W satisfy $(f_4.i)$, $(f_4.iv)$, $(f_4.v)$ with $\delta > o$ and $(f_4.vi)$.

In order to prove $(f_4.ii)$ it is sufficient to prove that all the eigenvectors of M belong to V or to W. Let $\bar{c}$ be an eigenvector of M and let $\bar{\lambda}$ be the corresponding eigenvalue. We have

$$L\,\bar{c} = -\,\omega^2 M\,\bar{c} = -\,\omega^2\,\bar{\lambda}\,\bar{c}$$

then $-\,\omega^2\,\bar{\lambda} \in \sigma(L)$. If M is positive semidefinite we have $-\,\omega^2\,\bar{\lambda} \leq o = \lambda_k$, then $\bar{c} \in W$. If $\nu \leq -\frac{\lambda_h}{\omega^2}$ we have $-\,\omega^2\,\bar{\lambda} \geq -\,\omega^2\,\nu \geq \lambda_h$, then $\bar{c} \in V$. By (I_6) it follows that $(f_4.ii)$ is verified.

In order to prove $(f_4.iii)$ we observe that if u is a constant and $f'(u) = o$, then $Mu = \nabla V(u)$. Hence by (I_7) and (1.1)

$$f(u) = -\frac{\omega^2}{2}\int_0^{2\pi} (Mu, u)\, dt + \omega^2 \int_0^{2\pi} V(u)\, dt \leq o.$$

Since $\delta > o$, then $(f_4.iii)$ is verified.

By the theorem 1.1 the conclusion of theorem 0.1 holds.

If the assumptions (I_4)-(I_7) are replaced by (I_4')-(I_7'), then we set $g(z) = - f(z)$ and (cf. definition (1.2))

$$V = H^-(s) \qquad W = H^+(k).$$

If $u \in V$ by (I_4') we get

$$g(u) = g''(o)[u,u] + o(||u||^2) =$$

$$= - \sum_{j=o}^{s} \lambda_j ||u_j||^2_{L^2} - \omega^2 \int_o^{2\pi} (V_{xx}(o)u,u)\, dt + o(||u||^2_H) \geqslant$$

(1.3)

$$\geqslant - \sum_{j=o}^{s} (\lambda_j + \omega^2 \mu) ||u_j||^2_{L^2} + o(||u||^2_H) \geqslant$$

$$\geqslant - (\lambda_s + \omega^2 \mu) ||u||^2_H + o(||u||^2_H) .$$

Then by (I_5') and (1.3) there exists $R > o$ such that

$$(1.4) \qquad g(u) \geqslant \delta > o \qquad \forall u \in V \cap S_R .$$

Moreover if $u \in W$

$$(1.5) \qquad g(u) = - \frac{1}{2} \sum_{j=k}^{\infty} (\lambda_j) ||u_j||^2_{L^2} - \omega^2 \int_o^{2\pi} V(u)\, dt \leqslant \gamma$$

Then by (1.4), (1.5), by same arguments used in the proof of theorem 0.1 and by theorem 1.1 we get the conclusion of theorem 0.2.

References

[1] P.BARTOLO - V.BENCI - D.FORTUNATO, Abstract critical point theorems and applications to some nonlinear problems with "strong resonance" at infinity, (to appear).

[2] V.BENCI - A.CAPOZZI - D.FORTUNATO, Periodic solutions of Hamiltonian systems with a prescribed period, (to appear).

[3] A.CAPOZZI - A.SALVATORE, Periodic solutions for nonlinear problems with strong resonance at infinity, to appear on Comm.Math.Univ. Carolinae.

[4] T.KATO, Perturbation theory for linear operators, Springer-Verlag, New York, 1966.

[5] P.H.RABINOWITZ, Periodic solutions of Hamiltonian systems: a survey, (to appear).

[6] K.THEWS, T-periodic solutions of time dipendent Hamiltonian systems with a potential vanishing at infinity, Manuscr. Math. 33, 327-338 (1981).

Author's address: Istituto di Matematica Applicata - Facoltà di Ingegneria
Via Re David, 200 - 70125 - BARI, (Italy).
S.I.S.S.A., Strada Costiera 11 - 34014 TRIESTE, (Italy).

STABILIZING EFFECTS OF DISSIPATION

C. M. DAFERMOS

Lefschetz Center for Dynamical Systems
Division of Applied Mathematics
Brown University
Providence, RI 02912 USA

1. Introduction

The balance laws of continuum physics in conjunction with the constitutive relations that characterize the type of material give rise to systems of evolution equations from which the thermokinetic processes of the material are to be determined. The nonlinear character of material response generally induces a destabilizing mechanism as a result of which acceleration waves are amplified and their amplitude may blow up in finite time thus generating shock waves. On the other hand, various dissipation mechanisms, such as viscosity, thermal diffusion, etc., have the opposite effect of damping out the amplitude of waves. The outcome of the contest between these competing mechanisms depends upon the nature of material response. Accordingly any material class may be classified into one of the following categories:

I. Dissipation is so powerful that it smoothens out instanteneously any discontinuity introduced by the initial conditions. Hence processes emanating from any initial data are smooth.

II. Dissipation is sufficiently strong to preserve the smoothness of smooth initial conditions but incapable to smoothen out discontinuities. Only processes emanating from smooth initial data are smooth.

III. Dissipation manages to preserve the smoothness of smooth initial data that are sufficiently close to equilibrium while processes emanating from smooth initial data that are far from equilibrium may develop discontinuities.

IV. Dissipation is very weak (or absent altogether) so even processes which emanate from smooth initial data near equilibrium may develop discontinuities.

Rigid bodies that conduct heat according to Fourier's law are typical members of Category I because their evolution is governed by the energy balance equation which is of parabolic type. On the opposite extreme, in Category IV, lie thermoelastic nonconductors of heat in which the balance laws of mass, momentum and energy constitute a system of (nonlinear) hyperbolic equations. The intermediate categories II and III constitute the territory of material classes in which internal dissipation is induced by heat conduction and/or viscosity. From the viewpoint of analysis, these models lead to systems of hyperbolic and parabolic equations that are coupled together. A number of such systems have been investigated recently in the literature (e.g. [1-10]).

As an illustration of the type of questions that arise in the above context, we shall discuss here in some detail a simple specific problem. We consider an incompressible Newtonian fluid, with viscosity μ that varies with temperature θ, which is sheared between two parallel plates occupying the planes $x=0$ and $x=1$. The flow is in the direction of the y-axis. We let v denote the y-component of velocity. Then the shearing stress is

(1.1) $$\sigma = \mu(\theta)v_x$$

Assuming that the density of the fluid is $\rho=1$ and upon identifying internal energy with temperature, the balance equations of momentum and energy take the form

(1.2) $$\begin{cases} v_t = \sigma_x = [\mu(\theta)v_x]_x \\ \theta_t = \sigma v_x = \mu(\theta)v_x^2 \end{cases} \qquad 0 \le x \le 1,\ 0 \le t < \infty,$$

while the corresponding boundary and initial conditions read

(1.3) $$v(0,t) = 0 \quad , \quad v(1,t) = 1, \qquad 0 \le t < \infty,$$

(1.4) $$v(x,0) = v_0(x), \quad \theta(x,0) = \theta_0(x), \qquad 0 \le x \le 1.$$

The important question here is whether the solution of (1.2),(1.3),(1.4) exists for all t and approaches asymptotically the uniform shearing flow

(1.5) $$v_x(x,t) = 1, \quad v_t(x,t) = 0, \quad \theta(x,t) = \Theta(t) \ ,$$

where

(1.6) $$\int_{\theta_0}^{\Theta(t)} \frac{d\xi}{\mu(\xi)} = t \quad ,$$

or else whether velocity gradient localizations may occur. The answer will depend upon the outcome of the contest between the destabilizing effect of stress power in the energy balance equation $(1.2)_2$ and the stabilizing effect of viscosity in the momentum balance equation $(1.2)_1$.

In a typical gas, viscosity is an increasing function of temperature while in liquids viscosity generally decreases with temperature. We will consider here the case of a liquid and for definiteness we will assume

(1.7) $$\mu(\theta) = \theta^{-\gamma} \quad , \quad 0 < \gamma < 1.$$

For viscosity of this type we will prove the following

<u>THEOREM</u>. Assume $v_0(x) \in W^{2,2}(0,1)$, $\theta_0(x) \in W^{1,2}(0,1)$, $v_0(0) = 0$, $v_0(1) = 1$, $\theta_0(x) > 0$, $0 \le x \le 1$. Then there is a unique solution $(v(x,t),\theta(x,t))$ of (1.2),(1.3),(1.4) on $[0,1]\times[0,\infty)$ and, as $t \to \infty$,

(1.8) $$v_x(x,t) = 1+O(t^{-\frac{1-\gamma}{1+\gamma}}),$$

(1.9) $$v_t(x,t) = O(t^{-1}) \ ,$$

(1.10) $$\int_{\theta_0(x)}^{\theta(x,t)} \frac{d\xi}{\mu(\xi)} = t + O(t^{\frac{2\gamma}{1+\gamma}}).$$

Thus, in the present situation, dissipation wins over and enforces asymptotic stability on the solution. Similar results for various types of $\mu(\theta)$ are obtained in [3].

2. Proof of Theorem

Let us assume that $(v(x,t),\theta(x,t))$ is a solution of (1.2),(1.3),(1.4) on $[0,1]\times[0,\infty)$ such that $v(\cdot,t), v_x(\cdot,t), v_t(\cdot,t), v_{xx}(\cdot,t), \theta(\cdot,t), \theta_x(\cdot,t)$ are all in $C^0([0,\infty);L^2(0,1))$ while $v_{xt}(\cdot,t)$ is in $C^0((0,\infty);L^2(0,1))$ and $v_{tt}(\cdot,t)$ is in $L^2_{loc}((0,\infty);L^2(0,1))$. We proceed to establish a priori estimates which will lead to the proof of the Theorem. Throughout this section K will stand for a generic constant which can be estimated from above solely in terms of γ and upper bounds of the $W^{2,2}(0,1)$ norm of $v_0(x)$ and the $W^{1,2}(0,1)$ norm of $\theta_0(x)$.

We rewrite (1.2) in the form

$$v_t = [\theta^{-\gamma}v_x]_x \quad , \tag{2.1}$$

$$\theta_t = \theta^{-\gamma}v_x^2 \quad . \tag{2.2}$$

Multiplying (2.1) by v_t and integrating over $[0,1]\times[0,t]$ we obtain, after two integrations by parts,

$$\int_0^t\int_0^1 v_t^2 dxd\tau + \frac{1}{2}\int_0^1 \theta^{-\gamma}(x,t)v_x^2(x,t)dx + \frac{\gamma}{2}\int_0^t\int_0^1 \theta^{-2\gamma-1}v_x^4 dxd\tau = \text{const.} \tag{2.3}$$

whence

$$\int_0^t\int_0^1 v_t^2 dxd\tau \le K, \qquad 0 \le t < \infty \; . \tag{2.4}$$

Next we multiply (2.2) by $\mu(\theta) = \theta^{-\gamma}$ and use (1.1), $(1.2)_1$ to get

$$\frac{1}{1-\gamma}\,[\theta^{1-\gamma}(x,t)]_t = \sigma^2(x,t) \tag{2.5}$$

$$= \int_0^1 \sigma^2(y,t)dy + 2\int_0^1\int_y^x \sigma(\xi,t)v_t(\xi,t)d\xi dy.$$

Hence, setting

$$\phi(t) = 1 + \int_0^t\int_0^1 \sigma^2 dxd\tau \quad , \quad 0 \le t < \infty \quad , \tag{2.6}$$

we deduce easily from (2.5) and (2.4)

$$\frac{1}{K}\,\phi^{\frac{1}{1-\gamma}}(t) \le \theta(x,t) \le K\phi^{\frac{1}{1-\gamma}}(t) \quad , \quad 0 \le x \le 1, \quad 0 \le t < \infty \; . \tag{2.7}$$

We proceed to estimate $\phi(t)$. By virtue of (2.7),

$$\int_0^1 \sigma^2(x,t)dx = \int_0^1 \theta^{-2\gamma}(x,t)v_x^2(x,t)dx \ge \frac{1}{K}\,\phi^{-\frac{2\gamma}{1-\gamma}}(t). \tag{2.8}$$

On the other hand, again by (2.7),

$$\int_0^1 \sigma^2(x,t)dx \le K\phi^{-\frac{\gamma}{1-\gamma}}(t)\int_0^1 \theta^{-\gamma}(x,t)v_x^2(x,t)dx. \tag{2.9}$$

In order to estimate the right-hand side of (2.9), we multiply (2.1) by tv_t, we integrate over $[0,1]\times[0,t]$ and integrate by parts thus obtaining

$$\int_0^t\int_0^1 \tau v_t^2 dxd\tau + \frac{t}{2}\int_0^1 \theta^{-\gamma}(x,t)v_x^2(x,t)dx + \frac{\gamma}{2}\int_0^t\int_0^1 \tau\theta^{-2\gamma-1}v_x^4 dxd\tau = \frac{1}{2}\int_0^t\int_0^1 \theta^{-\gamma}v_x^2 dxd\tau. \tag{2.10}$$

By account of (2.2),

$$\int_0^t\int_0^1 \theta^{-\gamma}v_x^2 dxd\tau = \int_0^1 \theta(x,t)dx - \int_0^1 \theta_0(x)dx. \tag{2.11}$$

Combining (2.9), (2.10) and (2.11),

$$\int_0^1 \sigma^2(x,t)dx \le K\,\phi^{-\frac{1+\gamma}{1-\gamma}}(t)\,\frac{1}{t}\{\int_0^1 \theta(x,t)dx\}^2. \tag{2.12}$$

We now multiply (2.1) by v and integrate over $[0,1]\times[0,t]$ to get, with the help of (2.11),

$$\frac{1}{2}\int_0^1 v^2(x,t)dx + \int_0^1 \theta(x,t)dx = \int_0^t \sigma(1,\tau)d\tau + \text{const.} \tag{2.13}$$

Therefore, by Schwarz's inequality,

$$[\int_0^1 \theta(x,t)dx]^2 \le t\int_0^t \sigma^2(1,\tau)d\tau + \text{const.} \tag{2.14}$$

$$\le 2t\int_0^t\int_0^1 \sigma^2 dxd\tau + t\int_0^t\int_0^1 \sigma_x^2\, dxd\tau + \text{const.}$$

Hence, using $(1.2)_1$,(2.4) and (2.6),

$$[\int_0^1 \theta(x,t)dx]^2 \le Kt\phi(t) + K \quad , \quad 0 \le t < \infty \;. \tag{2.15}$$

Combining (2.12) with (2.15),

$$\int_0^1 \sigma^2(x,t)dx \le K\phi^{-\frac{2\gamma}{1-\gamma}}(t) \quad , \quad 1 \le t < \infty \;. \tag{2.16}$$

From (2.8), (2.16) and since

$$\frac{d\phi(t)}{dt} = \int_0^1 \sigma^2(x,t)dx \quad , \tag{2.17}$$

we obtain

$$\frac{1}{K}t^{\frac{1-\gamma}{1+\gamma}} \le \phi(t) \le Kt^{\frac{1-\gamma}{1+\gamma}} \quad , \quad 1 \le t < \infty \quad , \tag{2.18}$$

and so (2.8),(2.16) and (2.7) yield

(2.19) $$\frac{1}{K} t^{-\frac{2\gamma}{1+\gamma}} \leq \int_0^1 \sigma^2 (x,t)dx \leq Kt^{-\frac{2\gamma}{1+\gamma}} , \quad 0 \leq t < \infty ,$$

(2.20) $$\frac{1}{K} t^{\frac{1}{1+\gamma}} \leq \theta(x,t) \leq Kt^{\frac{1}{1+\gamma}} , \quad 0 \leq x \leq 1 , \quad 1 \leq t < \infty .$$

Our next objective is to estimate the L^2 norm of v_t. To this end we differentiate (2.1) with respect to t,

(2.21) $$v_{tt} = [\theta^{-\gamma} v_{xt} - \gamma\theta^{-2\gamma-1} v_x^3]_x ,$$

we multiply (2.21) by v_t, we integrate over $[0,1]$ and we integrate by parts thus arriving at

(2.22) $$\frac{1}{2} \frac{d}{dt} \int_0^1 v_t^2(x,t)dx + \int_0^1 \theta^{-\gamma}(x,t) v_{xt}^2(x,t)dx$$
$$= \gamma \int_0^1 \theta^{-2\gamma-1}(x,t) v_x^3(x,t) v_{xt}(x,t)dx$$
$$= \gamma \int_0^1 \theta^{\gamma-1}(x,t)\sigma^3(x,t) v_{xt}(x,t)dx.$$

Applying the Cauchy-Schwarz inequality, (2.22) yields

(2.23) $$\frac{d}{dt} \int_0^1 v_t^2(x,t)dx + \int_0^1 \theta^{-\gamma}(x,t) v_{xt}^2(x,t)dx \leq \gamma^2 \int_0^1 \theta^{3\gamma-2}(x,t)\sigma^6(x,t)dx$$

or, using (2.20), (2.19) and $(1.2)_1$,

(2.24) $$\frac{d}{dt}\int_0^1 v_t^2(x,t)dx + \frac{1}{K_1} t^{-\frac{\gamma}{1+\gamma}} \int_0^1 v_{xt}^2(x,t)dx$$
$$\leq K_2 t^{\frac{3\gamma-2}{1+\gamma}} \{\max_{[0,1]} \sigma^2 (x,t)\}^2 \int_0^1 \sigma^2(x,t)dx$$
$$\leq K_3 t^{\frac{\gamma-2}{1+\gamma}} \{\int_0^1 \sigma^2 (x,t)dx + 2[\int_0^1 \sigma^2 (x,t)dx]^{3/4} [\int_0^1 \sigma_{xx}^2(x,t)dx]^{1/4}\}^2$$
$$\leq K_4 t^{-\frac{2+3\gamma}{1+\gamma}} + K_4 t^{-2} [\int_0^1 v_{xt}^2(x,t)dx]^{1/2}$$

whence

(2.25) $$\frac{d}{dt}\int_0^1 v_t^2(x,t)dx + \frac{1}{2K_1} t^{-\frac{\gamma}{1+\gamma}} \int_0^1 v_{xt}^2 (x,t)dx \leq K_5 t^{-\frac{2+3\gamma}{1+\gamma}} .$$

This yields the differential inequality

(2.26) $$\frac{d}{dt} \int_0^1 v_t^2(x,t)dx + \frac{1}{2K_1} t^{-\frac{\gamma}{1+\gamma}} \int_0^1 v_t^2(x,t)dx \leq K_5 t^{-\frac{2+3\gamma}{1+\gamma}}$$

upon integration of which we obtain

$$\int_0^1 v_t^2(x,t)dx \le Kt^{-2} \quad , \quad 0 < t < \infty \quad . \tag{2.27}$$

We proceed to estimate the L^1 norm of θ_x. We differentiate (2.5) with respect to x and use $(1.2)_1$ to get

$$[\theta^{-\gamma}(x,t)\theta_x(x,t)]_t = 2\sigma(x,t)\sigma_x(x,t) = 2\sigma(x,t)v_t(x,t). \tag{2.28}$$

Observing that on account of (2.19) and (2.27)

$$\left|\int_0^1 \sigma(x,t)v_t(x,t)dx\right| \le Kt^{-\frac{1+2\gamma}{1+\gamma}} \tag{2.29}$$

and using (2.20), we deduce easily from (2.28)

$$\int_0^1 |\theta_x(x,t)|dx \le Kt^{\frac{\gamma}{1+\gamma}} \quad , \quad 1 \le t < \infty \; . \tag{2.30}$$

We are now prepared to estimate the L^1 norm of v_{xx}.

Since $v_x = \sigma\theta^\gamma$, $\sigma_x = v_t$,

$$v_{xx} = v_t\theta^\gamma + \gamma\sigma\theta^{\gamma-1}\theta_x \; . \tag{2.31}$$

By (2.19) and (2.27),

$$\sigma^2(x,t) \le \int_0^1 \sigma^2(y,t)dy + 2\int_0^1 |\sigma(y,t)||v_t(y,t)|dy \le Kt^{-\frac{2\gamma}{1+\gamma}} \; ; \tag{2.32}$$

therefore, combining (2.31) with (2.20), (2.27), and (2.30) yields

$$\int_0^1 |v_{xx}(x,t)|dx \le Kt^{-\frac{1-\gamma}{1+\gamma}} \quad , \quad 0 < t < \infty \quad . \tag{2.33}$$

This estimate implies immediately (1.8) because

$$|v_x(x,t)-1| \le \int_0^1 |v_{xx}(y,t)|dy \quad , \quad 0 \le x \le 1 \quad , \quad 0 \le t < \infty. \tag{2.34}$$

Estimate (1.10) follows now easily from (1.8) since, by virtue of $(1.2)_2$,

$$\left|\int_{\theta_0(x)}^{\theta(x,t)} \frac{d\xi}{\mu(\xi)} - t\right| = \left|\int_0^t [v_x^2(x,\tau)-1]d\tau\right| \le Kt^{\frac{2\gamma}{1+\gamma}} \; . \tag{2.35}$$

It remains to show (1.9). To this end we first multiply (2.21) by t^2v_t, we integrate over $[0,1]\times[0,t]$ and integrate by parts to get

$$\frac{1}{2}t^2\int_0^1 v_t^2(x,t)dx - \int_0^t\int_0^1 \tau v_t^2 dxd\tau + \int_0^t\int_0^1 \tau^2\theta^{-\gamma}v_{xt}^2 dxd\tau \tag{2.36}$$

$$= \gamma \int_0^t \int_0^1 \tau^2 \theta^{-2\gamma-1} v_x^3 v_{xt} dx d\tau$$

which yields, with the help of (2.27),(2.20), (1.8), and Schwarz's inequality,

(2.37) $$\int_0^t \int_0^1 \tau^2 \theta^{-\gamma} v_{xt}^2 dx d\tau \le K t^{\frac{1}{1+\gamma}} \quad , \qquad 0 \le t < \infty \, .$$

Next we multiply (2.21) by $t^3 v_{tt}$, we integrate over $[0,1]\times[0,t]$ and perform a number of integrations by parts with respect to x and t thus obtaining

(2.38) $$\int_0^t \int_0^1 \tau^3 v_{tt}^2 dx d\tau + \frac{1}{2} t^3 \int_0^1 \theta^{-\gamma}(x,t) v_{xt}^2(x,t) dx$$

$$= \frac{3}{2} \int_0^t \int_0^1 \tau^2 \theta^{-\gamma} v_{xt}^2 dx d\tau - \frac{7}{2}\gamma \int_0^t \int_0^1 \tau^3 \theta^{-2\gamma-1} v_x^2 v_{xt}^2 dx d\tau$$

$$+ \gamma t^3 \int_0^1 \theta^{-2\gamma-1}(x,t) v_x^3(x,t) v_{xt}(x,t) dx$$

$$-3\gamma \int_0^t \int_0^1 \tau^2 \theta^{-2\gamma-1} v_x^3 v_{xt} dx d\tau + \gamma(2\gamma+1) \int_0^t \int_0^1 \tau^3 \theta^{-3\gamma-2} v_x^5 v_{xt} dx d\tau .$$

Combining (2.37),(1.8) and (2.20) we deduce from (2.38)

(2.39) $$\int_0^1 v_{xt}^2(x,t) dx \le K t^{-2} \quad , \qquad 0 < t < \infty ,$$

which, in conjunction with (2.27), yields (1.9).

Once the above a priori estimates have been derived, the existence of a globally defined solution of (1.2),(1.3),(1.4) can be established by a routine procedure, showing first that a local solution exists on a maximal time interval and then inferring from the estimates that this solution cannot escape in a finite time.

References

[1] DAFERMOS, C.M., Global smooth solutions to the initial-boundary value problem for the equations of one-dimensional nonlinear thermoviscoelasticity. SIAM J. Math. Analysis 13 (1982), 397-408.

[2] DAFERMOS, C.M. & HSIAO, L., Global smooth thermomechanical processes in one-dimensional nonlinear thermoviscoelasticity. J. Nonlinear Analysis 6 (1982), 435-454.

[3] DAFERMOS, C.M. & HSIAO, L., Adiabatic shearing of incompressible fluids with temperature dependent viscosity. Quart. Appl. Math. (to appear).

[4] DAFERMOS, C.M., & NOHEL, J.A., Energy methods for a class of nonlinear hyperbolic Volterra equations. Commun. in P.D.E. 4 (1979), 219-278.

[5] DAFERMOS, C.M. & NOHEL, J.A., A nonlinear hyperbolic Volterra equation in viscoelasticity. Am. J. Math. 1981 (suppl. dedicated to P. Hartman), pp. 87-116.

[6] HRUSA, W.J., A nonlinear functional differential equation in Banach space with applications to materials with fading memory. Arch. Rational Mech. Anal. (to appear).

[7] KAZHIKOV, A.M., & SHELUKHIN, V.V., Unique global solution with respect to time of initial-boundary value problems for one dimensional equations of a viscous gas. Appl. Math. Mech. 41 (1977), 273-282.

[8] MACCAMY, R.C., A model for one-dimensional, nonlinear viscoelasticity. Q. Appl. Math. 35 (1977), 21-33.

[9] MATSUMURA, A., & NISHIDA, T., The initial value problem for the equations of motion of viscous and heat-conductive gases. J. Math. Kyoto Univ. 20 (1980), 67-104.

[10] SLEMROD, M., Global existence, uniqueness and asymptotic stability of classical smooth solutions in one-dimensional non-linear thermoelasticity. Arch. Rational Mech. Anal. 76 (1981), 97-133.

Acknowledgment: This work was supported in part by NSF grant No. MCS 8205355 and in part by AROD contract No. DAAG-29-79-C-0161.

12/82

PERIODIC SOLUTIONS OF GENERALIZED LIENARD EQUATIONS WITH DELAY.*

E.De Pascale and R.Iannacci
Dipartimento di Matematica Università della Calabria
87030-Arcavacata di Rende (Cs) - Italia.

Abstract. We use classical Leray-Schauder techniques in order to derive the existence of periodic solutions of a generalized Liénard equation with delay.
Keywords. Delay, periodic solutions, Caratheodory's conditions.

1. INTRODUCTION

In this paper we prove the existence of 2π-periodic solutions for the following generalized Liénard equation with fixed delay $\tau \in [o,2\pi[$:

(1) $\ddot{x} + f(x)\dot{x} + g(t,x(t-\tau)) = e(t)$ a.e. for $t \in [o,2\pi]$,

where $f:R \to R$ is continuous, $g: [o,2\pi]\times R \to R$ verifies the Caratheodory's conditions and $e: [o,2\pi] \to R$ is integrable.

The unknown function $x: [o,2\pi] \to R$ in the equation (1) is defined for $o \leq t \leq \tau$ by the equality $x(t-\tau) = x(2\pi - (t-\tau))$ (i.e. in the standard way for the periodic case).

We will reach our main result in two steps:

1. We consider the auxiliary linear problem

(2) $\ddot{x} + f(t)\, x\,(t-\tau) = u(t)$, $x(o)-x(2\pi) = o = \dot{x}(0)-\dot{x}(2\pi)$.

For this problem, widely recognized as an interesting problem on its own, we derive uniqueness, existence and continuous dependence of its solution on u.

2. We apply Schaefer's fixed point theorem to a suitably defined map associated with equation (1).

We follow a technique due to J.Mawhin and J.R.Ward [7] in order to obtain the required a priori estimates.

We observe that the case of effective delay (i.e. $\tau \neq o$) is quite different, as it should be expected, from the case in which there is no delay (see Remark 1 - Example 1 - Proposition 2 - Remark 4).

Acknowledgement: We would like to thank Prof. M.Martelli for stimulating discussions on the subject matter.

2. NOTATIONS AND DEFINITIONS.

It is worth to point out once more that any real map defined on $[o,2\pi]$ has to be considered extended to all real axis by periodicity, possibly in a discontinuous fashion. In our notations a subscript P accounts, for the above assumption, in the basic spaces

---

* Work performed under the auspices of "Gruppo Nazionale per l'Analisi Funzionale ed Applicazioni" - Italy.

involved. Precisely:

1. $1 \le p < \infty$, $L^p_p[o,2\pi]$ are the usual Lebesgue spaces

2. $H^1_p[o,2\pi] = \{x$ absolutely continuous, $\dot{x} \in L^2_p[o,2\pi], x(o)-x(2\pi) = o = \dot{x}(o)-\dot{x}(2\pi)\}$ with the norm

$$||x||_{H^1} = \{(\frac{1}{2\pi} \int_0^{2\pi} x(t)dt)^2 + \frac{1}{2\pi} \int_0^{2\pi} |(\dot{x}(t))|^2 dt \}^{\frac{1}{2}}$$

3. $\tilde{H}^1_p[o,2\pi] = \{x \in H^1_p[o,2\pi], \bar{x} = \frac{1}{2\pi} \int_0^{2\pi} x(t)dt = o\}$

4. $W^{2,1}_p[o,2\pi] = \{x, \dot{x}$ absolutely continuous, $x(o)-x(2\pi) = o = \dot{x}(o)-\dot{x}(2\pi)\}$ with the norm $||x||_{W^{2,1}} = \frac{1}{2\pi} \sum_{i=o}^{2} \int_0^{2\pi} |x^{(i)}(t)|\, dt$

The conditions $x(o)-x(2\pi) = o = \dot{x}(o)-\dot{x}(2\pi)$ trivially ensure that x and $\dot{x}$, extended by periodicity over the real axis, are continuous.

We recall that every $x \in H^1_p[o,2\pi]$ can be written in the form $x = \tilde{x} + \bar{x}$, with $\tilde{x} \in \tilde{H}^1_p[o,2\pi]$ and $\bar{x} = \frac{1}{2\pi} \int_0^{2\pi} x(t)dt$.

A function $x \in W^{2,1}_p[o,2\pi]$ will be called a solution of (1), if it satisfies (1) almost everywhere on the real axis.

3. RESULTS: THE LINEAR CASE.

Lemma 1: (J.Mawhin - J.R.Ward [8])

If $\Gamma \in L^1_p[o,2\pi]$ is such that $\Gamma(t) \le 1$ for almost all $t \in [o,2\pi]$ and with strictly inequality on a subset of positive measure, then there exists $\delta(\Gamma) > o$ such that

$$B_\Gamma(\tilde{x}) = \frac{1}{2\pi} \int_0^{2\pi} |\dot{\tilde{x}}(t)^2 - \Gamma(t)\, \tilde{x}(t)^2|\, dt \ge \delta\, ||\tilde{x}||^2_{H^1} \quad \text{for any } \tilde{x} \in \tilde{H}^1_p[o,2\pi].$$

Theorem 1. Let Γ be as in Lemma 1. If $\Gamma(t) \ge 0$ for almost all $t \in [o,2\pi]$ and Γ is not identically zero then the equation

$$\ddot{x}(t) + \Gamma(t)\, x(t-\tau) = o \tag{3}$$

admits in $W^{2,1}_p[o,2\pi]$ only the trivial solution.

PROOF. If x is a possible solution of (3) we have, using the identity $-ab = \frac{(a-b)^2}{2} - \frac{a^2}{2} - \frac{b^2}{2}$:

$$0 = \frac{1}{2\pi} \int_0^{2\pi} [\bar{x} - \tilde{x}(t)]\, [\ddot{x}(t) + \Gamma(t)\, x(t-\tau)]dt =$$

$$= \frac{1}{2\pi} \int_0^{2\pi} [\bar{x} - \tilde{x}(t)]\, [\ddot{\tilde{x}}(t) + \Gamma(t)(\bar{x} + \tilde{x}(t-\tau))]\, dt =$$

$$= \frac{1}{2\pi} \int_0^{2\pi} [\dot{\tilde{x}}^2(t) - \frac{\Gamma(t)}{2} (\tilde{x}^2(t) + \tilde{x}^2(t-\tau))]\, dt +$$

$$+ \frac{1}{2\pi} \int_0^{2\pi} \frac{\Gamma(t)}{2} [(x(t-\tau) - \tilde{x}(t))^2 + \bar{x}^2]dt.$$

From the periodicity of x it follows that $\int_0^{2\pi} \tilde{x}^2(t)dt = \int_0^{2\pi} \tilde{x}^2(t-\tau)dt$.

Therefore, by the positivity of Γ, and by Lemma 1, we have

$$0 \geq \frac{1}{2}\{\frac{1}{2\pi}\int_0^{2\pi}[\dot{\tilde{x}}^2(t) - \Gamma(t)\,\tilde{x}^2(t)]dt + \frac{1}{2\pi}\int_0^{2\pi}[\dot{\tilde{x}}^2(t-\tau) - \Gamma(t)\tilde{x}^2(t-\tau)]dt\} \geq \delta\,||\tilde{x}||^2_{H^1}.$$

The last inequality implies $\tilde{x} = 0$ a.e. and $x = \bar{x}$ a.e.. But a constant map is a solution of (3) iff the constant is zero, because Γ is not identically zero. Q.E.D.

Theorem 2. Let $\Gamma \in L^1_P[0,2\pi]$ be such that $-1 \leq \Gamma(t) \leq 0$ a.e. with $-1 < \Gamma(t)$ on a subset of positive measure and assume that Γ be not identically zero. Then the equation (3) admits in $W^{2,1}_P[0,2\pi]$ only the trivial solution.

PROOF. If x is a possible solution of (3) we have:

$$0 = \frac{1}{2\pi}\int_0^{2\pi} - x(t)[\ddot{x}(t) + \Gamma(t)(\bar{x} + \tilde{x}(t-\tau))]\,dt =$$

$$= \frac{1}{2\pi}\int_0^{2\pi}[\dot{\tilde{x}}^2(t) + \frac{\Gamma(t)}{2}(\tilde{x}^2(t) + \tilde{x}^2(t-\tau))]dt - \frac{1}{2\pi}\int_0^{2\pi}\frac{\Gamma(t)}{2}[(x(t-\tau) + \tilde{x}(t))^2 + \bar{x}^2]dt \geq$$

$$\geq \delta\,||\tilde{x}||^2_{H^1}$$

where δ is the number associated by Lemma 1 to the function $-\Gamma$. The claim then follows as in Theorem 1. Q.E.D.

Corollary 1. Let Γ be as in Theorem 1 or 2. Then for every fixed $\tau \in [0,2\pi[$ and for every $u \in L^1[0,2\pi]$ the problem (2) admits in $W^{2,1}_P[0,2\pi]$ one and only one solution, which depends continuously on u.

PROOF. The operator $D: x \in W^{2,1}_P[0,2\pi] \to \ddot{x} \in L^1_P[0,2\pi]$ is Fredholm of index zero. The operator $F: x \in W^{2,1}_P[0,2\pi] \to \Gamma(t)x(t-\tau) \in L^1_P[0,2\pi]$ is completely continuous. Whence $D + F$ is Fredholm of index zero.

The existence and uniqueness follow from the fact that $\ker(D+F) = \{0\}$. The continuous dependence of the solution on u is a direct consequence of Banach Continuous Inverse Theorem.

Remark 1. It is well known that in the case $\tau = 0$ the assumption $\Gamma(t) \leq 0$, $\Gamma(t) \not\equiv 0$ ensures that (3) has only the trivial solution.

This is not true if $\tau \neq 0$. In fact, the equation

$$\ddot{x} - x(t-\pi) = 0$$

admits the periodic solutions $\sin t$ and $\cos t$. More precisely, it can be shown that the eigenvalues of the problem

$$\ddot{x} + \lambda x(t-\tau) = 0, \quad x(0) - x(2\pi) = 0 = x(0) - x(2\pi)$$

are

a) 0 if $\pi/\tau \in R \setminus Q$

b) $\lambda_n = (-1)^n(n\frac{\pi}{\tau})^2$ with $n \in N_0$ such that $n\frac{\pi}{\tau} \in N_0$, if $\frac{\pi}{\tau} \in Q$; in this case $\cos n\frac{\pi}{\tau}t$ and $\sin n\frac{\pi}{\tau}t$ are the eigenfunctions corresponding to λ_n.

In [8] and [9] J. Mawhin and J.R.Ward proved the following result.

(For other results concerning the existence of the only zero periodic solution see [5]).

Let $\Gamma \in L^1_P[0,2\pi]$ be such that $\Gamma(t) \leq 1$ with strict inequality on a subset of positive measure. Then the equation

$$\ddot{x} + \Gamma(t)x = 0$$

has only the trivial solution in $W_p^{2,1}[0,2\pi]$ provided that either one of the following conditions is verified

i) Γ has positive mean value, $\bar{\Gamma} > 0$

ii) Γ has mean value 0, $\bar{\Gamma} = 0$, and $\Gamma \neq 0$ on a set of positive measure.

The following example shows that under the above assumptions the Theorem no longer follows for $\tau \neq 0$.

Example 1. Let $b \in R$ be such that $|b| > 1$ and let $n \in N$. Consider the delay-differential equation

$$(*) \qquad \ddot{x} + \frac{n^2 \cos nt}{b + \cos n(t-\tau)} x(t-\tau) = 0$$

It is obvious that the function $x(t) = b + \cos nt$ is a non trivial 2π-periodic solution of $(*)$. Observe now that

$$\Gamma(t) = \frac{n^2 \cos nt}{b + \cos n(t-\tau)} \leq 1 \quad \text{for } |b| \geq n^2 + 1.$$

Moreover
$$\bar{\Gamma} = \frac{1}{2\pi}\int_0^{2\pi} \Gamma(t)dt = n^2\left(1 - \frac{|b|}{\sqrt{b^2-1}}\right)\cos n\tau.$$

Therefore, given $\tau \in (0,2\pi)$ we can choose n so as to obtain $\bar{\Gamma} > 0$ or $\bar{\Gamma} < 0$.
Moreover, if $\frac{\tau}{\pi}$ is a rational number of the type $\frac{\text{odd}}{\text{even}}$, we can also choose n so that $\bar{\Gamma} = 0$.

If $\tau = 0$ the above example shows that the assumption $\bar{\Gamma} \geq 0$ cannot be released in Mawhin-Ward Theorem. Nevertheless we have:

Proposition 1. Let $\Gamma(t) \leq 1$ with strict inequality on a subset of positive measure. Then the kernel of the linear operator $L : W_p^{2,1}[0,2\pi] \to L_p^1[0,2\pi]$ defined by $Lx = \ddot{x} + \Gamma(t)x$ is at most one dimensional.

PROOF. The result is obvious whenever $\bar{\Gamma} \geq 0$. Assume $\bar{\Gamma} < 0$ and let $x_1(t)$, $x_2(t)$ be linearly independent solutions of $Lx = 0$. Let

$$x_1(t) = r + \tilde{x}_1(t) \qquad\qquad x_2(t) = s + \tilde{x}_2(t)$$

Then $\alpha x_1(t) + \beta x_2(t)$ is also a solution and we can choose α,β so that $\alpha r + \beta s = 0$. Therefore the function $y(t) = \alpha x_1(t) + \beta x_2(t)$ will have mean value 0 with this choice of α and β, $y(t) = \tilde{y}(t)$.

We now have

$$\ddot{\tilde{y}}(t) + \Gamma(t)\tilde{y}(t) = 0$$

and

$$-\ddot{\tilde{y}}(t)\tilde{y}(t) - \Gamma(t)\tilde{y}^2(t) = 0$$

or

$$\int_0^{2\pi} \dot{\tilde{y}}^2(t)dt = \int_0^{2\pi} \Gamma(t)\tilde{y}^2(t)dt < \int_0^{2\pi} \tilde{y}^2(t)dt$$

which contradicts Wirtinger's inequality [2].

Hence $y \equiv 0$ or $\alpha\tilde{x}_1 + \beta\,\tilde{x}_2 = 0$. But then

$$x_2 = -\frac{\alpha}{\beta}\,x_1$$

Q.E.D.

Remark 1 shows that the assumption $\Gamma(t) \leq 1$ with inequality on a subset of positive measure does not ensure that the kernel of the operator $(Mx)(t) = \ddot{x}(t) + \Gamma(t)x(t-\tau)$, $\tau \neq 0$, is at most one dimensional. We have nevertheless the following result.

Proposition 2. Let $\Gamma \in L^1[0,2\pi]$ be such that $\Gamma \not\equiv 0$, $|\Gamma(t)| \leq 1$ almost everywhere, with $|\Gamma(t)| < 1$ on a set of positive measure. Then for every real number there is at most one function $\tilde{x}(t)$, periodic of period 2π and with mean value 0, such that $x(t) = r_0 + \tilde{x}(t)$ is a solution of (3).

PROOF. Assume that for some r_0 there are two functions $\tilde{x}(t)$ and $\tilde{y}(t)$ such that $r_0 + \tilde{x}(t)$ and $r_0 + \tilde{y}(t)$ are solutions of (3). This implies

$$\ddot{\tilde{x}}(t) + \Gamma(t)\,(r_0 + \tilde{x}(t-\tau)) = 0$$

$$\ddot{\tilde{y}}(t) + \Gamma(t)\,(r_0 + \tilde{y}(t-\tau)) = 0$$

Therefore $\tilde{z}(t) = \tilde{x}(t) - \tilde{y}(t)$ satisfies

$$\ddot{\tilde{z}}(t) + \Gamma(t)\tilde{z}(t-\tau) = 0$$

and has mean value 0.

We have

$$0 = -\int_0^{2\pi} \tilde{z}\,(\ddot{\tilde{z}} + \Gamma(t)\tilde{z}(t-\tau))dt = \int_0^{2\pi} \dot{\tilde{z}}^2(t)dt - \int_0^{2\pi} \Gamma(t)\tilde{z}(t)\tilde{z}(t-\tau)dt \geq$$

$$\geq \int_0^{2\pi} \dot{\tilde{z}}^2(t) - \int_0^{2\pi} |\Gamma(t)|\,|\tilde{z}(t)\,\tilde{z}(t-\tau)|dt > \int_0^{2\pi} \dot{\tilde{z}}^2(t)dt - \int_0^{2\pi} |\tilde{z}(t)\tilde{z}(t-\tau)|dt \geq$$

$$\geq \int_0^{2\pi} \dot{\tilde{z}}^2(t)dt - \int_0^{2\pi} \tilde{z}^2(t)dt \geq 0.$$

By Wirtinger's inequality, since $\tilde{z}(t)$ cannot be a linear combination of sint and cost, we obtain $\tilde{z}(t) \equiv 0$, i.e. $\tilde{x}(t) = \tilde{y}(t)$.

4. RESULTS: THE NONLINEAR CASE.

We begin with a lemma that enables us to get a priori estimates required to our results.

Lemma 2. Let $\Gamma \in L^1\,[0,2\pi]$ be such that $0 \leq \Gamma(t) \leq 1$ a.e. for $t \in [0,2\pi]$, with the strict inequalities on subsets of positive measure. Let $\delta > 0$ be associated to Γ by Lemma 1 and let $\varepsilon > 0$. Then for all $p \in L^1_P\,[0,2\pi]$ satisfying a.e. $0 \leq p(t) \leq \Gamma(t) + \varepsilon$ we have

$$\frac{1}{2\pi}\int_0^{2\pi} (\bar{x} - \tilde{x}(t))\,[\ddot{x}(t) + p(t)\,x(t-\tau)]\,dt \geq (\delta - \varepsilon)\,||\,\tilde{x}||^2_{H^1}$$

for every $\tilde{x} \in \tilde{H}^1_p[0,2\pi]$.

PROOF. Integrating by parts and using the identity $-ab = \frac{(a-b)^2}{2} - \frac{a^2}{2} - \frac{b^2}{2}$ we have:

$$\frac{1}{2\pi}\int_0^{2\pi} (\bar{x}-\tilde{x}(t)) \left[\ddot{x}(t) + p(t)x(t-\tau)\right]dt \geq \frac{1}{2\pi}\int_0^{2\pi}\left[\dot{\tilde{x}}^2(t) - \Gamma(t)\,\frac{\tilde{x}^2(t-\tau) + \tilde{x}^2(t)}{2}\right]dt +$$

$$-\varepsilon\cdot\frac{1}{2\pi}\int_0^{2\pi} \frac{\tilde{x}^2(t-\tau) + \tilde{x}^2(t)}{2}\,dt + \frac{1}{2\pi}\int_0^{2\pi} \frac{p(t)}{2}\left[(x(t-\tau) - \tilde{x}(t))^2 + \bar{x}^2\right]dt$$

In the last inequality the first addend is greater than or equal to $\delta||\tilde{x}||^2_{H^1}$ and the third addend is non negative. Moreover by Wirtinger's inequality we have

$$\frac{1}{2\pi}\int_0^{2\pi}\frac{\tilde{x}^2(t-\tau) + \tilde{x}^2(t)}{2}\,dt = \frac{1}{2\pi}\int_0^{2\pi}\tilde{x}^2(t)dt \leq \frac{1}{2\pi}\int_0^{2\pi}\dot{\tilde{x}}^2(t)dt = ||\tilde{x}||^2_{H^1}$$

and so we are done. Q.E.D.

We are now in a position to prove the following:

<u>Theorem 3. Let</u> $f: R \to R$ <u>be continuous and</u> $g: [0,2\pi] \times R \to R$ <u>a Caratheodory's function</u> (i.e. $g(.,x)$ <u>is measurable for each</u> $x \in R$ <u>and</u> $g(t,.)$ <u>is continuous for a.e.</u> $t \in R$).

<u>Assume that</u>

A - <u>There exists</u> $r > 0$ <u>such that for</u> $|x| \geq r$, $xg(t,x) \geq 0$

B - $\limsup_{|x|\to+\infty} \frac{g(t,x)}{x} \leq \Gamma(t)$ <u>uniformly a.e. for</u> $t \in [0,2\pi]$ with Γ as in <u>Lemma 2</u>.

C - <u>For every</u> $s > 0$ <u>there exists</u> $\gamma_s \in L^1_p[0,2\pi]$ <u>such that</u> $|g(t,x)| \leq \gamma_s(t)$ <u>for a.e.</u> t <u>and for all</u> $x \in [-s,s]$.

<u>Then for every</u> $\tau \in [0,2\pi[$, <u>the differential-delay equation</u>

$$\ddot{x} + f(x)\dot{x} + g(t,x(t-\tau)) = e(t)$$

has at least one 2π - periodic solution provided that $e \in L^1_p[0,2\pi]$, $\bar{e} = 0$.

PROOF. The operator $S: H^1_p[0,2\pi] \to L^1_p[0,2\pi]$ defined by $(Su(t) = \Gamma(t)\,u(t-\tau) + e(t) - g(t,u(t-\tau)) - f(u)\dot{u}$, is obviously continuous.

We denote by $T : L^1_p[0,2\pi] \to W^{2,1}_p[0,2\pi]$ the operator solution for the problem $\ddot{x} + \Gamma(t)x(t-\tau) = h(t)$, $x(0)-x(2\pi) = 0 = \dot{x}(0) - \dot{x}(2\pi)$, and by J the completely continuous embedding of $W^{2,1}_p[0,2\pi]$ in $H^1_p[0,2\pi]$. By Corollary 1, T is continuous and so the operator $J \circ T \circ S : H^1_p[0,2\pi] \to H^1_p[0,2\pi]$ is completely continuous and a fixed point of $J \circ T \circ S$ is a solution in $W^{2,1}_p[0,2\pi]$ of our equation.

Our aim will be attained by Shaefer's theorem [3], if we prove that the subset of $H^1_p[0,2\pi]$

$$\{x \in H^1_p[0,2\pi] : x = \lambda\, J \circ T \circ Sx,\ 0 < \lambda < 1\} = \{x \in H^1_p[0,2\pi] : \ddot{x} + (1-\lambda)x(t-\tau).\Gamma(t) +$$

$$+ \lambda f(x)\dot{x} + \lambda g(t,x(t-\tau)) - \lambda e(t) = 0 \ \text{ a.e. on } R,\quad 0 < \lambda < 1\}$$

is bounded.

Let $\delta > 0$ be associated to the function Γ by Lemma 1. By hypothesis B there exists

$m > 0$ such that

$0 \leq \frac{g(t,x)}{x} \leq \Gamma(t) + \frac{\delta}{2}$ a.e. for $t \in [0,2\pi]$ and $|x| \geq m$. Defined $\gamma : [0,2\pi] \times R \to R$ by

$$\gamma(t,x) = \begin{cases} g(t,x)/x & \text{if } |x| \geq m \\ g(t,m)/m & \text{if } 0 < x < m \\ \Gamma(t) & \text{if } x = 0 \\ -g(t,-m)/m & \text{if } -m < x < 0 \end{cases}$$

we have $0 \leq \gamma(t,x) \leq \Gamma(t) + \frac{\delta}{2\pi}$ a.e. for $t \in [0,2\pi]$ and $x \in R$. Moreover, the function $(t,x) \to \gamma(t,x)x$ satisfies Caratheodory's conditions and the function $h(t,x) \doteq g(t,x) - \gamma(t,x)x$ is such that for some $\alpha \in L^1_p [0,2\pi]$, $|h(t,x)| \leq \alpha(t)$ a.e. for $t \in [0,2\pi]$ and $x \in R$.

Let $\lambda \in]0,1[$ and $x \in H^1_p[0,2\pi]$ be such that

(5) $\ddot{x} + (1-\lambda)\,\Gamma(t)x(t-\tau) + \lambda f(x)\dot{x} + \lambda g(t,x(t-\tau)) - \lambda e(t) = 0$

or equivalently

(6) $\ddot{x} + [(1-\lambda)\,\Gamma(t) + \lambda\gamma(t,x(t-\tau))]\, x(t-\tau) + \lambda f(x)\dot{x} + \lambda[h(t,x(t-\tau)) - e(t)] = 0$

Using Lemma 2, with $p(t) = (1-\lambda)\,\Gamma(t) + \lambda\gamma(t,x(t-\tau))$ we have

$$0 = \frac{1}{2\pi}\int_0^{2\pi} (\bar{x}-\tilde{x}(t))\,[\ddot{\tilde{x}} + p(t)x(t-\tau)]\,dt + \frac{1}{2\pi}\int_0^{2\pi} \lambda f(x)\dot{x}(\bar{x}-\tilde{x}(t))dt +$$

$$+ \frac{\lambda}{2\pi}\int_0^{2\pi} (\bar{x}-\tilde{x}(t))\,[h(t,x(t-\tau))-e(t)]\,dt \geq \frac{\delta}{2}\,||\tilde{x}||^2_{H^1} - (|\bar{x}| + |\tilde{x}|_{C^0})(|\alpha|_{L^1} + |e|_{L^1})$$

Thus $||\tilde{x}||^2_{H^1} \leq \beta(|\bar{x}| + ||\tilde{x}||_{H^1})$

with $\beta \geq 0$ and independent of x. Integrating (5) over $[0,2\pi]$, we obtain

(7) $(1-\lambda)\int_0^{2\pi} \Gamma(t)x(t-\tau)\,dt = -\lambda\int_0^{2\pi} g(t,x(t-\tau)dt.$

If $x(t) \geq$ for all $t \in [0,2\pi]$, then (7) implies that $(1-\lambda)\bar{\Gamma}m \leq 0$, a contradiction with $\bar{\Gamma} > 0$.

Similarly if $x(t) \leq -m$ for all $t \in [0,2\pi]$, we reach a contradiction. Consequently, there exists $t_1 \in [0,2\pi]$ such that $|x,(t_1)| < m$.

Let t_2 be such that $\bar{x} = x(t_2)$, then $\bar{x} = x(t_2) = x(t_1) + \int_{t_2}^{t_1} \dot{x}(t)dt$; which implies

$|\bar{x}| \leq m + 2\pi||\tilde{x}||_{H^1}$ and consequently there exists $c > 0$ such that $||\tilde{x}||^2_{H^1} \leq c||\tilde{x}||_{H^1}$

Our goal is obtained, observing that $||x||_{H^1} \leq m + (2\pi + 1)c$. Q.E.D.

Corollary 1. Theorem 3 remains valid if we suppose, instead of A and $\bar{e} = 0$, that there exists three constants a,b.R such that

$$x.g(t,x) \geq b \text{ a.e. for } t \in [0,2\pi] \text{ and for } x > R,$$
$$x.g(t,x) \leq a \leq b \text{ a.e. for } t \in [0,2\pi] \text{ and for } x < -R,$$
$$a \leq \bar{e} \leq b.$$

PROOF. The equation (1) is equivalent to the equation

$$\ddot{x} + f(x)\dot{x} + g(t,x(t-\tau)) - \bar{e} = e(t) - \bar{e},$$

Observing that $e(t) - \bar{e}$ has mean value o and $g_*(t,x) = g(t,x) - \bar{e}$ verifies A, we are done.

Lemma 3. Let $\Gamma \in L^1_p[o,2\pi]$ be such that $1 < \Gamma(t) < 0$ a.e. for $t \in [o\ 2\pi]$ with the strict inequalities on subsets of positive measure. Let $\delta_1 > 0$ be associated to $-\Gamma$ by Lemma 1 and let $\varepsilon > 0$.

Then for all $p \in L^1_p\ [o,2\pi]$ satisfying a.e. $\Gamma(t)-\varepsilon \leq p(t) \leq 0$. we have

$$-\frac{1}{2\pi}\int_0^{2\pi} x(t)[\ddot{x}(t) + p(t)x(t\ \tau)]\ dt \geq (\delta_1-\varepsilon)\ ||\tilde{x}||^2_{H^1} \text{ for any } \tilde{x} \in \tilde{H}^1_p[o,2\pi].$$

PROOF. Integrating by parts and using the identity $ab = \frac{(a+b)^2}{2} - \frac{a^2}{2} - \frac{b^2}{2}$, we have

$$-\frac{1}{2\pi}\int_0^{2\pi} (\bar{x} + \tilde{x}(t))\ [\ddot{\tilde{x}}(t) + p(t)x(t-\tau\ )]\ dt \geq$$

$$\geq \frac{1}{2\pi}\int_0^{2\pi}\left[\dot{\tilde{x}}(t) + \Gamma(t)\ \frac{\tilde{x}^2(t) + \tilde{x}^2(t-\tau)}{2}\right]dt - \varepsilon\ .\ \frac{1}{2\pi}\int_0^{2\pi} \frac{\tilde{x}^2(t-\tau) + \tilde{x}^2(t)}{2}\ dt +$$

$$-\frac{1}{2\pi}\int_0^{2\pi}\frac{p(t)}{2}\left[\bar{x}^2 + (x(t-\tau) + \tilde{x}(t))^2\right]dt > (\delta_1 - \varepsilon)\ ||\tilde{x}||^2_{H^1_!}$$ Q.E.D.

Theorem 4. The thesis of theorem 3 holds also if the hypotheses A and B are substituted with the following ones:

A') there exists $m > 0$ such that for $|x| \geq m$ $xg(t,x) \leq 0$ a.e. for $t \in [o\ 2\pi]$.

B') $\liminf_{|x|\to+\infty} \frac{g(t,x)}{x} \geq \Gamma(t)$ uniformly a.e. for $t \in [o\ 2\pi]$ with Γ as in Lemma 3.

PROOF. The scheme of proof is similar to that of the proof in Theorem 3 and we omit it for sake of brevity. The required a priori estimates are obtained using Lemma 3.

Q.E.D.

Remark 4. If we consider the equation (1) without delay ($\tau = 0$) then the hypothesis B of Theorem 4 is superflous to the conclusion of Theorem 3. (see for example [1], [6] , [8]).

This is not so, also in linear case when $\tau \neq 0$. In fact the equation (see Remark 1)

$$\ddot{x} - x(t-\pi) = e(t)$$

admits a 2π-periodic solution iff

$$\int_0^{2\pi} e(t)dt = \int_0^{2\pi} e(t)\sin t dt = \int_0^{2\pi} e(t)\cos t dt = 0.$$

REFERENCES

1. J.BEBERNES, M.MARTELLI - Periodic solutions for Liénard Systems, Atti Equadiff 78, Firenze, 537-545.
2. E.F.BECKENBACH, R.BELLMANN - "Inequalities" Springer Verlag, Berlin 1961.
3. J.CRONIN - Fixed points and topological degree in nonlinear analysis, Amer. Math. Soc., Providence, R.I., 1964.
4. J.HALE - Functional differential equations, Springer Verlag Berlin, 1971.

5. A.LASOTA, Z.OPIAL Sur les solutions périodiques des equations différentielles ordinaires, Ann. Pol. Math. 16 (1964) 69-94.
6. M.MARTELLI - On forced nonlinear oscillations, J.Math.Anal.Appl. 69, 1979, 496-504.
7. J.MAWHIN - An extension of a theorem of A.C.Lazer on forced nonlinear oscillations, J.Math.Anal.Appl. 40, 1972, 20 29.
8. J.MAWHIN, J.R.WARD Jr. - Periodic Solutions of some forced Liénard equations at resonance (to appear in Arch. Math.).
9. J.MAWHIN, J.R.WARD Jr. - Nonuniform nonresonance conditions at the two first eigenvalues for periodic solutions of forced Liénard and Duffing equations. (to appear in Rocky Mountains Math. J. - 1982).
10. C.MIRANDA - Istituzioni di Analisi di Analisi Funzionale Lineare, Un.Mat.Ital., 1978.
11. R.REISSIG - Extension of some results concerning the generalized Liénard equation, Ann.Mat.Pura Appl., 1975, 269 281.
12. G.SANSONE, R.CONTI - Equazioni differenziali non lineari, Cremonese Roma, 1956.

ASYMPTOTIC AND STRONG ASYMPTOTIC EQUIVALENCE TO POLYNOMIALS FOR SOLUTIONS OF NONLINEAR DIFFERENTIAL EQUATIONS.

Allan L. Edelson
Department of Mathematics
University of California at Davis
Davis, California 95616

and

Jerry D. Schuur
Department of Mathematics
Michigan State University
East Lansing, Michigan 48824

This paper continues the study of [1] and [2] where we determined conditions under which the equations

$$(1_n) \qquad x^{(2n)} = xf(t,x), \quad f \text{ positive and continuous on } [\tau,\infty) \times (-\infty,\infty)$$

have a solution $x_m(t)$ asymptotic to t^m for some m, $0 \le m \le 2n-1$. (Much of the previous work we refer to studies the equation $(1_n)^{\pm}\ (r(t)x^n)^n = \pm\, xf(t,x)$, but for convenience we shall here let $r(t) = 1$ and just consider $+\, xf(t,x)$.)

Since such a solution is eventually positive it has been shown that it satisfies

$$(2) \qquad x_m^{(k)}(t) > 0,\ 0 \le k \le 2j\ ,\ (-1)^k\, x_m^{(k)}(t) > 0,\ 2j \le k \le 2n$$

on $[\sigma,\infty)$ for some $\sigma \ge \tau$, where j is such that $m = 2j - 1$ or $2j$.

A solution of (1_n) satisfying (2) is said to be of Type $2j$.

Our method depends on knowing when the equation

$$(3_n) \qquad x^{(2n)} = p(t)x,\ p \text{ positive and continuous on } [\tau,\infty),$$

has a solution $x_m(t)$ asymptotic to t^m.

Suppose we consider the fourth order equation (3_2) and ask for a solution $x_1(t)$ which satisfies $x_1(t) \sim t$ (i.e. $\lim_{t\to\infty} t^{-1}x_1(t) = c$). Either $x_1(t) > 0$ (for sufficiently large t) and $x_1'(t) < 0$ (i.e. x_1 is of Type $2j$ with $j = 0$) and hence $x_1''(t) > 0$, $\lim_{t\to\infty} x_1'(t)$ exists; or $x_1(t)$, $x_1'(t)$, and hence $x_1''(t) > 0$, $x_1'''(t) < 0$ (i.e. x_1 is of Type $2j$ with $j = 1$). In the latter case, using a theorem of Hardy [3], we may differentiate both sides of $x_1(t) \sim t$ and conclude that $x_1'(t) \sim 1$.

Using Taylor's theorem we have

$$x_1(t) = x_1(b) - x'(b)(b-t) + \frac{x''(b)}{2}(b-t)^2 - \frac{x'''(b)}{6}(b-t)^3 + \int_t^b \frac{(s-t)^3}{6}\, p(s)x_1(s)ds$$

hence

$$x_1'(t) = x'(b) - x''(b)(b-t) + \frac{x'''(b)}{3}(b-t)^2 - \int_t^b \frac{(s-t)^2}{3} p(s)x_1(s)ds.$$

If we knew that $\lim_{b\to\infty} x''(b)\, b = 0 = \lim_{b\to\infty} x'''(b)b^2$ (and this, in fact, is true), then we could let $b \to \infty$ and conclude $\int_t^\infty (s-t)^2 p(s)x_1(s)ds < \infty$, or

$$(4) \qquad \int_t^\infty s^3 p(s)ds < \infty .$$

Hence (4) is a necessary condition for the existence of $x_1(t) \sim t$.

On the other hand, if we define the mapping

$$(5) \qquad [Tx](t) = at + \int_\tau^t \frac{(s-t)^3}{6} p(s)x(s)ds$$

on the space of functions which are continuous on $[\tau,\infty)$ and satisfy $x(t)/t$ bounded, then by a careful use of the Schauder Fixed Point Theorem, see [4], we find that if (4) holds, then (5) has a fixed point $x(t)$ which is a solution of (3_2) and satisfies $x(t) \sim t$. Hence (4) is a necessary and sufficient condition for the existence of $x_1(t) \sim t$.

The growth conditions on $\lim_{b\to\infty} x^{(k)}(b)\, b^{k-1}$ are contained in [2]:

<u>Theorem 1</u>. If (3_n) has a solution x which satisfies $\lim_{t\to\infty} x^{(m)}(t) = A_m \geq 0$, then x satisfies $\lim_{t\to\infty} x^{(k)}(t)\, t^{k-m} = 0$ for $m + 1 \leq k \leq 2n-1$.

And the method just outlined can be used to prove:

<u>Theorem 2</u>. Equation (3_n) has a solution $x_m(t) \sim t^m$, $0 \leq m \leq 2n-1$, if and only if

$$(6) \qquad \int^\infty s^{2n-1} p(s)ds < \infty .$$

Theorem 2 may now be extended to the nonlinear equation by a fixed point method. See [5]. In the space of function continuous on $[\tau,\infty)$ with $x(t)/t^m$ bounded, choose $x_m(t) \sim t^m$. Impose conditions so that the linear equation $u^{(2n)} = uf(t,x_m(t))$ has a solution $u_m(t) \sim t_m$ and consider the mapping $x_m \to u_m$. A fixed point of this mapping will be a solution of (1_n), asymptotic to t^m. That is the outline of the proof of:

<u>Theorem 3</u>. In (3_n) assume that either f is increasing or decreasing with respect to x for all t, i.e. (3_n) is superlinear or sublinear. Then (3_n) has a solution $x_m(t) \sim t^m$, $0 \leq m \leq 2n-1$, if and only if

$$(7) \qquad \int^\infty t^{2n-1} f(t,ct^m)dt < \infty \text{ for some } c > 0 .$$

Remarks: This theorem, along with other results, may be found in : Kusano and Naito [4] - for $(1_2)^-$ with m = 0,3; Kreith [6] - for $(1_n)^-$ with m = 0, 2n-1 and r(t) = 1; Edelson and Schuur [1] - for $(1_n)^{\pm}$ with m = 0; and Edelson and Perri [7] - for $(1_n)^-$.

Returning to (3_n), instead of the mapping given by (5), the mapping

(8) $$[Ux](t) = at^m - \int_t^\infty \frac{(s-t)^{2n-1}}{(2n-1)!} p(s)x(s)ds \ , \ 0 \le m \le 2n-1,$$

and the space of functions which are continuous on $[\tau,\infty)$ and satisfy $x(t)/t^m$ bounded is often used, along with the Contraction Mapping Theorem. A fixed point of U will be a solution of (3_n) which satisfies $\lim_{t\to\infty} [x(t) - at^m] = 0$ - we call this strongly asymptotic. The result is:

Theorem 4. The following three conditions are equivalent:

(9) Equation (3_n) has a solution x_m, $0 \le m \le 2n-1$, of Type 2j (where j is such that m = 2j - 1 or 2j) satisfying

(9a) $$\lim_{t\to\infty} x_m(t) = A_m > 0 \text{ and}$$

(9b) $$\int_\tau^\infty s^{2n-1} p(s)x_m(s)ds < \infty .$$

(10) $$\int_\tau^\infty s^{2n-1+m} p(s)ds < \infty .$$

(11) Equation (3_n) has a solution x_m satisfying

$$x_m(t) = A_m t^m + \int_t^\infty \frac{(t-s)^{2n-1}}{(2n-1)!} p(s)x_m(s)ds, A_m > 0.$$

Further, if such an x_m exists, then

(12) $$\lim_{t\to\infty} \{x_m^{(k)}(t) - A_m t^{m-k}\} = 0 \text{ for } 0 \le k \le m.$$

The proof is given in [2]. Also an example is given to show that (12) does not imply (10) (or (11) or (9)). So the question of a necessary condition is not resolved.

Again the fixed point method can be used to extend Theorem 4 to (1_n), which we assume is either sublinear or superlinear.

Theorem 5. The following three conditions are equivalent:

(13) Equation (1_n) has a solution x_m, $0 \le m \le 2n-1$, of type 2j (where j is such that m = 2j-1 or 2j) satisfying

(13a) $$\lim_{t\to\infty} x^{(m)}(t) = A_m > 0 \text{ and}$$

(13b) $$\int_t^\infty s^{2n-1} f(s, cx_m(s))x_m(s)ds < \infty \text{ for some } c > 0$$

(14) $$\int_t^\infty s^{2n-1+m} f(s, cs^m)ds < \infty \text{ for some } c > 0$$

(15) Equation (1_n) has a solution x_m satisfying

$$x_m(t) = A_m t^m + \int_t^\infty \frac{(t-s)^{2n-1}}{(2n-1)!} f(s, cx_m(s))x_m(s)ds, \text{ for some } c > 0, A_m > 0.$$

Further, the higher derivitives of x_m satisfy (12).

Theorem 5 is also true for more general equations of the form $L_n[x] = x f(t,x)$, where L_n is an nth order, disconjugate linear differential operator. It is only necessary to introduce the corresponding linearly independent solutions, $R_0(t), \ldots, R_{n-1}(t)$, of the homogeneous equation $L_n[x] = 0$, and the generalized higher derivitives of the solutions $x_m(t)$ (see [2]).

References

1. Edelson, A. L. and J. D. Schuur, "Nonoscillatory solutions of $(rx^{(n)})^{(n)} \pm xf(t,x) = 0$", Pacific J. Math. (to appear).

2. Edelson, A. L. and J. D. Schuur, "Increasing solutions of $(r(t)x^{(n)})^{(n)} = xf(t,x)$", (preprint).

3. Hardy, G. H., "Divergent Series", Oxford University Press, London.

4. Kusano, T. and M. Naito, "Nonlinear oscillation of fourth order differential equations", Can. J. Math. XXVIII (1972), 840-852.

5. Schuur, J. D., "Qualitative behavior of ordinary differential equations of the quasilinear and related types," Proc. of International Conf. on Non-linear phenomena in abstract spaces (V. Lakshmikantham, Ed.) Univ. Texas-Arlington, 1980.

6. Kreith, K., "Extremal solutions for a class of nonlinear differential equations", Proc. Amer. Math. Soc. 79 (1980), 415-421.

7. Edelson, A. L. and E. Perri, "Asymptotic behaviour of nonoscillatory equations", (preprint).

8. Kamke, E., "Differentialgleichungen Lösungsmethoden und Lösungen", Chelsa Publishing Co., New York 1971.

ON SOME PARABOLIC INTEGRO-DIFFERENTIAL EQUATIONS: EXISTENCE AND ASYMPTOTICS OF SOLUTIONS

Hans Engler

Institut für Angewandte Mathematik
Universität Heidelberg
Im Neuenheimer Feld 293
6900 Heidelberg, West Germany

Mathematics Research Center
UW - Madison
610 Walnut Street
Madison, WI 53706 , USA

1. In this contribution we consider initial-boundary value problems of the type

$$\text{(I)}\qquad \begin{aligned} \partial_t u - \Delta_x u + a*g(u) &= f \quad \text{in } \Omega \times (0,T] \\ u &\equiv 0 \quad \text{on } \partial\Omega \times [0,T] \end{aligned}$$

together with an initial condition $u(\cdot,0) = u_0$. Here $\Omega \subset \mathbb{R}^n$ is a bounded domain, Δ_x is the Laplacian, and

$$a*g(u)(x,t) = \int_0^t a(t-s)g(u(x,s))\, ds .$$

Equation (I) can stand for a scalar equation or a system; we shall focus on the first case and only indicate how to handle the latter, as well as the related problem

$$\text{(II)}\qquad \begin{aligned} \partial_t u - \Delta_x u &= f \quad \text{in } \Omega \times (0,T], \\ \partial_\nu u + a*g(u) &= h \quad \text{on } \partial\Omega \times [0,T], \end{aligned}$$

where ∂_ν denotes the outer normal derivative.

A physical situation that leads to (I) or (II) is feed back heat control in the interior or at the boundary of some heat-conducting medium, where the control mechanism possesses some inertia, or a similar control situation for a reaction-diffusion problem. We want to show

--- existence results for solutions of (I) that are analogous to those for semilinear parabolic equations (section 2)

--- results on the global asymptotic behavior for some special cases of (I) (section 3).

In the scalar semilinear case ((I) with $a = \delta_0$) existence of classical solutions is known if, e.g.

$g'(r) \geq -K$ for all r (using comparison principles) or if

$\int_0^r g(s)\, ds \geq -K(r^2 + 1)$ and $g(r) \geq -K\cdot(|r|^\rho \cdot r + 1)$ for all r, $\rho < \frac{4}{n-2}$, if $n \geq 3$,

using a priori energy estimates and a hierarchy of test functions $|u|^q \cdot u$ ([7]).

Comparison principles only seem to be of value if g is sublinear in (I); hence we shall use the second approach and thus get dimesion dependent growth conditions for g (see Thm. 2.1).

In section 3 we give conditions for a and g such that solutions of (I) converge to a steady state solution and explicitly determine the decay rate, using a new technique for Volterra integral operators that may be of some interest of its own (Lemma 3.1). Related results on both the existence and convergence problem for (I) have been obtained in [1] and [2].

We shall state the results for the case of a scalar equation with homogeneous boundary conditions, sketch the proofs, and indicate how they can be generalized. The details will appear elsewhere.

The usual notation for Sobolev- and Hölder spaces is employed. Independent variables are sometimes omitted, where no confusion can arise. Various constants that can change from line to line are denoted by the same letter C.

2. In this section we want to show the existence of regular solutions of (I). By this we mean solutions that are continuous in $\overline{\Omega_T} = \overline{\Omega} \times [0,T]$ and for which all derivatives that appear in (I) exist in $\Omega_T = \Omega \times (0,T)$. Let $\Omega \subset \mathbb{R}^n$ be open and bounded with a sufficiently smooth (e.g., C^3-) boundary. We assume that $a \in W^{2,1}([0,T],\mathbb{R})$ with $a(0) = 1$ and that $g : \mathbb{R} \longrightarrow \mathbb{R}$ is Hölder-continuous.

Theorem 2.1: Let a and g be as above and for all r $\int_0^r g(s)\,ds \geq -C\cdot(r^2+1)$. For some $\alpha > 0$ let $u_0 \in (W_0^{1,2} \cap W^{2,2})(\Omega) \cap C^{\alpha}(\overline{\Omega})$ and $f \in W^{1,1}([0,T],L^2(\Omega)) \cap C^{\alpha}(\overline{\Omega_T})$. Then there exists a regular solution u of (I), $u(\cdot,0) = u_0$, if additionally

(i) $|g(r)| \leq C\cdot(|r|^{p(n)+1} + 1)$ for $n \geq 2$, or

(ii) $g'(r) \geq -C\cdot(|r|^{p(n)} + 1)$ in the sense of distributions for $n \geq 2$ and $|g(r)| \leq C\cdot(|r|^{q(n)+1} + 1)$ for $n \geq 4$.

Here $p(n) = \frac{4}{n-2}$ and $q(n) = \frac{4}{n-4}$ for $n > 2$ resp. $n > 4$; $p(2)$ and $q(4)$ can be any positive number.

Sketch of proof: Let $\bar{u}$ be the unique regular solution of (I) with a replaced by 0. For $v \in C(\overline{\Omega_T})$ and $0 \leq \sigma \leq 1$ define $y = K(\sigma,v)$ by

$$\partial_t y - \Delta_x y + a*(\sigma\cdot g(\bar{u} + v)) = 0 \quad \text{in } \Omega_T,$$
$$y(\cdot,0) \equiv 0,\ y \equiv 0 \ \text{ on } \partial\Omega \times [0,T].$$

Note that both $\bar{u}$ and $K(\sigma,v)$ will be in some $C^{\varepsilon}(\overline{\Omega_T})$, $\varepsilon > 0$, and in $L^{\infty}(0,T;W^{2,2}(\Omega))$, ε depending on α and g (cf. [6]). Then u is a regular solution of (I) iff it solves $u = \bar{u} + K(1,u-\bar{u})$. We want to show the existence of such a fixed point by means of a Leray-Schauder degree argument (cf. [5]).

Obviously 0 is the only solution of $u = \bar{u} + K(0,u-\bar{u})$. Also, by the standard regulatity theory, K maps $[0,1] \times C^{\varepsilon}(\overline{\Omega_T})$ into $C^{\varepsilon}(\overline{\Omega_T})$ and is completely continuous. Hence it remeains to show an a priori estimate for solutions of $u = \bar{u} + K(\sigma,u-\bar{u})$ that is independent of σ. We shallgive such an estimate for $\sigma = 1$:

Let $r \in W^{1,1}([0,T],\mathbb{R})$ be the resolvent kernel of a'; i.e. $r + a'*r + a' = 0$. Taking the convolution of (I) with r and adding it to (I) then gives the identity

$$\partial_t u + r(0)\cdot u + r'*u - \Delta_x u - r*\Delta_x u + 1*g(u) = f + r*f + r\cdot u_0 =: \bar{f}\,. \tag{2.1}$$

Take the t-derivative of (2.1) (formally, i.e. take backward difference qoutients), multiply by $\partial_t u$ and integrate over $\Omega \times [0,t]$. Using the boundedness assumptions on $\int g$, this gives after some integrations by parts and an application of Gronwall's Lemma the estimate

$$(2.2)\qquad \int_\Omega |\partial_t u(\cdot,t)|^2 + \int_0^t |\nabla_x \partial_t u(\cdot,s)|^2 ds \le C\cdot(\|u_0\|_{W^{2,2}} + \|f\|_{W^{1,1}([0,T],L^2(\Omega))})^2$$

for a.e. t. Now let assumption (i) hold. We employ the standard regularity theory for linear parabolic equations in $W_q := L^q(0,T;W^{2,q}(\Omega)) \cap W^{1,q}([0,T],L^q(\Omega))$ and get

$$\|u-\bar{u}\|_{W_q} \le C\cdot(\|a*g(u)\|_{L^q(\Omega_T)}).$$

Using the growth assumption on g and a standard calculus inequality:

$$(2.3)\qquad \|g(u(\cdot,t))\|_{L^q(\Omega)} \le (1+\|u-\bar{u}\|_{W^{2,q}(\Omega)})\cdot k(\|u(\cdot,t)\|_{W^{1,2}(\Omega)}),$$

with $k: \mathbb{R}^+ \longrightarrow \mathbb{R}^+$ locally bounded. A Gronwall - type argument then gives an estimate of $u-\bar{u}$ in W_q and hence in $C^\varepsilon(\overline{\Omega_T})$, if we choose q large enough.
If the assumptions (ii) hold, we first show that $\|u(\cdot,t)\|_{W^{2,2}}$ is essentially bounded: Differentiate (2.2) (formally), multiply with $-\Delta_x u$, and integrate over $\Omega \times [0,t]$. After some integrations by parts, one is left with an estimate

$$(2.4)\qquad \frac{1}{4}\int_\Omega |\Delta_x u(\cdot,t)|^2 - \iint_{0\Omega}^{t} g(u)\cdot\Delta_x u \le C\cdot(1+\int_0^t c(s)\cdot|\Delta_x u(\cdot,s)|^2 ds),$$

with $c \in L^1(0,T;\mathbb{R})$, c and C only depending on the data and on the quantities that have been estimated in (2.2). Without loss of generality $g(0) = 0$, thus

$$\int_\Omega g(u(\cdot,t))\cdot\Delta_x u(\cdot,t) \le C\cdot\int_\Omega (|u(\cdot,t)|^{p(n)}+1)\cdot|\nabla_x u(\cdot,t)|^2$$
$$\le C\cdot(\int_\Omega |\Delta_x u(\cdot,t)|^2 + 1)\cdot k(\|u(\cdot,t)\|_{W^{1,2}})$$

by the assumption on g' and a suitable imbedding theorem. Inserting this into (2.4) we can estimate $\|u(\cdot,t)\|_{W^{2,2}}$; modifying the estimate (2.3) in an obvious manner we then get a C^ε-estimate also under the weaker growth assumptions in (ii). The usual degree argument then completes the existence proof.

Uniqueness of the solution will follow if g is locally Lipschitz-continuous. The above argument can easily be modified to include the case of x - dependent kernels a and nonlinearities g, of nonhomogeneous boundary conditions and of a general second order elliptic operator in divergence form instead of the Laplacian. A further generalization gives an existence argument also in the case of an elliptic operator of order 2m.
We can also handle certain systems of the type (I) in a similar manner. Then u takes its values in $\mathbb{R}^N$, $-\Delta_x u$ is replaced by $-D\cdot\Delta_x u$, D a positive definite diagonal matrix. The kernel a is matrix-valued, and we assume a(0) and $D\cdot a(0)$ to be positive definite. The nonlinearity g has to be a gradient (Plus a perturbation term). The growth assumptions in Thm. 2.1 stay the same; the condition on g' in (ii) is replaced by a similar bound for the spectrum of the Jacobian Dg.
Finally it should be mentioned that it is also possible to handle the case where g is replaced by a quasilinear elliptic operator; see [4] where weak solutions are constructed.

3. We start this section with a general result for a class of nonlinear Volterra integral operators and then draw some consequences from it. Let $a \in W^{2,1}([0,T],\mathbb{R})$ be a scalar kernel, $a(0) = 1$, and let $g: \mathbb{R} \longrightarrow \mathbb{R}$ be continuous. Let G be a primitive of g.

Lemma 3.1: Let a be non-negative, $a' \leq 0$, and let $b : [0,T] \longrightarrow \mathbb{R}$ be absolutely continuous and non-negative such that for all $0 \leq s \leq s+\tau \leq T$

$$\frac{d}{ds}(a'(s)\cdot b(s+\tau)) \geq 0. \tag{3.1}$$

Let g be non-decreasing. Then for any absolutely continuous $u : [0,T] \longrightarrow \mathbb{R}$

$$\int_0^t b(s)\cdot u'(s)\cdot \frac{d}{ds}(a*g(u)(s))\,ds \geq a(t)\cdot b(t)\cdot G(u(t)) - b(0)\cdot G(u(0)) - \int_0^t (a(s)\cdot b(s))'\cdot G(u(s))\,ds. \tag{3.2}$$

Proof:

$$\int_0^t b(s)\cdot u'(s)\cdot \frac{d}{ds}(a*g(u)(s))\,ds = \int_0^t b(s)\cdot \frac{d}{ds}G(u(s))\,ds + \int_0^t b(s)\cdot u'(s)\cdot a'*g(u)(s)ds. \tag{3.3}$$

We turn to the second integral:

$$\ldots = \int_0^t \int_\tau^t b(s)u'(s)a'(s-\tau)ds\; g(u(\tau))\,d\tau$$

$$= \int_0^t \Big\{b(t)a'(t-\tau)u(t) - b(\tau)a'(0)u(\tau) - \int_\tau^t \frac{d}{ds}(b(s)a'(s-\tau))u(s)ds\Big\} g(u(\tau))d\tau.$$

The monotonicity of g implies that $g(y) \geq \dfrac{G(y) - G(y-h\cdot z)}{h}$ for all y,z,h; hence

$$\ldots \geq \int_0^t \frac{1}{h}\Big\{G(u(\tau)) - G((1+h\cdot a'(0)\cdot b(\tau))\cdot u(\tau) - h\cdot b(t)\cdot a'(t-\tau)\cdot u(t) + h\cdot \int_\tau^t \frac{d}{ds}(b(s)\cdot a'(s-\tau))\cdot u(s)ds\Big\}d\tau.$$

Choose h so small that $1 + h\cdot a'(0)\cdot b(\tau) \geq 0$ for all $\tau \in [0,t]$. Then the convexity of G and (3.1) imply

$$\ldots \geq \int_0^t \Big\{-a'(0)\cdot b(\tau)\cdot G(u(\tau)) + b(t)\cdot a'(t-\tau)\cdot G(u(t)) - \int_\tau^t \frac{d}{ds}(b(s)a'(s-\tau))G(u(s))ds\Big\}d\tau$$

$$= -\int_0^t a'(0)b(\tau)G(u(\tau))d\tau + (a(t)-1)b(t)G(u(t)) - \int_0^t \int_0^s \frac{d}{ds}(b(s)a'(s-\tau))d\tau G(u(s))ds.$$

Evaluating the last expression and combining it with the first integral in (3.3) then gives the result.

A Hilbert space version of this estimate (with a subdifferential $\partial\phi$ instead of g) can also be derived; for $b \equiv 1$ it has been used in [3].

We want to apply this result to the equation (I). Let A denote a linear elliptic operator in $L^2(\Omega)$ together with appropriate boundary conditions, A in divergence form, and let $k(\cdot,\cdot)$ denote the bilinear form defined by A. Let a,b,g be as in the above Lemma, and let $f \in L^2(\Omega)$ (t - independent for simplicity). Let u be a solution of

$$\partial_t u + Au + a*g(u) = f \quad \text{in } \Omega_T,\ u(\cdot,0) = u_0 . \tag{3.4}$$

How to construct such solutions has been indicated in the previous section. We differentiate (3.4) (formally, i.e. we take difference quotients), multiply with $b\cdot\partial_t u$ and integrate over $\Omega \times [0,t]$. Using Lemma 3.1 a.e. in Ω, we then get

$$(3.5)\qquad \frac{1}{2}\, b(t)\cdot\int_\Omega |\partial_t u(\cdot,t)|^2 - \int_0^t \frac{1}{2}\, b'(s)\int_\Omega |\partial_t u(\cdot,s)|^2 ds + \int_0^t b(s)k(\partial_t u(\cdot,s),\partial_t u(\cdot,s))ds$$

$$+ a(t)\cdot b(t)\cdot\int_\Omega G(u(\cdot,t)) - \int_0^t (a(s)b(s))'\int_\Omega G(u(\cdot,s))ds$$

$$\leq b(0)\Big(\int_\Omega G(u_0) + \frac{1}{2}\int_\Omega |Au_0 + \bar{f}|^2\Big).$$

If $k(v,v) \geq \lambda_0 \int_\Omega |v|^2$, $\lambda_0 > 0$, this implies

$$(3.6)\qquad \frac{1}{2}\, b(t)\cdot\int_\Omega |\partial_t u(\cdot,t)|^2 + a(t)b(t)\cdot\int_\Omega G(u(\cdot,t)) - \int_0^t (a(s)b(s))'\cdot\int_\Omega G(u(\cdot,s))ds$$

$$\leq \int_0^t \Big(\frac{1}{2}\, b'(s) - \lambda_0 b(s)\Big)\cdot\int_\Omega |\partial_t u(\cdot,s)|^2 ds + C(u_0,f).$$

Looking for suitable kernels b and thus singling out classes of kernels a, one can then deduce decay estimates. We formulate a result, if b is an exponential kernel and -A is the Laplacian with zero Dirichlet boundary conditions.

<u>Theorem 3.2:</u> Let a, g, u_0, f be as in Thm. 2.1, $f(x,t) = \bar{f}(x)$. Assume that g is non-decreasing, $a \geq 0$, $a' \leq 0$, and that

$$(3.7)\qquad \sup\{\,\varepsilon \mid a''(s) + \varepsilon\cdot a'(s) \geq 0 \text{ for almost every } s\,\} = \bar{\varepsilon} > 0 .$$

Let $\bar{\lambda} = \min\{2\lambda_0, \bar{\varepsilon}\}$, where λ_0 is the smallest eigenvalue of $-\Delta_x$ with zero Dirichlet boundary conditions.

Then for any solution u of (I) on $\Omega \times [0,\infty)$ and for all $\delta > 0$

$$\sup_{t>0}\int_\Omega |\partial_t u(\cdot,t)|^2 \leq C(u_0,\bar{f})\cdot e^{-\bar{\lambda}\cdot t},$$

$$\int_0^\infty e^{-\bar{\lambda}t}(1+t)^{1+\delta}\cdot\int_\Omega |\nabla_x\partial_t u(\cdot,t)|^2\, dt \leq C(u_0,f,\delta).$$

For the proof we take $b(t) = e^{\bar{\lambda}t}$ and employ the above argument. Noting that the assumptions on a imply that $(a(t)\cdot b(t))' \geq 0$, (3.6) then implies the boundedness of $b(t)\cdot\int|\partial_t u(\cdot,t)|^2$ which is the first estimate stated above. The second estimate can be derived from (3.5) and the first one:

$$\int_0^t e^{\bar{\lambda}s}\int_\Omega |\nabla_x\partial_t u(\cdot,s)|^2 ds \leq C + \frac{1}{2}\int_0^t b'(s)\cdot\int_\Omega |\partial_t u(\cdot,s)|^2 ds \leq C\cdot(1+t),$$

if we use Lemma 3.4 stated at the end of this section.

This result shows that certain integral operators do not slow down the "expected" decay of solutions. One can not expect any improvement of the decay for this general class of g's: This can be seen by choosing a to be a decreasing exponential and g to be an eigenfunction of $-\Delta_x$ for the eigenvalue λ_0; in this situation our estimates are sharp. Of course the above arguments are also valid, if g is x-dependent.

Theorem 3.2 can be applied to kernels that show "fast" (exponential) decay, including kernels with bounded support. However, kernels with algebraic decay are excluded. By choosing e.g. $b = a^{-1}$ we can still handle a certain class of such kernels.

Theorem 3.3: Let $a, g, u_o, f = \bar{f}$ be as in Theorem 2.1. Assume that g is non-decreasing, $a > 0$, $a' \leq 0$, and that log a is convex. Let

$$(3.7) \qquad a'(s) + 2\lambda_o a(s) \geq 0 \quad \text{for large } s,$$

where λ_o is the smallest eigenvalue of $-\Delta_x$ with zero Dirichlet boundary conditions. Then for any solution u of (I) on $\Omega \times [0,\infty)$ and for any $\delta > 0$

$$\int_\Omega |\partial_t u(\cdot,t)|^2 \leq a(t)\cdot C(u_o,\bar{f}) \quad \text{for all } t \geq 0$$

and

$$\int_0^\infty a(s)^{-1}(|\log a(s)| + 1)^{1+\delta}\cdot\int|\nabla_x\partial_t u(\cdot,s)|^2 \, ds \leq C(u_o,\bar{f},\delta).$$

For the proof we take $b = a^{-1}$. From the logarithmic convexity of a one then deduces that all the assumptions in Lemma 3.1 are satisfied and that $\frac{1}{2}b'(s) - \lambda_o\cdot b(s) \leq 0$ for large s. Then (3.6) immediately implies the first estimate. The second estimate follows just as above, using Lemma 3.4.

If a is log-convex and (3.7) does not hold (i.e. $a'(s) + 2\lambda_o\cdot a(s) < 0$ for all s), then from the log-convexity $a''(s) + 2\lambda_o\cdot a'(s) > 0$ for all s, and we are in the situation of Thm. 3.2.

From these explicit decay estimates one can deduce the convergence of solutions of (I) in $W^{1,2}(\Omega)$ to stationary solutions $\bar{u}$ of

$$-\Delta_x\bar{u} + \int_0^\infty a(s)ds\cdot g(\bar{u}) = \bar{f} \quad , \quad u|_{\partial\Omega} \equiv 0,$$

if the kernel a satisfies certain integrability properties. Again, x-dependence in a, g or in the differential operator does not change the argument significantly; it is also possible to handle t-dependent f's without much effort. If g is replaced by a quasilinear elliptic operator of the form $B(u) = -\mathrm{div}_x((\nabla_\xi G)(\nabla_x u))$, $G: \mathbb{R}^n \longrightarrow \mathbb{R}$ convex, then the Hilbert space version of Lemma 3.1 can be applied to give convergence of weak solutions to a solution of the associated quasilinear elliptic equation.

In the above arguments we have twice used the following

Lemma 3.4: Let $k, v \in L^1_{loc}([0,\infty),\mathbb{R}^+)$, $k \geq 2$ and nondecreasing. Let for all $t \geq 0$

$$\int_0^t k(s)\cdot v(s)\, ds \leq C_1\cdot\log k(t) \quad , \quad C_1 \text{ independent of } t.$$

Then for any $\delta > 0$ there exists $C_2(C_1,\delta)$ such that

$$\int_0^\infty k(s)\cdot(\log k(s))^{-1-\delta}\cdot v(s)\, ds \leq C_2(C_1,\delta).$$

Proof: If k is bounded from above, nothing has to be shown. Else define $(t_n)_{n\geq 0}$ by $t_o = 0$, $\log k(t_{n+1}) = 2\cdot\log k(t_n)$ for $n \geq 0$.

Let $t \geq 0$, $t_n \leq t \leq t_{n+1}$. Then

$$\int_0^t k(s)(\log k(s))^{-1-\delta}\cdot v(s)\,ds \leq \sum_{i=0}^{n} \int_{t_i}^{t_{i+1}} k(s)(\log k(s))^{-1-\delta} v(s)\,ds$$

$$\leq \sum_{i=0}^{n} (\log k(t_i))^{-1-\delta} \int_0^{t_{i+1}} k(s)\cdot v(s)\,ds \leq \sum_{i=0}^{n} (\log k(t_i))^{-\delta}\cdot 2\cdot C_1,$$

and this series converges geometrically by the choice of the t_n.

References:

1. V. Barbu: Nonlinear Volterra integro-differential equations in Hilbert spaces. Conf. del Sem. di Mat. Bari 143 (1976)
2. M.G. Crandall, S.-O. Londen, J.A. Nohel: An abstract nonlinear Volterra integro-differential equation. JMAA 64 (1978), 701-735.
3. H. Engler: A version of the chain rule and integrodifferential equations in Hilbert spaces. SIAM J. Math. Anal. 13 (1982), 801-810.
4. H. Engler, S. Luckhaus: Weak solution classes for parabolic integrodifferential equations. MRC-TSR, Madison, WI, 1982.
5. A.Friedman: Partial differential equations of parabolic type. Englewood Cliffs, N.J. 1964.
6. O.A. Ladyzenskaya, V.A. Solonnikov, N.N. Ural'tseva: Linear and qausilinear equations of parabolic type. Providence, Rh.I. 1968.
7. F. Rothe: Uniform bounds from bounded L^p- functionals in reaction-diffusion equations. JDE 45 (1982), 207-233.

Supported by Deutsche Forschungsgemeinschaft.
Sponsored by the United States Army under Contract No. DAAG29 -80-C-0041.

Oscillation and Nonoscillation Properties for Second Order Nonlinear Differential Equations

Lynn H. Erbe

Abstract: We survey oscillation and nonoscillation criteria for the generalized Emden-Fowler differential equation $y'' + q(x)y^{\gamma} = 0$, $q > 0$, $\gamma > 0$ with particular emphasis on the duality between the sublinear and superlinear cases.

1. Introduction: It is the purpose of this note to survey some of the recent (and not so recent) progress which has been made concerning the oscillation and nonoscillation properties of the nonlinear second order Emden-Fowler differential equation

$$(1.1) \qquad y'' + q(x)y^{\gamma} = 0$$

where $q(x)$ is positive, continuous and locally of bounded variation on $[a,\infty)$, and $\gamma > 0$ is the quotient of odd positive integers. Under these assumptions, all solutions of (1.1) exist on $[a,\infty)$ (cf, [6]). A nontrivial solution of (1.1) is said to be oscillatory if it has arbitrarily large zeros; otherwise it is said to be nonoscillatory. Equation (1.1) is said to be oscillatory (resp. nonoscillatory) in case all nontrivial solutions are oscillatory (resp. nonoscillatory). If $\gamma \neq 1$, then equation (1.1) allows the coexistence of both oscillatory and nonoscillatory solutions. Of primary concern will be a comparison of results in the two cases $0 < \gamma < 1$ (the so-called sublinear case) and $\gamma > 1$ (the superlinear case). It is possible to consider more general equations than (1.1) but for simplicity we confine attention to (1.1) and shall attempt to give the best results obtained thus far. In addition, we hope to illustrate to a certain degree the "duality principle" which exists between the sublinear and superlinear cases and also point out cases where the "dual results" have not yet been established. The duality principle was investigated in [8], [9], and [22]. For additional references to oscillation/nonoscillation criteria we refer to [7], [23].

2. Statements of the results: There are basically four types of oscillation-nonoscillation criteria which have been established for equation (1.1). These are criteria which guarantee that: 1) all solutions are oscillatory; 2) there exists a nonoscillatory solution; 3) there exists an oscillatory solution; 4)

all solutions are nonoscillatory. The first two cases are fairly easy to deal with while the last two offer substantially more difficulties. In fact, the former are completely answered in:

Oscillation criteria-existence of nonoscillatory solutions:

Theorem 1 ([1]): Let $\gamma > 1$. Then all solutions of (1.1) are oscillatory if and only if $\int_a^\infty tq(t)dt = +\infty$.

Theorem 1* ([2]) Let $0 < \gamma < 1$. Then all solutions of (1.1) are oscillatory if and only if $\int_a^\infty t^\gamma q(t)dt = +\infty$.

Existence of oscillatory solutions: In the criteria which follows we will use the notation $\phi(x) \equiv x^{\frac{\gamma+3}{2}} q(x)$.

Theorem 2: ([18],[20],[8]): Let $\gamma > 1$ and assume that $\phi(x)$ is nondecreasing for all (large) x. Then (1.1) has oscillatory solutions.

Theorem 2*: ([4],[8]): Let $0 < \gamma < 1$ and assume that $\phi(x)$ is nondecreasing for all (large) x. Then (1.1) has oscillatory solutions.

By means of energy function techniques along with appropriate changes of variables, the above results were extended in [10] to:

Theorem 3 ([10]): Let $\gamma > 1$ and assume that $\phi(x)(\ell n\ x)^\sigma$ is nondecreasing for some $\sigma \leq 0$. Then (1.1) has oscillatory solutions.

Theorem 3* ([10]): Let $0 < \gamma < 1$ and assume that $\phi(x)(\ell n\ x)^\sigma$ is nondecreasing for some $\sigma \leq 0$. Then (1.1) has oscillatory solutions.

In fact, Theorems 3 and 3* are special cases of the more general

Theorem 4: ([10]): Let $\gamma > 1$ and assume that $q(x)x^\delta(\log x)^\sigma = \pi(x) - \nu(x)$ where π, ν are nondecreasing and $\pi(x) - \nu(\infty) > 0$. Then (1.1) has oscillatory solutions if $\delta < \frac{\gamma+3}{2}$ or $\delta = \frac{\gamma+3}{2}$ and $\sigma \leq 0$.

Theorem 4* ([10]): Let $0 < \gamma < 1$ and assume that $q(x)x^\delta(\log x)^\sigma = \pi(x) - \nu(x)$, where π, ν are nondecreasing and $\pi(x) - \nu(\infty) > 0$. Then (1.1) has oscillatory solutions if $\delta < \frac{\gamma+3}{2}$ or $\delta = \frac{\gamma+3}{2}$ and $\sigma \geq 0$.

We note therefore that the above criteria for the existence of oscillatory solutions in Theorems 2,3,4 (resp. 2*,3*,4*) are the same for both the superlinear and sublinear case. The next result which is typical of several other criteria to follow, has as yet no sublinear analog.

Theorem 5 ([16]): Let $\gamma > 1$ and let $\psi = q(x)^{-\frac{1}{\gamma+3}}$. Assume that $\int_a^\infty \frac{dx}{\psi^2(x)} = \infty$ and $\int_a^\infty |\psi\psi''|dx < \infty$. Then (1.1) has oscillatory solutions.

Nonoscillation criteria: We present here the main results of the paper - results guaranteeing that all solutions have only finitely many zeros - and the results for which the analogs between the superlinear and sublinear cases have not, as yet, been estabished in all cases.

Theorem 6 ([19]): Let $\gamma > 1$ and assume that $\phi(x)x^\delta$ is nonincreasing for some $\delta > 0$. Then equation (1.1) is nonoscillatory.

Theorem 6* ([3]): Let $0 < \gamma < 1$ and assume that $\phi(x)x^\delta$ is nondecreasing and bounded above by a positive constant k, for some $0 < \delta < \frac{1-\gamma}{2}$. Then equation (1.1) is nonoscillatory.

Theorem 7 ([1]): Let $\gamma > 1$ and assume that $q(x)$ is nonincreasing and $\int_a^\infty t^\gamma q(t)dt < \infty$. Then equation (1.1) is nonoscillatory.

Theorem 7* ([14]): Let $0 < \gamma < 1$ and assume that $q(x)$ is nonincreasing and $\int_a^\infty tq(t)dt < \infty$. Then equation (1.1) is nonoscillatory.

Before continuing with the statements of the remaining results, we recall that the assumption that $q(x)$ is continuous and locally of bounded variation implies that $q(x)$ has the Jordan respresentation

$$q(x) = q_+(x) - q_-(x) \tag{2.1}$$

where $q_+(x)$, $q_-(x)$ are continuous and nondecreasing for $a \leq x < \infty$. We introduce the notation

$$Q_+(x) \equiv \exp\left(\int_a^x \frac{dq_+(t)}{q(t)}\right), \quad a < x < \infty$$

(2.2)

$$Q_-(x) \equiv \exp\left(\int_a^x \frac{dq_-(t)}{q(t)}\right), \quad a < x < \infty$$

so that we have the identity

(2.3) $$\frac{q(x)}{q(a)} \equiv \frac{Q_+(x)}{Q_-(x)}, \quad a < x < \infty.$$

Theorem 8 ([22]): Let $\gamma > 1$ and assume that $\lim_{x\to\infty} Q_+(x) < \infty$ and that $\lim_{x\to\infty} x^{\gamma+1} q(x) = 0$. Then equation (1.1) is nonoscillatory.

Theorem 8* ([27]): Let $0 < \gamma < 1$ and assume that $\lim_{x\to\infty} Q_+(x) < \infty$ and that $\lim_{x\to\infty} x^2 q(x) = 0$. Then equation (1.1) is nonoscillatory.

We may now state several results which improve some of the above results of Atkinson [1], Wong [22], Heidel [14] and others.

Theorem 9 ([12]): Let $\gamma > 1$ and assume that

$$\lim_{x\to\infty} x^{\gamma+1} q(x) \left(Q_+(x)\right)^{\frac{\gamma-1}{2}} = 0 .$$

Then equation (1.1) is nonoscillatory.

Theorem 9* ([12]): Let $0 < \gamma < 1$ and assume that

$$\lim_{x\to\infty} x^2 q(x) \left(Q_+(x)\right)^{\frac{1-\gamma}{1+\gamma}} = 0.$$

Then equation (1.1) is nonoscillatory.

Theorem 10 ([12]): Let $\gamma > 1$ and assume that

$$\int_a^\infty t^\gamma q(t)dt < \infty \quad \text{and} \quad \lim_{x\to\infty} \left(Q_+(x)\right)^{\frac{\gamma-1}{2}} \int_x^\infty t^\gamma q(t)dt = 0.$$

Then equation (1.1) is nonoscillatory.

Theorem 10* ([12]): Let $0 < \gamma < 1$ and assume that

$$\int_a^\infty t^\gamma q(t)dt < \infty \text{ and } \lim_{x\to\infty} \left(Q_-(x)\right)^{\frac{\gamma-1}{2}} \int_x^\infty t^\gamma q(t)dt = 0.$$

Then equation (1.1) is nonoscillatory.

It is interesting to note that if in Theorem 9 (or 10) one formally replaces $Q_+(x)$ by $\left(Q_-(x)\right)^{-1}$ then one obtains the corresponding result of Theorem 9* (or 10*), keeping in mind the identity (2.3). We note also that Theorems 9 and 9* include Theorems 8 and 8*, respectively, and Theorem 10 includes Theorem 7.

We continue this section with the statement of several results for the superlinear case for which sublinear analogs have not, as yet, been established.

Theorem 11 ([11]): Let $\gamma > 1$ and assume there exist $\beta > 0$ and $\eta > 0$ with $\frac{2\beta}{\gamma+3} + \frac{2\eta}{\gamma-1} > 1$ and such that $\phi(x)(\log x)^\beta$ is nonincreasing and $\phi(x)(\log x)^\eta$ is bounded. Then equation (1.1) is nonoscillatory.

Theorem 12 ([12]): Let $\gamma > 1$. Then equation (1.1) is nonoscillatory in case any of the following hold:

a) $\int_a^\infty t^{\gamma-1}q(t)\left(Q_+(t)\right)^{\frac{\gamma-1}{2}} dt < \infty$ and $\lim_{x\to\infty} x\int_x^\infty t^{\gamma-1}q(t)\left(Q_+(t)\right)^{\frac{\gamma-1}{2}} dt = 0$

b) $\int_a^\infty (q(t))^{\frac{2}{\gamma+1}}(Q_+(t))^{\frac{\gamma-1}{\gamma+1}}dt < \infty$ and $\lim_{x\to\infty} x \int_x^\infty q(t)^{\frac{2}{\gamma+1}}(Q_+(t))^{\frac{\gamma-1}{\gamma+1}}dt = 0$

c) $\int_a^\infty (q(t))^{\frac{1}{\gamma+1}}\left(Q_+(t)\right)^{\frac{\gamma}{\gamma+1}} dt < \infty$.

Finally, we state the following result for the sublinear case for which the analog in the superlinear case is not known.

Theorem 13* ([12]): Let $0 < \gamma < 1$ and assume

$$\int_a^\infty (q(t))^{\frac{1}{\gamma+1}} (Q_+(t))^{\frac{\gamma}{\gamma+1}}\, dt < \infty \text{ and}$$

$$\lim_{x\to\infty} (q(x))^{\frac{\gamma-1}{2(1+\gamma)}} (Q_+(x))^{\frac{1-3\gamma}{2(1+\gamma)}} \int_x^\infty (q(t))^{\frac{1}{\gamma+1}} (Q_+(t))^{\frac{\gamma}{\gamma+1}}\, dt = 0.$$

Then equation (1.1) is nonoscillatory.

3. Final comments and remarks on the proofs:

Theorem 1 (resp. Theorem 1*) is proved by showing that equation (1.1) has a bounded (resp., asymptotically linear) solution if and only if $\int_a^\infty tq(t)dt < \infty$ (resp., $\int_a^\infty t^\gamma q(t)dt < \infty$).

To establish Theorems 2-4 (and 2*-4*) one introduces the change of variables $x = e^t$, $y = t^\mu e^{(\frac{1}{2}+\nu)t}$ to obtain the equivalent equation

$$(3.1) \qquad (r(t)u')' + a(t,u)u = 0$$

where $r(t) = t^{2\mu}e^{2\nu t}$, $a(t,u) = r(t)(\sigma(t)u^{\gamma-1} - \lambda(t))$,

$$\sigma(t) = q(e^t)t^{\mu(\gamma-1)}\exp\left(\frac{\gamma+3}{2} + \nu(\gamma-1)\right)t, \quad \lambda(t) = \frac{1}{4} + \frac{\mu(1-\mu)}{t^2} - \nu^2 - \frac{2\mu\nu}{t}$$

The case $\nu = \mu = 0$ is the usual change of variable used in studying (1.1) for which the transformed equation is

$$(3.2) \qquad u'' + (p(e^t)e^{\frac{\gamma+3}{2}t}u^{\gamma-1} - \frac{1}{4})u = 0.$$

In (3.1) it follows that $a(t,u) > 0$ for large $|u|$ when $\gamma > 1$ and $a(t,u) > 0$ for small $|u|$ when $0 < \gamma < 1$. In conjunction with (3.1) for $\gamma > 1$ and for Theorem 3 one considers the energy function

$$(3.3) \qquad E(t) = (r(t)u'(t))^2 + 2r(t)\int_0^{u(t)} a(t,s)ds$$

and under the assumptions of the Theorem one may show that $E(t)$ is nondecreasing along solutions of (3.1) and that any solution of (3.1) with a zero is oscillatory. Similar considerations apply to Theorem 4 with a modified energy function. For Theorem 3* (and 4*) one shows that any solution with small enough initial conditions (i.e. $u'(t_0)^2 + u(t_0)^2$ small) is oscillatory.

Theorems 5, 6, and 6* are proved by change of variable techniques also, combined with certain integral inequalities. Theorem 7 and 7* were improved by Gollwitzer [13] who showed that the nonincreasing assumption could be replaced by the assumption that $\lim_{x\to\infty} Q_+(x) < \infty$. Theorems 9, 9* and 10, 10* as well as Theorems 12 and 13* are obtained by analysis of the energy functions

$$F(x) = \frac{(y')^2}{2} + \frac{1}{\gamma+1} q(x)y^{\gamma+1} \quad \text{and} \quad B(x) = \frac{F(x)}{q(x)}.$$

which satisfy the inequalities

$$\frac{Q_+(\xi)}{Q_+(x)} < \frac{F(\xi)}{F(x)} < \frac{Q_-(x)}{Q_-(\xi)} \quad \text{and}$$

$$\frac{Q_-(\xi)}{Q_-(x)} < \frac{B(\xi)}{B(x)} < \frac{Q_+(x)}{Q_+(\xi)}$$

for $\xi, x \in [a,\infty)$. Theorem 11, on the other hand, which improves and generalizes results of Nehari [21] and Chiou [5], is proved by the change of variable used in Theorem 3 along with certain geometric considerations.

In view of the above, it would be interesting to obtain the analogous results for Theorems 11, 12 and 13*.

References

[1] F. V. Atkinson, On second order nonlinear oscillation, Pacific J. Math. 5(1955), 643-647.

[2] S. Belohorec, Oscillatory solutions of certain nonlinear differential equations of second order, Mat.-Fyz. Casopis Sloven. Akad. Vied. 11 (1961), 250-255.

[3] S. Belohrec, On some properties of the equation $y''(x) + f(x)y^{\alpha}(x) = 0$, $0 < \alpha < 1$, Mat. Fyz. Caspois Sloven. Akad. Vied., 17(1967), 10-19.

[4] K. L. Chiou, The existence of oscillatory solutions for the equation $d^2y/dt^2 + q(t)y^{\gamma} = 0$, $0 < \gamma < 1$, Proc. Amer. Math. Soc. 35 (1972), 120-122.

[5] K.L. Chiou, A nonoscillation theorem for the superlinear case of second order differential equation $y'' + yF(y^2,x) = 0$, SIAM J. Appl. Math. 23 (1972), 456-459.

[6] C.V. Coffman and D.F. Ullrich, On the continuation of solutions of a certain nonlinear differential equation, Monatsh, Math. 71 (1967), 385-392.

[7] C.V. Coffman and J.S.W. Wong, On a second order nonlinear oscillation problem, Trans. Amer. Math. Soc. 147 (1970), 357-366.

[8] C.V. Coffman and J.S.W. Wong, Oscillation and nonoscillation of solutions of generalized Emden-Fowler equation, Trans. Amer. Math. Soc. 167 (1972), 399-434.

[9] C.V. Coffman and J.S.W. Wong, Oscillation and nonoscillation theorems for second order differential equations, Funkcialaj Ekvacioj 15 (1972), 119-130.

[10] L.H. Erbe and J.S. Muldowney, On the existence of oscillatory solutions to nonlinear differential equations, Ann. Math. Pura. Appl. 59 (1976), 23-37.

[11] L.H. Erbe and J.S. Muldowney, Nonoscillation results for second order nonlinear differential equations, Rocky Mountain Math. J., to appear.

[12] L.H. Erbe, Nonoscillation criteria for second order nonlinear differential equations, preprint.

[13] H.E. Gollwitzer, Nonoscillation theorems for a nonlinear differential equation, Proc. Amer. Math. Soc. 26 (1970), 78-84.

[14] J.W. Heidel, A nonoscillation theorem for a nonlinear second order differential equation, Proc. Amer. Math. Soc. 22 (1969) 485-488.

[15] J.W. Heidel and D.B. Hinton, Existence of oscillatory solutions for a nonlinear differential equation, Siam J. Math. Annal. 3(1972), 344-351.

[16] D.B. Hinton, An oscillation criterion for solutions of $(ry')' + qy^{\gamma} = 0$ Mich. Math. J. 16 (1969), 349-352.

[17] D.V. Izumova and I.T. Kiguradze, Some remarks on the solutions of the equation $u'' + a(t)f(u) = 0$, Differential Equations 4(1968), 589-605.

[18] M. Jasny, On the existence of an oscillatory solution of the nonlinear differential equation of the second order $y'' + f(x)y^{2n-1} = 0$, $f(x) > 0$. Casopis Pest. Mat. 85 (1960), 78-83.

[19] I.T. Kiguradze, On condition for oscillation of solutions of the equations $u'' + a(t)|u|^n \operatorname{sqn} u = 0$, Casopis Pest. Mat. 87 (1962), 492-495.

[20] J. Kurzweil, A note on oscillatory solutions of the equation $y'' + f(x)y^{2n-1} = 0$, Casopis Pest Mat. 85 (1960), 357-358.

[21] Z. Nehari, A nonlinear oscillation problem, J. Diff. Eqns. 5 (1969), 452-460.

[22] J.S.W. Wong, Remarks on nonoscillation theorems for a second order nonlinear differential equation, Proc. Amer. Math. Soc. 83 (1981),541-546.

[23] J.S.W. Wong, On the generalized Emden-Fowler equation, SIAM Review 17 (1975), 339-360.

Authors's Address:
Department of Mathematics
University of Alberta
Edmonton, Alberta Canada

GLOBAL HOPF BIFURCATION IN POROUS CATALYSTS

Bernold Fiedler
Universität Heidelberg, SFB 123
Im Neuenheimer Feld 293
D-6900 Heidelberg

0. Introduction

There is numerous experimental evidence for oscillatory reactions in porous catalysts [12, 14]. These oscillations are attributed to interaction of diffusion and reaction of the reactants inside the catalyst pellet Ω. Model equations for such systems are in [2] and read (in the simplest case) with some parameters L, β, Φ, R_1, $R_2>0$

$$
(0.1)\qquad
\begin{aligned}
d_t u &= \Delta u - \Phi^2\, r(u,v) \\
L\, d_t v &= \Delta v - \beta\Phi^2\, r(u,v)
\end{aligned}
\quad (\text{in } \Omega)
$$

$$
u = v = 1 \qquad (\text{on } \partial\Omega).
$$

For Neumann boundary conditions this system does not oscillate, of course (see combustion problems [13]). Oscillations for (0.1) were obtained numerically by Luss, Lee [9] and Uppal, Ray [15]. They used Langmuir-Hinshelwood kinetics

$$
(0.2)\qquad r(u,v) = \frac{uv}{(1+R_1 u+R_2 v)^2}
$$

rather than the non-oscillating kinetics [7]

$$
r(u,v) = uv\ .
$$

In our analysis $L\in]0,1]$ will be the bifurcation parameter. Note that the steady states $W = (U,V)$ of (0.1) are independent of L. But W does depend on $x\in\Omega$. Therefore it is hard to apply a standard local Hopf bifurcation theorem as in [4], [11]. There are difficulties with the transversality, multiplicity and non-resonance conditions. Choosing L as a parameter even kicks (0.1) out of the class of problems handled in [4], [11].

In section 2 we will analyze stability of a steady state W. We obtain a net change of stability as L decreases from 1 to 0. This occurs as some eigenvalues cross the imaginary axis. Any local analysis of these crossings is avoided. Instead, a global Hopf bifurcation theorem from section 1 is applied and we get a global continuum of periodic orbits. Unfortunately the stability analysis still requires some (easy) numerical computations which are given in section 3.

The abstract bifurcation theorem of section 1 is in the author's dissertation [6] which was supervised by Prof. W. Jäger. The author wishes to thank Dr. H. Ederer,

Prof. H. Ray and Prof. W. Jäger for advice and encouragement.

1. Global Hopf Bifurcation

We consider parabolic systems of the form

$$(1.1) \qquad d_t w + A(L)w + f(L,w) = 0$$

in the following abstract setting :

(1.2) Assumption

X is a real Hilbert space with norm $|\cdot|$; A(L) is closed, densely defined for all L in]0, 1] with domain $X_\alpha := \mathcal{D}(A(L)^\alpha) \subset X$ independent of L for $0 \leq \alpha < 2$; -A(L) generates an analytic semigroup on X by an appropriate L-independent estimate for the compact resolvent. Concerning differentiability we require C^4 for the maps

$$A(\cdot) : \quad]0, 1] \longrightarrow \mathcal{L}(X_{1+\alpha}, X_\alpha)$$

$$f \; : \quad]0, 1] \times X_\alpha \longrightarrow X$$

and some fixed $\alpha \in [0, 1[$. We further need a growth condition

$$|D_w^k f(L,w)| \leq C(L)\,(1 + |w|_\alpha)$$

for all k=0,...,4 with C depending continuously on L.

(1.3) Assumption

The steady states W of (1.1), i.e. the solutions of

$$A(L)W + f(L,W) = 0 \;,$$

are independent of L. The linearization at W

$$\Lambda_L := -A(L) - D_w f(L,W)$$

is analytic in L with (uniformly) bounded inverse.

(1.4) Definition

A steady state (L,W) of (1.1) is called Hopf point of f, if its linearization Λ_L has a purely imaginary nonzero eigenvalue. Let H_f denote the set of all Hopf points of f.

(1.5) Assumption

The set H_f of Hopf points is contained in a closed bounded subset of $]0, 1[\times X_\alpha$.

We now introduce some notation to analyze stability properties of a steady state W as the parameter L varies. Let E(L,W) be the number of eigenvalues of Λ_L with positive real part, counting multiplicities. By assumption 1.3, E(L,W) changes by even integers as L varies. Define

$$\sigma(W) := (-1)^{E(1,W)}$$

and the net change of stability as L decreases from 1 to 0

$$\chi(W) := \lim_{\varepsilon \downarrow 0} (E(1-\varepsilon,W) - E(\varepsilon,W))/2$$

(1.6) Definition (Hopf index)

$$H(f) \;:=\; \sum_{W} \alpha(W)\,\chi(W)$$

is called Hopf index of f. The sum extends over all steady states W. The sum is finite because the steady states are isolated by assumption 1.3 and because assumption 1.5 guarantees $\chi(W)$ to be zero for large $|W|_\alpha$. Note that A has compact resolvent. If the steady state W is unique we have

$$H(f) \;=\; \pm\chi(W).$$

(1.7) Definition (virtual period)

$\tau \in R^+$ is called virtual period of a periodic solution $(L,w(t))$ of (1.1) with primitive period p, if either $\tau = p$ or for some integer m

$$\tau \;=\; mp$$

and $(L,w(t))$ has a primitive m-th root of unity as a Floquet multiplier.

We can now state our abstract result.

(1.8) Theorem (global Hopf bifurcation)

Let assumptions (1.2),(1.3) and (1.5) be satisfied. In addition assume the Hopf index to be nonzero:

(1.9) $$H(f) \neq 0 \,.$$

Then there exists a continuum $Z \subset \,]0,\,1] \times X_\alpha$ of periodic solutions and Hopf points of (1.1), such that

- Z is not contained in a closed bounded subset of $]0,\,1[\times X_\alpha$ (Z is "unbounded") or
- the virtual periods of the periodic solutions in Z are unbounded.

(1.10) Remarks:

The virtual periods need not vary continuously along the continuum Z. Virtual periods were introduced by Alligood and Yorke [1] in an ODE setting. The proof of the theorem uses generic bifurcation pictures and approximation arguments. The generic case for ODEs was treated by Mallet-Paret and Yorke [10]. A corresponding generic bifurcation picture for PDEs was given by the author in [6]; a proof of the theorem can be found there. It relies heavily on compactness of the flow (induced by compactness of the resolvent) and transversality theory.

In an ODE setting J. Ize has a related result for multiparameter problems. Note that crossing of double eigenvalues and crossing at resonance are cases covered by the theorem.

2. Application

We apply our abstract bifurcation result to the catalyst problem (0.1) and reduce our assumptions to some simple numerical computations which are given in section 3.

For simplicity let $\Omega =]-1,+1[$ and choose $R_1=1$, $R_2=30$, $\beta=1.2$ (cf. [9], [15]). Define $X := (H_0^1(\Omega))^2$, $w(x) := (u(x)-1, v(x)-1)$ to get normalized boundary conditions and

$$(2.1) \qquad A(L) := -\begin{pmatrix} 1 & 0 \\ 0 & 1/L \end{pmatrix} \Delta$$

$$(2.2) \qquad f(L,w)(x) := \begin{pmatrix} 1 \\ \beta/L \end{pmatrix} \cdot \phi^2 \cdot r(u(x),v(x)) \quad ;$$

the domain of A is $\mathcal{D}(A) = (H_0^3(\Omega))^2$ and we let $\alpha = 0$.

Check our technical assumptions 1.2 first: A(L) obviously fits into the framework presented there. The growth estimate on f is immediate by $r \in BC^\infty(R^+ \times R^+, R^+)$ and the embedding $H_0^1(\Omega) \longrightarrow C(\overline{\Omega})$. Note that $]0,1] \times]0,1]$ is an invariant set under (0.1) by maximum principle arguments [3]; therefore we restrict discussions to positive u and v.

Now we study assumptions 1.3 and 1.5 . Section 3, Lemma 3.2 tells us that (0.1) has a unique steady state $W = W(\phi)$ for any fixed ϕ. $W = (U-1, V-1)$ is trivially independent of L and the linearization

$$(2.3) \quad \Lambda_L := -A(L) - D_w f(L,W) = \begin{pmatrix} 1 & 0 \\ 0 & 1/L \end{pmatrix} \cdot \left(\Delta - \begin{pmatrix} 1 \\ \beta \end{pmatrix} \cdot \phi^2 \cdot D_{(u,v)} r(U,V)\right)$$

depends analytically on L. Its inverse is bounded iff zero is not an eigenvalue of

$$(2.4) \qquad \Lambda_1 = \Delta - \begin{pmatrix} 1 \\ \beta \end{pmatrix} \cdot \phi^2 \cdot D_{(u,v)} r(U,V) \ .$$

To discuss eigenvalues of (2.3) we define

$$(2.5) \qquad \begin{aligned} a(x) &:= \phi^2 \cdot D_u r(U(x),V(x)) \\ b(x) &:= \phi^2 \cdot D_v r(U(x),V(x)) \\ c(x) &:= a(x) + \beta\, b(x). \end{aligned}$$

The eigenvalue equation for Λ_L then reads

$$(2.6)_L \qquad \begin{aligned} \lambda u &= \Delta u - a(x)u - b(x)v \\ L\lambda v &= \Delta v - \beta a(x)u - \beta b(x)v \end{aligned} \quad ; \quad u(\pm 1) = v(\pm 1) = 0$$

(2.7) Lemma

Suppose

$$(2.8) \qquad |c(x)|_\infty < \pi^2/4 \ .$$

Then assumption 1.3 is satisfied and W is stable as a steady state of equation (0.1) for L = 1 .

If in addition $(\Delta - \beta b)^{-1}$ exists then assumption 1.5 is satisfied, i.e. the set of Hopf points is "bounded".

Proof:

For L = 1 one obtains from equations $(2.6)_1$ that $\lambda<0$ or $\beta u - v \equiv 0$. In the latter case we are left with

$$\lambda u = \Delta u - c(x)u .$$

By (2.8) assumption 1.3 is thus proved, along with stability of the steady state W for L = 1 .

The proof of assumption 1.5 is indirect. Assume there is a sequence (L_n,λ_n,u_n,v_n) of nontrivial solutions of $(2.6)_{L_n}$ with $L_n \downarrow 0$, $\lambda_n \in \mathrm{Im}^+$ and $|v_n| = 1$. We consider two cases.

Case 1: $\lim |\lambda_n| = \infty$

We obtain a contradiction by (2.6) and the resolvent estimate for $\Delta - a$:

$$1 = |v_n| = |(\Delta - \beta b - L_n\lambda_n)^{-1}(\beta a u_n)| \leq C\,|u_n| =$$

$$= C\,|(\Delta - a - \lambda_n)^{-1}(bv_n)| \leq C\,|v_n|/|\lambda_n| \longrightarrow 0 .$$

Without loss of generality we are left with

Case 2: $\lim \lambda_n = \lambda \in \mathrm{Im}^+$ exists.

By compactness of the resolvents we may assume that $u = \lim u_n$ and $v = \lim v_n$ exist and

$$u = (\Delta - a - \lambda)^{-1}\, b\, (\Delta - \beta b)^{-1}\, \beta a u \; ; \qquad u \neq 0 .$$

With $E := (\Delta - a - \beta b(\Delta-\beta b)^{-1}a)$ this is equivalent to the eigenvalue condition

$$\lambda u = Eu$$

which in turn can be written as

$$(2.9) \qquad \lambda(\Delta - c)\Delta^{-1}u = (\Delta - c)\,(\Delta - \beta b)^{-1}\,(\Delta - c)u .$$

We test equation (2.9) by u in L^2 to obtain a contradiction. Note that the right hand side of (2.9) defines a selfadjoint operator, hence

$$\overline{\lambda}\cdot(\,u\,,\,(\Delta - c)\Delta^{-1}u\,) \in \mathbb{R}\,, \qquad \lambda \in \mathrm{Im}^+, \quad \text{and}$$

$$0 = \mathrm{Re}\,(\,u\,,\,(\Delta - c)\Delta^{-1}u\,) = |u|^2_{L^2} - \mathrm{Re}\,(\,u\,,\,c\Delta^{-1}u\,) \geq$$

$$\geq |u|^2\cdot(1 - |c|_\infty\cdot|\Delta^{-1}|_{L^2,L^2}\,) = |u|^2\cdot(1 - |c|_\infty\cdot 4/\pi^2) > 0 .$$

□□ □

We have found the unique steady state W being stable at L = 1 . To show

$$H(f) \neq 0$$

we only need W to be unstable for sufficiently small positive L.

(2.10) Lemma (L = 0)

If $\Delta - \beta b$ has a positive eigenvalue, then W is unstable for sufficiently small po- positive L.

Proof:

Let $\mu := L\lambda$ and rewrite (2.6) as an eigenvalue equation for an L-dependent operator.

$$\mu u = L(\Delta u - au - bv)$$
$$\mu v = \Delta v - \beta au - \beta bv \,.$$

Given an eigenvalue $\mu_0 > 0$ for $L = 0$ we want a solution of

$$\mu v = \Delta v - \beta\cdot[a\cdot(L\Delta - La - \nu)^{-1}L + 1]\cdot bv$$

with $|v| = 1$ and $\mu = \nu$ such that μ is close to μ_0. Then μ is an eigenvalue of the system above.

By the resolvent estimate for $(\Delta - a - \nu/L)^{-1}$ we get analytic maps [8]

$$\Psi_L : \quad \nu \longmapsto \mu$$

in a small neighborhood of μ_0 in $\mathbb{C}$ - we may regard the terms involving L as a small perturbation. Ψ_L depends continuously on L for small $L \geq 0$ and, by a degree homotopy argument, Ψ_L has a fixed point $\mu(L)$ near μ_0; note that μ_0 has fixed point index 1 as a fixed point of the constant map $\Psi_0(\nu) \equiv \mu_0$. Then $\lambda(L) = \mu(L)/L$ is the eigenvalue which we looked for; Re $\lambda(L)$ is positive because μ_0 is positive.

□ □ □

3. Numerical Computations

In the last section we have reduced the assumptions of our bifurcation theorem to some easy computations. Once these are done (in Lemma 3.2) we get

(3.1) Theorem

The following conclusion holds for system (0.1) with $\Omega =]-1,+1[$, $R_1=1$, $R_2=30$, $\beta=1.2$ and $\Phi \in [30.81, 34.22]$:

- there are periodic solutions for arbitrarily small positive L or
- there are periodic solutions of arbitrarily large virtual period.

The proof is by theorem 1.8 and the following lemma which checks the remaining assumptions. Note that, by invariance of $]0, 1] \times]0, 1]$ under (0.1), periodic solutions cannot blow up in amplitude, i.e. stay bounded in X-space.

(3.2) Lemma (computations)

(i) Equation (0.1) has a unique steady state $W = W(\Phi)$ for any $\Phi \geq 0$.

(ii) $|c(x)|_\infty < \pi^2/4$,
i.e. (2.8) is satisfied, for all Φ less than 34.22 .

(iii) $\Delta - \beta b$ has exactly one simple positive eigenvalue and no zero eigenvalue for all Φ between 30.81 and 34.88 . The ranges of Φ are respectively maximal to the last digit.

Proof:

The steady states $W = (U-1, V-1)$ satisfy the linear relation

$$\beta U + V = \beta + 1 \quad .$$

We are therefore left with a simple boundary value problem (to be solved numerically)

$$0 = U_{xx} - \phi_0^2 \cdot r(U, \beta+1-\beta U)$$

$$U_x(0) = 0 \; ; \quad U(\phi/\phi_0) = 1$$

where we utilized the symmetry of U. The solutions for varying ϕ can be parametrized over $U(0) \in]0,1]$. We followed the unique solution branch by solving initial value problems for varying U(0) to determine $\phi = \phi(U(0))$. Inequality (2.8) was checked directly along the solution. To check (iii) we solved the initial value problem

$$0 = v_{xx} - \beta\phi_0^2 D_v r(U(x), V(x)) \cdot v$$

$$v_x(0) = 0 \; ; \quad v(0) = 1 .$$

We then used Sturm oscillation theory to see:

$\Delta - \beta b$ has a positive eigenvalue

$\Leftrightarrow$ v has a zero in $]0, \phi/\phi_0[$.

This last condition is easily verified along the solution of the initial value problem.

The computations were done on the IBM 370/168 at URZ Heidelberg and used the program DIFEX 1 by Deuflhard [5]. The relative error bounds were prescribed as $1/2 \cdot 10^{-4}$ with maximum step size 10^{-2} and the scaling factor $\phi_0 = 35$.

□ □ □

References:

[1] K. T. ALLIGOOD, J. A. YORKE: Families of periodic orbits: virtual periods and global continuability. Preprint.

[2] R. ARIS: The mathematical theory of diffusion and reaction in permeable catalysts, Vol. I, II. Oxford: Clarendon Press 1975.

[3] K. N. CHUEH, C. C. CONLEY, J. A. SMOLLER: Positively invariant regions for systems of nonlinear diffusion equations. Ind. Univ. Math. J. 26 (1977), 373

[4] M. CRANDALL, P. H. RABINOWITZ: The Hopf bifurcation theorem in infinite dimensions. Arch. Rat. Mech. Anal. 67 (1977), 53

[5] P. DEUFLHARD: Order and stepsize control in extrapolation methods. SFB 123 preprint 93 (1980) Heidelberg.

[6] B. FIEDLER: Stabilitätswechsel und globale Hopf-Verzweigung. Dissertation, Heidelberg 1982.

[7] C. S. KAHANE: On a system of nonlinear parabolic equations arising in chemical engineering. J. Math. Anal. Appl. 53 (1976), 343

[8] T. KATO: Perturbation theory for linear operators. Berlin, Heidelberg, New York: Springer Verlag 1980

[9] D. LUSS, J. C. M. LEE: Stability of an isothermal reaction with complex rate expression. Chem. Eng. Sci. 26 (1971), 1433

[10] J. MALLET-PARET, J. A. YORKE: Snakes: oriented families of periodic orbits, their sources, sinks, and continuation. Preprint.

[11] J. E. MARSDEN, M. McCRACKEN: The Hopf bifurcation theorem and its applications. Berlin, Heidelberg New York: Springer Verlag 1976

[12] H. RAY: Bifurcation phenomena in chemically reacting systems. Applications of bifurcation theory, Rabinowitz (ed.), Madison. Academic Press 1977, 285.

[13] D. H. SATTINGER: A nonlinear parabolic system in the theory of combustion. Q. Appl. Math. 33 (1975), 47

[14] M. SHEINTUCH, R. A. SCHMITZ: Oscillations in catalytic reactions. Catalysis Reviews 15 (1977), 107

[15] A. UPPAL, H. RAY: On the steady state and dynamic behaviour of permeable isothermal catalysts. Chem. Eng. Sci. 32 (1977), 649.

Weakly Nonlinear Systems and Bifurcation of Higher Dimensional Tori

Dietrich Flockerzi
Mathematisches Institut
Universität Würzburg
Am Hubland
D-8700 Würzburg, F.R.G.

I. INTRODUCTION

We consider the situation where a k-dimensional torus $\tau^k(\alpha)$ that is invariant for a sufficiently smooth one-parameter family of ordinary differential equations

$$\dot{\xi} = F(\xi,\alpha) \ , \quad \xi \in \mathbb{R}^n , \quad \alpha \in (-\alpha_o,\alpha_o) \subset \mathbb{R} , \tag{1.1}$$

encounters a change of stability as α crosses o. It is well known that this may lead to the bifurcation of a (k+1)-dimensional invariant torus from the given k-dimensional one. Let us only mention the result in [9] for k=o and the ones in [13],[12] or [1o,Ch.III] for k=1. In 1979 Sell[15] and Iooss[1o,Ch.VI] presented sufficient conditions for this bifurcation phenomenon to occur also for larger values of k. It is our goal to prove Sell's result under weaker assumptions on the non-linear terms in (1.1) and in doing so to generalize the results in [3], [6],[1o,Ch.III] or [11] to higher dimensions k. The main tools for the construction of the bifurcating torus are contained in section III. Although Lemma 3.1 about the persistence of an invariant manifold is not new we would like to present a simple proof of it which we have not found in the literature. Also Lemma 3.1 may help to clarify the construction of the invariant torus for System (4.11) in [15,pp.216-219].

II. THE BIFURCATION PROBLEM

We suppose that there exist curvilinear coordinates (x,y,z) so that in a neighborhood of $\tau^k(\alpha)$ System (1.1) takes the form

$$\begin{aligned}
\dot{x} &= A(y,\alpha)x + \alpha A_2(y,\alpha)z + X(x,y,z,\alpha), && x \in \mathbb{R}^2, \\
(2.1) \qquad \dot{y} &= \omega_1 \quad + Y(x,y,z,\alpha), && y \in T^k, \\
\dot{z} &= B(y,\alpha)z + \alpha B_1(y,\alpha)x + Z(x,y,z,\alpha), && z \in \mathbb{R}^{n-k-2},
\end{aligned}$$

where the capital letters denote smooth mappings between the obvious spaces. The y-dependent terms are supposed to be periodic in y with vector period 2π, the nonlinearities X and Z are to be of the form $O(|x|^2+|z|^2)$ as $(x,z) \to (o,o)$ uniformly in (y,α). We assume $Y(0,y,0,\alpha)$ to be of the form $O(|\alpha|)$ and consider the unperturbed linear system

(2.2) (A) $\dot{x} = A(\omega_1 t+y,o)x$ (B) $\dot{z} = B(\omega_1 t+y,o)z$

where y is any element of the k-dimensional standard torus T^k. We impose the following three conditions on (2.1):

A1) Spectral Conditions:

i) $A(y,o) = \begin{pmatrix} 0 & -\omega_o \\ \omega_o & 0 \end{pmatrix}$ for all $y \in T^k$, $\omega_o \neq o$.

ii) The realpart of the spectrum of $A(y,\alpha)$ has a first order contact with the imaginary axis (see (2.5) in C3 below).

iii) The fundamental matrix solution $\mathcal{B}(t,s,y)$ of (2.2B) with $\mathcal{B}(s,s,y) = I$ satisfies

$$\| \mathcal{B}(t,s,y) \| \leq \kappa_o e^{-\kappa(t-s)} \quad \text{for } t \geq s$$

with constants $\kappa_o \geq 1$ and $\kappa > o$ (uniformly in y).

A2) Nonresonance Condition:

For $\omega = \begin{pmatrix} \omega_o \\ \omega_1 \end{pmatrix}$ there exist positive constants κ_1 and κ_2 such that for all $\nu \in Z^{k+1} \setminus \{o\}$ one has

$$| \nu \cdot \omega | \geq \kappa_1 |\nu|_\infty^{-\kappa_2} .$$

A3) Vague Attractor Condition:

$T^k(o)$ is a vague attractor or repellor of order m (see C3 and (2.9))

Comments:

C1) A1i) suggests the introduction of cylindrical coordinates

(2.3) $x=(r\cos\theta,\ r\sin\theta)^T$, $\varphi = (\theta,y)^T$, $z = z$

in which System (2.1) takes the form

$$(2.4)_\alpha \qquad \begin{aligned} \dot r &= \alpha A_1^*(\varphi,\alpha)r + \alpha A_2^*(\varphi,\alpha)z + X^*(r,\varphi,z,\alpha)\,, \\ \dot\varphi &= \omega + \Phi(r,\varphi,z,\alpha)\,, \\ \dot z &= B(\varphi,o)z + \alpha B_1^*(\varphi,\alpha)r + \alpha B_2^*(\varphi,\alpha)z + Z^*(r,\varphi,z,\alpha) \end{aligned}$$

with $(r,\varphi,z) \in \mathbb{R}\times T^{k+1}\times\mathbb{R}^{n-k-2}$. The first order contact condition now reads:

(2.5) The mean value of $\alpha A_1^*(\varphi,\alpha)$ with respect to φ has a first order zero at $\alpha=o$ and is thus of the form

$$\alpha A_1^*(\varphi,\alpha) = \alpha K_1 + O(\alpha^2)\,, \quad K_1 \neq o\,.$$

In [7] we deal with the case of a higher order contact of the spectrum of $A(y,\alpha)$ with the imaginary axis.

C2) A2 is superfluous in case k is o and can be considerably weakened in case k is 1 (see e.g. [1o], [13]).

C3) A3 generalizes the notion of an m-th order focus or of m-asymptotic stability in case k is o (see [1, §24] or [11]). What is required in A3 is that the radial equation in $(2.4)_o$ can be put in the form

$$\dot r = r^{2m+1}(K_o + O(r))\,, \quad K_o \neq o\,.$$

THEOREM

Under the above assumptions there exists a unique family of (k+1)-dimensional invariant tori $\tau^{k+1}(\alpha)$ for (2.1) near $\alpha=o$ bifurcating from $\tau^k(\alpha)$. In case $K_1 > o > K_o$ ($K_1 < o, K_o < o$) the bifurcation occurs supercritically (subcritically) and $\tau^{k+1}(\alpha)$ is asymptotically stable. In case $K_1 < o < K_o$ ($K_1 > o, K_o > o$) the bifurcation occurs supercritically (subcritically) and $\tau^{k+1}(\alpha)$ is unstable. In any case in terms of System $(2.4)_\alpha$ the bifurcating torus is for small $|\alpha|$ of the form

$$(2.6) \qquad \{(r,\varphi,z,\alpha):\ r=r(\varphi,\alpha)=\left|\frac{K_1\alpha}{K_o}\right|^{1/2m}(1+o(1)),\ z=z(\varphi,\alpha)=o(|\alpha|^{1/2m})\}.$$

Proof:

A preliminary scaling $r\to\varepsilon r$, $z\to\varepsilon z$, $\alpha=\varepsilon\beta$ with $\varepsilon > o$ in $(2.4)_\alpha$ leads to a system of the form

(2.7a) $$\dot r = \varepsilon\beta A_1^*(\varphi,o)r + \varepsilon\beta A_2^*(\varphi,o)z + O(|\varepsilon^2\beta|) + O(\varepsilon\beta^2) + \sum_{i=1}^{M} \varepsilon^i \sum_{\ell+|\lambda|_1=i+1} R_{\ell,\lambda}(\varphi)r^\ell z^\lambda + O(\varepsilon^{M+1}) \quad ,$$

(2.7b) $$\dot\varphi = \omega + O(\varepsilon) ,$$

(2.7c) $$\dot z = B(\varphi,o)z + O(\varepsilon) \quad ,$$

where $|\lambda|_1$ denotes the sum of the components of the nonnegative multiindex $\lambda \in Z^{n-k-2}$ and $M \in \mathbb{N}$ is a sufficiently large number. We now average successively the coefficients of $\varepsilon\beta$ and ε^i for $i=1,\dots,M$ in (2.7a) so that the z-independent terms are replaced by their mean values and the z-dependent ones by o . Transformations of the type

$$\bar r = r + \varepsilon\beta L_1(\varphi)r + \varepsilon\beta L_2(\varphi)z$$

for the linear terms or

(2.8)$_i$ $$\bar r = r + \varepsilon^i \sum_{\ell+|\lambda|_1=i+1} u_{\ell,\lambda}(\varphi)r^\ell z^\lambda , \quad i = 1,\dots,M,$$

for the nonlinear terms lead in the well known way to partial differential equations which can be solved uniquely because of A1iii) and A2 (cf.[15],[2,§3o]). After the i-th change of variable (2.8)$_i$ we are left with a system for $(\bar r,\varphi,z)$ having again the form (2.7). We denote the new coefficients by $R^{(i+1)}_{\ell,\lambda}$,call $\bar r$ again by r and apply (2.8) successively for i+1,i+2,....,M. Since these averaging transformations do not destroy the symmetry properties with respect to θ (cf.(2.3)) in (2.7) a nonzero mean value $\bar R^{(i)}_{i+1,o}$ corresponds necessarily to an odd power of r . Hence the condition A3 amounts to the following:

(2.9) There exists an $m \in \mathbb{N}$ so that $\bar R^{(i)}_{i+1,o} = \begin{cases} o \text{ for } i=1,\dots,2m-1, \\ K_o \neq o \text{ for } i=2m \end{cases}$.

For a detailed description of the averaging procedure we refer to [7]. In summing up we arrive at the following equation for r :

$$\dot r = \varepsilon\beta K_1 r + \varepsilon^{2m} K_o r^{2m+1} + O(|\varepsilon^2\beta|) + O(\varepsilon\beta^2) + O(\varepsilon^{2m+1}).$$

To obtain a small (k+1)-dimensional torus we need to choose $\beta=\pm\varepsilon^{2m-1}$ for $\mp K_oK_1 > o$. In case $K_oK_1 < o$ we define

$$\rho \equiv \left|\frac{K_1}{K_o}\right|^{1/2m} \qquad \text{and} \qquad P \equiv \frac{\partial}{\partial r}(K_1 r + K_o r^{2m+1})\Big|_{r=\rho} \quad .$$

The averaging transformations and the subsequent translation $r \to r+\rho$ thus lead from (2.7) to the normal form

$$
\begin{aligned}
\dot{r} &= \varepsilon^{2m} Pr + \varepsilon^{2m} P_1(r) r^2 + \varepsilon^{2m+1} P_2(r,\varphi,z,\varepsilon) \ , \\
(2.10)_\varepsilon \qquad \dot{\varphi} &= \omega + O(\varepsilon) \ , \\
\dot{z} &= B(\varphi, o) z + O(\varepsilon)
\end{aligned}
$$

with some polynomial $P_1(r)$ and some smooth function P_2.

In the next section we will show that $(2.10)_\varepsilon$ possesses a (k+1)-dimensional invariant toroidal manifold of the form

$$\{(r,\varphi,z,\varepsilon): r=r(\varphi,\varepsilon)=O(\varepsilon),\ z=z(\varphi,\varepsilon)=O(\varepsilon)\} \ .$$

Tracing back all the substitutions we then will have shown that $(2.4)_\alpha$ possesses a family of (k+1)-dimensional invariant tori of the form (2.6) bifurcating from $\tau^k(\alpha)$. The case $K_1 K_o > o$ is treated in the analogous way. That all (k+1)-dimensional invariant tori of the original bifurcation problem are obtained this way can be seen as in [3].

III. WEAKLY NONLINEAR SYSTEMS

In this section we show that the invariant manifold $\{(\psi,z): z=o\}$ of a system of the type

$$(3.1) \qquad \dot{\psi} = \Omega \ , \quad \dot{z} = B(\psi) z \ , \qquad (\psi,z) \in \mathbb{R}^p \times \mathbb{R}^q \ ,$$

persists under small perturbations. Here Ω is a fixed vector in $\mathbb{R}^p$ and the matrix $B(\psi)$ is such that the fundamental matrix solution $\mathcal{B}(t,s,\psi)$ of

$$(3.2) \qquad \dot{z} = B(\Omega t + \psi) z \ , \quad \psi \in \mathbb{R}^p \ ,$$

with $\mathcal{B}(s,s,\psi) = I$ satisfies

$$(3.3) \qquad \|\mathcal{B}(t,s,\psi)\| \leq M e^{-\kappa(t-s)} \quad \text{for} \quad t \geq s$$

with uniform constants $M \geq 1$ and $\kappa > o$. To be more precise we state the following lemma which can be shown by the results in [5] or [14]. We prefer to present a rather simple way of proving it whereby we follow the ideas in [4,Ch.1].

LEMMA 3.1

Given

$$(3.4)\qquad \begin{aligned} \dot\psi &= \Omega + \Psi(\psi,z,\varepsilon), & \psi &\in \mathbb{R}^p, \\ \dot z &= B(\psi)z + Z(\psi,z,\varepsilon), & z &\in \mathbb{R}^q, \end{aligned}$$

for $0 < \varepsilon \leq \varepsilon_o$, $z \in U=\{z: |z| \leq \rho_o\}$ we assume:

H1) Ψ, Z are continuous in (ψ,z,ε), lipschitzian with respect to ψ and z with Lipschitz constants of order $O(\varepsilon)$ and $|\Psi|, |Z|$ are bounded by εK in $\mathbb{R}^p \times U \times (0,\varepsilon_o]$.

H2) B is bounded and possesses the Lipschitz constant L.

H3) The fundamental matrix solution $\mathcal{B}$ of (3.2) satisfies (3.3).

Then there exist an $\varepsilon_1 \in (0,\varepsilon_o]$ and functions $\rho(\varepsilon)=O(\varepsilon)$, $\lambda(\varepsilon)=O(\varepsilon)$ and $f(\psi,\varepsilon)$ such that

$$\{(\psi,z,\varepsilon): z = f(\psi,\varepsilon)\}, \quad 0 < \varepsilon \leq \varepsilon_1,$$

is an invariant manifold for (3.4) and f is the only such function in

$$S_{\rho(\varepsilon),\lambda(\varepsilon)} = \left\{ f: \mathbb{R}^{p+1} \to \mathbb{R}^q : \begin{array}{l} f \text{ continuous}, \ \|f\| \leq \rho(\varepsilon), \\ |f(\psi_1,\varepsilon)-f(\psi_2,\varepsilon)| \leq \lambda(\varepsilon)|\psi_1-\psi_2| \end{array} \right\}$$

If the right-hand side of (3.4) is periodic in ψ then f is periodic in ψ with the same period.

Proof:

Let $\psi(t,\tau) = \psi(t,\tau,\xi,\varepsilon)$ denote the solution of

$$(3.5)\qquad \dot\psi = \Omega + \Psi(\psi,f(\psi,\varepsilon),\varepsilon), \quad \psi(\tau,\tau) = \xi,$$

with $f \in S_{\rho,\lambda}$ and let $\mathcal{B}(t,s) = \mathcal{B}(t,s,\tau,\xi,\varepsilon)$ be the fundamental matrix solution of

$$(3.6)\qquad \dot z = B(\psi(t,\tau))z$$

with $\mathcal{B}(s,s) = I$. We fix $s_o \in \mathbb{R}$ and show that for small $\varepsilon > 0$ $\mathcal{B}(t,s)$ satisfies

$$(3.7)\qquad \|\mathcal{B}(t,s)\| \leq Me^{-\kappa'(t-s)} \quad \text{for } t \geq s$$

with uniform constants $M \geq 1$ and $\kappa' \in (0,\kappa_o]$. A thorough analysis of Theorem 2.1 of [8,Ch.VII] shows that the hypotheses of Lemma 3.1 together with the estimate (3.7) suffice to ensure the existence of an invariant manifold as claimed above.

Let $y_0(t,s_0) = y_0(t,s_0,\tau,\xi,\varepsilon)$ be the solution of $\dot{\psi} = \Omega$ with initial value $y_0(s_0,s_0) = \psi(s_0,\tau)$. For the solution $z(t,s_0)$ of (3.6), i.e. of

$$\dot{z} = B(y_0(t,s_0))z + [B(\psi(t,\tau)) - B(y_0(t,s_0))]z ,$$

we obtain by the variation of constants formula

$$|z(t,s_0)| \leq Me^{-\kappa(t-s_0)}|z(s_0,s_0)| + \int_{s_0}^{t} Me^{-\kappa(t-u)} L\varepsilon K|u-s_0||z(u,s_0)|du$$

and thus by Gronwall's inequality

$$|z(t,s_0)| \leq M|z(s_0,s_0)|\exp[-\kappa(t-s_0) + \frac{\varepsilon MLK}{2}(t-s_0)^2] \quad ,t \geq s_0.$$

With

$$h \equiv \left(\frac{2\log M}{\varepsilon MLK}\right)^{1/2} \quad \text{and} \quad \gamma \equiv h^{-1}\log M$$

we reach the estimate

$$|z(t,s_0)| \leq M|z(s_0,s_0)|\exp[(-\kappa+\gamma)(t-s_0)] \quad \text{for} \quad s_0 \leq t \leq s_0+h.$$

We repeat the above proceeding with $s_n \equiv s_0+nh$ and with the solution $y_n(t,s_n) = y_n(t,s_n,\tau,\xi,\varepsilon)$ of

$$\dot{\psi} = \Omega \quad , \quad y_n(s_n,s_n) = \psi(s_n,\tau) \quad ,$$

instead of s_0 and $y_0(t,s_0)$ for $n = 1,2,\ldots$. This leads to the estimate

$$\begin{aligned} |z(t,s_n)| &\leq M|z(s_n,s_0)|\exp[(-\kappa+\gamma)(t-s_n)] \\ &\leq M^{n+1}|z(s_0,s_0)|\exp[(-\kappa+\gamma)(t-s_0)] \\ &\leq M|z(s_0,s_0)|\exp[(-\kappa+2\gamma)(t-s_0)] \quad \text{for} \quad s_n \leq t \leq s_{n+1} \end{aligned}$$

and for all $n = 0,1,2,\ldots$. Since γ is of the form $O(\sqrt{\varepsilon})$ we can choose an $\varepsilon_1 > 0$ and a κ' such that (3.7) holds.

□

We now return to System $(2.10)_\varepsilon$. If its right-hand side is modified outside a small neighborhood of $(r,z)=(o,o)$ such that the local properties are globally valid then Lemma 3.1 allows us to reduce the dimension of $(2.10)_\varepsilon$ to k+2. The reduced system is then given by

$$\text{(3.8)} \qquad \begin{aligned} \dot{r} &= \varepsilon^{2m}Pr + \varepsilon^{2m}P_1(r)r^2 + \varepsilon^{2m+1}P_2(r,\varphi,f(r,\varphi,\varepsilon),\varepsilon) \quad , \\ \dot{\varphi} &= \omega + O(\varepsilon) \end{aligned}$$

for sufficiently small $\varepsilon > 0$.

LEMMA 3.2

System (3.8) possesses a unique periodic invariant manifold of the form

$$\{(r,\varphi,\varepsilon): r = r(\varphi,\varepsilon) = O(\varepsilon)\}$$

for small $\varepsilon > o$ which is asymptotically stable for $P < o$ and unstable for $P > o$.

Proof:

We just outline the proof of Lemma 3.2 and refer to [7] for the details. Since the coupling between the two equations in (3.8) is generally not weak enough for an immediate application of an invariant manifold theorem we perform further averaging transformations of the type (2.8) for the (r,φ)-variables in (2.7). The subsequent reduction via Lemma 3.1 then yields a System (3.8) in the following normal form:

$$\dot{r} = \varepsilon^{2m}Pr + \varepsilon^{2m}P_1(r)r^2 + \varepsilon^{2m+1}P_2(r,\varepsilon) + O(\varepsilon^{4m+1}),$$

$$\dot{\varphi} = \omega + \varepsilon\,\Phi_1(r,\varepsilon) + O(\varepsilon^{2m+1})$$

with smooth functions P_2 and Φ_1. Now the various Lipschitz constants are just of the right order in ε to guarantee the existence of an invariant manifold as claimed in the Lemma (cf. [8,Ch.VII]). It is clear that the sign of P determines the stability properties of this manifold.

□

With the results of this section all the statements of the Theorem are now easily proven.

VI. REFERENCES

[1] Andronov A.A. et al.: *Theory of Bifurcations of Dynamic Systems on a Plane*, Israel Program for Scientific Translations, Jerusalem (1971).

[2] Bogoljubov N.N. et al.: *Methods of Accelerated Convergence in Nonlinear Mechanics*, Springer Verlag Berlin (1976).

[3] Chow S.N. & Mallet-Paret J.: Integral Averaging and Bifurcation, JDE 26, No.1, 112-159 (1977).

[4] Coppel W.A.:*Dichotomies in Stability Theory*, Lecture Notes in Mathematics Vol.629,Springer Verlag Berlin(1978).

[5] Fenichel N.:Persistence and Smoothness of Invariant Manifolds for Flows,Indiana Univ.Math.Journal 21,No.3,193-226(1971).

[6] Flockerzi D.:Bifurcation Formulas for ODE's in $\mathbb{R}^n$,Journal Nonlinear Analysis 5,No.3,249-263(1981).

[7] Flockerzi D.: Generalized Bifurcation of Higher Dimensional Tori, Univ.Würzburg preprint No.9o(1982).

[8] Hale J.K.:*Ordinary Differential Equations*,Wiley-Interscience , New York(1969).

[9] Hopf E.:Abzweigung einer periodischen Lösung von einer stationären Lösung eines Differentialsystems,Berichte Math.-Phys.Klasse der Sächsischen Akademie der Wissenschaften (Leipzig) 94,1-22(1942).

[1o] Iooss G.:*Bifurcation of Maps and Applications*, Math.Studies 36, North Holland,Amsterdam (1979).

[11] Negrini P.&Salvadori L.:Attractivity and Hopf Bifurcation,Journal Nonlinear Analysis 3,No.1,87-99(1979).

[12] Ruelle D.&Takens F.:On the Nature of Turbulence,Comm.Math.Phys.2o, 167-192(1971).

[13] Sacker R.J.:On Invariant Surfaces and Bifurcation of Periodic Solutions of Ordinary Differential Equations,IMM-NYU 333(1964).

[14] Sacker R.J.&Sell G.R.:A Spectral Theory for Linear Differential Systems,JDE 27,32o-358(1978).

[15] Sell G.R.:Bifurcation of Higher Dimensional Tori,Archive Rat.Mech. Anal.69,199-23o(1979).

PERIODIC SOLUTIONS OF FUNCTIONAL DIFFERENTIAL EQUATIONS

Tetsuo Furumochi

1. Introduction.

Among various methods to show the existence of periodic solutions of functional differential equations, one of the fundamental ideas is to find a fixed point of Poincaré mapping T which maps a compact convex set S into itself. Recently, Grimmer [4] defined a convex Liapunov function (or functional) to choose a suitable set S, and by using this function the existence of periodic solutions of functional differential equations with finite delay has been discussed by Grimmer [4] and Furumochi [2]. In this paper, we shall consider the existence of periodic solutions for an ω-periodic functional differential equation with a delay r which is large comparing with the period ω. For such an ω-periodic functional differential equation in a very special form, it is known that the existence of ω-periodic solutions can be reduced to an auxiliary equation with a delay not greater than ω. For example, the existence of ω-periodic solutions of the ω-periodic difference-differential equation $\dot{x}(t) = f(t, x(t), x(t-r))$ with a finite delay r can be reduced to the auxiliary equation $\dot{x}(t) = f(t, x(t), x(t-\tau))$ with $\tau = r-k\omega$, where k is an integer and $0 \leqq \tau < \omega$, refer to Krasnosel'skii [6, p. 278]. The same reduction is known even for some integral equations with infinite delay, refer to Grimmer [4], Leitman and Mizel [7]. But such a reduction is not obvious for a general equation, and in many theorems the condition $r \leqq \omega$ is posed as an important assumption (see Halanay [5] and Yoshizawa [9]), which fails to hold for any $\omega > 0$ when $r = \infty$. On the other hand, several efforts have been made for functional differential equations with infinite delay, see Chow and Hale

[1]. We shall show that the reduction in the above is always possible for a general ω-periodic functional differential equation with any delay greater than ω. Moreover, we shall obtain a Razumikhin type theorem concerning the existence of ω-periodic solutions of functional differential equations by using a strongly convex Liapunov function.

Let R^+ and R denote the intervals $0 \leqq t < \infty$ and $-\infty < t < \infty$, respectively. For a given r, $0 < r \leqq \infty$, C_r denotes the Banach space of continuous and bounded functions defined by

$$C_r = \{ \phi : [-r, 0] \to R^n, \text{ continuous}\}, \quad 0 < r < \infty,$$

or

$$C_\infty = \{ \phi : (-\infty, 0] \to R^n, \text{ continuous and bounded}\},$$

with the uniform norm $\|\phi\| = \sup \{ |\phi(\theta)| : -r < \theta \leqq 0 \}$, where $|\cdot|$ denotes the Euclidean norm in R^n. For a given continuous function $x(s)$, the symbol x_t will denote the element of C_r such that $x_t(\theta) = x(t+\theta)$, $-r < \theta \leqq 0$.

Consider the functional differential equation

$$\dot{x}(t) = f(t, x_t). \tag{1}$$

The superposed dot denotes the right-hand derivative and $f(t, \phi) : R \times C_r \to R^n$ is a completely continuous function which is ω-periodic in t. We shall denote by $x(t_0, \phi)$ a solution of Equation (1) such that $x_{t_0}(t_0, \phi) = \phi$ and denote by $x(t, t_0, \phi)$ the value at t of $x(t_0, \phi)$.

2. Reduction of Equation (1).

The reduction of the existence of ω-periodic solutions of Equation (1) with $r > \omega$ to an auxiliary equation with the delay ω is done by using the following mapping. Let $\sigma(t,\psi) : R \times C_\omega \to C_r$ be a mapping such that $\sigma(t,\psi)$ is ω-periodic in t, continuous on $R \times C_\omega$, takes bounded sets in $R \times C_\omega$ into bounded sets, $\sigma(t,\psi)(\theta)$ is ω-periodic in θ on $(-r, 0]$ if $\psi(0) = \psi(-\omega)$, and that $\sigma(t,\psi)(\theta)$ is ω-periodic in θ on $(-r, -\omega]$ if $r > 2\omega$. A simple example is :

$$(2) \quad \sigma(\psi)(\theta) = \begin{cases} \psi(\theta), & -\omega \leqq \theta \leqq 0, \\ \psi(\theta + k\omega) - \dfrac{\theta + k\omega + \omega}{\omega}(\psi(0) - \psi(-\omega)), & -\min\{r, k\omega + \omega\} \leqq \theta < -k\omega, \ k = 1, 2, \cdots. \end{cases}$$

For such a mapping $\sigma(t,\psi)$, let $g(t,\psi) : R \times C_\omega \to R^n$ be a function defined by

$$g(t, \psi) = f(t, \sigma(t, \psi)).$$

Clearly $g(t,\psi)$ is completely continuous, and ω-periodic in t. For this function $g(t,\psi)$, consider an auxiliary equation

$$(3) \qquad \dot{x}(t) = g(t, x_t).$$

Then this equation has a solution for the initial value problem, while Equation (1) may fail to have a solution for some initial value problem (see Seifert [8]). However, we have the following theorem.

<u>Theorem 1</u>. An ω-periodic solution $x(t)$ $(t \in R)$ of Equation (1) is an ω-periodic solution of Equation (3), and vice versa.

The proof of this theorem is clear from the definition of $g(t,\psi)$ and the properties of the mapping $\sigma(t,\psi)$.

3. Existence of periodic solutions of Equation (1).

We shall discuss the existence of ω-periodic solutions of Equation (1). A function $V(t, x) : R \times R^n \to R$ is said to be a Liapunov function if $V(t, x)$ is continuous on $R \times R^n$ and satisfies $V(t, x) \geq a(|x|)$ for a continuous function $a(u)$ such that $a(u) \to \infty$ as $u \to \infty$. We shall say a Liapunov function $V(t, x)$ is convex if for each fixed $t \in R$ the set $\{x \in R^n : V(t, x) \leq k\}$ is convex in R^n. Moreover, a convex Liapunov function $V(t, x)$ is said to be strongly convex if for each fixed $t \in R$ the set $\{x \in R^n : V(t, x) = k\}$ is the boundary of $\{x \in R^n : V(t, x) \leq k\}$. We define the derivative of V along the solutions of Equation (1) by

$$V'_{(1)}(t, \phi) = \limsup_{\tau \to 0+} \frac{1}{\tau}\{V(t+\tau, x(t+\tau, t, \phi)) - V(t, \phi(0))\}.$$

Clearly we have

$$V'_{(1)}(t, \phi) = \limsup_{\tau \to 0+} \frac{1}{\tau}\{V(t+\tau, \phi(0)+\tau f(t, \phi)) - V(t, \phi(0))\} \tag{4}$$

if $V(t, x)$ is locally Lipschitzian with respect to x.

Consider the scalar equation

$$\dot{u} = h(t, u), \tag{5}$$

where $h(t, u) : R \times R \to R$ is continuous and locally Lipschitzian with respect to u. For Equation (1) with finite delay r, the following theorem is obtained in [2].

Theorem 2. Let $V(t, x) : R \times R^n \to R^+$ be a continuous, ω-periodic in t, convex Liapunov function, and let $L : R \to R$ be a nondecreasing continuous function such that $L(u) > u$ for all $u > 0$ or $L(u) \equiv u$. Suppose that Equation (5) has a solution on $[0, \omega]$ such that $u(0) \geq u(\omega)$, $u(t) \geq$

$V(t, 0)$ for $0 \leq t \leq \omega$, $\max_{0\leq t\leq\omega} u(t) \leq L(\min_{0\leq t\leq\omega} u(t))$, and that we have

$$V'_{(1)}(t, \psi) \leq h(t, V(t, \psi(0)))$$

for all functions $\psi \varepsilon C_r$ with the property that

(6) $V(t, \psi(0)) \geq u(t)$, $V(t+\theta, \psi(\theta)) \leq L(V(t, \psi(0)))$ for $-r \leq \theta \leq 0$.

Then Equation (1) with finite delay r has an ω-periodic solution $x(t)$ such that $V(t, x(t)) \leq u(t)$ for $0 \leq t \leq \omega$.

This theorem can be proved by employing Schauder's fixed point theorem. For the details, see [2]. Now we can prove the following theorem concerning the existence of an ω-periodic solution of Equation (1) with $r = \infty$, by using Theorem 1 and Theorem 2.

<u>Theorem 3</u>. Let $V(t, x) : R \times R^n \to R^+$ be a continuous, ω-periodic in t, strongly convex Liapunov function which is locally Lipschitzian with respect to x, and let $L : R \to R$ be a nondecreasing continuous function such that $L(u) > u$ for all $u > 0$ or $L(u) \equiv u$. Suppose that Equation (5) has a solution $u(t)$ on $[0, \omega]$ such that $u(0) \geq u(\omega)$, $u(t) > V(t, 0)$ for $0 \leq t \leq \omega$, $\max_{0\leq t\leq\omega} u(t) \leq L(\min_{0\leq t\leq\omega} u(t))$, and that we have

(7) $$V'_{(1)}(t, \phi) \leq h(t, V(t, \phi(0)))$$

for all functions $\phi \varepsilon C_\infty$ with the property that

(8) $\phi \varepsilon P_\infty$, $V(t, \phi(0)) \geq u(t)$, $V(t+\theta, \phi(\theta)) \leq L(V(t, \phi(0)))$ for $-\infty<\theta\leq 0$,

where $P_\infty = \{\phi \varepsilon C_\infty : \phi(\theta)$ is ω-periodic on $(-\infty, -\omega]\}$. Then Equation (1) with $r = \infty$ has an ω-periodic solution $x(t)$ such that $V(t, x(t)) \leq u(t)$ for $0 \leq t \leq \omega$.

Proof. First, we define a function $\bar{\sigma}(t,\psi): R\times C_\omega \to C_\infty$, which is similar to $\sigma(t,\psi)$ in Section 2, by using the strongly convex Liapunov function $V(t,x)$. For any $(t,\psi)\in R\times C_\omega$, let $v = \max_{-\omega\leq\theta\leq 0} V(t+\theta,\psi(\theta))$ and $\bar{v} = \max\{v, L(\min_{0\leq t\leq\omega} u(t))\}$. Define a set Σ by

$$\Sigma = \{(s,x)\in R\times R^n : V(s,x)\leq \bar{v}\}.$$

Let $\sigma(\psi)$ be a function defined by (2). For $\theta\leq 0$ such that $(t+\theta,\sigma(\psi)(\theta))\in\Sigma$, define $\bar{\sigma}(t,\psi)(\theta)$ by $\bar{\sigma}(t,\psi)(\theta) = \sigma(\psi)(\theta)$. For $\theta < -\omega$ such that $(t+\theta,\sigma(\psi)(\theta))\notin\Sigma$, define $\bar{\sigma}(t,\psi)(\theta)$ by $\bar{\sigma}(t,\psi)(\theta) = \lambda(\theta)\sigma(\psi)(\theta)$, where $\lambda(\theta)$ is a uniquely determined number such that $0<\lambda(\theta)<1$ and $V(t+\theta,\lambda(\theta)\sigma(\psi)(\theta)) = \bar{v}$. It is easy to see that the function $\bar{\sigma}(t,\psi)(\theta)$ has the same properties with $\sigma(t,\psi)$ in Section 2. Moreover, we have $V(t+\theta,\bar{\sigma}(t,\psi)(\theta))\leq\bar{v}$ for $\theta\leq 0$ by the definition of $\bar{\sigma}(t,\psi)$. For the function $\bar{\sigma}(t,\psi)$, define an auxiliary function $\bar{g}(t,\psi): R\times C_\omega\to R^n$ by

$$\bar{g}(t,\psi) = f(t,\bar{\sigma}(t,\psi)), \tag{9}$$

and consider the following auxiliary equation

$$\dot{x}(t) = \bar{g}(t,x_t). \tag{10}$$

If we show that

$$V'_{(1)}(t,\psi) \leq h(t, V(t,\psi(0))) \tag{11}$$

holds for (t,ψ) which satisfies (6) with $r=\omega$, then we can apply Theorem 2 to Equation (10). Let (t,ψ) satisfy (6) with $r=\omega$, and take $\phi = \bar{\sigma}(t,\psi)$. Then we obtain

$$V(t+\theta,\phi(\theta))\leq\bar{v}\leq L(V(t,\phi(0))) \quad \text{for } -\infty<\theta\leq 0.$$

Thus (6) with $r=\omega$ implies (8), and we have (7), which is not different from (11) by (4) and (9), since $V(t, x)$ is locally Lipschitzian with respect to x. Therefore all assumptions of Theorem 2 are satisfied with this $u(t)$, and Equation (10) has an ω-periodic solution. Thus we can obtain the conclusion of this theorem by Theorem 1.

References

[1] Shui-Nee Chow and J. K. Hale, Periodic solutions of autonomous equations, J. Math. Anal. Appl. 66(1978), 495-506.

[2] T. Furumochi, Periodic solutions of periodic functional differential equations, Funkcialaj Ekvacioj 24(1981), 247-258.

[3] T. Furumochi, Periodic solutions of functional differential equations with large delays, Funkcialaj Ekvacioj 25(1982), 33-42.

[4] R. Grimmer, Existence of periodic solutions of functional differential equations, J. Math. Anal. Appl. 72(1979), 666-673.

[5] A. Halanay, "Differential Equations, Stability, Oscillations, Time Lags," Academic Press, 1966.

[6] M. A. Krasnosel'skii, "Translation along Trajectories of Differential Equations," Amer. Math. Soc., 1968.

[7] M. J. Leitman and V. J. Mizel, Asymptotic stability and the periodic solutions of $x(t) + \int_{-\infty}^{t} a(t-s)g(s, x(s))ds = f(t)$, J. Math. Anal. Appl. 66(1978), 606-625.

[8] G. Seifert, Positively invariant closed sets for systems of delay differential equations, J. Differential Equations 22(1976), 292-304.

[9] T. Yoshizawa, "Stability Theory by Liapunov's Second Method," Math. Soc. Japan, 1966.

Author's Address :

Department of Mathematics

Iwate University

Morioka, Japan 020

SUBHARMONIC AND CHAOTIC SOLUTIONS OF THE FORCED VAN DER POL RELAXATION OSCILLATOR

by

J. Grasman

Department of Applied Mathematics
Mathematical Centre, Amsterdam,
The Netherlands

1. INTRODUCTION

In electronic circuits it has been observed that periodically forced nonlinear oscillators may behave chaotically in certain parameter ranges [2]. In this paper we present an asymptotic analysis of such a system, the forced Van der Pol oscillator,

$$\frac{d^2x}{dt^2} + \nu(x^2-1)\frac{dx}{dt} + x = (\alpha\nu+\beta)k\cos kt, \qquad 0 < \alpha < 2/3 \tag{1}$$

for large values of ν. Eventually, we construct a difference equation that contains all necessary quantitative and qualitative information for describing the possible solutions of this system. Besides the well-known stable solutions of period $T = 2\pi(2n-1)$, also irregular types of solutions are analyzed. Existence of such solutions was expected by LITTLEWOOD [8] and made possible by LEVINSON [7] in a study of a related piecewise linear equation. The horseshoe mapping created by Smale turned out to be an important tool in establishing the existence of these irregular solutions. LEVI [6] used this concept and symbolic dynamics in his study of a modified version of (1) which comes close to the piece-wise linear variant of Levinson. Our results agree qualitatively with those of Levi. Furthermore, (1) has been solved numerically by FLAHERTY and HOPPENSTEADT [1]. A comparison shows that there is also a good agreement between the outcome of their work and that of our asymptotic investigations.

2. OUTLINE OF THE ASYMPTOTIC METHOD

A solution of (1) has a behaviour that is characteristic for singular perturbation problems. Locally, the solution exhibits a bounary layer type of action like one meets in problems of fluid mechanics. On the other hand it passes a large time interval, where a two time scales expansion is applied. Finally, we distinguish a sequence of points, determined by the intersections with the lines $x = \pm 1$, where the local behaviour of the solution is analyzed by a stretching procedure in both the dependent and independent variable. For a picture of the different regions which are successively crossed by a subharmonic solution, we refer to Fig. 1. The method of matched asymptotic expansions yields formal local asymptotic solutions in which the integration constants are determined by averaging and by matching pairs of local solutions of adjacent regions. These computations have been carried out in [3] and a summary is given in [5]. Here we give only the leading terms of the local solutions.

For the region A of Fig. 1 the solution behaves asymptotically as

$$x_0(t,\tau) = 2\cos[\tfrac{1}{3}\arccos\{\tfrac{3}{2}\alpha\sin kt + \tfrac{3}{2}C_0(\tau)\}],$$

(2) $$\frac{\partial C_0}{\partial\tau} = \frac{-k}{2\pi}\int_{\tau\nu}^{\tau\nu+2\pi/k} x_0(t,\tau)dt, \quad C_0(0) = 2/3-\alpha,$$

$$\tau = (t-t^*)/\nu, \quad t^* = t_{1/2}, \quad t_r = (-\pi/2+2\pi r)/k.$$

The solution will leave the region A at a time t_m, when x_0 approaches the line $x = 1$, which is the case if C_0 reaches the value $\alpha - 2/3$. In the slow time scale this will be for

(3) $$T(\alpha) = \frac{-2\pi}{k}\int_{2/3-\alpha}^{-2/3+\alpha}\{\int_0^{2\pi/k} x_0(t;C_0)dt\}^{-1}dC_0.$$

For a region A_m, which is entered by the solution in a neighbourhood of $(x,t) = (1,t_{m-1})$ and left near $(x,t) = (1,t_m)$, the local asymptotic solution reads

(4) $$x_0^{(m)} = 2\cos\{\tfrac{1}{3}\arccos(\tfrac{3}{2}\alpha\sin kt + \tfrac{3}{2}\alpha - 1)\}.$$

For the $\nu^{-1/2}$-neighbourhood of $(x,t) = (1,t_m)$, that is region B_m, we have

(5) $$x \approx 1+\nu^{-1/2}v_0^{(m)}(\xi), \quad \xi = (t-t_m)\nu^{1/2},$$

(6) $$v_0^{(m)}(\xi) = -aD'_{K_m/a^2}(-a\xi)/D_{K_m/a^2}(-a\xi), \quad a = \sqrt[4]{2\alpha k^2},$$

(7) $$K_{m+1} = K_m + 2\pi I, \quad I = \frac{1}{2\pi}\int_{t_{m-1}}^{t_m} x_0^{(m)}(t)dt,$$

where $D_\mu(z)$ denotes the parabolic cylinder function of order μ. For $\mu \le 0$, this function has no zero's. Let $K_m < 0$, then for some m, say $m = n$, $K_{n-1} < 0 < K_n$. Similarly, near $x = -1$, $K_{-1/2} < 0 < K_{1/2}$. Since the asymptotic solution (5)-(6) is singular in region $B_{1/2}$ (and B_n) at $\xi = \xi_0$, where the denominator vanishes at a zero of the parabolic cylinder function, a boundary layer type of solution occurs at that point.

(8) $$x \approx W_0(\eta), \quad \eta = (t-t_{1/2}-\xi_0\nu^{-1/2})\nu,$$

(9) $$(1-W_0)^{-1} + 1/3\log\{(W_0+2)(1-W_0)^{-1}\} = -\eta + W_0.$$

Notice that $W_0 \to 2$ as $\eta \to \infty$. Apparently, the boundary layer solution matches the solution for region A given by (2).

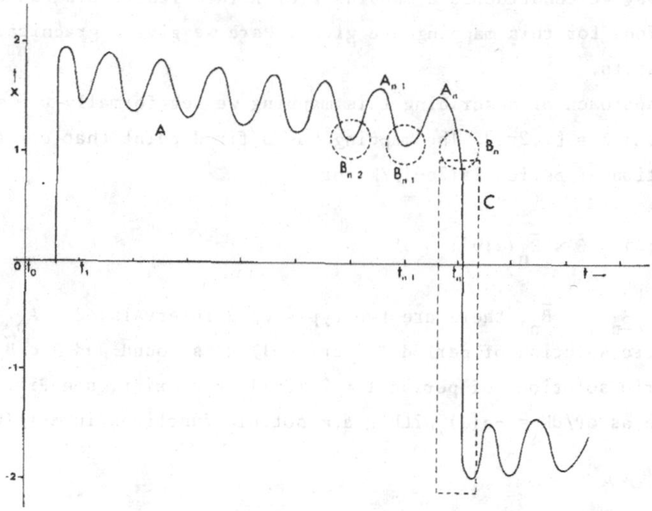

Fig. 1: Regions with local solutions in the subharmonic case

If K_n is $O(e^{-d\nu})$ with $d > 0$, then a local behaviour near the line $x = 1$ arises which we indicate by dips and slidings of the solution. For a complete analysis of those phenomena we refer to [4]. In Fig. 2 we sketch the regions that are crossed and for which local asymptotic solutions similar to the above ones can be computed.

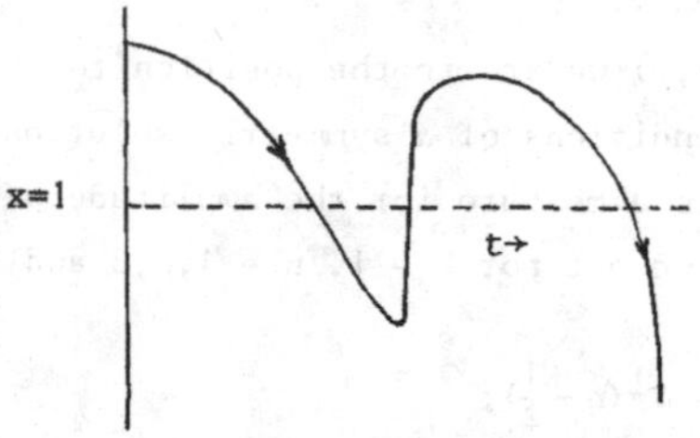

Fig. 2a: Local behaviour of a dipping solution K_n is negative and and of order $O(e^{-d\nu})$

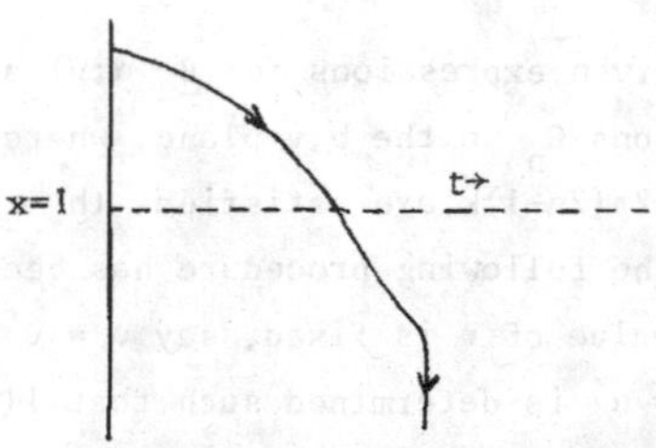

Fig. 2b: Local behaviour of a sliding solution K_n is positive and of order $O(e^{-d\nu})$

3. A MAPPING ON AN INTERVAL

Using the matched local asymptotic solutions of the foregoing section, including higher order terms of the local asymptotic expansions, one can relate the value of K_n of the region B_n with the value of $K_{1/2}$. We restrict $K_{1/2}$ to a compact interval X of length $2\pi I$ and take for n the first positive integer for which K_n returns in this interval. In this way we constructed a mapping P of X into itself. In [5] explicit asymptotic expressions for this mapping are given. Here we give a graphical representation of these results.

In a first approach of describing this mapping we let formally $\nu \to \infty$ and consider P for the interval $X = [0, 2\pi I]$. The mapping has a fixed point that corresponds with a symmetric solution of period $2\pi(2n-1)/k$ for

$$(10) \qquad \underline{\beta}_n(\alpha;\nu) < \beta < \bar{\beta}_n(\alpha;\nu).$$

Since $\underline{\beta}_n < \bar{\beta}_{n+1} < \underline{\beta}_{n-1} < \bar{\beta}_n$, there are two types of β intervals: $\beta \in \tilde{A}_n = (\bar{\beta}_{n+1}, \underline{\beta}_{n-1})$, where one symmetric solution of period $T = 2\pi(2n-1)/k$ is found and $\beta \in \tilde{B}_n = (\underline{\beta}_n, \bar{\beta}_{n+1})$, where two symmetric solutions of period $T = 2\pi(2n\pm 1)/k$ coexist, see Fig. 3. All fixed points are stable as $dP/dK = -g(\alpha)$ with g a monotonic functions in α ($g(0) = 1/2$, $g(2/3) = 1$).

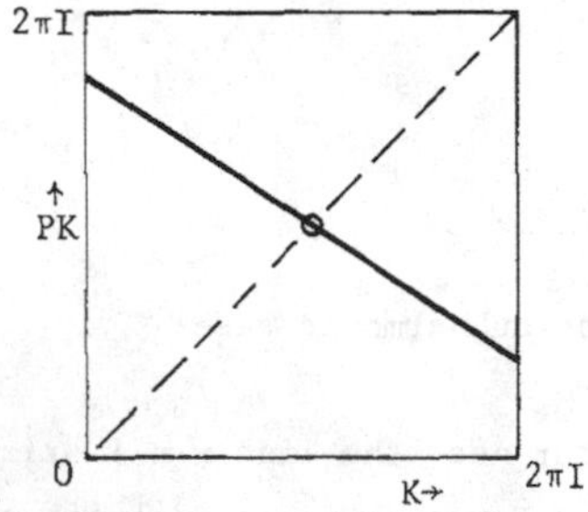

Fig. 3a: one fixed point $\beta \in \tilde{A}_n$

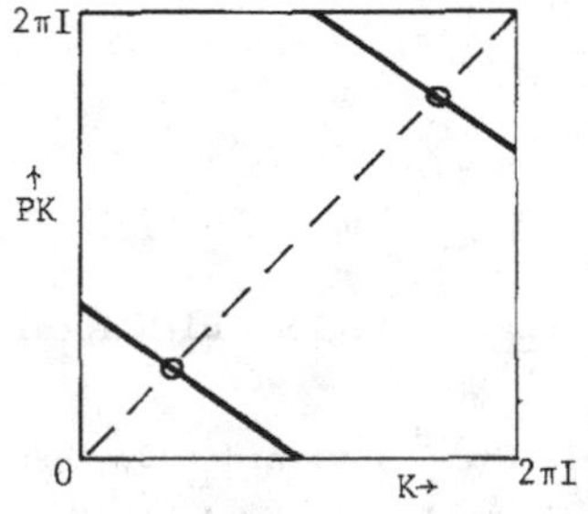

Fig. 3b: two fixed points $\beta \in \tilde{B}_n$

Using the given expressions for $\underline{\beta}_n(\alpha;\nu)$ and $\bar{\beta}_n(\alpha;\nu)$ we are in the position to construct regions Ω_n in the b,ν-plane, where the conditions of a symmetric solution of period $T = 2\pi(2n-1)k$ are satisfied ($b = \alpha+\beta/\nu$ is a measure for the amplitude of the forcing). The following procedure has been carried out for k = 1, n = 1,2,3 and 4:

Step 1. a value of ν is fixed, say $\nu = \nu^*$;

Step 2. $\alpha = \alpha^*$ is determined such that $T(\alpha^*)\nu^* = 2\pi(n-\frac{1}{2})$;

Step 3. the corresponding values of $\underline{\beta}_n$ and $\bar{\beta}_n$ are computed;

Step 4. the line $\nu = \nu^*$ is within Ω_n for $\alpha^* + \underline{\beta}_n/\nu^* < b < \alpha^* + \bar{\beta}_n/\nu^*$.

The shape of Ω_n follows from a sufficient number of values ν^*, see Fig. 4, which is in good agreement with the results of FLAHERTY and HOPPENSTEADT [1].

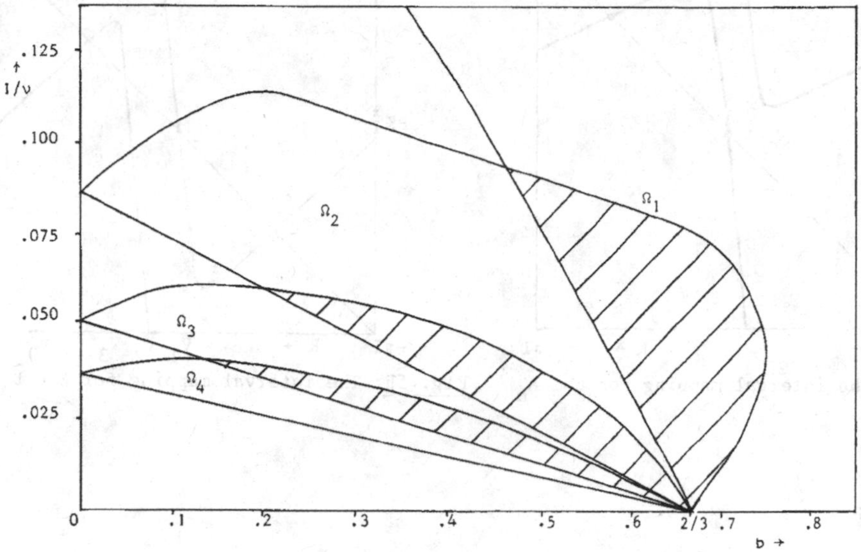

Fig. 4: Domains Ω_n with periodic solution $T = 2\pi(2n-1)$

In the above approximation for $\beta \in \tilde{A}_n$ we only found the stable fixed point; the unstable one is situated in the boundary layer, as is seen in Fig. 5. Note that we shifted the X-interval in order to have K = 0 on the interior of the interval. For the case $\beta \in \tilde{B}_n$ there are two unstable fixed points within the boundary layer, see Fig. 5b. Besides these two unstable points there is Cantor set of points which do not belong to the domain of attraction of the stable fixed points. In order to describe this set we use symbolic dynamics. As given in Fig. 5b we consider subintervals V_i ($i = 0, 1, 2$ and 3) and keep track of the mapping of points remaining in $\cup V_i$ by the transition matrix

$$M = \begin{pmatrix} 0 & 1 & 1 & 1 \\ 0 & 1 & 1 & 1 \\ 1 & 0 & 0 & 0 \\ 0 & 1 & 1 & 1 \end{pmatrix} . \qquad (11)$$

If $M_{ij} = 1$ a point of V_i is mapped in V_j, while for $M_{ij} = 0$ such a mapping is not possible. We introduce the topological space Σ_M consisting of all biinfinite sequences of the symbols 0, 1, 2 and 3 allowing only combinations ij for which $M_{ij} = 1$. Then every element of Σ_M corresponds with a solution of (1). The two unstable fixed points are represented by sequences of just the symbol 1 or just the symbol 3. Furthermore, any irregular solution satisfying the transition rule (11) is possible.

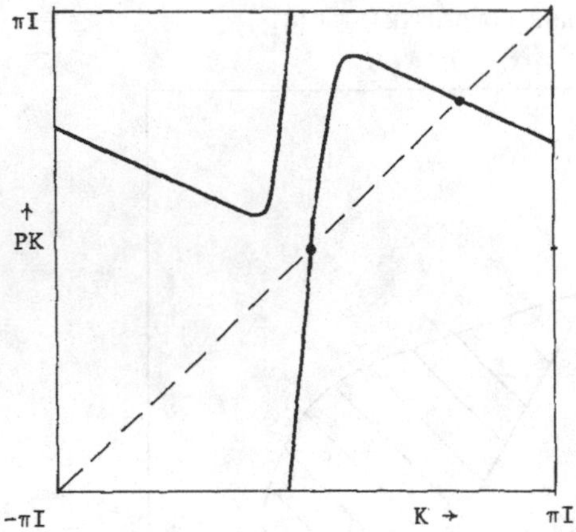

Fig. 5a: The interval mapping for $\beta \in \tilde{A}_n$

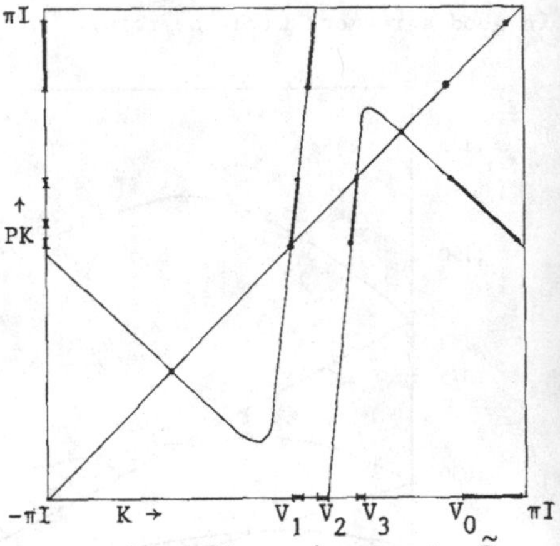

Fig. 5b: The interval mapping for $\beta \in \tilde{B}_n$

REFERENCES

[1] FLAHERTY, J.E. & F.C. HOPPENSTEADT, *Frequency entrainment of a forced Van der Pol oscillator*, Stud. Appl. Math. 18 (1978), 5-15.

[2] GOLLUB, J.P., T.O. BRUNNER & B.G. DANLY, *Periodicity and chaos in coupled nonlinear oscillators* Science 200 (1978), 48-50.

[3] GRASMAN, J., *On the Van der Pol relaxation oscillator with a sinusoidal forcing term*, Mathematisch Centrum, Amsterdam, Report TW 207 (1980).

[4] GRASMAN, J., *Dips and slidings of the forced Van der Pol relaxation oscillator*, Mathematisch Centrum, Amsterdam, Report TW 214 (1981).

[5] GRASMAN, J., H. NIJMEIJER & E.J. VELING, *Singular perturbations and a mapping on an interval for the forced Van der Pol relaxation oscillator*, Mathematisch Centrum, Amsterdam, Report TW 221 (1982).

[6] LEVI, M., *Qualitative analysis of the periodically forced relaxation oscillations*, Mem. Amer. Math. Soc. 244 (1981).

[7] LEVINSON, N., *A second order differential equation with singular solutions*, Ann. of Math. 50 (1949), 127-153.

[8] LITTLEWOOD, J.E., *On nonlinear differential equations of the second order: III The equation* $\ddot{y} - k(1-y^2)\dot{y} + y = b\mu k \cos(\mu t+\alpha)$ *for large* k, *and its generalizations*, Acta. Math. 97 (1957), 267-308.

FUNCTIONAL DIFFERENTIAL EQUATIONS WITH INFINITE DELAY ON THE SPACE C_γ

Thomas HAGEMANN and Toshiki NAITO

The variation-of-constants formula is well known for linear functional differential equations with finite delay [cf. 1, 2, 3]. It is a representation of the solution in terms of the fundamental matrix, the function of forcing and the initial function. For equations with infinite delay on the space C_γ defined below, the corresponding formula contains an additional term described by the "exponential limit at $-\infty$" of the initial function.

For arbitrary $\gamma \in R$ we consider the space C_γ of all continuous functions $\phi:(-\infty, 0] \to C^n$, such that the limit $\tilde{\phi}(-\infty) = \lim_{\theta \to -\infty} e^{-\gamma\theta}\phi(\theta)$ exists in C^n, with the norm $|\phi| := \sup\{e^{-\gamma\theta}|\phi(\theta)| : \theta \leqq 0\}$. The limit $\tilde{\phi}(-\infty)$ is called the exponential limit of ϕ at $-\infty$.

For the Lemma to follow we need a further definition. Let J be either the interval $(-\infty, 0]$ or $[-1, 0]$ and $\xi:J \to M$ a function, where M is the space of complex $n \times n$-matrices with the operator norm induced by the usual vector norm on C^n. We say ξ is normalized, if $\xi(0) = 0$ and ξ is continuous to the left on the interior of J.

Lemma. Let $L:R \times C_\gamma \to C^n$ be a continuous function such that $L(t, \cdot):C_\gamma \to C^n$ is linear for all $t \in R$. Then there are functions $\Lambda:R \to M$ and $\eta:R \times (-\infty, 0] \to M$ such that

(i) Λ and η are Borel measurable on R and $R \times (-\infty, 0]$ respectively, and for fixed $t \in R$ the function $\eta(t, \cdot):(-\infty, 0] \to M$ is normalized and locally of bounded variation;

(ii) for $t \in R$ and $a < b \leqq 0$

$$|\Lambda(t)| \leqq c|L(t, \cdot)|$$

$$\mathrm{Var}(\eta(t, \cdot), [a, b]) \leqq c \max\{e^{-\gamma a}, e^{-\gamma b}\}|L(t, \cdot)|,$$

where $\mathrm{Var}(\eta(t, \cdot), [a, b])$ is the variation of $\eta(t, \cdot)$ on $[a, b]$

and c is a real constant depending only on the dimension n;

(iii) for all $(t, \phi) \in R \times C_\gamma$

$$L(t, \phi) = \Lambda(t)\tilde{\phi}(-\infty) + \lim_{R\to\infty} \int_{-R}^{0} d_\theta[\eta(t, \theta)]\phi(\theta).$$

Furthermore the following uniqueness proposition holds:

(iv) The functions $\Lambda: R \to M$ and $\eta: R \times (-\infty, 0] \to M$, such that $\eta(t, \cdot)$ is normalized and locally of bounded variation on $(-\infty, 0]$ for all $t \in R$ and (iii) is satisfied, are uniquely determined by L.

<u>Proof.</u> First we observe that C_γ and the space C of C^n-valued continuous functions on $[-1, 0]$ with the supremum-norm are isometric; the mapping $i: C_\gamma \to C$

$$i(\phi)(s) := \begin{cases} e^{-\gamma s/(1+s)}\phi(s/(1+s)), & s \in (-1, 0] \\ \tilde{\phi}(-\infty) = \lim_{\theta\to-\infty} e^{-\gamma\theta}\phi(\theta), & s = -1 \end{cases}, \quad \phi \in C_\gamma,$$

is linear, bijective and norm-preserving: $|\phi| = |i(\phi)|$. The function $M: R \times C \to C^n$, $M := L\circ(id_R \times i)^{-1}$ is continuous with $M(t, \cdot): C \to C^n$ linear for all $t \in R$. By the Riesz-Representation-Theorem there is a function $\mu: R \times [-1, 0] \to M$ such that $\mu(t, \cdot)$ is normalized and of bounded variation on $[-1, 0]$ with

$$M(t, \psi) = \int_{-1}^{0} d_s[\mu(t, s)]\psi(s) \quad \text{for all} \quad (t, \psi) \in R \times C.$$

Defining $\Lambda(t) := \mu(t, -1 + 0) - \mu(t, -1)$, $t \in R$, one has for $(t, \psi) \in R \times C$ (cf. [5, Theorem 5c, Chapter I])

$$M(t, \psi) = \Lambda(t)\psi(-1) + \lim_{r\to 1-} \int_{-r}^{0} d_s[\mu(t, s)]\psi(s).$$

Since $L(t, \phi) = M(t, i(\phi))$, this relation is rewritten as (cf. [5, Theorem 5c, Chapter I])

$$L(t, \phi) = \Lambda(t)\tilde{\phi}(-\infty) + \lim_{r\to 1-} \int_{-r}^{0} d_s[\mu(t, s)]\, e^{-\gamma s/(1+s)}\phi(s/(1+s)).$$

Setting $\zeta(t, \theta) := \mu(t, \theta/(1 - \theta))$ for $(t, \theta) \in R \times (-\infty, 0]$,

$$\eta(t, \theta) := \int_{0}^{\theta} e^{-\gamma\alpha} d_\alpha \zeta(t, \alpha) \quad \text{for} \quad (t, \theta) \in R \times (-\infty, 0),$$

and $\eta(t, 0) = 0$, one has

$$L(t, \phi) = \Lambda(t)\tilde{\phi}(-\infty) + \lim_{R\to\infty} \int_{-R}^{0} d_\theta[\zeta(t, \theta)]e^{-\gamma\theta}\phi(\theta)$$
$$= \Lambda(t)\tilde{\phi}(-\infty) + \lim_{R\to\infty} \int_{-R}^{0} d_\theta[\eta(t, \theta)]\phi(\theta)$$

for all $(t, \phi) \in R \times C_\gamma$ (cf. [5, Theorem 6b, Chapter I]), which proves the statement (iii).

Next we prove (i). Since $\mu(t, \cdot)$, $t \in R$, is normalized and of bounded variation on $[-1, 0]$, $\zeta(t, \cdot)$ is clearly normalized and of bounded variation on $(-\infty, 0]$, and hence $\eta(t, \cdot)$ is normalized and locally of bounded variation on $(-\infty, 0]$ (cf. [5, Theorem 5c, Chapter I]). If one can prove the Borel measurability of μ on $R \times [-1, 0]$, one has at once the Borel measurability of Λ on R and of ζ on $R \times (-\infty, 0]$; therefore $\eta(t, \cdot)$ is continuous to the left for all $t \in R$ and $\eta(\cdot, \theta)$ is Borel measurable for all $\theta \in (-\infty, 0]$, which together shows the Borel measurability of η on $R \times (-\infty, 0]$.

To prove the Borel measurability of μ we define, for all natural numbers $m \geq 1$, $\psi_m:[-1, 0] \to C([-1, 0], M)$ by $[\psi_m(s)](r) := -I, 0,$ and $-(1 - m(s-r))I$ for $r \in [s, 0]$, $[-1, \max(-1, s-1/m))$ and $[\max(-1, s-1/m), s)$ respectively, where $s \in [-1, 0]$. For $s \in (-1, 0)$ and $-1 < s - 1/m$, integration by parts implies that

$$M(t, \psi_m(s)) = \left\{\int_{s-\frac{1}{m}}^{s} + \int_{s}^{0}\right\} d_r\mu(t, r)[\psi_m(s)](r)$$
$$= \mu(t, s) - \int_{s-\frac{1}{m}}^{s} \mu(t, r)(-m)\, dr - \mu(t, s)$$
$$= m\int_{s-\frac{1}{m}}^{s} \mu(t, r)\, dr$$

Since $\mu(t, r)$ is continuous to the left in $(-1, 0)$, the last integral tends to $\mu(t, s)$ as $m \to \infty$, namely,

$$\mu(t, s) = \lim_{m\to\infty} M(t, \psi_m(s)) \quad \text{for} \quad (t, s) \in R \times (-1, 0).$$

Since ψ_m is easily to be seen as continuous function, the function $[(t, s) \mapsto M(t, \psi_m(s))]:R \times [-1, 0] \to M$ is continuous. Therefore $\mu(t, s)$ is Borel measurable for $(t, s) \in R \times (-1, 0)$ as a limit function of a sequence of continuous functions $M(t,\psi_m(s))$. Since $\mu(t, 0) = 0$ and $\mu(t, -1) = M(t, \psi_m(-1))$ for any $m \geq 1$, we know that $\mu(t, s)$ is Borel measurable for $(t, s) \in R \times [-1, 0]$.

From the Riesz-Representation-Theorem, we have $\mathrm{Var}(\mu(t, \cdot), [1, 0]) \leq |M(t, \cdot)|$ in the case $n = 1$; in the general case notice that all norms in C^n are equivalent, so that there is a real constant c depending only on the dimension n such that

$$\mathrm{Var}(\mu(t, \cdot), [-1, 0]) \leqq c|M(t, \cdot)|.$$

By this inequality, one has

$$|\Lambda(t)| = |\mu(t, -1+0) - \mu(t, -1)| \leqq \mathrm{Var}(\mu(t, \cdot), [-1, 0])$$

$$\leqq c|M(t, \cdot)| = c|L(t, \cdot)|;$$

it is clear that $|M(t, \cdot)| = |L(t, \cdot)|$ since the isomorphism i^{-1} is norm-preserving. Next we have

$$(1) \qquad \mathrm{Var}(\zeta(t, \cdot), [a, b]) \leqq \mathrm{Var}(\mu(t, \cdot), [a/(1-a), b/(1-b)])$$

$$\leqq \mathrm{Var}(\mu(t, \cdot), [-1, 0]) \leqq c|L(t, \cdot)|.$$

Since

$$\mathrm{Var}(\eta(t, \cdot), [a, b]) \leqq \int_a^b e^{-\gamma\alpha}|d_\alpha\zeta(t, \alpha)|$$

(cf. [5, Theorem 5c, Chapter I]), we have

$$(2) \quad \mathrm{Var}(\eta(t, \cdot), [a,b]) \leqq \max\{e^{-\gamma a}, e^{-\gamma b}\}\ \mathrm{Var}(\zeta(t, \cdot), [a, b])$$

$$\leqq c\cdot\max\{e^{-\gamma a}, e^{-\gamma b}\}\ |L(t, \cdot)|,$$

which is (ii).

It remains to prove (iv). We show that, if there is a complex $n \times n$-matrix Λ and a normalized function $\eta:(-\infty, 0] \to M$ locally of bounded variation with

$$\Lambda\tilde{\phi}(-\infty) + \lim_{R\to\infty} \int_{-R}^0 d_\theta[\eta(\theta)]\phi(\theta) = 0 \quad \text{for all} \quad \phi \in C_\gamma,$$

then $\Lambda = 0$ and $\eta = 0$. Let C_γ^0 be the space of all $\phi \in C_\gamma$ with $\tilde{\phi}(-\infty) = 0$; we have

$$\lim_{R\to\infty} \int_{-R}^0 d_\theta[\eta(\theta)]\phi(\theta) = 0 \quad \text{for all} \quad \phi \in C_\gamma^0.$$

Suppose there is a point $\sigma < 0$ of continuity of η with $\eta(\sigma) \neq 0$, and choose a $v \in C^n$ with $\eta(\sigma)v \neq 0$. By [5, Theorem 3b, Chapter I], one can find an $\varepsilon > 0$ with $\mathrm{Var}(\eta, [\sigma - \varepsilon, \sigma]) < |\eta(\sigma)v|/|v|$. Define $\phi \in C^0_\gamma$ by $\phi(\theta) := 0$ if $\theta \leqq \sigma - \varepsilon$, $\phi(\theta) := v$ if $\theta \in [\sigma, 0]$ and linear on $[\sigma - \varepsilon, \sigma]$ such that ϕ becomes continuous. For large $R > |\sigma - \varepsilon|$ one has therefore $\int_{-R}^0 d_\theta[\eta(\theta)]\phi(\theta) = 0$. On the other hand

$$\begin{aligned}
\left|\int_{-R}^{0} d_\theta[\eta(\theta)]\phi(\theta)\right| &= \left|\int_{\sigma-\varepsilon}^{\sigma} d_\theta[\eta(\theta)]\phi(\theta) + \int_{\sigma}^{0} d_\theta[\eta(\theta)]v\right| \\
&= \left|\int_{\sigma-\varepsilon}^{\sigma} d_\theta[\eta(\theta)]\phi(\theta) - \eta(\sigma)v\right| \\
&\geqq |\eta(\sigma)v| - \left|\int_{\sigma-\varepsilon}^{\sigma} d_\theta[\eta(\theta)]\phi(\theta)\right| \\
&\geqq |\eta(\sigma)v| - \mathrm{Var}(\eta, [\sigma-\varepsilon, \sigma]) \sup\{|\phi(\theta)| : \theta \leqq 0\} \\
&> 0,
\end{aligned}$$

which is a contradiction. Therefore $\eta(\theta) = 0$ at all points $\theta < 0$ of continuity of η and thus $\eta = 0$, since η is normalized. It follows $\lim_{R\to\infty} \int_{-R}^0 d_\theta[\eta(\theta)]\phi(\theta) = 0$ for all $\phi \in C_\gamma$, which leads to $\Lambda\tilde{\phi}(-\infty) = 0$ for all $\phi \in C_\gamma$; but this means $\Lambda = 0$.

For a function $x:(-\infty, A) \to C^n$, let $x_t:(-\infty, 0] \to C^n$, $t < A$, be defined by $x_t(\theta) = x(t + \theta)$, $\theta \leqq 0$. For every $(\sigma, \phi) \in R \times C_\gamma$, we consider a linear system of equations

$$(3) \qquad x'(t) = L(t, x_t) + h(t) \quad \text{for} \quad t \geqq \sigma, \quad x_\sigma = \phi,$$

where L satisfies the assumption in Lemma and $h:[\sigma, \infty) \to C^n$ is locally integrable. In the following, the functions $\eta(t, \theta)$ and $\zeta(t, \theta)$ are always assumed to be the same as in the Lemma and its Proof.

Now let us introduce some result, the proof of which will be published elsewhere under more general conditions (see [4] for autonomous systems). Under the above assumption on L, there exists a unique $n \times n$ matrix $X(t, \sigma)$, called the fundamental matrix, locally absolutely continuous in $t \in [\sigma, \infty)$ for each σ, locally of B.V. in $\sigma \in (-\infty, t]$ for each t, such that

$$(\partial X/\partial t)(t, \sigma) = \int_{\sigma-t}^{0} [d_\theta \eta(t, \theta)]X(t + \theta, \sigma) \quad \text{a.e. in } t \geqq \sigma,$$

$$X(\sigma, \sigma) = I, \quad X(t, \sigma) = 0 \quad \text{for} \quad t < \sigma.$$

Furthermore it satisfies the relation

(4) $$X(t, \sigma) + \int_\sigma^t X(t, \alpha)\eta(\alpha, \sigma - \alpha)\, d\alpha = I \quad \text{for} \quad \sigma \leqq t.$$

For $t \geqq 0$, let $S(t): C_\gamma \to C_\gamma$ be defined by $[S(t)\phi](\theta) = \phi(t + \theta)$ for $t + \theta \leqq 0$ and $[S(t)\phi](\theta) = \phi(0)$ for $t + \theta \geqq 0$, $\phi \in C_\gamma$. Then the function defined by $x(t,\sigma,\phi,h) = \phi(t - \sigma)$ for $t \leqq \sigma$ and

(5) $$x(t,\sigma,\phi,h) = \phi(0) + \int_\sigma^t X(t, s)L(s, S(s - \sigma)\phi)\, ds + \int_\sigma^t X(t, s)h(s)\, ds \quad \text{for} \quad t \geqq \sigma$$

is the unique solution of System (3).

<u>Theorem.</u> Suppose $L(t, \phi)$ satisfies the assumption in Lemma and $h:[\sigma, \infty) \to C^n$ is locally integrable. Then the solution $x(t,\sigma,\phi,h)$ of System (3) is represented as

(6) $$\begin{aligned} x(t,\sigma,\phi,h) = X(t, \sigma)\phi(0) &+ [\int_\sigma^t X(t, s)\Lambda(s)e^{\gamma(s-\sigma)}\, ds]\tilde{\phi}(-\infty) \\ &+ \lim_{R\to\infty} \int_{-R}^0 d_\theta[\int_\sigma^t X(t, s)\eta(s, \sigma + \theta - s)\, ds]\phi(\theta) \\ &+ \int_\sigma^t X(t, s)h(s)\, ds \quad \text{for} \quad t \geqq \sigma, \end{aligned}$$

where $\eta(t, \theta)$ is normalized in $\theta \in (-\infty, 0]$.

<u>Proof.</u> It suffices to show that Relation (5) can be rewritten as Relation (6). From the trivial relation $[S(t)\phi]^\sim(-\infty) = e^{\gamma t}\tilde{\phi}(-\infty)$, the second term of the right hand side of (5) becomes

(7) $$\begin{aligned} &\int_\sigma^t X(t, s)\Lambda(s)e^{\gamma(s-\sigma)}ds\ \tilde{\phi}(-\infty) \\ &\quad + \int_\sigma^t X(t, s)\{\lim_{R\to\infty} \int_{-R}^0 [d_\theta\eta(s, \theta)][S(s - \sigma)\phi](\theta)\}ds. \end{aligned}$$

Thus it appears the second term of the right hand side of Relation (6).

Observe the second term of (7). Suppose $R > s - \sigma$ and devide the integration interval of the integral in the braces as $[-R, 0] = [-R, \sigma - s] \cup [\sigma - s, 0]$. The integral on the second subinterval becomes $[\eta(s, 0) - \eta(s, \sigma - s)]\phi(0) = -\eta(s, \sigma - s)\phi(0)$ since $[S(s - \sigma)\phi](\theta) = \phi(0)$ for $s - \sigma + \theta \geqq 0$. This result and Relation (4) lead to

$$(8)\quad \int_{\sigma}^{t} X(t, s)\int_{\sigma-s}^{0} [d_{\theta}\eta(s, \theta)][S(s - \sigma)\phi](\theta)\, ds = [X(t, \sigma) - I]\phi(0).$$

Since the limit operation in the brackets of Relation (7) acts only on the integral on $[-R, \sigma - s]$, the sum of $\phi(0)$ and the left hand side of (8) yields the term $X(t, \sigma)\phi(0)$ in Relation (6).

Since $[S(s - \sigma)\phi](\theta) = \phi(s - \sigma + \theta)$ for $\theta \leqq \sigma - s$, the above limit operation is rewritten as

$$(9)\quad \begin{aligned} &\int_{\sigma}^{t} X(t, s)\{\lim_{R\to\infty} \int_{-R}^{\sigma-s} [d_{\theta}\eta(s, \theta)][S(s - \sigma)\phi](\theta)\}\, ds \\ &= \int_{\sigma}^{t} X(t, s)\ \{\lim_{R\to\infty} \int_{s-\sigma-R}^{0} [d_{\xi}\eta(s, \sigma + \xi - s)]\phi(\xi)\}\, ds \\ &= \int_{\sigma}^{t} X(t, s)\ \{\lim_{R\to\infty} \int_{-R}^{0} [d_{\xi}\eta(s, \sigma + \xi - s)]\phi(\xi)\}\, ds. \end{aligned}$$

The integral $I_R(s)$ in the last bracket is the limit of the sequence of Riemann sums

$$A_P(s) = \sum_{j=1}^{N} [\eta(s, \sigma + \xi_j - s) - \eta(s, \sigma + \xi_{j-1} - s)]\phi(\tau_j),$$

as $m(P) \to 0$, where $P{:}\,-R = \xi_0 < \xi_1 < \dots < \xi_N = 0$, $\xi_{j-1} \leqq \tau_j \leqq \xi_j$, $j = 1, \dots, N$, and $m(P) = \max\{|\xi_j - \xi_{j-1}|: j = 1, \dots, N\}$. Since $\eta(t, \theta)$ is Borel measurable in (t, θ), the function $\eta(s, \sigma + \xi - s)$ is Borel measurable in $s \in [\sigma, t]$ for each $\xi \in [-R, 0]$. Thus the integral $I_R(s)$ is a Borel measurable function in $s \in [\sigma, t]$ as a limit of the sequence of the Borel measurable functions $A_P(s)$. Furthermore, Relation (1) implies

$$\begin{aligned} |I_R(s)| &\leqq \int_{-R}^{0} |d_{\xi}\zeta(s, \sigma + \xi - s)|e^{-\gamma(\sigma+\xi-s)}|\phi(\xi)| \\ &\leqq e^{\gamma(s-\sigma)}|\phi|c|L(s, \cdot)| \quad \text{for} \quad \sigma \leqq s \leqq t, \quad R > 0. \end{aligned}$$

From the dominated convergence theorem it follows that

$$(10)\quad \begin{aligned} &\int_{\sigma}^{t} X(t, s)\{\lim_{R\to\infty} \int_{-R}^{0} [d_{\xi}\eta(s, \sigma + \xi - s)]\phi(\xi)\}\, ds \\ &= \lim_{R\to\infty} \int_{\sigma}^{t} X(t, s)\{\int_{-R}^{0} [d_{\xi}\eta(s, \sigma + \xi - s]\phi(\xi)\}\, ds. \end{aligned}$$

Again from Relations (2), one knows that

$$(11)\quad \mathrm{Var}(\eta(s, \sigma + \cdot - s), [-R, 0]) = \mathrm{Var}(\eta(s, \cdot), [\sigma - s - R, \sigma - s])$$
$$\leqq c\, e^{\gamma(s-\sigma)} \max(1, e^{\gamma R})\, |L(s, \cdot)|.$$

This implies that, for any fixed $R > 0$, $|A_P(s)|$ is uniformly bounded in $s \in [\sigma, t]$. From the bounded convergence theorem and the definition of the Stieltjes integral, it follows that

$$(12)\quad \int_\sigma^t X(t, s)\{\int_{-R}^0 [d_\xi \eta(s, \sigma + \xi - s)]\phi(\xi)\}\, ds$$
$$= \int_\sigma^t X(t, s)\{\lim_{m(P)\to 0} A_P(s)\}\, ds = \lim_{m(P)\to 0} \int_\sigma^t X(t, s)A_P(s)\, ds$$
$$= \lim_{m(P)\to 0} \sum_{j=1}^N [\int_\sigma^t X(t,s)\eta(s,\sigma+\xi_j-s)ds - \int_\sigma^t X(t,s)\eta(s,\sigma+\xi_{j-1}-s)ds]\phi(\tau_j)$$
$$= \int_{-R}^0 d_\xi[\int_\sigma^t X(t, s)\eta(s, \sigma + \xi - s)\, ds]\, \phi(\xi).$$

From Estimate (11), we claim that the integral in the last bracket is of bounded variation in $\xi \in [-R, 0]$. Summarizing (9, 10, 12) and replacing ξ by θ, we obtain the third term in the right hand side of Relation (6).

REFERENCES

[1] H. T. Banks, Representation for solutions of linear functional differential equations, J. Differential Equations, 5(1969), 399-409.

[2] A. Halanay, Differential Equations Stability, Oscillations, Time Lags, Academic Press, 1966.

[3] J. K. Hale, Theory of Functional Differential Equations, Springer-Verlag, 1977.

[4] T. Naito, Fundamental matrices of linear autonomous retarded equations with infinite delay, Tohoku Math. J., 32(1980), 539-556.

[5] D. V. Widder, The Laplace Transform, Princeton University Press, Princeton, 1941.

Authors;

Thomas Hagemann
Department of Mathematics
Keio University
Hiyoshi-cho, Kohoku-ku
Yokohama 223, Japan

Toshiki Naito
The University of Electro-Communications
1-5-1, Chofugaoka, Chofu
Tokyo 182, Japan

DETERMINISTIC AND STOCHASTIC MODELS FOR THE DYNAMICS OF ANIMAL POPULATIONS

U. HALBACH
Ecology Group, Department of Biology
Johann Wolfgang Goethe - Universität
D-6000 Frankfurt am Main
West Germany

Population ecology is a discipline of Biology, which deals with the distribution of the individuals in space and the population density pattern in time. Goal is the elucidation of the causal relationships.

Some of the main problems studied are:
- Long-term regulation of a population
- Reasons for accidental population outbreaks such as pest calamities
- Conditions for the final extinction of a population
- The relation between the individual life histories, the age structure and the population dynamics
- Optimal harvesting strategy for predators and hunters

In order to get basic insights into the complicated matter of population dynamics complex organisms such as vertebrates are not as suitable as simple model organisms. After understanding their population dynamics we can try to extend our studies to the complex organisms step by step. For these reasons I have studied the population dynamics of the rotifer Brachionus calyciflorus PALLAS (Fig.1) for many years; ecotoxicological experiments have been made with the related species Brachionus rubens EHRENBERG (Fig.1).

Many features of these organisms make them ideal tools for our puposes:
1. With a body length of about 300 µm they are among the smallest metazoans.
2. They are cosmopolitan.
3. They have a relatively simple life history and behaviour pattern.
4. They are easy to rear under controlled laboratory conditions.
5. They have a short generation time of only a few days.
6. Due to their predominant parthenogenetic reproduction they can be cloned easily in order to get genetically pure lines.
7. They have a relatively schematic life table, e.g. nearly rectangular survivorship curves. The reason is the cell constancy of these organisms, which is consequence of evolutionary selection for small body size of these plankton organisms and of a strictly deterministic embryological development. As there are no more cell divisions after hatching, the tissues do not regenerate like in higher organisms.

These animals pass their life until the physiological exhaustion - and then die.

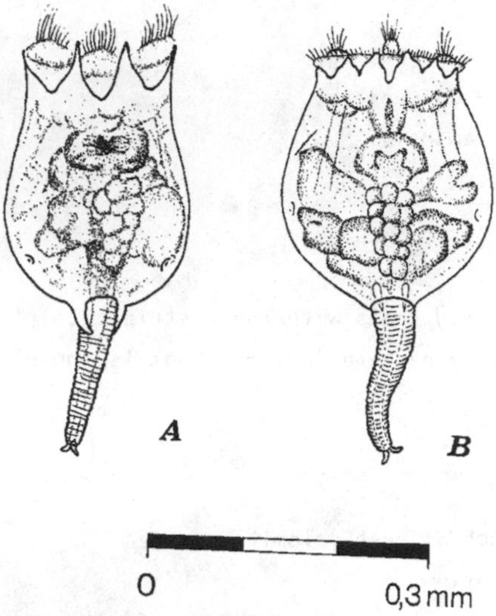

Fig.1. Two representatives of the rotifer genus Brachionus. These aquatic organisms, which are very common in eutrophic and mesotrophic ponds and lakes, belong to the smallest metazoans; but their body size of about 300 µm length is not larger than that of many unicellular ciliates (after Halbach 1979a).

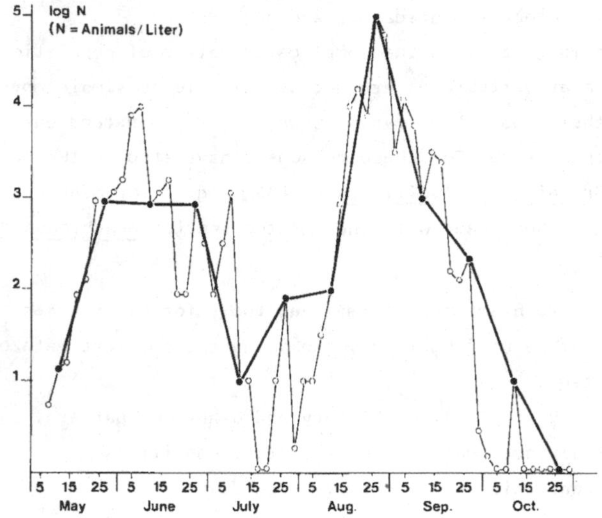

Fig.2. Population dynamics of the rotifer Brachionus calyciflorus in a pond of the Botanical Garden in Würzburg, registered with a sampling distance of 14 days (solid symbols and thick line) and of 3 days (open symbols and thin line). With the coarse resolution we lose information: the real lapse of the population dynamics is unrecognizable. The actual sizes of maxima and minima are incorrectly estimated; they differ from the real value up to the factor of ten (after Halbach 1975).

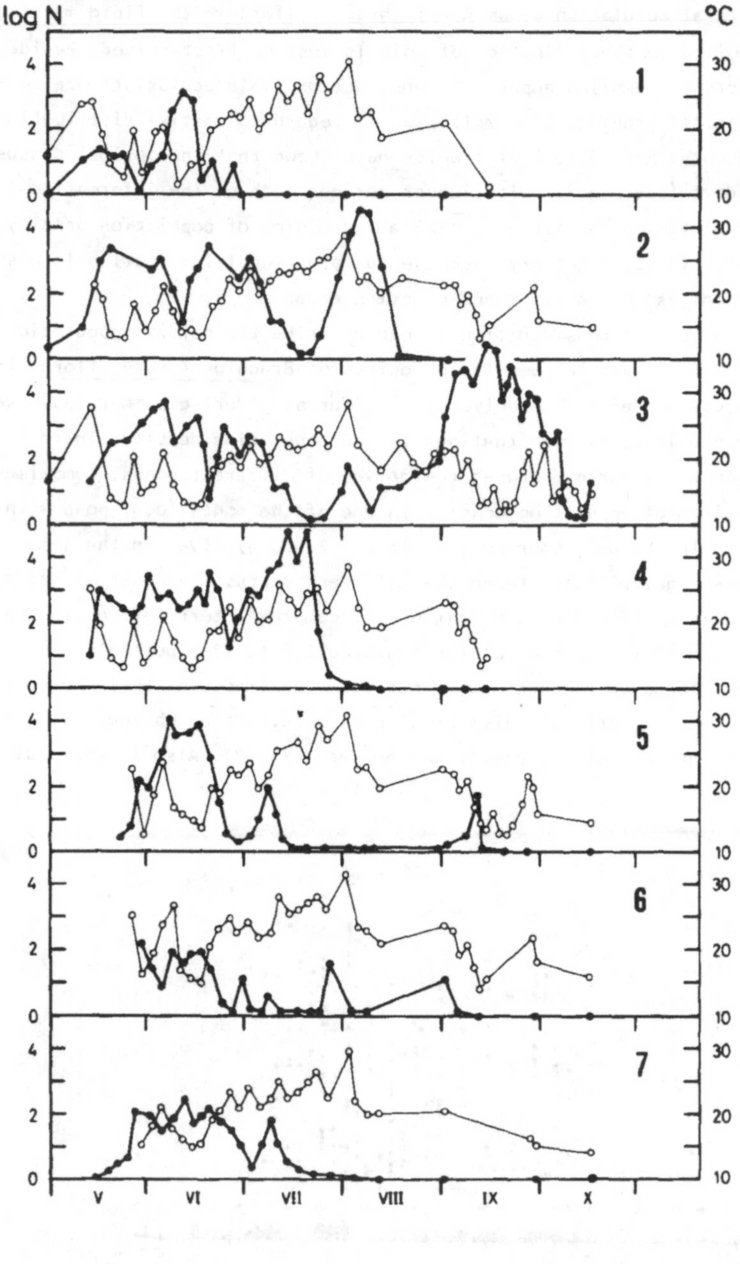

Fig.3. Changes in temperature (open symbols, thin line) and population dy= namics of Brachionus calyciflorus (solid symbols, thick line) in 7 ponds of the Botanical Garden in Würzburg during summer 1967. Note the very irregu= lar fluctuations without any obvious correlation (after Halbach 1970).

To describe the natural population dynamics of these rotifers in the field methodological problems involved in the collection of animals must be first solved. Besides the technical aspects of sampling apparatus construction field ecologists are confronted with statistical problems of sample size in regard to spatial distribution and of census frequency. Methodological studies have shown that the optimal census distance is three days (see Fig.2): With longer periods much of the information is lost, e.g. the actual timing and sizes of maxima and minima of population density. Higher census density is not necessary, because the additional information is masked by the background 'noises' of sampling and counting error.

Taken these experiences into consideration we can describe the natural population dynamics adequately. Fig.3 gives the population curves of Brachionus calyciflorus in 7 ponds in the Botanical Garden of the University of Würzburg during summer 1967. We observe very large and irregular fluctuations in the population density. There is no apparent correlation or synchrony between the curves of different ponds. Sometimes there is even a maximum of population density in one of the ponds (e.g. pond 4 in July) during a minimum in the neighbouring ones (ponds 2 and 3). Even in the same pond there are no obvious regularities between the different years. The observed natural population dynamics in the field result from highly complex interrelationships between many abiotic and biotic ecological factors. Looking to find possible causal relationships we first focus on temperature, since it is an important environmental parameter, which influences nearly all biological processes. But no obvious correlation between temperature and population density can be seen (Fig.3). Also in the scatter

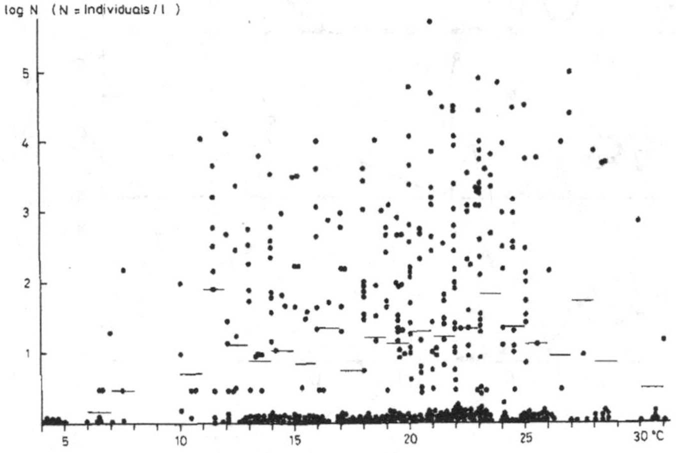

Fig.4. Measured population density of Brachionus calyciflorus (log N with N = animals/liter) in relation to the instantaneous temperature in 19 ponds during 1967. The small cross-lines represent the averages of the tempera= ture classes (after Halbach 1970).

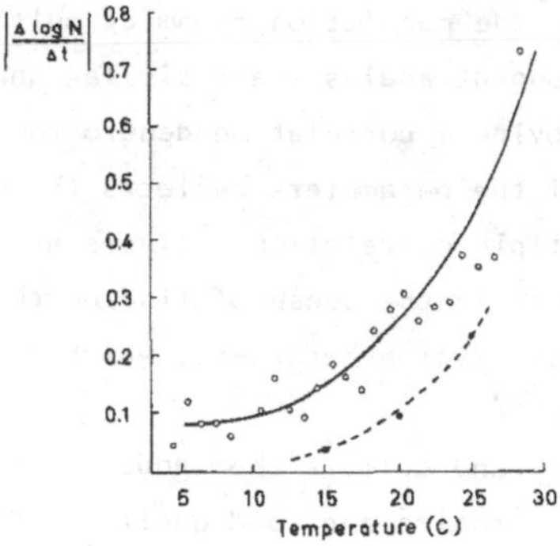

Fig.5. The relationship between absolute slope of the population fluctuations of Brachionus calyciflorus and temperature. The slopes increase nearly exponentially. Open circles and continuous line: field data. Solid circles and broken line: laboratory experimental data (after Halbach 1973).

diagram (Fig.4) there is no concrete relationship recognizable and distribution seems to be random. However, using statistical analysis we find indeed a relationship between temperature and population dynamics. Fig.5 demonstrates that the absolute value of the slope of the population curve depends on the temperature. Fluctuations increase with temperature and are also greater in field data; the causes for this discrepancy are not known.

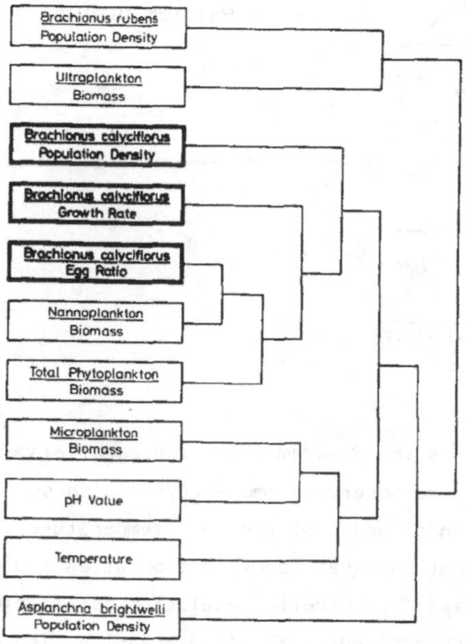

Fig.6. Dendrogram: Relationship between population density of Brachionus calyciflorus and diverse ecological factors: result of a statistical correlation analysis (after Halbach 1978).

Since many further environmental factors may influence the population dynamics multivariate analyses, correlation matrices, principal component analyses and similar analyses are used in field ecology. These methods can provide a correlation dendrogram (Fig.6). The degree of linkage or the neighbourhood of the parameters reflects the degree of correlation. These and similar methods of multiple correlation analysis are widely used, but the results are difficult to interprete in the sense of finding the biologically causal basis. However, they can be used to develop hypotheses which can be tested in laboratory experiments.

For this reason we take the animals into the laboratory and culture them under controlled conditions: constant volume, temperature, light regime and food quality and quantity.

Fig.7 demonstrates the population dynamics at three temperatures: 15°C, 20°C and 25°.

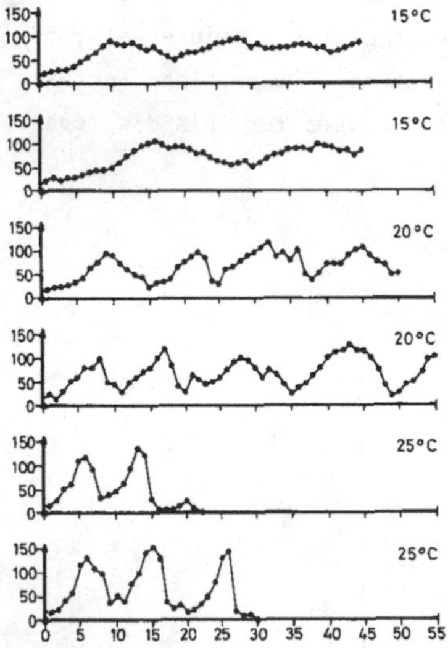

Fig.7. Experimental population dynamics of Brachionus calyciflorus at three diffe= rent temperatures (one couple of replicates each). With increasing temperature frequen= cies and amplitudes of the oscillations rise (after Halbach 1970).

For each of these temperatures the diagram shows two parallels, which indicate the differences between replicates. We observe sigmoid curves with subsequent oscillations around an equilibrium. Frequency and amplitude are temperature dependent; both increase with temperature. At 25°C that may lead to extinction after two or three peaks. The observed periodic fluctuations are cybernetic regulation oscillations around the carrying capacity (C), which is determined by the food quantity. The oscillations are caused by a time lag between food uptake and the resulting offspring production, because the food must be filtered, ingested, assimilated, transformed to yolk and deposited into eggs; eggs must be formed, extruded and attached to the body wall; finally another 1/2 to 1 day is necessary for the ripening of the eggs and the hatching of the offspring depending on temperature. During this time lag the nutrition situation for the individuals is changed by the growing population. Thus, the food dose per individual may be lowered drastically, resulting in an overshoot and subsequent regulation oscillations around the food-determined carrying capacity.

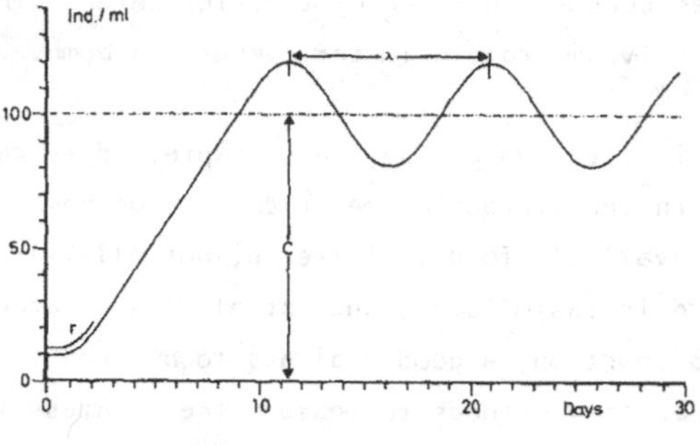

Fig.8. Schematical representation of the population dynamics under limited conditions. Note the sigmoid growth followed by oscillations around an equilibrium (C). This curve is characterized by three population parameters: r (intrinsic rate of natural increase), C (carrying capacity), λ (wavelength of oscillations). The curve is described by the logistic growth function with simple time lag (after Halbach 1975).

This population curve can be described by the logistic growth function with simple time lag,

$$\frac{dN_{(t)}}{dt} = r \cdot N_{(t)} \quad \frac{C - N_{(t-T)}}{C} \qquad (1),$$

where N = population density, t = time, r = intrinsic rate of natural increase, C = carrying capacity, and T = time lag.

This curve (Fig.8) is determined by the three population parameters r, C and T; T can be substituted by the frequency f (= 1/λ), as there is a close connection between both (3).

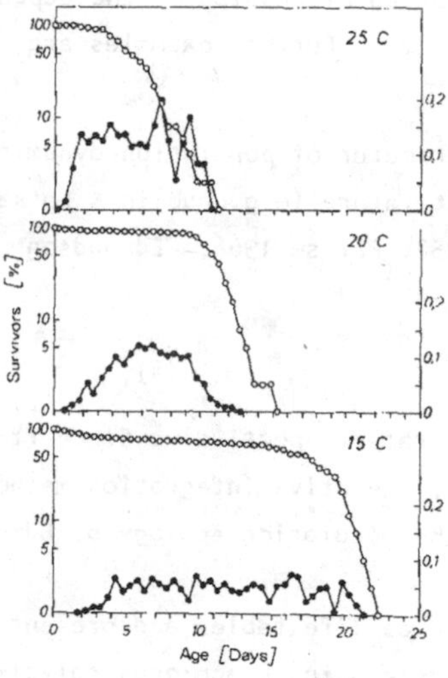

Fig.9. Life data of Brachionus calyciflorus at 3 different temperatures. Survivorship curve (open symbols), fertility curve (solid symbols). The lifespan increases with decreasing temperature. Duration of reproductive period and fertility rate differ at different temperatures. The data have been measured with individually cultured animals (100 specimens per experimental run) (after Halbach 1970).

Biologically most important is that these parameters can be calculated by measurements made on single isolated individuals.

C:

$$C = \frac{p \cdot F \cdot L}{B} \qquad (2)$$

B = biomass of an adult animal expressed as combustion value or caloric content (in Joule) and measured with a highly sensitive microcalorimeter combustion bomb.

L = average life span (in days).

F = food dose; the inflow of algae per milliliter and per day, also expressed as caloric content (in Joule) and measured in the microcalorimeter combustion bomb.

p = efficiency of foraging, since not all available food is filtered, not all filtered food is digested, not all digested food is assimilated, and not all assimilated food is used for body growth and egg production; a good deal has to be used for respiration. We use very sensitive laboratory methods to measure these rates of isolated individuals, e.g. radioactive traced algae and liquid scintillation counter for filtration rate (Haney et al. 1982) and Kartesian diver method for respiration rates (Leimeroth 1980). From these we calculate p, which is in an order of approximately 0.3 (30%).

This formula has been developed for food limited populations. It is based on the simple assumption that in the equilibrium for every dying individual one newborn enters the population.

f:

$$f = \frac{1}{T \cdot \pi \cdot \sqrt{2}} \qquad (3)$$

T = time delay. In a small range around the stable oscillations the frequency depends only on the time lag, which lies between food uptake and the resulting offspring production (Cook 1965). May (1974) gives a more detailed discussion of the dependency of T and confirms his results by numerical studies. Further examples are given in May (1976).

r: The intrinsic rate of natural increase is a central parameter of population dynamics. On this parameter there exists an immense amount of literature (e.g. Dublin & Lotka 1925, Birch 1948, Leslie & Park 1949, Evens & Smith 1952, Parise 1966, Edmondson 1968). It is not possible to express r explicitely:

$$\int_0^\infty l_x m_x e^{-rx} dx = 1 \qquad (4),$$

where l_x and m_x = age specific fecundity and mortality rate respectively. Since it is complicated to determine r from the life table data, iterative integration methods must be applied or graphical methods introduced into the population ecology by Edmondson (1968) can be used.

The age specific natality and mortality rates are measured as life tables and presented as survivorship and fertility curves. Fig.9 gives an example using Brachionus calyciflorus at three different temperatures indicating the relatively rectilinear physiological survivorship curves. The life expectations increase with decreasing temperature. Duration of the reproductive period and the fertility rate differ at different temperatures. Reproduction is prolonged but retarded at low temperatures. The data are measured with individually cultured animals (100 per experimental run).

More complicated is the influence of food dose on the life tables (Fig.10). The highest

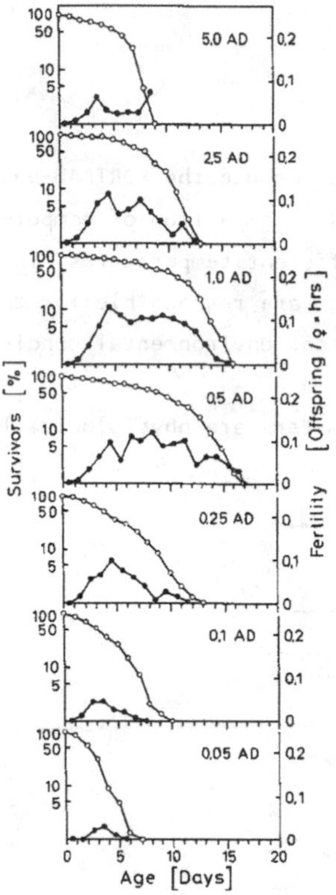

Fig.10. Survivorship curves (open circles) and fertility curves (solid circles) at different food quantities. An algal dose of AD 1.0 means $1.0 \cdot 10^6$ Chlorella-cells/ml/12h. Survivorship and fertility have a maximum between AD 0.5 and 1.0 (after Halbach and Halbach-Keup 1974).

vitality is at moderate algal doses (of 0.5 - $1.0 \cdot 10^6$ Chlorella-cells/ml/12 hours). The influence of algae is twofold. At low algal doses the animals die soon by starvation and at high doses by toxic metabolites. This is the cause of the "animal exclusion principle" in limnology, which means the exclusion of zooplankton organisms during algal blooms in lakes.

From the life table data the intrinsic rate of natural increase r can be calculated. Hence we are able to present the complete causal chain in population dynamics (Fig.11): Ecological factors such as temperature, nutrition or chemical substances influence physiological characters (like filtration rate, ingestion rate, assimilation rate, and respiration rate). These influence life table data (such as age specific natality and mortality), which determine the population parameters r, C and T (2-4). From these the population dynamics can be simulated using deterministic models, e.g. the logistic

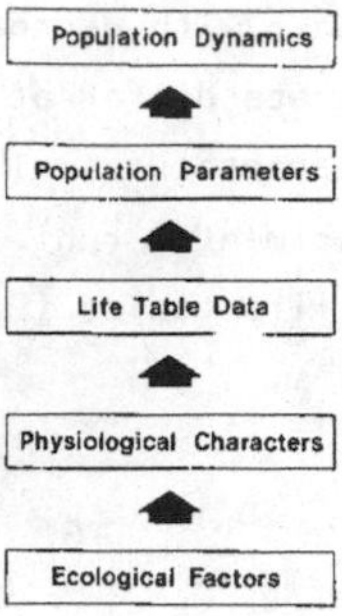

Fig.11. The causal chain of population dynamics.

growth function with time lag. For the numerical integration we use the FORTRAN-based algorithm CSMP including RUNGE-KUTTA-methods. Fig.12 gives a comparison of computer simulations with empirical population dynamics at three different temperatures. The similarity between both types is evidence that the time lags are responsible for the periodic oscillations of the population density under constant environmental conditions.

Since r is dependent on birth and death rate and both parameters are physiologically

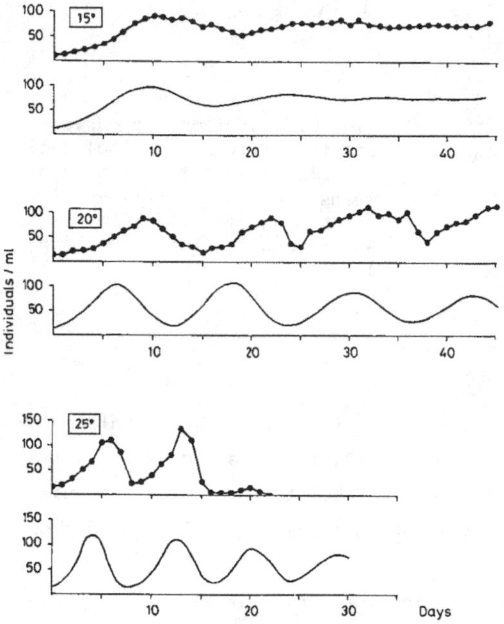

Fig.12. Comparison of empirical population dynamics of Brachionus calyciflorus (upper curves) with computer simulations (lower curves) at three different temperatures. The simulations are carried out using the logistic growth function with simple time lag (after Halbach 1974).

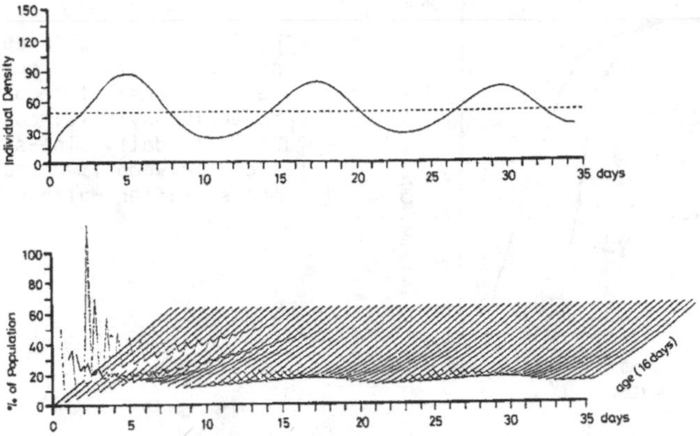

Fig.13. Computer simulation of the population dynamics of Brachionus calyciflorus with separate time lags for birth and death rate. The lower diagram represents a computer plot of the age structure of the population (after Halbach 1979b and Girke and Halbach 1982).

distinct, we also developed models with two time lags. Fig.13 gives a computer plot of a simulated 20°C-population with separate time lags for natality and mortality including the age structure.

Another complication involves the use of variable time lags. In contrast to the constant temperature of our laboratory cultures (with a deviation of ± 1°C) nearly nowhere in nature do constant temperatures occur. More frequent are diurnal oscillations of the temperature (Fig.14). Therefore we also used sine-shaped temperature oscillations in

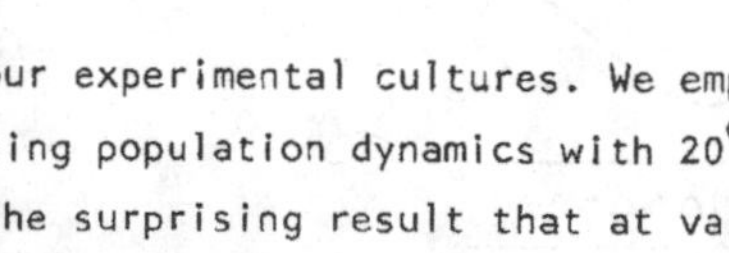

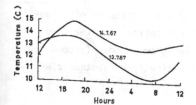

Fig.14. The course of temperature in a pond in the Botanical Garden of the University of Würzburg on two consecutive days with different weather conditions (after Halbach 1973).

our experimental cultures. We employed a range of 15°C to 25°C and compared the resulting population dynamics with 20°C-constant-temperature reactions (Fig.15). We found the surprising result that at varying temperature the animals have a higher vitality than at constant temperature; particularly, the life expectation is lengthened. Using these data for simulation leads to fairly good results (Fig.16). It can be seen that the population oscillations do not simply reflect the temperature waves; both curves have clearly distinct periodicities. A comparison of empirical population curves with constant and with changing temperature is given in Fig.17.

These simple deterministic models allow qualitative conclusions. It seems the logistic

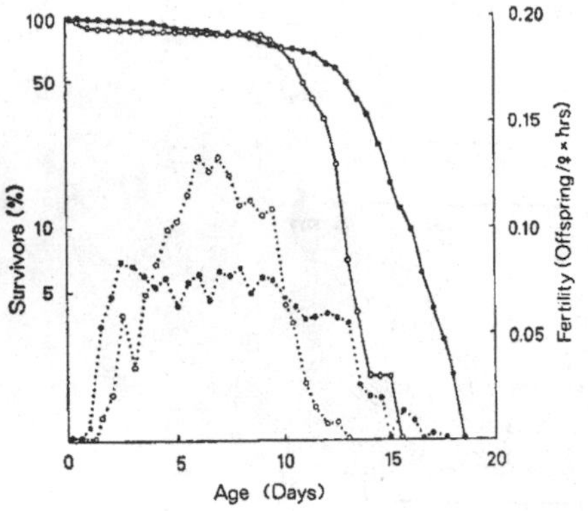

Fig.15. Survivorship curves (solid lines) and fertility curves (dotted lines) at constant temperature of 20°C (white symbols) abd at tempera= tures with daily sine-shaped oscil= lations between 15°C and 25°C (black symbols) (after Halbach 1973).

growth function can be used to determine the optimum harvesting strategy, e.g. in fishing and hunting (Halbach 1979c), to predict the final carrying capacity of growing populations of animals (Halbach and Friz 1978), or even of human (Halbach 1974): Pearl (1936) made prognoses using the logistic growth function and known census values of the past. For example, he predicted an equilibrium number of human beings of $2.64 \cdot 10^9$ would be reached in the year 2100 (Fig.18). His prognoses has been discounted, because the population explosion continued. The obvious fault was, that the logistic function does not fit to the human population. Instead of continually increasing resistance from zero there should be threshold values in vertebrates, which is indicated by experimental studies of mice and tupajas (von Holst 1973, Halbach 1979d). In the future we need to incorporate these threshold values into the deterministic models, but first they must be determined experimentally.

Another biological complication is the change of life habits, for which yeast (Saccharomyces cerevisiae) can be a simple example. They display the phenomenon of 'diauxy' (Fig.19). Fed with sugar or with acetate they grow by fermentation with the production of alcohol. After this food supply is exhausted population growth stops (second lag phase in Fig.19). After a certain time they adapt physiologically by synthesizing the

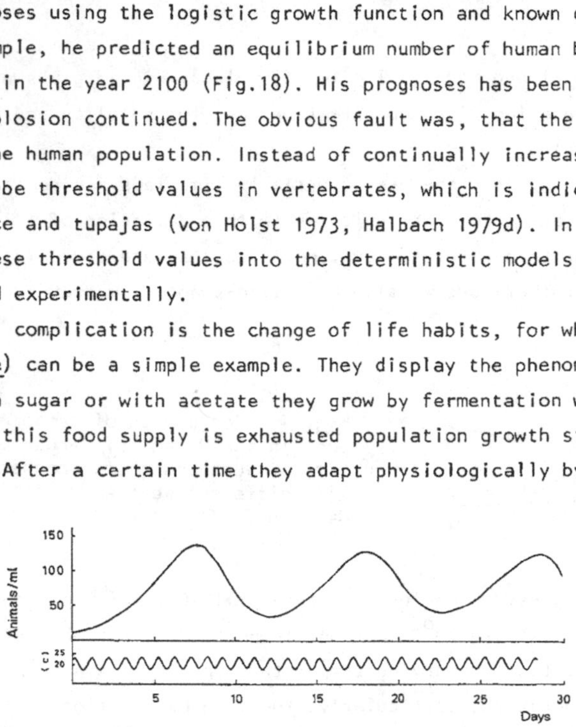

Fig.16. Computer simulation of the experimental population dynamics with changing temperatures (15-25°C - s.lower diagram) using the logistic growth function with simple time lag $dN_{(t)}/dt = rN_{(t)}(C-N_{(t-T)})/C$ and the empirically determined parameters: r=0.63, C=89.o, and T=3.0 (upper diagram) (after Halbach 1973).

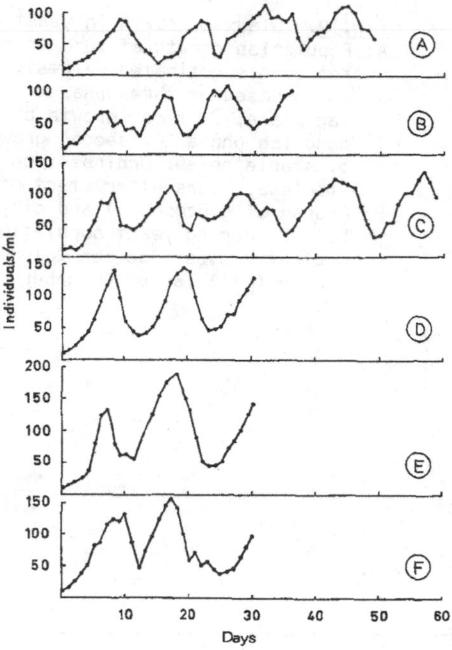

Fig.17. Experimental population dynamics at constant temperature of 20°C (A,B,C) and at a temperature changing between 15°C and 25°C (D,E,F) (after Halbach 1973).

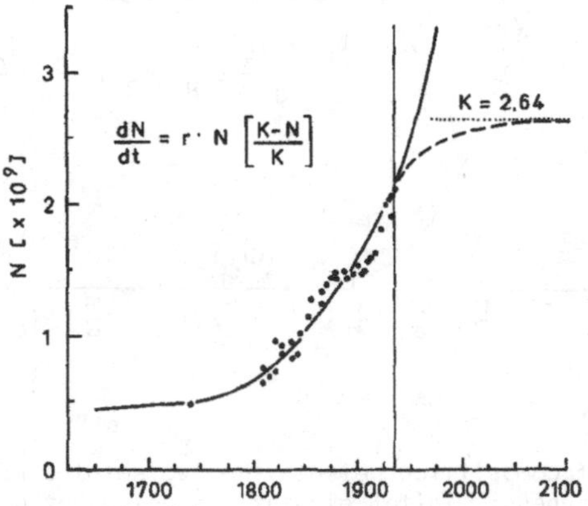

Fig.18. Predictions of the human population dynamics (situation of 1936) using the logistic growth function (dashed line). The final number of $2.64 \cdot 10^9$ should be reached at about 2100 A.D. This prognosis has been completely wrong, as we transgressed just the 4.10^9-threshold (solid line) (after Pearl and Gould 1936 and Halbach 1974).

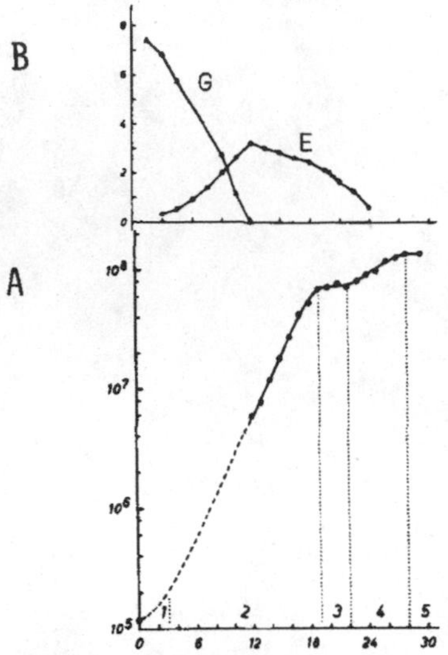

Fig.19. Diauxic growth in yeast cultures.
A: Population growth at 20°C (dashed line
   represents estimated values). The curve
   is composed of three phases: 1. first
   lag phase, 2. first growth phase, 3. se=
   cond lag phase, 4. second growth phase,
   5. stable phase. Ordinate: cells/ml.
   Abscissa: hours after start of culture.
B: Course of glucose (G) and ethyl alco=
   hol (E) during yeast population growth.
   Ordinate: substance per volume culture
   medium (g/l) (after Ehrenberg et al.
   1973).

$$\frac{dN_C}{dt} = r_C \cdot N_C \cdot \left[\frac{K_C - N_C - \alpha \cdot N_R}{K_C}\right]$$

$$\frac{dN_R}{dt} = r_R \cdot N_R \cdot \left[\frac{K_R - N_R - \beta \cdot N_C}{K_R}\right]$$

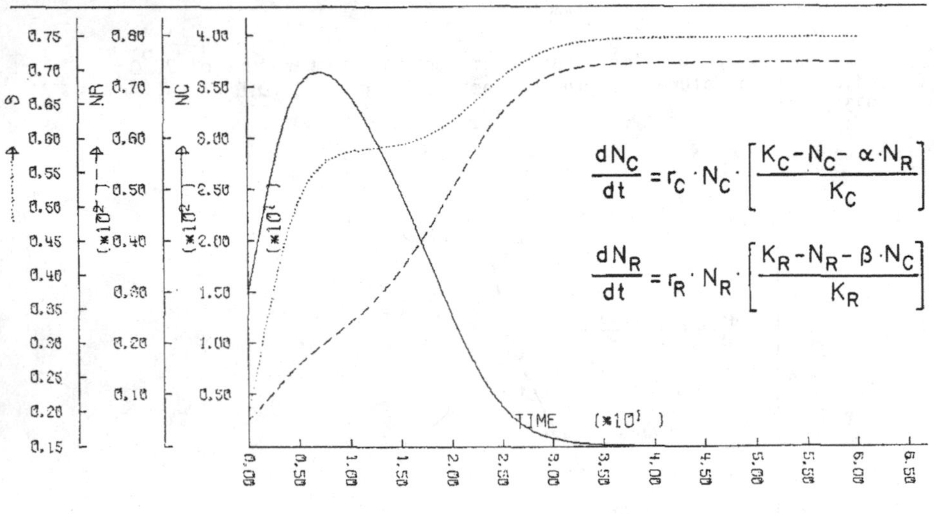

Fig.20. Computer simulation of competition between the two seleted rotifers
Brachionus calyciflorus and B.rubens: Digital simulations using Lotka-Volterra-
equations and the algorithm DSL/90 (computer-plot).
Ordinate: solid line: log (individual density) of B.calyciflorus (NC);
           dashed line: log (individual density) of B.rubens (NB);
           dotted line: log (individual density) of sum of both species (S).
Abscissa: Time course of mixed population experimenet (days).
Note the different scales of the ordinate; note also that the symbol C refers
to B.calyciflorus in this case, the carrying capacity is represented by the
symbol K (after Halbach 1974).

requisite enzyme make-up. Now they are able to respire the alcohol which they produced before. After this time delay they start a second growth phase ('4' in Fig.19) which stops when the alcohol supply is exhausted (stable phase '5' in Fig.19). This type of response can also be incorporated into the function.

There is another advantage of this type of deterministic model which is that it can be used for multi-species systems (with one differential equation for each species), because the numerical integration of the whole set of equations can be made synchronously by the computer.

As an example, the LOTKA-VOLTERRA-system for interspecific competition is given. Fig.20 shows the equations (with α and β as competition coefficients) and a computer plot of the competition between the two related rotifers Brachionus calyciflorus and Brachionus rubens (see Fig.1). B.calyciflorus dies out after 35 days. In mixed experimental cultures B.rubens is indeed always the winner - independent of the original proportions (Fig.21). Through additional experimentation we found that B.rubens has a better foraging efficiency, but also produces toxic metabolites into the medium, which lower the vitality of its competitors (an interspecific activity, which is calles 'interference').

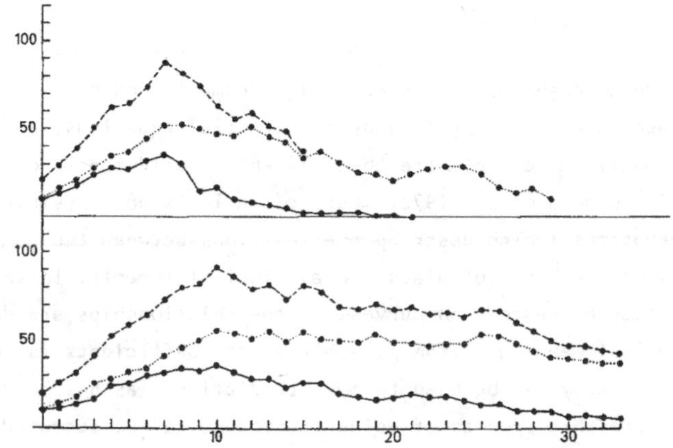

Fig.21. Results of interspecific competition. Course of individual density in mixed populations of Brachionus calyciflorus (solid line) and B.rubens (dotted line) with identical start density of 10 animals/ml. Dashed line: sum of both. Upper diagram: unrenewed culture medium; lower diagram: daily renewed culture medium. Ordinate: Individual density (animals/ml). Abscissa: experimental time (days) (after Halbach 1979a).

Similar equations can be used for predator-prey relationships (formulas 5 and 6). They lead to the known oscillations with phase delay of the predator (Fig.22).

In this way whole ecosystems can be simulated, where the compartments represent species or groups of species with identical or similar ecological functions. They are

$$\frac{dN_B}{dt} = r_B \cdot N_B \ (1 - N_B/C_B) \ - \ N_R \cdot k \ (1 - \exp \ (-c \cdot N_B^2 \cdot N_R^{1-b})) \qquad (5)$$

$$\frac{dN_R}{dt} = N_R \ (\beta \cdot N_B N_R \ - \ b \cdot N_R) \qquad (6)$$

(see May 1976).

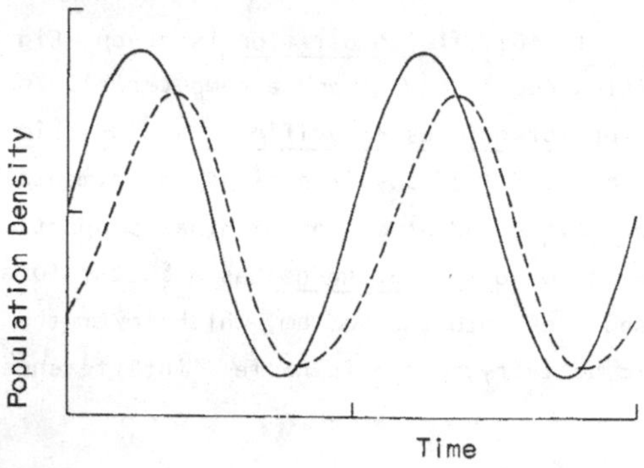

Fig.22. Predator-prey cycles (fic= tive example). The oscillations of the predator (dashed line) follow those of the prey (continuous line) with a phase delay (simulated by an analogue computer; original).

connected by biotic relationships such as predation, competition or symbiosis. In all cases these relationships can be expressed by differential equations, which can can be integrated simultaneously. An example are the well known world models of the Club of Rome (Forrester 1971, Meadows et al. 1972). Very often it is not possible to construct the differential equations (which describe the relations between the compartments of the model) logically on the basis of plausible a priori statements. In these cases the measured data are fitted by regression curves. If the relationships are described by polynomes, the biological meaning of the parameters and coefficients is, of course, completely unknown. But they can be used to make predictions (as has be done by the world models). Therefore this program of using the models can be accepted for application in applied systems. However, there is a more severe disadvantage of this type of deterministic model; it handles the population like a homogeneous distribution, or at least random distribution, which is unrealistic. In nature we normally find a more or less heterogeneous or clustered distribution pattern (Halbach 1975). Furthermore many animals migrate; Brachionus calyciflorus shows, like many other plankton organisms, the phenomenon of diurnal vertical migration (Halbach 1975). During the night most of the population is located near the surface while during the day they move to the deeper water, probably in order to avoid high doses of radiation . To account for these heterogenities in the pattern of density of the organisms in space and time we have to introduce partial differential equations or diffusion equations. Some preliminary work in this direction has be done (Halbach 1975).

There is another disadvantage of the deterministic model. The variance of the biological parameters is completely suppressed by using averages. However, this variability is an inherent characteristic of life which is necessary for processes such as evolution. In order to incorporate this variation we should use stochastic models. Fig.23 gives the flow chart for the stochastic simulation of the population dynamics of Brachionus. Using the ALGOL-based algorithm SIMULA the computer synthesizes the life history of each newborn animal by MONTE-CARLO-methods. The synthesized population consists of artificially composed animals, which we call "Frankenstein rotifers". Fig.24 demonstrates three stochastic simulations of Brachionus calyciflorus at 20°C. These curves seem more

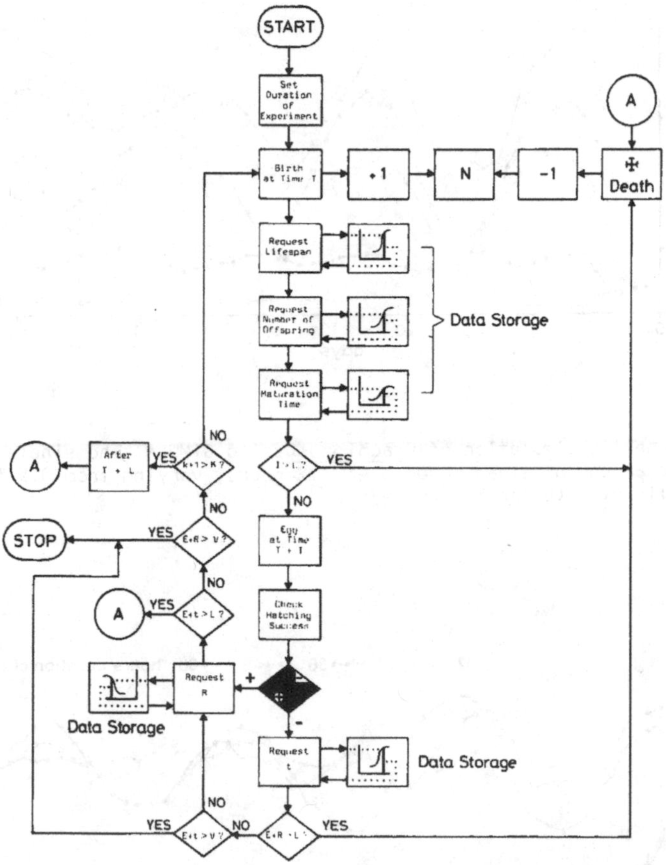

Fig.23. Flow chart of the stochastic simulation of the population dynamics of Brachionus calyciflorus using Monte-Carlo methods. N = individual density, L = lifespan, K = number of offspring, I = maturation time, t = time span between two egg extrusions, T = moment of birth, E = moment of egg extrusion, R = development time of eggs, V = experimental time (after Halbach 1978).

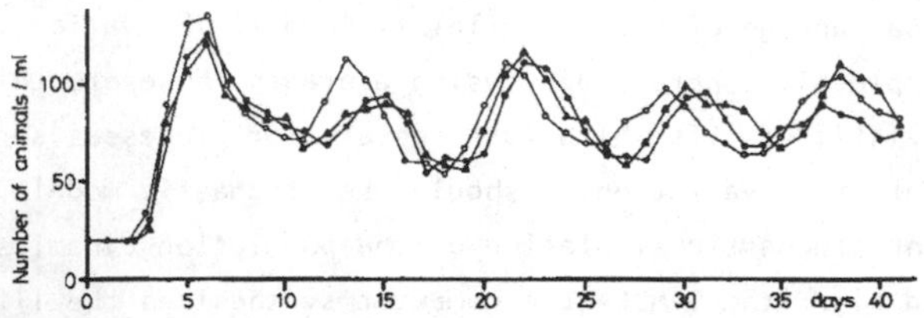

Fig.24. Stochastic simulation of population dynamics of Brachionus calyciflorus at 20°C using SIMULA as algorithm (after Halbach 1978).

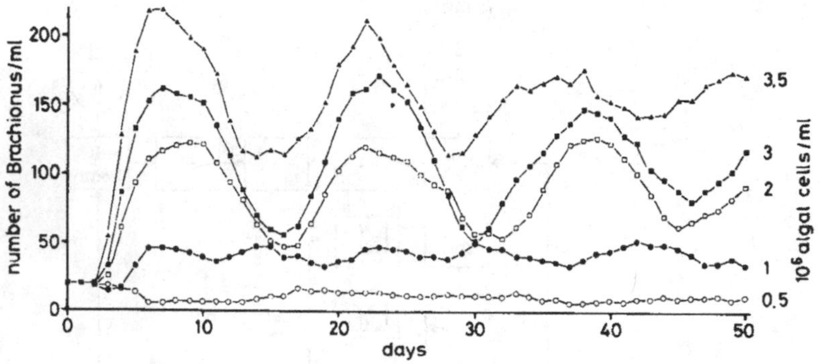

Fig.25. Stochastic simulation of Brachionus using SIMULA. The single simulations differ in the amount of algae added every 12 hours, varying from 0.5 to $3.5 \cdot 10^6$ Chlorella-cells ml^{-1} (after Kaiser 1975).

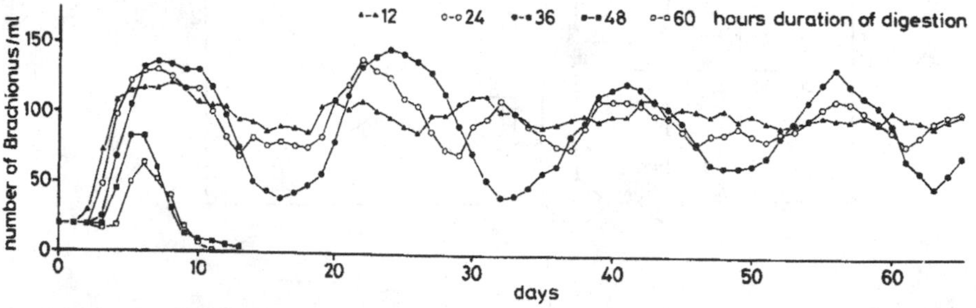

Fig.26. Stochastic simulation of the population dynamics of Brachionus. The single simulations differ in the duration of digestion (time lag!), varying from 12 to 60 hours. The algal dose was $1.7 \cdot 10^5$ Chlorella-cells ml^{-1} h^{-1} (after Kaiser 1975).

realistic because they are more similar to the experimental populations (see Fig.7). The simulations can be substituted for experiments. In the models we are able to change parameters as in experiments, but they can be treated completely independently and varied through a broader range. We can then look for the interesting correlations which should be tested experimentally. This type of deductive research can save time and money.

For example, higher food doses in the simulations (Fig.25) lead to higher carrying capacities as well as increased oscillation amplitudes. Both results are plausible. But the frequency did not change greatly. Since this finding was unexpected, we re-examined these results and have been able to confirm them (Halbach 1982).

In a similar way we have been able to vary parameters which cannot be changed in the experiments without altering the animals' physiologies. As an example, the change of the time delay in simulations is presented (Fig.26). It can be seen that unrealistic high delays of more than 36 hours leads to extinction. By extrapolating the known tem-

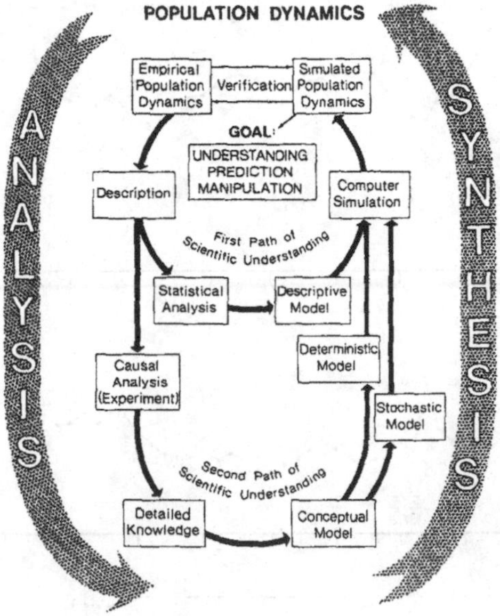

Fig.27. Schematic representation of the two pathways of scientific under= standing of the natural population dynamics and predictions, especially of manipulated populations. We can reach these goals by five steps: desrip= tion, analysis, synthesis, verification, prediction. If the analysis is superficial we can look for patterns by statistical analysis and use them to construct descriptive models (first path of scientific understanding). A more detailed analysis by experiments leads to delailed knowledge, which can be used for constructing conceptional models. The simulations can be done using deterministic or stochastic techniques (second path of scienti= fic understanding). For more details see text (after Halbach 1980).

perature-dependent time-lags between food uptake and offspring production we get a lower temperature border for survival of *Brachionus calyciflorus* which lies at 5°C. This is exactly identical to the experimentally determined tolerance (Halbach, in prep.). Manipulating simulations allows the test of evolutionary strategies, which are technically existing, but are not realized in nature (see Rechenberg 1973). Stochastic models are biologically very satisfying, because they are realistic, precise and give a vivid impression of the dynamic processes. On the other hand they are laborious and need a huge amount of computer capacity.

If there is inadequate time to construct such a detailed realistic model simpler and more pragmatic solutions must be sought. In another approach (see Fig.27) we do not look for causal relationships, but rather use statistical methods to describe the basic pattern of the population dynamics, which then can be used to make predictions by extrapolation of the pattern (Halbach 1978, 1979b). Fig.27 is a schematic representation of the different heuristic approaches. In contrast to deterministic and stochastik

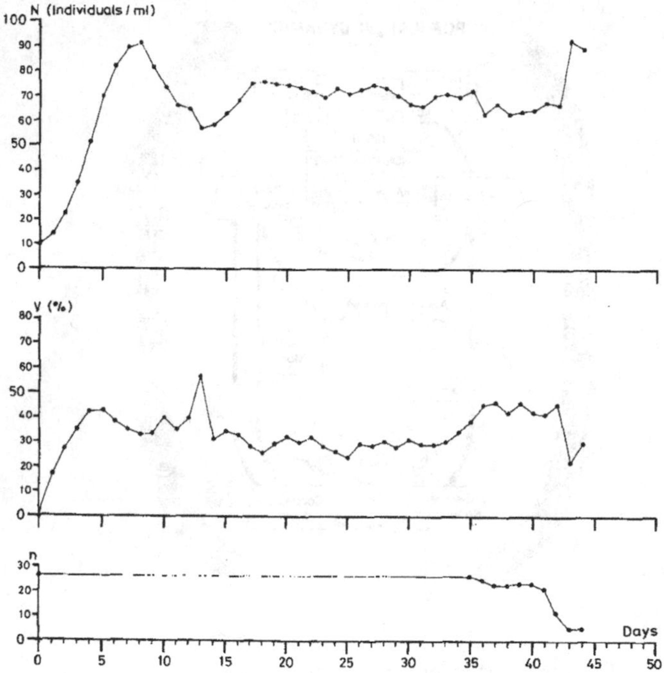

Fig.28. Averaging of population dynamics of *Brachionus calyciflorus*. The curve at the top shows the average of 26 population dynamics at 20°C. By this type of generalization we lose the typical oscillations (see Figs. 4,5,14). The curve in the middle represents the coefficient of variation. The bottom diagram demonstrates the number of parallel experiments resp. replicates (after Halbach 1980).

models descriptive models based on statistical analysis represent an abbreviated path of scientific understanding.

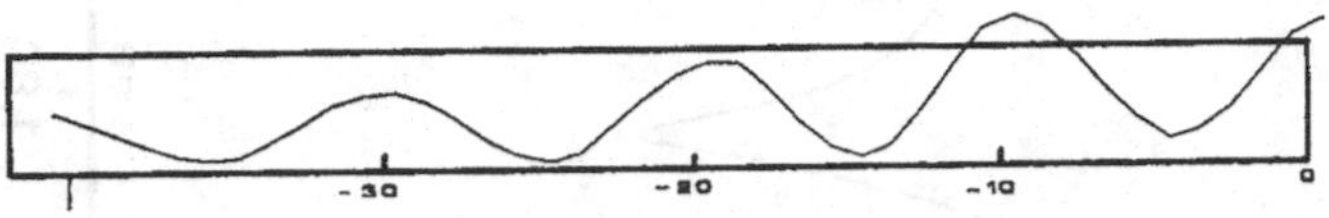

Fig.29. Crosscorrelation function between diverse populations of Brachionus calyciflorus in parallel experiments with a periodicity of 10 days (abscissa). Calculations made by Beuter (after Halbach 1978).

Averaging is in this case an inadequate method for detecting the underlying deterministic pattern in the population dynamics, since it levels out the oscillations (Fig.28). Cross-correlations can be used to eliminate the irregularities, which are caused by uncontrolled random effects (Fig.29). After suppressing such 'noise' we find indeed at 20°C a periodicity with a wave length of about 10 days. However, power spectra and FOURIER-analysis indicate that there is a more complicated underlying pattern (Fig.30), which now can be used to make predictions by extrapolations including stochastic simulation techniques (Fig.31).

I have demonstrated that the population dynamics of the rotifers can be simulated from life table characteristics of thousands of animals. Even very small reductions of vitality (e.g. by low doses of new invented chemical substances) can give rise to tiny sublethal effects such as statistical decrease in lifespan or fertility. Small effects of chronic poisoning, which can be hardly detected in isolated animals, can lead to projections into the next higher integration level, i.e. population dynamics. In this way, population dynamics can act as a 'magnifying glass' enabling the detection of small sublethal ecotoxicological effects, so that it can be used as a very sensitive bioassay. We developed this method to be used as a standard test in order to check new chemical compounds before distribution. As an example, the life tables of Brachionus rubens under the influence of the pesticide pentachlorophenol (PCP) are shown inFig.32. As the determination of survivorship and fertility (using 100 individual cultures each) is too complicated for routine work, we used the population curves in the culture medium with different concentrations of the substance (Fig.33). At high concentrations, the toxic effect of the test-substance is obvious: extinction at 0.20 ppm PCP after 5 days; lowered carrying capacity at 0.15 ppm PCP. By superficial consideration it cannot be decided whether the observed differences at low concentrations of the substance are significant. For this decision we use autocorrelation as a statistical method (Fig.34). The frequency can be measured by the autocorrelation curves much more accurately than in the original curves. I have introduced another parameter into ecotoxicology: the 'pregnancy' (p) of the oscillations, which is the level distance between the first minimum

contd. on p. 242

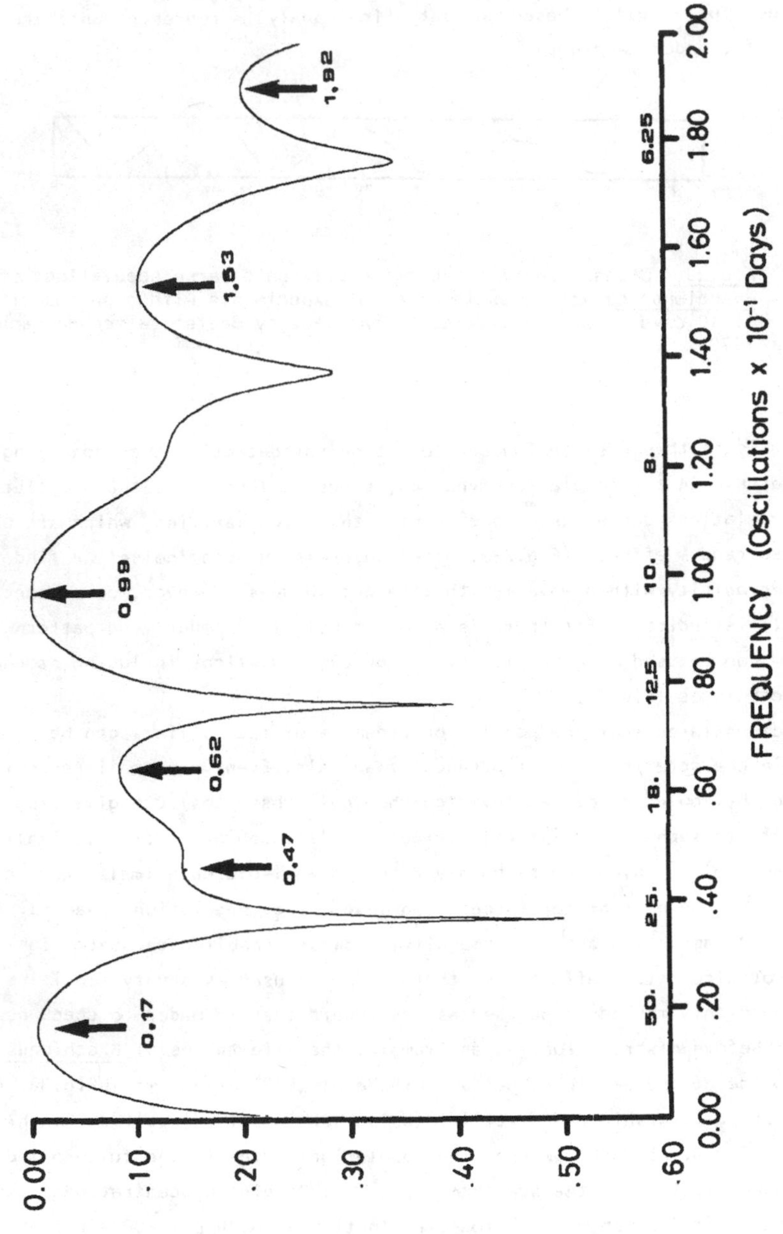

Fig. 30. Fourier-analysis of population dynamics of Brachionus calyci=florus at 20°C. Calculations made by Beuter (after Halbach 1978).

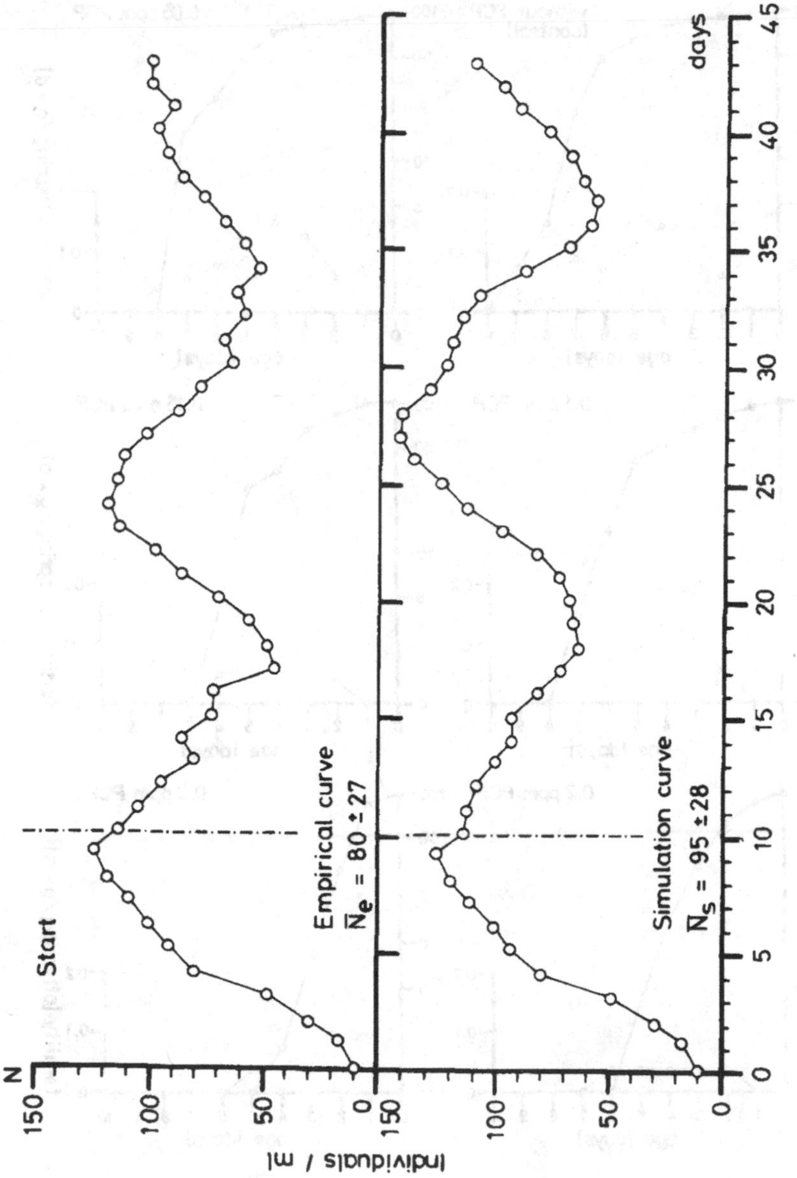

Fig.31. Predictions of population dynamics of Brachionus calyciflorus at 20°C, The curve at the top is empirical. The curve at the bottom is identical during the first 10 days (starting phase). The rest is pre=dicted using the information of the first 10 days (after Halbach 1978).

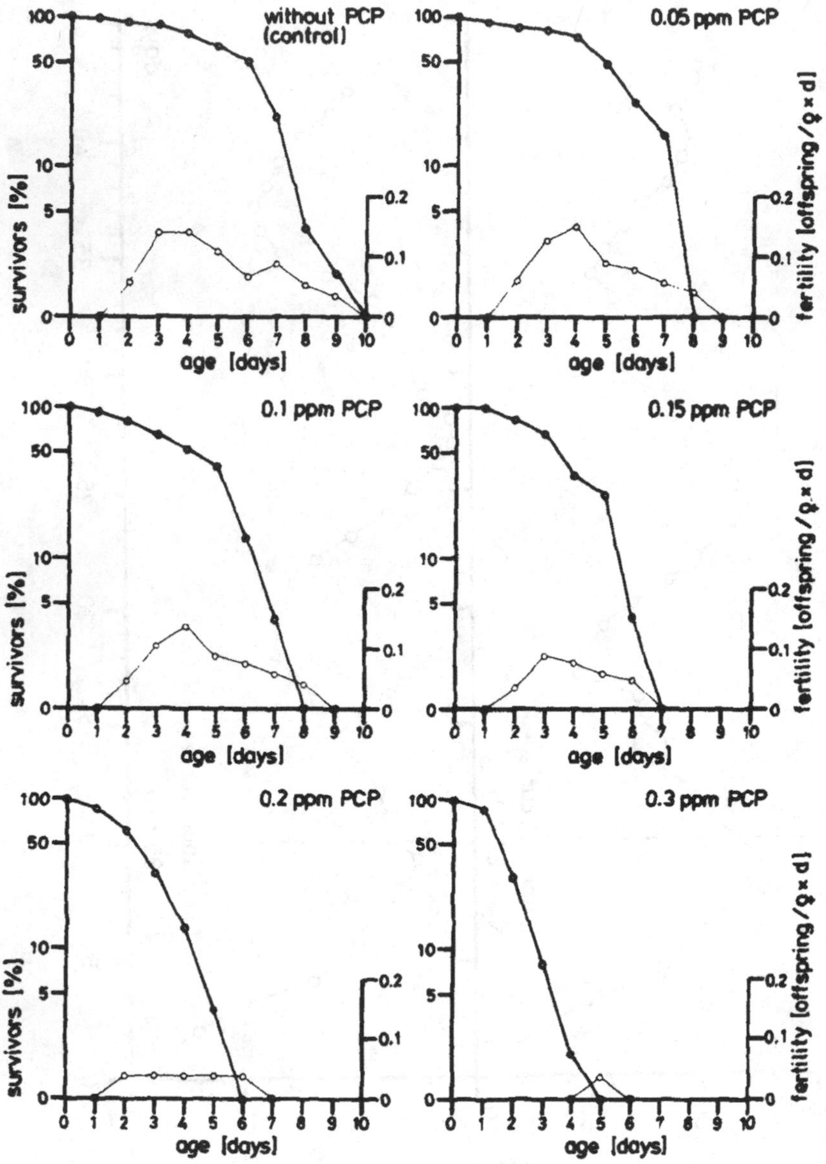

<u>Fig.32.</u> Survivorship curves (circles) and fertility curves (triangles) of <u>Brachinus rubens</u> in pure culture without PCP (control) and at dif= ferent concentrations of PCP. The abscissa represents the age of the animals in days (birth in the origin) (after Halbach et al. 1981).

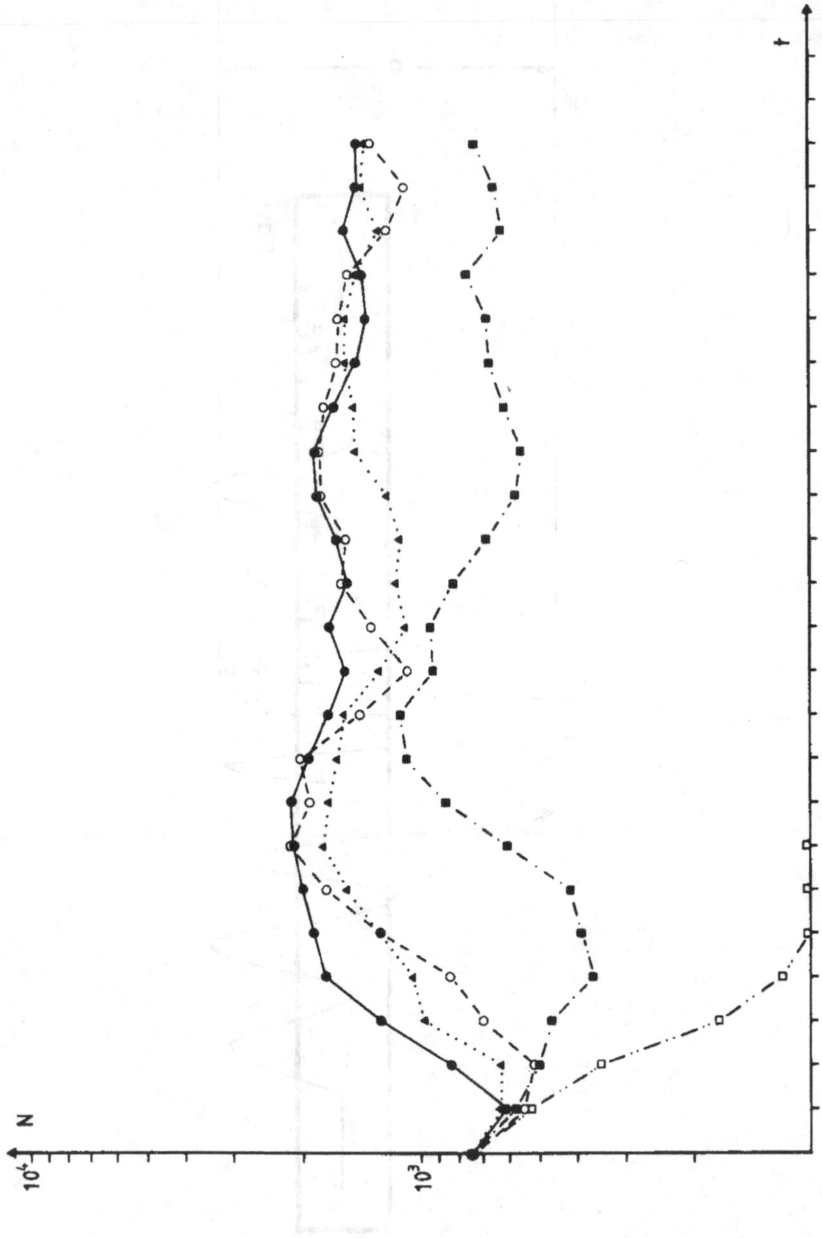

Fig.33. Population dynamics of Brachionus rubens at different concen= tration of pentachlorophenol (PCP). The curves are averages of 6 paral= lels each. Abscissa: time in days; ordinate: individuals/15 ml (log.).

●——● 0 ppm PCP, ○- - -○ 0.05 ppm PCP, ▲·······▲ 0.10 ppm PCP, ■-·-·-■ 0.15 ppm PCP, □-···-□ 0.20 ppm PCP. (after Halbach et al.1981).

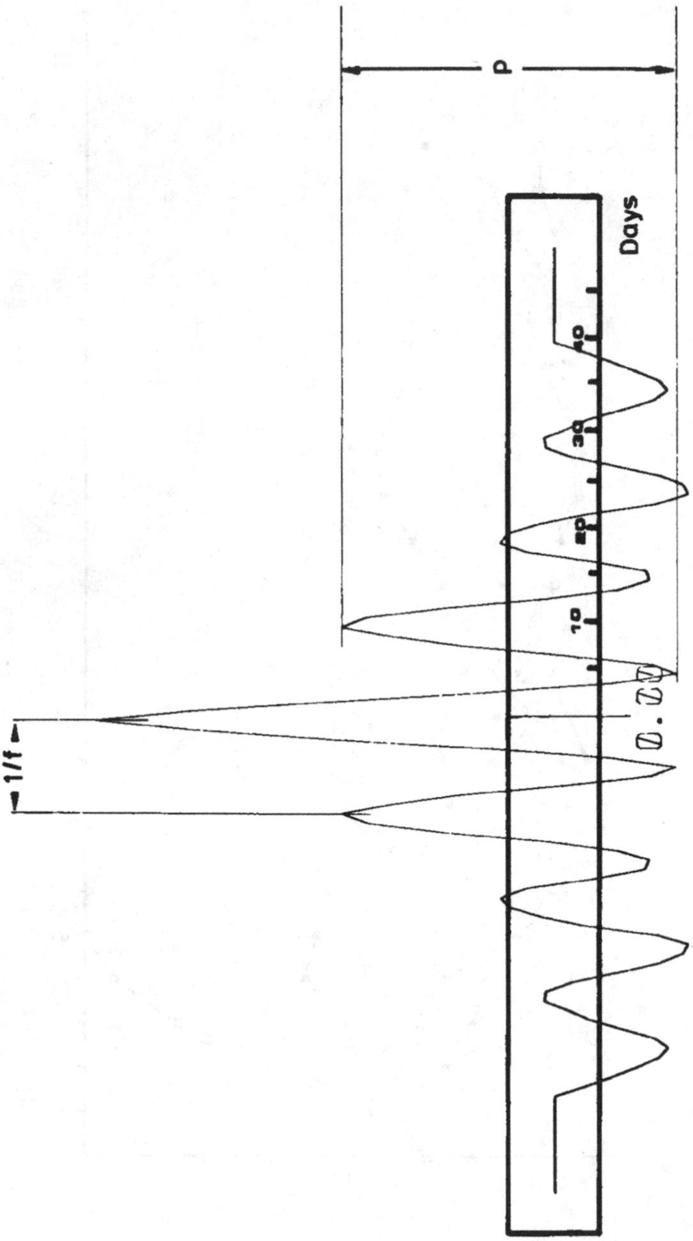

Fig. 34. Autocorrelation function of population dynamics of Brachionus at 20°C. The statistically determined duration of one period is slightly more than 10 days. Calculations made by Beuter (after Halbach 1978).

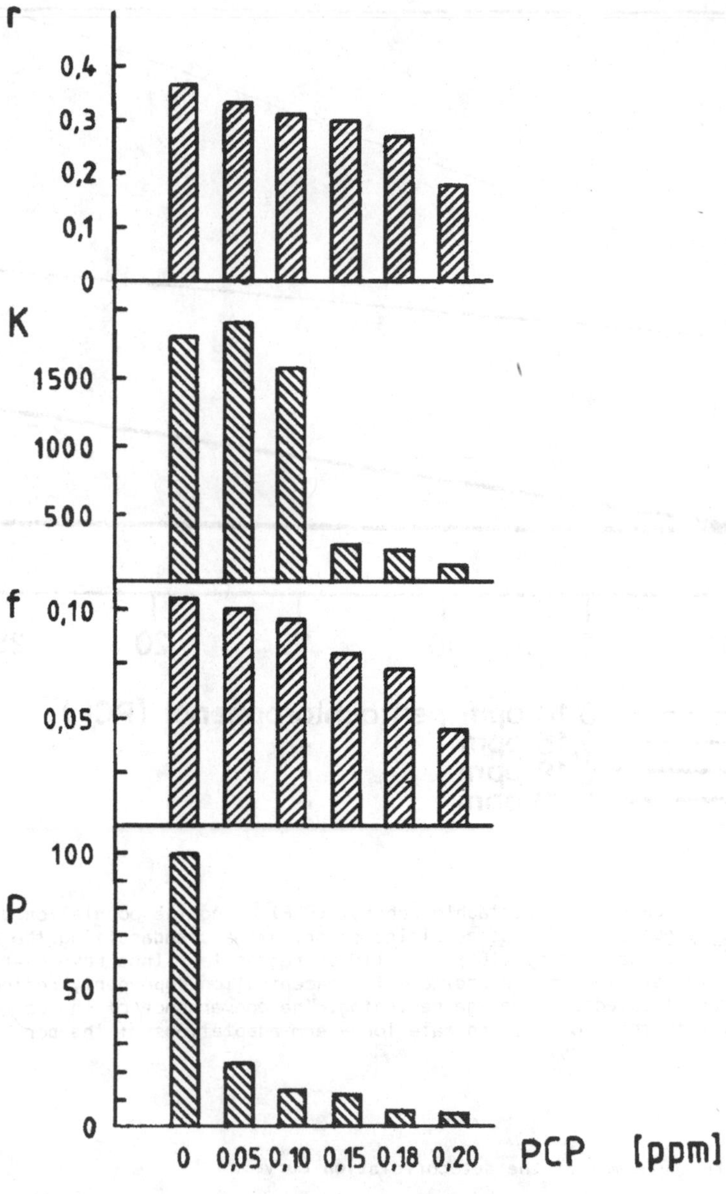

Fig.35. The four population parameters of Brachionus rubens which serve as bio-indicator for sublethal to= xic effects exemplified with PCP in 6 different con= centrations.
r: intrinsic rate of natural increase (offspr./♀ · h),
C: carrying capacity (animals/15 ml),
f: frequency of the oscillations (day^{-1}),
p: pregnancy of the oscillations (%).
For details see text (after Halbach et al.1981).

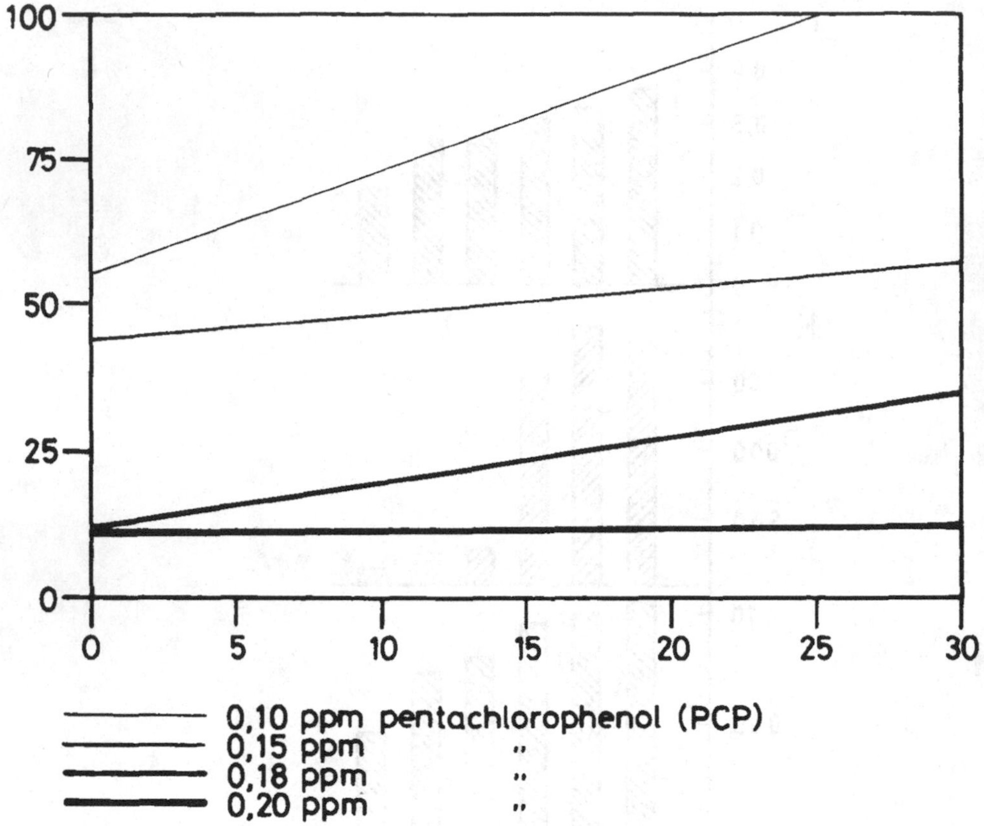

Fig.36. Bioassay for pentachlorophenol (PCP) using the population dynamics of Brachionus calyciflorus as bioindicator. After standardizing the carrying capacity to the control (100%) the linear regression lines have been calcu= lated. The section of the ordinate is concentration dependent, reflecting detrimental effects since the beginning. The convergence of the carrying ca= pacity with the control indicate long-term adaptations of the population (after Halbach 1982b).

and the first maximum of the autocorrelation curve.

For practical purposes we use four parameters as bioindicators: the intrinsic rate of natural increase (r), the carrying capacity (C), the frequency (f), and the pregnancy (p) of the oscillations (Fig.35).

The highest spreading power (steepest slope) of the four parameters lies at different concentrations of the toxic substances. We can thus use the different parameters specifically for different concentrations of the substance.

Sometimes we observe long-term effects in our ecotoxicological studies, such as an increase in carrying capacity, which can be detected by standardizing the population curves to the control and calculating the linear regression line for the carrying capacity. The converging curves (Fig.36) indicate an adaptation to the poisoning substances over many generations. For example, with 4-chloroanilin a long-term chronic effect has been observed, which can be explained as an accumulation effect over many generations (Halbach et al. 1982). The biological basis for these long-term phenomena is still under study.

Thus, the mathematical models of population dynamics can provide valuable insights into the physiological basis of population dynamics. Also, they can be useful in applied studies as in ecotoxicological bioassays.

REFERENCES

1. G.A. BECUS, (1980): Stochastic predator-prey relationships. Lect. Notes Pure Appl. Math. 58, 171-195.
2. K. BEUTER, C. WISSEL & U. HALBACH, (1981): Correlation and spectral analyses of the dynamics of a controlled rotifer population. In: D.G. CHAPMAN & V.F. GALUCCI (Eds.): Quantitative Population Dynamics (Statistical Ecology Series 13), 61-82.
3. C. BIRCH, (1948): The intrinsic rate of natural increase of an insect population. J. Anim. Ecol. 17, 15-26.
4. L.M. COOK, (1965): Oscillations in the simple logistic growth model. Nature (Lond.), 207, 316.
5. L.J. DUBLIN & A.J. LOTKA, (1925): On the true rate of natural increase. J. Amer. Statist. Ass. 20, 305-339.
6. W.T. EDMONDSON, (1968): A graphical model for evaluating the use of egg ratio for measuring birth and death rates. Oecologia 1, 1-37.
7. F.C. EVANS & E.F. SMITH, (1952): The intrinsic rate of natural increase of the human louse Pediculus humanus L. Amer. Nat. 86, 229-310.
8. J.W. FORRESTER, (1971): World Dynamics. Cambridge, Mass.
9. D. GIRKE & U. HALBACH, (1982): A new population model: Computer simulation using different time lags for birth and death rate. Ber. Ökol. Aussenst. Schlüchtern 11 (in press).
10. B.S. GOH, (1980): Stability of some multispecies population models. Lect. Notes Pure Appl. Math. 58, 209-216.
11. U. HALBACH, (1970): Influence of temperature on the population dynamics of the rotifer Brachionus calyciflorus PALLAS. Oecologia 4, 176-207.
12. U. HALBACH, (1973): Life table data and population dynamics of the rotifer Brachionus calyciflorus PALLAS as influenced by periodically oscillating temperature. In: W. WIESER (Ed.): Effects of Temperature on Ectothermic Organisms. Springer-Verlag, Berlin - Heidelberg - New York, 217-228.
13. U. HALBACH, (1974): Modelle in der Biologie. Naturwiss. Rundschau 27, 3-15.
14. U. HALBACH, (1975): Methoden der Populationsökologie. Verh. Ges. Ökol., Erlangen 1974, 1-24.
15. U. HALBACH, (1976): Populations- und synökologische Modelle in der Ornithologie. J. Ornithol. 117, 279-296.

16. U. HALBACH, (1978a): Problems of ecosystem research as exemplified by limnology. Verh. Dtsch.Zool. Ges. 1977, 41-66.

17. U. HALBACH, (1978b): Populationdynamik planktischer Rotatorien. Verh. Ges. Ökol. Kiel 1977, 173-183.

18. U. HALBACH, (1979a): The ecological niche and derived concepts. Abh. Geb. Vogelkunde 6, 53-65.

19. U. HALBACH, (1979b): Introductory remarks: Strategies in population research exemplified by rotifer population dynamics. In: U. HALBACH & J. JACOBS (Eds.): Population Ecology. Fortschr. Zool. 25, 1-27.

20. U. HALBACH, (1979c): Modelle und Modellvorstellungen in der Biologie. Handbuch d. prakt. und exper. Schulbiologie I/1, 61-112.

21. U. HALBACH, (1979d): Computer sagt Bevölkerungsentwicklung voraus. Mathematische Modelle für Schwankungen der Individuendichte. Umschau 79(11), 341-346.

22. U. HALBACH, (1982a): Population dynamics of rotifers and its consequences for ecotoxicology. Hydrobiologia (in press).

23. U. HALBACH, (1982b): Population ecology of rotifers as a bioassay tool for ecotoxicological tests in aquatic environments. Ecotoxicology and Environmental Safety (in press).

24. U. HALBACH & H.-J. BURKHARDT (1972): Are simple time-lags responsible for cyclic variation of population density? A comparison of laboratory population dynamics with computer simulations. Oecologia 9, 215-222.

25. U. HALBACH & I. FRIZ, (1978): Bei welcher Individuendichte stoppt eine Bevölkerungsexplosion? Ber. Ökol. Aussenstelle Schlüchtern 1, 107-127.

26. U. HALBACH & G. HALBACH-KEUP, (1974): Quantitative relations between phytoplankton and the population dynamics of the rotifer Brachionus calyciflorus PALLAS. Results of laboratory experiments and field studies. Arch. Hydrobiol. 73, 273-309.

27. U. HALBACH et al. (1981a): Population dynamics of rotifers as bioassay tool for toxic effects of organic pollutants. Verh. Intern. Verein. Limnol. 21, 1147-1152.

28. U. HALBACH et al. (1981b): The population dynamics of rotifers as bioassay for sublethal ecotoxicological effects exemplified with pentachlorophenol (PCP). Verh. Ges. Ökol., Berlin 1980, 261-267.

29. T.G. HALLAM, (1980): Persistence in Lotka-Volterra models of food chains and competition. Lect. Notes Pure Appl. Math. 58, 1-12.

30. J.F. HANEY, M. BRAUER & G. NÜRNBERG, (1982): Cerenkov Counting: A useful method for determining feeding, egestion, and excretion rates of zooplankton. Limnol. & Oceanogr. (in press).

31. A. HASTINGS, (1980): Population dynamics in patchy environments. Lect. Notes Pure Appl. Math. 58, 217-224.

32. D. VON HOLST, (1974): Sozialer Stress bei Tier und Mensch. Verh. Ges. Ökol., Saarbrücken 1973, 97-106.

33. E. HUTCHINSON, (1948): Circular causal systems in ecology. Ann. N.Y. Acad. Sci. 50, 221-246.

34. E. HUTCHINSON, (1954): Theoretical notes on oscillating populations. J. Wildlife Mgmt. 18, 107-109.

35. H. KAUSER, (1975): Dynamics of populations and properties of single individuals. Verh. Ges. Ökol., Erlangen 1974, 25-38.

36. N. LEIMEROTH, (1980): Respiration of different stages and energy budgets of juvenile Brachionus calyciflorus. Hydrobiologia 73, 195-197.

37. P.H. LESLIE & T. PARK, (1949): The intrinsic rate of natural increase of Tribolium castaneum HERBST. Ecology 30, 469-477.

38. R.M. MAY, (1974): Stability and complexity in model ecosystems. Monographs in Population Biology 6, 2nd Edition. Princeton, New Jersey.

39. R.M. MAY, (1976): Theoretical Ecology - Principals and Applications. Oxford.

40. D.H. MEADOWS et al., (1972): The Limits of Growth. New York.

41. A.J. NICHOLSON, (1954): An outline of the dynamics of animal populations. Austr. J. Zool. 2, 9-65.

42. A. PARISE, (1966): Ciclo sessuale e dinamica popolazioni di Euchlanis (Rotatoria) in condizioni sperimentali. Arch. Oceanogr. Limnol. 16, 387-411.

43. R. PEARL & S.A. GOULD, (1936): Human Biology 8, 399-511.

44. I. RECHENBERG, (1973): Evolutionsstrategie. Stuttgart - Bad Cannstatt.

45. C. RORRES, (1980): Optimal age-specific harvesting policy for a continuous time-population model. Lect. Notes Pure Appl. Math. 58,239-254.

46. A. SEITZ & U. HALBACH, (1973): How is the population density regulated? Experimental studies on rotifers and computer simulations. Naturwiss. 60, 51.

47. P.J. WANGERSKY & W.J. CUNNINGHAM, (1957): Time lag in population models. Cold Spr. Harb. Symp. Quant. Biol. 22, 329-338.

48. C. WISSEL, K. BEUTER & U. HALBACH, (1981): Correlation functions for the evaluation of repeated time series with fluctuations. ISEM Journal 3, 11-29.

49. D.J. WOLLKIND, A. HASTINGS & J.A. LOGAN, (1980): Models involving differential equations appropriate for describing a temperature dependent predator-prey mite ecosystem on apples. Lect. Notes Pure Appl. Math. 58,255-277.

STABILITY PROPERTIES FOR FUNCTIONAL DIFFERENTIAL EQUATIONS WITH INFINITE DELAY

Yoshiyuki HINO
Department of Mathematics
Chiba University
Chiba, 260, JAPAN

1. Introduction. For ordinary differential equations, there are many results with respect to stability properties and limiting equations by using the techniques of topological dynamics (cf. see [6, 7]). Clearly, these results required the uniqueness of solutions of a given equations and its limiting equations for initial conditions (regularity).

Recently, Hale and Kato [2] have extended Sell's result (Theorem 6 in [6], Theorem VIII.9 in [7]) to functional differential equations with infinite delay without assuming regularity.

In this note, we also discuss functional differential equations with infinite delay and extend some results given for ordinary differential equations and functional differential equations by using Hale and Kato's result, that is, we show that some stability properties of a bounded solution of given equation follows from stability properties of its limiting equations. We also do not assume regularity. In particular, our results contain the fact that for periodic systems, if a bounded solution is uniformly asymptotically stable, then it is totally stable.

2. Phase space B, notations and definitions. First, we shall give the space B discussed by Kato [5] (also, refer [1, 2, 3]). Let $|x|$ be any norm of x in R^n. Let B be a real linear vector space of functions mapping $(-\infty, 0]$ into R^n with a semi-norm $|\cdot|_B$. For any elements ϕ and ψ in B, $\phi = \psi$ means $\phi(t) = \psi(t)$ for all $t \in (-\infty, 0]$. If x is a function defined on $(-\infty, a)$, then for each $t \in (-\infty, a)$ we define the function x_t by the relation $x_t(s) = x(t+s)$, $-\infty < s \leqq 0$. The space B is assumed to have the following properties:

(I) If x(t) is defined on $(-\infty, a)$, continuous on $[\sigma, a)$, $\sigma < a$, and $x_\sigma \in B$, then for $t \in [\sigma, a)$

(I.1) $x_t \in B$,

(I.2) x_t is continuous in t with respect to $|\cdot|_B$,

(I.3) there are a $K > 0$ and a positive continuous function $M(\beta)$, $M(\beta) \to 0$ as $\beta \to \infty$, such that $|x_t|_B \leqq K \sup_{\sigma \leqq \theta \leqq t} |x(\theta)| + M(t-\sigma)|x_\sigma|_B$.

(II) $|\phi(0)| \leqq M_1|\phi|_B$ for $M_1 > 0$.

Remark 1. In [2], Hale and Kato have given hypotheses on the space B in a slightly different way. However, in our present context, there is no difference between the two.

Let S be a compact subset in B and let $\alpha > 0$ and $\beta > 0$. Define $X^*(S, \alpha, \beta)$ by

$$X^*(S, \alpha, \beta) = \{x(\cdot) \mid x(\theta) = \phi(\theta) \text{ for } \theta \in (-\infty, 0],\ \phi \in S,\ x(\theta) \text{ is continuous on } [0, \infty),\ |x(\theta)| \leqq \alpha \text{ for } \theta \in [0, \infty) \text{ and } |x(\theta^1) - x(\theta^2)| \leqq \beta|\theta^1 - \theta^2| \text{ for } \theta^1, \theta^2 \in [0, \infty)\}.$$

Furtheremore, define $X(S, \alpha, \beta)$ by

$$X(S, \alpha, \beta) = \{x_t \mid t \geqq 0,\ x(\cdot) \in X^*(S, \alpha, \beta)\}.$$

Lemma 1 (Corollary 3.2 in [2]). $\overline{X(S, \alpha, \beta)}$ is compact in B, where $\overline{X(S, \alpha, \beta)}$ is the closure of $X(S, \alpha, \beta)$.

Suppose that $F(t, \phi)$ is an R^n-valued function, continuous on $R \times B$, $R = (-\infty, \infty)$, and for any compact set $S \subset B$, $F(t, \phi)$ is bounded and uniformly continuous on $R \times S$. Then, it is known that for any compact set $W \subset R \times B$, any sequence $\{t_n\}$, $t_n \geqq 0$, contains a subsequence $\{t_{n_j}\}$ such that $\{F(t+t_{n_j}, \phi)\}$ converges uniformly for $(t, \phi) \in W$ (cf. see [2, 4]). The hull $H(F)$ $(H^+(F))$ denotes the set of pairs (G, Ω), $\Omega \subset R \times B$, such that there exists a sequence $\{t_n\}$, $t_n \geqq 0$, $(t_n \geqq 0$ and $t_n \to \infty$ as $n \to \infty)$ such that $F(t+t_n, \phi)$ converges to $G(t, \phi)$ for $(t, \phi) \in \Omega$.

Remark 2. We shall note that if $(G, \Omega) \in H(F)$, then for any compact set $S \subset B$, there exists a $(G^*, \Omega^*) \in H(F)$ such that $\Omega^* \supset \Omega \cup \{I \times S\}$ and $G^*(t, \phi) = G(t, \phi)$ on Ω, where $I = [0, \infty)$, because $I \times S$ is separable (refer [2, 4]).

Consider the system of functional differential equations

$$\dot{x}(t) = F(t, x_t), \tag{1}$$

where $\dot{x}(t)$ denotes the right hand derivative of a given continuous function $x(t)$. Let $x(t, F)$ be a solution of (1). In particular, let $x(t, s, \phi^o, F)$ be a solution of (1) through (s, ϕ^o). We assume that System (1) has a bounded solution $u(t)$ defined on I and $L = \sup\{|F(t, \phi)|| t \geqq 0, |\phi|_B \leqq 2H\} < \infty$, where H is some constant and satisfies $|u_t|_B \leqq H$ for $t \geqq 0$. We shall define $H(u, F)$ $(H^+(u, F))$ such that for $(v, G, \Omega) \in H(u, F)$ $(H^+(u, F))$, there exists a sequence $\{t_n\}$, $t_n \geqq 0$ ($t_n \geqq 0$, $t_n \to \infty$ as $n \to \infty$), such that $F(t+t_n, \phi)$ converges to $G(t, \phi)$ for $(t, \phi) \in \Omega$ and $u(t+t_n) \to v(t)$ as $n \to \infty$ uniformly on any compact interval in I. By noting that $\overline{\{u_t | t \geqq 0\}}$ is compact by Lemma 1, we may assume that for any $(G, \Omega) \in H(F)$, $\Omega \supset I \times \overline{\{u_t | t \geqq 0\}}$, by Remark 2. Hence it is easily shown that $v(t)$ is a solution of

$$\dot{x}(t) = G(t, x_t) \tag{2}$$

defined on I and $|u_{t+t_n} - v_t|_B \to 0$ as $n \to \infty$ uniformly on any compact interval in I, if $(v, G, \Omega) \in H(u, F)$ (cf. see [2]). We shall denote the set of such solutions $v(t)$ by $H(u)$ and call that System (1) is regular, if for any $(G, \Omega) \in H(F)$, every solution of (2) is unique for the initial conditions.

We shall give some definitions of stabilities.

<u>Definition 1</u>. The solution $u(t)$ is uniformly stable, if for any $\varepsilon > 0$ there exists a $\delta(\varepsilon) > 0$ such that if $s \geqq 0$ and $|\phi^o - u_s|_B < \delta(\varepsilon)$, then $|x_t(s, \phi^o, F) - u_t|_B < \varepsilon$ for all $t \geqq s$. Furthermore, the solution $u(t)$ is uniformly asymptotically stable, if it is uniformly stable and if there exists a $\delta_o > 0$ and for any $\varepsilon > 0$ there exists a $T(\varepsilon) > 0$ such that if $s \geqq 0$ and $|\phi^o - u_s|_B < \delta_o$, then $|x_t(s, \phi^o, F) - u_t|_B < \varepsilon$ for $t \geqq s + T(\varepsilon)$.

<u>Definition 2</u>. The solution $u(t)$ is weakly uniformly asymptotically stable, if it is uniformly stable and if there exists a $\delta_o > 0$ such that for any $s \in I$, $|\phi^o - u_s|_B < \delta_o$ implies $|x_t(s, \phi^o, F) - u_t|_B \to 0$ as $t \to \infty$.

<u>Definition 3</u>. The solution $u(t)$ is uniformly stable in $H(F)$ $(H^+(F))$, if for any $\varepsilon > 0$, there exists a $\delta(\varepsilon) > 0$ such that for any $s \in I$ and $(v, G, \Omega) \in H(u, F)$ $(H^+(u, F))$, $|\phi^o - v_s|_B < \delta(\varepsilon)$ implies $|x_t(s, \phi^o, G) - v_t|_B < \varepsilon$ for all $t \geqq s$.

Definition 4. The solution $u(t)$ is attracting in $H(F)(H^+(F))$, if there exists a $\delta_o > 0$ such that for any $(v, G, \Omega) \in H(u, F)(H^+(u, F))$, $|\phi^o - v_0|_B < \delta_o$ implies $|x_t(0, \phi^o, G) - v_t|_B \to 0$ as $t \to \infty$. Furthermore, the solution $u(t)$ is weakly uniformly asymptotically stable in $H(F)$ $(H^+(F))$, if it is uniformly stable in $H(F)(H^+(F))$ and attracting in $H(F)(H^+(F))$.

Definition 5. The solution $u(t)$ is uniformly asymptotically stable in $H(F)(H^+(F))$, if it is uniformly stable in $H(F)(H^+(F))$ and if there exists a $\delta_o > 0$ and for any $\varepsilon > 0$ there exists a $T(\varepsilon) > 0$ such that for any $s \in I$ and $(v, G, \Omega) \in H(u, F)(H^+(u, F))$, $|\phi^o - v_s|_B < \delta_o$ implies $|x_t(s, \phi^o, G) - v_t|_B < \varepsilon$ for $t \geqq s + T(\varepsilon)$.

Definition 6. The solution $u(t)$ is totally stable, if for any $\varepsilon > 0$ there exists a $\delta(\varepsilon) > 0$ such that if $g(t)$ is continuous and satisfies $|g(t)| < \delta(\varepsilon)$ on $[s, \infty)$ for an $s \in I$ and if $|\phi^o - u_s|_B < \delta(\varepsilon)$, then $|x_t(s, \phi^o, F+g) - u_t|_B < \varepsilon$ for all $t \geqq s$, where $x(t, s, \phi^o, F+g)$ is a solution of

$$\dot{x}(t) = F(t, x_t) + g(t) \tag{3}$$

through (s, ϕ^o).

Remark 3. In the above concepts, if the semi-norm $|\cdot|_B$ can be replaced by R^n-norm, then the concepts of the stabilities in R^n will be obtained. However, it is known that the concepts of stabilities and stabilities in R^n are equivalent under hypotheses (I) and (II) (see Theorem 5 in [5]).

We have following propositions that are well known for ordinary differential equations and functional differential equations (cf. [9]).

Proposition 1. Assume that System (1) is regular. If the solution $u(t)$ is uniformly stable (uniformly asymptotically stable), then it is uniformly stable in $H(F)$ (uniformly asymptotically stable in $H(F)$).

Proof. If $G(t, \phi) = F(t+\sigma, \phi)$ for some $\sigma > 0$ and the solution $u(t)$ is uniformly stable (uniformly asymptotically stable), then the solution $v(t) = u(t+\sigma)$ of (2) is also uniformly stable (uniformly asymptotically stable). Hence we shall consider only the case where $(v, G, \Omega) \in H^+(u, F)$, for which there exists a sequence $\{t'_n\}$, $t'_n \to \infty$ as $n \to \infty$, such that $F(t+t'_n, \phi)$ converges to $G(t, \phi)$ for $(t, \phi) \in \Omega$ and $u(t+t'_n) \to v(t)$ as $n \to \infty$ uniformly on any compact interval in I.

Let $0 \leqq s < r$, $0 < \varepsilon < M_1H$ and $|\phi^0 - v_s|_B < \delta(\varepsilon/2)/2$, where $\delta(\cdot)$ is the same one given for uniform stability of u(t). Since $[s, r] \times X(\{\phi^0, u_0\}, 2M_1H, 2L)$ is compact by Lemma 1, there exists a subsequence $\{t_n\}$ of $\{t'_n\}$ such that $F(t+t_n, \phi) \to G(t, \phi)$ as $n \to \infty$ uniformly on $[s, r] \times X(\{\phi^0, u_0\}, 2M_1H, 2L)$. Let x(t) be a solution of (1) such that $x_{s+t_n} = \phi^0$. Then $x^n(t) = x(t+t_n)$ is the solution of

$$\dot{x}(t) = F(t+t_n, x_t) \tag{4}$$

through $x^n_s = \phi^0$.

We shall show only that v(t) is uniformly stable, if u(t) is uniformly stable, because the remaining parts can be shown by using parallel arguments as in the proof of Theorem 13.3 in [9]. Since $u^n(t) = u(t+t_n)$ also is uniformly stable with the same pair $(\varepsilon, \delta(\varepsilon))$ as the one for u(t) and $|u^n_s - x^n_s|_B \leqq |u^n_s - v_s|_B + |v_s - x^n_s|_B < \delta(\varepsilon/2)$ for all large n, we have

$$|u^n_t - x^n_t|_B < \varepsilon/2 \text{ for all } t \geqq s. \tag{5}$$

Hence $x^n_t \in X(\phi^0, 2M_1H, 2L)$ for all $t \geqq s$, and therefore $\{x^n(t)\}$ converges to the solution y(t) of (2) through (s, ϕ^0), which is uniquely determined, uniformly on [s, r]. Thus, if n is sufficiently large, we have

$$|x^n_t - y_t|_B < \varepsilon/4 \text{ and } |u^n_t - v_t|_B < \varepsilon/4 \text{ on } [s, r]. \tag{6}$$

It follows from (5) and (6) that $|v_t - y_t|_B < \varepsilon$ on [s, r]. Since r is arbitrary, $|v_t - y_t|_B < \varepsilon$ for all $t \geqq s$, if $|v_s - \phi^0|_B < \delta(\varepsilon/2)/2$.

The following proposition can be proved by using the parallel arguments as in the proof of Theorem 13.2 in [9].

Proposition 2. If $F(t, \phi)$ is periodic in t with period $\omega > 0$, then the solution u(t) is uniformly stable in H(F) (weakly uniformly asymptotically stable in H(F), uniformly asymptotically stable in H(F)), if it is uniformly stable (weakly uniformly asymptotically stable, uniformly asymptotically stable).

Proposition 3 (Theorem in [3]). If $F(t, \phi)$ is linear in ϕ, then the null solution of (1) is totally stable if and only if it is uniformly asymptotically stable.

3. Main theorem and related results. Hale and Kato's theorem is the following:

Theorem A (Theorem 6.2 in [2]). If $u(t)$ is uniformly stable and attracting in $H^+(F)$, then it is uniformly asymptotically stable.

Now we shall give our theorem. The proof of Theorem will be given in Section 4.

Theorem. If the solution $u(t)$ is unique for the initial conditions and weakly uniformly asymptotically stable in $H^+(F)$, then it is uniformly asymptotically stable in $H(F)$ and totally stable.

By Propositions 1 and 2 and Theorem, we have the following corollaries that are well known for ordinary differential equations (cf. [9]).

Corollary 1. Assume that System (1) is regular. If $u(t)$ is uniformly asymptotically stable, then it is totally stable.

Corollary 2. Assume that $F(t, \phi)$ is periodic in t. Then $u(t)$ is weakly uniformly asymptotically stable if and only if it is uniformly asymptotically stable.

Corollary 3. Assume that $F(t, \phi)$ is periodic in t. If $u(t)$ is uniformly asymptotically stable, then it is totally stable.

4. Proof of Theorem. We shall use the following lemmas to prove Theorem.

Lemma 2. Assume that the solution $u(t)$ is attracting in $H^+(F)$.
Let S be a compact subset of B and let $\{x^n(t, F+g^n)\}$ be a sequence of solutions of $\dot{x}(t) = F(t, x_t) + g^n(t)$ defined on $[s_n, r_n]$ and satisfies $x^n_{s_n}(F+g^n) \in S$ for $n = 1, 2, 3, \cdots$, and

(7) $\min\{H, \delta_0\} > \delta_1 \geq |x^n_t(F+g^n) - u_t|_B \geq \varepsilon$ on $s_n \leq t \leq r_n$

for some sequences $\{\varepsilon_n\}$, $\varepsilon_n \to 0$ as $n \to \infty$, $\{s_n\}$, $\{r_n\}$, $r_n > s_n \geq 0$ and $\{g^n(t)\}$, $|g^n(t)| < \varepsilon_n$ on $[s_n, \infty)$, and for some constants ε and δ_1, $0 < \varepsilon < \delta_1$, where δ_0 is the one given for attractor in $H^+(F)$ of $u(t)$ and $g^n(t)$ is continuous on $[s_n, \infty)$.

Then the sequence $\{r_n - s_n\}$ is bounded.

<u>Proof</u>. Put $q_n = r_n - s_n$ and suppose that $q_n \to \infty$ as $n \to \infty$. Set $t_n = s_n + (q_n/2)$ and $y^n(t) = x^n(t+t_n, F+g^n)$, then we have $y^n{}_t \in \overline{X(S, 2M_1H, 2L)}$ for $t \in [0, q_n/2]$, because $y^n{}_{-q_n/2} = x^n{}_{s_n}(F+g^n) \in S$, $|y^n{}_t|_B = |x^n{}_{t+t_n}(F+g^n)|_B \leq |x^n{}_{t+t_n}(F+g^n) - u_{t+t_n}|_B + |u_{t+t_n}|_B < \delta_1 + H < 2H$ for $t \in [-q_n/2, q_n/2]$ by (7), and hence $|y^n(t)| \leq 2M_1H$ by Hypothesis (II), $|\dot{y}^n(t)| \leq |F(t, y^n{}_t)| + |g^n(t)| \leq L + \varepsilon_n < 2L$ for $t \in [-q_n/2, q_n/2]$ and $\overline{X(S, 2M_1H, 2L)} \supset \overline{X(\overline{X(S, 2M_1H, 2L)}, 2M_1H, 2L)}$. Clearly, $y^n(t)$ and $u(t+t_n)$ are solutions of $\dot{x}(t) = F(t+t_n, x_t) + g(t+t_n)$ and $\dot{x}(t) = F(t+t_n, x_t)$, respectively, and hence we can easily show that there are a function $G(t, \phi)$ and solutions $y(t)$ and $v(t)$ of (2) defined on I such that $F(t+t_n, \phi) + g^n(t+t_n) \to G(t, \phi)$ uniformly on any compact subset of I and uniformly on $\overline{X(\{S, u_0\}, 2M_1H, 2L)}$ as $n \to \infty$ and $y^n(t) \to y(t)$ and $u(t+t_n) \to v(t)$ uniformly on any compact subset of I as $n \to \infty$, taking a subsequence, if necessary, because $q_n \to \infty$ as $n \to \infty$ and $\overline{X(\{S, u_0\}, 2M_1H, 2L)}$ is the compact subset of B by Lemma 1. For a fixed $t > 0$, there exists an $n_o > 0$ such that for every $n \geq n_0$, $r_n - t_n = r_n - s_n - (q_n/2) = q_n/2 > t$, because $q_n \to \infty$ as $n \to \infty$. Hence for $n \geq n_0$, we have $s_n < t+t_n < r_n$. Thus for $n \geq n_0$,

$$(8) \qquad |y^n{}_t - u_{t+t_n}|_B \geq \varepsilon$$

by (7). There exists an $n_1 > n_0$ such that for every $n \geq n_1$

$$(9) \qquad |y^n{}_t - y_t|_B \leq \varepsilon/4 \text{ and } |u_{t+t_n} - v_t|_B \leq \varepsilon/4.$$

By (8) and (9), for every $n \geq n_1$

$$(10) \quad |y_t - v_t|_B \geq |y^n{}_t - u_{t+t_n}|_B - |u_{t+t_n} - v_t|_B - |y^n{}_t - y_t|_B \geq \varepsilon/2.$$

However, $|y_0 - v_0|_B < \delta_0$ implies $|y_t - v_t|_B \to 0$ as $t \to \infty$, which is a contradiction of (10).

The following lemma can be proved by using the parallel arguments as in the proof of Lemma 2 in [10].

Lemma 3. Assume that any solution in $H(u)$ is unique for the initial conditions. Let $T > 0$. Then for any $\varepsilon > 0$, there exists a $\delta(\varepsilon) > 0$ such that for any $s \in I$, if $|\phi^o - u_s|_B < \delta(\varepsilon)$ and $|g(t)| < \delta(\varepsilon)$ on $[s, s+T]$, then $|x_t(s, \phi^o, F+g) - u_t|_B < \varepsilon$ for $t \in [s, s+T]$.

Proof of Theorem. We shall show only that the solution $u(t)$ is totally stable, because remaining parts follow from Theorem A, immediately. Suppose not. Then there exist sequences $\{t_n\}$, $t_n \geqq 0$, $\{r_n\}$, $r_n > 0$, $\{\varepsilon_n\}$, $\varepsilon_n \to 0$ as $n \to \infty$, $\{g^n(t)\}$ and $\{x^n(t, F+g^n)\}$ and a constant δ_1, $\delta_1 < \min\{H, \delta_o/2\}$, such that

(11) $|x^n_{t_n}(F+g^n) - u_{t_n}|_B < \varepsilon_n$ and $|g^n(t)| < \varepsilon_n$ on $[t_n, \infty)$

and

(12) $|x^n_{t_n+r_n}(F+g^n) - u_{t_n+r_n}|_B = \delta_1$ and $|x^n_t(F+g^n) - u_t|_B < \delta_1$ on $[t_n, t_n+r_n)$,

where δ_o is the one given for attractor in $H^+(F)$ of $u(t)$. There exists a sequence $\{q_n\}$, $0 < q_n < r_n$, such that

(13) $$|x^n_{t_n+q_n}(F+g^n) - u_{t_n+q_n}|_B = \delta(\delta_1/2)/2$$

and

(14) $$\delta(\delta_1/2)/2 \leqq |x^n_t(F+g^n) - u_t|_B \leqq \delta_1 \text{ on } [t_n+q_n, t_n+r_n],$$

by (11) and (12), where $\delta(\cdot)$ is the one given for uniform stability in $H^+(F)$ of $u(t)$. We can show that $q_n \to \infty$ as $n \to \infty$. Suppose that there exists a subsequence of $\{q_n\}$, which we shall denote by $\{q_n\}$ again, such that q_n converges to some q, $q \in I$. Since any solution in $H(u)$ is unique for initial conditions, it follows from (11) that there exists an $n_o > 0$ such that for any $n \geqq n_o$, $q+1 \geqq q_n \geqq 0$ and $|x^n_{t_n+t}(F+g^n) - u_{t_n+t}|_B < \delta(\delta_1/4)/4$ for $t \in [0, q+1]$ by Lemma 3, which contradicts to (13).

Set $p_n = r_n - q_n$, then it follows from (11) and (14) that $\{p_n\}$ is bounded by Lemma 2, because $x^n_{t_n+q_n}(F+g^n) \in \overline{X(S, 2M_1H, 2L)}$, where $S = \overline{\{x^n_{t_n}(F+g^n)\ n = 1, 2, 3, \cdots\}}$. Hence we may assume that p_n converges to p, $p \in I$, as $n \to \infty$ and $0 \leqq p_n < p+1$ for all n. Set

$$y^n(t) = \begin{cases} x^n(t+t_n+q_n,\ F+g^n), & t \in (-\infty,\ p_n), \\ x^n(t_n+r_n,\ F+g^n), & t \in [p_n,\ p+1]. \end{cases}$$

Then we can assume that $F(t+t_n+q_n, \phi) + g^n(t+t_n+q_n) \to G(t, \phi)$, $G \in H^+(F)$, uniformly on $[0, p] \times \overline{X(\{S, u_0\}, 2M_1H, 2L)}$ as $n \to \infty$ and that $y^n(t)$ and $u(t+t_n+q_n)$ converge to solutions $y(t)$ and $v(t)$ of (2) as $n \to \infty$ uniformly on $[0, p]$, respectively. Since $|y_0 - v_0|_B = \delta(\delta_1/2)/2$ by (13), we have $|y_{p_n} - v_p|_{B_n} < \delta_1/2$. However, we have a contradiction by (12), because $|y^n_{p_n} - y^n_p|_B = |x^n_{p_n+t_n+q_n}(F+g^n) - x^n_{p+t_n+q_n}(F+g^n)|_B \leqq K \sup_{-(p_n+q_n) \leqq \theta \leqq 0} |x^n(p_n+t_n+q_n+\theta,\ F+g^n) - x^n(p+t_n+q_n+\theta,\ F+g^n)| + M(p_n+q_n)|x^n_{t_n}(F+g^n) - x^n_{t_n+p-p_n}(F+g^n)|_B \leqq 2KL|p_n - p| + M(p_n+q_n)\{|x^n_{t_n}(F+g^n)|_B + |x^n_{t_n+p-p_n}(F+g^n)|_B\} \leqq 2KL|p_n - p| + M(p_n+q_n)\{\varepsilon_n+\delta_1\}$, and hence we have $|y_p - v_p|_B \geqq |x^n_{t_n+r_n}(F+g^n) - u_{t_n+r_n}|_B - |y_p - y^n_p|_B - |y^n_p - y^n_{p_n}|_B - |y^n_{p_n} - x^n_{t_n+r_n}(F+g^n)|_B - |u_{t_n+r_n} - v_{p_n}|_B - |v_{p_n} - v_p|_B \geqq \delta_1/2$ for all large n. Thus the solution u(t) is totally stable.

References

[1] J.K.Hale, Dynamical systems and stability, J. Math. Anal. Appl., 26(1969), 39 - 69.

[2] J.K.Hale and J.Kato, Phase space for retarded equations with infinite delay, Funkcial. Ekvac., 21(1978), 11 - 41.

[3] Y.Hino, Total stability and uniformly asymptotic stability for linear functional differential equations with infinite delay, Funkcial. Ekvac., 24(1981), 345 - 349.

[4] J.Kato, Uniformly asymptotic stability and total stability, Tohoku Math. J., 22(1970), 254 - 269.

[5] J.Kato, Stability problem in functional differential equations with infinite delay, Funkcial. Ekvac., 21(1978), 63 - 80.

[6] G.R.Sell, Nonautonomous differential equations and topological dynamics, I, II., Trans. Amer. Math. Soc., 127(1967), 241 - 262, 263-283.

[7] G.R.Sell, Topological Dynamics and Ordinary Differential Equations, Van Nostrand Reinhold Company, London, 1971.

[8] T.Yoshizawa, Asymptotically almost periodic solutions of an almost periodic system, Funkcial. Ekvac., 12(1969), 23 - 40.

[9] T.Yoshizawa, Stability Theory and the Existence of Periodic Sol-

utions and Almost Periodic Solutions, Appl. Math. Sci., 14, Springer-Verlag, 1975.

[10] J.Kato and T.Yoshizawa, A relationship between uniformly asymptotic stability and total stability, Funkcial. Ekvac., 12(1969), 233-238.

AN AVERAGING METHOD FOR VOLTERRA INTEGRAL EQUATIONS WITH APPLICATIONS TO PHASE-LOCKED FEEDBACK SYSTEMS

F. C. Hoppensteadt*
Department of Mathematics
University of Utah
Salt Lake City, Utah 84112

ABSTRACT

A method of averaging is derived to study systems of Volterra integral and integro-differential equations. The results are applied to phase locked loops arising in synchronous control mechanisms in electrical and biological systems.

INTRODUCTION.

A straightforward method is needed to model and to analyze phase locking phenomena observed in physical and biological systems. The method derived here meets these needs for systems of Volterra integro-differential equations that model low gain feedback control systems and weakly coupled Hamiltonian systems.

Phase locked loops are important for the design of synchronous control systems and the tracking of carrier signals. The example presented here is for a bimodal phase-locked loop, but many other systems that have appeared in the engineering literature [1] are also included in the general class of models analyzed here. In our example, an input signal is estimated by the loop circuitry and then demodulated using the estimate. This technique is widely used in computer communications and frequency control of radios and for control of electric motors and power systems.

Biological systems also exhibit phase locking behavior. Examples are synchronization of biological clocks, such as ultradian metabolic and circadian hormonal cycles [2,3], and synchronization of SA and AV node firing in heart muscle contraction [4]. Such behavior is recognizable, although less well understood, in crowd behavior (e.g., rock concerts and cheering fans at sporting events), in synchronization of insect signals (e.g., firefly blinking and cicada buzzing), and in population fertility cycles (e.g., tidal animal's fertility synchronized by lunar periods). All of these phenomena are believed to be governed by coupled nonlinear oscillators, but it is unknown in most cases what the basic oscillators are and how they are coupled.

Section 1 presents the mathematical model studied here and the main result that is used later (in Section 2) to analyze a bimodal phase locked loop circuit. The

*This work was supported in part by NSF Grant No. MCS 80 15359 and was in part completed while the author was visiting the Forschungsschwerpunkt "Dynamische Systeme" at the University of Bremen

example studied in Section 2 describes a typical feedback loop where the averaging method can be used effectively. Proofs and details of the analysis are presented in Section 3, and for more general problems in [5].

1. THE MAIN RESULTS.

The synchronous control systems studied here are modelled by Volterra integro-differential equations of the form

$$(1) \qquad dx/dt = f(t,\varepsilon) + \varepsilon \int_0^t k(t-s)F(x(s),\varepsilon)ds, \quad x(0) = x_0$$

where $x, f, F \in E^N$ and $k \in E^{N\times N}$. Here ε is a small, positive parameter that (roughly) measures coupling intensities and the amplitude of noise in the system. The variables $x_1, x_2, \ldots, x_N$, the components of x, are phase-like variables, and this equation turns out in applications to be analogous to the phase equations in perturbed Hamiltonian systems. F is a periodic function of the components of x.

In many applications, the kernel k corresponds to a weak filter, and so it is approximately the Dirac delta function. In particular, if $k(t-s) = \delta(t-s)$, then (1) becomes

$$(2) \qquad dx/dt = f(t,\varepsilon) + \varepsilon F(x(t),\varepsilon),$$

and this system can be studied directly using Bogoliuboff's averaging method [6]. When $k(t-s) = \delta(t-s-T)$, then rather than (2) we get

$$(3) \qquad dx/dt = f(t,\varepsilon) + \varepsilon F(x(t-T,\varepsilon),\varepsilon)$$

and an averaging method has been developed for such differential-delay equations [7]. A rigorous demonstration of the relationship between (1) and (2) or (3) was given in [8] in a more general setting. Thus, some connections can be made between (1) and more familiar ordinary and delay-differential equations for which averaging procedures have been established.

<u>Assumptions</u>. The following assumptions, denoted by H1, H2, etc., are made about the data in equation (1).

H1: (Smoothness of the data)
$f \in C^2([0,\infty) \times [0,\varepsilon_0); E^N)$,
$F \in C^2(E^N \times [0,\varepsilon_0); E^N)$,
$k \in C^1([0,\infty); E^{N\times N})$, and there are positive constants K_0 and b such that $|k(t)| < K_0 e^{-bt}$ where ε_0 is some positive (small) constant.

Here $C^K(X;Y)$ denotes the functions from space X to space Y having K continuous derivatives.

H2: F is 2π periodic in each component of x. In particular, F has a Fourier expansion that is absolutely and uniformly convergent:

$$F(x,\varepsilon) = \sum_{|j|=-\infty}^{\infty} p_j(\varepsilon)\exp[i(j\cdot x)]$$

where $j = (j_1,\dots,j_N)$ is a multi-index of integers, $i = \sqrt{-1}$, $|j| = j_1 + \dots + j_N$ and $j \cdot x = j_1x_1 + \dots + j_Nx_N$.

From H1 we have that for each j, $p_j \in C^2([0,\varepsilon_0);E^N)$. Next, it is assumed that the system is completely resonant:

H3: The forcing function f has the form

$$f(t,\varepsilon) = w_0 + \varepsilon w_1 + O(\varepsilon^2)$$

uniformly for $0 \leqslant t < \infty$. Here w_0 is a vector of *integers*.

Preliminary transformation. We now perform a preliminary change of variables. A new basis for E^N is made up of w_0 and vectors $W_2, \dots, W_N \in E^N$ that have integer components and are orthogonal to w_0. These are taken to be pairwise orthogonal:

$$W_i\cdot W_j = W_i^2\delta_{i,j} \quad \text{and} \quad W_i\cdot w_0 = 0 \quad \text{for} \quad i, j = 1, \dots, N.$$

Here $W_i^2 = W_i\cdot W_i$, etc., and $\delta_{i,j} = 1$ if $i = j$ and is otherwise zero. If W^{tr} denotes the $(N-1)\times N$ matrix whose j-th row is W_j/W_j^2 and $u = \mathrm{col}(u_2,\dots,u_N)$, then

$$x(t) = (v(t)/w_0^2)w_0 + Wu(t)$$

Obviously, $u_j = (W_j \cdot x)/W_j^2$.

Using these projections, we get the new system

$$\begin{aligned} dv/dt &= w_0^2 + \varepsilon\{w_0\cdot w_1 + O(\varepsilon) + \int_0^t w_0\cdot k(t-s)F(x(s),\varepsilon)ds\} \\ v(0) &= w_0\cdot x_0 \\ du/dt &= \varepsilon\{W^{tr}w_1 + O(\varepsilon) + \int_0^t W^{tr}k(t-s)F(x(s),\varepsilon)ds\}^1 \\ u(0) &= W^{tr}x_0. \end{aligned} \tag{4}$$

We define

---

[1] Here and below W^{tr} denotes the transpose of W suitably normalized so $W^{tr}W = I$.

$$\overline{F}(U,\varepsilon) = \sum_{(j\cdot w_0)=0} p_j(\varepsilon)\exp[i\, j \cdot WU].$$

This amounts to averaging the function $F(x)$ with respect to v. We define the remainder by $g(v,u,\varepsilon) = F(x,\varepsilon) - \overline{F}(U,\varepsilon)$. In addition to (4), we consider the averaged equation

$$\text{(5)} \qquad dU/dt = \varepsilon[W^{tr} w_1 + \int_0^t W^{tr} k(t-s)\overline{F}(U(s),0)ds]$$

$$U(0) = W^{tr} x_0.$$

We suppose that there is an exponentially stable (asymptotic) static state for this system:

H4: There exists $U^* \in E^{N-1}$ such that

$$(\int_0^\infty k(t)dt)\overline{F}(U^*,0) = -w_1,$$

and there is a positive constant α such that all roots of the characteristic equation

$$\det[pI_{N-1} - \varepsilon W^{tr}\tilde{k}(p)\overline{F}_U(U^*,0)] = 0$$

satisfy

$$\mathrm{Re}\ p \leqslant -\alpha\varepsilon < 0.$$

Here $\overline{F}_U(U^*,0)$ denotes the Jacobian matrix of F evaluated at $U = U^*$, $\varepsilon = 0$, and $\tilde{k}(p)$ denotes the Laplace transform of k.

With these conditions, we state the main result.

THEOREM. Let conditions H1-H4 be satisfied. Then there are positive constants β and γ such that if

$$|U(0) - U^*| < \beta \quad \text{and} \quad |\varepsilon| < \gamma,$$

then

$$\lim(t \to \infty)x_1(t) : x_2(t) : \ldots : x_N(t) = w_{0,1} : \ldots : w_{0,N}$$

where $w_0 = (w_{0,1}, w_{0,2}, \ldots, w_{0,N})$. (Proof presented in Section 3.)

The averaging result given here is by no means the most general one possible, but it is tailored to treat the example in the next section. A detailed development of averaging for more general integro-differential equations than (1) as well as a proof of this theorem are given in [5]. This theorem is an extension of the study of phase locking for perturbed Hamiltonian systems developed in [9].

The same limiting value of the ratios results if w_1 and the function F are changed slightly in such a way that H1-H4 still hold. In this case, the relative frequencies of the components remain (asymptotically) the same, and the system (1) is said to be in *phase lock*. Thus, all of the oscillators remain synchronized relative to each other, under slight changes in w_1 and F although their actual frequencies might change.

2. A BIMODAL PHASE LOCKED LOOP.

An input signal (say, a voltage) of the form $A_I \cos x_I$ is applied to the circuit. Its amplitude A_I is assumed to stay constant, but the phase $x_I(t)$ will vary, say $x_I(t) = (\nu_0 + \varepsilon_1\nu_1)t$. x_I is to be tracked by the circuit and demodulated. A brief description of the circuit elements follows:

Filter: A filter is denoted by

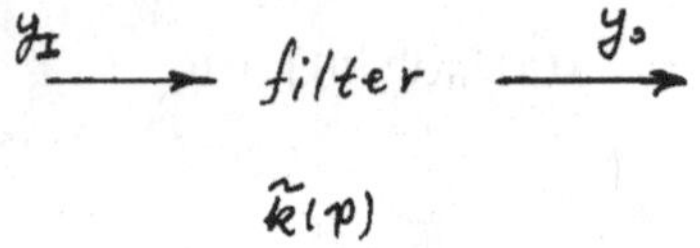

where y_I and y_0 are the input and output signals, respectively, and $\tilde{k}(p)$ is the filter's transfer function. Then the equation

$$y_0 = \tilde{k}(p)y_I$$

describes the filter. Here p is a Laplace transform variable, and this equation has the interpretation that

$$y_0(t) = \int_0^t k(t-s)y_I(s)ds.$$

For example, an RC-filter has $k(t) = (1/RC)\exp[-t/RC]$ and so $\tilde{k}(p) = 1/(RCp + 1)$. Note that if $RC \to 0$, then $k(t) \to \delta(t)$. Thus, in that case, $y_I = y_0$. This follows from applying Laplace's method to the convolution integral.

We suppose that $|k(t)| < K_0 e^{-bt}$ for some positive constants K_0 and b.

Acquisition voltage: The loop dynamics are facilitated by introducing an acquisition voltage, say

$$e(t) = \mu_0 + \varepsilon A_A \cos(\mu_A t)$$

where μ_0 is the direct current bias.

Voltage controlled oscillator: This is the basic element of the circuit. It is described by

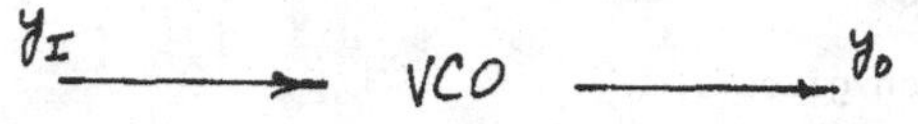

where $y_0 = A_{VCO}\cos(\nu_0 t + \int^t y_I(t')dt')$. A_{VCO} is called the VCO gain and $\nu_0/2\pi$ is its resonant frequency.

Phase detector: A phase detector is denoted by

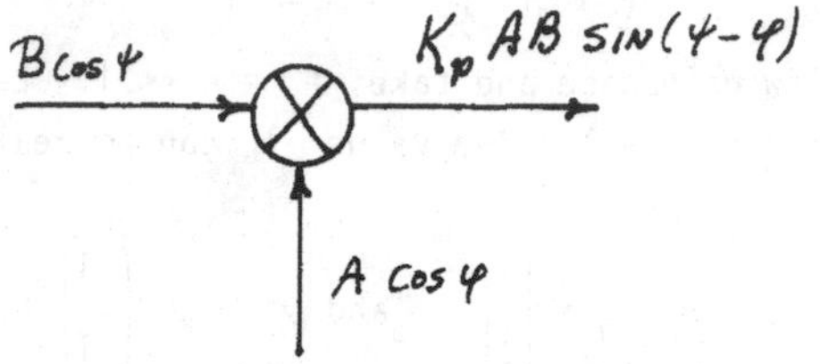

It processes two input signals and puts out a signal $z(t) = K_p AB\sin(\psi-\phi)$ where K_p is the gain and $\psi-\phi$ is the phase deviation or error.

Frequency multiplier: This simply doubles the frequency of an harmonic input signal.

The bimodal phase-locked loop circuit is depicted in Figure 1.

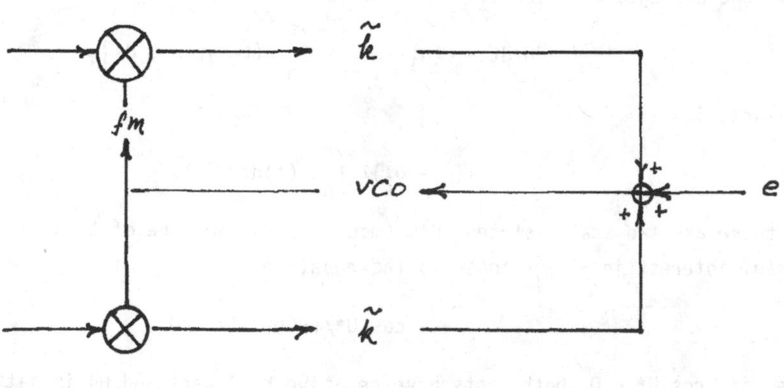

Figure 1: Bimodal Phase-Locked Loop.

A model for this feedback system is easily derived. The VCO (phase) output is

$$x_0 = \nu_0 t + \int_0^t [e(t') + \tilde{k}(p)A_c[\sin(2x_0-x_I) + \sin(x_0-x_I)]]dt'.$$

Differentiating this equation gives

$$dx_0/dt = \nu_0 + \mu_0 + \varepsilon\mu_1(t) + \int_0^t k(t-s)A_c[\sin(2x_0-x_I) + \sin(x_0-x_I)]ds$$

$$dx_I/dt = \nu_1 + \varepsilon\nu_2$$

where the second equation describes the input signal: ν_1 is the input center frequency and $\varepsilon\nu_2$ is noise. This system is of the form (1) where

$$w_0 = \begin{pmatrix} \nu_0 + \mu_0 \\ \nu_1 \end{pmatrix}, \quad w_1 = \begin{pmatrix} \mu_1 \\ \mu_2 \end{pmatrix}, \text{ etc.}$$

We consider the low gain case and take $A_c = \varepsilon << 1$. Let us first tune the circuit so that $\nu_0 + \mu_0 = \mu_1 = 1$. The value 1 can be realized by changing the time scale. Then

$$w_0 = \begin{pmatrix} 1 \\ 1 \end{pmatrix} \quad \text{and} \quad W_2 = \begin{pmatrix} 1 \\ -1 \end{pmatrix}$$

and $v(t) = x_0 + x_I$ and $u(t) = x_0 - x_I$. The equation becomes

$$dv/dt = 2 + \varepsilon[\mu_1 + \nu_1 + \int_0^t k(t-s)[\sin(\tfrac{3u}{2} + v) + \sin u]ds$$

$$du/dt = \varepsilon[\mu_1 - \nu_2 + \int_0^t k(t-s)[\sin(\tfrac{3u}{2} + v) + \sin u]ds$$

and so the averaged equation is

(5) $$dU/dt = \varepsilon[\mu_1 - \nu_2 + \int_0^t k(t-s)\sin U\, ds].$$

Therefore, if

(6) $$|(\mu_1 - \nu_2)/\int_0^\infty k(t)dt| < 1,$$

then there are two static states U^* (modulo 2π), and one of them is stable. The roots of interest in H4 are those to the equation

$$p - \varepsilon \cos U^*/(RCp + 1) = 0.$$

Hence, if $\cos U^* < 0$ both roots have negative real part and H4 is satisfied.

Therefore, $x_0 : x_I \to 1:1$ as $t \to \infty$, no matter what μ_1, ν_2 and k are as long as (6) is satisfied. The system is in 1:1 phase lock.

Next, let us increase the acquisition voltage bias so that

$$(\nu_0 + \mu_0) = 1 = \nu_1$$

Now,

$$w_0 = \begin{pmatrix} 1 \\ 2 \end{pmatrix}, \quad W_2 = \begin{pmatrix} 2 \\ -1 \end{pmatrix},$$

and $v(t) = x_0 + 2x_I$ and $u(t) = 2x_0 - x_I$. The averaged equation is again (5). Thus, if (6) is satisfied,

$$\lim(t \to \infty) x_0 : x_I = 1:2,$$

and the system is 1:2 phase lock.

This model reduces to the one in [9] when $\tilde{k}(p) = 1$ (no filter). That model was derived to simulate rhythm splitting behavior in the biological rhythms of smal mammals.

3. PROOF OF THE THEOREM.

First, we set $v = t + y$. Then

$$dy/dt = \varepsilon[w_0 \cdot w_1 + w_0 O(\varepsilon) + \int_0^t w_0 \cdot k(t-s) F \, ds]/w_0^2$$

$$du/dt = \varepsilon[W^{tr} w_1 + W^{tr} O(\varepsilon) + \int_0^t W^{tr} k(t-s) F \, ds].$$

We will show that if ε is sufficiently small, then $|u(t)|$ remains bounded uniformly for $0 \leqslant t < \infty$.

Let $z = u - U$, $K(t) = W^{tr} k(t)$, and define

$$H(z,U) = \overline{F}(z+U,0) - \overline{F}(U,0) - \overline{F}_U(U,0)z.$$

It follows that $H(z,U) = O(|z|^2)$ uniformly for $U \in E^{N-1}$. Then

$$\text{(7)} \qquad dz/dt - \varepsilon \int_0^t K(t-s) F_U(U^*,0) z(s) ds = O(\varepsilon^2) + $$
$$+ \varepsilon \int_0^t K(t-s)\{[F_U(U,0) - F_U(U^*,0)]z + H(U,z(s))\} ds$$
$$+ \varepsilon \int_0^t K(t-s) g(s+y, z + U, \varepsilon) ds.$$

Now consider the linear equation

$$dY/dt - \varepsilon \int_0^t K(t-s) F_U(U^*,0) Y(s) ds = h(t), \ Y(0) \text{ given.}$$

Because of hypothesis H4,

$$Y(t) = R_0(t) + \int_0^t R(t-s)h(s)ds$$

where $R_0(0) = Y(0)$, $|R_0(t)| < C_1 e^{-\varepsilon t\alpha}$, and $|R(t)| < C_2 e^{-\varepsilon t\alpha}$ for $0 \leq t < \infty$.

Applying this calculation to (7), we have

$$(8)\quad z(t) = O(\varepsilon) + \varepsilon \int_0^t L(t-s)[\overline{F}_U(U,0) - \overline{F}_U(U^*,0)]ds + \varepsilon \int_0^t L(t-s)H(z,U)ds + \varepsilon G(t,\varepsilon)$$

where $L(t) = \int_0^t R(t-s)K(s)ds$ and $G(t,\varepsilon) = \int_0^t L(t-s)g(s+y,\ z+U,\varepsilon)ds.$

It follows from H1 and H4 that $|L(t)| < C_3 e^{-\varepsilon\alpha t}$ for $0 \leq t < \infty$. Estimates of the remaining terms follow from the next two lemmas.

LEMMA 1: There is a constant C_4 such that

$$|G(t,\varepsilon)| < C_4$$

for $0 \leq t < \infty$ and all sufficiently small ε.

Proof of Lemma 1. We first observe that $L(t) = e^{-\varepsilon\alpha t} L^*(t)$ where L^* is bounded and $dL^*/dt = O(\varepsilon)$ uniformly for $0 \leq t < \infty$. Then

$$G(t,\varepsilon) = \sum_j e^{-\varepsilon\alpha t} \int_0^t e^{(ij\cdot w_0+\alpha\varepsilon)s} L^*(t-s)p_j(\varepsilon)\exp\{i(y\, j\cdot w_0 + j\cdot WU\}ds.$$

Integrating these integrals by parts where the first exponential is integrated and the remaining terms $L^*(t-s)p_j\ldots$ are differentiated, establishes the result.

LEMMA 2. If $|U(0) - U^*|$ is sufficiently small, then

$$|U(t) - U^*| < C_0\, e^{-a_0\varepsilon t}$$

where C_0 and a_0 are some positive constants.

The proof of Lemma 2 follows from the stability hypotheses H1 and H4 and Gronwall's inequality [10].

Using these results, we have from (8) that

$$e^{\varepsilon\alpha t}|z(t)| \leq C\varepsilon\, e^{\varepsilon\alpha t} + \varepsilon \int_0^t C_3C_0\, e^{\varepsilon(a_2-a_0)s} ds + \varepsilon\, C_3\, \mu \int_0^t e^{\varepsilon a_2 s}|z(s)|ds$$

provided that $z(t)$ is small, say $|z(t)| \leq \delta(\mu)$. It follows from Liapunov's argument [10, Ch. 13] that $z(t)$ is uniformly bounded for $0 \leq t < \infty$.

Thus, $|u| \leq |u - U| + |U| < M_1$ for $0 \leq t < \infty$ and some positive constant M_1. Since $t + y \to \infty$ as $t \to \infty$, we have that

$$x(t) = (y + t)w_0 + Wu$$

implies that

$$x(t)/(y + t) \to w_0 \quad \text{as} \quad t \to \infty.$$

This completes the proof of the theorem.

BIBLIOGRAPHY

1. W. C. Linsey, Synchronization systems in communications and control, Pretice-Hall, Engelwood, 1972.

2. A. T. Winfree, the geometry of biological time, Springer-Verlag, 1980.

3. C. Rowsesmitt, et. al., Photo-periodic induction of diurnal locomotor activity in Microtus montanus, the montane vole, Canad. J. Zoo., 60(1982), 2798-2803.

4. J. P. Keener, On cardiac arrythmias: AV conduction block (preprint).

5. F. C. Hoppensteadt, Singular perturbation methods for systems of Volterra integro-differential equations, in preparation.

6. V. I. Arnold, Chaiptres supplementaires de la theorie des equations differentielles ordinaires, Editions MIR, Moscow, 1978.

7. Jack Hale, Theory of functional differential equations, Springer-Verlag, New York, 1976.

8. F. C. Hoppensteadt, An algorithm for approximate solutions to weakly filtered synchronous control systems and nonlinear renewal processes, SIAM J. Appl. Math. (in press).

9. F. C. Hoppensteadt, J. P. Keener, Phase locking in biological clocks, J. Math. Biology, 15(1982), 339-349.

10. E. A. Coddington, N. Levinson, Theory of Ordinary Differential Equations, McGraw-Hill, New York, 1955.

THE ASYMPTOTIC ANALYSIS OF BOUNDARY VALUE
PROBLEMS BY THE METHOD OF ORDER REDUCTION

F. A. Howes
Department of Mathematics
University of California, Davis
Davis, CA. 95616
U.S.A.

1. INTRODUCTION

Consider the simple singularly perturbed boundary value problem

$$(E_1)\qquad \begin{aligned} &\varepsilon y'' = -y'+y,\ 0 < x < 1, \\ &y(0,\varepsilon) = \alpha,\ y(1,\varepsilon) = \beta, \end{aligned}$$

where $\varepsilon > 0$ is a small parameter. Although the exact solution of (E_1) is readily available, it is perhaps more instructive to proceed as follows. First of all, since ε is small, we begin by setting ε equal to zero and solving the reduced problem $u' = u$, $u(1) = \beta$. The solution $u = u(x) = \beta e^{x-1}$ is our candidate for the limit of the solution $y(x,\varepsilon)$ of (E_1) in the sense that

$$\lim_{\varepsilon \to 0^+} y(x,\varepsilon) = \beta e^{x-1}$$

in "most" of $[0,1]$. (Note that the solution of theother reduced problem $u' = u$, $u(0) = \alpha$, is <u>not</u> the limit of $y(x,\varepsilon)$. This follows from various geometric or analytic conditions; cf. [2; Chapter 2] and [3,5].) Since $u(0) = \beta e^{-1} \neq \alpha$, in general, the function u must be supplemented near $x = 0$ with a boundary layer corrector term that makes up for this deficiency. Obviously the determination of the boundary layer corrector must involve the second-order derivative in (E_1) since the neglect of this term led to the reduced solution u. Once again, there are several analytic techniques (cf. [2; Chapter 2] and [3,5]) for showing that the corrector $Y(x,\varepsilon)$ is the solution of the "boundary layer" problem

$$\begin{gathered} \varepsilon y'' = -y', \\ y(0,\varepsilon) = \alpha-\beta e^{-1},\ \lim_{\xi \to \infty} y(\xi,\varepsilon) = 0\ (\xi = x/\varepsilon), \end{gathered}$$

that is, $Y(x,\varepsilon) = (\alpha-\beta e^{-1})e^{-x/\varepsilon}$. Combining our results, we have that the solution of (E_1) as $\varepsilon \to 0^+$ is represented in $[0,1]$ by

(1)
$$\begin{aligned} y(x,\varepsilon) &= u(x) + Y(x,\varepsilon) + \mathcal{O}(\varepsilon) \\ &= \beta e^{x-1} + (\alpha - \beta e^{-1})e^{-x/\varepsilon} + \mathcal{O}(\varepsilon), \end{aligned}$$

and so

(2)
$$\lim_{\varepsilon \to 0^+} y(x,\varepsilon) = u(x) \quad \text{for} \quad 0 < \delta \leq x \leq 1$$

$(0 < \delta < 1)$.

Either from our arguments above or by examining the exact solution we see that the solution of (E_1) is approximated in $[0,1]$ (with the exception of an asymptotically small neighborhood of $x = 0$) by the solution of an even simpler first-order problem. We can now reconsider (E_1) in this light and attempt what may be called an "asymptotic reduction of order". The idea is to replace the differential equation in (E_1) with the approximate equation $\varepsilon y'' = -y' + u(x)$, integrate once and so obtain the first-order initial value problem

(E_1')
$$\begin{aligned} &\varepsilon z' = -z + \beta e^{x-1}, \quad 0 < x \leq 1, \\ &z(0,\varepsilon) = \alpha. \end{aligned}$$

Note that the reduced equation obtained by setting ε equal to zero in (E_1') has the solution $z \doteq \beta e^{x-1}$ which satisfies $z(1) = \beta$, and so the right-hand boundary condition in (E_1) is not really lost in passing to (E_1'). Now (E_1') is a first-order problem, and hence, it is easier to solve than the original second-order one. Its solution clearly satisfies in $[0,1]$

$$z(x,\varepsilon) = \beta e^{x-1} + (\alpha - \beta e^{-1})e^{-x/\varepsilon} + \mathcal{O}(\varepsilon),$$

in agreement with the previous relation (1). Thus the technique of replacing a second-order problem with an asymptotically equivalent first-order one affords another means of attacking various classes of singular perturbation problems.

2. AN APPLICATION

As an illustration of these remarks, let us consider the quasilinear problem (cf. [1] and [3,5])

(E_2)
$$\begin{aligned} &\varepsilon y'' = f(y)y' + g(y), \quad 0 < x < 1, \\ &y(0,\varepsilon) = \alpha, \; y(1,\varepsilon) = \beta, \end{aligned}$$

in which f and g are continuously differentiable functions, and let us assume that the reduced problem

$$f(u)u' + g(u) = 0, \quad 0 < x < 1,$$
$$u(1) = \beta,$$

has a twice continuously differentiable solution $u = u(x)$. Then the theory tells us that if there exists a positive constant k such that

(3) $$f(u(x)) \leq -k < 0 \text{ in } [0,1]$$

and

(4) $$f(\lambda) \leq -k < 0 \text{ for } \lambda \text{ between } u(0) \text{ and } \alpha,$$

the problem (E_2) has a solution $y = y(x,\epsilon)$ as $\epsilon \to 0^+$ which satisfies in $[0,1]$ (cf. (1))

$$y(x,\epsilon) = u(x) + (\alpha - u(0))e^{-kx/\epsilon} + \mathcal{O}(\epsilon).$$

Therefore, the limiting relation (2) for the simple linear problem (E_1) is seen to obtain for the general nonlinear problem (E_2) under the conditions (3) and (4). It is perhaps helpful to mention that the condition (4) may be replaced with the less restrictive integral condition (cf. [1], [3])

(5) $$(u(0)-\alpha)\int_{\eta}^{u(0)} f(s)ds < 0$$

for all values of η strictly between $u(0)$ and α.

We can recapture these results by passing from (E_2) to the asymptotically equivalent initial value problem

(E_2') $$\epsilon z' = F(z) - F(u(x)) = H(x,z), \quad 0 < x \leq 1,$$
$$z(0,\epsilon) = \alpha,$$

where F is an antiderivative of f. Note that $H(x,u(x)) \equiv 0$ and that $(\partial H/\partial z)(x,u(x)) \leq -k < 0$ in $[0,1]$ by virtue of condition (3). Consequently, the theory of singularly perturbed initial value problems (cf. [6] and [4]) allows us to assert that under conditions (3) and (4) (or (5)) the solution $z = z(x,\epsilon)$ of (E_2') satisfies

$$\lim_{\epsilon \to 0^+} z(x,\epsilon) = u(x) \text{ for } 0 < \delta \leq x \leq 1.$$

In view of the asymptotic equivalence of (E_2) and (E_2'), we can assert that the solution of (E_2) satisfies this same limiting relation. Note that in terms of

(E_2') the integral condition (5) is translated into the requirement that

$$(u(0)-\alpha)\,H(0,\lambda) > 0$$

for all values of λ strictly between $u(0)$ and α.

3. HIGHER ORDER REDUCTION

The suggestions given above for solving second-order perturbed problems apply mutatis mutandis to scalar problems of order higher than two and to systems. Let us illustrate this by considering the third-order problem

$$(E_3)\qquad \begin{aligned} &\varepsilon^2 y''' = f(y)y' + g(y), \quad a<x<b,\\ &y(a,\varepsilon)=\alpha,\ y'(a,\varepsilon)=\gamma,\ y(b,\varepsilon)=\beta, \end{aligned}$$

in which f and g are twice continuously differentiable functions. Problems of this type arise, among other places, in the study of travelling wave solutions of dispersive equations such as the Korteweg-de Vries equation

$$(E_4)\qquad v_t + vv_x = -\varepsilon^2 v_{xxx}$$

(cf. for instance [8]). (To see this, simply make the change of variable $\xi = x-ct$ which converts (E_4) to an ordinary differential equation in ξ.) In order to solve (E_3) asymptotically as $\varepsilon \to 0^+$ we proceed as above by first setting ε equal to zero and solving the reduced problem

$$\begin{aligned} &f(u)u' + g(u) = 0, \quad a<x<b,\\ &u(a)=\alpha. \end{aligned}$$

We then use this solution $u=u(x)$ to replace the third-order problem with the asymptotically equivalent <u>second</u>-order one

$$(E_3')\qquad \begin{aligned} &\varepsilon^2 z'' = F(z)-F(u(x)) = H(x,z), \ a<x<b,\\ &z'(a,\varepsilon)=\gamma,\ z(b,\varepsilon)=\beta, \end{aligned}$$

where F is an antiderivative of f. As before, we have that $H(x,u(x)) \equiv 0$, and so the boundary condition at $x=a$ is not lost in passing from (E_3) to (E_3'). The problem (E_3') is second-order and fortunately, there is a great deal known about such problems (cf. [5]). For instance, if there exists a positive constant m such that

$$(\partial H/\partial z)(x,u(x)) = f(u(x)) \geq m^2 > 0 \text{ in } [a,b]$$

and

$$f(\lambda) \geq m^2 > 0$$

for all values of λ between $u(b)$ and β, then it is known that the problem (E_3') has a solution $z = z(x,\varepsilon)$ as $\varepsilon \to 0^+$ which satisfies in $[a,b]$

$$z(x,\varepsilon) = u(x) + (\varepsilon/m)(u'(a)-\gamma)e^{-m(x-a)/\varepsilon} + (\beta-u(b))e^{-m(b-x)/\varepsilon} + \mathcal{O}(\varepsilon^2). \tag{6}$$

Consequently, the original problem (E_3) has a solution $y = y(x,\varepsilon)$ as $\varepsilon \to 0^+$ satisfying (6) as well.

We conclude by noting that the equation (E_4) may also possess so-called solitary wave solutions which represent a transition from one limiting state to another (cf. [8]); this means that we seek solutions of (E_4) satisfying $\lim_{\xi \to -\infty} v(\xi,\varepsilon) = v_-$ and $\lim_{\xi \to +\infty} v(\xi,\varepsilon) = v_+$ for appropriate constants v_-, v_+ and $\xi = x - ct$. In terms of the problem (E_3) a solitary wave is translated into a solution $y = y(x,\varepsilon)$ which displays shock layer behavior as $\varepsilon \to 0^+$, that is,

$$\lim_{\varepsilon \to 0^+} y(x,\varepsilon) = \begin{cases} u_-(x), & a \leq x \leq x_0 - \delta, \\ u_+(x), & x_0 + \delta \leq x \leq b \end{cases}$$

$(0 < \delta < b-a)$. Here u_- and u_+ $(u_- \neq u_+)$ are solutions of the reduced equation $f(u)u' + g(u) = 0$ which satisfy $u_-(a) = \alpha$ and $u_+(b) = \beta$. The location of the transition point x_0 is determined by passing to the asymptotically equivalent problem (E_3') and requiring that $J[x_0] = 0$ and $J'[x_0] \neq 0$ for $J[x] = \int_{u_-(x)}^{u_+(x)} H(x,s)ds$ (cf. [3]). Finally, if the function H is independent of x (whence, $J[x] \equiv \text{const.}$), then we can apply the theory of O'Malley [7] to the autonomous problem (E_3') and thereby study solutions of (E_4) (and related problems) which exhibit sharp spike-like behavior.

In conclusion, let us hasten to point out that our results are formal, in the sense that having obtained an approximate solution, we must also verify that it is indeed a valid approximation. This can be accomplished by means of differential inequality and/or integral equation techniques. The construction of the a'priori approximation is themore difficult part of the analysis, and for this reason, it is the aspect of the problem we have addressed here.

ACKNOWLEDGMENT

The author wishes to thank the typist, Mrs. Karen E. Bigbee, for her assistance in preparing this camera-ready copy.

REFERENCES

1. E.A. Coddington and N. Levinson, A Boundary Value Problem for a Nonlinear Differential Equation with a Small Parameter, Proc. Amer. Math. Soc. 3(1952), 73-81.

2. J.D. Cole, Perturbation Methods in Applied Mathematics, Ginn-Blaisdell, Waltham, Mass., 1968.

3. F.A. Howes, Boundary-Interior Layer Interactions in Nonlinear Singular Perturbation Theory, Memoirs Amer. Math. Soc., vol. 203, 1978.

4. __________, An Improved Boundary Layer Estimate for a Singularly Perturbed Initial Value Problem, Math. Z. 165(1979), 135-142.

5. __________, Some Old and New Results on Singularly Perturbed Nonlinear Boundary Value Problems, in Singular Perturbations and Asymptotics, ed. by R.E. Meyer and S.V. Parter, Academic Press, New York, 1980, pp. 41-85.

6. M. Nagumo, Über das Verhalten der Integrale von $\lambda y''+f(x,y,y',\lambda)=0$ für $\lambda \to 0$, Proc. Phys. Math. Soc. Japan 27(1939), 529-534.

7. R.E. O'Malley, Jr., Phase-Plane Solutions to Some Singular Perturbation Problems, J. Math. Anal. Appl. 54(1976), 449-466.

8. A.C. Scott, F.Y.F. Chu and D.W. McLaughlin, The Soliton: A New Concept in Applied Science, Proc. IEEE 61(1973), 1443-1483.

AN OBSTRUCTION APPROACH TO MULTIPARAMETER HOPF BIFURCATION

Jorge IZE

IIMAS - UNAM
A.P. 20-726
MEXICO 20, D.F.

Several topological methods have been used in the study of the Hopf bifurcation [A.Y], [I.1], [C.M.Y.]. The basic idea of this paper is to consider the existence of a periodic solution to an autonomous differential equation as an obstruction to the construction of an extension for equivariant mappings.

In the Hopf bifurcation problem one looks for periodic solutions to the equation

$$(*)\quad \frac{dX}{dt} = L(\mu)X + g(\mu,X) \equiv f(\mu,X)$$

where $X \in \mathbb{R}^M$, $\mu \in \mathbb{R}^{k-1}$, $k \geq 2$, $L(\mu)$ is a continuous family of M×M matrices, g is C^1 in X and continuous in μ.

Furthermore the following hypotheses hold

H.1 : $g(\mu,0)=0$, $g(\mu,X)=o(X)$ as X tends to 0, $L(0)$ is invertible.

H.2 : there is a $\beta > 0$ and integers $1=m_1 < m_2 < \cdots < m_\ell$ such that $\pm i\beta, \pm im_2\beta, \ldots, \pm im_\ell\beta$ belong to the spectrum of $L(0)$ and no other eigenvalue of $L(0)$ has the form $m\beta$, m an integer.

H.3 : Let d_j be the algebraic multiplicity of $im_j\beta$ and let
$\alpha_{j,k_0}(\mu)+i(m_j\beta + \gamma_{j,k_0}(\mu))$
with $\alpha_{j,k_0}(0)= \gamma_{j,k_0}(0)= 0$, $k_0=1,\ldots,d_j; j=1,\ldots,\ell$,
be the corresponding eigenvalues of $L(\mu)$. Assume that $\alpha_{j,k_0}(\mu)\neq 0$ for μ small, $\mu \neq 0$.

For the case of a one parameter problem (k=2) let $n_j^{\pm}$ be the number of eigenvalues crossing from left to right (respectively from right to left) the imaginary axis at $im_j\beta$ as μ passes through 0. Set $n_j=n_j^+-n_j^-$.

If k=2 and $|\Sigma n_j|$ is odd, Alexander and Yorke proved that there is a continuum of periodic solutions, periods and values of μ, starting from $(0,2\pi/\beta,0)$ which is either unbounded or going to a different stationary point. Their proof used cohomology theory. Another proof was given in [I.1] using homotopy arguments and their result was later improved by Chow, Mallet-Paret and Yorke, using the Fuller's degree; it is enough to have $\Sigma\, n_j/m_j \neq 0$ to have the same alternative.

The present paper is an attempt to give another proof to this last result, treating it as an example of an equivariant problem.

Theorem 1) If $2 \leq k \leq 2M$, there are integers $n_j(k)$, $j=1,\ldots,\ell$, such that if $n_j(k) \neq 0$ then there is a continuum in $C^0(\mathbb{R}^M) \times \mathbb{R}^k \times \mathbb{R}^+$ of periodic solutions of period T branching off $(0,0,2\pi/m_j\beta)$ which is either unbounded or goes to another stationary point. $n_j(k)$ is the class of $im_j\nu I - L(\mu)$, for $\|\mu\|+|\nu-\beta|=\rho$ small, in the stable group $\Pi_{k-1}(GL(\mathbb{C}^M))$. This group is Z if k is even, 0 if k is odd (Bott's periodicity theorem) and $n_j(2)$ is the above n_j.

2) If the continuum is bounded and each stationary point on it (X_p,μ_p,T_p), p in a finite set I, is such that the linearization of $f(\mu,X)$ at (μ_p,X_p) has the form of $(*)$ with $L(\mu)$ replaced by $L_p(\mu)$ and satisfies H_1,H_2,H_3, then the following formula holds

$$0= \Sigma_I \text{ Sign det } L_p(\mu_p)\Sigma_{j=1,\ldots,\ell_p}\, n_j(k,p)/(m_j(p))^{k/2}$$

where $n_j(k,p)$, $m_j(p)$ are the corresponding integers for (X_p,μ_p) and all eigenvalues of $L_p(\mu_p)$ of the form $2\pi m_j(p)/T_p$.

In next section the local bifurcation will be proved and in the following part the global bifurcation will be considered. Due to the length of the proof and to space limitations some parts of the proof will be sketchy, a detailed proof will be published elsewhere. However the main ideas will be presented here.

This research was done while the author was visiting the IMSP, Université de Nice, during the winter semester 1981-82.

I) LOCAL BIFURCATION

I.1) The bifurcation equations

The bifurcation problem will be studied in the space of 2π- periodic functions where the group action is explicitely given in terms of Fourier series.

Fix the period to 2π by leting $\tau=\nu t$, so that $(*)$ becomes

$(*)_\nu$ $\quad \nu\,\dot{X}= L(\mu)X + g(\mu,X)$

2π periodic solutions of $(*)_\nu$ are $2\pi/\nu$ periodic solutions of $(*)$. Secondly the reals mod 2π are identified to S^1 and the Sobolev spaces $W^{\varepsilon,2}(S^1)$, $\varepsilon=0,1$, are identified with the spaces of Fourier series $\Sigma_{-\infty}^{\infty}\, x_n\, e^{in\tau}$, with $x_{-n}=\bar{x}_n$ and $\Sigma\, (1+\varepsilon\, n^2)|x_n|^2 < \infty$.

$(*)_\nu$ defines a C^1 map from $\mathbb{R}^k\times W^{1,2}(S^1)^M$ into $L^2(S^1)^M$, $\nu\,\dot{X} - L(\mu)X$ is a Fredholm operator of index 0 and $(*)_\nu$ is equivalent to the system

$$(*)_n \quad (i\,\nu\,n\,I - L(\mu))X_n - g_n(\mu, X_0, X_1, \ldots) = 0 \qquad n \geqslant 0$$

where $X(\tau) = \Sigma\ X_n\ e^{in\tau}$, $X_n \in \mathbb{C}^M$, $X_{-n} = \bar{X}_n$ and $g_n(\mu, X_0, \ldots)$ is the Fourier coefficient of $g(\mu, X(\tau))$.

For ν close to β and μ close to 0, the linear part of $(*)_n$ will be singular only for $\nu=\beta$, $\mu=0$, $n=m_1,\ldots,m_\ell$, so that one can reduce locally the system to a finite dimensional bifurcation equation: if $X(\tau)$ is written as

$$X(\tau) = \Sigma_{j=-\ell}^{\ell}\ X_j\ e^{im_j\tau} + Y(\tau)$$

with $m_{-j} = -m_j$, $X_{-j} = \bar{X}_j$, $X_0 = 0$ (note the change in the indices), then $Y(\tau)$ satisfies $(*)_n$ for $n \neq m_1,\ldots,m_\ell$ which has a unique solution $Y(X_1,\ldots,X_\ell,\mu,\nu,\tau)$ which is C^1 in $X_1,\ldots,X_\ell$, continuous in (μ,ν) and

$$\|Y\|_{W^{1,2}(S^1)} = o\ (\ \|(X_1,\ldots,X_\ell)\|\ ).$$

The bifurcation equation

$$(*)_b \quad (i\,\nu\,m_j I - L(\mu))\ X_j - g_j\ (\mu,\nu,X_1,\ldots,X_\ell) = 0 \quad j = 1,\ldots,\ell$$

or else $B_j(\lambda)\ X_j - g_j(\lambda, X) = 0$

with $\lambda = (\mu, \nu-\beta)$ in $\mathbb{R}^k$, has the following properties

a) $B_j(\lambda)$ belong to $GL(\mathbb{C}^M)$ for $\lambda \neq 0$ small.

b) $g_j(\lambda, X) = o\ (\|X\|)$ for X small

c) for each ρ small enough, there is an r such that $(*)_b$ has no solution with $\|\lambda\|=\rho$, $0 < \|X\| \leqslant r$

d) $g_j(\lambda, e^{i\phi}X_1, e^{im_2\phi}X_2, \ldots, e^{im_\ell\phi}X_\ell) = e^{im_j\phi} g_j(\lambda, X_1, \ldots, X_\ell)$

this comes from the fact that g_j is a Fourier coefficient and that, from the uniqueness, $Y(e^{i\phi}X_1,\ldots,e^{im_\ell\phi}X_\ell,\mu,\nu,\tau) = Y(X_1,\ldots,X_\ell,\mu,\nu,\tau+\phi)$.

Choosing ρ and r as in property (c), one wishes to find solutions to $(*)_b$ in the set $B^k \times S^{2\ell M-1}$ of (λ, X) with $\|\lambda\| \leqslant \rho, \|X\| = r$, knowing that $(*)_b$ maps $S^{k-1} \times S^{2\ell M-1}$ into $C^{\ell M} - \{0\}$. From the topological point of view, one would like a negative answer to the stronger question: does there exist an equivariant extension of $(*)_b$ from $S^{k-1} \times S^{2\ell M-1}$ to $B^k \times S^{2\ell M-1}$ into $\mathbb{C}^{\ell M} - \{0\}$?

Now continuous (real) extensions are treated using obstruction theory; in order to take account of the action of the group, that is of the phase translation along orbits, it is necessary to look at the graph map

$$S^{k-1} \times S^{2\ell M-1} \longrightarrow S^{k-1} \times S^{2\ell M-1} \times S^{2\ell M-1} \quad \text{given by}$$

$$(\lambda, X) \longrightarrow (\lambda, X, (*)_b (\lambda, X))$$

where $(*)_b$ has been normalized. Then if one defines the group action as $e^{i\phi}X \equiv (e^{i\phi} X_1, \ldots, e^{im_\ell\phi} X_\ell)$ it is clear that $(\lambda, e^{i\phi}X)$ is sent into $(\lambda, e^{i\phi}X, e^{i\phi}(*)_b(\lambda,X))$ and that it is enough to know the value of this map at one point of the orbit. This implies that, taking the quotient on both sides by the action, one will get a continuous map and conversely. However, due to the fact that the action is not free, the quotient space $S^{2\ell M-1}/S^1$ is a stratified set and obstruction theory is not readily applicable, so that the free action case will be first studied ($\ell=1$) and then the general case will be reduced to the free case by a trick which will partly explain the change in the Fuller's degree.

(.2) The free action case.

The quotient of S^{2M-1} by the standard action is the complex projective space $\mathbb{C}P^{M-1}$ and the quotient of $S^{2M-1} \times S^{2M-1}$ by the free action $e^{i\phi}(X,Y)=(e^{i\phi}X, e^{i\phi}Y)$ is denoted by $S^{2M-1} \times_{S^1} S^{2M-1}$. Now if one considers the sphere bundle

$$S^{2M-1} \longrightarrow S^{k-1} \times S^{2M-1} \times_{S^1} S^{2M-1} \xrightarrow{P} S^{k-1} \times \mathbb{C}P^{M-1}$$

with unitary bundle group and $P(\lambda,[X,Y]) = (\lambda,[X])$ it is easy to see that cross-sections of this bundle, i. e. maps $\psi(\lambda,[X])=(\lambda,[X,F(\lambda,X)])$, are in one to one correspondance with equivariant maps $F(\lambda,X)$ from $S^{k-1} \times S^{2M-1}$ into S^{2M-1} (see [Br. chap. II, 2.6]). Then one has to extend the map induced by $(*)_b$ to a continuous cross-section of the bundle with S^{k-1} replaced by B^k, a situation in which one may apply obstruction theory (see [St] and [Hu] for the basic facts about obstruction theory).

Eventual obstructions will be elements of the cohomology groups $H^{q+1}(B^k \times \mathbb{C}P^{M-1}, S^{k-1} \times \mathbb{C}P^{M-1}; \Pi_q(S^{2M-1}))$. Now $\mathbb{C}P^{M-1}$ has the following cell structure: $\mathbb{C}P^{M-1} = B^0 \cup B^2 \cup \ldots \cup B^{2M-2} = \mathbb{C}P^{M-2} \cup B^{2M-2}$ where the cell $B^{2M-2} = \{(x_1,\ldots,x_{M-1}) \in \mathbb{C}^{M-1} / \Sigma |x_i|^2 \leq r^2\}$ is sent into $[x_1,\ldots,x_{M-1}, \sqrt{(r^2 - \Sigma |x_i|^2)}]$ (see [Sp. p 146]). This implies that cells in $B^k \times \mathbb{C}P^{M-1} - S^{k-1} \times \mathbb{C}P^{M-1}$ are of the form $B^k \times B^{2P}$, $p=0,\ldots,M-1$, and so the only non-trivial cohomology groups are $H^{k+2P}(B^k \times \mathbb{C}P^{M-1}, S^{k-1} \times \mathbb{C}P^{M-1}) = Z$.

The primary obstruction, for $2 \leq k \leq 2M$, will be for $p=(2M-k)/2$ if k is even or for $p=(2M+1-k)/2$ if k is odd. However the primary obstruction is unique and is the same for maps which are homotopic on

$S^{k-1} \times \mathbb{C}P^{M-1}$, i. e. for equivariant homotopic maps from $S^{k-1} \times S^{2M-1}$ into $\mathbb{C}^M-\{o\}$. Thus $(*)_b$ and $B(\lambda)X$ will have the same primary equivariant obstruction, as well as $D(\lambda)X$ for any $D(\lambda)$ in the class of $B(\lambda)$ in $\Pi_{k-1}(GL(\mathbb{C}^M))$.

<u>Theorem</u> (Bott) If $k \leqslant 2M$, $\Pi_{k-1}(GL(\mathbb{C}^M))$ is 0 if k is odd and Z if k is even. If k is odd any matrix $B(\lambda)$ is deformable to I and if k is even there is an integer, degree $B(\lambda)$, such that $B(\lambda)$ is homotopic to I if and only if this degree is 0. In fact from the exact sequence

$$\Pi_k(S^{2M-1}) \xrightarrow{\delta} \Pi_{k-1}(U_{M-1}) \xrightarrow{i_*} \Pi_{k-1}(U_M) \xrightarrow{P_*} \Pi_{k-1}(S^{2M-1})$$

for the fiber bundle of unitary groups, i_* is an isomorphism for $k < 2M-1$ and, for $k = 2s \leqslant 2M$, $B(\lambda)$ is deformable to

$$\begin{pmatrix} I & 0 \\ 0 & C(\lambda) \end{pmatrix}$$

with $C(\lambda)$ in $GL(\mathbb{C}^s)$. P_* sends the generator of $\Pi_{2s-1}(U_s)$ into an element of degree $(s-1)!$ in $\Pi_{2s-1}(S^{2s-1})$ and degree $B(\lambda) \equiv (-1)^{s-1} \times$ degree $P_* C(\lambda)/(s-1)!$ [At.]. If $k = 2$, $B(\lambda)$ is deformable to $D(\lambda)$ if and only if $\det B(\lambda)$ is deformable to $\det D(\lambda)$ as mappings from S^1 into $\mathbb{C}-\{o\}$ and degree $B(\lambda)$ = degree $\det B(\lambda)$.

In order to compute the obstruction, let $X = \tilde{X} \oplus X_o$ with $\tilde{X}$ in $\mathbb{C}^p$, $p = M-s$, $k = 2s \leqslant 2M$ (if k is odd, $(*)_b$ is deformable to X which is clearly extendable) and consider $\mathbb{C}P^{M-1} = \mathbb{C}P^{p-1} \cup B^{2p} \cup \ldots \cup B^{2M-2}$ with $\mathbb{C}P^{p-1} = \{ [\tilde{X}, 0, \ldots, 0] \} \subset \mathbb{C}P^{M-1}$. Then the section $(\lambda, [\tilde{X}, X_o]) \longrightarrow (\lambda, [\tilde{X}, X_o, \tilde{X}, C(\lambda)X_o])$ is already defined on the $k+2p-1$ skeleton of $B^k \times \mathbb{C}P^{M-1}_o$ and on the single cell $B^k \times B^{2p}$, attached by $\{(\lambda, \tilde{X}) \,/\, \|\lambda\| \leqslant \rho,\ \|\tilde{X}\| \leqslant r\} \longrightarrow (\lambda, [\tilde{X}, (r^2 - \|\tilde{X}\|^2)^{1/2}, 0, \ldots, 0])$, the section looks like $(\lambda, [\tilde{X}, (r^2-\|\tilde{X}\|^2)^{1/2}, 0, \ldots, \tilde{X}, C(\lambda)((r^2-\|\tilde{X}\|^2)^{1/2}, 0, \ldots, 0)^T)])$. After local trivialization of the bundle, the obstruction is the class of the map $(\lambda, \tilde{X}) \longrightarrow (\tilde{X}, C(\lambda)((r^2-\|\tilde{X}\|^2)^{1/2}, 0 \ldots, 0)^T)$ as a mapping from $\partial(B^k \times B^{2p})$ into S^{2M-1}. Now this map is deformable to $(\tilde{X}, C(\lambda)(1, 0, \ldots, 0)^T)$ and so the degree of this map is, by the product theorem, the degree of $C(\lambda)(1, 0, \ldots, 0)^T$, i.e. $(-1)^{s-1}(s-1)!$ degree $B(\lambda)$.

The numbers $n_j(k)$ in the statement of the theorem are just the degrees of $i\nu m_j I - L(\mu)$. To complete the proof of the local result (the existence of the continuum will be a consequence of the global result) it remains to compute $n(2)$ = degree $\det(B(\lambda) = \nu i I - L(\mu))$. $\det B(\lambda) = \Pi(\nu i - e_j(\mu))$, where $e_j(\mu)$ are the eigenvalues of $L(\mu)$, $j = 1, \ldots, 2M$. The eigenvalues not passing through $i\beta$ give terms which are easily deformed to 1, for $|\nu-\beta|^2 + |\mu|^2 = \rho^2$, ρ small enough. The degree of $\Pi(-\alpha_j(\mu)$

$+ (\nu-\beta-\gamma_j(\mu))i)$ is the same as the degree of $((\nu-\beta)i-\mu)^{n+}((\nu-\beta)i+\mu)^{n-}$ after deforming $\gamma_j(\mu)$ to 0, $\alpha_j(\mu)$ to sign $\alpha_j(\mu)$, if it doesn't change sign, or to $\pm\mu$ if $\alpha_j(\mu)$ changes sign as $\pm\mu$. $(\nu-\beta)i-\mu$ contributes a winding number 1 and $(\nu-\beta)i+\mu$ a winding number -1, so that the degree is $n_+ - n_- = n$.

Remarks 1) If $B(\lambda)$ is deformable to I via $B(\lambda,t)$ for $\|\lambda\|=\rho$, then $B(\lambda\rho/\|\lambda\|, \|\lambda\|/\rho)\ (\|\lambda\|/\rho + \|X\|^2)\ X$ is an equivariant extension of $B(\lambda)= B(\lambda,1)$, for $\|\lambda\|=\rho$, without nontrivial zeroes.

2) Examples of matrices $L(\mu)$ giving any degree for $k \geq 4$ and any permissible degree (between- $[M/2]$ and $[M/2]$ for $k = 2$ can easily be constructed. One may also apply this procedure to higher order differential equations.

3) One may also consider the non stable case ($k > 2M$). Then one may have several obstructions.

I.3) The non free action case.

The fact that the action has fixed points implies a stratification on the set of solutions of $(*)_b$ which has to be taken into account for the construction of equivariant extensions. However one can reduce the non free case to a free action problem by considering the following trick:

If Z is a vector in $\mathbb{C}^M$ denote $Z^s \equiv (Z_1^s,\ldots,Z_M^s)$ for any positive integer s. Let $S^{2\ell M-1} \equiv \{(\eta_1,\ldots,\eta_\ell),\ \eta_j \in \mathbb{C}^M,\ \Sigma\,|\eta_j|^2=1\}$ be the unit sphere in $\mathbb{C}^{\ell M}$ and consider the system

$$(\tilde{*})\quad [B_j(\lambda)\,\tilde{r}\,\eta_j^{m_j} - g_j(\lambda,\tilde{r}\,\eta_1^{m_1},\tilde{r}\,\eta_2^{m_2},\ldots,\tilde{r}\eta_\ell^{m_\ell})]^{\Pi m_i}_{i\neq j} = 0,$$

$$j = 1,\ldots,\ell, \text{ with } \tilde{r} = r(\,\Sigma|\eta_j|^{2m_j})^{-1/2}.$$

Since the transformation $\eta \equiv (\eta_1,\ldots,\eta_\ell) \longrightarrow X_j = \tilde{r}\,\eta_j^{m_j}$ is onto, zeroes of $(\tilde{*})$ will give zeroes of $(*)_b$ and each zero of $(*)_b$ will give many roots of $(\tilde{*})$; extensions without zeroes of $(*)_b$ to $B^k \times S^{2\ell M-1}$ will give extensions without zeroes of $(\tilde{*})$ but <u>not</u> conversely, so that if the obstruction sets for $(\tilde{*})$ do not contain 0 then $(*)_b$ will have a solution.

Now if one has the standard action on $S^{2\ell M-1}$ then $(\tilde{*})\,(e^{i\phi}\eta) = e^{\Pi m_i\phi}(\tilde{*})(\eta)$, one gets a cross-section of the fiber bundle and one is back to the previous situation. The primary obstruction being unique, one may deform g_j to 0, $\tilde{r}$ to 1 and $B_j(\lambda)$ to any other matrix in its class in $\Pi_{k-1}(GL(\mathbb{C}^M))$.

a) Case k=2

Since this case is easier it will be treated before. $B_j(\lambda)$ is deformable to the suspension of det $B_j(\lambda)$ and the obstruction sets for $(\tilde{*})$ are the same for

$$\left[\begin{pmatrix} I & \\ & \det B_1(\lambda)\end{pmatrix}_{\eta_1}\right]^{\Pi m_i} = \begin{pmatrix} I & \\ & \det B_1(\lambda)^{\Pi m_i}\end{pmatrix}_{\eta_1}^{\Pi m_i}$$

$$\vdots$$

$$\left[\begin{pmatrix} I & \\ & \det B_\ell(\lambda)\end{pmatrix}_{\eta_\ell}^{m_\ell}\right]^{\prod_{i\neq\ell} m_i} = \begin{pmatrix} I & \\ & \det B_\ell(\lambda)^{\prod_{i\neq\ell} m_i}\end{pmatrix}_{\eta_\ell}^{\Pi m_i} .$$

The above matrix (having the same determinant) may be equivariantly deformed to

$$\begin{pmatrix} I & \\ & \det \Pi B_j(\lambda)^{\prod_{i\neq j} m_i}\end{pmatrix}_{\eta}^{\Pi m_i} .$$

The first obstruction (and the only one) is then the degreee of the

mapping $\qquad (\lambda,\tilde{\eta}) \longrightarrow (\tilde{\eta}^{\Pi m_i}, \det \Pi B_j(\lambda)^{\prod_{i\neq j} m_i})$

for $(\lambda,\tilde{\eta})$ on $\partial(B^2 \times B^{2\ell M-2})$. This degree is $(\Pi m_i)^{M\ell}\, \Sigma\, n_j(2)/m_j$. Thus the obstruction for extending $(\tilde{*})$ to $B^2 \times S^{2\ell M-1}$ is precisely the change in the Fuller's degree given in [C.M.Y.].

b) Case $2 < k \leqslant 2M$

If k is odd each $B_j(\lambda)$ is deformable to I giving a system which is extendable to $B^k \times S^{2\ell M-1}$. Suppose then that k is even, $k = 2s$, and deform $B_j(\lambda)$ to the suspension of $C_j(\lambda)$, an element of $GL(\mathbb{C}^s)$. One has to look at the map

$(\lambda,\eta) \longrightarrow (\tilde\eta^{\Pi m_i}, (C_\ell(\lambda)\,\xi_\ell^{m_\ell})^{\Pi_{i\neq\ell} m_i}, \ldots, (C_1(\lambda)\xi_1)^{\Pi m_i})$, with ξ_j in

^{s} and $\tilde\eta$ in $\mathbb{C}^{\ell(M-s)}$. Extending $C_j(\lambda)$ by $\rho^{-1}|\lambda|C_j(\lambda\rho/|\lambda|)$ and puting

$=\rho^2-|\lambda|^2$ one gets an equivariant extension to the 2 $M\ell$-1 skeleton of

$^k \times \mathbb{C}P^{M\ell-1}$ by $(\tilde\eta^{\Pi m_i}, (C_\ell(\lambda)\xi_\ell^{m_\ell})^{\Pi_{i\neq\ell} m_i}, (C_{\ell-1}(\lambda)\xi_{\ell-1}^{m_{\ell-1}})^{\Pi_{i\neq\ell-1} m_i} + t\,\xi_\ell^{\Pi m_i}, \ldots$

$\ldots, (C_1(\lambda)\xi_1^{m_1})^{\Pi_{i\neq 1} m_i} + t\,\xi_2^{\Pi m_i})$, i. e. for points $(\lambda,[\eta])$ with

$|\lambda| \leq \rho$, $[\tilde\eta, \xi_\ell, \ldots, \xi_2, 0]$.

The first obstruction is the degree of the above mapping, where ξ_1 has been replaced by $(1,0,\ldots,0)^T$, from the boundary of the ball

$|\lambda| \leq \rho$, $\|(\tilde\eta,\xi_\ell,\ldots,\xi_2)\| \leq 1$ into $\mathbb{C}^{\ell M}-\{0\}$.

The map $(P_* C(\lambda))^m = (C(\lambda)(1,0,\ldots,0)^T)^m$, from S^{2s-1} into itself, is the composition of $P_*C(\lambda)$ with the application $\xi \longrightarrow \xi^m$, of degree m^s, and hence has a degree m^s degree $P_*C(\lambda)=$ degree $P_* C(\lambda)^{m^s}$, so that $(P_* C(\lambda))^m$ is homotopic to $P_* C(\lambda)^{m^s}$. Using this homotopy one may replace the former expression by the later and get an equivariant extension

for the map $(\tilde\eta^{\Pi m_i}, (C_\ell(\lambda)\xi_\ell^{m_\ell})^{\Pi_{i\neq\ell} m_i}, \ldots, C_1(\lambda)^{\Pi m_i^s}\,\xi_1^{\Pi m_i})$. Conmuting ξ_1 and ξ_2 and the last two elements of this map, one has an equivariant

homotopy to the map $(\tilde\eta^{\Pi m_i}, (C(\lambda)\,\xi_\ell^{m_\ell})^{\Pi_{i\neq\ell} m_i}, \ldots, C_1(\lambda)^{m_i^s}\,\xi_2^{\Pi m_i}$,

$(C_2(\lambda)\,\xi_1^{m_2})^{\Pi_{i\neq 2} m_i})$.

Repeating the previous argument the obstruction (here one is using strongly the fact that the primary obstruction is unique and depends only on the equivariant homotopy class of maps on $S^{k-1} \times S^{2\ell M-1}$) is the same for the initial map and for the map

$(\tilde\eta^{\Pi m_i}, (C_\ell(\lambda)\xi_\ell^{m_\ell})^{\Pi_{i\neq\ell} m_i}, \ldots, C_2(\lambda)^{\Pi_{i\neq 2} m_i^s}\,\xi_2^{\Pi m_i}, C_1(\lambda)^{\Pi m_i^s}\,\xi_1^{\Pi m_i})$,

and, after repetition of the process, for the map $(\tilde\eta^{\Pi m_i}, C_\ell(\lambda)^{\Pi_{i\neq\ell} m_i^s}\,\xi_\ell^{\Pi m_i}, \ldots$

$\ldots, C_1(\lambda)^{\Pi m_i^s}\, \xi_1^{\Pi m_i})$.

Now if A and B belong to $GL(\mathbb{C}^M)$ the deformation

$$\begin{bmatrix} 1-\tau & \tau \\ -\tau & 1-\tau \end{bmatrix} \begin{bmatrix} (1-\tau)A & -\tau\, A\, B \\ \tau\, I & (1-\tau)B \end{bmatrix} \text{ is valid and}$$

so one obtains an equivariant homotopy to $(\tilde{\eta}^{\Pi m_i}, \xi_\ell^{\Pi m_i}, \ldots, \xi_2^{\Pi m_i},$

$\Pi_j\, C_j(\lambda)^{\Pi_{i\neq j} m_i^s}\, \xi_1^{\Pi m_i})$. The obstruction is then the degree of the above map,

with ξ_1 replaced by $(1,0,\ldots,0)^T$, that is $(\Pi\, m_i)^{\ell M}(k/2-1)!\, \Sigma_j\, n_j(k)/m_j^{k/2}$,

where $n_j(k)$ is the class of $C_j(\lambda)$ in $\Gamma_{k-1}(GL(\mathbb{C}^{k/2}))$.

Remarks: 1) It can be shown that, for the case $k > 2M$, the primary obstruction is always zero, due to the fact that $B_j(\lambda) = i\nu m_j I - L(\mu)$ has a non-trivial range for $\lambda = 0$ and hence is always deformable to a suspension. However a local reduction to the kernels of $B_j(0)$ is possible giving non trivial results but too long to develop here.

2) If M_{j_0} is the set of all j's in $\{1,\ldots,\ell\}$ such that m_j is a multiple of m_{j_0}, then if one takes $X_j = 0$ for all j's in $M_1 - M_{j_0}$ it is easy to see that the corresponding $g_j(\lambda,X)$ are zero. Equivalently one may look for solutions of $(*)_b$ with ν close to $m_{j_0}\beta$. Thus if $n_{j_0}(k)$ is non zero one may look at the largest j in M_{j_0} with $n_j(k)\neq 0$ and reduce the above obstruction to $n_j(k)$, proving the local result. It can be seen, by constructing examples, that this does not imply that the minimal period is $2\pi/m_j\beta$.

II) GLOBAL BIFURCATION.

In this part the equation $(*)_\nu$ will be considered in the space of 2π-periodic continuous functions $C^0(S^1)^M$. $W^{1,2}(S^1)^M$ is compactly contained (Sobolev embedding theorem) in it and, in turn, $C^0(S^1)$ is contained in $L^2(S^1)$. Fourier series will be used for all three spaces. Let C be the connected component of $(0,0,2\pi/\beta)$ in the set $P \cup \{(0,0,2\pi/\beta)\}$, where P is the set of $(X(\tau), \mu, T = 2\pi/\nu)$ solutions of $(*)_\nu$ in $C^0(S^1)^M \times \mathbb{R}^{k-1} \times \mathbb{R}^+$ of non trivial periodic solutions, and let S be the set of stationary solutions. It is easy to see that limit points of P, which are not in P, belong to S ([A.M.Y.], [I.1]) and are bifurcation points.

II.1) Proof of the first part of the theorem

Suppose C is bounded and C does not contain any other stationary point but $(0,0,2\pi/\beta)$. From last remark, the sum of the degrees of $B_j(\lambda)/\ m_j^{k/2}$ is assumed to be non-zero, $k=2s \leq 2M$ (in all other cases it can be proved that the global invariant is zero).

Construct an open bounded set Ω is $C^o(S^1)^M \times \mathbb{R}^k$ with the following properties:

1) $C \subset \Omega$

2) The only stationary points in $\bar{\Omega}$ are of the form $(0,\mu,T)$ with (μ,T) close to $(0,2\pi/\beta)$, (this comes from the fact that the intersection of S with any bounded set is compact).

3) If $X(\tau) = X_o + Y(\tau)$, $X_o \in \mathbb{R}^M$, $Y(\tau)$ L^2-orthogonal to the constants, $\lambda = (\mu,\nu-\beta)$, $\nu = 2\pi/T, \bar{\Omega}$ close to $(0,0,2\pi/\beta)$ is a ball $\|Y\| \leq \varepsilon_1$, $\|X_o\| \leq \varepsilon_o$, $\|\lambda\| \leq \varepsilon_k$, where the local reduction done in the first part is valid and where the only solution with $\|\lambda\| = \varepsilon_k$ is $X = 0$.

4) There are constants c, $C > 0$ such that $o < c < \nu < C$ on Ω (this comes from the fact that if T tends to 0 on C then one must approach a stationary point [I.2 Remark II.4 p. 1328, Lemma II.2 p. 1348], [M.Y. Appendix].

5) If $\|Y\| + \|X_o\| \leq \varepsilon_o + \varepsilon_1$, then (X,μ,T) belongs to the ball in (3) (if not one would have a stationary point $(X=0,\mu_1,T_1)$ on $\bar{C}$).

6) If $Z(\tau)$ is in Ω, so is $Z(\tau+\phi)$ (see the proof of Whyburn's lemma).

7) $(*)_\nu$ has no solutions on $\partial\Omega$ but points of the form $X=0$, (μ,T) close to $(0,2\pi/\beta)$.

The construction of such a set Ω is a straightforward adaptation of the proof in [I.1], [C.M.Y] etc...

Consider the equivalent equation $\nu X - Kf(\mu,X) = 0$ where K is the compact operator

$$C^o(S^1)^M \longrightarrow L^2(S^1)^M \xrightarrow{\tilde{K}} W^{1,2}(S^1)^M \longrightarrow C^o(S^1)^M$$

and $\tilde{K}(Y_o + \Sigma_{n\neq o} Y_n e^{in\tau}) = Y_o + \Sigma_{n\neq o} Y_n e^{in\tau}/(in)$. Let

$P_N(\Sigma Y_n e^{in\tau}) = \Sigma_{-N}^{N} Y_n e^{in\tau}$, it is then a standard argument (from the compactness of K) to show that, since $\nu X - Kf(\mu,X)$ is non zero on

$\partial\Omega \cap \{X_o, Y,\mu,T)/ \|X_o\| + \|Y\| \geq \varepsilon_o + \varepsilon_1\}$, $\nu X - P_n Kf(\mu,X)$ is non-zero, on the same set, for $n \geq N$ sufficiently large. N will be taken larger than m_ℓ.

Let then Ω_N be $\Omega \cap P_N (C^0(S^1)^M) \times \mathbb{R}^k$, thus $\nu P_N X - P_N Kf(\mu,P_N X)$ is non zero for $(P_N X,\mu,T)$ in $\partial\Omega_N$ and $\|X_o\| + \|Y\| \geq \varepsilon_o + \varepsilon_1$.

If $X(\tau) = \Sigma_{-N}^{N} X_n e^{in\tau}$, $X_{-n} = \bar{X}_n$ and Y denotes the vector $(X_1,\ldots,X_N)$ in $\mathbb{C}^{MN}$, look at the following triple in Ω_N (in fact in a diffeomorphic set), for any $\varepsilon > 0$,

$$F_\varepsilon(X_o,Y,\lambda) = \left| \begin{array}{ll} i\nu X_n - f_n(X_o,Y,\lambda) & 0 < n \leq N \\ f_o(X_o,Y,\lambda) & \\ \|X_o\| + \|Y\| - \varepsilon & \end{array} \right.$$

where $f_j(X_o,Y,\lambda)$ is the Fourier coefficient of $f(X(\tau),\lambda)$.

From the above argument and property (3) of Ω, the triple in non zero on $\partial\Omega_N$. Since Ω_N is bounded, let B be a big ball $\{\|Y\| \leq R_1, \|X_o\| \leq R_o, \|\lambda\| \leq R_k\}$ containing it, and on $\mathbb{C}^{MN} \times \mathbb{R}^M \times \mathbb{R}^k$ put the action $e^{i\phi}(Y,X_o,\lambda) = (e^{i\phi}Y_1, e^{2i\phi}Y_2,\ldots,e^{iN\phi}Y_N,X_o,\lambda)$. Ω_N is then invariant under this action, as well as B since the action is an isometry.

<u>Definition</u>: Let X be a closed, bounded, invariant subset of $\mathbb{C}^{MN} \times \mathbb{R}^M \times \mathbb{R}^k$ and let $F(Y,X_o,\lambda) \equiv (F_1,\ldots,F_N,F_o)$ be a mapping from X into $\mathbb{C}^{MN} \times \mathbb{R}^{M+1} - \{0\}$ which is equivariant, i. e. $F_j(e^{i\phi}(Y,X_o,\lambda)) = e^{ij\phi}F_j(Y,X_o,\lambda)$ $j = 0,\ldots,N$ (F_j belong to $\mathbb{C}^M$, if $j \geq 1$, and to $\mathbb{R}^{M+1}$, if $j=0$). Then, if Y is a closed, bounded, invariant subset containing X, F is called <u>S^1-inessential with respect to</u> Y if and only if there is an equivariant non zero extension of F to Y.

<u>Lemma</u> 1) If F_o and F_1 are equivariantly homotopic on X and F_1 is S^1-inessential with respect to Y, then so is F_o.

2) F is S^1 inessential with respect to any Y (or with respect to a ball B) if and only if F is equivariantly deformable to $(0,\ldots,0,C)$, C a constant in $\mathbb{R}^{M+1}$.

Proof It is enough to adapt the standard proof (see [I.4]) by replacing Tietze's extension lemma by Gleason's and constructing an invariant separating function [Pa].

Here, for E large enough, $F_E|_{\partial\Omega}$ is S^1-inessential with respect to $\bar{B}$ and equivariantly deformable to $(0,0,-E)$ on $\partial\bar{B}$. Use this deformation to extend radially $F_E|_{\partial B}$ to a slightly larger ball, which will be also called B, as in[I.4]. Then $F_E|_{\partial\Omega} \cup (0,0,-E)|_{\partial B}$ is S^1-inessential with respect to $\bar{B}$.

Fix then $\varepsilon \leqslant \varepsilon_1$. $F_\varepsilon|_{\partial\Omega} \cup (0,0,-E)|_{\partial B}$ is S^1-homotopic to the preceding map and hence also S^1-inessential with respect to $\bar{B}$. Let $F(Y,X_0,\lambda) = (F_1,\ldots,F_N,F_0,F_r)$, with F_j in $\mathbb{C}^M$, $j=1,\ldots,N$, F_0 in $\mathbb{R}^M$, F_r in $\mathbb{R}$, be defined on $\bar{B}$ as the extension of $F\varepsilon|_{\partial\Omega} \cup (0,0,-E)|_{\partial B}$ on $\bar{B}-\Omega$ and F_ε on Ω. Then $F(Y,X_0,\lambda) \neq 0$ on $\bar{B} - B_{\varepsilon_1,\varepsilon_0,\varepsilon_k}$ since if $\|X_0\| + \|Y\| = \varepsilon < \varepsilon_1$ and F_ε has a solution it must be in the small ball $B_{\varepsilon_1,\varepsilon_0,\varepsilon_k}$.

Consider the unit sphere $\{(\eta_1,\ldots,\eta_N)/ \Sigma |\eta_j|^2=1\}$ and the mapping $\tilde{F}(\eta_1,\ldots,\eta_n,r,X_0,\lambda)$ from $S^{2MN-1} \times B^1_{R_1} \times B^M_{R_o} \times B^k_{R_k}$ into $\mathbb{C}^{MN} \times \mathbb{R}^M \times \mathbb{R}$ defined by $([F_n(\tilde{r}\eta_1, \tilde{r}\eta_2^2,\ldots,\tilde{r}\eta_N^N,X_0,\lambda)]^{N!/n}$, $n=1,\ldots,N$; $F_0(\tilde{r}\eta_1,\ldots,\tilde{r}\eta_N^N, X_0,\lambda)$, $F_r(\tilde{r}\eta_1,\ldots,\lambda))$ with $\tilde{r} = r(\Sigma|\eta_j|^{2j})^{-1/2}$.

Then $\tilde{F} \neq 0$ on $S^{2MN-1} \times (B^1_{R_1} \times B^M_{R_o} \times B^k_{R_k} - B^1_{\varepsilon_1} \times B^M_{\varepsilon_o} \times B^k_{\varepsilon_k})$, since if $\tilde{F}$ has a zero, then F has a zero for $Y_j = \tilde{r}\eta_j^{m_j}$ with $\|Y\| = r$. But then an inversion with respect to the boundary of $B^1_{\varepsilon_1} \times B^M_{\varepsilon_o} \times B^k_{\varepsilon_k}$ gives an extension to this ball(the value at the origin is $(0,0,-E)$, value of the map on the boundary of the big ball).

This means that the mapping

$\{((\text{inv}-L(\mu))\tilde{r}\eta_n^n - g_n(\tilde{r}\eta_1,\ldots,\lambda))^{N!/n}, L(\mu)X_0 - g_0(\tilde{r}\eta_1,\ldots,\lambda), r+\|X_0\|-\varepsilon\}$

has an equivariant extension from $S^{2MN-1} \times S^{k+M}$ to $S^{2MN-1} \times B^{k+M+1}$, and a corresponding extension for the cross-section of the bundle

$$S^{2MN+M} \longrightarrow B^{k+M+1} \times S^{2MN-1} \times_{S^1} S^{2MN+M} \longrightarrow B^{k+M+1} \times \mathbb{C}P^{MN-1}$$

(the only difference with the previous case is that, having 2MN+M+1 equations, the dimension of the fiber is different). The eventual obstructions are in $H^{k+M+1+2p}(\mathbb{C}P^{MN-1} \times B^{M+k+1}, \mathbb{C}P^{MN-1} \times S^{M+k}; \Pi_{k+2M+2p}(S^{2MN+M}))$ and here they vanish since there is an extension. For k odd or $k > 2M$ the stable classes are trivial, so, for $k = 2s \leq 2M$, the primary obstruction is for $p = MN-s$ and is the degree of the map $(\tilde{\eta} \in \mathbb{C}^{MN-s}, r, X_o, \lambda) \longrightarrow (\tilde{\eta}^{N!}, r + \|X_0\| - \varepsilon, L(\mu)X_0, \Pi C_j(\lambda)^{N!/m_j}(1,0,\ldots,0)^T)$ as it is now clear from the previous calculation.

Since $L(0)$ is invertible, this degree is $(N!)^{MN}(s-1)!$ sign det $L(o) \times \Sigma$ degree $B_j(\lambda)/m_j^s$, leading to a contradiction and finishing the proof of this part of the theorem.

II.2) Proof of the second part of the Theorem

Suppose again that C is bounded and that $\bar{C}$ connects $(0,0,2\pi/\beta)$ to stationary points (X_p,μ_p,T_p), with $f(\mu_p,X_p)=0$, $D_X f(\mu_p,X_p)$ invertible, so that the stationary points near (μ_p,X_p) are of the form $X_p(\mu)$, $f(\mu,X)= L_p(\mu)\,(X-X_p(\mu)) + g_p(\mu,X)$, $g_p(\mu, X) = o(\|X-X_p(\mu)\|)$.

From [I.2 lemma II.2 p. 1348] or [M. Y. Appendix] it follows that $T_p > 0$, $T_p = 2\pi q(p)/\beta(p)$ where $q(p)$ is an integer ($q(p)=1$ if one is considering least periods), $i\beta(p)$ belongs to the spectrum of $L_p(\mu_p)$. Assume that $L_p(\mu)$ has the corresponding properties H.2, H.3, with $m_j(p)$, $\ell(p)$ and the classes $n_j(k,p)$ for $i\nu m_j(p)I-L_p(\mu)$ (k even $\leq 2M$). It is then clear that the number of such bifurcation points in a bounded set is finite: $p \in I$ a finite set.

Construct Ω as before with the obvious changes (in (3) the ball is $\|Y\| \leq \varepsilon_1$, $\|X_0-X_p\| \leq \varepsilon$, $\|\mu-\mu_p\|+|\nu-\beta(p)/q(p)| \leq \varepsilon_k$; in (5) $\|Y\| + \|X_0\|$ has to be replaced by distance$(X,\mu;S)$). The reduction to a finite dimensional system goes through and one has to look at the mapping F_ε where the last equation has been replaced by $d(X,\mu;S) - \varepsilon$. Again F_ε is inessential with respect to B (the large ball) and for a fixed $\varepsilon \leqslant \varepsilon_1$ one gets the extension $F(Y,X_0,\lambda)$ on $\overline{B} - \bigcup_I B_{\varepsilon_1,\varepsilon_0,\varepsilon_k}(p)$, the above balls. Defining $\tilde{F}(\eta_1,\ldots,\eta_N,r,X_0,\lambda)$ as before, one gets an extension of the cross section on $L \equiv \mathbb{C}P^{MN-1} \times \partial(B^1_{R_1} \times B^M_{R_o} \times B^k_{R_k} - \bigcup_I B^1_{\varepsilon_1} \times B^M_{\varepsilon_o} \times B^k_{\varepsilon_k}(p))$ to $K \equiv \mathbb{C}P^{MN-1} \times (B^1_{r_1} \times B^M_{R_o} \times B^k_{R_k} - \bigcup_I B^1_{\varepsilon_1} \times B^M_{\varepsilon_o} \times B^k_{\varepsilon_k}(p))$.

Then all obstructions in $H^{k+M+1+2p}(k, L; \Pi_{k+M+2p}(S^{2MN+M}))$ vanish. It is not difficult to show, but tedious, that the algebraic manipulations done for cohomotopy theory in [I.4] work also for cohomology and so, using the fact that primary obstructions are given by characteristic elements in $H^{2MN+M}(L)$ (see [Hu p. 189]), the sum of the local degrees has to be zero.

However locally $d(X,\mu;S) = \|Y\| + \|X_0 - X_p(\mu)\|$, only the n's with $n = m_j(p)$ will make the corresponding matrices to be singular ($q(p)$ is chosen to be the smallest integer with that property), so that the local degree will be $(N!)^{MN}(s-1)!$ sign det $L_p(\mu_p) \times \Sigma\, n_j(k,p)/(m_j(p))^{k/2}$, proving thus the theorem.

REFERENCES

[A.Y.] Alexander J. C., Yorke J. A.: Global bifurcation of periodic orbits. Amer. J. Math. 100 (1978), 263-292.

[At.] Atiyah M. F. : Algebraic topology and elliptic operators. Comm. Pure and Applied Math. XX (1967),237-249.

[Br.] Bredon G. E.: Introduction to compact transformation groups. Academic Press, New-York, 1972.

[C.M.Y.] Chow S. N., Mallet-Paret J., Yorke J. A. : Global Hopf bifurcation from a multiple eigenvalue. Nonlinear Anal. 2 (1978), 753-763.

[F.R.] Fadell F. R., Rabinowitz P. H. : Generalized cohomological index theories for Lie group actions with an application to bifurcation questions for Hamiltonian systems. Invent. Math. 45 (1978), 134-174.

[Hu.] Hu S. T. : Homotopy theory. Academic Press. New-York (1959).

[I.1] Ize J. : Bifurcation theory for Fredholm operators. Memoirs.

A. M. S. 7, 174, (1976).

[I.2] Ize J. : Periodic solutions of nonlinear parabolic equations. Comm. in P. D. E. 4, 12 (1979), 1299-1387.

[I.3] Ize J. : Le problème de bifurcation de Hopf. Séminaire Brezis-Lions 1975. A paraître.

[I.4] Ize J.: Introduction to bifurcation theory. In Proceedings of the 1st. Latin-American School of Differential Equations, Springer Lecture Notes in Mathematics 957 (1982), 145-202.

[Ko.] Kosnioswski C. : Equivariant cohomology and stable cohomotopy. Math. Ann. 210, (1974), 83-104.

[M.Y.] Mallet-Paret J., Yorke J. A. : Snakes: oriented families of periodic orbits, their sources, sinks and continuation. University of Maryland preprint (1979).

[Pa.] Palais R. S. : The classification of G-spaces. Memoirs A. M. S. 36, (1960).

[Sp.] Spanier E. : Algebraic topology. Mc. Graw-Hill, (1966).

[St.] Steenrod N. : Topology of fiber bundles, Princeton University Press, Princeton, New-Jersey (1965).

[To] Toda H. : Composition methods in homotopy groups of spheres, Annals of Math. Studies, No. 49, Princeton University, (1962).

FINITE DIMENSIONAL APPROXIMATION

TO SYSTEMS WITH INFINITE DIMENSIONAL STATE SPACE

F. Kappel
Institute of Mathematics
University of Graz
A-8010 Graz, Austria

In recent years an increasing number of papers in the applied mathematics literature is concerned with effective computational methods for control systems with infinite dimensional state space. In this paper we concentrate on an approach originally developed for delay systems. We first present the general idea and comment on some relevant papers. Instead of giving a complete survey we prefer to present in a preliminary setting two new results.

1. THE GENERAL APPROACH

Let Z be a real Banach space and $S(t)$, $t \geq 0$, a C_0-semigroup of bounded linear operators on Z with infinitesimal generator A. Furthermore, let B be a bounded linear operator $\mathbb{R}^m \to Z$. We assume that the state $z(t) = z(t;z_0,u)$ of our system evolves according to

$$z(t;z_0,u) = S(t)z_0 + \int_0^t S(t-s)Bu(s)ds, \quad t \geq 0, \tag{1.1}$$

where $z_0 \in Z$ and $u \in L^2_{loc}(0,\infty;\mathbb{R}^m)$. The following result shows that in order to approximate $z(t;z_0,u)$ one has to find a good approximation for the homogeneous part $S(t)z_0$.

Theorem 1.1. Assume that $S^N(t)$, $t \geq 0$, $N = 1,2,\dots$, is a sequence of C_0-semigroups on Z such that for constants $M \geq 1$, $\omega \in \mathbb{R}$

$$||S^N(t)|| \leq Me^{\omega t}, \quad t \geq 0, \tag{1.2}$$

and for all $z \in Z$

$$\lim_{N\to\infty} S^N(t)z = S(t)z \tag{1.3}$$

uniformly on bounded t-intervals. Furthermore, let B^N, $N = 1,2,\dots$, be a sequence of bounded linear operators $\mathbb{R}^m \to Z$ such that

$$\lim_{N\to\infty} B^N z = Bz \quad \text{for all } z \in Z. \tag{1.4}$$

Then for all $z_0 \in Z$ and all $t_1 > 0$

$$\lim_{N\to\infty} z^N(t;z_0,u) = z(t;z_0,u) \tag{1.5}$$

uniformly for $t \in [0,t_1]$ and $u \in$ bounded subsets of $L^2(0,t_1;\mathbb{R}^m)$, where $z^N(t;z_0,u)$ is defined by

$$z^N(t;z_0,u) = S^N(t)z_0 + \int_0^t S^N(t-s)B^N u(s)ds, \quad t \geq 0. \tag{1.6}$$

Proof. The proof of this result is already given in [3]. Define the families $T(t)$, $T^N(t)$ of bounded linear operators $\mathbb{R}^m \to Z$ by

$$T(t)\xi = S(t)B\xi, \quad T^N(t)\xi = S^N(t)B^N\xi, \quad t \geq 0, \ \xi \in \mathbb{R}^m.$$

From (1.2), (1.3), (1.4) and an application of the triangel inequality we see that, for all $\xi \in \mathbb{R}^m$, $T^N(t)\xi \to T(t)\xi$ uniformly on $[0,t_1]$, which implies

$$T^N(t) \to T(t) \quad \text{uniformly on } [0,t_1]$$

in the uniform operator topology. Similarly one sees that the maps $t \to T(t)$ and $t \to T^N(t)$ are continuous with respect to the uniform operator topology. Therefore

$$\lim_{N\to\infty} \int_0^{t_1} ||T(t) - T^N(t)||^2 dt = 0.$$

Then the result follows from

$$\begin{aligned} |z(t;z_0,u) - z^N(t;z_0,u)| &\leq |S(t)z_0 - S^N(t)z_0| + \int_0^{t_1} ||T(t-s) - T^N(t-s)||\,|u(s)|ds \\ &\leq |S(t)z_0 - S^N(t)z_0| + \left(\int_0^{t_1} ||T(s) - T^N(s)||^2 ds\right)^{1/2} |u|_{L^2(0,t_1,\mathbb{R}^m)}, \quad t \geq 0. \end{aligned}$$

Remarks. 1. Since $\sup_{[0,t_1]} ||T(t) - T^N(t)|| \to 0$ it is clear that (1.5) is also uniform for $u \in$ bounded subsets of $L^p(0,t_1;\mathbb{R}^m)$, $1 \leq p \leq \infty$.

2. If $S(t)$, $S^N(t)$ are strongly continous semigroups of nonlinear globally Lipschitzean operators with Lipschitz constants $Me^{\omega t}$ for all N and (1.3), (1.4) are true, then (1.5) holds uniformly for $t \in [0,t_1]$ and $u \in$ bounded subsets of $L^\infty(0,t_1;\mathbb{R}^m)$. The proof is similar as for Theorem 1.1 using compactness of closed bounded sets in $\mathbb{R}^m$.

3. In concrete cases the semigroup $S^N(t)$ is constructed such that a (finite dimensional) subspace Z^N of Z is invariant, i.e. $S^N(t)Z^N \subset Z^N$, $t \geq 0$. Assume that $P^N: Z \to Z^N$, $N = 1,2,\ldots$, is a uniformly bounded sequence of projections converging strongly to I.

Define $B^N = P^N B$, $N = 1,2,\dots$, and

$$\tilde{z}^N(t;z_0,u) = S^N(t)P^N z_0 + \int_0^t S^N(t-s)P^N Bu(s)ds, \quad t \geq 0.$$

Then for all $z_0 \in Z$

$$\lim_{N\to\infty} \tilde{z}^N(t;z_0,u) = z(t;z_0,u)$$

uniformly for $t \in [0,t_1]$ and $u \in$ bounded subsets of $L^2(0,t_1;\mathbb{R}^m)$. This is clear from

$$|z^N(t;z_0,u) - \tilde{z}^N(t;z_0,u)| \leq |S^N(t)z_0 - S^N(t)P^N z_0| \leq Me^{\omega t_1}|z_0 - P^N z_0|.$$

If $\dim Z^N < \infty$, then $\tilde{z}^N(t;z_0,u)$ is the unique solution of the ordinary differential equation on Z^N

$$\dot{w}(t) = \tilde{A}^N w(t) + P^N Bu(t), \quad t \geq 0,$$

$$w(0) = P^N z_0,$$

where $\tilde{A}^N$ is the restriction to Z^N of the infinitesimal generator of $S^N(t)$.

In order to get properties (1.2) and (1.3) for the sequence of semigroups $S^N(t)$, $N = 1,2,\dots$, we impose the following conditions. Let A^N, $N = 1,2,\dots$, be the infinitesimal generator of $S^N(t)$. For a subset $D \subset Z$ define $E(t_1) = \bigcup_{0\leq t\leq t_1} S(t)D$, t_1

(H1) There is a constant $\omega \in \mathbb{R}$ such that $A^N - \omega I$ is dissipative for all N.

(H2) There is a subset $D \subset \operatorname{dom} A \cap \bigcap_N \operatorname{dom} A^N$ and a $t_1 > 0$ such that D is dense in Z and

$$\lim_{N\to\infty} A^N z = Az \quad \text{for all } z \in E(t_1).$$

(H3) Let $z(t;z_0) = S(t)z_0$, $0 \leq t \leq t_1$, $z_0 \in D$. Then there exists a function $m \in L^1(0,t_1;\mathbb{R})$ such that for all N

$$|A^N z(t;z_0)| \leq m(t) \quad \text{a.e. on } [0,t_1].$$

<u>Theorem 1.2.</u> Assume (H1), (H2) and (H3). Then for all $z_0 \in Z$

$$\lim_{N\to\infty} z^N(t,z_0) = z(t;z_0)$$

uniformly for $t \in [0,t_1]$.

<u>Proof.</u> Consider first $z_0 \in D$. Then $z(t) = S(t)z_0$ and $z^N(t) = S^N(t)z_0$ are strong solutions of

$$\dot{z} = Az, \quad z(0) = z_0,$$

and

$$\dot{z}^N = A^N z^N, \quad z^N(0) = z_0,$$

respectively, i.e. $z(t) = z_0 + \int_0^t Az(s)ds$ and $z^N(t) = z_0 + \int_0^t A^N z^N(s)ds$, $t \geq 0$. Since the derivative $\frac{d}{dt}\Delta^N(t)$ of $\Delta^N(t) = z(t) - z^N(t)$ exists for all $t > 0$, the left-hand derivative of $|\Delta^N(t)|$ exist for all $t > 0$ is given by (cf.[29],p.228)

$$\frac{d^-}{dt}|\Delta^N(t)| = \tau_-(\Delta^N(t), \dot{\Delta}^N(t)),$$

where $\tau_\pm(x,y) = \lim_{h\to 0\pm} \frac{1}{h}(||x+hy||-||x||)$ (cf. [29] for properties of $\tau_\pm$). Using the fact that (H1) is equivalent to $\tau_-(z, A^N z) \leq \omega|z|$ for all $z \in \mathrm{dom}\, A^N$ and furthermore the estimates $|\tau_-(x,y)| \leq |y|$ and $\tau_-(x, y_1+y_2) \leq \tau_-(x,y_1) + \tau_-(x,y_2)$ we get

$$\frac{d^-}{dt}|\Delta^N(t)| \leq |Az(t) - A^N z(t)| + \omega|\Delta^N(t)| \quad \text{for all } t \geq 0.$$

This inequality implies

$$|\Delta^N(t)| \leq \int_0^t |Az(s) - A^N z(s)| e^{\omega(t-s)} ds, \quad t \geq 0.$$

Assumptions (H2) and (H3) imply that we can use the dominated convergence theorem in order to get $\int_0^{t_1} |Az(s) - A^N(s)| e^{\omega(t-s)} ds \to 0$. This proves the theorem for $z_0 \in D$. For $z_0 \in Z$ we choose a sequence (z_{on}) in D such that $z_{on} \to z_0$ and get for some constant M

$$\begin{aligned} |z(t;z_0) - z^N(t;z_0)| &\leq |z(t;z_0) - z(t;z_{on})| + |z(t;z_{on}) - z^N(t;z_{on})| \\ &\quad + |z^N(t;z_{on}) - z^N(t;z_0)| \\ &\leq 2Me^{\omega t}|z_0 - z_{on}| + |z(t;z_{on}) - z^N(t;z_{on})|, \quad t \geq 0, \end{aligned}$$

which implies the assertion of the theorem z_0.

As we shall see the following modification of Theorem 1.2 is quite useful for applications. Let (Z^N) be a sequence of finite dimensional subspaces of Z and let $P^N: Z \to Z^N$ be projections such that $||P^N|| \leq \pi$ for all N, π some constant, and $\lim_{N\to\infty} P^N z = z$ for all $z \in Z$. Furthermore, let $S^N(t)$ be a C_0-semigroup on Z^N with infinitesimal generator A^N. Note, that A^N is a bounded linear operator $Z^N \to Z^N$ and $S^N(t) = e^{A^N t}$. We extend the definition of A^N to all of Z by putting $A^N z = A^N P^N z$, $z \in Z$. Instead of (H1) we shall need

(H1*) There is a constant $\omega \in \mathbb{R}$ such that $A^N - \omega I$ is dissipative on Z^N for all N, i.e. $\tau_-(z, A^N z) \le \omega|z|$ for $z \in Z^N$ and all N.

Theorem 1.3. Assume (H1*), (H2) and (H3). Then for all $z_0 \in Z$

$$\lim_{N\to\infty} z^N(t;P^N z_0) = z(t;z_0)$$

uniformly for $t \in [0,t_1]$.

Proof. For $z_0 \in D$ we define $\Delta^N(t) = P^N z(t) - z^N(t)$, where $z(t) = z_0 + \int_0^t Az(s)ds$ and $z^N(t) = P^N z_0 + \int_0^t A^N z^N(s)ds$, $t \ge 0$. Similarly as in the proof of Theorem 1.2 by using (H1*) we get the inequality

$$\frac{d^-}{dt} |\Delta^N(t)| \le |P^N Az(t) - A^N z(t)| + \omega|P^N z(t) - z^N(t)|, \quad t > 0.$$

Here it is important to note that $A^N z(t) = A^N P^N z(t)$. This inequality implies

$$|\Delta^N(t)| \le \int_0^t e^{\omega(t-s)} |P^N Az(s) - A^N z(s)|ds, \quad t \ge 0.$$

If we observe $|P^N Az(s) - A^N z(s)| \le \pi|Az(s) - A^N z(s)|$ we get $|\Delta^N(t)| \to 0$ uniformly for $t \in [0,t_1]$ by using (H2), (H3). Uniform boundedness of $||P^N||$ and a compactness argument show $|P^N z(t) - z(t)| \to 0$ uniformly for $t \in [0,t_1]$. This implies $|z(t) - z^N(t)| \to 0$ uniformly for $t \in [0,t_1]$. The proof for arbitrary $z_0 \in Z$ is as in case of Theorem 1.2 using $|z(t;z_0) - z^N(t;P^N z_0)| \le |z(t;z_0) - z(t;z_{on})| + |z(t;z_{on}) - z^N(t;P^N z_{on})| + |z^N(t;P^N z_{on}) - z^N(t;P^N z_0)| \le (M+\pi)e^{\omega t}|z_0 - z_{on}| + |z(t;z_{on}) - z^N(t;P^N z_{on})|$.

Remark. If we assume $Z^N \subset \mathrm{dom}\, A$ the operators A^N can be defined by $A^N = P^N A P^N$. If in addition $||P^N|| = 1$ for all N then (H1*) is a consequence of

(H1**) There exists an $\omega \in \mathbb{R}$ such that $A - \omega I$ is dissipative.

This is immediate from $\frac{1}{h}(|z + hA^N z| - |z|) = \frac{1}{h}(|P^N z + hP^N Az| - |z|) \le \frac{1}{h}(|z + hAz| - |z|)$ for $z \in Z^N$.

A condition of type (H1) is usually called a stability condition for the approximating scheme, whereas conditions of type (H2) together with (H3) are known as consistency conditions for the scheme. The classical Lax-Richtmyer equivalence theorem states that a consistent scheme is convergent if and only if it is stable. The proof for Theorem 1.2 and 1.3 is essentially the classical one for the Lax-Richtmyer theorem (cf. for instance [16]).

The idea to relate an abstract Cauchy problem to a delay system and then to use an approximation scheme for the abstract problems appears possibly for the first time in

[26]. The first ones to use this approach in full generality for autonomous linear delay systems were Banks and Burns in [3]. A complete presentation of the results announced in [3] was given in [4], a paper which we also especially recommend for an extensive discussion of the relevant literature up to 1976. The state space used in [4] is the Hilbert space $\mathbb{R}^n \times L^2(-r,0;\mathbb{R}^n)$. The subspaces Z^N are subspaces of step functions with jumps at points of an equally spaced mesh and the projections P^N are the orthogonal projections $Z \to Z^N$, usually called averaging projections in this context. In order to get dissipativity of A or A^N one has to introduce an equivalent weighted norm on $L^2(-r,0;\mathbb{R}^n)$, an idea which goes back to G. Webb [33]. In order to overcome the at most first order rate of convergence for the scheme of averaging projections in [9] the elements of the spaces Z^N were chosen to be splines (first order or cubic) with knots at an equally spaced mesh. The projections are orthogonal projections and the scheme is a classical Galerkin procedure, i.e. the approximating generators are defined as $A^N = P^N A P^N$. The scheme of averaging projections was generalized to nonlinear delay equations in [24] and [1]. Linear autonomous equations of neutral type have been considered in [27], [17] (averaging projections) and in [22], [19] (spline approximation). Spline approximation for nonlinear equations was considered in [21] and [2] (see also [20]). In most of these papers versions of the Trotter-Kato theorem are used in order to prove convergence of the approximating semigroups. This theorem not only gives convergence of the approximating semigroups, but assures that the limiting operator is an infinitesimal generator, i.e. we at the same time get well-posedness of the limiting abstract Cauchy problem. In the case of linear autonomous equations this is not crucial, but for nonlinear and/or nonautonomous equations this poses unnecessary restrictions, since existence of solutions is in most cases already established by other means. [2] is the first paper wherefor delay systems the more direct approach presented also in this paper was used. This approach can be easily modified for nonlinear and nonautonomous systems and was used for instance in [8] and [13] for nonautonomous and nonlinear delay systems and PDE-systems. Recently also infinite delay systems have been considered successfully using this approach [23]. The main purpose for developing approximation schemes of the type discussed in this paper is to get numerically efficient algorithms for the solution of control and identification problems for delay and distributed parameter systems. The interested reader is referred to [6], [10], [7], [5], [11], [28], [15] for details, further references and many numerical examples. Finally we want to mention a few papers where also discretization with respect to time is involved [30], [31], [12].

2. APPROXIMATION OF DELAY SYSTEMS IN THE STATE SPACE C

In all of the papers discussed at the end of Section 1 the state space was a Hilbert-space. [25] is the only paper where the state space C is used (for nonlinear and nonautonomous delay equations). There the nonlinear equation is considered as a perturbation of the trivial equation $\dot{x} \equiv 0$ and is transformed to an equivalent integral equation which involves the solution semigroup S(t) of $\dot{x} \equiv 0$. The integral equation is just the variation of parameters formula. The approximation is done by a sequence of integral equations involving approximations $S^N(t)$ of S(t). Main tool is a fixed point theorem which assures continuous dependence of fixed points on parameters. However, for the construction of the approximating semigroups $S^N(t)$ L^2-methods are used. In this section we use Theorem 1.3 in order to get an approximation scheme entirely working in the state space $C = C(-r,0;\mathbb{R}^n)$, $r > 0$.

For simplicity of the presentation we consider a very simple situation here. A more complete version of the results will appear elsewhere.

The equation considered is

$$\begin{aligned} \dot{x}(t) &= A_0x(t) + A_1x(t-r), \\ x_0 &= \varphi \in C. \end{aligned} \tag{2.1}$$

The semigroup S(t), $t \geq 0$, is defined by $S(t)\varphi = x_t(\varphi)$, where $x(t;\varphi)$ is the solution of (2.1) ($x_t(\varphi)$ denotes the function $x_t(\varphi)(s) = x(t+s;\varphi)$, $s \in [-r,0]$). The infinitesimal generator A of S(t) is given by (see [18], for instance).

$$\text{dom } A = \{\varphi \in C^1(-r,0;\mathbb{R}^n) \,|\, \dot{\varphi}(0) = A_0\varphi(0) + A_1\varphi(-r)\},$$

$$A\varphi = \dot{\varphi} \quad \text{for} \quad \varphi \in \text{dom } A.$$

For N = 1,2,... , we define

$$t_j^N = -j\frac{r}{N}, \quad j = 0,\ldots,N,$$

$$Z^N = \{\varphi \,|\, \varphi \text{ is a spline of first order with knots at the points } t_j^N\}$$

and the projections $P^N\colon C \to Z^N$ by

$$(P^N\varphi)(t_j^N) = \varphi(t_j^N), \quad j = 0,\ldots,N,$$

i.e. $P^N\varphi$ is the interpolating first order spline for φ with respect to the mesh $\{t_j^N\}$. It is clear that $||P^N|| = 1$ for all N and $P^N\varphi \to \varphi$ for all $\varphi \in C$. (For results on spline functions which we are using here and in the sequel see for instance [32] or [14].)

For $\varphi \in Z^N$ we define $A^N\varphi$ by

$$(A^N\varphi)(t_j^N) = \frac{N}{r}\,(\varphi(t_{j-1}^N) - \varphi(t_j^N)), \quad j = 1,\dots,N,$$
$$(A^N\varphi)(0) = A_o\varphi(0) + A_1\varphi(-r). \tag{2.2}$$

We have to prove uniform dissipativeness of the operators A^N on Z^N. As usual we define the duality map F by

$$F(\varphi) = \{x' \in C' \,|\, ||x'||^2 = ||\varphi||^2 = x'(\varphi)\},\ \varphi \in C.$$

Then dissipativeness of $A^N - \omega I$ on Z^N is equivalent to

$$j(A^N\varphi) \leq \omega|\varphi|^2 \quad \text{for all } \varphi \in Z^N,$$

for at least one $j \in F(\varphi)$ (see for instance [29]). It is not difficult to prove that one $j_\varphi \in F(\varphi)$, $\varphi \in C$, is always given by

$$j_\varphi(\chi) = \int_{-r}^{0} [d\psi^T(\theta)\,\chi(\theta)],$$

where $\psi \in BV(-r,0,\mathbb{R}^n)$ is given by

$$\psi(\theta) = \begin{cases} 0 & \text{for } \theta_o \leq \theta \leq 0, \\ -\varphi(\theta_o) & \text{for } -r \leq \theta < \theta_0 \end{cases}$$

and $\theta_0 = \max\{\theta\in[-r,0] \,|\, |\varphi(\theta)| = |\varphi|\}$. For $\theta_0 = -r$ we put $\psi(\theta) = 0$ for $-r < \theta \leq 0$ and $\psi(-r) = -\varphi(-r)$. For $\varphi \in Z^N$ θ_0 is always one of the points t_j^N. Assume $\theta_0 = -j\frac{r}{N}$ with $j > 0$. Then

$$j_\varphi(A^N\varphi) = \varphi(t_j^N)^T(A^N\varphi)(t_j^N) = \varphi(t_j^N)^T\,\frac{N}{r}\,[\varphi(t_{j-1}^N) - \varphi(t_j^N)] = \frac{d^+}{d\theta}\,|\varphi(\theta)|^2\Big|_{\theta=t_j^N}.$$

Since $|\varphi(\theta)|^2 \leq |\varphi(\theta_0)|^2$ for $\theta \geq \theta_0$, we get

$$j_\varphi(A^N\varphi) \leq 0.$$

If $\theta_0 = 0$ then

$$j_\varphi(A^N\varphi) = \varphi(0)^T\,[A_o\varphi(0) + A_1\varphi(-r)]$$
$$\leq (|A_0| + |A_1|)|\varphi|^2.$$

This proves (H1*) with $\omega = |A_0| + |A_1|$.

It is not difficult to see that $A^N - \omega I$ is not uniformly dissipative on C for any $\omega \in \mathbb{R}$. In order to verify (H2) and (H3) we choose

$$D = \operatorname{dom} A.$$

Since D is invariant for S(t), $t \geq 0$, we have $E(t_1) = D$ for all $t_1 > 0$. Fix $\varphi \in D$ and $\theta \in [-r,0)$. For N sufficiently large θ is in $[t^N_{j+1}, t^N_j]$ with $j \geq 1$. Then

$$(A^N\varphi)(\theta) - (A\varphi)(\theta)$$

$$= [1 - \frac{N}{r}(\theta - t^N_{j+1})]\frac{N}{r}[\varphi(t^N_j) - \varphi(t^N_{j+1})] + \frac{N}{r}(\theta - t^N_{j+1})\frac{N}{r}[\varphi(t^N_{j-1}) - \varphi(t^N_j)] - \dot{\varphi}(\theta)$$

$$= [1 - \frac{N}{r}(\theta - t^N_{j+1})](\dot{\varphi}(\xi_1) - \dot{\varphi}(\theta)) + \frac{N}{r}(\theta - t^N_{j+1})(\dot{\varphi}(\xi_0) - \dot{\varphi}(\theta)),$$

where $\xi_0 \in (t_j, t^N_{j-1})$, $\xi_1 \in (t^N_{j+1}, t^N_j)$. Since $(A^N\varphi)(0) - (A\varphi)(0) = A_0\varphi(0) + A_1\varphi(-r) - \dot{\varphi}(0) = 0$, we immediately get

$$|A^N\varphi - A\varphi| \leq \sup\{|\dot{\varphi}(\xi) - \dot{\varphi}(\eta)| \,\big|\, |\xi - \eta| < 2\frac{r}{N},\ \xi, \eta \in [-r,0]\},$$

i.e. (H2) is satisfied.

For $\varphi \in D$ the solution of (2.1) is continuously differentiable on $[-r,\infty)$. Since a first order spline attains its maximum always at a mesh-point, we have

$$|A^N S(t)\varphi| = |A^N x_t(\varphi)| = |A^N x_t(\varphi)(t^N_j)|$$

for a $j \geq 0$. Then the estimates

$$|(A^N x_t(\varphi))(t^N_j)| = \frac{N}{r}|x_t(\varphi)(t^N_{j-1}) - x_t(\varphi)(t^N_j)|$$

$$= \frac{N}{r}|x(t + t^N_{j-1};\varphi) - x(t + t^N_j;\varphi)| = |\dot{x}(\xi;\varphi)|$$

with $\xi \in [-r,t_1]$ for $t \in [0,t_1]$, in case $j \geq 1$, and

$$|(A^N x_t(\varphi))(0)| = |A_0 x(t;\varphi) + A_1 x(t-r;\varphi)| = |\dot{x}(t;\varphi)|,$$

in case $j = 0$, show

$$|A^N S(t)\varphi| \leq \sup_{-r \leq \xi \leq t_1} |x(\xi;\varphi)| \quad \text{for all } t \in [0,t_1] \text{ and all } N,$$

i.e. (H3) is also satisfied.

An application of Theorem 1.3 gives

<u>Theorem 2.1.</u> For any $\varphi \in C$

$$\lim_{N\to\infty} e^{A^N t}\varphi = x_t(\varphi)$$

uniformly for t in bounded intervals, where A^N is defined by (2.2).

3. SPLINE APPROXIMATION FOR THE DUAL SEMIGROUP

In [15] Gibson considered the linear quadratic optimal control problem for linear autonomous delay systems and proved that the solution of the corresponding infinite dimensional Riccati equation can be approximated by solutions of Riccati matrix equations in a very strong sence (i.e. with respect to the trace-norm). Basic for this result is that in case of averaging projections the solution semigroup $S(t)$ and the dual semigroup $S^*(t)$ are approximated in the strong operator topology by $S^N(t)$ and $S^N(t)^*$, respectively. For the spline schemes presented in [9] we can only show that $S^N(t)^*$ converges to $S^*(t)$ in the weak operator topology. The following spline scheme was developed recently by D.Salamon and the author of this paper. In this section we state the convergence result for the simplest case. Proofs, more general situations and the implications for the Riccati equation are considered in a forthcomming paper.

We consider again equation (2.1) but now in the state space $Z = \mathbb{R}^n \times L^2(-r,0;\mathbb{R}^n)$. The solution semigroup $T(t)$ is defined by $T(t)(z^1, z^2) = (x(t;z), x_t(z))$ for $t \geq 0$ and $z = (z^1, z^2) \in Z$, where $x(t;z)$ is the unique solution of (2.1) corresponding to the initial condition $x(0) = z^1$, $x(s) = z^2(s)$ a.e. on $[-r,0)$. The subspaces Z^N are given by

$$Z^N = \mathbb{R}^n \times X^N,$$

where X^N is the subspace of $L^2(-r,0;\mathbb{R}^n)$ consisting of all first order splines with knots at the points $t_j^N = -j\frac{r}{N}$, $j = 0,\dots,N$. P^N is the orthogonal projection onto Z^N and the matrix representation $[A^N]$ of A^N: $Z^N \to Z^N$ with respect to the basis $(I,0),(0,e_j^N)$, $j = 0,\dots,N$, e_j^N the spline with $e_j^N(t_i^N) = \delta_{ij}$, is given by

$$[A^N] = (Q^N)^{-1}H^N.$$

The matrices Q^N and H^N are

$$Q^N = \frac{r}{N}\begin{pmatrix} \frac{N}{r} & 0 & \cdots & \cdots & \cdots & 0 \\ 0 & \frac{1}{3} & \frac{1}{6} & 0 & \cdots & 0 \\ \vdots & \frac{1}{6} & \frac{2}{3} & \frac{1}{6} & \ddots & \vdots \\ \vdots & 0 & \ddots & \ddots & \ddots & 0 \\ \vdots & \vdots & \ddots & \frac{1}{6} & \frac{2}{3} & \frac{1}{6} \\ 0 & 0 & \cdots & 0 & \frac{1}{6} & \frac{1}{3} \end{pmatrix} \otimes I,$$

$$H^N = \begin{pmatrix} A_0 & 0 & \cdots & \cdots & 0 & A_1 \\ I & -\frac{1}{2}I & -\frac{1}{2}I & 0 & \cdots & 0 \\ 0 & \frac{1}{2}I & 0 & \ddots & \ddots & \vdots \\ \vdots & 0 & \ddots & \ddots & \ddots & 0 \\ \vdots & \vdots & \ddots & \ddots & 0 & -\frac{1}{2}I \\ 0 & 0 & \cdots & 0 & \frac{1}{2}I & -\frac{1}{2}I \end{pmatrix}$$

The semigroups $S^N(t)$ on Z^N are given by $S^N(t) = e^{A^N t}$. The adjoint semigroups $S^N(t)^*$ on Z^N are given by $S^N(t)^* = e^{(A^N)^* t}$, where the matrix representation of $(A^N)^*$ with respect to the basis chosen above for Z^N is

$$[(A^N)^*] = (Q^N)^{-1}(H^N)^T.$$

It should be noted, that the infinitesimal generator A^* of $S^*(t)$ is given by

$$\text{dom } A^* = \{(z^1, z^2) \in Z \mid z^2 \in W^{1,2}(-r,0,\mathbb{R}^n),\ z^2(-r) = A_1^T z^1\},$$

$$A^*(z^1, z^2) = (z^2(0) + A_0^T z^1, \dot{z}^2).$$

Therefore, neither $Z^N \subset \text{dom } A$ nor $Z^N \subset \text{dom } A^*$.

The fundamental result is given by

Theorem 2.1. For all $z \in Z$

$$\lim_{N\to\infty} S^N(t)P^N z = S(t)z$$

and

$$\lim_{N\to\infty} S^N(t)^* P^N z = S^*(t)z$$

uniformly on bounded t-intervals.

REFERENCES

[1] Banks, H.T.: Approximation of nonlinear functional differential equation control systems, J. Optimization Theory Appl. 29 (1979), 383-408.

[2] Banks, H.T.: Identification of nonlinear delay systems using spline methods, in Proc. Int. Conf. Nonl. Phenomena in Math. Sci., Arlington, Texas, June 1980, to appear.

[3] Banks, H.T., J.A. Burns: An abstract framework for approximate solutions to optimal control problems governed by hereditary systems, in Proc. Int. Conf. Diff. Eqs., Univ. So.Calif., Sept. 74 (H.A. Antosiewicz, Ed.) pp. 10-25, Academic Press, New York 1975.

[4] Banks, H.T., J.A. Burns: Hereditary control problems: Numerical methods based on averaging approximations, SIAM J. Control Optimization 16 (1978), 169-208.

[5] Banks, H.T., J.A. Burns, E.M. Cliff: A comparison of numerical methods for identification and optimization problems involving control systems with delays, LCDS Techn. Report 79-7, Brown University, Providence, R.I.

[6] Banks, H.T., J.A. Burns, E.M. Cliff: Parameter estimation and identification for control systems with delays, SIAM J. Control and Optimization 19 (1981), 791-829.

[7] Banks, H.T., J.M. Crowley, K. Kunisch: Cubic spline approximation techniques for parameter estimation in distributed systems, to appear.

[8] Banks, H.T., P.L. Daniel: Parameter estimation of nonlinear nonautonomous distributed systems, in Proc. 20th IEEE Conf. Decision and Control, San Diego, December 1981, 228-232.

[9] Banks, H.T., F. Kappel: Spline approximation for functional differential equations, J. Differential Eqs. 34 (1979), 496-522.

[10] Banks, H.T., K. Kunisch: An approximation theory for nonlinear partial differential equations with applications to identification and control, LCDS Techn. Report 81-7, Brown Univ., Providence, R.I.

[11] Burns, J.A., E.M. Cliff: Methods for approximating solutions to linear hereditary quadratic optimal control problems, IEEE Trans. AC 23 (1978), 21-36.

[12] Burns, J.A., P.D. Hirsch: A difference equation approach to parameter estimation for differential-delay equations, Appl. Math. and Computation 7 (1980), 281-311.

[13] Daniel, P.L.: Spline approximations for nonlinear hereditary control systems, J. Optimization Theory and Appl., submitted.

[14] De Boor, C.: A Practical Guide to Splines, Springer Verlag, New York 1978.

[15] Gibson, J.S.: Linear quadratic optimal control of hereditary differential systems: infinite dimensional Riccati equations and numerical approximations, SIAM J. Control and Optimization, to appear.

[16] Gottlieb, D., S.A. Orszag: Numerical Analysis of Spectral Methods: Theory and Applications, CBMS-NSF Regional Conference Series in Applied Mathematics, SIAM 1977.

[17] Halanay, A., Vl. Rasvan: Approximation of delays by ordinary differential equations, in Proc. Intern. Conf. Recent Adv. Diff. Eqs., Trieste, August 1978, (R. Conti, Ed.), pp. 155-197, Academic Press 1981.

[18] Hale, J.K.: Theory of Functional Differential Equations, Springer-Verlag, New York 1977.

[19] Kappel, F.: Approximation of neutral functional differential equations in the state space $\mathbb{R}^n \times L^2$, in Colloquia Mathematica Societatis Jànos Bolyai Vol. 30: Qualitative Theory of Differential Equations, Szeged 1979 (Farkas, Ed.), pp. 463-506, North Holland, Amsterdam 1981.

[20] Kappel, F.: An approximation scheme for delay equations, in Proc. Int. Conf. Nonl. Phenomena in Math. Sci., Arlington, Texas, June 1980, to appear.

[21] Kappel, F.: Spline approximation for autonomous nonlinear functional differential equations, J. Nonlinear Analysis: Theory, Methods and Appl., to appear.

[22] Kappel, F., K. Kunisch: Spline approximations for neutral functional differential equations, SIAM J. Numer. Analysis 18 (1981), 1058-1080.

[23] Kappel, F., K. Kunisch: Approximation of the state of infinite delay and Volterra-type equations, Preprint No. 8-1982, Math. Inst.,Tech. Univ. and University of Graz, 1982.

[24] Kappel, F., W. Schappacher: Autonomous nonlinear functional differential equations and averaging approximations, J. Nonlinear Analysis: Theory, Methods and Applications 2 (1978), 391-422.

[25] Kappel, F., W. Schappacher: Non-linear functional differential equations and abstract integral equations, Proc. Royal Soc. Edinburgh 84A (1979), 71-91.

[26] Krasovskij, N.N.: The approximation of a problem of analytic design of controls in a system with time-lag, J. Appl. Math. Mech. 28 (1964), 876-885.

[27] Kunisch, K.: Neutral functional differential equations in L^p-spaces and averaging approximations, J. Nonlinear Analysis: Theory, Methods and Appl. 3 (1979), 419-448.

[28] Kunisch, K.: Approximation schemes for the linear quadratic optimal control problem associated with delay equations, SIAM J. Control and Optimization, to appear.

[29] Martin, Jr., R.H.: Nonlinear Operators and Differential Equations in Banach Spaces, J. Wiley, New York 1976.

[30] Reber, D.C.: A finite difference technique for solving optimization problems governed by linear functional differential equations, J. Differential Eqs. 32 (1979), 192-232.

[31] Rosen, I.G.: A discrete approximation framework for hereditary systems, J. Differential Eqs. 40 (1981), 377-449.

[32] Schultz, M.H.: Spline Analysis, Prentice Hall, Englewood Cliffs, N.J., 1973.

[33] Webb, G.F.: Functional differential equations and nonlinear semigroups in L^p-spaces, J. Differential Eqs. 20 (1976), 71-89.

ASYMPTOTIC BEHAVIOR IN FUNCTIONAL DIFFERENTIAL EQUATIONS WITH INFINITE DELAY

Junji Kato
Tohoku University, Sendai, Japan

The relationship between the stability (including the boundedness) of a system and that of the limiting equations has been discussed by many authors. For a part of references, see [17]. Recently we have added some remarks including several illustrative examples which show that supplement conditions can not be dropped in order that the stability property is inherited to the limiting equations [17].

In this report, we shall show how we can extend those results to functional differential equations, though the situation is not so obvious as shown by the example given in [16]. This will be done in Section 2.

In Section 3, we shall review the various developments in the Liapunov second method for functional differential equations based on the ideas inspired by Razumikhin [19], by Barnea [1] and by Burton [5].

1. Phase Space.

Consider the functional differential equation with infinite delay

$$\dot{x}(t) = f(t, x_t) \tag{E}$$

defined on the domain $I \times X$, where $I = [0,\infty)$, $x_t(s) := x(t+s)$ for $s \leqq 0$ and $(X, \|\cdot\|)$ is an admissible phase space (see [11]), that is, a semi-normed space which satisfies:

(H1) $x_t \in X$,

(H2) x_t is continuous in t,

(H3) $|x(t)| \leqq \|x_t\| \leqq K(t-\tau)\sup_{\tau \leqq s \leqq t}|x(s)| + M(t-\tau)\|x_\tau\|$

for positive continuous function $K(s)$ and $M(s)$, whenever $x_\tau \in X$, $x(t)$ is continuous on $[\tau, \tau+\alpha)$ and $t \in [\tau, \tau+\alpha)$. Here, $f(t,\phi)$ is assumed to be in $C(I \times X, R^n)$, the space of continuous functions of (t,ϕ), and hence it takes a common value on each equivalent class, which enables us to indentify the elemement of X and its equivalent class.

In [11] we have shown that these hypotheses supply adequate features to the phase space to discuss the local properties of solutions and that the global properties such as the stability can be treated in a systematical way under a fading memory (H3*), which means that in (H3) $K(s)$ is independent of s and $M(s) \to 0$ as $s \to \infty$.

The following will be easily established.

Lemma 1. Suppose that X has a fading memory (H3*). Then the sequence $\{x^k_{t_k}\}$ contains a convergent subsequence if $x^k \in S(\alpha,L)$ for positive constants α and L and if $t_k \to \infty$, where $S(\alpha,L)$ is the collection of the functions $x(t)$ on $(-\infty,\infty)$ which satisfy $\|x_0\| \leqq \alpha$, $|x(t)| \leqq \alpha$ and $|x(t) - x(s)| \leqq L|t-s|$ on $[0,\infty)$.

Typical examples of such phase spaces are the space M_γ of measurable functions ϕ with finite $\|\phi\| := |\phi(0)| + \int_{-\infty}^{0} e^{\gamma s}|\phi(s)|ds$ and the space C_γ of ϕ such that $e^{\gamma s}\phi(s)$ is bounded and uniformly continuous on $(-\infty,0]$ with $\|\phi\| := \sup\{e^{\gamma s}|\phi(s)| : s \leqq 0\}$, where $\gamma \geqq 0$ is a constant ($\gamma>0$ for(H3*)). For the axiomatic investigations of the phase space, see [6], [11], [12], [20] etc.

The hypothesis (H2) excludes the space of bounded continuous functions with the uniform norm, but the example due to Seifert [22] shows that this exclusion is not altogether without reason.

2. Limiting Equation.

For the equation (E) suppose that X is separable and that $H(f)$ is a compact subset of $C(I \times X,R^n)$ endowed with the compact open topology, where $H(f)$ is the hull, that is, the closure of the set $T(f) := \{f_\tau : \tau \in I\}$, $f_\tau(t,\phi) := f(t+\tau,\phi)$, in $C(I \times X,R^n)$. Here we note that $C(I \times X,R^n)$ is not complete in this topology but that $H(f)$ is compact if, and only if, $f(t,\phi)$ is bounded and uniformly continuous on $I \times K$ for any compact set K of X (for the proof, refer to [17] also to [23]). Then we have the following:

Lemma 2. (i) $g \in \Omega(f) := \bigcap_{t>0} H(f_t)$ if, and only if, there exists a sequence $\{t_k\}$, $t_k \to \infty$, for which $\{f_{t_k}\}$ converges to g uniformly on any compact set of $I \times X$;

(ii) $H(f) = T(f) \cup \Omega(f)$;

(iii) Both of $H(f)$ and $\Omega(f)$ are invariant, namely, if $g \in H(f)$ [resp. $\Omega(f)$], then $H(g) \subset H(f)$ [resp. $\Omega(f)$];

(iv) If $f(t,\phi)$ is almost periodic in t uniformly for ϕ in each compact set of X, then

(1) $\Omega(g) = H(f)$ for any $g \in H(f)$.

Remark 1. The converse of the statement (iv) is not true. Consider the example $f(t) := F(t,0)$, where

$$F(t,\alpha) := \begin{cases} \sigma(k,\alpha) := \operatorname{sgn} \cos(2\pi(k\theta+\alpha)) & \text{if } t = k \in Z \\ \left(\dfrac{\sin 2\pi t}{2\pi}\right)^2 \displaystyle\sum_{k=-\infty}^{\infty} \dfrac{\sigma(k,\alpha)}{(t-k)^2} & \text{if } t \notin Z \end{cases}$$

for a given irrational number θ, given by Veech [24] to show that an almost automorphic function needs not be almost periodic (for the definition, see [24] or [3]). As stated by Veech, $f(t)$ is not almost periodic. However, it is not difficult to see that $H(f) = \{F(t+\tau,\alpha): (\tau,\alpha) \in [0,1) \times [0,1)\}$ and that for a given $\alpha \in [0,1)$ there is a sequence of integers $\{n_k\}$, $n_k \to \infty$, such that $\{F(t+n_k,\alpha)\}$ converges to $F(t,0)$ uniformly on any compact interval. The crucial part of the proof of these facts is based on Kronecker's Lemma (for this lemma, refer to [2]).

For any $g \in \Omega(f)$, the equation

(LE) $\dot{x}(t) = g(t,x_t)$

is called <u>a limiting equation</u> of (E), and (E) is said to be <u>regular</u> if the solution of (LE) is unique for the initial value problem and continuable to the right as long as it remains bounded whatever $g \in \Omega(f)$ is. A property of solutions of (E) is said to be an <u>inherited</u> property for (E) if the solutions of each limiting equation preserve the same property, refer to [8].

Remark 2. We note that neither of the uniqueness nor the continuability is an inherited property. A counter-example for the uniqueness is

$$\dot{x}(t) = (x^2 + a(t))^{1/3}$$

given in [23], where $a(t)$ is an almost periodic continuous function, due to Sibuya, which statisfies

(i) $a(t) > 0$ for all $t \in (-\infty,\infty)$,

(ii) there is a $b \in \Omega(a)$ such that $b(t) \equiv 0$ on $[0,\ell]$ for a preasigned constant $\ell > 0$,

while a counter-example for the continuability is

$$\dot{x}(t) = f(x_{t-a(t)}),$$

where $a(t)$ is the one in the above and $f(\phi)$ is the function given by Yorke [25] so that $x(t) = \sin 1/t$, $t < 0$, is a solution of

$$\dot{x}(t) = f(x_t).$$

A sufficient condition for the uniqueness to be inherited is that $f(t,\phi)$ satisfies a uniform Lipschitz condition on $I \times K$ for each compact set $K \subset X$. On the other hand, a sufficient condition for the continuability is that

(2) $\qquad f(t,\phi)$ is uniformly bounded on $I \times B$

for each bounded set $B \subset X$.

Finally we shall say that the solutions of (E) are <u>interval-bounded</u> if for given $\alpha > 0$ and $T \geqq 0$ there exists a $\beta > 0$ such that $|x(t)| \leqq \beta$ on $[\tau,\tau+T]$ for any solution x of (E) if $\|x_\tau\| \leqq \alpha$ for a $\tau \in I$.

Other definitions of boundedness such as the uniform boundedness and the uniform ultimate boundedness can be defined as in [17].

For ordinary differential equations we have the following. For the proof, see [17].

Proposition 1. If every solution of (LE) for any $g \in H(f)$ is continuable up to $t = \infty$, then the solutions of (E) are interval-bounded, and the converse is true when (E) is regular.

Proposition 2. If the solutions of (LE) are uniformly ultimately bounded for any $g \in H(f)$, then the solutions of (E) are uniformly bounded.

However, the example in [16] tells us that the same assertions are no more true for functional differential equations even if f is autonomous, of finite delay, locally Lipschitzian and completely continuous.

This example also shows that the uniform ultimate boundedness does not necessarily imply the uniform boundedness even for autonomous systems, which prevents us from extending the results given for ordinary differential equations to functional differential equations in a general way.

For functional differential equations we can only state the following obvious theorems, based on the following lemma (the second part):

Lemma 3. Suppose that $\{f_{t_k}\}$ converges to a $g \in \Omega(f)$, and let $x^k(t)$ be a noncontinuable solution of (E) such that $x^k_{t_k}$ tends to a $\xi \in X$. Then, the sequence $\{x^k(t+t_k)\}$ contains a subsequence which converges to a solution $x(t)$ of (LE) through ξ at $t = 0$ uniformly on every compact interval of the domain of $x(t)$. Furthermore, if $x(t)$ is such a unique solution, then $\{x^k(t+t_k)\}$ itself must converge to $x(t)$.

Theorem 1. If the solutions of (E) are interval-bounded and uniformly ultimately bounded, then they are uniformly bounded.

Theorem 2. If (E) is regular, then the following are inherited properties for (E):

(a) the interval-boundedness,
(b) the uniform (asymptotic) stability of the zero solution,
(c) the uniform boundedness,
(d) the uniform ultimate boundedness under the interval-boundedness.
(e) the global uniform asymptotic stability of the zero solution under the interval-boundedness.

Remark 3. It was shown by examples [17] that even for almost periodic ordinary differential equations the regularity can not be omitted in the above. Similarly the interval-boundedness can not be deleted in Theorem 2 (d) and (e). However, in the latter we can delete this condition if the solutions of every limiting equation are continuable up to $t = \infty$, in spite of the remark after Propositions 1 and 2.

If (E) lacks the regularity, the situation becomes more complicate. For example, we have the following theorem.

Theorem 3. Suppose that X has a fading memory (H3*), that f satisfies the conditions (1) and (2) and that solutions of every limiting equation of (E) are uniformly bounded. Then, the uniform ultimate boundedness is an inherited property for (E).

Proof. Suppose that there is a $g \in H(f)$ for which the solutions of (LE) are not uniformly ultimately bounded. Then, there exist a constant $\alpha > 0$, sequences $\{\tau_k\}$, $\{t_k\}$ and $\{x^k(t)\}$, solutions of (LE), such that $\tau_k \geqq 0$, $t_k - \tau_k (=2s_k) \geqq k$, $\|x^k_{\tau_k}\| \leqq \alpha$ and $\|x^k_t\| \geqq K(B+1)+1$ on $[\tau_k, t_k]$, because otherwise we will have $|x(t)| \leqq \gamma(K(B+1)+1)$ for all $t \geqq \tau + \sigma$ and some $\sigma > 0$ if $\|x_\tau\| \leqq \alpha$, where B is a bound for the uniform ultimate boundedness of (E) and $\gamma(\alpha)$ is the number which is associated with the definition of the uniform boundedness of (LE). Here, note that $g \in H(f) = \Omega(f)$ by Lemma 2. Since $|x^k(t)| \leqq \gamma(\alpha)$ for all $t \geqq \tau_k$, the sequence $\{y^k\}$ defined by $y^k(t) = x^k(t+\tau_k)$ belongs to $S(B(\alpha), L(\alpha))$ under (H3*), where $L(\alpha)$ is a bound for $|f(t,\phi)|$ on $I \times \{\phi : \|\phi\| \leqq B(\alpha)\}$ and $B(\alpha) = \max\{\alpha, \gamma(\alpha)\}$. Thus, by Lemma 1 $\{y^k_{s_k}\}$ contains a convergent subsequence, or simply, we may assume that $\{y^k_{s_k}\}$ is convergent to a $\xi \in X$. On the other hand, by applying Lemma 3 we may also assume that $x^k(t + \tau_k + s_k) = y^k(t + s_k)$ converges to an $x(t)$ uniformly on any compact interval of $[0,\infty)$ and that $\{g_{\tau_k+s_k}\}$ converges to an $h \in \Omega(g)$, where $x(t)$ is a solution of

$$\dot{x}(t) = h(t, x_t)$$

through ξ at $t = 0$. Clearly $\|x^k_{t+\tau_k+s_k}\| \geqq K(B+1)+1$ on $[0, t_k - \tau_k - s_k] = [0, s_k]$ implies that $\|x_t\| \geqq K(B+1)+1$ on $[0,\infty)$. Since $h \in \Omega(g) \subset H(f) = \Omega(f)$ by Lemma 2, we have $f \in \Omega(h)$ by (1) and there is a sequence $\{r_k\}$ for which $\{h_{r_k}\}$ converges to the $f \in \Omega(h)$. Thus, applying Lemma 3 again we can see that $\{x(t + r_k)\}$ converges to a solution $y(t)$ of (E), and $\|y_t\| \geqq K(B+1)+1$ on $[0,\infty)$. Suppose that there is a $T \geqq 0$ for which $|y(t)| \leqq B+1$ on $[T,\infty)$. Then, under the condition (H3*) we can find a T_1 so that $M(t)\gamma(\alpha) < 1$ for all $t \geqq T_1$, and hence $\|y_t\| < K(B+1)+1$ for $t \geqq T + T_1$, a contradiction. Thus, we have $|y(t)| \geqq B+1$ for an arbitrarily large t, again a contradiction.

3. Liapunov's Second Method.

For simplicity we only discuss about a continuous (Liapunov) function $V(t,\phi)$ defined on $I \times \{\phi \in X : \|\phi\| < H\}$ which provides a sufficient condition for the uniform asymptotic stability of the zero solution of (E), namely (refer to [9]), for the fact

$$|x(t,\tau,\xi)| \leqq \sigma(t - \tau, \|\xi\|), \tag{3}$$

where $x(t,\tau,\xi)$ denotes a solution of (E) through (τ,ξ) and $\sigma(t,\varepsilon)$ is a continuous function on $[0,\infty) \times [0,H_o]$, $H_o < H$, approaching to 0 uniformly as $t \to \infty$ or as $\varepsilon \to 0$.

When (E) is ordinary, namely, when $X \cong R^n$, a typical theorem says that the triple of the conditions

(A) $\quad a(|\phi(0)|) \leqq V(t,\phi)$,

(B) $\quad V(t,\phi) \leqq b(|\phi(0)|)$

(C) $\quad \dot{V}_{(E)}(t,\phi) \leqq -c(V(t,\phi))$

is sufficient for this purpose, where $a(r)$, $b(r)$, $c(r)$ are continuous functions which are positive for $r > 0$, $b(0) = 0$, and $\dot{V}_{(E)}(t,\phi)$ is defined by

$$\dot{V}_{(E)}(t,\phi) = \overline{\lim_{h\to+0}} \frac{1}{h}\{V(t+h,x_{t+h}) - V(t,\phi)\}$$

for solutions $x(s)$ of (E) through (t,ϕ). Clearly, under the conditions (A) and (B), the condition (C) is equivalent to requiring

(C^*) $\quad \dot{V}_{(E)}(t,\phi) \leqq -c^*(|\phi(0)|)$

(the function $c^*(r)$ is similar to $c(r)$ in (C)). It is also known that the condition (A)-(B)-(C) is necessary under a smooth condition on f.

The sufficiency of (A)-(B)-(C) is also valid for delay equations such as

$$\dot{x}(t) = -(1 + x(t-1)^2)x(t)$$

with $V(t,\phi) = \phi(0)^2$, but this is far from being a necessary condition. A legitimate generalization for delay equations may be (A_Δ)-(B_Δ)-(C) with

(A_Δ) $\quad a(\|\phi\|) \leqq V(t,\phi)$,

(B_Δ) $\quad V(t,\phi) \leqq b(\|\phi\|)$,

and this condition or (A)-(B_Δ)-(C) or even (A)-(B_Δ)-(C^*_Δ) are sufficient and also necessary under a smooth condition on f, where

(C^*_Δ) $\quad \dot{V}_{(E)}(t,\phi) \leqq -c^*(\|\phi\|)$,

which follows from (C) under (A_Δ) while implies (C) under (B_Δ). However t is quite difficult to obtain a suitable Liapunov function satisfying these conditions for a practical equations.

Thus, there are several attempts to find a sufficient condition such that it is easier to construct a Liapunov function endowed with the condition. Among them the following ideas are specifically notable. (The statements are not exact as in the literatures, since our aim is to sketch their ideas). For the moment we shall consider the case of a finite delay h, that is, the case where $X \cong C([-h,0],R^n)$.

(I). Krasovski [18; Th. 31.1] and Yoshizawa [26: Th. 33.3] show that the condition $(A)-(B_\Delta)-(C^*)$ is sufficient if f is uniformly bounded. This can be thought as a sort of generalized LaSalle's invariant principle.

(II). Burton [4] also shows that $(A)-(B_\Delta)-(C^*)$ is sufficient without the boundedness condition on f if $X \cong M_0([-h,0],R^n)$ or if the norm appeared in (B_Δ) is given by

$$\|\phi\| = |\phi(0)| + \int_{-h}^{0} |\phi(s)|ds. \tag{4}$$

Here, we note that the stability in $M_0([-h,0],R^n)$ implies that in $C([-h,0],R^n)$ since

$$|\phi(0)| + \int_{-h}^{0} |\phi(s)|ds \leq (1 + h)\sup_{-h\leq s\leq 0}|\phi(s)|.$$

(III). Razumikhin [19] (also, see [18], [7]) presents an important idea: $(A)-(B_\Delta)-(C_F)$ is sufficient, that is, the condition (C) suffices to hold if only ϕ satisfies

(F) $V(t+s,\phi_s) \leq F(V(t,\phi))$ for $s \in [-p,0]$,

where $p = h$ and $F(r)$ is a continuous function which satisfies $F(r) > r$ for $r > 0$.

This idea allows us to use such a simple function as $V(t,\phi) = \phi(0)^2$, and it turns out that this is very useful for many practical equations though in theoretical we can construct a Liapunov function satisfying $(A)-(B_\Delta)-(C)$ based on the Liapunov function with $(A)-(B_\Delta)-(C_F)$ (see [13]).

For example, consider the equation

$$\dot{x}(t) = - ax(t) + bx(t - h), \ |b| < a, \tag{5}$$

Then, $V(t,\phi) = \phi(0)^2$ satisfies $(A)-(B_\Delta)-(C_F)$ with $a(r) = b(r) = r^2$, $c(r) = 2(a - \lambda|b|)r$ and $F(r) = \lambda^2 r$ for a λ, $1 < \lambda < a/|b|$. Thus, the zero solution of (5) is uniformly asymptotically stable. On the other hand, we can see that

$$(6) \qquad W(t,\phi) := \sup_{-h\leq s\leq 0} V(t+s,\phi_s)\exp\left[\frac{s}{h}\log\frac{V(t+s,\phi_s)}{F^{-1}(V(t+s,\phi_s))}\right]$$
$$= \sup_{-h\leq s\leq 0} |\phi(s)|^2 \exp\left[\left(\frac{2}{h}\log\lambda\right)s\right]$$

satisfies $(A_\Delta)-(B_\Delta)-(C)$ with $a(r) = r^2/\lambda^2$, $b(r) = r^2$, $c(r) = \min\{2(a - \lambda|b|)r, (2r\log\lambda)/h\}$. Another way to construct a Liapunov function for (5) is

$$V(t,\phi) = \phi(0)^2 + \mu\int_{-h}^{0}\phi(s)^2 ds$$

with a $\mu > 0$ satisfying $(2a - \mu)\mu > b^2$ (see [10]) under the condition $|b| < a$. We should note that this function satisfies $(A)-(B_\Delta)-(C^*)$ with $a(r) = r^2$, $b(r) = (1 + \mu h)r^2$, $c^*(r) = \{2a - \mu - b^2/\mu\}r^2$ but does not satisfy $(A)-(B_\Delta)-(C)$, though to see the stability the both of (I) and (II) are applicable.

(IV). Another idea has been presented by Barnea [1] (also, see [13]). Extending his idea we can see that $(A)-(B_\Delta)-(C_{FD})$ is sufficient, namely, the condition (C) needs not hold unless we have (F) and

(D) $\phi = x_t$ for x being a solution of (E) on the duration $[t-p,t]$

with $p = kh$ for an integer k, where it is assumed that f satisfies a uniform Lipschitz condition, or at least the relation (3) holds for a $\delta(t,\varepsilon)$ which tends to 0 as $\varepsilon \to 0$ uniformly in $t \in [0,T]$ for each $T > 0$.

This assertion makes it possible to show that the zero solution of

$$\dot{x}(t) = -bx(t-h)$$

is uniformly asymptotically stable if $0 < bh < 3/2$ by using the simple function $V(t,\phi) = \phi(0)^2$ and setting $k = 2$ (see [13], [14]). Here we note that by a similar way as in (6) the condition $(A)-(B_\Delta)-(C_{FD})$ guarantees the existence of a Liapunov function which satisfies $(A)-(B_\Delta)-(C_D)$.

In (IV) the Lipschitz condition on $f(t,\phi)$ is crucial. For example, consider the equation

$$\dot{x}(t) = a(t)x(g(t)), \tag{7}$$

where $a(t)$ is a continuous function defined on $(-\infty,\infty)$ and satisfying

(i) $a(t) = 0$ except for $t \in \bigcup_{k=-\infty}^{\infty} [s_k, s_k+2\sigma_k]$

(ii) $\int_{s_k}^{s_k+2\sigma_k} a(s)ds = -1$

Suppose that $g(t) = t - h$ and that $s_k = 2kh$, $\sigma_k = h/2$ for integers k. Then, the only element satisfying (D) with $p = 5h$ is the zero function. Clearly, the zero solution of (7) is not uniformly stable if $a(t)$ satisfies

$$\int_{s_k}^{s_k+\sigma_k} a(s)ds \geqq k.$$

(V). Recently, Burton [5] has constructed a Liapunov function for a Volterra integral differential equations. Along his idea we can say that the condition (A_{F*})-(B_Δ)-(C) is sufficient, namely, the Liapunov function $V(t,\phi)$ suffices to satisfy the condition (A) if only ϕ satisfies

(F^*) $\quad |\phi(s)| \leqq F(|\phi(0)|) \qquad$ for $s \in [-p,0]$

with $p = h$ and F as in (F).

The extension of these results to the infinite delay case are not immediate, though the sufficiency of (A)-(B_Δ)-(C) is obvious.

The result (I) is valid for the infinite delay case if the phase space admits the fading memory $(H3^*)$, while so is (II) if the norm (4) is replaced by

$$\|\phi\| = |\phi(0)| + \int_{-\infty}^{0} e^{\gamma s}|\phi(s)|ds$$

for a $\gamma > 0$, see [15].

It is obvious that we cannot let $p = \infty$ in the results (III), (IV) and (V). Here we note that if $\phi = x_t$ does not satisfy (F^*) for all $t \geqq \tau$ then we have

$$|x(t)| \leqq (F^{-1})^k(\|x_\tau\|) \qquad \text{for } t \geqq \tau + kp,$$

which guarantees that $x(t)$ tends to zero uniformly as $t \to \infty$ but that if $p = \infty$ in (F^*) then the best possible conclusion is

$$\overline{\lim_{t\to\infty}}|x(t)| < F^{-1}(\sup_{s\leq\tau}|x(s)|).$$

Actually, for the equation

$$\dot{x}(t) = -3x(t) + x(0)$$

the function $V(t,\phi) = \phi(0)^2$ satisfies (A)-(B_Δ)-(C_F) with $F(r) = 2r$ and $p = \infty$. However, any solution $x(t)$ can not approach to zero unless $x(0) = 0$. On the other hand, again consider the equation (7), and suppose that

$$g(t) = \begin{cases} t - 2 & \text{if } t \leqq 4 \\ t/2 & \text{if } t \geqq 4 \end{cases}$$

and that $s_k = 4k$, $\sigma_k = 1$ for nonpositive integer k and $s_k = 2\cdot 3^k$ $\sigma_k = 3^k$ for positive integer k. Then the only solution defined on $(-\infty,t]$ is the zero solution, that is, the only element satisfying (D) with $p = \infty$ is the zero function. Moreover, we may assume that $a(t)$ is bounded but satisfies

$$\int_{s_k}^{s_k+\sigma_k} a(s)ds \geqq 3^k,$$

which prevents the zero solution of (7) from being uniformly stable.

Therefore, it will be substantial to assume some restriction on p. It is shown that if $p = p(V(t,\phi))$ in (F) and (D) and if $p = p(\|\phi\|)$ in (F^*) for a continuous function $p(r)$ of $r > 0$ ($p(r)$ may tend to ∞ as $r \to +0$), then the results (III), (IV) and (V) are valid even for infinite delay equations on a phase space with the fading memory $(H3^*)$, see [14], [15], [21] for (III) and (IV).

Following the idea of Professor Haddock [27] in this meeting, we may conjecture that for the infinite delay version of (III) and (V) the restriction (F) could be replaced by

(F_Δ) by setting $v(t+s) = V(t+s,\phi_s)$ for given (t,ϕ), $v_t \in X(R)$ and $\|v_t\| \leqq F(V(t,\phi))$,

where $X(R)$ is an admissible phase space of scalar functions with a fading memory, and (F^*) by

$$(F_\Delta^*) \quad \|\phi\| \leq F(|\phi(0)|).$$

Also, refer to [28; Theorem 8.2.2].

References.

[1]. B. I. Barnea, A method and new results for stability and instability of autonomous functional equations, SIAM J. Appl. Math., 17 (1969), 681-697.

[2]. A. S. Bescovitch, Almost Periodic Functions, Cambridge Univ. Press, Cambridge, 1932.

[3]. S. Bochner, A new approach to almost periodicity, Proc. Nat Acad. Sci. U. S., 48(1962), 2039-2043.

[4]. T. A. Burton, Uniform asymptotic stability in functional differential equations, Proc. Amer. Math. Soc., 68(1978), 195-199.

[5]. T. A. Burton, Perturbed Volterra equations, J. Differential Eq., 43(1982), 168-183.

[6]. B. D. Coleman and V. J. Mizel, On the stability of solutions of functional differential equations, Arch. Rational Mech. Anal., 30(1968), 178-196.

[7]. R. D. Driver, Existence and stability of solutions of a delay-differential system, Arch. Rational Mech. Anal., 10(1962), 401-426

[8]. A. M. Fink. Almost Periodic Differential Equations, Lec. Note in Math. 377, Springer-Verlag, Berlin-Heidelberg-New York, 1974.

[9]. W. Hahn, Stability of Motion, GMWE. 138 Springer-Verlag, Berlin-Heidelberg-New York, 1967.

[10]. J. K. Hale, Theory of Functional Differential Equations, Appl. Math. Sci. 3, Springer-Verlag, Berlin-Heidelberg-New York, 1977.

[11]. J. K. Hale and J. Kato, Phase space for retarded equations with infinite delay, Funkcialaj Ekvacioj, 21(1978), 11-41.

[12]. F. Kappel and W. Schappacher, Some considerations to the fundamental theory of infinite delay equations, J. Differential Eq., 37 (1980), 141-183.

[13]. J. Kato, On Liapunov-Razumikhin type theorems for functional differential equations, Funkcialaj Ekvacioj, 16(1973), 225-239.

[14]. J. Kato, Stability in functional differential equations, Proc. on Functional Differential Equations and Bifurcation, São Carlos, Brazil 1979 (Lec. Note in Math. 799, Springer-Verlag 1980).

[15]. J. Kato, Liapunov's second method in functional differential equations, Tohoku Math. J., 32(1980) 487-497.

[16]. J. Kato, An autonomous system whose solutions are uniformly ultimately bounded but not uniformly bounded, Tohoku Math. J., 32 (1980), 499-504.

[17]. J. Kato and T. Yoshizawa, Remarks on global properties in limiting equations, Funkcialaj Ekvacioj, 24(1981), 363-371.

[18]. N. N. Krasovskii, Stability of Motion, Standford Univ. Press, Standford, 1963.

[19]. R. S. Razumikhin, On the stability of systems with a delay, Prikl. Mat. Meh., 20(1956), 500-512.

[20]. K. Schumacher, Existence and continuous dependence for functional differential equations with infinite delay, Arch. Rational Mech. Anal., 67(1978), 315-334.

[21]. G. Seifert, Liapunov-Razumikhin conditions for asymptotic stability in functional differential equations of Volterra type, J. Differential Eq., 16(1974), 289-297.

[22]. G. Seifert, Positively invariant closed sets for systems of delay differential equations, J. Differential Eq., 22(1976), 292-304.
[23]. G. R. Sell, Nonautonomous differential equations and topological dynamic I. The basic theory, Trans. Amer. Math. Soc., 127(1967), 241-262; II. Limiting equations, ibd., 127(1967), 263-283.
[24]. W. A. Veech, Almost automorphic functions on groups, Amer. J. Math., 87(1965), 719-751.
[25]. J. A. Yorke, Noncontinuable solutions of differential delay equations, Proc. Amer. Math. Soc., 21(1969), 648-652.
[26]. T. Yoshizawa, Stability Theory by Liapunov's Second Method, Publication 9, Math. Soc. of Japan, 1966.
[27]. J. R. Haddock, Invariance principles for autonomous functional differential equations, EQUADIFF 82, Würzburg Aug. 23, 1982.
[28]. V. Lakshmikantham and S. Leela, Differential and Integral Inequalities, II, Academic Press, New York, London, 1969.

Uniqueness and nonexistence of limit cycles for the FitzHugh equation

E. Kaumann and U. Staude

1. Introduction

In [1], [2] R.FitzHugh proposed a system of ordinary differential equations as an approximation for the Hodgkin-Huxley model of the squid giant axon. This system of differential equations is equivalent to

$$\begin{aligned} \dot{x} &= y - \frac{x^3}{3} + x + \mu \\ \dot{y} &= \rho(a - x - by) \quad , \end{aligned} \tag{1}$$

$b \in (0,1)$, $a \in \mathbb{R}$, $\rho > 0$, [10].

In these equations x is the negative of the membrane potential, y is the quantity of refractoriness and μ is the magnitude of stimulating current.

Using numerical methods FitzHugh found periodic solutions for special values of the parameters.

System (1) has exactly one stationary point $(x(\mu),y(\mu))$ for every $\mu \in \mathbb{R}$

Let us take $\eta := x(\mu)$ as a new parameter. By the transformation $x - \eta \to x$, $y - \frac{a}{b} + \frac{\eta}{b} \to y$ system (1) can be transformed to

$$\begin{aligned} \dot{x} &= y - (\frac{x^3}{3} + \eta x^2 + (\eta^2 - 1)x) = y - H(x,\eta) \\ \dot{y} &= \rho(-x - by) \quad , \end{aligned} \tag{2}$$

where the origin is the only stationary point.

The stationary point is asymptotically stable for

$$|\eta| > \eta_o := \sqrt{1 - \rho b} \tag{3}$$

and unstable for $|\eta| < \eta_o$, [10]. For $\rho b > 1$ the stationary point is asymptotically stable for all η, [3].

In [3] - [6], [10] the Hopf bifurcation theory was applied to this system. It was shown that periodic solutions bifurcate from the stationary point when η is crossing $\pm\eta_o$. If

$$(b,\rho) \in A := \{(b,\rho) \mid b \in (0,1),\ \rho > 0,\ \rho b^2 - 2b + 1 \geq 0\} \tag{4a}$$

these periodic solutions are asymptotically stable, and they exist for $\eta_o - \delta < |\eta| < \eta_o$. In the case

$$(b,\rho) \in B := \{(b,\rho) \mid b \in (0,1),\ \rho > 0,\ \rho b^2 - 2b + 1 < 0\} \tag{4b}$$

the periodic solutions are unstable, and they exist for $\eta_o < |\eta| < \eta_o + \delta$.

In [3] the existence of at least one stable periodic solution for $|\eta| < \eta_o$ is proved.

2. Uniqueness

Our main result is the following

Theorem: $b \in (0,1)$, $a \in \mathbb{R}$, $\rho > 0$, $0 < \rho b < 1$.

i) For $|\eta| \leq \frac{1}{2}\eta_o$ system (1) has exactly one (asymptotically stable) limit cycle.

ii) For $(b,\rho) \in B_2 := \{(b,\rho) \mid b \in (0,1),\ \rho > 0,\ \rho b^2 - 7b + 6 < 0\}$ (5)
system (1) has exactly one (asymptotically stable) limit cycle for $|\eta| < \eta_o$, $(B_2 \subsetneqq B)$.

From [3] we know that system (2) can be transformed to a generalized Liênard system. However, it is not necessary to differentiate $H(x,\eta)$. By the transformation $y + \rho bx \to y$ we find

$$\begin{aligned} \dot{x} &= y - \left(\frac{x^3}{3} + \eta x^2 + (\eta^2 - \eta_o^2)x\right) = y - F(x,\eta) \\ \dot{y} &= -\rho b\left(\frac{x^3}{3} + \eta x^2 + (\eta^2 + \frac{1}{b} - 1)x\right) = -g(x,\eta) \quad , \end{aligned} \tag{6}$$

where $F(x,\eta) = \frac{x^3}{3} + \eta x^2 + (\eta^2 - \eta_o^2)x$. (7)

Further we have $xg(x,\eta) = \frac{1}{3}\rho bx^2\left[(x + \frac{3}{2}\eta)^2 + \frac{3}{4}\eta^2 + 3(\frac{1}{b} - 1)\right] > 0$ for $x \neq 0$.

Thus we can apply the Conti transformation ([7] , p.156) to system (6). Put

$$G(x,\eta) := \int_0^x g(s,\eta)ds = \frac{\rho b}{12}(x^4 + 4\eta x^3 + 6(\eta^2 + \frac{1}{b} - 1)x^2), \tag{8}$$

$z(x,\eta) := \sqrt{2G(x,\eta)}\ \mathrm{sgn}x$ and $x(z,\eta)$ the inverse of this function and $\Phi(z,\eta) := F(x(z,\eta),\eta)$. Then system (6) is equivalent to the ordinary Liênard system

$$\begin{aligned} \dot{z} &= y - \Phi(z,\eta) \\ \dot{y} &= -z \quad . \end{aligned} \tag{9}$$

For $|\eta| < \eta_o$ we have $xF(x,\eta) > 0$, $x \neq 0$, in a neighborhood of the origin and $\lim_{x\to\pm\infty} F(x,\eta) = \pm\infty$.

The function $\Phi(z,\eta)$ has corresponding properties. Therefore there exists at least one

$$z^*(\eta) > 0 \text{ such that } \Phi(z^*(\eta),\eta) = \Phi(-z^*(\eta),\eta) \ . \tag{10}$$

By virtue of these properties of $\Phi(z,\eta)$ we can apply Filippov's existence theorem for periodic solutions of the Liênard system (cf. [7], p.156), and we find at least one (asymptotically stable) limit cycle for $|\eta| < \eta_o$.

There existsa large number of uniqueness results for periodic solutions of the Liênard system (see [9]). Except for the case $\eta = 0$, only the theorem in [8] is applicable to system (9).

In order to apply this theorem we have to verify:

There exists exactly one $z^* > 0$ such that

a) $\Phi(z) < \Phi(-z)$ for $0 < z < z^*$,

b) $\Phi(z) < \Phi(-z)$ for $z > z^*$, (11)

c) $\Phi(z)$ is non decreasing for $|z| > z^*$.

Then the theorem yields the desired uniqueness of the limit cycle.

However, there is a difficulty in verifying properties (11) in the present case. This difficulty stems from the fact that the functions $x(z,\eta)$ and $\Phi(z,\eta)$ are not explicitly known. Thus we shall derive a sufficient condition for (11) in the sequel.

Observe that (6) resp. (9) remain unchanged by the transformation $y \to -y$, $\eta \to -\eta$, $x \to -x$ resp. $z \to -z$. Therefore it suffices to give proofs only for the case $\eta \leqq 0$.

The graph of $F(x,\eta)$ is N-shaped for every $\eta \in \mathbb{R}$. More precisely, for $\eta \in [-\eta_o, 0]$ there exist exactly one $-x_1(\eta)$ and one $x_2(\eta)$, $(x_1(\eta), x_2(\eta) \geqq 0)$, such that

$F(x,\eta)$ is monotone increasing for $x \notin [-x_1(\eta), x_2(\eta)]$,

$F(x,\eta)$ is monotone decreasing for $x \in (-x_1(\eta), x_2(\eta))$.

Put $z_1(\eta) := \sqrt{2G(-x_1(\eta),\eta)}$, $z_2(\eta) := \sqrt{2G(x_2(\eta),\eta)}$. Then

$\Phi(z,\eta)$ is monotone increasing for $z \notin [-z_1(\eta), z_2(\eta)]$,

$\Phi(z,\eta)$ is monotone decreasing for $z \in (-z_1(\eta), z_2(\eta))$. (12)

From (7) we find $x_1(\eta) = \eta_o + \eta$, $x_2(\eta) = \eta_o - \eta$.

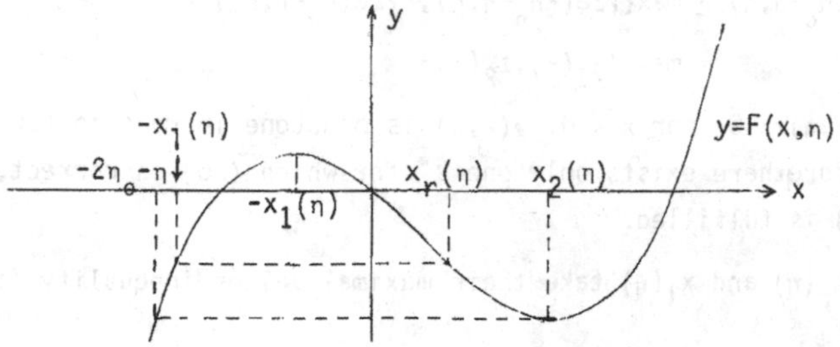

Further, we have $F(-2\eta_o-\eta,\eta) = F(\eta_o-\eta,\eta)$. Therefore for every $\eta \in [-\eta_o,0]$ and

$$x_r(\eta) \in (0,\eta_o-\eta] \text{ there exists exactly one } x_1(\eta) \in (0,2\eta_o+\eta] \tag{13}$$

such that

$$F(-x_1(\eta),\eta) = F(x_r(\eta),\eta). \tag{14}$$

In particular, we define

$$x_r(-\eta_o) = :X_r \in (0,2\eta_o] \ , \quad x_1(-\eta_o) = :X_1 \in (0,\eta_o] \ , \tag{15}$$

and then we have

$$x_r(\eta) = X_r - \eta_o - \eta \text{ and } x_1(\eta) = X_1 + \eta_o + \eta \ . \tag{16}$$

Now let us show that properties (11) are satisfied when

$$G(x_r(\eta),\eta) < G(-x_1(\eta),\eta) \tag{17}$$

for all $x_r(\eta)$, $x_1(\eta)$ fulfilling (13), (14).

Note that

$$G(x_r(\eta),\eta) < G(-x_1(\eta),\eta) \quad \text{and} \quad F(x_r(\eta),\eta) = F(-x_1(\eta),\eta)$$

is equivalent to

$$z_r(\eta) < z_1(\eta) \text{ and } \Phi(z_r(\eta),\eta) = \Phi(-z_1(\eta),\eta) \tag{18}$$

where

$$z_r(\eta): = \sqrt{2G(x_r(\eta),\eta)} \quad \text{and} \quad z_1(\eta): = \sqrt{2G(-x_1(\eta),\eta)} \ .$$

By (12) $\Phi(z,\eta)$ is monotone decreasing for $0 < z < \sqrt{2G(\eta_o-\eta,\eta)}$ and hence it follows that

$$\Phi(-z_1(\eta),\eta) = \Phi(z_r(\eta),\eta) > \Phi(z_1(\eta),\eta). \tag{19}$$

Finally, from (10) we have that there exists at least one $z^*(\eta) > 0$ such that

$$\Phi(z^*(\eta),\eta) = \Phi(-z^*(\eta),\eta). \tag{20}$$

From (18), (19) we find that

$$z^*(\eta) \geq \sqrt{2G(-2\eta_o-\eta,\eta)} \geq \max\{\sqrt{2G(-\eta_o-\eta,\eta)}, \sqrt{2G(\eta_o-\eta,\eta)}\} =$$
$$= \max\ \{z_1(\eta),z_2(\eta)\}$$

since $G'(x,\eta) = g(x,\eta) < 0$ for $x < 0$. $\Phi(z,\eta)$ is monotone increasing for $|z| > z^*$, and therefore there exists only one z^* for which (20) is correct, and for $z > z^*$ (11b) is fulfilled.

For the case where $x_r(\eta)$ and $x_1(\eta)$ take their maximal values inequality (17) takes the form

$$G(\eta_o-\eta,\eta) < G(2\eta_o-\eta,\eta) \tag{21}$$

for $\eta \in [-\eta_0, 0]$, where we have used (13).

Put $\varepsilon := \frac{-\eta}{\eta_0}$, $p := (\frac{1}{b} - 1)\frac{1}{\eta_0^2} > 0$ then (21) is equivalent to $\rho b \eta_0^4 h(\varepsilon) < 0$,

where $h(\varepsilon) := \varepsilon^3 + 3p(\varepsilon - \frac{1}{2}) - \frac{5}{4}$. Denote by ε_0 the smallest positive zero of $h(\varepsilon)$. Then we have $\varepsilon_0 > \frac{1}{2}$, and for $\eta \in (-\varepsilon_0 \eta_0, 0]$ (21) is correct.

The three equivalent conditions $p < \frac{1}{6}$,

$$6(\frac{1}{b} - 1) < \eta_0^2 \tag{22}$$

$$\text{and } \rho b^2 - 7b + 6 < 0 \tag{23}$$

thus imply that $h(1) < 0$. Moreover, for (b,ρ) fulfilling (23) we have that (21) is correct for every $\eta \in [-\eta_0, 0]$.

But $\rho b^2 - 2b + 1 < \rho b^2 - 7b + 6$, and therefore (23) is satisfied only if $(b,\rho) \in B_2 \subset B$.

Now let $x_r(\eta)$, $x_l(\eta)$ be arbitrary. From $\eta \in [-\frac{\eta_0}{2}, 0]$ we find $0 \leqq x_r(\eta) < x_l(\eta)$ since $F(-x,\eta) > F(x,\eta)$ for $x \in (0, \eta_0 - \eta)$.

By $\eta < 0$ we have

$$G(-x_l(\eta),\eta) = \frac{\rho b}{12}(x_l^4 - 4\eta x_l^3 + 6(\eta^2 + \frac{1}{b} - 1)x_l^2) >$$

$$> \frac{\rho b}{12}(x_r^4 + 4\eta x_r^3 + 6(\eta^2 + \frac{1}{b} - 1)x_r^2) = G(x_r(\eta),\eta).$$

Thus we have proved the first part of the theorem.

To prove the second part we consider the two functions

$$\begin{aligned} P(x_l(\eta),\eta) :&= \frac{12}{\rho b}G(-x_l(\eta),\eta) - 12\eta F(-x_l(\eta),\eta) = \\ &= x_l^4 + 6(\frac{1}{b} - 1 - \eta^2)x_l^2 + 12\eta(\eta^2 - \eta_0^2)x_l) , \\ Q(x_r(\eta),\eta) :&= \frac{12}{\rho b}G(x_r(\eta),\eta) - 12\eta F(x_r(\eta),\eta) = \\ &= x_r^4 + 6(\frac{1}{b} - 1 - \eta^2)x_r^2 - 12\eta(\eta^2 - \eta_0^2)x_r . \end{aligned} \tag{24}$$

From (14) we see that inequality (17) is equivalent to

$$P(x_l(\eta),\eta) > Q(x_r(\eta),\eta) . \tag{25}$$

Taking $\eta = -\eta_0$, $x_l(\eta) = x_l(-\eta_o) = X_l$, $x_r(\eta) = x_r(-\eta_0) = X_r$, (where $X_l < X_r$), (25) is reduced to

$$S(X_l) := X_l^4 + 6(\frac{1}{b} - 1 - \eta_0^2)X_l^2 > X_r^4 + 6(\frac{1}{b} - 1 - \eta_0^2)X_r^2 =: S(X_r) . \tag{26}$$

Since $(b,\rho) \in B_2$ it follows from (22) that the coefficients of the quadratic terms are negative. The function $S(x)$ is monotone decreasing for

$$0 < x < \sqrt{3(\eta_0^2 + 1 - \frac{1}{b})} \tag{27a}$$

and monotone increasing for

$$x > \sqrt{3(\eta_o^2 + 1 - \frac{1}{b})} \ . \tag{27b}$$

Therefore $S(X_1)$ is monotone decreasing for all X_1 since for $X_1 = \eta_o$ the inequalities (27a) and (23) are equivalent.

From (13) we have $X_r < 2\eta_o$, and by (22) we have

$$S(2\eta_o) - S(\eta_o) = 3\eta_o^2(6(\frac{1}{b} - 1) - \eta_o^2) < 0.$$

In the case $X_r \leq \sqrt{3(\eta_o^2 + 1 - \frac{1}{b})}$ we clearly have $S(X_1) > S(X_r)$ for $X_1 < X_r$.

On the other hand, when $\sqrt{3(\eta_o^2 + 1 - \frac{1}{b})} < X_r < 2\eta_o$ then

$S(X_1) > S(\eta_o) > S(2\eta_o) > S(X_r)$ is satisfied.

Thus we have

$$G(-X_1,-\eta_o) > G(X_r,-\eta_o). \tag{28}$$

Further we find

$$\begin{aligned} \frac{d}{d\eta}\Big[G(-x_1(\eta),\eta) - G(x_r(\eta),\eta)\Big] &= \frac{d}{d\eta}\Big[G(-X_1-\eta_o-\eta,\eta) - G(X_r-\eta_o-\eta,\eta)\Big] = \\ &= 2\rho b(\eta^2 + \frac{1}{b} - 1)(X_1 + X_r) > 0. \end{aligned} \tag{29}$$

From (28) and (29) we find $G(-x_1(\eta),\eta) > G(x_r(\eta),\eta)$ for all $\eta \in [-\eta_o,0]$, $(b,\rho) \in B_2$. Thus we have proved part ii) of the theorem.

3. Nonexistence

In [3] it was shown that system (1) has no closed orbits for $|\eta| > 2\sqrt{1 + \frac{1}{b}}$. We shall give a stronger result, proving the

Theorem: $b \in (0,1)$, $a \in \mathbb{R}$, $\rho > 0$, $0 < \rho b < 1$.

i) If $(b,\rho) \in A$ then system (1) has no limit cycle for $|\eta| \geq \eta_o$.

ii) If $(b,\rho) \in B$ then system (1) has no limit cycle for $|\eta| > \sqrt{2\eta_o^2 - (\frac{1}{b} - 1)}$.

For the proof of the theorem we use the following

Lemma: a) $\Phi(z) \in \mathrm{Lip}(\mathbb{R})$, $\Phi(0) = 0$,

b) $\Phi(-z) \leq \Phi(z)$ for $z > 0$, and $\Phi(-z) \neq \Phi(z)$ for $0 < z < d$, (30)

c) $\frac{\Phi(z)}{z} > -k$, $0 < k < 2$, for $|z| > c$.

Then for system (9) the origin is globally asymptotically stable, and therefore system (9) has no periodic solution.

Proof of the lemma: We can find a function $H(z)$ with the following properties:

$$H(z) \in \mathrm{Lip}(\mathbb{R}),\ H(0) = 0,$$
$$H(-z) = H(z), \tag{31}$$
$$\frac{H(z)}{z} > -k,\quad 0 < k < 2,\quad \text{for } |z| > \varepsilon, \tag{32}$$
$$\Phi(-z) \geqq H(z) \geqq \Phi(z),\quad \text{for } z > 0. \tag{33}$$

Now we consider the system

$$\begin{aligned} \dot{z} &= y - H(z) \\ \dot{y} &= -z \end{aligned} \tag{34}$$

From property (31) it follows that all orbits are symmetric with respect to the y-axis. Condition (32) ensures that every orbit starting in $(0,y)$, $y > 0$, will either meet the half-axis $z = 0$, $y < 0$, in one point or will tend to the origin for $t \to +\infty$. Thus every orbit in the phase portrait of system (34) is either closed or such that its union with the origin is a closed curve. Applying the comparision theorem in [8] , from the condition (33) we find that all orbits of system (9) cross the orbits of system (34) from the left hand side to the right hand side. Therefore the origin is globally asymptotically stable, with respect to system (9).

Proof of the theorem: By the properties of $\Phi(z,\eta)$ given before (10) conditions a) and c) of the lemma are fulfilled. Thus we only have to consider condition b) of the lemma.

For $(b,\rho) \in B$ we have

$$\eta_o^2 > \frac{1}{b} - 1\ , \tag{35}$$

and therefore under the assumption of the second part of the theorem we have $|\eta| > \eta_o$, too.

Again we prove the theorem only for $\eta \leqq 0$.

In the case $\eta \leqq -2\eta_o$ the proof of the theorem is very short. We see at once that $xF(x,\eta) > 0$ for $x \neq 0$, and thus we have $\Phi(-z,\eta) < 0 < \Phi(z,\eta)$ for $z > 0$, and therefore condition b) of the lemma is fulfilled.

Now let $\eta \in [-2\eta_o,-\eta_o]$. Analogously to (13), (14), for every η in this interval and every $x_l(\eta) \in (0,2\eta_o+\eta]$ there exists exactly one $x_r(\eta) \in [0,\eta_o-\eta]$ such that $F(-x_l(\eta),\eta) = F(x_r(\eta),\eta)$, where

$$x_l(\eta) < x_r(\eta). \tag{36}$$

We shall verify this inequality for all $x_1(\eta)$, $x_r(\eta)$ fulfilling condition (36). Therefore we can omit the argument η in $x_1(\eta)$ and $x_r(\eta)$.

We again consider the functions defined in (24), and we show that

$$P(x_1,\eta) < Q(x_r,\eta). \tag{37}$$

In the case $(b,\rho) \in A$ we have $\frac{1}{b} - 1 \geq \eta_o^2$, and thus for $\eta = -\eta_o$ and $x_1 < x_r$ we find

$$P(x_1,-\eta_o) = x_1^4 + 6(\tfrac{1}{b} - 1 - \eta_o^2)x_1^2 <$$

$$< x_r^4 + 6(\tfrac{1}{b} - 1 - \eta_o^2)x_r^2 = Q(x_r,-\eta_o),$$

$$\frac{\partial P}{\partial \eta}(x_1,\eta) = -12\eta x_1^2 + 24\eta^2 x_1 + 12(\eta^2 - \eta_o^2)x_1 \geq 0$$

and

$$\frac{\partial Q}{\partial \eta}(x_r,\eta) = -12(\eta x_r + 2\eta^2)x_r - 12(\eta^2 - \eta_o^2)x_r \leq 0.$$

From this it follows that

$$P(x_1,\eta) \leq P(x_1,-\eta_o) < Q(x_r,-\eta_o) \leq Q(x_r,\eta).$$

Thus (37) is correct,and part i) of the theorem is proved.

In the case $(b,\rho) \in B$ we have $\frac{1}{b} - 1 < \eta_o^2 < \eta^2$. We find the equality

$$x_1\frac{\partial P}{\partial x_1}(x_1,\eta) = 4P(x_1,\eta) + 12(\eta^2 - \tfrac{1}{b} + 1)x_1^2 - 36\eta(\eta^2 - \eta_o^2)x_1,$$

which implies:

$$\text{If } P(x_1,\eta) \geq 0 \quad \text{then} \quad \frac{\partial P}{\partial x_1}(x_1,\eta) \geq 0\ . \tag{38}$$

On the other hand we have $Q(x_r,\eta) = x_r q(x_r,\eta)$, where

$$q(x_r,\eta) = x_r^3 - 6(\eta^2 - \tfrac{1}{b} + 1)x_r - 12(\eta^2 - \eta_o^2).$$

The function $q(x_r,\eta)$ has at least one negative zero. From a theorem of elementary algebra it follows that $q(x_r,\eta)$ has exactly one real zero iff

$$36\eta^2(\eta^2 - \eta_o^2)^2 > 8(\eta^2 - \tfrac{1}{b} + 1)^3\ . \tag{39}$$

Since $\eta^2 > \eta^2 - (\frac{1}{b} - 1)$ inequality (39) is fulfilled if

$$36(\eta^2 - \eta_o^2)^2 > 9(\eta^2 - \tfrac{1}{b} + 1)^2, \text{or equivalently, } 2(\eta^2 - \eta_o^2) > \eta^2 - \tfrac{1}{b} + 1.$$

Thus we have

$$Q(x_r,\eta) > 0 \quad \text{for } x_r > 0 \quad \text{if } \eta^2 > 2\eta_o^2 - (\tfrac{1}{b} - 1). \tag{40}$$

Further from (24) we find

$$P(x_r,\eta) < Q(x_r,\eta)\ . \tag{41}$$

Now we have either $P(x_1,\eta) < 0$, in which case (40) implies (37), or $0 < P(x_1,\eta)$, and in this case, using (38) and (41) we conclude

$$P(x_1,\eta) \leq P(x_r,\eta) < Q(x_r,\eta).$$

Thus (37) is again fulfilled. This proves the second part of the theorem.

Literature:

[1] FITZHUGH,R.: Thresholds and plateaus in the Hodgkin-Huxley nerve equations. J. Gen. Physiology 43, 867 - 896 (1960).

[2] FITZHUGH,R.: Impulses and physiological states in theoretical models of nerve membrane. Biophys. J. 1, 445 - 466 (1961).

[3] HADELER,K.P., AN DER HEIDEN,U., SCHUMACHER,K.: Generation of the nervous impulse and periodic oscillations. Biol. Cybernetics 23, 211 - 218 (1976).

[4] HSÜ,J.D., KAZARINOFF,N.D.: An applicable Hopf bifurcation formula and and instability of small periodic solutions of the Field-Noyes model. J. Math. Anal. Appl. 55, 61 - 89 (1976).

[5] HSÜ,J.D.: A high-order Hopf bifurcation formula and its application to FitzHugh's nerve conduction equations. J. Math. Anal. Appl. 60, 47 - 57 (1977).

[6] NEGRINI,P.,SALVADORI,L.: Attractivity and Hopf bifurcation. Nonlinear Anal. 3, 87 - 99 (1979).

[7] REISSIG,R., SANSONE,G., CONTI,R.: Qualitative Theorie nichtlinearer Differentialgleichungen. Rom, 1961.

[8] STAUDE,U.: Ein Eindeutigkeitssatz für periodische Lösungen der Liénard-Gleichung. VII Internationale Konferenz über nichtlineare Schwingungen, Berlin, 1975, Bd. I.2, 295 - 302, Akademie-Verlag, Berlin 1977.

[9] STAUDE,U.: Uniqueness of periodic solutions of the Liénard equation. in Conti,R.(ed.): Recent advances in differential equations, 421 - 429, Academic Press, New York, 1981.

[10] TROY,W.C.: Bifurcation phenomena in FitzHugh's nerve conduction equations. J. Math. Anal. Appl. 54, 678 - 690 (1976).

Mathematisches Institut
der Universität Mainz
Saarstraße 21
D-65 Mainz

Periodic Solutions of Nonlinear Heat Equations under Discontinuous Boundary Conditions

B. Kawohl & R. Rühl
Institut für Angewandte Mathematik
Universität Erlangen-Nürnberg
Martensstr. 3
D 852o Erlangen, W.Germany

Introduction

This paper is concerned with the existence, uniqueness and regularity of T-periodic solutions to the parabolic boundary value problem

$$
(P)\quad
\begin{aligned}
u_t(t,x) - \Delta u(t,x) + \beta_0(u(t,x)) &\ni f(t,x) && \text{in } (o,\infty) \times \Omega ,\\
-\frac{\partial u}{\partial n}(t,x) &\in \beta_1(u(t,x)) && \text{on } (0,s]\cup(T,T+s]\cup\ldots\times\partial\Omega,\\
-\frac{\partial u}{\partial n}(t,x) &\in \beta_2(u(t,x)) && \text{on } (s,T]\cup(T+s,2T]\cup\ldots\times\partial\Omega,
\end{aligned}
$$

where $\Omega \subset \mathbb{R}^n$ is a bounded domain with sufficiently smooth boundary, and where $\beta_i : \mathbb{R} \supset D(\beta_i) \to 2^{\mathbb{R}}$ (i=o,1,2) are maximal monotone mappings. f is T-periodic in T.

The problem is motivated by applications in the thermostat-control of heat conducting media, which occurs in material testing, where e.g. a sample of steel is exposed to periodically changing boundary conditions. Examples of such boundary conditions are: the Dirichlet-, Neumann- and so-called radiation condition, the Signorini condition $u \geq a$, $\frac{\partial u}{\partial n} \geq 0$, $(u-a)\frac{\partial u}{\partial n} = 0$, or the nonlinear Stefan Boltzmann law of heat radiation. Further examples and applications are given in [5].

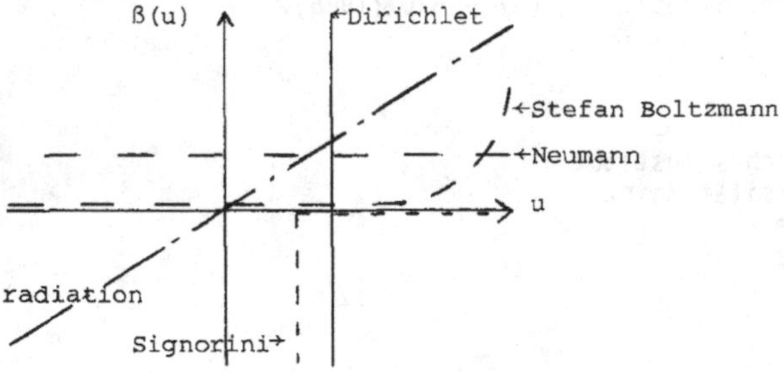

Intuitively one would expect that any periodic solution shows up after prescribing suitable initial data and waiting for a sufficiently long time. Therefore we first study the initial value problem (P) plus fictitious initial data $u(0,x) = u_0(x)$. In §2 we prove the existence of periodic solutions, and §3 and 4 are devoted to the questions of uniqueness and regularity.

1. Initial Value Problems

First we consider the initial value problem

$$(1)\qquad \begin{aligned} u_t(t,x) - \Delta u(t,x) + \beta_0(u(t,x)) &\ni f(t,x) && \text{in } (o,s)\times\Omega, \\ -\frac{\partial u}{\partial n}(t,x) &\in \beta_1(u(t,x)) && \text{on } (o,s)\times\partial\Omega, \\ u(o,x) &= u_o(x) && \text{in } \Omega. \end{aligned}$$

Problems of this type are frequently treated in the literature, e.g. in [1,3]. It can be shown that the initial value problem (1) is equivalent to the following abstract evolution equation in Hilbertspace $H = L^2(\Omega)$

$$(2)\qquad \begin{aligned} \frac{du}{dt} + \partial\Phi_1(u) &\ni f(t) && \text{in } (o,s), \\ u(o) &= u_o . \end{aligned}$$

Here $\partial\Phi_1$ is a maximal monotone Operator in $L^2(\Omega)$. $\partial\Phi_{1(2)}$ is the subdifferential of the functional $\Phi_{1(2)}: L^2(\Omega) \to (-\infty,+\infty]$ defined by

$$(3)\qquad \Phi_{1(2)}(u) := \begin{cases} \int_\Omega (\frac{1}{2}|\nabla u|^2 + j_o(u))dx + \int_{\partial\Omega} j_{1(2)}(u)ds, & \text{if the integrals exist,} \\ +\infty & \text{, otherwise,} \end{cases}$$

and $\partial j_i = \beta_i$ $(i=o,1,2)$, i.e. the j_i's are the primitives to the β_i's.

We have to explain what we mean by a solution of problem (2).

<u>Definition:</u>

Let $H = L^2(\Omega)$ and let $\partial\Phi_1 : H \supset D(\partial\Phi_1) \to 2^H$ be given by (3). Suppose $f \in L^2(o,s;H)$. A function $u \in C([o,s],H)$ is called a solution of problem (2) iff

a) $u: [o,s] \to H$ is absolutely continuous on every compact $K \subset (o,s)$,

b) $u(t) \in D(\partial\Phi_1)$ for a.e. $t \in (o,s)$,

c) $\frac{du}{dt}(t) + \partial\Phi_1(u(t)) \ni f(t)$ for a.e. $t \in (o,s)$.

The following result can be found in [1,3]:

<u>Proposition 1:</u>

i) For $u_o \in \overline{D(\Phi_1)}$ and $f \in L^2(o,s;H)$ the initial value problem (2) has a unique solution $u \in C([o,s];H)$ with

$u(t) \in D(\Phi_1)$ for $t \in (o,s]$,

$u(t) \in D(\partial\Phi_1)$ for a.e. $t \in (o,s)$,

$\sqrt{t}\,\frac{du}{dt} \in L^2(o,s;H)$.

ii) If in addition $u_o \in D(\Phi_1)$ then $\frac{du}{dt} \in L^2(o,s;H)$.

iii) The mapping $u(o) \to u(s)$ is nonexpansive.

Until now we know that there exists a solution up to tim s of the following sequence of subsequent initial value problems ($k \in \mathbb{N}_o$):

$$(4)\qquad \begin{aligned} &\frac{du}{dt} + \partial\Phi_1(u(t)) \in f(t) \qquad \text{for } t \in (kT, kT+s), \\ &\qquad u(kT) = u_{kT}, \\ &\frac{du}{dt} + \partial\Phi_2(u(t)) \in f(t) \qquad \text{for } t \in (kT+s, kT+T), \\ &\qquad u(kT+s) = u_{kT+s}. \end{aligned}$$

Here u_s is $\lim_{t \to s} u(t)$ and the other Cauchy data are defined analogously. Observe that

$$\overline{D(\Phi_1)} = \overline{\{ u \in H^1(\Omega) \mid \int_\Omega j_o(u)\, dx < \infty \}}^{L^2(\Omega)} = \overline{D(\Phi_2)}$$

so that fortunately $u_s \in D(\Phi_2)$ and hence, starting at t=s, there exists a solution up to time T and so forth. Therefore Proposition 1 leads to

Corollary 2:

If we prescribe $u(o,x) = u_o(x) \in D(\Phi_1)$, then the sequence of initial boundary value problems (4) has a global solution in the sense of Proposition 1i).

Remark 1:

Observe that the time gerivative of the solution becomes singular at the switching times s, T, T+s, 2T etc. More regularity can be expected for $f \in W^{1,1}(o,T;H)$ or under compatibility conditions of the type $j_k(r) \leq c(1 + j_h(r))$, $r \in \mathbb{R}$; $k,h \in \{1,2\}$, $k \neq h$. For details we refer to [5,8].

2. Existence of Periodic Solutions

As mentioned in the introduction, once we have the existence of a global solution to the initial value problem, we can prove the existence of a periodic solution. Note that the mapping K: $u(o) \to u(T)$ is nonexpansive. Any periodic solution to (P) will be a fixed point of K and vice versa. Therefore our method of proof will rest on the following fixed point result [4,7].

Proposition 3:

Let $C \subset H$ be a closed convex (but not necessarily bounded) subset of a Hilbertspace H. Let K: $C \to C$ be nonexpansive and suppose there exists $\tilde{x} \in C$ such that $\| K^n \tilde{x} \|$ is uniformly bounded for $n \in \mathbb{N}$. Then K has a fixed point.

Generalizations of this result were given in [9]. The assumptions of Proposition 3 are satified if we manage to show that the sequence u(o), u(T), u(2T),... obtained in Corollary 2 is uniformly bounded in $L^2(\Omega)$. Intuitively this means that the energy Φ of a solution cannot blow up.

Theorem 4:

Suppose the operators $\partial\Phi_i$ (i=1,2) defined in (3) are coercive, i.e. there exists an $x_i \in D(\partial\Phi_i)$ such that

$$\lim_{\|x\|\to\infty} \frac{(x-x_i,y)}{\|x\|} = +\infty \text{ for any } x \in D(\partial\Phi_i),\ y \in \partial\Phi_i(x).$$

Then the sequence $\{u(nT)\}_{n\in\mathbb{N}}$ is uniformly bounded.

The proof of this theorem is a nontrivial extension of an idea of Benilan and Brezis [2,3] and will be sketched below. A sufficient criterion for the coerciveness of $\partial\Phi_i$ can be given in terms of the nonlinearities as follows:

Lemma 5:

If $\lim_{\substack{|r|\to\infty \\ r\in\mathbb{R}}} \{ |\Omega| \frac{j_o(r)}{|r|} + |\partial\Omega| \frac{j_i(r)}{|r|} \} = +\infty$, then $\partial\Phi_i$ is coercive.

For the proof of this lemma see [6].

Corollary 6:

Under the assumptions of Lemma 5 there exists a periodic solution of the problem (P).

Proof of Theorem 4:

The basic idea is to derive suitable estimates for u(t) at the switching times t=nT+s and t=nT. Set $v_n(\tau) := u(t)$ for $nT+\tau := t \in (nT,nT+T]$ and observe that the coerciveness of Φ_i implies the existence of a constant $R \in \mathbb{R}$ and of $x_{oi} \in D(\partial\Phi_i)$ such that for every $x_i \in D(\partial\Phi_i)$ and $y_i \in \partial\Phi_i(x_i)$ with $\|x_i\| \geq R$ the inequality

$$(5) \qquad (y_i, x_i - x_{oi}) \geq L\ \|x_i - x_{oi}\| \qquad (i=1,2)$$

holds. Here (.,.) denotes the scalar product in $L^2(\Omega)$, $L > \frac{1}{T}(C_1 + C_2 + 4C_o)$ with $C_o := \max\{\|x_{o1}\|, \|x_{o2}\|\}$, $C_1 := \int_o^T \|f(t)\|\,dt$ and $C_2 \geq \|u(T)-u(o)\| \geq \|u(nT)-u(nT+T)\|$.

Step 1:

We show that the sequence v_n is bounded in at least one point $t_o \in (o,T]$. If this were not the case then $\|v_n(\tau)\| > R$ for any $\tau \in (o,T)$ and for a sufficiently large n. Due to (5) we would get

$$(f - \frac{dv_n}{d\tau}, \frac{v_n - x_{oi}}{\|v_n - x_{oi}\|}) \geq L\ \|v_n(\tau) - x_{oi}\| \quad \text{, and integration over } \tau \text{ would yield}$$

$$LT \leq \int_o^s (f - \frac{dv_n}{d\tau}, \frac{v_n - x_{oi}}{\|v_n - x_{oi}\|})\,d\tau + \int_s^T (f - \frac{dv_n}{d\tau}, \frac{v_n - x_{oi}}{\|v_n - x_{oi}\|})\,d\tau$$

$$\leq C_1 - \int_o^s \frac{d}{d\tau}\|v_n(\tau) - x_{o1}\|\,d\tau - \int_s^T \frac{d}{d\tau}\|v_n(\tau) - x_{o2}\|\,d\tau$$

$$= C_1 - \|v_n(s) - x_{o1}\| + \|v_n(o) - x_{o1}\| - \|v_n(T) - x_{o2}\| + \|v_n(s) - x_{o2}\| \leq$$

$\leq C_1 + \| v_n(o) - v_n(T) \| + 2 \| x_{o1} \| + 2 \| x_{o2} \|$

$\leq C_1 + C_2 + 4 C_o$. But this contradicts the definition of L.

Step 2:

Now we extend the estimate for $v_n(t_o)$ to the whole interval (o,T]. We have to distinguish the two cases $t_o \in (s,T]$ and $t_o \in (o,s]$. Suppose $t_o \in (s,T]$. Then $\| v_n(t_o) \| \leq R$ and for any $x_2 \in D(\partial\Phi_2)$, $y_2 \in \partial\Phi_2(x_2)$ and $t \in [t_o,T]$ we obtain

$$\| v_n(t) - x_2 \| - \| v_n(t_o) - x_2 \| = \int_{t_o}^{t} \frac{d}{d\tau} \| v_n(\tau) - x_2 \| \, d\tau$$

(6)

$$= \int_{t_o}^{t} \left(\frac{dv_n}{d\tau}, \frac{v_n(\tau) - x_2}{\| v_n(\tau) - x_2 \|} \right) d\tau \leq \int_{t_o}^{t} \| f(\tau) - y_2 \| \, d\tau .$$

Hence $\| v_n(T) \| \leq R + 2 \| x_2 \| + C_1 + T \| y_2 \|$ and

$$\| v_n(o) \| \leq \| v_n(T) \| + C_2 \leq R + 2 \| x_2 \| + C_1 + C_2 + \| y_2 \| ,$$

i.e. $v_n(o)$ is bounded, too. Integration as in (6) yields

$$\| v_n(t) - x_1 \| \leq \| v_n(o) - x_1 \| + \int_{o}^{t} \| f(\tau) - y_1 \| \, d\tau \quad \text{for } t \in (o,s],$$

i.e. $v_n(s)$ is bounded. This and the estimate

$$\| v_n(t) - x_2 \| \leq \| v_n(s) - x_2 \| + \int_{s}^{t} \| f(\tau) - y_2 \| \, d\tau \quad \text{for } t \in (s,T]$$

lead to the uniform boundedness of $v_n(t)$ for every $t \in [o,T]$ and $n \in \mathbb{N}$:

$$\| v_n(t) \| \leq R + 2 \| x_1 \| + 4 \| x_2 \| + 3C_1 + C_2 + 2T \| y_2 \| + s \| y_1 \| .$$

The second case $t_o \in (o,s]$ is treated analogously.

3. UNIQUENESS

The following counterexample shows that in general one cannot expect uniqueness. Let $f = \beta_o \equiv o$, let $j_1 = I_{(-\infty,1]}$ and $j_2 = I_{[o,\infty)}$, where I_A denotes the indicator function of $A \subset \mathbb{R}$. Then any constant between zero and one is a stationary and hence periodic solution to problem (P).

Additional assumptions are needed to obtain uniqueness. Suppose that one of the operators $\partial\Phi_i$ (i=1,2) is strictly coercive, i.e. there exists a constant $\omega_i > o$ such that for any $x, \xi \in D(\partial\Phi_i)$ and $y \in \partial\Phi_i(x)$, $\eta \in \partial\Phi_i(\xi)$

(7) $$(y - \eta , x - \xi) \geq \omega_i \| x - \xi \|^2 .$$

Then the mapping $K: u(o) \to u(T)$ is a (strict) contraction and Banach's fixed point theorem implies uniqueness of the periodic solution. Moreover the solution of the initial value problem (4) converges to the periodic solution strongly in $L^2(\Omega)$ as t goes to infinity.

Remark 2:

The strict coerciveness condition (7) can again be expressed in terms of the nonlinearities. As in Lemma 5 one can show that a strictly coercive j_i (i=o,1,2), i.e. one which grows at least quadratically, leads to the contractivity of K. This is for instance the case if one of the boundary conditions is the Dirichlet condition or if one deals with the equation $u_t - \Delta u + u = f$.

4. Regularity

The regularity of periodic solutions is described in Proposition 1 and Remark 1. Due to the smoothing behaviour of parabolic operators one might expect more regularity for feriodic solutions; and for the degenerate situation s=T this is indeed the case [1,3]. In general, however, any smoothing that has been built up is destroyed at the switching times.

References

[1] BARBU, V. Nonlinear semigroups and differential equations in Banach spaces. Leyden: Noordhoff 1976.

[2] BENILAN,Ph., BREZIS,H. Solutions faibles d'évolution d'équation dans les espaces de Hilbert. Ann. Inst. Fourier, 22 (197o) p. 311-329.

[3] BREZIS,H. Operateurs maximaux monotones et semigroupes de contractions dans les espaces de Hilbert. Amsterdam: North Holland 1973.

[4] BROWDER, F., PETRYSHYN, W.V. The solution by iteration of nonlinear functional equations in Banach spaces. Bull. Amer. Math. Soc. 72 (1966) p. 571-575.

[5] KAWOHL, B. On nonlinear parabolic equations with abruptly changing boundary conditions. Nonlin. Analysis TMA 5 (1981) p. 1141-1153.

[6] KAWOHL, B. A singular perturbation approach to nonlinear elliptic boundary value problems. to appear.

[7] KIRK, W.A. A fixed point theorem for mappings which do not increase distances. Amer. Math. Monthly 72 (1965) p. 1oo4-1oo6.

[8] RÜHL, R. Periodische Lösungen einer nichtlinearen Wärmeleitungsgleichung unter zeitlich stückweise konstanten Randbedingungen. Thesis, Erlangen 1981.

[9] WEBB, J.R.L. Approximation solvability of nonlinear equations. to appear in: Proc. Spring School on Nonlinear Analysis II, Pisek ČSSR, May 24-28,1982

Homoclinic Bifurcation of Perturbed Reversible Systems

by

Klaus Kirchgässner+)

Abstract

In this contribution certain types of semilinear second order equations are studied in a Hilbert space. The vector field has a quasi accretive linearization, i. e. it is strictly accretive up to a finite dimensional part. These equations include elliptic boundary value problems in cylindrical domains as well as reaction-diffusion systems. Bounded solutions are sought, in particular those which join two saddle nodes (heteroclinic solutions) or a saddle node with itself (homoclinic solution). The reversible case is studied by using a center manifold reduction. Global existence and persistence under nonreversible perturbations is shown for homoclinic solutions.

+) Mathematisches Institut A, Universität Stuttgart,
Pfaffenwaldring 57, D-7000 Stuttgart 80

1. Introduction

In this contribution we describe a general framework to recent results about bifurcation of homoclinic or heteroclinic solutions for certain types of partial differential equations. These equations include reaction-diffusion systems as well as the Euler equations for inviscid fluid motion. Numerous examples of homoclinic bifurcation from an equilibrium have been published in the literature. We mention in particular the systematic work for finite-dimensional problems by Kopell and Howard [9], [14] and the work of Renardy [15], [16] which is applicable for "reversible" systems generally and thus could be used to treat the equations studied in section 3 of this paper. As an example for reaction-diffusion systems we quote the formal work of Cohen, Hoppenstaedt and Miura [5], which has been studied with the method applied here in [12]. For a good discussion of further applications see [16].

Our main interest is in homoclinic solutions, although most of our results carry over to heteroclinic solutions as well. Here we mean by a homoclinic solution a path (in infinite-dimensional space) joining a saddle point with itself. The typical application we have in mind is a semilinear elliptic equation of any order over a domain which extends to infinity in one direction. If the equations are autonomous in the unbounded variables (say t) and covariant to reflecting t then we have a special situation of reversibility. We develop a dynamic approach for these problems as it was done for special equations in [11], [12] and for general elliptic systems by G. Fischer in [7].

Homoclinic solutions describe e.g. solitary waves in inviscid fluids. The well known local theory of these waves could in many cases be understood as a bifurcation of homoclinic solutions from a trivial equilibrium [11]. The breaking of these solutions through the influence of outer forces into a transverse homoclinic point is an open question, to which we will comment in this contribution. Another problem having attracted recent interest is the global behavior of these solutions. Amick and Toland [1], [2] have treated this problem in the framework of their remarkable work on the Stokes conjecture. We mention also some work of Esteban-Lions [6] for second order equations and Renardy's recent paper on reaction-diffusion equations [16].

In presenting our results we are not striving for the greatest generality and sacrifice some obvious extensions to technical relieve. In order to describe the problem treated we freely make use of notations which are defined at the end of this introduction. Let H be a real Hilbert-space with scalar product $(\cdot,\cdot)$ and norm $|\cdot|$ and let $\mathbb{H}$ be its complexification. $\Lambda \subset \mathbb{R}$ denotes an open parameter set. $T(\lambda)$ is a family of closed operators in H with dense domain $D(T) \subset H$, independent of $\lambda \in \Lambda$. We assume that, with the possible exception of a finite-dimensional part, $T(\lambda)$ is strictly accretive ([10], p. 279), i.e. there is a T-invariant decomposition $H = H_o \oplus H_1$, $\dim H_o < \infty$, $TH_j \subset H_j$, $j = 0,1$, and a positive constant δ such that

$$\mathrm{re}(\mathbb{T}_1 u\,,\,u) \geqslant \delta |u|^2\,,\quad u \in D(\mathbb{T})$$

where $\mathbb{T}$ denotes the natural extension of T to $\mathbb{H}$. δ is called a lower bound of T_1. Moreover 0 should be in the resolvent set of T_1. It is well known that T_1 possesses a unique strictly accretive square root S_1 with lower bound $\delta^{1/2} = \delta_s$ and $D(T_1) \subset D(S_1) \subset H$. If $\dim H_0 = 0$ we set $T = T_1$, $S = S_1$, if not, define $D(S) = H_0 \oplus D(S_1)$.

Let us now consider the nonlinear mappings $f(\lambda,u)$ and $F(\lambda,t,u)$ having the following properties

$$\begin{aligned}
&f \in C^r(\Lambda \times D(S)\,,H)\,, \quad r \geqslant 2 \\
&f(\lambda,0) = 0\,, \quad D_u f(\lambda,0) = 0\,, \quad \lambda \in \Lambda \\
(1.1) \qquad &F \in C^r(\Lambda \times \mathbb{R} \times D(S)\,,H) \\
&\sup_{|Su| \leqslant \rho} |D_u F(\lambda,t,u)| \leqslant C(\rho)\,, \quad \lambda \in \Lambda\,, \quad t \in \mathbb{R}\,.
\end{aligned}$$

We study the following equation in H

$$(1.2)_\varepsilon \qquad \frac{d^2u}{dt^2} - T(\lambda)u + f(\lambda,u) = \varepsilon F(\lambda,t,u)$$

for $\varepsilon \in \mathbb{R}$ near 0. Solutions are sought in

$$X = C_b^2(\mathbb{R},H) \cap C_b^1(\mathbb{R},D(S)) \cap C_b^0(\mathbb{R},D(T))$$

where D(S) and D(T) are endowed with their graph norms. Observe that $u = 0$ is a solution for all λ and ε. The linear part about O has infinite spectrum on both sides of the imaginary axis, however symmetrically distributed.

First, we investigate the case $\varepsilon = 0$. Assume that for $\lambda_o \in \Lambda$, all eigenvalues of $T_o(\lambda)$ vanish. Thus, if T depends "smoothly" on λ, $|\Sigma T_o(\lambda)|$ is small, if Λ is a sufficiently small neighborhood of λ_o. Hence, for fixed r and $|\mathrm{re}\,\Sigma S_1(\lambda)| > \alpha > 0$, independently of λ, we can construct a center-manifold to reduce $(1.2)_o$ to an ordinary differential equation of order 2n. $(n = \dim H_o)$; i.e. we show, for certain neighborhoods U_o, U_1 of O, in H_o resp. H_1, and $\Lambda_o \subset \Lambda$ of λ_o, the existence of a C^{r-1}-function $h(\lambda, u_o, u_o')$ with values in $D(S_1)$ such that $h = O((|u_o| + |u_o'|)^2)$, and every solution $u \in U_o \oplus U_1$ of $(1.2)_o$ satisfies $u_1(t) = h(\lambda, u_o, u_o')(t)$, $(u_o' = d\,u_o/d\,t)$. The function h preserves reversibility, i.e. $h(\lambda, u_o, -u_o')$ $= h(\lambda, u_o, u_o')$.

Theorems of this type are well known in finite dimensions (c.f. [3]) and for partial differential equations of parabolic type (c.f. [8]). For the case under consideration it was proved first for a certain elliptic equation in [11] and for general elliptic systems in [7]. For the use in reaction-diffusion systems a particular result was given in [12]. The present version contains all known examples.

This reduction principle is used then to treat in detail the case $\dim H_o = 1$, i.e. $T_o(\lambda)$ has a simple eigenvalue $\sigma_o(\lambda)$ with $\sigma_o(\lambda_o) = 0$, $\sigma_o'(\lambda_o) < 0$. If the leading nonlinear term in $f(\lambda, u)$ is of even power then, $u = 0$, $\lambda = \lambda_o$ is the bifurcation point of homoclinic solutions $p(\cdot,\lambda) \in X$. They have exponential decay at $t = \pm\infty$. The precise formulations can be found in section 3.

In section 4 we treat the case $\varepsilon \neq 0$ and restrict attention to the example described above. For fixed $\lambda < \lambda_0$, $u = 0$ is a saddle-point and $p(t,\lambda)$ decays exponentially to 0 as $t \to \pm\infty$. The subsequent analysis follows closely the work of Chow - Hale and Mallet-Paret [4]. We show that $u = 0$ has a unique continuation $u^*(\cdot,\varepsilon)$ in X for small $|\varepsilon|$. Local stable and unstable manifolds exist. However, the globalization of this manifold does not yield a manifold anymore, since time reversion is not allowed. Nevertheless we construct conditions under which a bounded solution exists near $p(\cdot,\lambda)$. For the example in section 3 I owe the proof to G. Fischer. The global behaviour of these solutions is studied.

In section 3 and 4 we demonstrate the applicability of our results by treating the existence of solitary waves in channels with a variable temperature distribution and an inviscid, incompressible fluid. We show also the continuability of these solitary waves with respect to perturbations. In section 2 we present the necessary material for the linear equation, construct explicit formulae for the stable and unstable manifolds, etc. According to the spirit of this contribution, details of straightforward proofs are omitted.

<u>Notations</u>:

H real Hilbert-space, scalar product $(\cdot,\cdot)$, norm $|\cdot|$

$\mathbb{H}$ its complexification

$\Lambda \subset \mathbb{R}$ open neighborhood of λ_o

$T(\lambda) : D(T) \to H$, $D(T)$ independent of λ , dense in H , $T(\lambda)$ closed

$H = H_o \oplus H_1$, $T(\lambda)H_j \subset H_j$, $j = 0,1$

$T(\lambda) = T_o \oplus T_1$

$T(\lambda)$ strictly accretive, i.e. T_1^{-1} exists and is bounded, $\mathrm{re}(\mathbb{T}_1 u,u) \geq \delta|u|^2$, $\delta > 0$ (re = Real part, $\mathbb{T}_1$ complexification of T_1)

$S_1(\lambda)$ strictly accretive square root with lower bound $\delta^{1/2} = \delta_s$ and domain $D(S_1)$

$S(\lambda) = S_1(\lambda)$ if $\dim H_o = 0$

$D(S) = H_o \oplus D(S_1)$

$C^r(A,B)$ all mappings from A to B which are r-times continuously differentiable

$C_b^r(A,B)$ subspace of $C^r(A,B)$ containing those mappings whose derivatives are bounded

$C_\delta^r(A,B) \subset C_b^r(A,B)$, all derivatives are bounded by δ

$C_\eta^{Lip}(A,B) \subset C_\eta^o(A,B)$, mappings are Lipschitz continuous with constant η

$C_{\eta,o}^{Lip}(A,B) = \left\{ f \in C_\eta^{Lip} \;/\; f(0) = 0 \right\}$

$X = C_b^2(\mathbb{R},H) \cap C_b^1(\mathbb{R},D(S)) \cap C_b^o(\mathbb{R},D(T))$ with the norm

$$\|u\|_X = |u''|_o + |Su'|_o + |Tu|_o + |u|_o$$

$$|u|_o = \sup_{t\in\mathbb{R}} |u(t)| , \quad u'' = \frac{d^2u}{dt^2}$$

$X(I)$, I interval in $\mathbb{R}$, replace $\mathbb{R}$ by I in the definition of X . Norm as above when $|u|_o$ is replaced by $|u|_{o,I}$ etc.

$$|u|_{o,I} = \sup_{t\in I} |u(t)|$$

$X^o = C_b^1(\mathbb{R},H) \cap C_b^o(\mathbb{R},D(S))$

$X^1 = C_b^1(\mathbb{R},D(S))$

2. Linear accretive equations

The strictly accretive Operator T with lower bound $\delta > 0$ has a unique strictly accretive square root S with lower bound $\delta^{1/2} = \delta_s$. Its numerical range is contained in $\left\{z \in \mathbb{C} \,/\, |\arg z| \leq \pi/4 \,,\ \operatorname{re} z \geq \delta_s\right\}$ (c.f. [10], S. 279).

Lemma 2.1

- S generates a holomorphic semigroup $\exp(-S\,t)$, $t \geq 0$. The following estimates hold for any $\alpha < \delta_s$

$$|S^k \exp(-S\,t)| \leq \frac{c(\alpha)}{t^k} e^{-\alpha t} \,, \quad t > 0 \,,\quad k = 0,1 \,.$$

Proof: Consider

$$(S + \zeta)u = v \quad \text{for} \quad \zeta = |\zeta| e^{i\theta}$$

then we obtain, using the accretivity of S,

$$|\zeta| \sin(\Theta + \tfrac{\pi}{4}) \; |u| \leq |v| \,.$$

Hence ζ belongs to the resolvent set of $-S$ if $\sin(\Theta + \frac{\pi}{4}) \geq \eta > 0$ holds. The resolvent satisfies

$$|(S+\zeta)^{-1}| \leq \frac{1}{\eta|\zeta|} \,.$$

This however suffices to define the semigroup via the Laplace integral ([10], p. 487), and the remaining part of the proof is standard.

Observe that, if T is real, then S is real as well. This follows e.g. from the representation

$$Su = \frac{1}{\pi} \int_0^\infty \zeta^{-1/2} (S+\zeta)^{-1} Tu \, d\zeta$$

which is valid for $u \in D(T)$, $D(T)$ being a core of S ([10], p. 281). Hence $\exp(-St)$ is real and the estimates of Lemma 2.1 remain true.

Now we consider the linear inhomogeneous differential equation

$$(2.1) \qquad \frac{d^2u}{dt^2} - Tu = f(t)$$

and the corresponding system

$$(2.2) \qquad \begin{aligned} \frac{du}{dt} &= S\,v \\ \frac{dv}{dt} &= S\,u + S^{-1} f(t) \ . \end{aligned}$$

<u>Lemma 2.2</u>

Let I be an interval in $\mathbb{R}$, $f \in C^o(I,H)$.
$u \in X = C^2(I,H) \cap C^1(I,D(S)) \cap C^o(I,D(T))$ solves (2.1) if and only if $u \in C^1(I,D(S))$, $v \in C^1(I,D(S))$ solve (2.2) .

The proof is trivial and therefore omitted. For bounded f the bounded solutions can be computed. We first show a weak version for (2.2) and then extend it to (2.1) by stronger assumptions for f .

<u>Lemma 2.3</u>

Let $I = [t_o,\infty)$, $f \in C_b^o(I,H)$, $w \in D(S)$. Then there exists a unique solution of (2.2) in $C_b^1(I,H) \cap C_b^o(I,D(S))$. It is given for $t \geq t_o$ by the following integral representations ($u_1 = u$, $u_2 = v$):

$$(2.3)\qquad u_j(t) = (-1)^{j-1}\exp(-S(t-t_o))w + \int_{t_o}^{\infty} K_j(t-s)S^{-1}f(s)ds$$

with

$$K_1(t) = -\frac{1}{2}\exp(-S|t|)$$
$$K_2(t) = -\frac{1}{2}\,\mathrm{sign}\,t\exp(-S|t|)\ .$$

For the proof set $U = u+v$, $V = u-v$ and obtain

$$U_t = SU + S^{-1}f(t)$$
$$V_t = -SV - S^{-1}f(t)\ .$$

Obviously, the second equation can be solved for arbitrary $V(t_o) = 2w$. The only solution of the first equation is given by

$$U(t) = \int_t^{\infty}\exp(-S(s-t))S^{-1}f(s)ds\ .$$

For V we obtain

$$V(t) = 2\exp(-S(t-t_o)w - \int_{t_o}^{t}\exp(S(t-s))S^{-1}f(s)ds$$

whence (2.3) follows by elementary calculations. The regularity is easy to prove, since S commutes with the integral.

If $f \in C_b^1(I,H)$ one obtains u and v in $C_b^1(I,D(S))$ and thus a solution of (2.1) in X. The arguments are less trivial but well known in semigroup circles.

<u>Corollary 2.4</u>

If $f \in C_b^1(I,H)$, then the solutions u and v of the preceding Lemma belong to $C^1(I,D(S))$.

Let us write the equations (2.3) in the form

$$\underline{u}(t) = \underline{K}^+(f,w)\ ,\quad \underline{u} = (u,v)\ .$$

Fix $w \in D(S)$ and consider the set

$$(2.4)\qquad M_{\beta,C} = \left\{ f \in C_b^o(I,H)\ /\ \sup_{t\in I} |e^{\beta|t|}\ f(t)| \leq C \right\}$$

where β is some positive constant, $\beta < \alpha < \delta_s$. It is easy to see that

$$(2.5)\qquad |S\,u(t)| + |S\,v(t)| \leq c_1\,C e^{-\beta|t|}\ ,\quad t \in I$$

where c_1 depends on α and β . In view of (2.2), the set $\underline{K}^+(M_{\beta,C},w)$ is bounded and equicontinuous and hence relatively compact on every compact subinterval of I . Inequality (2.5) shows that it is relatively compact in $C_b^o(I,H)$.

Lemma 2.5

Given $w \in D(S)$, $C > 0$, $\beta \in (0,\alpha)$, $\alpha < \delta_s$. Then $\underline{K}^+(M_{\beta,C},w)$ is relatively compact in $C_b^o(I,H)$.

Similar propositions hold for solutions bounded on $(-\infty,t_0]$. While, in the preceding situation, $(u-v)(t_0)$ could be chosen arbitrarily in $D(S)$, we can now prescribe $(u+v)(t_0)$. We formulate these results without proof in the following Lemma.

Lemma 2.6

Let be $I = (-\infty, t_o]$, $C > 0$ and $\beta \in (0,\alpha)$, $\alpha < \delta_s$, $w \in D(S)$.

1. If $f \in C_b^o(I,H)$ then there exists a unique solution of (2.2) in $C_b^1(I,H) \cap C_b^o(I,D(S))$. This solution is given by the following integral representation ($u_2 = v$, $j = 1,2$)

$$(2.6) \qquad u_j(t) = \exp(-S(t_o-t))w + \int_{-\infty}^{t_o} K_j(t-s)\, S^{-1} f(s)\, ds .$$

2. If f belongs to $C_b^1(I,H)$ then u and v belong to $C_b^1(I,D(S))$ and thus u solves (2.1).

3. Denote the mapping defined by (2.6) by $\underline{K}^-(f,w)$. Define $M_{\beta,C}$ as in (2.4). Then, for given $w \in D(S)$, $\underline{K}^-(M_{\beta,C},w)$ is relatively compact in $C_b^o(I,H)$.

It is clear that for $I = R$ one obtains similar assertions for bounded solutions. Observe that $w = (u+v)(t_o)/2$ in Lemma 2.6 and $w = (u-v)(t_o)/2$ in Lemma 2.3.

Theorem 2.7

Given positive constants C, β and α with $\beta < \alpha < \delta_s$. If $f \in C_b^k(\mathbb{R},H)$, $k = 0,1$, then there exists a unique solution of (2.2) in X^k. This solution has the representation ($u_2 = v$, $j = 1,2$)

$$(2.7) \qquad u_j(t) = \int_{-\infty}^{\infty} K_j(t-s)\, S^{-1} f(s)\, ds .$$

The thus defined mapping $\underline{u} = \underline{K} f$ is continuous. Moreover, the image of $M_{\beta,c}$ defined in (2.4) for $I = \mathbb{R}$, is relatively compact in $C_b^o(\mathbb{R},H)$.

For later use we extend the compactness results to $L_2(I,H)$, the space of H-valued square integrable functions. We formulate the result for $I = \mathbb{R}$. Similar propositions hold for $I = [t_o,\infty)$ or $I = (-\infty,t_o]$. Define the set

$$(2.8) \qquad N_{\beta,c} = \left\{ f \in L_2(I,H) \;/\; \| e^{\beta|t|} f \|_{L_2} \leq C \right\}$$

with

$$\| f \|_{L_2} = \left\{ \int_{\mathbb{R}} e^{2\beta|t|} |f(t)|^2 \, dt \right\}^{1/2} .$$

Observe that Lemma 2.1, together with the Cauchy - Schwarz inequality, implies

$$|u'(t)| + |S\,u(t)| \leq c(\alpha,\beta)\, C\, e^{-\beta|t|} .$$

Thus one obtains by integration

$$\| S\,u \|_{L_2} \leq \frac{2c(\alpha,\beta)C}{\beta} .$$

<u>Remark 2.8</u>

Given C and β as in Theorem 2.7. Then the image of $N_{\beta,c}$ unter (2.7) is relatively compact in $L_2(\mathbb{R},H)$.

3. Bifurcation of Homoclinic Solutions

In this section we formulate a reduction principle for $(1.2)_0$ using a center manifold approach. Then we proceed to study the case where $T_0(\lambda)$ has a simple eigenvalue O at λ_0. Under generic conditions, $u = 0$, $\lambda = \lambda_0$ is a bifurcation point of homoclinic (heteroclinic) solutions. An application to solitary wave solutions of Euler's equation is given.

We decompose H into two T-invariant subspaces which may depend on λ

$$H = H_0 \oplus H_1 , \quad T(\lambda) = T_0(\lambda) \oplus T_1(\lambda)$$
$$\dim H_0 = n .$$

A 1: Assume that the projection $\Pi_0(\lambda) : H \to H_0$ is a C^r-map in the uniform topology. Suppose further that the function $(\beta_o > 0)$

$$\beta_o^2(\lambda) = \sup \left\{ |\sigma| / \sigma \in \Sigma T_0(\lambda) \right\}$$

approaches O as λ tends to λ_0.

A 2: The operator $T_1(\lambda)$ is strictly accretive with square-root $S_1(\lambda) = S_1$ having an exponential dichotomy of the form given in Lemma 2.1 with

$$\alpha < \inf \left\{ \operatorname{re} \sigma / \sigma \in \Sigma S_1(\lambda) \right\}$$

independent of $\lambda \in \Lambda$.

A 3: Suppose $f \in C^r(\Lambda \times D(S), H)$ for some $r \geq 2$.

Moreover

$$f(\lambda,0) = 0\,, \quad D_u f(\lambda,0) = 0\,, \quad \lambda \in \Lambda\,.$$

Define a cutoff function $\chi(u)$, $\chi \in C^\infty(D(S); H)$ such that for some $\delta > 0$

$$\chi(u) = \begin{cases} 1 & \text{for } |u| + |S\,u| \leq \delta \\ 0 & \text{for } |u| + |S\,u| \geq 2\delta \end{cases}$$

holds. Then

$$g(\lambda,u) = \chi(u)\, f(\lambda,u)$$

coincides with f in some neighborhood of 0. Moreover we have

$$g \in C_b^r(\Lambda \times D(S), H) \cap C_\eta^1$$

where we used the notation

$$C_\eta^1 = \left\{ g \in C^1(\Lambda \times D(S); H) \,/\, \|g\|_1 \leq \eta \right\}$$

$$\|g\|_1 = \sup_{|\alpha| \leq 1} \; \sup_{\Lambda \times D(S)} |D^\alpha g(u,\lambda)|\,.$$

We study the localized equation $(1.2)_0$ which reads

$$(3.1) \qquad \frac{d^2u}{dt^2} - T(\lambda)\, u + g(\lambda,u) = 0\,.$$

Obviously, $u = 0$ is a solution for all $\lambda \in \Lambda$. Theorem 2.7 immediately implies that $u = 0$ is an isolated solution in X as long as $\Sigma\, T(\lambda) \neq 0$. Hence, to obtain nontrivial solutions in X, we have to look near $\lambda = \lambda_0$. We construct a function $h(\lambda, u_0, u_0')$ with values in $D(S_1)$ such that every solution $u \in X$ of (3.1) satisfies

$$u_1(t) = h(\lambda, u_o(t), u_o'(t))$$

thus reducing (3.1) to an ordinary differential equation of order $2n$.

To construct h we follow the well known device for the finite-dimensional case (c.f. [3]). The proof is a straightforward extension, after Theorem 2.7 had been established. First solve ($g_o = \Pi_o(\lambda)g$)

$$(3.2) \quad \begin{aligned} & u_o'' - T_o(\lambda)\, u_o + g_o(\lambda, u_o + h(u_o, u_o')) = 0 \\ & u_o(0) = \xi_o, \quad u_o'(0) = \xi_1 \end{aligned}$$

obtaining a solution $u_o(t; \xi, \lambda, h)$ which exists for all $t \in \mathbb{R}$. On bounded t-intervals it is a C^r-function of $\lambda \in \Lambda$, $\xi \in H_o \times H_o$ and $h \in C_b^k(\Lambda \times H_o \times H_o, D(S_1)) \cap C_\kappa^1$ for some $\kappa > 0$ and $h(\lambda, 0, 0) = 0$, $D_u h(\lambda, 0, 0) = 0$, $1 \leqslant k < r$. Moreover, for any $\beta > \beta_o(\lambda)$ we obtain the estimate

$$|u_o(t)| \leqslant c_o(|\xi_o| + |\xi_1||t|)\, e^{\beta|t|}$$

and, if D^γ denotes any derivative with respect to ξ, λ or h of order $|\gamma| > 0$

$$|D^\gamma u_o(t)| \leqslant c_o\, e^{|\gamma|\beta|t|}, \qquad |\gamma| \leqslant k .$$

Similar estimates hold for $u_o' = d u_o/dt$.

Now consider (3.1) in H_1 and apply Theorem 2.7. Any solution in X must satisfy (2.7). Setting $t = 0$ in this representation we define

$$(3.3) \quad Z\,h(\lambda,\xi) = \int_{-\infty}^{\infty} K_1(-s)\,S_1^{-1}\,f_1(\lambda,u_0+h(u_0,u_0'))(s;\lambda,\xi,h)\,ds\,.$$

Here K_1 is given as in (2.3), only S is replaced by S_1. Observe that, if η and κ are sufficiently small positive numbers, Z is a strict contraction of $C^{Lip}_{\eta,o}$ into itself. Its unique fixed point, which we call $h(\lambda,\xi)$ again, is k-times continuously differentiable, provided that $k\beta < \alpha$ holds, moreover $D_\xi h(\lambda,0) = 0$. Tus Λ has to be a sufficiently small neighborhood of λ_0.

<u>Theorem 3.1</u>

Suppose that the assumptions A 1, A 2, A 3 are valid. Then for every integer k, $1 \leq k < r$, there exists a neighborhood Λ of λ_0 and positive numbers η_1, κ_1 such that for every $\eta \in (0,\eta_1)$, $\kappa \in (0,\kappa_1)$ there is a unique function $h \in C^k_b(\Lambda \times H_0 \times H_0, D(S_1)) \cap C^{Lip}_\kappa$ with the following properties

1. $h(\lambda,0) = 0$, $D_\xi h(\lambda,0) = 0$
2. $M_\lambda = \left\{ u \,/\, u = u_0 + h(\lambda,u_0,u_0') \right\}$ is invariant under (3.1)
3. All solutions of (3.1) which belong to X belong to M_λ.

Observe further that $h(\lambda,u_0,-u_0') = h(\lambda,u_0,u_0')$ since, with $u(t)$, $u(-t)$ is a solution of (3.1). This is a special case of reversibility, which can be used when f depends on du/dt as well (c.f. [11], [13]).

The above result is applied to the case where H_o is 1-dimensional, i.e. $T_o(\lambda)$ has a simple eigenvalue $\sigma_o(\lambda)$. We suppose

$$(3.4) \qquad \sigma_o(\lambda_o) = 0\,, \qquad \sigma_o'(\lambda_o) < 0\,.$$

Thus $\sigma_o(\lambda)$ is positive for $\lambda < \lambda_o$ and negative for $\lambda > \lambda_o$. Let f have the form

$$(3.5) \qquad f(\lambda,u) = f^{(k)}(\lambda,u) + O((|u| + |S\,u|)^{k+1})$$

for some $k \geqslant 2$, where f is homogeneous of degree k in u. Since $h = O(|u_o|^2 + |u_o'|^2)$ we obtain

$$(3.6) \qquad u_o'' - \sigma_o(\lambda)u_o + \gamma(\lambda)u_o^k + O(|u_o|^{k+1} + |u_o'|^{k+1}) = 0$$

with

$$\gamma(\lambda) = f_o^{(k)}(\lambda,u_o) \neq 0\,.$$

Equation (3.6) is readily discussed. First truncate by neglecting the O-terms. For k even, homoclinic solutions exist on both sides of λ_o satisfying

$$\lim_{t\to\pm\infty} u_o(t) = \begin{cases} 0 & \text{for } \lambda < \lambda_o \\ \left(\dfrac{\sigma_o(\lambda)}{\gamma(\lambda)}\right)^{\frac{1}{k-1}}, & \text{for } \lambda > \lambda_o \end{cases}\,.$$

The equilibrium points being approached at infinity are both saddle nodes. Hence there is a unique equilibrium $u_\infty(\lambda)$ of (3.4) such that

$$u_\infty(\lambda) = \begin{cases} 0 & \text{for } \lambda < \lambda_o \\ \left(\dfrac{\sigma_o(\lambda)}{\gamma(\lambda)}\right)^{\frac{1}{k-1}} + O(|\sigma_o|) & \text{for } \lambda > \lambda_o \end{cases}$$

Moreover, the homoclinic orbits of the truncated equation are continuable to those of the full equation (c.f. [11]) in view of the reversibility.

If k is odd, the situation is somewhat different. If $\gamma(\lambda_o)$ is positive, then two homoclinic solutions exist for $\lambda < \lambda_o$ with

$$\lim_{t\to\pm\infty} u_o(t) = 0 .$$

If $\gamma(\lambda_o)$ is negative then two heteroclinic solutions exist for $\lambda > \lambda_o$ with

$$\lim_{t\to\pm\infty} u_o(t) = u_{\pm\infty}(\lambda)$$

$$u_{\pm\infty}(\lambda) = \pm\left(\frac{\sigma_o(\lambda)}{\gamma(\lambda)}\right)^{\frac{1}{k-1}} + O(|\sigma_o(\lambda)|) .$$

Using the center manifold we obtain the corresponding solutions for the full equations (3.1) resp. $(1.2)_o$. The analysis developed above gives the complete picture. In particular uniqueness holds, up to translations of the t-axis, in the class of solutions having limits for $t \to \pm\infty$.

<u>Proposition 3.2</u>

In addition to A 1 , A 2 , A 3 assume that dim $H_o = 1$ and that the simple eigenvalue $\sigma_o(\lambda)$ of $T(\lambda)$ satisfies (3.4) . Suppose that (3.5) holds for f with $\gamma(\lambda_o) \neq 0$.

1. If k is even then there exists, for $\lambda > \lambda_o$, a unique t-independent nontrivial solution $U_\infty(\lambda) \in D(T)$. Moreover for each λ in a suitable neighborhood Λ of λ_o, $\lambda \neq \lambda_o$, there exists a homoclinic solution $p(\cdot,\lambda)$ of (3.1), unique up to translations in t in some X-neighborhood of O. This solution satisfies

$$\lim_{t\to\pm\infty} p(t,\lambda) = \begin{cases} 0 & \text{for } \lambda < \lambda_o \\ U_\infty(\lambda) & \text{for } \lambda > \lambda_o \end{cases}$$

2. If k is odd, $\gamma(\lambda_o) > 0$, there are exactly two homoclinic solutions for $\lambda < \lambda_o$. They satisfy

$$\lim_{t\to\pm\infty} p(t,\lambda) = 0 .$$

3. If k is odd, $\gamma(\lambda_o) < 0$, then there are two nontrivial, t-independent solutions $U_{\pm\infty}(\lambda)$ near O. Moreover there exist exactly two heteroclinic solutions p_1, p_2 such that

$$\lim_{t\to\pm\infty} |p_j(t;\lambda) - U_{\pm\infty}(\lambda)| = 0, \quad j = 1,2 .$$

All these solution bifurcate in $\Lambda \times X$ from $(\lambda_o, 0)$.

For the rest of this paper we concentrate on case 1 for $\lambda < \lambda_o$, i.e. $\sigma_o(\lambda) = \omega^2 > 0$. Then $T(\lambda)$ is strictly accretive with lower bound ω^2. Let $p(t;\lambda)$ denote the homoclinic solution, set

$$(3.7)\qquad \begin{aligned} p_0(t;\lambda) &= \omega^2 P_0(\omega t) \\ p_1(t;\lambda) &= \omega^4 P_1(\omega t) \end{aligned}$$

then it is not hard to see that P_0 and P_1 are uniformly bounded in t [15]. In particular P_0 satisfies

$$P_0'' - P_0 + \gamma(\lambda) P_0^2 + O(\omega^2) = 0 .$$

Moreover the following estimate is valid

$$(3.8)\qquad |S\,p(t,\lambda)| \leq c_1 e^{-\beta|t|} \quad \text{for } \beta < |\omega|$$

An example: We consider an inviscid fluid in the two-dimensional strip $(x,y) \in \Omega = (0,h) \times \mathbb{R}$ with varying density $\rho = \rho(\theta)$, θ = temperature, under the influence of gravity g (acting in negative y-direction). $\underline{u} = (u,v)$, denotes the velocity vector, p the pressure. We search for solitary wave solutions, i.e. waves of permanent shape travelling with constant speed $c > 0$ along the x-axis and vanishing at infinity. Introducing a coordinate system moving with the wave, we obtain from the (dimensionless) Euler equations (c.f. [11])

$$(3.9)\qquad \begin{aligned} \rho(\underline{u} \cdot \nabla \underline{u}) + \nabla p &= (-\lambda\rho + \mu\rho\theta)\,\underline{e}_2 \\ \underline{u} \cdot \nabla \rho + \rho \nabla \cdot \underline{u} &= 0 \\ \underline{u} \cdot \nabla \theta &= 0 \end{aligned}$$

We have neglected thermal conductivity. The following notations have been used

$$\lambda = \frac{g\,h}{c^2} , \qquad \mu = \lambda\,\beta\,\Delta T$$

where β is the coefficient of thermal expansion. ΔT the temperature difference between upper and lower boundary and $\underline{e}_2 = (0,1)$. The boundary condition reads ($\underline{e}_1 = (1,0)$)

$$\underline{u} \circ \underline{n}|_{\partial\Omega} = 0, \quad \lim_{|x|\to\infty} u(x,y) = c\underline{e}_1$$

(3.10)

$$\theta(0) = 0, \quad \theta(1) = 1 .$$

We conclude from $\rho = \rho(\theta)$ and (3.9) that $\underline{u} \circ \nabla\rho = 0$ (Nondiffusivity). Introducing the streamfunction ψ by

$$\psi_y = \sqrt{\rho}\, u, \qquad \psi_x = -\sqrt{\rho}\, v$$

the temperature becomes a function of ψ alone. Bernoulli's equation

$$\frac{\rho}{2}(u^2 + v^2) + p + \rho\lambda y = H(\psi) + \mu\rho\theta y$$

yields ($\rho(\psi) = \rho(\theta(\psi))$), together with (3.9)

$$\Delta\psi + \lambda y \rho' = H'(\psi) + \mu(\rho\theta)' y$$

(3.11)

$$\psi(x,0) = \psi_o, \quad \psi(x,1) = \psi_1$$

$$\lim_{|x|\to\infty} \psi(x,y) = \psi_o(y) = \psi_o + c\int_0^y \sqrt{q(\eta)}\, d\eta$$

where q is the density distribution at infinity. For $\mu = 0$ this is the well known Long - Yih equation (c.f. [11]). Obviously, $\psi = \psi_o(y)$ is a solution. Setting $\psi = \psi_o + \phi$ we obtain ($\Delta = \partial^2/\partial x^2 + \partial^2/\partial y^2$)

$$\Delta\phi + a(\lambda,y)\phi + b(\lambda,y)\phi^2 + r(\lambda,y,\phi) = \mu F(y,\phi)$$

(3.12)

$$\phi(x,0) = \phi(x,1) = 0, \quad \lim_{|x|\to\infty} \phi(x,y) = 0$$

where a and b have been calculated in [11], e.g.

$$a(\lambda,y) = -\lambda s' - \frac{s'^2}{4} - \frac{s''}{2}, \qquad s = \log q .$$

Moreover we have

$$r(\lambda,y,\phi) = O(\phi^3) \text{ uniformly in } y \text{ and } \lambda$$

$$\begin{aligned} F(y,\phi) &= y(\rho\,\theta)'(\psi_o + \phi) \\ &= \frac{1}{c\sqrt{q}} \left\{ (q\,\theta_o)_y + \frac{\phi}{c^2}\left(\frac{q_y\,\theta_{oy}}{q}\right)_y + O(\phi^2) \right\} \end{aligned}$$

$$\rho(\psi_o(y)) = q(y)$$

$$\theta(\psi_o(y)) = \theta_o(y) .$$

Now we make the following identifications

$$H = L_2(0,1), \qquad T(\lambda) = -\frac{\partial^2}{\partial y^2} - a(\lambda,y)$$

$$f(\lambda,\phi) = b(\lambda,y)\,\phi^2 + \dots, \qquad x = t .$$

If $q' < 0$ holds, then $T(\lambda)$ has a smallest value λ_o for which $0 \in \Sigma\, T(\lambda_o)$. This eigenvalue is simple and proposition 3.2 applies (for $\mu = 0$).

Proposition 3.3:

Set $\mu = 0$. Assume $q'(y) < 0$ for $y \in (0,1)$ and denote the eigenfunction to $\sigma_o(\lambda)$ of $T(\lambda)$ by $\varphi_o(\lambda,y)$. Suppose that

$$\tilde{b}(\lambda) = \int_0^1 b(\lambda,y)\,\varphi_o^3(\lambda,y)\,dy \neq 0 .$$

Then there is, for $\lambda < \lambda_o$, a solution of (3.12), unique up to translation in x in some $L_2(0,1)$-neighborhood of 0.

4. Nonreversible Perturbations

In the situation described by Proposition 3.2, no. 1, to which we restrict our attention now, $\lambda = \lambda_o$ is a bifurcation point of homoclinic solutions. For $\lambda < \lambda_o$, these solutions decay exponentially to 0 at infinity (c.f. (3.8)). The trivial solution is a saddle point in H. In this section we study the question, whether homoclinic points persist under nonreversible perturbations. We follow the work of Chow, Hale and Mallet-Paret [4].

Consider the equation $((1.2)_\varepsilon)$

$$\frac{d^2u}{dt^2} - T(\lambda)\, u + f(\lambda,u) = \varepsilon\, F(t,\lambda,u) \tag{4.1}$$

with the assumptions formulated in section 1. First we show, that, for fixed $\lambda < \lambda_o$, $u = 0$ has a unique continuation in X^1 for $\varepsilon \neq 0$; call it $u^*(\cdot,\varepsilon)$. After having established good integral representations in section 2 this is a straightforward step. The same is true for the construction of local stable and unstable manifolds.

However the globalization to invariant manifolds, as it is common for ordinary differential equations, does not work here, since backward in t-integration is not possible in general. Therefore we restrict ourselves to the construction of homoclinic solutions u near p, for which

$$\lim_{|t|\to\infty} |S\,u(t) - S\,u^*(t,\varepsilon)| = 0$$

holds. Moreover we show how global existence follows and discuss an example, which extends the one of the last section.

Lemma 4.1

Make the assumptions of Proposition 3.2 and fix $\lambda < \lambda_o$. Then there is a positive number ε_1 and a X-neighborhood U(O) of O such that, for every ε with $|\varepsilon| < \varepsilon_1$, there is a unique C^r-solution $u^*(\circ,\varepsilon)$ in U(O) which solves (4.1).

Proof: Apply Theorem 2.7. Set $\underline{u} = (u,v)$ and define

$$G(\varepsilon,\underline{u}) = \underline{u} - \underline{K}(-f(\lambda,u) + \varepsilon F(\cdot,\lambda,u)) .$$

Observe that $G(0,0) = 0$ and $G \in C^r(\mathbb{R} \times X^1 \times X^1, H)$ holds. Moreover, $D_u G(0,0)$ is the idendity. Hence, the implicit function theorem yields the assertion in $X^1 \times X^1$. The regularity follows by applying Corollary 2.4 and Lemma 2.6

It is rather clear how to construct stable and unstable manifolds for u^*. One sets $u = u^* + U$ and applies the Lemmas 2.3 and 2.6. Since we do not really use these manifolds we omit the detailed formulation. Let us simply mention that for any $(t_o,\underline{u}_o)$ on the stable manifold we have

$$|S\,U(t;t_o,\underline{u}_o)| \leq c\,e^{-\beta'(t-t_o)}$$

for $t \geq t_o$. Here $\beta' < \omega(\lambda)$ and c depends on the choice of β'. A similar estimate holds for the unstable manifold and $t \leq t_o$. The neighborhood of u^* in $X^1 \times X^1$ where these manifolds exist can be chosen independently of t_o, the solutions constructed below eventually enter them.

For the subsequent analysis it is appropriate to work in $L_2(\mathbb{R},H)$. Denote by $p(\cdot,\lambda)$ the homoclinic solution constructed in Prop. 3.2, part 1 for $\lambda < \lambda_o$. In particular we use (3.8). Let us investigate the linear inhomogeneous differential equation

$$\frac{d^2u}{dt^2} - T(\lambda)\,u + f_u(p)\,u = F \tag{4.2}$$

where $f_u(p) = D_u f(\lambda,p(\cdot,\lambda))$. From Theorem 2.7 we obtain ($K=K_1$)

$$u(t) = \int_{-\infty}^{\infty} K(t-s)\,S^{-1} f_u(p(s))\,u(s)\,ds + \int_{-\infty}^{\infty} K(t-s)\,S^{-1} F(s)\,ds \tag{4.3}$$

which we write formally

$$u = \mathcal{K}_p u + \mathcal{F} \tag{4.3'}$$

Lemma 4.2

$\mathcal{K}_p$ is a compact linear operator in $L_2(\mathbb{R},H)$.

Proof: Observe that

$$|f_u(p(t))| \leq c\,e^{-\beta|t|}$$

holds and thus, by Remark 2.8, $\mathcal{K}_p$ maps bounded sets in $L_2(\mathbb{R},H)$ into relatively compact sets.

Introduce the notation

$$(4.4)\qquad [u,v] = \int_{-\infty}^{\infty} (u,v)(t)\,dt$$

for the scalar product in $L_2(T,H)$. Moreover assume

$$(4.5)\qquad |S^{-1} f_u(p)\,u| \leq c_p|u|$$

or all $u \in H$.

<u>Lemma 4.3</u>

$\mathcal{K}_p$ has the simple eigenvalue 1 .

Proof: Obviously 1 is an eigenvalue with eigenfunction p_t . Assume $u \neq 0$ to be another independent eigenfunction. Set $u(t) = \omega^2 U(\omega t)$, use the decomposition $S = S_0 \oplus S_1$, now in $L_2(\mathbb{R},H)$, and (3.7) to obtain

$$U_0(\tau) = -\frac{1}{2}\int_{-\infty}^{\infty} e^{-|\tau-\sigma|} f_u^0(P(\sigma))\,U(\sigma)\,d\sigma$$

$$U_1(\tau) = -\frac{\omega}{2}\int_{-\infty}^{\infty} e^{-S_1|\tau-\sigma|/\omega} S_1^{-1} f_u^1(P(\sigma))\,U(\sigma)\,d\sigma .$$

Hence we have

$$(4.6)\qquad \begin{aligned} U_0(\tau) &= -\frac{1}{2}\int_{-\infty}^{\infty} e^{-|\tau-\sigma|} f_u^0(P_0(\sigma))\,U_0(\sigma)\,d\sigma + O(\omega^2) \\ U_1(\tau) &= O(\omega^2) \end{aligned}$$

where the rest terms tend to 0 with ω^2 in $L_2(\mathbb{R},D(S_1))$. Setting $\omega = 0$ one obtains the following second order equation for U_o :

$$U_o'' - U_o + f_u^{o'}(P_o)\, U_o = 0 .$$

$U_o = P_o' V_o$ yields V_o = const. A simple implicit function argument now yields that the dimension of the kermel $\ker(\mathrm{id} - \mathcal{K}_p)$ is one.

It remains to be shown that the adjoint equation has a solution $q \in L_2(\mathbb{R},H)$ satisfying $[q,\dot{p}] = 0$. Denote by "~" the adjointness in H , und by "*" in $L_2(\mathbb{R},H)$. Observe that (4.5) implies that $\widetilde{S^{-1} f_u(p)}$ is defined in all of H . Then q must satisfy

$$q(t) = \widetilde{S^{-1} f_u(p(t))} \int_{-\infty}^{\infty} \tilde{K}(s-t)\, q(s)\, ds . \tag{4.7}$$

Define

$$q(t) = \widetilde{S^{-1} f_u(p(t))}\, Q(t)$$

then the following equation holds for Q

$$Q(t) = \int_{-\infty}^{\infty} \tilde{K}(s-t)\, \tilde{S}^{-1} \tilde{f}_u(p(s))\, Q(s)\, ds .$$

In view of (3.7) we see that

$$Q_1 = \omega^2 \mathcal{L}_1(\omega)\, Q_o$$

where $\mathcal{L}_1(\omega)$ is a linear compact operator depending continuously on ω . To obtain an equation for Q_o set $\tau = \omega t$, $\sigma = \omega s$, $Q_o(t) = \omega^2 R_o(\omega t)$

$$\begin{aligned} R_o &= \mathcal{L}_o f_u^o(P)\, R_o - \omega^2 \mathcal{L}_o f_u^o(P)\, \mathcal{L}_1(\omega)\, R_o \\ \mathcal{L}_o f(\tau) &= -\frac{1}{2} \int_{-\infty}^{\infty} e^{-|\tau-\sigma|} f(\sigma)\, d\sigma . \end{aligned} \tag{4.8}$$

Observe that $\widetilde{f_u^o}(P) = f_u^o(P)$. For $\omega = 0$ we have

$$R_o = \mathscr{L}_o f_o'(P_o) R_o .$$

This is the same equation as (4.6) for $\omega = 0$. Therefore 1 is a simple eigenvalue of this equation. Standard perturbation arguments yield the existence of simple eigenvalue $\rho(\omega)$ with $\rho(0) = 1$ for (4.8). Moreover we have $R_o = P_{ot} + O(\omega^2)$ and thus

$$[q,p_t] = [Q, f_u p_t] = [p_{ott}, p_{ott}] + O(\omega^2) .$$

For sufficiently small $|\omega|$ the product $[q, p_t]$ does not vanish. Observe that q satisfies (4.7) with $\rho(\omega)$ as factor on the right side. Multiply by p_t to obtain

$$[q, p_t] = \rho[q, p_t]$$

whence $\rho \equiv 1$ follows, q.e.d.

The solvability condition for (4.3') reads now

$$[q,\mathscr{F}] = [\tilde{S}^{-1} \int_{-\infty}^{\infty} \tilde{K}(s - \cdot)\, q(s)\, ds , F]$$

$$= [Q, F] = 0 .$$

To treat the equation (4.2) we need stronger assumptions on f and F as formulated in (1.1). Denote by $W^k(\mathbb{R},H)$ the Sobolev-space $W^{2,k}$ of order k , thus $W^o(\mathbb{R},H) = L_2(\mathbb{R},H)$.

A 4 : Assume $f \in W^r(\Lambda \times H, H)$ for some $r \geq 2$, $f(\lambda,0) = 0$, $D_u f(\lambda,0) = 0$. $F \in W^r(\Lambda \times \mathbb{R} \times H, H)$ with

$$\sup_{|u| \leq \rho} |D_u F(\lambda,t,u)| \leq C(\rho), \; \lambda \in \Lambda, \; t \in \mathbb{R}.$$

Moreover we introduce the space

$$Y = W^2(\mathbb{R},H) \cap W^1(\mathbb{R},D(S)) \cap W^0(\mathbb{R},D(T))$$

$$Y_\perp = \left\{ z \in Y \; / \; [z, p_t] = 0 \right\}$$

$$W^1_\perp = \left\{ F \in W^1(\mathbb{R},H) \; / \; [Q, F] = 0 \right\}.$$

Lemma 4.4

Given Prop. 3.2, 1 for $\lambda < \lambda_o$ with the principal eigenvalue $\omega^2(\lambda)$. For sufficiently small ω^2(λ close to λ_o) , (4.2) defines an isomorphism between $F \in W^1_\perp$ and $Y_\perp$.

The proof follows from Lemmas 4.2 and 4.3 and a straightforward extension of the results in section 2 to the spaces considered here.

A solution of the nonlinear equation (4.1) is sought in the form

$$(4.9) \qquad u(\tau) = p(\tau+\alpha) + z(\tau+\alpha), \quad t = \tau + \alpha .$$

Then z has to solve the equation

$$(4.10) \qquad z_{tt} - T(\lambda)\, z + f_u(p)\, z = G(\varepsilon,\lambda,t-\alpha,z)$$

where

$$G(\varepsilon,\lambda,t,z) = \varepsilon F(t,p+z) - r(\lambda,t)$$

$$r(\lambda,z) = f(\lambda,p+z) - f(\lambda,p) - f_u(p)\,z\ .$$

According to the preceding analysis one can solve (4.10) uniquely in $Y_\perp$ if the right side is replaced by

$$G(\varepsilon,\lambda,t-\alpha,z+\gamma\,p_t) - [Q,G]\,p_t$$

where Q has been normalized by $[Q,p_t]=1$. Denote by $\tilde{z}(\varepsilon,\alpha,\gamma)$ the unique solution for fixed α . Then, the solvability of (4.1) in some Y-neighborhood of O is equivalent to (suppress λ)

$$(4.11)\qquad 0 = [Q,\, G(\varepsilon,\cdot-\alpha,\ \tilde{z}+\gamma\,p_t)] = g(\varepsilon,\alpha,\gamma)\ .$$

<u>Theorem 4.5</u>

Assume Prop. 3.2,1 and fix $\lambda < \lambda_o$ such that the preceding Lemmas hold. Suppose further the validity of A 4 . Then there exists an $\varepsilon_1 > 0$ and an Y-neighborhood U(O) of O such that, for every ε with $|\varepsilon| < \varepsilon_1$, the solvability of (4.1) in U(O) is equivalent with the equation (4.11) for $\alpha,\gamma \in \mathbb{R}$ via (4.9) .

The solution u in (4.9) eventually enters the stable and unstable manifold of $u^*(\cdot,\varepsilon)$. It would be of interest to study the consequences of a "transverse" intersection in u . Let us finally make some remarks about the existence in the large. We keep $\lambda = \lambda_1 < \lambda_o$ fixed.

If F decays sufficiently fast at infinity and $F(\cdot,\lambda,0) = 0$ then

$$(4.11) \qquad \mathcal{K}_p z + K\, G(\varepsilon,\lambda,\cdot - \alpha, z)$$

defines a completely continuous mapping in some space $L_{2,\beta}(\mathbb{R},H)$ with exponential decay. Since 1 is a simple eigenvalue of $\mathcal{K}_p$ for $\varepsilon = 0$ we just need a transversality condition to establish global existence of (4.10) in $\mathbb{R} \times Y$ for given α and λ. Assume that F satisfies

$$\sup_{|u| \leq \rho} |F(t,\lambda,u)| \leq C_\rho\, e^{-\beta|t|}$$

for some $\beta > 0$ and define

$$L_{2,\beta} = \left\{ u : \mathbb{R} \to H \;/\; e^{\beta|t|}\, u \in L_2(\mathbb{R},H) \right\} .$$

Then it is easy to see that (4.11) defines a completely continuous operator in $L_{2,\beta}$ with the simple eigenvalue 1 for $\varepsilon = 0$. Moreover, if $(p = p(\cdot,\lambda_1))$

$$[\, q,\, F_u(\cdot,\lambda,p_t)\, p_t \,] \neq 0$$

holds, we can apply the well known theorem of Rabinowitz and obtain global existence of solutions of (4.1).

Example: We apply the results to the problem (3.11), now for $\mu \neq 0$. Obviously this equation is still reversible (' denotes derivative with respect to ψ). θ is an arbitrary function of ψ satisfying the boundary conditions. Set $\psi = \psi_0 + \phi$ and define $T(\lambda)$ as in section 3. The nonlinearity does not satisfy A 4. However, if we introduce $\Psi = S\psi$, then the right side of (4.3), after S has been applied, defines a completely continuous map in $L_2(0,1)$. ($D(S)$ lies compactly in $L_2(0,1)$). Therefore, our above analysis applies and we obtain solutions of (3.11) if

$$[y(\rho\theta)'(\psi_0 + p(\cdot,\lambda_1) + z) - \tilde{r}(\lambda_1,z)\,,\, p_x(\cdot,\lambda_1)] = 0$$

with the notation (see (3.12))

$$\begin{aligned}\tilde{r}(\lambda_1,z) &= b(\lambda_1,\cdot)((p+z)^2 - 2pz)\\ &\quad + r(\lambda_1,\cdot,p+z) - r(\lambda_1,p)\\ &\quad - D_u r(\lambda_1,p)z\end{aligned}$$

$p(\cdot,\lambda_1)$ denotes the homoclinic solution of Prop. 3.3.

Choose $p(\cdot,\lambda_1)$ and z to be even in x, then the above equality is always fulfilled. Thus, given any temperature distribution $\theta(\psi)$ we obtain, for sufficiently small $\mu \neq 0$, a solution of (3.11). We could apply these arguments also to a nonreversible situation, e.g. when heat sources are present in the strip. However the analysis is too technical and we suppress it here.

References

[1] C.J. Amick and J.F. Toland, On solitary water waves of finite amplitude, Arch. Rat. Mech. Anal. 76 (1981), 9-95.

[2] C.J. Amick and J.F. Toland, Nonlinear elliptic eigenvalue problems on an infinite strip - global theory of bifurcation and asymptotic bifurcation, manuscript.

[3] S. Chow and J.K. Hale, Methods of bifurcation theory, Grundlehren der math. Wissenschaften Nr. 251, Springer-Verlag, 1982.

[4] S. Chow, J.K. Hale, and J. Mallet-Paret, An example of bifurcation to homoclinic orbits, J. Diff. Equ. 37 (1980), 351-373.

[5] D.S. Cohen, F.C. Hoppenstaedt, and R.M. Miura, Slowly modulated oscillations in nonlinear diffusion processes, SIAM J. Appl. Math. 33 (1977), 217-229.

[6] M.J. Esteban and P.L. Lions, Existence and nonexistence results for semilinear elliptic problems in unbounded domains, to appear in Proc. Roy. Soc. Edinburgh.

[7] G. Fischer, Zentrumsmannigfaltigkeiten bei elliptischen Differentialgleichungen, to appear in Math. Nachrichten.

[8] D. Henry, Geometric theory of semilinear parabolic equations, Lect. Notes in Math., Nr. 840, Springer-Verlag, 1981.

[9] L.N. Howard and N. Kopell, Slowly varying waves and shock structures in reaction-diffusion equations, Studies Appl. Math. 56 (1977), 95-145.

[10] T. Kato, Perturbation theory for linear operators, Grundlehren der math. Wissenschaften, Nr. 132, Springer-Verlag, 1966.

[11] K. Kirchgässner, Wave-Solutions of reversible systems and applications, J. Diff. Equ. 45 (1982), 113-127.

[12] K. Kirchgässner, Waves in weakly-coupled parabolic media, manuscript, to appear.

[13] K. Kirchgässner and J. Scheurle, On the bounded solutions of a semilinear elliptic equation in a strip, J. Diff. Equ. 32 (1979), 119-148.

[14] N. Kopell and L.N. Howard, Bifurcations and trajectories joining critical points, Adv. Math. 18 (1975), 306-358.

[15] M. Renardy, Bifurcation of singular and transient solutions. Spatially nonperiodic patterns for chemical reaction models in infinitely extended domains, in "Recent Contributions to Nonlinear Partial Differential Equations", (H. Berestycki and H. Brezis, Eds.), Pitman, Boston, London, Melbourne, 1981.

[16] M. Renardy, Bifurcation of singular solutions in reversible systems and applications to reaction-diffusion equations, Adv. in Appl. Math. 3 (1982), 384-406.

ON LINEAR DIFFERENTIAL EQUATIONS WITH ALMOSTPERIODIC COEFFICIENTS AND THE PROPERTY THAT THE UNIT SPHERE IS INVARIANT

J. Kurzweil, A. Vencovská
Mathematical Institute
of Czechoslovak Academy of Sciences
Žitná 25, 115 67 Praha 1,
Czechoslovakia

Denote by Matr (n) the set of $n \times n$-matrices with complex or real entries, by F^* the adjoint of F for $F \in$ Matr (n) (F^* is the transpose of F in the real case), by I the identity matrix. Let $A : \mathbb{R} \to$ Matr (n) be continuous and let $X : \mathbb{R} \to$ Matr (n) be the matrix solution of

(1) $$\dot{x} = A(t)x$$

fulfilling $X(0) = I$. It is an easy exercise to prove that the following conditions are equivalent:

(2) $A(t) + A^*(t) = 0$ for $t \in \mathbb{R}$.

(3) If x is a solution of (1), then $\|x(t)\| = (x(t), x(t))^{1/2}$ does not depend on t.

(4) $X(t) \in U(n)$ for $t \in \mathbb{R}$ in the complex case, $U(n)$ being the set of unitary matrices of order n ($X(t) \in O(n)$ for $t \in \mathbb{R}$ in the real case, $O(n)$ being the set of orthonormal matrices of order n).

Let $AP(n)$ be the set of uniformly almostperiodic functions $A : \mathbb{R} \to$ Matr (n) that fulfil (2), let $AP_{sol}(n)$ be the set of such functions $A \in AP(n)$ that X is uniformly almostperiodic.

Problem 1. Is $AP_{sol}(n)$ dense in $AP(n)$ (in the topology of uniform convergence) ?

In the complex case every $A \in AP(1)$ is a uniformly almostperiodic function with purely imaginary values. It can be approximated by a trigonometrical polynomial T with purely imaginary values. It follows from $X(t) = \exp \int_0^t T(s)ds$ that X is uniformly almostperiodic and the answer to Problem 1 is affirmative. In the real case every $A \in AP(2)$ can be written in the form

$$A(t) = \begin{pmatrix} 0, & a(t) \\ -a(t), & 0 \end{pmatrix}$$

with $a : \mathbb{R} \to \mathbb{R}$ uniformly almostperiodic and by an analogous argument the answer to Problem 1 is affirmative. Therefore we shall assume $n > 1$ in the complex case and $n > 2$ in the real case.

A function $A : \mathbb{R} \to \mathrm{Matr}(n)$ will be called quasiperiodic with at most $r+1$ frequencies, if there exist such $\omega_0 \geq 0$, $\omega_1 \geq 0$, ..., $\omega_r \geq 0$ and such a continuous function $B : \mathbb{R}^{r+1} \to \mathrm{Matr}(n)$ that

(5) $\quad A(t) = B(\omega_0 t, \omega_1 t, \dots, \omega_r t)$, $\quad t \in \mathbb{R}$,

(6) $\quad B(s_0, \dots, s_j+1, \dots, s_r) = B(s_0, \dots, s_j, \dots, s_r)$
for $s_0, \dots, s_r \in \mathbb{R}$, $j = 0, 1, \dots, r$.

Let $QP(n,r)$ be the set of such quasiperiodic functions $A : \mathbb{R} \to \mathrm{Matr}(n)$ with at most $r+1$ frequencies that (2) holds. Let $QP_{sol}(n,r)$ be the set of such $A \in QP(n,r)$ that X is quasiperiodic with at most $r+1$ frequencies.

<u>Problem 2.</u> Is $QP_{sol}(n,r)$ dense in $QP(n,r)$?

An affirmative answer to Problem 2 would obviously imply an affirmative answer to Problem 1. However, the results on Problem 2 are far from being complete. In general it can be proved that the answer to Problem 2 is affirmative provided that the manifolds $SU(n)$, $n=2,3,\dots$ (special unitary matrices of order n, i.e. unitary matrices with determinant equal to one) and the manifolds $SO(n)$, $n=3,4,\dots$ (special orthonormal matrices of order n) have certain estimation properties with respect to homotopies (cf. Proposition below).

Let M be a connected Riemannian manifold, $r=1,2,\dots$. It will be said that M has the estimation property of order r (with respect to homotopies) and written $M \in EP(r)$, if there exists such a $c = c(M,r) > 0$ that the following situation takes place:

Assume that the following conditions are fulfilled:

$$m \in M, \ L \geq 1, \ g : J = \langle 0,1 \rangle^r \to M,$$
$$g(\alpha) \in M \quad \text{for} \quad \alpha \in \partial J,$$
$$g \ \text{is of class} \ C^{(1)}, \quad \left\| \frac{\partial g}{\partial \alpha_j} \right\| \leq L, \quad j=1,2,\dots,r,$$

g is homotopic to g_0, $g_0(\alpha) = m$ for $\alpha \in J$.
Then there exists such a homotopy

$$h : J \times \langle 0,1 \rangle \longrightarrow M \quad \text{that}$$

$$h(\alpha,1) = g(\alpha),\ h(\alpha,0) = g_0(\alpha) \quad \text{for} \quad \alpha \in J,$$

$$h(\alpha,\beta) = m \quad \text{for} \quad \alpha \in \partial J,\ \beta \in \langle 0,1 \rangle,$$

$h, \dfrac{\partial h}{\partial \alpha_j}, \dfrac{\partial h}{\partial \beta}, \dfrac{\partial^2 h}{\partial \alpha_j \partial \beta}, \dfrac{\partial^2 h}{\partial \beta^2}$ are continuous and

$$\left\| \frac{\partial h}{\partial \beta} \right\| \leq c,\ \left\| \frac{\partial h}{\partial \alpha_j} \right\|,\ \left\| \frac{\partial^2 h}{\partial \alpha_j \partial \beta} \right\| \leq cL,\ \left\| \frac{\partial^2 h}{\partial \beta^2} \right\| \leq cL^2 .$$

<u>Proposition.</u> If $SU(n) \subset EP(1) \cap EP(2) \cap \ldots \cap EP(r)$ in the complex case (if $SO(n) \subset EP(1) \cap EP(2) \cap \ldots \cap EP(r)$ in the real case), then $QP_{sol}(n,r)$ is dense in $QP(n,r)$.

It is known (cf. [2]) that

$$\pi_1(SU(n)) = 0,\ \pi_2(SU(n)) = 0,\ n=2,3,\ldots$$

$$\pi_1(SO(n)) = Z/2,\ \pi_2(SO(n)) = 0,\ n=3,4,\ldots .$$

The fact that $\pi_1(SU(n))$, $\pi_2(SU(n))$, $\pi_2(SO(n))$ are trivial and that the structure of $\pi_1(SO(n))$ is very simple can be used to prove that

$$SU(n) \in EP(1) \cap EP(2),\quad SO(n) \in EP(1) \cap EP(2) .$$

However it cannot be proved in a similar way that $SU(n)$ or $SO(n) \in EP(3)$, for $\pi_3(SU(n)) = Z$, $\pi_3(SO(n)) = Z$.

Thus we obtain from Proposition

<u>Theorem.</u> $QP_{sol}(n,1)$ is dense in $QP(n,1)$, $QP_{sol}(n,2)$ is dense in $QP(n,2)$,

$n=2,3,\ldots$ in the complex case,

$n=3,4,\ldots$ in the real case.

In order to indicate some steps which are to be made to obtain the above results, put $\omega_0 = 1$ and start from equation

(7) $$\dot{x} = B(t, \omega_1 t, \ldots, \omega_r t)x$$

with $B : \mathbb{R}^{r+1} \to \mathrm{Matr}(n)$ of class $C^{(1)}$ fulfilling (6),

(8) $$B(s_0,s_1,\ldots,s_r) + B^*(s_0,s_1,\ldots,s_r) = 0$$

for $s_0,s_1,\ldots,s_r \in \mathbb{R}$

and

(9) $$\left\| \frac{\partial B}{\partial \alpha_j} \right\| \leq L , \quad j=0,1,\dots,r .$$

Assume that

(10) $1, \omega_1, \dots, \omega_r$ are rationally independent .

Find such integers $p_1,\dots,p_r,q$, q sufficiently large, that

(11) $$\left| \omega_j - \frac{p_j}{q} \right| \leq \frac{1}{q^{1+1/r}} .$$

If $SU(n) \in EP(1) \cap \dots \cap EP(r)$ in the complex case (or if $SO(n) \in EP(1) \cap \dots \cap EP(r)$ in the real case), then such a $\hat{B} : \mathbb{R}^{r+1} \to \mathrm{Matr}(n)$ of class $C^{(1)}$ can be constructed that $\hat{B}$ fulfils (6), (8),

(12) $\left\| B(s_0,\dots,s_r) - \hat{B}(s_0,\dots,s_r) \right\|$ is small for $s_0,\dots,s_r \in \mathbb{R}^{r+1}$,

(13) $$\left\| \frac{\partial \hat{B}}{\partial \alpha_j} \right\| \leq \varkappa L , \quad j=0,1,\dots,r ,$$

$\varkappa$ being independent of $B,p_1,\dots,p_r,q$,

(14) $\hat{X}(q,\alpha_1,\dots,\alpha_r) = I$ for $\alpha_1,\dots,\alpha_r \in \mathbb{R}$,
$\hat{X}(\cdot,\alpha_1,\dots,\alpha_r)$ being the matrix solution of

$$\dot{x} = \hat{B}(t,\frac{p_1}{q}t+\alpha_1,\dots,\frac{p_r}{q}t+\alpha_r)x , \quad \hat{X}(0,\alpha_1,\dots,\alpha_r) = I .$$

Put $Y(t) = \hat{X}(t,(\omega_1-\frac{p_1}{q})t,\dots,(\omega_r-\frac{p_r}{q})t)$ and calculate $D(t)$ so that $\dot{Y}(t) = D(t)Y(t)$. We have

(15) $$D(t) = \hat{B}(t,\omega_1 t,\dots,\omega_r t) + \sum_{j=1}^{r} (\omega_j - \frac{p_j}{q}) \frac{\partial \hat{X}}{\partial \alpha_j}(\ )\hat{X}^*(\ ) .$$

It is not difficult to find that $\left\| \frac{\partial \hat{X}}{\partial \alpha_j}(\ ) \right\| \leq \varkappa Lq$ (by (13) and (14)) so that $\left\| D(t) - B(t,\omega_1 t,\dots,\omega_r t) \right\|$ is small by (12) and (11).

References

[1] J. Kurzweil, A. Vencovská: On a problem in the theory of linear differential equations with quasiperiodic coefficients, Proceedings of IX International Conference on Nonlinear Oscilations held in Kiev, August 30 - September 6, 1981

[2] D. Husemoller: Fibre Bundles, McGraw-Hill Book Company, New York, 1966

ON SOME CLASSES OF NONLINEAR HYPERBOLIC EQUATIONS

N.A.Lar'kin

The Institute of Theoretical & Applied Mechanics
The Siberian Branch of the USSR Academy of Sciences
Novosibirsk 630090 USSR

Introduction.

In this paper we consider nonlinear evolution equations of the form

$$Lu = A(t,u)u_{tt} + B(t)u_t + C(t)u + G(t,u,u_t) = f(t) \qquad (E)$$

Properties of the operators $A(t,u)$, $B(t)$, $C(t)$, $G(t,u,u_t)$ will be described below. If $A(t,u) = A(t)$ is a linear positively defined operator, $C(t) = -\Delta_x = \sum_{i=1}^{n} \partial^2/\partial x_i^2$, then initial-boundary value problems for (E) were studied by Visik [1] , Lions, Strauss [2] , Lions [3] . Evolution equations with a nonnegative linear operator A(t) were investigated by Showalter [4] , Egorov [5] , Belov, Savvateev [6] . If A(t) is a nonnegative function of variables (x,t), see papers of Tersenov [7] , Vragov [8] , Lar'kin [9] , Medeiros [10] . In the case of A(t,u) is a nonnegative function, $C(t) = -\Delta_x$, B(t) is a positive constant, $G(t,u,u_t) = |u_t|^p u_t$, initial-boundary value problems and the characteristic Darboux and Goursat problems for hyperbolic-parabolic equation (E) were studied by Lar'kin [11] . Here, results obtained in [11] are generalized for nonlinear operator equations. It should be noted that a global solvability of the Cauchy problem for (E) is due to the dissipative nonlinear operator $G(t,u,u_t)$. Moreover, there are no smallness conditions for the initial data and the right-hand side of (E). In connection with this notion we have to add that in the case of a linear dissipation global solvability theorems for nonlinear hyperbolic equations with small initial conditions and small right-hand side were obtained by Matsumura [12] , Yamada [13] .

Notations. Assumptions.

All functional spaces will be real. Let V, V_1, H be Hilbert spaces. V^* is a dual space of V with respect to an inner product in H, $H^* = H$. $V \subseteq V_1 \subseteq H \subseteq V_1^* \subseteq V^*$ and the inclusion mapping is continuous. Let W, W_1 be Banach spaces contained in H and the inclusion mapping is continuous. Moreover, $V \cap W \cap W_1$ is dense in H. The inner product in H at a fixed $t \in (0,T)$ is denoted by $(u,v)(t)$. The space of functions $g(t)$ which are L_p over $(0,T)$ with values in the Banach space X we denote by $L_p(0,T;X)$.

$$\|g\|_{L_p(0,T;X)} = \left(\int_0^T \|g(t)\|_X^p dt \right)^{1/p},$$

$$\|g\|_{L_\infty(0,T;X)} = \sup_{t \in (0,T)} \text{ess} \|g(t)\|_X .$$

As usually, the space of linear operators from V to V^* is denoted by $L(V,V^*)$ and by $C^k(S)$ we denote the space of continuous in S functions which have k continuous derivatives in S.

Assumptions I.

The operator $C(t) \in L(V,V^*)$ generates a continuous bilinear form on $V \times V$ with following properties:

I.1. $(C(t)u,v) = c(t;u,v) \quad (u,v \in V)$.

I.2. $c(t;u,v) = c(t;v,u)$, $c(t;u,u) \geq \alpha \|u\|_V^2$ for any $u \in V$, $t \in [0,T]$ α is a positive constant.

I.3. For every u,v V the function $t \to c(t;u,v) \in C^2[0,T]$; $c'(t;u,v) = \partial c(t;u,v)/\partial t$.

Assumptions II.

The nonlinear operator $G(t,u,v)$ generates a continuous linear form on $W \cap W_1$:

II.1. $(G(t,u,v),w) = g(t;u,v,w) \quad (w \in W \cap W_1)$,
where u,v are fixed elements from $W \cap W_1$ and W respectively.

II.2. There exists a continuous linear form on W: $w \to g_1(t;v,w)$ $(w \in W)$, which is a jointly continuous function of $t \in [0,T]$ and $v \in W$. For every $v \in W$ $g_1(t;v,v) \geq 0 \quad (t \in [0,T))$.

II.3. There exists a continuous linear form on W_1: $z \to g_2(t;u,z)$ $(z \in W_1)$, which is a jointly continuous function of $t \in [0,T]$ and $u \in W_1$. For every $u \in W_1$ $g_2(t;u,u) \geq 0 \quad (t \in [0,T])$.

II.4. For an arbitrary function $t \to u(t)$ over $(0,T)$ with values in $W \cap W_1$ and such that $u_t = \lim_{h \to 0} h^{-1}(u(t+h) - u(t)) \in W$, functions

$g(t;u(t),u_t(t),u_t(t))$, $g_1(t;u_t(t),u_t(t))$, $g_2(t;u(t),u(t))$ are measurable of t. For every $t \in [0,T]$ following integrals exist:

$$\int_0^t g(s;u(s),u_t(s),u_t(s))ds, \quad \int_0^t g_1(s;u_t(s),u_t(s))ds;$$

$g_2(t;u(t),u(t)) < \infty$, and the inequality is satisfied

$$\int_0^t g(s;u(s),u_t(s),u_t(s))ds \geq C_1 \int_0^t g_1(s;u_t(s),u_t(s))ds +$$
$$C_2 g_2(t;u(t),u(t)) - C_3 \int_0^t g_2(s;u(s),u(s))ds.$$

Here positive constants C_1, C_2, C_3 do not depend on t.

II.5. For all $u,w \in W \cap W_1 \cap V$, $v \in W \cap V$ the function $t \to g(t;u,v,w)$ has the derivative $g'(t;u,v,w)$, which is the continuous function of u,v,w,t.

II.6. For every $t \in [0,T]$, $u,h,w \in W_1 \cap W \cap V$; $z,v \in W \cap V$ the following limit exists: $\lim_{s\to 0} s^{-1}(g(t;u + sh,v,w) - g(t;u,v,w)) + \lim_{q\to 0} q^{-1}(g(t;u,v + qz,w) - g(t;u,v,w)) \equiv g^*(t;u,h,v,z,w)$, that is a continuous linear function of h,z,w depending continuously on t,u,v, when two of the latter belong to every finite-dimensional subspaces of $W \cap W_1 \cap V$ and $W \cap V$ respectively.

Assumptions III.

III.1. The operator $A(t,u)$ can be written as $A(t,u) = A_1(t) + A_2(t,u)$.

III.2. For fixed $t \in [0,T]$ $A_1(t) \in L(V_1,V_1^*)$ generates a continuous bilinear form on $V_1 \times V_1$: $u,v \to a_1(t;u,v) = (A_1(t)u,v)$, $a_1(t;u,v) = a_1(t;v,u)$ $(u,v \in V_1)$ and $a_1(t;u,u) \geq 0$ for every $u \in V_1$.

III.3. The function $t \to a_1(t;u,v) \in C^1[0,T]$ for all $u,v \in V_1$.

III.4. For fixed $t \in [0,T]$ the nonlinear operator $A_2(t,u)$ generates a continuous linear form on W: $w \to a_2(t;u,v,w) = (A_2(t,u)v,w)$ $(w \in W)$ that is a continuous function of $t, u \in W \cap W_1$ and a continuous linear one of $v \in W$.

III.5. $a_2(t;u,v,w) = a_2(t;u,w,v)$, $a_2(t;u,v,v) \geq 0$ for every $u \in W \cap W_1$, $v \in W$.

III.6. For every $u,h \in W \cap W_1$, $v,w \in W$ we assume the existence of the limit: $a_{2u}(t;u,h,v,w) \equiv \lim_{s\to 0} s^{-1}(a_2(t;u + sh,v,w) - a_2(t;u,v,w))$, which is a linear continuous form in h,v,w and is a continuous function of t,u. $a_{2u}(t;u,h,v,w) = a_{2u}(t;u,v,h,w) = a_{2u}(t;u,h,w,v)$.

III.7. For all $u \in W \cap W_1$, $w \in W$ the function $t \to a_2(t;u,v,w) \in C^1[0,T]$.

III.8. For an arbitrary function $t \to u(t)$ over $(0,T)$ with values in $W \cap W_1$ and such that $u_t(t) \in W$ the functions $a_2(t;u(t),u_t(t),u_t(t))$,

$a_2'(t;u(t),u_t(t),u_t(t))$, $a_{2u}(t;u(t),u_t(t),u_t(t),u_t(t))$ are measurable of t. For every t (O,T) following integrals exist:

$$\int_0^t a_2'(s;u(s),u_t(s),u_t(s))ds, \quad \int_0^t a_{2u}(s;u(s),u_t(s),u_t(s),u_t(s))ds,$$

$a_2(t;u(t),u_t(t),u_t(t)) < \infty$ and the inequality is valid:

$$-\int_0^t |a_{2u}(s;u(s),u_t(s),u_t(s),u_t(s)) + a_2'(s;u(s),u_t(s),u_t(s))|\,ds + \int_0^t g(s;u(s),u_t(s),u_t(s))ds \geqslant C_1 \int_0^t g_1(s;u_t(s),u_t(s))ds + C_2 g_2(t;u(t),u(t)) - C_3\Big(\int_0^t g_2(s;u(s),u(s))ds + \int_0^t a_2(s;u(s),u_t(s),u_t(s))ds\Big),$$

where positive constants C_1, C_2, C_3 do not depend on $t \in (O,T)$.

III.9. For every $u,w \in W \cap W_1 \cap V$, for a.e. $t \in (O,T)$ the inequality is valid:

$$-|a_2'(t;u,w,w) + a_{2u}(t;u,v,w,w)| + g^*(t;u,v,v,w,w) + g'(t;u,v,w) \geqslant C_1 a_2(t;u,w,w) - C_2(g_1(t;v,v) + g_2(t;u,u) + \|u\|_V \|w\|_{V_1}),$$

where positive constants C_1, C_2, C_3 do not depend on t.

Assumptions IV.

IV.1. The linear operator B(t) may be written as $B(t) = B_1(t) + B_2(t)$.
IV.2. For fixed $t \in [0,T]$ $B_1(t) \in L(V,V_1^*)$ and for every $u \in V$ the identity $2(B_1(t)u,u) = (B_{11}(t)u,u)$ is valid, where $B_{11}(t) \in L(V_1,V_1^*)$.
IV.3. The operator $B_1(t)$ has a strong derivative $B_1'(t)$ and for all $u,v \in V$ the inequality holds: $|(B_1'(t)u,v)| \leq C\,\|u\|_V \|v\|_{V_1}$, where a constant does not depend on $t \in [0,T]$.
IV.4. For fixed t the operator $B_2(t) \in L(V_1,V_1^*)$ has a strong derivative $B_2'(t)$, which satisfies the inequality: $|(B_2'(t)u,v)| \leqslant C\,\|u\|_{V_1} \|v\|_{V_1}$ $(u,v \in V_1)$.
IV.5. For every $t \in [0,T]$ and $u \in V_1$ one has the inequality:

$$((2B_2(t) + B_{11}(t))u,u) - a_1'(t;u,u) \geqslant 2\delta\|u\|_{V_1}^2,$$

where δ is a positive constant.

Assumption V.

If u_t, $u \in L_\infty(O,T;V)$, $u(t) \in W \cap W_1$, $u_t(t) \in W$ for a.e. $t \in (O,T)$ and assumptions III.4, III.8 are fulfilled, then for every $w \in W \cap W_1 \cap V$ functions $t \rightarrow$: $g(t;u(t),u_t(t),w)$, $a_2(t;u(t),u_t(t),w)$, $a_{2u}(t;u(t),$

$u_t(t),u_t(t),w)$, $a_2'(t;u(t),u_t(t),w)$ are integrable over $(0,T)$.

Assumption VI.

Let a sequence of functions $\{u^N(t)\}$ continuous with values in $W\cap W_1\cap V$ be given such that

$u^N \to u$ weak star in $L_\infty(0,T;V)$,

$u_t^N \to u_t$ weak star in $L_\infty(0,T;V)$,

$u_{tt}^N \to u_{tt}$ weakly in $L_2(0,T;V)$, $\int_0^T g_1(t;u_t^N(t),u_t^N(t))dt \leq C$;

$a_2(t;u^N(t),u_t^N(t),u_t^N(t)) + g_2(t;u^N(t),u^N(t)) \leq C$ and the constant C does not depend on N. Then $u(t) \in W\cap W_1$, $u_t(t) \in W$ for a.e. $t \in (0,T)$; $g(t;u(t),u_t(t),u_t(t))$, $a_2(t;u(t),u_t(t),u_t(t))$, $a_2'(t;u(t),u_t(t),u_t(t))$, $a_{2u}(t;u(t),u_t(t),u_t(t),u_t(t))$, $g_1(t;u_t(t),u_t(t))$, $g_2(t;u(t),u(t))$ are measurable functions of t.

$$\int_0^T g_1(t;u_t(t),u_t(t))dt < \infty, \quad \int_0^T g(t;u(t),u_t(t),u_t(t))dt < \infty \quad ;$$

$$g_2(t;u(t),u(t)) + a_2(t;u(t),u_t(t),u_t(t)) < \infty \quad \text{for a.e. } t \in (0,T).$$

A subsequence $\{u^l\}$ of $\{u^N\}$ may be extracted such that

$g(t;u^l(t),u_t^l(t),v) \to g(t;u(t),u_t(t),v)$,

$a_2(t;u^l(t),u_t^l(t),v) \to a_2(t;u(t),u_t(t),v)$,

$a_2'(t;u^l(t),u_t^l(t),v) \to a_2'(t;u(t),u_t(t),v)$,

$a_{2u}(t;u^l(t),u_t^l(t),u_t^l(t),v) \to a_{2u}(t;u(t),u_t(t),u_t(t),v)$

as a distribution over $(0,T)$ for every $v \in W\cap W_1\cap V$.

Assumptions VII.

We define $D(C(t))$ as a set of all $u \in V$ such that $C(t)u \in H$ and we define $D(G(t))$ as a set of all (u,v): $u \in W_1\cap W\cap V$, $v \in W\cap V$ such that $G(t,u,v) \in H$.

VII.1. $W\cap W_1\cap V$ is separable.

VII.2. $D(C(0))\cap W\cap W_1$ is dense in $D(C(0))$.

1. Existence Theorem.

Theorem 1.1. Let spaces V, V_1, W, W_1, H and operators $A(t,u)$, $B(t)$, $C(t)$, $G(t,u,v)$ be given, satisfying assumptions I - VII. Known functions $f(t)$, u_0, u_1 are given such that

$f \in L_2(0,T;H)$, $f_t \in L_2(0,T;H)$, $u_0 \in D(C(0))$, $(u_0, u_1) \in D(G(0))$. Besides,

$$B(0)u_1 + C(0)u_0 + G(0,u_0,u_1) = f(0). \tag{1.1}$$

Then there exists at least one function $u(t)$:

$u \in L_\infty(0,T;V)$, $u_t \in L_\infty(0,T;V)$, $u(t) \in W \cap W_1$, $u_t(t) \in W$ for a.e. $t \in (0,T)$; $u_{tt} \in L_2(0,T;V_1)$. $g(t;u(t),u_t(t),u_t(t))$, $a_2(t;u(t),u_t(t),u_t(t))$, $a_2'(t;u(t),u_t(t),u_t(t))$, $a_{2u}(t;u(t),u_t(t),u_t(t),u_t(t))$ are measurable functions of t;

$$\int_0^T g(t;u(t),u_t(t),u_t(t))dt < \infty \ , \quad \int_0^T g_1(t;u_t(t),u_t(t))dt < \infty \ ,$$

$$a_2(t;u(t),u_t(t),u_t(t)) + g_2(t;u(t),u(t)) < \infty \text{ for a.e. } t \in (0,T).$$

Function $u(t)$ satisfies the initial data

$$u(0) = u_0, \quad u_t(0) = u_1 \tag{1.2}$$

and the integral identity

$$\int_0^T a_1(t;u_{tt},v)dt - \int_0^T a_2(t;u,u_t,v_t)dt + \int_0^T ((B(t)u_t,v) + c(t;u,v) + g(t;u,u_t,v))dt - \int_0^T (a_2'(t;u,u_t,v) + a_{2u}(t;u,u_t,u_t,v))dt = \int_0^T (f,v)(t)dt + a_2(0;u_0,u_1,v(0)). \tag{1.3}$$

Here $v(t)$ is an arbitrary function from $L_2(0,T;V)$ having the following properties: $v(t) \in W \cap W_1$, $v_t \in W$ for a.e. $t \in (0,T)$; $v(T) = 0$.

Proof. First of all we regularize (E) by nondegenerate equation

$$L_h u_h = hIu_{htt} + Lu_h, \tag{1.4}$$

where I is the identity operator, h is a positive constant. Problem (1.4), (1.2) will be solved by Galerkin's method (see Lions, Strauss [2]). Let $\{w_j\}$ be a countable set which is dense everywhere in $W \cap W_1 \cap V$. P_N is the orthogonal projection in H onto the space generated by $w_1, \dots, w_N$. Let us denote by $u_h^N(t)$ a solution of the non-linear ordinary differential equations system:

$$h(u^N_{htt},w_j)(t) + a_1(t;u^N_{htt},w_j) + a_2(t;u^N_h,u^N_{htt},w_j) + (B(t)u^N_{ht},w_j) +$$

$$c(t;u^N_h,w_j) + g(t;u^N_h,u^N_{ht},w_j) = (f,w_j)(t) \quad (j = 1,\ldots,N) \qquad (1.5)$$

$$u^N_h(0) = P_N u_0, \quad u^N_{ht}(0) = P_N u_1 . \qquad (1.6)$$

For any $h > 0$ system (1.5) is nondegenerate because of assumptions III.2, III.5. Since II.5, II.6, III.3, III.6, III.7 are valid the solution $u^N_h(t)$ exists in some interval $(0,t_N)$. If necessary energy estimates are available, one may show that this solution exists in the whole interval $(0,T)$. Below the index h will be omitted in solutions of (1.5).

The first estimate is obtained by replacing w_j in (1.5) by u^N_t, by integrating the result with respect to s from 0 to t and taking into account assumptions I, III.1 - III.7, IV:

$$h\,(u^N_t,u^N_t)(t) + a_1(t;u^N_t,u^N_t) + a_2(t;u^N,u^N_t,u^N_t) + c(t;u^N,u^N) + \int_0^t ((2B_2(s)+$$

$$B_{11}(s))u^N_t,u^N_t)ds - \int_0^t (a_1'(s;u^N_t,u^N_t) + c'(s;u^N,u^N) + a_2'(s;u^N,u^N_t,u^N_t) +$$

$$a_{2u}(s;u^N,u^N_t,u^N_t,u^N_t))ds + 2\int_0^t g(s;u^N,u^N_t,u^N_t)ds = 2\int_0^t (f,u^N_t)(s)ds +$$

$$h(u^N_t,u^N_t)(0) + a_1(0;u^N_t(0),u^N_t(0)) + a_2(0;u^N(0),u^N_t(0),u^N_t(0)) + c(0;$$

$$u^N(0),u^N(0)).$$

Hence, making use of assumptions I.2, II.1 - II.4, III.1 - III.8, IV and (1.6) we come to the inequality

$$a_1(t;u^N_t,u^N_t) + a_2(t;u^N,u^N_t,u^N_t) + c(t;u^N,u^N) + \delta\|u^N_t\|^2_{L_2(0,t;V_1)} + C_1 g_2(t;$$

$$u^N,u^N) + C_2\int_0^t g_1(s;u^N_t,u^N_t)ds \leqslant C\int_0^t (a_2(s;u^N,u^N_t,u^N_t) + g_2(s;u^N,u^N) +$$

$$c(s;u^N,u^N))ds + K(f,u_0,u_1),$$

where positive constants C, K, C_1, C_2 do not depend on h, N, $t \in (0,T)$. Gronwall's Lemma gives us the first estimate which is uniform in h, N:

$$a_1(t;u^N_{ht},u^N_{ht}) + a_2(t;u^N_h,u^N_{ht},u^N_{ht}) + c(t;u^N_h,u^N_h) + g_2(t;u^N_h,u^N_h) \leqslant C,$$

$$\int_0^T g_1(t;u^N_{ht},u^N_{ht})dt + \|u^N_{ht}\|^2_{L_2(0,T;V_1)} \leqslant C. \qquad (1.7)$$

Before to prove the second a priori estimate we must note that from (1.1), assumptions III.2, III.5 and from (1.5) one may find the trace of $u^N_{htt}(t)$ at $t = 0$:

$$u^N_{htt}(0) = 0. \tag{1.8}$$

Differentiating (1.5) with respect to t, replacing w_j by u^N_{htt} **and** suppressing indices h,N, with the aid of assumptions II.5, II.6, III.6, III.7 we obtain

$$h(u_{ttt},u_{tt})(t) + a_1(t;u_{ttt},u_{tt}) + a_1'(t;u_{tt},u_{tt}) + a_2(t;u,u_{ttt},u_{tt}) +$$

$$a_2'(t;u,u_{tt},u_{tt}) + a_{2u}(t;u,u_t,u_{tt},u_{tt}) + (B(t)u_{tt},u_{tt}) + (B'(t)u_t,u_{tt}) +$$

$$c(t;u_t,u_{tt}) + c'(t;u,u_{tt}) + g'(t;u,u_t,u_{tt}) + g^*(t;u, u_t,u_t,u_{tt},u_{tt}) =$$

$$(f_t,u_{tt})(t).$$

Integrating this equality over $(0,t)$, making use of (1.8) and assumptions I.3, II.2 - II.4, III.9, IV after some transformations we have:

$$a_1(t;u_{tt},u_{tt}) + h(u_{tt},u_{tt})(t) + a_2(t;u,u_{tt},u_{tt}) + c(t;u_t,u_t) +$$

$$\delta\|u_{tt}\|^2_{L_2(0,t;V_1)} \leq C_1\int_0^t (a_2(s;u,u_{tt}u_{tt}) + c(s;u_t,u_t))ds +$$

$$C_2\int_0^t (g_1(s;u\ ,u\ ) + g_2(s;u,u))ds + C_3\|f_t\|^2_{L_2(0,T;H)} + C_4,$$

where constants in the right-hand side of the inequality do not depend on h, N, $t \in (0,T)$. Gronwall's Lemma gives us the second estimate uniform in h, N:

$$h(u^N_{htt},u^N_{htt})(t) + a_1(t;u^N_{htt},u^N_{htt}) + a_2(t;u^N_h,u^N_{htt},u^N_{htt}) \leq C,$$

$$\|u^N_{htt}\|_{L_2(0,T;V_1)} + \|u^N_{ht}\|_{L_\infty(0,T;V\ )} \leq C. \tag{1.9}$$

Estimates (1.7), (1.9) imply that as $h \to 0$, $N \to \infty$, a subsequence $\{u^l\}$ of $\{u^N_h\}$ can be extracted such that

$u^l \to u$ weakly star in $L_\infty(0,T;V)$,

$u^l_t \to u_t$ weakly star in $L_\infty(0,T;V)$,

$u^l_{tt} \to u_{tt}$ weakly in $L_2(0,T;V_1)$,

$u^l_{htt}h \to 0$ weakly in $L_2(0,T;H)$. (1.10)

Assumptions VI make us sure that $u(t)\in W\cap W_1$, $u_t(t)\in W$ for a.e. $t\in (0,T)$; $\int_0^T g_1(t;u_t(t),u_t(t))dt<\infty$, $\int_0^T g(t;u(t),u_t(t),u_t(t))dt<\infty$, $g_2(t;u(t),u(t)) + a_2(t;u(t),u_t(t),u_t(t))<\infty$ for a.e. $t\in(0,T)$.

$$g(t;u^1(t),u_t^1(t),v) \to g(t;u(t),u_t(t),v),$$

$$a_2(t;u^1(t),u_t^1(t),v) \to a_2(t;u(t),u_t(t),v),$$

$$a_2'(t;u^1(t),u_t^1(t),v) \to a_2'(t;u(t),u_t(t),v),$$

$$a_{2u}(t;u^1(t),u_t^1(t),u_t^1(t),v) \to a_{2u}(t;u(t),u_t(t),u_t(t),v)$$

in the sense of distributions over $(0,T)$ for every $v\in W\cap W_1\cap V$. It remains to prove that $u(t)$ satisfies integral identity (1.3).

Let $\{g_j(t)\}$ be smooth functions, $g_j(T) = 0$ for all j. Multiplying (1.5) by $g_j(t)$ and integrating the result over $(0,T)$ we get:

$$\int_0^T ((u_{htt}^N,g_jw_j)(t)h + a_1(t;u_{htt}^N,g_jw_j) + c(t;u_h^N,g_jw_j) + (B(t)u_{ht}^N, g_jw_j) + g(t;u_h^N,u_{ht}^N,g_jw_j) - a_2'(t;u_h^N,u_{ht}^N,g_jw_j) - a_{2u}(t;u_h^N,u_{ht}^N,u_{ht}^N,g_jw_j) - a_2(t;u_h^N,u_{ht}^N,g_{jt}w_j))dt = \int_0^T (f,g_jw_j)(t)dt - a_2(0;P_Nu_0,P_Nu_1,g_j(0)w_j).$$

Recalling (1.10), (1.11) we may tend $N\to\infty$, $h\to 0$ at fixed j. Then making use of a fact that linear combinations of $g_j(t)w_j$ at fixed t are dense in $W\cap W_1\cap V$ we complete the demonstration of Theorem 1.1.

2. Initial-Boundary value Problems for Differential Equations.

Example 1. A parabolic-hyperbolic equation.

Let S be a domain in R^n; $x \in S$, $t \in (0,T)$, $D = D_x$. We take $V_1 = H = L_2(S)$. $V = \overset{\circ}{W}{}^1_2(S)$ is the Sobolev space (see Sobolev [14] about notations $\overset{\circ}{W}{}^1_p(S)$), $W = L_2(S)$, $W_1 = L_p(S)$, $p>2$, $Q = S \times (0,T)$. In Q consider the equation:

$$Lu = K(x,t)u_{tt} - \sum_{i,j=1}^{n} (a_{ij}(x,t)u_{xi})_{xj} + \sum_{i=1}^{n} b_i(x,t)u_{xit} + au_t + |u|^{p-2}u = f(x,t). \tag{2.1}$$

The following conditions are imposed on coefficients of (2.1):

$$0 \leq K(x,t) \in C^1(\bar{Q}),\ a_{ij}(x,t) \in C^2(\bar{Q}),\ b_i(x,t) \in C^1(\bar{Q}),\ 0 < p-2 < 2/(n-2),$$

$$a_{ij}s_is_j \geqslant \alpha |s|^2, \tag{2.2}$$

Here a, α are positive constants. The inequality is fulfilled

$$2a - |K_t| - \sum_{i=1}^{n} b_{ixi} \geqslant 2\delta > 0 . \tag{2.3}$$

If $K(x,t) > 0$ in $\bar{Q}$, then (2.3) may be omitted. In the case $p = 2$ parabolic-hyperbolic equations were studied by Tersenov [7], Vragov [8], Egorov [15]. For $p>2$ see papers of Lar'kin [9], Medeiros [10]. Now we must check the validity of assumptions of Theorem 1.1 by putting $A_1(t) = K(x,t)$, $A_2(t,u) = 0$, $B_1(t) = \sum_{i=1}^{n} b_i \partial / \partial x_i$, $B_2(t) = a$, $C(t)u = -\sum_{i,j=1}^{n} (a_{ij}u_{xi})_{xj}$, $G(t,u,u_t) = |u|^{p-2}u$.

The Mixed Problem. Find a generalized solution u(t) of (2.1) such that $u \in L_\infty(0,T;\overset{\circ}{W}{}^1_2(S) \cap L_p(S))$, $u_t \in L_\infty(0,T;\overset{\circ}{W}{}^1_2(S))$, $u_{tt} \in L_2(Q)$, $K^{1/2}u_{tt} \in L_\infty(0,T;L_2(S))$. u(t) satisfies the initial data $u(0) = 0$, $u_t(0) = 0$. To fulfill the compatibility condition (1.1) it suffices that $f(0) = 0$. Assumptions I are valid since (2.2). Assumptions II are fulfilled if one takes

$$g_2(t;u,v) = \int_S |u|^{p-2}uv\, dx, \quad g_1(t;v,w) = 0,\ C_1 = p^{-1},\ C_3 = 0,$$

$$g^*(t;u,h,v,z,w) = (p-1)\int_S |u|^{p-2}hv dx.$$

Assumptions III are evidently fulfilled. Assumptions IV take place in virtue of (2.3). Assumptions V are valid because for $u \in L_p(S)$ we have $g_2(t;u,u) = \int_S |u|^p dx < \infty$, and $t \to g_2(t;u(t),u(t))$ is a measurable function of t. To check assumptions VI consider a sequence of functions $\{u^N\}$ such that

$u^N \to u$ weak star in $L_\infty(0,T;\overset{\circ}{W}{}_2^1(S) \cap L_p(S))$,

$u_t^N \to u_t$ weak star in $L_\infty(0,T;\overset{\circ}{W}{}_2^1(S))$,

$K^{1/2}u_{tt}^N \to K^{1/2}u_{tt}$ weak star in $L_\infty(0,T;L_2(S))$,

$u_{tt}^N \to u_{tt}$ weakly in $L_2(Q)$.

It follows from here that $\{u^N\}$ converges weakly in $W_2^1(Q)$ thereby one can extract a subsequence of $\{u^N\}$ which converges strongly in $L_2(Q')$, (there Q' is any compact in Q), and a.e. in Q'. Since $g^N = |u^N|^{p-2}u^N \in L_{p/(p-1)}(Q)$, then $g^N \to |u|^{p-2}u$ weakly in $L_{p/(p-1)}(Q)$. (See Lions [3]). Assumptions VII also take place if ∂S is sufficiently smooth. Then Theorem 1.1 gives the following.

Let f, f_t $L_2(Q)$, $f(0) = 0$, then there exists at least one generalized solution of the Mixed Problem for (2.1).

In fact, a stronger assertion is got because this generalized solution is regular one, i.e. it satisfies (2.1) a.e. in Q. Indeed, from (2.1) it follows that

$$-\sum_{i,j=1}^{n}(a_{ij}u_{xi})_{xj} \in L_\infty(0,T;L_2(S)).$$

Since the operator $(a_{ij}u_{xi})_{xj}$ is elliptical and ∂S is smooth, then $u \in L_\infty(0,T;\overset{\circ}{W}{}_2^1(S) \cap W_2^2(S))$. Moreover, the solution is unique. To prove that let us consider for the difference of two possible solutions $z = u_1 - u_2$ the integral :

$$2\int_0^t\int_S \Big(Kz_{tt}z_t + \sum_{i,j=1}^{n} a_{ij}z_{xi}z_{xjt} + \big\{|u_1|^{p-2}u_1 - |u_2|^{p-2}u_2\big\}z_t + az_t^2 - \sum_{i=1}^{n} b_{xi}z_t^2\Big)dx\,dt_1 = \int_S \Big(K(t)z_t^2 + \sum_{i,j=1}^{n} a_{ij}(t)z_{xi}z_{xj}\Big)dx + \int_0^t\int_S \Big(\big(2a - K_t - \sum_{i=1}^{n} b_{xi}\big)z_t^2 - \sum_{i,j=1}^{n} a_{ijt}z_{xi}z_{xj} - 2\,\big||u_1|^{p-2}u_1 - |u_2|^{p-2}u_2\big||z_t|\Big)dx\,dt_1 \le 0.$$

From this inequality one can easily get $\|z_t\|_{L_2(Q)} = 0$. Since $z(0,x) = 0$, then $z(x,t) = 0$ in Q, which proves the uniqueness.

Example 2. Evolution equation of a composite type.

Let S be a domain in R^n. $V_1 = \mathring{W}_2^1(S)$, $V = \mathring{W}_2^m(S)$, $H = L_2(S)$, $W = W_1 = L_p(S)$, $p > 2$; l, m, n are positive entire numbers; $l \leqslant m$. We take $A(t,u) = A_1(t)$, $a_1(t;u,v) = \sum_{|i|,|j|\leq l} \int_S K_{ij}(x,t) D^i u D^j v dx$, $a_1(t;u,u)$ 0. $B(t) = B_2(t)$, $(B_2(t)u,u) = \sum_{|i|,|j|\leq l} \int_S B_{ij}(x,t) D^i u D^j v dx$.

Let the inequality holds

$2(B_2(t)u,u) - a_1(t;u,u) \geqslant 2\delta \|u\|^2_{\mathring{W}_2^l(S)}$, $u \in \mathring{W}_2^l(S)$. Besides,

$c(t;u,v) = \sum_{|i|,|j|\leq m} \int_S C_{ij}(x,t) D^i u D^j v dx$, $c(t;u,u) \geqslant a\|u\|^2_{\mathring{W}_2^m(S)}$,

where δ, a are positive constants. Take $g(t;u,v,w) = g_1(t;v,w) = \int_S |v|^{p-2} vw dx$. It is easy to check that assumptions I -VII hold, recalling that $\mathring{W}_2^l(S)$ $L_2(S)$ and putting $g_2(t;u,v) = 0$. Then Theo - rem 1.1 leads to the assertion.

Let $f, f_t \in L_2(Q)$, $u_0 \in W_{2m}^2(S) \cap \mathring{W}_m^2(S) \cap L_{2(p-1)}(S)$; $u_1 \in \mathring{W}_2^m(S) \cap L_{2(p-1)}(S)$ and the equation

$$B(0)u_1 + C(0)u_0 + |u_1|^{p-2}u_1 = f(0)$$

takes place. Then there exists at least one function $u(t)$: $u \in L_\infty(0,T;\mathring{W}_2^m(S) \cap L_p(S))$, $u_t \in L_\infty(0,T;\mathring{W}_2^m(S)) \cap L_p(Q)$, $u_{tt} \in L_2(0,T;\mathring{W}_2^l(S))$, $a_1(t;u_{tt},u_{tt}) \in L_\infty(0,T)$,

which satisfies the boundary value problem:

$$Lu = \sum_{|i|,|j|\leq l} (-1)^i D^i(K_{ij}(x,t) D^j u_{tt}) + \sum_{|i|,|j|\leq m} (-1)^i D^i(C_{ij}(x,t) D^j u) +$$
$$\sum_{|i|,|j|\leq l} (-1)^i D^i(B_{ij}(x,t) D^j u_t) + |u_t|^{p-2}u_t = f(x,t).$$
$$u(0) = u_0, \quad u_t(0) = u_1.$$

As in example 1, one can easily prove uniqueness of the solution described above. Particularly, taking $B_2 = -\Delta + 1$, $n = 1$, $A_1(t) =$

0, $C(t) = -\Delta$, we get the equation

$$u_t - u_{xtx} - u_{xx} + |u_t|^{p-2}u_t = f.$$

This equation was studied by many authors. (See Benjamin, Bona, Ma - honey [17], Kozanov, Lar'kin, Janenko [18], Tsutsumi, Matahashi [19]).

Example 3. A quasilinear hyperbolic equation.

Let D be a bounded domain in R^n. $x \in D$, $t \in (0,T)$, $Q = D \times (0,T)$. Put $H = V_1 = L_2(D)$, $V = \overset{\circ}{W}{}_2^1(D)$, $W_1 = W = L_{p+2}(D)$, $p > 1$; $A(t,u) = K(t,u)$ is a jointly continuous function of arguments and has first derivatives satisfying the inequalities:

$$K(t,u) \geqslant K_0 > 0,$$

$$|K_t(t,u)| + |K_u(t,u)|^{p/(p-1)} \geqslant a + C(a)K(t,u) \qquad (2.4)$$

(2.4) imply that the growth of $K(t,u)$ function in u does not exceed the p-power. We put $A_1(t) = 0$, $C(t) = -\Delta$, $B(t) = 0$, $G(t,u,u_t) = |u_t|^p u_t$, $g_1(t;v,w) = \int_D |v|^p vwdx$, $g_2(t;u,v) = 0$. Assumptions I, II evidently are valid. From assumptions III it suffices to check III.5-III.9. Since $u(t) \in L_p(D)$ for a.e. $t \in (0,T)$, then (2.4) gives

$$\int_D K(u)vwdx = \int_D K(u)wvdx = a_2(t;u,v,w), \quad \int_D K(u)u_t^2dx \geqslant 0.$$

So, III.5 holds. Furthermore,

$$a_{2u}(t;u,h,v,w) = \int_D K_u(t,u)hvwdx = \int_D K_u(t,u)vhwdx = \int_D K_u(t,u)hwvdx$$

Hence, III.6 is fulfilled. Assumption III.7 evidently holds. To check III.8 we can use Young's inequality to estimate the expression:

$$I = -\int_0^t \int_D (K_u(s,u)u_t^3 + K_t(s,u)u_t^2 - |u_t|^{p+2})dxds \geqslant -$$

$$\int_0^t \int_D K_t(s,u)\, u_t^2dxds + \int_0^t \int_D (u_t^2(|u_t|^p - h|u_t|^p - C(h)|K_u(s,u)|^{p/(p-1)})dx\times$$

ds, where h is an arbitrary positive constant. Choosing it sufficiently small in virtue of (2.4) we arrive at the inequality

$$I \geqslant C \int_0^t \int_D |u_t|^{p+2}dxds - C\int_0^t \int_D (1 + K(s,u))u_t^2dxds.$$

Since $K(t,u) > 0$, the latter inequality may be written in the form

$$I \geqslant C \int_0^t \int_D |u_t|^{p+2} dxds - C \int_0^t \int_D (K(s,u)u_t^2 dxds,$$

that approves the validity of the assumption III.8. To check III.9 let us consider an expression

$$I_1 = - |K_t(t,u)| u_{tt}^2 - |K_u(t,u)u_t| u_{tt}^2 + |u_t|^p u_{tt}^2 .$$

By Young's inequality rewrite it in the manner

$$I_1 \geqslant u_{tt}^2 (|u_t|^p - h|u_t|^p - C(h)|K_u(t,u)|^{p/(p-1)} - |K_t(t,u)|).$$

As above, we get, choosing $h > 0$ small enough

$$I_1 \geqslant C_1 u_{tt}^2 |u_t|^p - C_2 K(t,u) u_{tt}^2 ,$$

that proves the validity of III.9. By the first of conditions (2.4) we have $K(t,u) > 0$, hence the operator $B(t)$ may be null operator. Since $u_t \in L_{p+2}(Q)$, $K(t,u) \leqslant C(1 + |u|^p)$, then $K(t,u)u_t \in L_{(p+2)/(p+1)}(Q)$, $|u_t|^{p+1} \in L_{(p+2)/(p+1)}(Q)$, i.e. for every $w \in L_{p+2}(Q)$ there exist integrals: $\int_Q K(t,u)u_t w dQ$, $\int_Q |u_t|^p u_t w dQ$, $\int_Q (.) dQ = \int_0^T \int_D (.) dxdt$. Therefore, the assumption V is also valid. Now check the assumption VI. Let a sequence $\{u^N\}$ be given such that

(1) $\quad u^N \to u$ weak star in $L_\infty(0,T;\mathring{W}_2^1(D))$,

(2) $\quad u_t^N \to u_t$ weak star in $L_\infty(0,T;\mathring{W}_2^1(D))$,

(3) $\quad u_{tt}^N \to u_{tt}$ weakly in $L_2(Q)$,

(4) $\quad \int_Q |u_t^N|^{p+2} dQ \leqslant C$,

(5) $\quad \int_D (K(t,u^N)u_t^{N2} + K(t,u^N)u_{tt}^{N2}) dx \leqslant C$ uniformly in N.

From (2), (3), (4) we have: $u_t^N \in L_{p+2}(Q)$, $u_t^N \in W_2^1(Q)$, hence $|u_t^N|^p u_t^N \to |u_t|^p u_t$ weakly in $L_{(p+2)/(p+1)}(Q)$. Restrictions on $K(t,u)$ gives us: $K_t(t,u^N)u_t^N \in L_{(p+2)/(p+1)}(Q)$, $K_u(t,u^N)u_t^{N2} \in L_{(p+2)/(p+1)}(Q)$. Since $u_t^N \to u_t$ a.e. in Q, then $u^N \to u$ also a.e. in Q, therefore,

$$K_t(u^N)u_t^N \to K_t(u)u_t \quad \text{weakly in } L_{(p+2)/(p+1)}(Q).$$

$K(u^N)u_t^N \rightharpoonup K(u)u_t$ weakly in $L_{(p+2)/(p+1)}(Q)$,

(6) $K_u(u^N)u_t^{N2} \rightharpoonup K_u(u)u_t^2$ weakly in $L_{(p+2)/(p+1)}(Q)$.

Moreover, from (5) and from the first inequality of (2.4) we get

$K^{1/2}(u^N)u_{tt}^N \in L_\infty(0,T;L_2(D))$, $u_{tt}^N \in L_\infty(0,T;L_2(D))$. From the second inequality of (2.4) $K^{1/2}(u^N) \in L_{2(p+2)/p}(Q)$, hence $K(u^N)u_{tt}^N \in L_{(p+2)/(p+1)}(Q)$. Those imply existence of some $z \in L_{(p+2)/(p+1)}(Q)$ such that

$K(u^N)u_{tt}^N \rightharpoonup z$ weakly in $L_{(p+2)/(p+1)}(Q)$,

$u_{tt}^N \rightharpoonup u_{tt}$ weakly star in $L_\infty(0,T;L_2(D))$.

On the other hand,

$$K(t,u)u_{tt} = (K(t,u)u_t)_t - K_t(t,u)u_t - K_u(t,u)u_t^2.$$

From here and from (6) we derive: $K(u^N)u_{tt}^N \rightharpoonup K(u)u_{tt}$ as a distribution over (Q). This fact approves the validity of the assumption VI. At last, if ∂D is smooth enough and $u_0 \in W_2^2(D) \cap \mathring{W}_2^1(D) \cap L_{2(p+1)}(D)$, $u_1 \in \mathring{W}_2^1(D) \cap L_{2(p+1)}(D)$, so assumptions VII hold, and Theorem 1.1 gives.

<u>Let</u> $f, f_t \in L_2(Q)$, $p > 1$, <u>and conditions</u> (2.4) <u>be fulfilled, then there exists at least one function</u> $u(t)$:

$u \in L_\infty(0,T;\mathring{W}_2^1(D) \cap L_{p+2}(D))$, $u_t \in L_\infty(0,T;\mathring{W}_2^1(D)) \cap L_{p+2}(Q)$, $K^{1/2}(t,u)u_{tt} \in L_\infty(0,T;L_2(D))$, $u_{tt} \in L_\infty(0,T;L_2(D))$,

<u>which is the solution of the mixed problem</u>:

$$Lu = K(t,u)u_{tt} - \Delta u + |u_t|^p u_t = f(x,t),$$

$$u(0) = u_0, \quad u_t(0) = u_1, \quad u\big|_{\partial D} = 0.$$

In case $n = 1$, this solution is unique. In fact, imbedding theorems of Sobolev [14] and (1) - (6) imply that $u \in C(\bar{Q})$, $u_t \in C(\bar{Q})$. For the difference of two possible solutions, $z = u_1 - u_2$, we can get

$$2\int_0^t \int_D (Lu_1 - Lu_2)z_t\,dxds = 0; \int_D (K(t,u_1)z_t^2 + z_x^2)dx - \int_0^t \int_D ((K_t(s,u_1) + K_u(s,u_1)u_{1t})z_t^2 + 2(K(s,u_2) - K(s,u_1))u_{2tt}z_t)dxds \leq 0.$$

After some transformations we come to $\|z_t\|_{L_2(Q)} = 0$. Since $z(x,0)=$

0, then $z(x,t) = 0$, that proves the uniqueness. Making use of the solvability of the Mixed Problem, one can study the characteristic Goursat and Darboux Problems, when the data are given on a surface of a characteristic cone of a hyperbolic equation. For more details see Lar'kin [11, 16].

Example 4. A quasilinear evolution equation of a composite type.

As in example 3, by the same methods one can investigate the solvability of initial-boundary value problems for the equation

$$Lu = -\sum_{i=1}^{n}((1+|u_{xi}|^{p})u_{xitt})_{xi} + \Delta^2 u - \sum_{i=1}^{n}(|u_{xit}|^{p}u_{xit})_{xi} = f. \quad (2.5)$$

Here $V_1 = \overset{\circ}{W}{}_2^1(D)$, $H = L_2(D)$, $V = \overset{\circ}{W}{}_2^2(D)$, $p > 1$, $W = W_1 = \overset{\circ}{W}{}_{p+2}^1(D)$. The validity of assumptions I - VII may be easily proved. Therefore, Theorem 1.1 asserts.

Let $f, f_t \in L_2(Q)$, then there exists at least one solution of (2.5) $u(t)$: $u \in L_\infty(0,T;\overset{\circ}{W}{}_2^2(D) \cap \overset{\circ}{W}{}_{p+2}^1(D))$, $u_t \in L_\infty(0,T;\overset{\circ}{W}{}_2^2(D)) \cap L_{p+2}(0,T;\overset{\circ}{W}{}_{p+2}^1(D))$, $u_{tt} \in L_\infty(0,T;\overset{\circ}{W}{}_2^1(D))$, which satisfies the initial data: $u(0) = u_0 = D(\Delta^2) \cap \overset{\circ}{W}{}_{2(p+1)}^1(D)$, $u_t(0) = u_1 \in \overset{\circ}{W}{}_2^2(D) \cap \overset{\circ}{W}{}_{2(p+1)}^1(D)$.

Remark. One can replace the Laplace operator by any elliptic operator and in an appropriate manner modify spaces V, V_1, W, W_1.

Literature.

1. Visik M. The Cauchy Problem for equations with operator coefficients; mixed boundary value Problems for systems of differential equations and approximation methods for their solution. Math. USSR Sb., 39 (81), 51-148, (1956).
2. Lions J.-L., Strauss W.A. Some nonlinear evolution equations. Bull. Soc. Math. France. 93, 1, 43-96, (1965).
3. Lions J.-L. Quelques methodes de resolition des problemes aux limites non lineaires. Dunod, Paris (1969).
4. Showalter R.E. Nonlinear degenerate evolution equations and partial differential equations of mixed type. SIAM J. Math. Anal., 6, 25-42, (1975).
5. Egorov I.E. On a Cauchy Problem for a second-order degenerate operator equation. Sib. Math. J., 20, 5 (Russian), 1015-1021, (1979).
6. Belov Y.J., Savvateev E.G. On approximation of composite type systems. Chislennye metody mehaniki sploshnoi sredy. Collect. works. USSR, Novosibirsk, 9, 6 (Russian), 12-24, (1978).
7. Tersenov S.A. Introduction in the theory of equations degenerating on a boundary. USSR, Novosibirsk, Novosibirskij state university. (Russian), (1973).
8. Vragov V.N. On a mixed Problem for a class of Hyperbolic-Parabo-

lic Equations. Soviet Math. Dokl., 16, 1179-1183, (1975).
9. Lar'kin N.A. Mixed Problem for a class of Hyperbolic Equations. Sib. Math. J., 18, 1414-1419, (1977).
10. Medeiros L.A. Non linear hyperbolic-parabolic partial differential equations. Funkc. Ekvacioj., Ser. Int., 23, 151-158, (1979).
11. Lar'kin N.A. Global solvability of Boundary value Problems for a Class of Quasilinear Hyperbolic Equations. Sib. Math. J., 22, 1, 82-111, (1981).
12. Matsumura A. Global existence and asymptotics of the solution of the second-order quasilinear hyperbolic equations with the first-order dissipation. Publ. RIMS, Kyoto Univ., 13, 349-379, (1979).
13. Yamada Y. Quasilinear wave equations and related nonlinear evolution equations. Nagoya Math. J., 84, 31-83, (1981).
14. Sobolev S.L. Applications of functional analysis in mathematical physics. Transl. Math. Monograph. 7. A.M.S. (1963).
15. Egorov I.E. On a mixed Problem for one hyperbolic-parabolic equation. Mat. zametki. 23, 389-400, (1978).
16. Lar'kin N.A. On a class of quasilinear hyperbolic equations having global solutions. Soviet Math. Dokl., 20, 1, 28-31, (1979).
17. Benjamin T.B., Bona J.L., Mahoney J.J. Model equations for long waves in nonlinear dispersive systems. Philos. Trans. Roy. Soc., London, ser. A 272, 47-78, (1972).
18. Kozanov A.I., Lar'kin N.A., Janenko N.N. On a regularization of equations of variable type. Sov. Math. Dokl., 21, 758-761, (1980).
19. Tsutsumi M., Matahashi T. On some nonlinear Pseudo-parabolic equations. J. Differ. Equations, 32, 65-75, (1979).

STATISTICAL STABILITY OF DETERMINISTIC SYSTEMS

A.Lasota
Institute of Mathematics
Silesian University
40-007 Katowice , Poland

Introduction.

It is well known that dynamical systems with an extremely irregular (chaotic) behavior of trajectories are quite regular from the statistical point of view. More precisely, if simultaneously with a semidynamical system $\{S_t\}$ acting on a measure space $(X, \mathcal{A}, m)$ we consider the corresponding stochastic semigroup

$$P^t f = \frac{d}{dm}(m_f \circ S_t^{-1}) \qquad dm_f = f\, dm$$

acting on $L^1(X)$ (see Section 4 for the details), then the "irregular" behavior of $\{S_t\}$ such as mixing or exactness is equivalent to the asymptotical stability of $\{P^t\}$.

The purpose of this paper is to prove a simple necessary and sufficient condition for the stability of stochastic semigroups. This criterion was proved in [9] in the special case for Markov operators generated by discrete time processes which turns out to be irrelevant. The proof given here is based on a different technique and works equally well for discrete and continuous time processes. Because of a special role played in this condition by a "lower function" it will be called the l-condition.

The main advantage of the l-condition is its applicability to different classes of semigroups. For example using this condition it is easy to prove the classical ergodic theorem for Markov chains, the existence of an invariant measure for point transformations and the asymptotical stability of solutions of some transport equations (e.g., the linear Boltzmann equation).

The paper is divided into nine sections. In Section 1 we formulate and prove the l-condition. Then, in Section 2, we show some simple sufficient conditions for the existence of a lower function. Section 3 contains an application of the l-condition to Markov chains which allows to compare it with the classical Markov type conditions. In Section 4 we show some details related with the construction of stochastic semigroups corresponding to "deterministic" semidynamical systems and the relationship between the asymptotical stability and exactness. Sections 5, 6 and 7 are devoted to applications of the l-condition to different classes of discrete time semidynamical systems (on manifolds, on the unit interval and on the real line). Section 8 contains preliminary remarks concerning the linear Boltzmann equation in the Tjon-Wu representation. The asymptotical stability of this equation is proved in Section 9.

The results stated in Sections 1, 2, 6 and 9 are unpublished. The remaining applications of the l-condition are new only from the methodological point of view. No attempt is made to present the results in the most general form since our primary concern is to indicate the variety of problems to which the l-condition is applicable.

1. The l-condition

Let $(X,\mathcal{A},m)$ be a measure space with a nonnegative σ-finite measure m. A linear mapping $P : L^1 \to L^1$ $(L^1 = L^1(X,\mathcal{A},m))$ will be called a Markov operator if it satisfies the following two conditions

(a) $Pf \geq 0$ for $f \geq 0$, $f \in L^1$;

(b) $\|Pf\| = \|f\|$ for $f \geq 0$, $f \in L^1$

where $\|\cdot\|$ stands for the norm in L^1 (cf.[3]).

From conditions (a) and (b) it is easy to derive the following well known properties of Markov operators:

(c) $|Pf| \leq P|f|$ and $\|Pf\| \leq \|f\|$ for $f \in L^1$;

(d) $(Pf)^+ \leq Pf^+$ and $(Pf)^- \leq Pf^-$ for $f \in L^1$

where $f^+ = \max(f,0)$ and $f^- = \max(-f,0)$;

(e) $Pf = f \Rightarrow Pf^+ = f^+$ and $Pf^- = f^-$ for $f \in L^1$.

Now let T be a nontrivial semigroup of real nonnegative numbers, i.e. $T \neq \{0\}$ and $t_1 \pm t_2 \in T$ for every $t_1 \geq t_2 \in T$. A family of Markov operators $\{P^t\}_{t \in T}$ will be called a <u>stochastic semigroup</u> if

$$P^{t_1 + t_2} = P^{t_1} P^{t_2} \qquad \text{for } t_1, t_2 \in T.$$

By $D = D(X, \mathcal{A}, m)$ we shall denote the set of all (normalized) densities on X, i.e.

$$D = \{ f \in L^1 : f \geq 0 \text{ and } \| f \| = 1 \}.$$

A nonnegative function $h \in L^1$ will be called a <u>lower function</u> for a semigroup $\{P^t\}$ if

(1.1) $\lim_{t \to \infty} \| (P^t f - h)^- \| = 0$ for <u>every</u> $f \in D$.

A nonnegative lower function will be called <u>nontrivial</u> if $\|h\| > 0$.

A stochastic semigroup $\{P^t\}$ will be called <u>asymptotically stable</u> if there is a unique $f_* \in D$ such that

(1.2) $P^t f_* = f_*$ for $t \in T$

and

(1.3) $\lim_{t \to \infty} \| P^t f - f_* \| = 0$ for <u>every</u> $f \in D$.

<u>Theorem 1.1.</u> A stochastic semigroup is asymptotically stable if and only if it has a nontrivial lower function.

<u>Proof.</u> The if part is obvious, since (1.3) implies (1.1) with $h = f_*$. The proof of the only if part will be given in two steps. First we are going to show that

$$(1.4) \qquad \lim_{t\to\infty} \| P^t(f_1 - f_2)\| = 0 \qquad \text{for} \quad f_1,f_2 \in D.$$

Then we shall construct the desired function f_*.

Step I. For every fixed pair $f_1,f_2 \in D$ the function

$$t \longrightarrow \| P^t(f_1 - f_2)\|$$

is decreasing. In fact according to (c) every Markov operator is contractive and consequently

$$\| P^{t+t'}(f_1 - f_2)\| = \|P^{t'} P^t(f_1 - f_2)\| \leq \| P^t(f_1 - f_2)\| .$$

Now write

$$g = f_1 - f_2 \quad , \qquad c = \|g^+\| = \| g^-\| = \frac{1}{2} \| g \|$$

and assume that $c > 0$. We have $g = g^+ - g^-$ and

$$\| P^t g \| = c \, \| (P^t(\tfrac{1}{c} g^+) - h) - (P^t(\tfrac{1}{c} g^-) - h)\| .$$

Since the functions g^+/c and g^-/c belong to D, there exists, according to (1.1), a number $t_1 \in T$ such that

$$\| (P^t(\tfrac{1}{c} g^+) - h)^- \| \leq \frac{1}{4} \| h \|$$

and

$$\| (P^t(\tfrac{1}{c} g^- - h)^- \| \leq \frac{1}{4} \| h \| \qquad \text{for} \quad t > t_1 .$$

This in turn implies

$$\| P^t(\tfrac{1}{c} g^+) - h \| \leq 1 - \| h \| + \frac{2}{4} \| h \|$$

and

$$\| P^t(\tfrac{1}{c} g^-) - h \| \leq 1 - \| h \| + \frac{2}{4} \| h \| .$$

Finally

$$(1.5) \qquad \| P^t g \| \leq c(2 - \| h \|) = \| g \| (1 - \frac{1}{2} \| h \|).$$

For $c = \frac{1}{2}\|g\| = 0$ this inequality is obvious. Thus for any $f_1, f_2 \in D$ we can find a time t_1 such that

$$\|P^{t_1}(f_1 - f_2)\| \leqslant \|f_1 - f_2\|(1 - \tfrac{1}{2}\|h\|).$$

Applying this argument to the pair $P^{t_1}f_1$, $P^{t_1}f_2$ we may find a time t_2 such that

$$\begin{aligned}\|P^{t_1+t_2}(f_1 - f_2)\| &\leqslant \|P^{t}(f_1 - f_2)\|(1 - \tfrac{1}{2}\|h\|) \\ &\leqslant \|f_1 - f_2\|(1 - \tfrac{1}{2}\|h\|)^2\end{aligned}$$

and after n steps

$$\|P^{t_1+\dots+t_n}(f_1 - f_2)\| \leqslant \|f_1 - f_2\|(1 - \tfrac{1}{2}\|h\|)^n.$$

Since $t \to \|P^t(f_1 - f_2)\|$ is decreasing, this implies (1.4).

<u>Step II.</u> We shall construct the maximal lower function. Write

$$r = \sup\{\|h\| \; : h \text{ is a lower function}\}.$$

It is evident that $0 < r \leqslant 1$. Now observe that for any two lower functions h_1 and h_2 the function $h = \max(h_1, h_2)$ is also a lower function. In fact

$$\|(P^t f - h)^-\| \leqslant \|(P^t f - h_1)^-\| + \|(P^t f - h_2)^-\|.$$

Choose a sequence $\{h_n\}$ of lower functions such that $\|h_n\| \to r$. Replacing, if necessary, h_n by $\max(h_1, \dots, h_n)$ we may assume that $\{h_n\}$ is an increasing sequence of lower functions. The limiting function

$$h_* = \lim_{n \to \infty} h_n$$

is also a lower function because

$$\|(P^t f - h_*)^-\| \leqslant \|(P^t f - h_n)^-\| + \|h_n - h_*\|$$

and according to the Lebesgue convergence theorem $\|h_n - h_*\| \to 0$. The function h_* is the largest lower function. In fact, for any other lower function h_* the $\max(h,h_*)$ is also a lower function and

$$\|\max(h,h_*)\| \leq r = \|h_*\|$$

which implies $h \leq h_*$. Now observe that according to (d)

$$\|(P^t f - P^{t'} h_*)^-\| \leq \|P^{t'}(P^{t-t'} f - h_*)^-\|$$
$$\leq \|(P^{t-t'} f - h_*)^-\| \qquad \text{for} \quad t > t'$$

which implies that for any $t' \in T$ the function $P^{t'} h_*$ is a lower function. Thus $P^{t'} h_* \leq h_*$ and, since $P^{t'}$ preserves the integral $P^{t'} h_* = h_*$. The function $f_* = h_* / \|h_*\|$ is a density satisfying (1.2).

Now, according to (1.4) we have

$$\lim_{t \to \infty} \|P^t f - f_*\| = \lim_{t \to \infty} \|P^t f - P^t f_*\| = 0 \quad \text{for} \quad f \in D$$

which implies (1.3). In turn, condition (1.3) implies that f_* is a unique function in D satisfying (1.2). This completes the proof.

A function $f \in D$ satisfying condition $P^t f = f$ for all $t \in T$ will be called a stationary density. The statement concerning the uniqueness of stationary densities for asymptotically stable stochastic semigroups can be strengthened as follows.

Proposition 1.1. Let $\{P^t\}$ be an asymptotically stable stochastic semigroup and let f_* be the unique stationary density. Then for every normalized $f \in L^1$ ($\|f\| = 1$) and every $t' \in T$ condition

(1.6) $$P^{t'} f = f, \quad t' > 0$$

implies that either $f = f_*$ or $f = -f_*$.

Proof. Condition (1.6) implies that f^+ and f^- are fixed points of $P^{t'}$ (property (e)). Assume that $\|f^+\| > 0$. Then $\tilde{f} = f^+ / \|f^+\|$ is a normalized density and $P^{t'} \tilde{f} = \tilde{f}$. By induction we obtain

$$P^{nt'}\tilde{f} = \tilde{f} \qquad \text{for } n=1,2,\ldots$$

and , according to (1.3), $\lim_n P^{nt'}\tilde{f} = f_*$. Thus $\tilde{f} = f_*$ which implies

$$f^+ = f_* \, \|f^+\| .$$

This equality is also evidently true for $\|f^+\| = 0$.
Analogously

$$f^- = f_* \, \|f^-\| .$$

Thus

$$f = f^+ - f^- = (\|f^+\| - \|f^-\|)\, f_* = \alpha f_* .$$

Since $\|f\| = \|f_*\|$, we have $|\alpha| = 1$ and the proof is completed.

In many applications it is easier to verify condition (1.1) for some special elements in D. This justifies the following

Proposition 1.2. Let D_o be such that $\overline{\text{conv}}\, D_o = D$ ($\overline{\text{conv}}$ = = convex closed hull). Then (1.1) is equivalent to

(1.7) $$\lim_{t\to\infty} \|(P^t f - h)^-\| = 0 \qquad \text{for every } f \in D_o .$$

The proof follows from the fact that the operators P^t are linear and uniformly continuous.

2. Existence of lower functions.

From Theorem 1.1 it follows that in order to prove the stability of $\{P^t\}$ it is sufficient to find an arbitrary nontrivial lower function. In this section we shall show some simple proofs of the existence of such functions. We shall consider semigroups $\{P^t\}$ on different spaces $L^1(X)$ and we shall write $\{P^t;\, L^1(X)\}$ to underleine the role of X.

We shall use the following notation: By o_t we denote an arbitrary function from T into L^1 such that

$$\lim_{t\to\infty} \|o_t\| = 0.$$

Using this notation we may rewrite condition (1.1) in the form

$$P^t f \geqslant h + o_t \qquad \text{for} \quad f \in D\,, \quad t \in T.$$

We shall assume that for every $f \in D_0(\overline{\text{conv}}\ D_0 = D(X))$ the trajectory $P^t f$ may be written in the form

$$(2.1) \qquad P^t f = f_t + o_t \qquad \text{for} \quad t \in T\,, \quad t \geqslant t_0(f)$$

where $f_t : X \to [0,\infty)$ satisfy some additional conditions.

Our first criterion for the existence of a lower function will be formulated in the special case when $X = (a,b)$ is an interval on the real line (a,b finite or not) with the usual Lebesgue measure. We shall use some standard notions from the theory of differential inequalities. A function $f : (a,b) \to R$ is called lower semicontinuous if

$$\liminf_{\varepsilon \to 0} f(x - \varepsilon) \geqslant f(x) \qquad \text{for} \quad x \in (a,b).$$

It is left lower semicontinuous if

$$\liminf_{\substack{\varepsilon \to 0 \\ \varepsilon > 0}} f(x + \varepsilon) \geqslant f(x) \qquad \text{for} \quad x \in (a,b).$$

For any function $f : (a,b) \to R$ we may define its right lower derivative by setting

$$\frac{d_+ f(x)}{dx} = \liminf_{\substack{\varepsilon \to 0 \\ \varepsilon > 0}} \frac{1}{\varepsilon}(f(x + \varepsilon) - f(x)) \quad \text{for} \quad x \in (a,b).$$

It is well known that every left lower semicontinuous function $f : (a,b) \to R$ satisfying

$$\frac{d_+ f(x)}{dx} \leqslant 0 \qquad \text{for} \quad x \in (a,b)$$

is nonincreasing on (a,b). (The same is true for functions defined on a half closed interval [a,b)).

By 1_A we denote the characteristic function of the set A.

<u>Proposition 2.1.</u> Let a stochastic semigroup $\{P^t\ ;\ L^1((a,b))\}$ be given. Assume that there exist a nonnegative function $g \in L^1((a,b))$ and a constant $k \geq 0$ such that for each $f \in D_0$ the functions f_t in (2.1) are left lower semicontinuous and satisfy the following conditions

$$f_t(x) \leq g(x) \qquad \text{a.e. in } (a,b) \tag{2.2}$$

$$\frac{d_+ f_t(x)}{dx} \leq k\, f_t(x) \qquad \text{for all } x \text{ in } (a,b). \tag{2.3}$$

Then there exists an interval $\Delta \subset (a,b)$ and an $\varepsilon > 0$ such that $h = \varepsilon 1_\Delta$ is a lower function for $\{P_t\}$.

Proof. Let $x_0 < x_1 < x_2$ be chosen in (a,b) such that

$$\int_a^{x_1} g(x)\,dx < \frac{1}{4} \qquad \text{and} \qquad \int_{x_2}^b g(x)\,dx < \frac{1}{4}. \tag{2.4}$$

Set

$$\varepsilon = \min(x_1 - x_0\ ,\ M(x_2 - x_0)^{-1}), \qquad M = \frac{1}{4}\, e^{-k(x_2 - x_0)}.$$

Since $\|P^t f\| = 1$, condition (2.1) implies

$$\int_a^b f_t(x)\,dx > \frac{3}{4} \tag{2.5}$$

for sufficiently large t (say $t > t_1(f)$). Now we are going to show that $h = \varepsilon 1_{(x_0,x_1)}$ is a lower function. Suppose not. Then there is $t' > t_1$ and $y \in (x_0, x_1)$ such that $f_{t'}(y) < h(y) = \varepsilon$. Integrating inequality (2.3) we obtain

$$f_{t'}(x) \leq f_{t'}(y) e^{k(x-y)} \leq \frac{\varepsilon}{4M} \qquad \text{for } x \in [y, x_2]. \tag{2.6}$$

Furthemore, since $f_{t'} \leq g$, we have

$$\int_a^b f_{t'}(x)\,dx \leq \int_a^{x_1} g(x)\,dx + \int_y^{x_2} f_{t'}(x)\,dx + \int_{x_2}^b g(x)\,dx.$$

Applying inequalities (2.4) and (2.6) we obtain finally

$$\int_a^b f_{t'}(x)\, dx \leqslant \frac{1}{4} + (x_2 - y)\,\frac{\varepsilon}{4M} + \frac{1}{4} \leqslant \frac{3}{4}$$

which contradicts (2.5).

Remark 2.1. In the proof of Proposition 2.1 the left lower semicontinuity of f_t and inequality (2.3) were used only to obtain the evaluation

$$f_t(x) \leqslant f_t(y)e^{k(x-y)} \qquad \text{for } x \geqslant y.$$

Therefore Proposition 2.1 remains true under this condition; for example it is true if all f_t are nonincreasing.

It is obvious that in Proposition 2.1 we may replace (2.3) by $d_- f_t/dx \geqslant -k f_t$ and assume f_t right lower continuous (or assume f_t nondecreasing in Remark 2.1). In the case of a bounded interval we may omit condition (2.2) and replace (2.3) by a two-side inequality. This fact is summarized in the following

Proposition 2.2. Let (a,b) denote a bounded interval and let $\{P^t ; L^1((a,b))\}$ be a stochastic semigroup. Assume that for each $f \in D_0$ the functions f_t in (2.1) are lower semicontinuous and satify inequality

$$\left| \frac{d_+ f_t(x)}{dx} \right| \leqslant k f_t(x) \qquad \text{for all } x \in (a,b) \tag{2.7}$$

where $k \geqslant 0$ is a constant independent on f. Then there exists an $\varepsilon > 0$ such that $h = \varepsilon 1_{(a,b)}$ is a lower function.

Proof. As in the previous proof we assume inequality (2.5). Set

$$\varepsilon = \frac{1}{2(b-a)}\, e^{-k(b-a)}.$$

Now it is easy to show that $f_t \geqslant h$ for $t > t_1$. If not, then $f_{t'}(y) < \varepsilon$ for some $y \in (a,b)$ and $t' > t_1$. Consequently by (2.7)

$$f_{t'}(x) \leq f_{t'}(y)e^{k|x-y|} \leq \frac{1}{2(b-a)} \quad .$$

This evidently contradicts (2.5). The inequality $f_t \geq h$ completes the proof.

Analogous results may be formulated for stochastic semigroups on R^n and on manifolds. Thus assume now that $X = M$ is an finite dimensional compact connected smooth (C^∞) manifold equipped with a Riemannian metric $|\cdot|$. The metric induces on M the natural (Borel) measure m and the distance ϱ. A function $f : M \rightarrow R$ is called Lipschitzean if there exists a constant $c > 0$ such that

$$|f(x) - f(y)| \leq c\,\varrho(x,y) \qquad \text{for } x,y \in M.$$

For any Lipschitzean f the gradient of f is defined almost everywhere and we denote by $|f'(x)|$ the length of the gradient at the point x.

<u>Proposition 2.3.</u> Let a stochastic semigroup $\{P^t\,;\,L^1(M)\}$ be given. Assume that for every $f \in D_o$ the functions f_t in (2.1) are Lipschitzean and satisfy inequality

$$|f_t'(x)| \leq k\,f_t(x) \qquad \text{a.e. in } M$$

where $k \geq 0$ is a constant independent on f. Then there exists an $\varepsilon > 0$ such that $h = \varepsilon 1_M$ is a lower function.

<u>Proof.</u> The proof is almost the same as the previous one. As before assume inequality $\|f_t\| > \frac{3}{4}$ for $t > t_1$. Set

$$\varepsilon = \frac{1}{2m(M)}\,e^{-kr} \qquad \text{where} \quad r = \sup_{x,y \in M} \varrho(x,y)$$

Let $\gamma(s)$ $(0 \leq s \leq 1)$ be a smooth arc joining the points $y = \gamma(0)$ and $x = \gamma(1)$. The differentiation of $f_t \circ \gamma$ gives

$$\frac{d}{ds}\,f_t(\gamma(s)) = \langle f_t(\gamma'(s)),\gamma'(s)\rangle \leq k\,|\gamma'(s)|\,f_t(\gamma(s))$$

and consequently

$$f_t(x) \leqslant f_t(y) \exp\Big(k \int_0^1 |\gamma'(s)|\,ds\Big) \leqslant f_t(y) e^{kr}.$$

Now suppose that $f_{t'}(y) < \varepsilon$ for some $t' > t_1$ and $y \in M$. Then

$$f_{t'}(x) \leqslant \varepsilon e^{kr} = \frac{1}{2m(M)} \qquad \text{for} \quad x \in M$$

which contradicts (2.5). Again we have $f_t \geqslant h$ for $t > t_1$ which completes the proof.

3. Markov chains.

Our first application of Theorem 1.1 (the 1-condition) is motivated by some historical and methodological purposes only. Let us come back to the notation of Section 1, so let $(X, \mathcal{A}, m)$ be an arbitrary (σ-finite) measure space and T a nontrivial semigroup of nonnegative reals.

Consider a family of stochastic kernels $K_t : X \times X \to R$ $(t \in T)$. Thus we assume that K_t are measurable and satisfy the following conditions

(i) $K_t(x,y) \geqslant 0$ a.e. in X^2 for $t \in T$;

(ii) $\int_X K_t(x,y)m(dx) = 1$ a.e. in X for $t \in T$;

(iii) $K_{t_1+t_2}(x,y) = \int_X K_{t_1}(x,z) K_{t_2}(z,y)m(dz)$ a.e. in X for $t_1, t_2 \in T$.

Given K_t we may define a stochastic semigroup by setting

$$(3.1) \quad P^t f(x) = \int_X K_t(x,y) f(y) m(dy) \qquad \text{for} \quad f \in L^1(X), \ t \in T.$$

In the case $T = N$ (positive integers) the family $\{K_n\}_{n \in N}$ is fully defined if K_1 is given. If, in addition, $X = 1,\ldots,s$ is a finite space and m is uniformly distributed on X $(m(\{j\}) = s^{-1}$; $j = 1,\ldots,s)$, then K_n represents simply the n-th power of the $s \times s$-matrix K_1 and (3.1) is a homogeneous Markov chain.

Now let us come back to the general case. For any $t, t_0 \in T$ $(t > t_0)$

we have

$$K_t(x,y) = \int_X K_{t_0}(x,z)\, K_{t-t_0}(z,y)\, m(dz)$$

$$\geqslant \inf_z K_{t_0}(x,z) \int_X K_{t-t_0}(z,y)\, m(dz) = \inf_z K_{t_0}(x,z)$$

and consequently for any normalized density f we obtain

$$P^t f(x) = \int_X K_t(x,y)\, f(y)\, m(dy)$$

$$\geqslant \inf_z K_{t_0}(x,z) \int_X f(y)\, m(dy) = \inf_z K_{t_0}(x,z).$$

Thus $h_0(x) = \inf_z K_{t_0}(x,z)$ is a lower function for the semigroup (3.1) and from Theorem 1.1 we obtain the following classical result [11]

<u>Corollary 3.1.</u> If $\{K_t\}$ is a family of stochastic kernels such that

(3.2) $$\int_X \inf_z K_{t_0}(x,z)\, m(dx) > 0$$

for some $t_0 \in T$, then the semigroup (3.1) is asymptotically stable.

Let us observe that condition (3.2) is far from to be necessary for the asymptotical stability of (3.1). It is much stronger and implies that

$$\| P^{nt_0} f - f_* \| \leqslant 2(1 - \|h_0\|)^n$$

uniformly for all $f \in D$ (which can be directly verified). On the other hand no kind of uniform convergence (with respect to f) is required in our definition of asymptotical stability and in the l-condition. This is the main difference between the l-condition and classical conditions like (3.2). It also makes the l-condition applicable to the stability problems for stochastic semigroups generated by deterministic systems where the convergence to equilibrium is seldom uniform.

<u>4. Stochastic semigroups of deterministic systems.</u>

Stochastic semigroups appear mainly in pure probabilistic problems such as random walks, stochastic differential equations and many others. It is of great importance that they can all be generated by "deterministic" semidynamical systems.

As before let $(X,\mathcal{A},m)$ be a σ-finite measure space and T a nontrivial semigroup of nonnegative reals. A family of transformations $S_t : X \to X$ $(t \in T)$ will be called a semidynamical system if it satisfies the following two conditions

(a) S_t are double measurable, that is

$$S_t^{-1}(A) \in \mathcal{A} \qquad \text{and} \qquad S_t(A) \in \mathcal{A} \qquad \text{for } A \in \mathcal{A},\ t \in T;$$

(b) $$S_{t_1+t_2} = S_{t_1} \circ S_{t_2} \qquad \text{for } t_1, t_2 \in T.$$

A semidynamical system will be called nonsingular if in addition

(c) $$\left.\begin{array}{l} A \in \mathcal{A} \\ m(A) = 0 \end{array}\right\} \Longrightarrow \left\{\begin{array}{l} m(S_t(A)) = 0 \\ m(S_t^{-1}(A)) = 0 \end{array}\right. \quad \text{for } t \in T.$$

Given a nonsingular semidynamical system $\{S_t\}$ we may define a family of operators $P_S^t : L^1 \to L^1$ by setting

$$\int_A P_S^t f(x)\, m(dx) = \int_{S_t^{-1}(A)} f(x)\, m(dx) \qquad \text{for } f \in L^1,\ A \in \mathcal{A},\ t \in T. \tag{4.1}$$

Due to the nonsingularity of $\{S_t\}$ the integrals on the right-hand side of (4.1) are absolutely continuous with respect to m. Therefore, according to Radon-Nikodym theorem, condition (4.1) defines $\{P_S^t\}$ in a unique way.

It is easy to verify that $\{P_S^t\}$ is a stochastic semigroup. It has an additional important property, namely

$$\operatorname{supp}(P_S^t f) \subset S_t(\operatorname{supp} f) \qquad \text{for } f \in T \tag{4.2}$$

where $\operatorname{supp} f = \{x : f(x) \neq 0\}$. In fact setting $A = \operatorname{supp} f$ we have

$$\int_{X\setminus S_t(A)} |P_S^t f(x)|\, m(dx) \leq \int_{X\setminus S_t(A)} P_S^t |f(x)|\, m(dx)$$

$$\leq \int_{X\setminus S_t^{-1}(S_t(A))} |f(x)|\, m(dx) \leq \int_{X\setminus A} |f(x)|\, m(dx) = 0$$

which proves that $P_S^t f(x) = 0$ for $x \notin S_t(A)$.

The semigroup $\{P_S^t\}$ has a simple probabilistic interpretation. Namely, if x is a random variable with a probability density function f, then for each $t \in T$ the variable $S_t(x)$ has the probability density function $P_S^t f$.

The behavior of $\{P_S^t\}$ allows to determine many properties of the semi-dynamical system $\{S_t\}$ such as preservation of a measure, ergodicity, mixing and exactness. We shall concentrate here only on the first and the last problem.

Recall that $\{S_t\}$ preserves a measure m_o (equivalently m_o is invariant under $\{S_t\}$) if

$$m_o(S_t^{-1}(A)) = m_o(A) \qquad \text{for } A \in \mathcal{A},\ t \in T. \tag{4.3}$$

Assume now that a measure m_o is normalized $(m_o(X) = 1)$ and invariant under $\{S_t\}$. The quadruple $(X, \mathcal{A}, S_t, m_o)$ is called exact (shortly $\{S_t\}$ with the measure m_o is exact, cf.[13]) if

$$\left.\begin{array}{l} A \in \mathcal{A} \\ m_o(A) > 0 \end{array}\right\} \Longrightarrow \lim_{t\to\infty} m_o(S_t(A)) = 1. \tag{4.4}$$

Comparing (4.1) and (4.3) we obtain immediately the following

<u>Proposition 4.1.</u> Let $\{S_t\}$ be a nonsingular dynamical system and let $f \in L^1(X, \mathcal{A}, m)$. Then the measure

$$m_f(A) = \int_A f(x)\, m(dx) \qquad (A \in \mathcal{A})$$

is invariant under $\{S_t\}$ if and only if $P_S^t f = f$ for all $t \in T$.

The discovery that the exactness of a semidynamical system $\{S_t\}$ may be characterized by the asymptotical behavior of $\{P_S^t\}$ is due to M. Lin [10] . The following proposition is close to one of his results.

Proposition 4.2. Let $\{S_t\}$ be a nonsingular semidynamical system. If the stochastic semigroup $\{P_S^t\}$ is asymptotically stable and f_* is its unique stationary density, then the system $\{S_t\}$ with the measure

$$m_*(A) = \int_A f_*(x)\, m(dx) \qquad \text{for } A \in \mathcal{A}$$

is exact. Moreover m_* is the unique absolutely continuous normalized (nonnegative) measure invariant under $\{S_t\}$.

Proof. From Propositions 1.1 and 4.1 it follows immediately that m_* is an invariant measure and that it is unique. Thus it remains to prove condition (4.4) for m_*. Assume that $m_*(A) > 0$ and define

$$f_A(x) = \frac{1}{m_*(A)} f_*(x)\, 1_A(x) \qquad \text{for } x \in X.$$

Of course $f_A \in D(X, \mathcal{A}, m)$ and

$$r_t = \| P_S^t f_A - f_* \| \to 0 \quad \text{as} \quad t \to \infty .$$

From the definition of m_* we have

$$m_*(S_t(A)) = \int_{S_t(A)} f_*(x)\, m(dx) \tag{4.5}$$

$$\geq \int_{S_t(A)} P_S^t f_A(x)\, m(dx) - r_t .$$

According to (4.2) $P_S^t f_A$ is supported on $S_t(A)$ and consequently

$$\int_{S_t(A)} P_S^t f_A(x)\, m(dx) = \int_X P_S^t f_A(x)\, m(dx) = 1,$$

substituting this into (4.5) we complete the proof.

In general, Proposition 4.2 is not inver-tible. The asymptotical stability of $\{P_S^t\}$ implies the existence of a unique invariant measure m_* and the exactness but not vice-versa. The inverse implication may be formulated and proved in the case when the initial measure m is invariant (cf. [10]).

We admit the following definition. A nonsingular semidynamical system $\{S_t\}$ will be called <u>statistically stable</u> if the corresponding stochastic semigroup is asymptotically stable.

At the end let us consider a special case when $T = N$ (positive integers) and the semidynamical system $\{S^n\}_{n\in N}$ consists of the iterates of a (nonsingular) transformation S. Then the semigroup $\{P_S^n\}_{n\in N}$ is given by a unique operator

$$(4.6) \qquad P_S f = \frac{dm_f}{dx} \qquad \text{where} \quad m_f(A) = \int_{S^{-1}(A)} f(x)\, m(dx)\,.$$

Following S.Ulam, P_S is called the Frobenius-Perron operator corresponding to S (cf. 16 , VI.4).

5. Expanding mappings on manifolds.

From condition (4.2) it follows that any exact semidynamical system is in some sense expanding. It is not easy to express this "expansivness" in terms of differential properties of transformations $\{S_t\}$. This problem is relatively simple for some local diffeomorphisms on compact manifolds without boundary.

Let M be a (finite dimensional) compact connected smooth (C^∞) manifold equipped with a Riemannian metric $|\cdot|$ and let m be the corresponding Borel measure. A C^1 mapping $S : M \to M$ is called expanding if there exists a constant $\lambda > 1$ such that at each point $x \in M$ the differential $d\,S(x)$ satisfies

$$(5.1) \qquad |d\,S(x)\xi| \geq \lambda|\xi|$$

for each tangent vector ξ.

Using this definition K.Krzyżewski and W.Szlenk [6], [7] were able to prove the existence of a unique absolutely continuous normalized measure invariant under S and to establish many properties of this measure. A large part of their results is summarized in the following

Theorem 5.1. Assume that $S : M \to M$ is an expanding mapping of class C^2. Then the semidynamical system $\{S^n\}_{n\in N}$ is statistically stable.

Proof. (cf.[8]). Condition (5.1) implies that $dS(x)$ is a nonsingular mapping for every $x \in M$. Thus for every x there exists a neighbourhood U of x such that $S^{-1}(U)$ can be written as a union of disjoint open sets $V_1,\dots,V_q$ and S restricted to V_i $(i=1,\dots,q)$ is a homeomorphism from V_i onto U. Thus, on U the Frobenius-Perron operator P_S corresponding to S can be written in the form

$$P_S f(x) = \sum_i |\det dg_i(x)| \, f(g_i(x))$$

where g_i denotes the inverse function to $S_{|V_i}$. Now let $D_0 \subset D(M)$ be the set of all C^1 strictly positive densities. For $f \in D_0$ the differentiation of $P_S f$ gives

$$\frac{|(P_S f)'|}{P_S f} \leqslant \frac{\sum |J_i'|(f\circ g_i)}{\sum J_i (f\circ g_i)} + \frac{\sum J_i |f'\circ g_i| |dg_i|}{\sum J_i (f\circ g_i)}$$

$$\leqslant \max_i \frac{|J_i'|}{J_i} + \max_i \frac{|f'\circ g_i| |dg_i|}{(f\circ g_i)}$$

where $J_i(x) = |\det dg_i(x)|$. From (5.1) it follows that $|dg_i| \leqslant 1/\lambda$. Thus

$$\sup \frac{|(P_S f)'|}{P_S f} \leqslant c + \frac{1}{\lambda} \sup \frac{|f'|}{f}$$

where $c = \sup_{i,x} |J_i'(x)|/J_i(x)$. Consequently by induction

$$\sup \frac{|(P_S^n f)'|}{P_S^n f} \leqslant \frac{c\lambda}{\lambda - 1} + \frac{1}{\lambda^n} \sup \frac{|f'|}{f} \quad \text{for } n=1,2,\ldots$$

Choose a real $k > c/(\lambda - 1)$. Then

$$(5.2) \qquad \sup \frac{|(P_S^n f)'|}{P_S^n f} \leqslant k$$

for sufficiently large $n(n > n_o(f))$. A straightforward application of Proposition 2.3 and Theorem 1.1 completes the proof.

Remark 5.1. From inequality (5.2) it follows that the functions $\{P_S^n f\}$ are uniformly bounded and equicontinuous. Therefore the convergence of $\{P_S^n f(x)\}$ to the stationary density $f_*(x)$ is uniform in x for every $f \in D_o$. Moreover f_* is continuous (in fact C^1) and satisfies

$$\frac{1}{m(M)} e^{-kr} \leqslant f_*(x) \leqslant \frac{1}{m(M)} e^{kr} \qquad \text{for } x \in M$$

where $r = \sup\{ \rho(x,y) : x,y \in M \}$.

6. Rényi transformations.

In the special case, when M is the unit circle, Theorem 5.1 follows from classical results of Rényi (existence of an invariant measure [12]) and Rochlin (exactness [13]). Actually Rényi and Rochlin were considering two classes of mappings, namely

$$(6.1) \qquad S(x) = \varphi(x) (\text{mod } 1) \qquad \text{for } 0 \leqslant x \leqslant 1$$

where φ is a given smooth (e.g. C^2) function such that $\inf \varphi' > 1$, $\varphi(0) = 0$, $\varphi(1)$ is an integer and

$$(6.2) \qquad S(x) = r\,x \ (\text{mod } 1) \qquad \text{for } 0 \leqslant x \leqslant 1$$

where $r > 1$ is a real constant. The first class consists of expanding mappings on the unit circle (if in addition $\varphi'(0) = \varphi'(1)$) but the second is quite different. When r is not an integer then the stationary density f_* is a piecewise constant function with a finite (or countable) set of points of discontinuity.

Using the 1-condition it is easy to prove the statistical stability of S^n for a large class of transformations $S : [0,1] \to [0,1]$ which contains (6.1) and (6.2) as special cases.

Consider a mapping $S : [0,1] \to [0,1]$ which satisfies the following conditions

(a) There is a partition $0 = a_0 < \dots < a_r = 1$ of the unit interval such that for each integer $i=1,\dots,r$ the restriction of S to the interval $[a_{i-1}, a_i)$ is a C^2 function.

(b) $S(a_i) = 0$ for $i = 0,\dots r-1$.

(c) There is $\lambda > 1$ such that $S'(x) \geq \lambda$ for $0 \leq x < 1$ ($S'(a_i)$ denotes the right derivative).

(d) There is a real c such that

$$(6.3)\qquad \frac{-S''(x)}{(S'(x))^2} \leq c \qquad \text{for } x \neq a_i \ (i=0,\dots,r)$$

<u>Theorem 6.1.</u> If $S : [0,1] \to [0,1]$ satisfies conditions (a) - (d), then the semidynamical system $\{S^n\}_{n\in N}$ is statistically stable.

<u>Proof.</u> Using (4.6) it is easy to write an explicit formula for the Frobenius-Perron operator corresponding to S ; namely

$$(6.4)\qquad P_S f(x) = \sum_{i=1}^{r} g_i'(x)\, f(g_i(x)) \qquad \text{for } 0 \leq x < 1$$

where

$$g_i(x) = \begin{cases} S^{-1}_{(i)}(x) & \text{for } 0 \leq x < b_i , \\ a_i & \text{for } b_i \leq x < 1. \end{cases}$$

In this formula $S_{(i)}$ denotes the restriction of S to the interval

$[a_{i-1}, a_i)$ and $b_i = \lim_{x \to a_i} S_i(x)$. If $b_i < 1$, then $g_i'(b_i)$ denotes the right derivative. Therefore $g_i'(x) = 0$ for $b_i \leq x < 1$ and all g_i' are left lower semicontinuous.

Denote by D_o the subset of $D([0,1])$ consisting all functions which on the interval $[0,1)$ are bounded, left lower semicontinuous and satisfy inequality

(6.5) $$f_+'(x) = \frac{d_+ f(x)}{dx} \leq k_f\, f(x)$$

where k_f is a constant which depends on f. (The values of f's at $x = 1$ are irrelevant.) For any $f \in D_o$ the function $P_S f$ given by formula (6.4) is again bounded and left lower semicontinuous.

For $f \in D_o$ the differentiation of $P_S f$ gives

$$(P_S f)_+' = \sum_{i=1}^{r} (g_i')_+' \,(f \circ g_i) + \sum_{i=1}^{r} (g_i')^2 (f_+' \circ g_i).$$

A standard application of the implicit function theorem yields

$$g_i' \leq \sup \frac{1}{S'} \leq \frac{1}{\lambda}, \qquad \frac{(g_i')_+'}{g_i'} \leq \sup \frac{-S''}{(S')^2} \leq c$$

and consequently

$$(P_S f)_+' \leq c \sum_{i=1}^{r} g_i'(f \circ g_i) + \frac{1}{\lambda} \sum_{i=1}^{r} g_i'(f_+' \circ g_i).$$

Using (6.5) we obtain

$$(P_S f)_+' \leq \left(c + \frac{k_f}{\lambda}\right) P_S f .$$

Now write $f_n = P_S^n f$. An induction argument shows that

$$(f_n)_+' \leq \left(\frac{c\lambda}{\lambda - 1} + \frac{k_f}{\lambda^n}\right) f_n .$$

Choose a real $k > c\lambda/(\lambda - 1)$. Then

(6.6) $$(f_n)_+' \leq k\, f_n$$

for sufficiently large n (say $n > n_0(f)$).

Thus, the condition (2.3) in Proposition 2.1 is satisfied. Now we are going to show that f_n are bounded. From (6.4) we obtain

$$(6.7) \qquad f_{n+1}(0) \leqslant \frac{1}{\lambda} f_n(0) + \frac{1}{\lambda} \sum_{i=2}^{r} f_n(a_{i-1}).$$

From (6.6) it follows that

$$f_n(a_i) \leqslant f_n(x)\, e^k \qquad \text{for } x \leqslant a_i$$

and consequently

$$1 \geqslant \int_0^{a_i} f_n(x)dx \geqslant e^{-k} f_n(a_i)a_i \qquad \text{for } i=1,\dots,n.$$

Thus $f_n(a_i) \leqslant e^k/a_i$ and (6.7) implies

$$f_{n+1}(0) \leqslant \frac{1}{\lambda} f_n(0) + \frac{L}{\lambda} \qquad \text{where} \qquad L = \sum_{i=2}^{r} \frac{e^k}{a_{i-1}} \quad .$$

Again by a simple induction argument it follows that

$$f_n(0) \leqslant \frac{1}{\lambda^n} + \frac{L}{\lambda - 1}$$

and consequently

$$f_n(0) \leqslant (1 + \frac{L}{\lambda - 1})$$

for large n $(n > n_1(f))$. Using this and the differential inequality (6.6) we obtain finally

$$(6.8) \qquad f_n(x) \leqslant (1 + \frac{L}{\lambda - 1})e^k \qquad \text{for } 0 \leqslant x < 1 \ , \ n > n_1.$$

According to (6.6) and (6.8) all assumptions of Proposition 2.1 are satisfied. The proof is completed.

7. Transformations on the real line.

All of the transformations considered in the previous sections were defined on compact spaces. Transformations on unbounded regions, in particular on the real line, may have some specific properties. Thus, for example, the requirement that $|S'| \geq \lambda > 1$ for $S : R \to R$ is not sufficient for the statistical stability. This is amply illustrated by the behavior of the semidynamical system S^n with $S(x) = 2x$. Thus

$$P_S^n f(x) = \frac{1}{2^n} f\left(\frac{x}{2^n}\right) \qquad \text{for } x \in R$$

and for every bounded f the sequence $\{P_S^n f\}$ conveges uniformly to zero and does not converge in L^1 to any normalized density.

A typical example of a statistically stable system on R is given by S^n with $S(x) = a \tan(bx + c)$, $|ab| > 1$. This section will treat a class of such transformations. Theorem 7.1 summarized results of J.H.B Kemperman [5], F.Schweiger [14], M.Jabłoński - A.Lasota [4] and P.Bugiel [2].

Consider a mapping $S : R \to R$ which satisfies the following conditions:

(a) There is a partition $\dots a_{-1} < a_0 < a_1 \dots$ of the real line such that for each integer $i = 0, \pm 1, \dots$ the restriction $S_{(i)}$ of S to the interval (a_{i-1}, a_i) is a C^2 function.

(b) $S((a_{i-1}, a_i)) = R$ for $i = 0 \pm 1, \dots$

(c) There is a constant $\lambda > 1$ such that $|S'(x)| > \lambda$ for $x \neq a_i$ $(i = 0 \pm 1, \dots)$.

(d) There is a constant $L \geq 0$ and a function $q \in L^1(R)$ such that

$$(7.1) \qquad a_i - a_{i-1} \leq L, \qquad |g_i'(x)| \leq q(x)(a_i - a_{i-1})$$

where $g_i = S_{(i)}^{-1}$, $i = 0, \pm 1, \dots$

(e) There is a real c such that

$$(7.2) \qquad \frac{|S''(x)|}{(S'(x))^2} \leq c \qquad \text{for } x \neq a_i \ (i=0,\pm 1,\dots).$$

<u>Theorem 7.1.</u> If $S : R \to R$ satisfies conditions (a) - (d), then

the dynamical system $\{S^n\}_{n\in N}$ is statistically stable.

Proof. The explicit formula for P_S admits the form

$$P_S f(x) = \sum_i q_i(x)\, f(g_i(x)) \qquad \text{for } x \in R \tag{7.3}$$

where $q_i = |g_i'|$. Denote by $D_0 \subset D(R)$ the set of all densities of bounded variation on R which are positive, continuously differentiable and satisfy the inequality

$$|f'(x)| \leqslant k_f f(x) \qquad \text{for } x \in R \tag{7.4}$$

with a constant k_f which depends upon f.

In order to examine the properties of $P^n f$ $(f \in D_0)$, the variation of $P_S f$ will be required. Thus

$$\overset{+\infty}{\underset{-\infty}{V}} P_S f = \sum_i \overset{+\infty}{\underset{-\infty}{V}} q_i (f \circ g_i)$$

$$\sum_i \left(\frac{1}{\lambda} \overset{+\infty}{\underset{-\infty}{V}} (f \circ g_i) + \int_{-\infty}^{+\infty} |q_i'(x)|\, f(g_i(x))dx \right).$$

Further using $|q_i'| \leqslant c\, q_i$ (which follows from (7.2)) and substituting $y = g_i(x)$ we obtain

$$\overset{+\infty}{\underset{-\infty}{V}} P_S f \leqslant \sum_i \left(\frac{1}{\lambda} \overset{a_i}{\underset{a_{i-1}}{V}} f + c \int_{a_{i-1}}^{a_i} f(y)dy \right)$$

$$= \frac{1}{\lambda} \overset{+\infty}{\underset{-\infty}{V}} f + c .$$

Thus by an induction argument

$$\overset{+\infty}{\underset{-\infty}{V}} P_S^n f \leqslant \frac{1}{\lambda^n} \overset{+\infty}{\underset{-\infty}{V}} f + \frac{\lambda c}{\lambda - 1} .$$

Choose a real $\alpha > c/(\lambda - 1)$. Then

$$(7.5) \qquad \bigvee_{-\infty}^{+\infty} P_S^n f \leq \alpha$$

for sufficiently large n $(n > n_o(f))$.

Now we are in a position to evaluate $P_S^n f$ by an integrable function. From (7.1) and (7.3) it follows that

$$(7.6) \qquad P_S f(x) \leq q(x) \sum_i f(g_i(x))(a_i - a_{i-1}).$$

In every interval (a_{i-1}, a_i) pick a point z_i such that

$$(a_i - a_{i-1}) f(z_i) \leq \int_{a_{i-1}}^{a_i} f(x)dx \qquad \text{for } i=0,\pm 1,\ldots$$

From (7.6) and (7.1) we obtain

$$P_S f(x) \leq q(x) \sum_i \left(L(f(g_i(x)) - f(z_i)) + \int_{a_{i-1}}^{a_i} f(x)dx \right)$$

$$\leq L\, q(x) \bigvee_{-\infty}^{+\infty} f + q(x) \int_{-\infty}^{+\infty} f(x)dx = q(x)(L \bigvee_{-\infty}^{+\infty} f + 1).$$

Substituting $P^{n-1} f$ for f and using (7.5) we have

$$(7.7) \qquad P_S^n f(x) \leq q(x)(\alpha L + 1) \qquad \text{for } x \in R, \quad n > n_o(f) + 1.$$

Further the differentiation of (7.3) gives

$$\frac{(P_S f)'}{P_S f} \leq \frac{\sum_i |q_i'|(f \circ g_i)}{\sum_i q_i (f \circ g_i)} + \frac{\sum_i (q_i)^2 (f' \circ g_i)}{\sum_i q_i (f \circ g_i)}$$

$$\leq c + \frac{1}{\lambda} \sup \frac{|f'|}{f}$$

and by induction

$$\frac{|(P_S^n f)'|}{P_S^n f} \leq \frac{c\lambda}{\lambda - 1} + \frac{1}{\lambda^n} \sup \frac{|f'|}{f} .$$

Choose $k > \lambda c/(\lambda - 1)$. For sufficiently large n $(n > n_1(f))$ we have

$$(7.8) \qquad |(P_S^n f)'| \leq k P_S^n f .$$

From (7.7) and (7.8) it follows that the sequence $f_n = P_S^n f$ satisfies all conditions required by Proposition 2.1. This completes the proof.

Remark 7.1. In the special case when S is periodic with period $\omega = a_i - a_{i-1}$, condition (d) is automatically satisfied. The remaining conditions simply generalize the properties of the function $S(x) = a \tan (bx + c)$ with $|ab| > 1$.

Let us observe that the proofs of Theorems 5.1, 6.1 and 7.1 are similar and quite elementary. On the other hand these theorems cover (and even extend) a large part of known results concerning exact dynamical systems which were originally proved by different and sophisticated methods. This unification shows the power of the probabilistic approach offered by the 1-condition.

8. A linear Boltzmann equation.

From the analytical point of view the classical Boltzmann equation is quite complicated. However, under some additional conditions (absence of external forces, spacial homogeneity of the gas ...) it may be reduced to the following simple form

$$(8.1) \qquad \frac{\partial u(t,x)}{\partial t} + u(t,x) = \int_x^\infty \frac{dy}{y} \int_0^y u(t,y-z)\, u(t,z)dz \quad (x > 0,\ t \geq 0)$$

which is known as the Tjon-Wu representation [1] , [15] . In this equation $u(t,x)$ denotes the density function of the distribution of molecules with respect to the energy x.

It is easy to see that the function $u = e^{-x}$ is a stationary solution of (8.1). Replacing, on the right-hand side of (8.1), $u(t,z)$ by e^{-z} we obtain a linear equation

$$(8.2) \qquad \frac{\partial u(t,x)}{\partial x} + u(t,x) = \int_x^\infty \frac{dy}{y} \int_0^y e^{-z}\, u(t,y-z)dz \quad (x > 0,\ t \geqslant 0)$$

which has the following physical interpretation. Assume that a large part of a gas is in the equilibrium stage corresponding to the density function e^{-x} and that a relatively small part is distributed according to a given density f. Then equation (8.2) with the initial condition

$$u(0,x) = f(x) \qquad \text{for} \quad x > 0$$

describes the evolution of the distribution of energy in the small part of the gas.

Using the definition of the exponential integral

$$- \mathrm{Ei}(-x) = \int_x^\infty \frac{e^{-y}}{y}\, dy \qquad \text{for} \quad x > 0$$

equation (8.2) may be rewritten as

$$(8.3) \qquad \frac{\partial u(t,x)}{\partial x} + u(t,x) = \int_0^\infty b(x,y)u(t,y)dy \qquad \text{for} \quad x > 0$$

where

$$(8.4) \qquad b(x,y) = \begin{cases} -e^{y}\ \mathrm{Ei}(-y) & \text{for} \quad x \leqslant y, \\ -e^{y}\ \mathrm{Ei}(-x) & \text{for} \quad x > y. \end{cases}$$

It is easy to verify that the operator

$$(8.5) \qquad Bv(x) = \int_0^\infty b(x,y)\ v(y)dy \qquad \text{for} \quad x > 0$$

maps $L^1((0,\infty))$ into itself and is Markovian. Thus we may avoide the difficulties related with the classical solutions and consider (8.4) as the evolution equation

(8.6) $$\frac{du}{dt} + u = Bu \quad \text{for} \quad t \geqslant 0, \quad (u(0) = u_o)$$

in the space $L^1((0,\infty))$.

From this moment we shall admit a more general point of view. Namely we shall consider (8.6) with the operator B given by an arbitrary kernel $b(x,y)$ in (8.5) satisfying the following conditions:

(i) The function $b : (0,\infty)^2 \to R$ is measurable and stochastic, that is

$$b(x,y) \geqslant 0, \qquad \int_0^\infty b(x,y)dy = 1 \qquad \text{for} \quad x>0, \ y>0.$$

(ii) For every fixed $y > 0$ the function $x \to b(x,y)$ is non-increasing in x.

(iii) There is a function $g \in L^1((0,\infty))$ such that

$$\frac{b(x,y)}{1+y} \leqslant g(x) \qquad \text{for} \quad x>0, \quad y>0.$$

(iv) There are constants $\alpha \geqslant 0$, $0 \leqslant \beta < 1$ such that

$$\int_0^\infty x\, b(x,y)dx \leqslant \alpha + \beta y \qquad \text{for} \quad y>0.$$

It is easy to verify that the kernel $b(x,y)$ given by (8.4) satisfies condition (i) - (iv) with

$$g(x) = -\frac{e^x}{1+x} \operatorname{Ei}(-x)$$

and constants $\alpha = \beta = \frac{1}{2}$.

Proposition 8.1. If $b(x,y)$ satisfies condition (i), then the unique solution of equation (8.6) with the initial condition $u_o \in L^1((0,\infty))$ is given by the formula

(8.7) $$u(t) = e^{(B-I)t}u_o = e^{-t}\sum_{n=0}^{\infty} \frac{t^n}{n!} B^n u_o \qquad \text{for } t \geqslant 0.$$

and the family of operators $\{e^{(B-I)t}\}_{t \geq 0}$ is a stochastic semigroup.

The proof of Proposition 8.1 is straightforward. It should be underlined, however, an important property of formula (8.7). Namely for every $t \geq 0$ the sum of the coefficient on the right-hand side is equal to one. This is the reason that the operators $e^{(B-I)t}$ preserve the integral.

9. Asymptotic stability of the Linear Boltzmann Equation.

In order to examin the behavior of solutions of equation (8.6) we need an additional notion. By M we denote the subspace of $L^1 = L^1((0,\infty))$ consisting of all functions v for which the value

$$(9.1) \qquad \|v\|_M = \int_0^\infty (1+x)|v(x)|\,dx = \|v\| + \int_0^\infty x|v(x)|\,dx$$

is finite. The space M with the norm $\|\cdot\|_M$ is a Banach space. The last integral on the right-hand side of (9.1) has a simple physical interpretation; it is equal to the mean energy of molecules distributed according to the density v $(v \geq 0)$.

Proposition 9.1. Assume that the kernel $b(x,y)$ satisfies conditions (i) - (iv). If $u_0 \in M$, then the solution of (8.6) satisfies

$$(9.2) \qquad \|u(t)\|_M \leq \sigma\|u_0\| + e^{-(1-\beta)t}\|u_0\|_M$$

where $\sigma = (1+\alpha-\beta)/(1-\beta)$ and

$$(9.3) \qquad |u(t)(x)| \leq g(x)(\sigma\|u_0\| + \|u_0\|_M) + e^{-t}|u_0(x)|$$

for all $t \geq 0$.

Proof. Let $v \in M$. We have

$$\|Bv\|_M = \int_0^\infty (1 + x)\,|Bv(x)|dx$$

$$\leqslant \int_0^\infty (1 + x) \int_0^\infty b(x,y)|v(y)|\,dy$$

$$= \|v\| + \int_0^\infty |v(y)|\,dy \int_0^\infty xb(x,y)dx.$$

Using (iv) we obtain

$$\|Bv\|_M \leqslant \|v\| + \int_0^\infty |v(y)|\,(\alpha + \beta y)dy$$

$$= (1 + \alpha - \beta)\|v\| + \beta\|v\|_M .$$

Consequently by an induction argument we obtain

$$\|B^n v\|_M \leqslant \sigma\|v\| + \beta^n \|v\|_M .$$

From this and formula (8.7) we have

$$\|u(t)\|_M \leqslant e^{-t}\sum_{n=0}^{\infty} \frac{t^n}{n!}\,(\sigma\|u_0\| + \beta^n\|u_0\|_M) \quad \text{for} \quad t \geqslant 0$$

which is equivalent to (9.2). Further

$$|Bv(x)| \leqslant \int_0^\infty b(x,y)|v(y)|\,dy$$

$$\leqslant \int_0^\infty \frac{b(x,y)}{1 + y}(1 + y)\,|v(y)|dy \leqslant g(x)\|v\|_M$$

and consequently, for $n \geqslant 1$

$$|B^n v(x)| \leqslant g(x)\|B^{n-1} v\|_M \leqslant g(x)(\sigma\|v\| + \beta^{n-1}\|v\|_M).$$

Again from this and formula (8.7) we obtain

$$|u(t)(x)| \leqslant g(x)e^{-t} \sum_{n=1}^{\infty} \frac{t^n}{n!} (\sigma \|u_0\| + \beta^{n-1} \|u_0\|_M) + e^{-t}|u_0(x)|$$

which implies (9.3) The proof is completed.

The inequalities (9.2) and (9.3) allow us to prove the following

<u>Theorem 9.1.</u> If the kernel $b(x,y)$ satisfies conditions (i) - (iv), then the semigroup $\{e^{(B-I)t}\}_{t \geqslant 0}$ generated by equation (8.6) is asymptotically stable.

<u>Proof.</u> Define $D_0 = D((0,\infty)) \cap M$. Choose an arbitrary $f \in D_0$ and consider

$$u(t) = e^{(B-I)t} f \qquad \text{for} \quad t \geqslant 0.$$

Applying inequality (9.3) to the function $u(t + t_0)$ we obtain

(9.4) $$u(t + t_0) \leqslant g(x)(\sigma \|u(t_0)\| + \|u(t_0)\|_M) + e^{-t} u(t_0)(x)$$

Since $f \in D$ and the semigroup $\{e^{(B-I)t}\}_{t \geqslant 0}$ is stochastic, $\|u(t_0)\| = 1$. Further, according to (9.2)

$$\|u(t_0)\|_M \leqslant \sigma + e^{-(1-\beta)t_0} \|f\|_M .$$

Since $\beta < 1$, we may admit that $\|u(t_0)\|_M < 2\sigma$ for sufficiently large t (say $t > t_0(f)$). Thus, from (9.4) it follows that

$$u(t + t_0)(x) \leqslant 3\sigma g(x) + e^{-t} u(t_0)(x)$$

or

$$u(t)(x) \leqslant 3\sigma g(x) + e^{-(t-t_0)} u(t_0)(x) \qquad \text{for} \quad t > t_0.$$

Now setting

$$f_t = u(t) - e^{-(t-t_0)} u(t_0)$$

we have

$$u(t) = f_t + o_t , \qquad f_t \leqslant 3\sigma g \qquad \text{for} \quad t > t_0.$$

Applying formula (8.7) to the function $u((t-t_0)+t_0)$ we obtain

$$f_t = u((t-t_0)+t_0) - e^{-(t-t_0)} u(t_0)$$

$$= e^{-(t-t_0)} \sum_{n=1}^{\infty} \frac{(t-t_0)^n}{n!} B^n u(t_0) \quad \text{for} \quad t \geq t_0 .$$

According to (ii) every term $B^n u(t_0)$ $(n \geq 1)$ is a nonincreasing function and the same property has f_t. Thus, in virtue of Remark 2.1 the semigroup has a nontrivial lower function. An application of Theorem 1.1 completes the proof.

Remark 9.1. The family of operators $\{e^{(B-I)t}\}_{t \in R}$ is of course a group but in general it is not stochastic for $t < 0$.

Remark 9.2. In the special case when $b(x,y)$ is given by formula (8.4) every solution $u(t)$ of equation (8.6) with the initial value $u_0 \in D((0,\infty))$ satisfies

$$\lim_{t \to \infty} u(t)(x) = e^{-x} \quad (\text{in } L^1 \text{ norm}).$$

In fact in this case $u_* = e^{-x}$ is a fixed point of B and a stationary density.

The proof of Theorem 9.1 was quite complicated and the conditions (iii) and (iv) do not look natural. They can be replaced be a "more natural" condition

(v) There is a function $g \in L^1((0,\infty))$ such that

$$b(x,y) \leq g(x) \qquad \text{for } x > 0, \quad y > 0.$$

Unfortunately condition (v) is not satisfied by the kernel (8.4) in the Tjon-Wu representation. Nevertheless it is easy to prove the following

<u>Theorem 9.2.</u> If the kernel $b(x,y)$ satisfies conditions (i),(ii) and (v) then the semigroup $\{e^{(B-I)t}\}_{t\geq 0}$ is asymptotically stable.

<u>Proof.</u> Choose an $f \in D((0,\infty))$ and define

$$f_t = e^{-t} \sum_{n=1}^{\infty} \frac{t^n}{n!} B^n f \ , \qquad o_t = e^{-t} f \qquad \text{for} \quad t \geq 0.$$

Then

$$e^{(B-I)t} f = f_t + o_t \qquad \text{for} \quad t \geq 0$$

and according to conditions (ii) and (v) the functions f_t are non-increasing and bounded by g. Again an application of Remark 2.1 and Theorem 1.1 completes the proof.

<u>References</u>

[1] M.F.Barnsley and G.Turchetti, On the Abel Transformation and the Nonlinear Boltzmann Equation, Phys. Letters 72A (1979), 417-419.

[2] P.Bugiel, Approximation for the measures of ergodic transformation on the real line, Z. Wahrscheinlichkeitstheorie und verw. Gebiete, 59 (1982), 27-38

[3] S.R.Foguel, The Ergodic Theory of Markov Processes, Van Nostrand Reinhold Company, 1969.

[4] M.Jabłoński and A.Lasota, Absolutely continuous invariant measures for transformations on the real line, Zeszyty Nauk. Uniw. Jagiello. Prace Mat. 22 (1981), 7-13.

[5] J.H.B.Kempermann, The ergodic behavior of a class of real transformations, In: Proceeding of the summer res. Inst. on Statist. Inference for Stochastic Processes, Academic Press, 1975.

[6] K.Krzyżewski, Some results on expanding Mappings, Société Mathématique de France, Astérisque 50 (1977), 205-218.

[7] K.Krzyżewski and W.Szlenk, On invariant measures for expanding differentiable mappings, Studia Math. 33 (1969), 83-92.

[8] A.Lasota, A fixed point theorem and its application in ergodic theory, Tohoku Math. J. 32 (1980), 567-575.

[9] A.Lasota and J.Yorke, Exact dynamical systems and the Frobenius-

-Perron operator, Trans. Amer. Math. Soc. (to appear)

[10] M.Lin, Mixing for Markov operators Z.Wahrscheinlichkeitstheorie und verw. Gebiete 16 (1971), 231-242.

[11] A.A.Markov, Wahrscheinlichkeitsrechnung, B.G. Teubner, Leipzig 1912.

[12] A.Rényi, Representation for real numbers and their ergodic properties, Acta Math. Acad. Sci. Hungar. 8 (1957), 477-493.

[13] V.A.Rochlin, Exact endomorphisms of Lebesgue spaces, Izv. Akad. Nauk SSSR Ser. Math. 25 (1961), 499-530 (Amer. Math. Soc. Transl. (2) 39 (1964), 1-36).

[14] F.Schweiger, tan x is ergodic, Proc. Amer. Math. Soc. 1 (1978), 54-56.

[15] J.A.Tjon and T.T.Wu, Numerical aspects of the approach to a Maxwellian Distribution, Phys. Rev. A. 19 (1979), 883-888.

[16] S.M.Ulam, A Collection of Mathematical Problems, (Interscience Tracts in Pure and Appl. Math., No 8). Wiley 1960.

RECENT DEVELOPMENTS IN STABILITY AND ERROR ANALYSIS OF NUMERICAL METHODS FOR ORDINARY DIFFERENTIAL EQUATIONS

Werner Liniger

IBM Thomas J. Watson Research Center
Yorktown Heights, New York 10598

Abstract: We survey recent theoretical work on four types of integration methods for ordinary differential equations: multistep-, one-leg-, Runge-Kutta-, and extrapolation methods. Rigorous stability results and error bounds were obtained for such methods as applied to the linear test equation with constant or variable coefficient and/or certain classes of nonlinear systems, notably dissipative (monotone negative) ones. Investigations were carried out both for constant and variable steps.

1. Introduction

A linear multistep formula ($\Sigma := \sum_{j=0}^{k}$),

$$\Sigma\alpha_j x_{n+j} - h\Sigma\beta_j \dot{x}_{n+j} = 0, \tag{1.1}$$

applied to the test equation (TE)

$$\dot{x} = \lambda x \tag{1.2}$$

on an equidistant grid $\{t_n\}$, $t_n = nh$, $n = 0,1,2,\ldots$, $h>0$, is said to be stable at $q = h\lambda$ if all solutions $\{x_n\}$ of the difference equation

$$\Sigma(\alpha_j - q\beta_j)x_{n+j} = 0 \tag{1.3}$$

are bounded as $n \to \infty$, which is the case iff the characteristic equation

$$\rho(\zeta) - q\sigma(\zeta) = 0 \tag{1.4}$$

satisfies the "root condition", i.e., the condition that all of its roots ζ_i, $i = 1, \ldots, k$, satisfy $|\zeta_i| \leq 1$ and $|\zeta_i| = 1 \Rightarrow \zeta_i$ is simple. Here

$$\rho(\zeta) := \Sigma\alpha_j\zeta^j, \quad \sigma(\zeta) := \Sigma\beta_j\zeta^j. \tag{1.5}$$

The set S of all q's at which the formula is stable is called the stability region and the formula (1.1) is said to be A-stable [1] if S contains the entire left half plane $\mathbb{C}_- := \{q, Re\, q \leq 0\}$. A-stability is the most important linear stability concept for identifying formulas which are efficient for solving stiff problems. It was shown [1] that the order of an A-stable linear multistep formula cannot exceed two and that such a formula cannot be explicit. The Trapezoidal Rule (TR)

$$x_{n+1}-x_n-h\left(\frac{1}{2}\dot{x}_{n+1}+\frac{1}{2}\dot{x}_n\right)=0, \tag{1.6}$$

the two-step Backward Differentiation Formula (BD2)

$$\frac{3}{2}x_{n+2}-\frac{4}{2}x_{n+1}+\frac{1}{2}x_n-h\dot{x}_{n+2}=0, \tag{1.7}$$

the Adams-type formula

$$x_{n+2}-x_{n+1}-h\left(\frac{3}{4}\dot{x}_{n+2}+\frac{1}{4}\dot{x}_n\right)=0, \tag{1.8}$$

and the formula

$$\frac{5}{6}x_{n+2}-\frac{4}{6}x_{n+1}-\frac{1}{6}x_n-h\left(\frac{5}{9}\dot{x}_{n+2}+\frac{2}{9}\dot{x}_{n+1}+\frac{2}{9}\dot{x}_n\right)=0 \tag{1.9}$$

are all examples of second-order A-stable formulas.

When applied to a more general problem than (1.2), e.g. to the variable coefficient test equation (VTE)

$$\dot{x}=\lambda(t)x \tag{1.10}$$

or to a nonlinear system

$$\dot{x}=f(t,x), \tag{1.11}$$

satisfying the root condition in general does not guarantee boundedness of the numerical solutions. The stability properties depend first of all on how the formula (1.1) is implemented. Two different methods can be associated with any given formula (1.1), namely the familiar Multistep (MS) Method

$$\Sigma\alpha_j x_{n+j}-h\Sigma\beta_j f(t_{n+j},x_{n+j})=0 \tag{1.12}$$

and its One-leg (OL) "twin" which was introduced for theoretical purposes in [2] and, subject to the normalization

$$\Sigma\beta_j=1, \tag{1.13}$$

is defined by

$$\Sigma\alpha_j x_{n+j} - hf(\Sigma\beta_j t_{n+j}, \Sigma\beta_j x_{n+j}) = 0. \tag{1.14}$$

With respect to the TE, (1.12) and (1.14) are identical and their stability properties are thus the same. In general, however, they are different. For example [3], the MS implementation of the TR is unstable ($x_{2n} = x_0(-2)^n$) for the VTE with $\lambda(t_{2n}) = 0$, $\lambda(t_{2n+1}) = -1$, $n = 0,1,...$, when applied with variable steps $h_{2n} = 7$, $h_{2n+1} = \frac{1}{2}$, $n = 1,2,...$; here $h_n := t_n - t_{n-1}$. By contrast, the Implicit Midpoint Rule (the OL-implementation of the TR) is stable with respect to the VTE for any $\lambda(t)$, $Re\lambda(t) \leq 0$ and any step sequence $\{h_n\}$. In [4] it was observed that also for multistep formulas in the strict sense ($k \geq 2$), some variable-step OL-methods are "more stable" than their MS-counterparts and it was suggested therefore that those OL-methods could advantageously be used for computation. For example, the variable-step MS-versions of (1.8) and (1.9) applied to the VTE can be unstable [4 – 6]. By contrast, the variable-step OL-methods associated with (1.8) and (1.9) are stable for the VTE for any step sequence $\{h_n\}$ and any $\lambda(t) \leq 0$ [4], respectively any $\lambda(t)$ with $Re\lambda(t) \leq 0$ [5,6]. (Note that for the uniform step case the stability properties of the MS- and OL-method associated with the same formula are the same, even for a class of nonlinear problems [2]). However, not all OL-methods are that strongly stable. For example, the OL- and MS-methods associated with BD2 (or any other Backward Differentiation Formula) are identical. For the TE and increasing steps a contractivity condition (see Section 2 hereafter) becomes violated for step ratios $r > 1 + \sqrt{2}$ with $\lambda < 0$ [7], and for the VTE with a certain pure imaginary $\lambda(t)$ and exponentially increasing steps the BD2 method is unstable for any $r > 1$ [5,6]. (Recall that for these problems the variable-step OL-methods (1.8), respectively (1.9), are stable for any step sequence $\{h_n\}$).

For variable coefficient problems or nonlinear problems and for variable steps it is in general not possible to give necessary and sufficient conditions for stability. One approach for obtaining sufficient conditions for stability and global error bounds for OL-methods is to analyze the *contractivity* of the integration method, as discussed in Section 2 hereafter. Roughly speaking, a method is called contractive if the numerical solutions are non-increasing in some norm. A functional analytic technique for obtaining error bounds for MS-solutions of nonlinear problems with constant steps is summarized in Section 3 of this paper. Linear stability and nonlinear contractivity is discussed for Runge-Kutta methods in Section 4, and for extrapolation methods in Section 5.

2. Contractivity Analysis for One-leg Methods

A. Linear contractivity

The formula (1.1) is said to be contractive [8] at $q = h\lambda$ with respect to any given norm $\| \bullet \|$ in $\mathbb{C}_k$ if for all solutions $\{x_n\}$ of the TE we have $\|X_n\| \geq \|X_{n+1}\|$, $n = 0,1,\ldots$, where $X_n := (x_n, x_{n+1}, \ldots, x_{n+k-1})$. The contractivity region K (with respect to $\| \bullet \|$) is the set of all q at which the formula is contractive. The formula is called A-contractive if $\mathbb{C}_- \subseteq K$. Examples of formulas which are A-contractive w.r. to $\| \bullet \|_\infty$ are (1.6) and (1.9).

Clearly, contractivity at q implies stability at q and thus, in any norm, $K \subseteq S$. A-contractivity (in any norm) implies A-stability.

Although contractivity is defined for the constant coefficient TE, we immediately get results for OL-methods applied to the VTE as well. First we note that for constant steps $\{q_n := h\lambda(\Sigma\beta_j t_{n+j})\} \subseteq K$ is sufficient for stability of an OL-method as applied to the VTE. For variable steps, the formula coefficients in general depend on n (i.e., $\alpha_j = \alpha_{j,n}$, $\beta_j = \beta_{j,n}$, and thus $K = K_n$. In this case stability for any OL-method applied to the VTE is assured if $q_n := h_n\lambda(\Sigma\beta_{j,n}t_{n+j}) \epsilon K_n$ for all n. In particular, if at every step an A-contractive OL-method is used (i.e., $\mathbb{C}_- \subseteq K_n$ for all n), then we get stability with variable steps for the VTE with any $\lambda(t)$, $Re\lambda(t) \leq 0$ (for then $q_n \epsilon \mathbb{C}_- \subseteq K_n$). The unconditional stability result stated above for the variable-step OL-method associated with (1.9) is based on this observation.

Contractivity in the ℓ_∞-norm as defined above was first introduced in [9] and studied for classical Adams- and BD-methods with constant steps and for $\lambda(t) \leq 0$. Contractivity of variable-step BD-methods in certain "polygonal norms" (i.e., norms whose "unit-ball" is a polygon) was investigated in [7,10]. A rather complete theory of ℓ_∞-norm-contractivity for fixed steps is given in [8,4], as well as a boundedness result for diagonally dominant (but not necessarily monotone) nonlinear systems. This latter result was subsequently strengthened somewhat and its proof was simplified [6]. In [5], A_0-contractivity (the analogue of A_0-stability) and A-contractivity in the ℓ_∞-norm were studied for the two-parameter family of all two-step, second-order formulas with variable steps. It was shown that for any step ratio $r = r_n := h_{n+1}/h_n$ there exists a one-parameter subfamily: the set of all A-contractive formulas (containing a particular formula which for $r = 1$ reduces to (1.9)). This result was generalized in [11] where it was shown by explicit construction that the $(2k - 2)$-parameter family of all second-order k-step formulas with arbitrary variable steps contains a $(k - 1)$-parameter subfamily: the set of all formulas which are A-contractive in the ℓ_∞-norm.

B. Nonlinear contractivity

Consider a dissipative nonlinear system (1.11), i.e. one which satisfies

$$Re<x-y, f(t,x)-f(t,y)>\leq \mu\,|x-y|^2,\ \mu\leq 0 \tag{2.1}$$

in some scalar product $<,>$; here $|x|^2 = <x,x>$. For differentiable f one can replace (2.1) by

$$Re<v, Jv>\leq \mu\,|v|^2 \tag{2.2}$$

with suitable arguments of $J := f_x$. Condition (2.2) states that the numerical range of the Jacobian matrix J lies in a half-plane $Re\, q\leq \mu$ and thus we say that we are dealing with a "half-plane-bounded" nonlinearity. It is well known that for the difference $v(t)$ between any two solutions $x(t)$ and $x(t)+v(t)$ of a system satisfying (2.1) we have $|v(t)|\leq |v(0)|\,e^{\mu t}$, i.e. $|v(t)|$ is bounded or decaying. It is then natural to ask whether for certain integration methods this property carries over to the numerical solution?

If we let $\rho := \rho(E)$, $\sigma := \sigma(E)$, where $\rho(\zeta)$ and $\sigma(\zeta)$ are defined by (1.5) and where E is the displacement operator, then we can write the MS-method in operator form:

$$\rho x_n - h\sigma f(t_n,x_n) = 0, \tag{2.3}$$

and the OL-method as

$$\rho x_n - hf(\sigma t_n,\sigma x_n) = 0. \tag{2.4}$$

If $\{x_n\}$ and $\{x_n+v_n\}$ are any two OL-solutions then $\{v_n\}$ satisfies

$$\rho v_n = h[f(\sigma t_n,\sigma x_n+\sigma v_n) - f(\sigma t_n, \sigma x_n)] \tag{2.5}$$

and it follows from (2.1) and (2.5) that

$$Re<\sigma v_n,\rho v_n>\leq h\mu\,|\sigma v_n|^2\leq 0. \tag{2.6}$$

Definition: A multistep formula is said to be G-contractive (formerly called G-stable [2]) if there exists a positive definite symmetric matrix $G = (g_{ij})$, $i,j = 0,1,\ldots,k-1$, such that for any sequence $\{y_n\}$ we have

$$G(Y_{n+1}) - G(Y_n)\leq 2Re<\sigma y_n,\rho y_n>; \tag{2.7}$$

here Y_n is the composite vector $(y_n,y_{n+1},\ldots,y_{n+k-1})^+$ and $G(Y_n) := \sum_{i,j=0}^{k-1} g_{ij}<y_{n+i},y_{n+j}>$. □

Note that for a G-contractive OL-method it follows form (2.6) and (2.7) that

$$G(V_{n+1}) - G(V_n)\leq 2Re<\sigma v_n,\rho r_n>\leq 2h\mu\,|\sigma v_n|^2\leq 0, \tag{2.8}$$

i.e. we have a contraction in the weighted ℓ_2-norm associated with G. Trivially, G-contractivity implies A-stability (since the TE with $Re\lambda \leq 0$ is dissipative). But is was shown [12] that the converse holds as well, i.e. G-contractivity and A-stability are equivalent. (Note however that his is true only for constant steps [5]).

For $k = 2$, G-contractivity was analyzed in [5] for arbitrary variable steps. In this case, although by the equivalence theorem of [12] a G-norm can be constructed for every formally A-stable method (i.e. every formula whose coefficients $\alpha_{j,n}$, $\beta_{j,n}$ are such that $\rho_n(\zeta) - q\sigma_n(\zeta)$ satisfies the root condition for all q, $Re\, q \leq 0$), that norm will depend on n and from its existence we cannot necessarily infer boundedness. The only second-order two-step formulas for constant steps which can be extended to arbitrary variable steps in such a way that they are G-contractive in a *fixed* G-norm are those which are A-contractive in the ℓ_∞-norm (e.g. (1.9)) [5]. For the latter, stability is thus assured for the variable-step OL-solutions of any nonlinear dissipative systems.

3. Nonlinear Stability of Multistep Methods

The following stability result for MS-methods with constant steps applied to dissipative nonlinear systems was derived independently in [13] and in [14] by functional analytic techniques: Let $\{x_n\}$ be any MS-solution of a dissipative system generated by an A_∞-stable formula (i.e. a formula for which $q = \infty$ is an interior point of S and thus the root-locus curve $\{q \mid q = \rho(e^{i\theta})/\sigma(e^{i\theta}), 0 \leq \theta \leq 2\pi\}$ is bounded). For such a formula there exist quantities $c := \max_\theta [1/\Sigma\beta_j e^{i(k-j)\theta}]$ and b where $-b := \min_\theta Re[\rho(e^{i\theta})/\sigma(e^{i\theta})]$, and we have "input-output" stability,

$$\|\{e_n\}\| \leq m \|\{\ell_n\}\|, \tag{3.1}$$

if

$$\mu h < -b. \tag{3.2}$$

Here $\{\ell_n\}$ is the local truncation error sequence, with $\ell_n := \rho x(t_n) - h\sigma\dot{x}(t_n)$ and $x(t)$ is the exact solution, $\{e_n\}$ is the global error sequence ($e_n := x_n - x(t_n)$), $\|\{e_n\}\| := (\sum_{n=0}^{N} |e_n|^2)^{1/2}$ and $m := c[(-h\mu) - b]^{-1}$, independently of N. Pointwise (rather than ℓ_2-)estimates can be given under slightly stronger assumptions. For $b = 0$ the above result can be viewed as a nonlinear A-stability result since it is then valid for any negative μ arbitrarily small in absolute value, i.e. for any system with an arbitrarily small amount of dissipation.

Generalizations of the above result were given in [15] for circle-bounded nonlinearities, i.e.

problems satisfying

$$(1+2a\mu h)Re\langle x-y,f(t,x)-f(t,y)\rangle \leq \mu\,|x-y|^2 + ah(1+a\mu h)\,|f(t,x)-f(t,y)|^2, \tag{3.3}$$

and in [16] by the multiplier theory for angle-bounded nonlinearities as discussed hereafter.

We summarize the idea of the multiplier theory for the autonomous case, writing $x := x(t_n)$, $e := e_n$ etc. Let $\{\gamma_n\}$ be such that $\rho(\zeta)/\sigma(\zeta) = \sum_{n=0}^{\infty} \gamma_n \zeta^{-n}$. The global MS-error satisfies

$$\rho e - h\sigma[f(x+e) - f(x)] = \ell. \tag{3.4}$$

Upon convolution $(*)$ with $\{\gamma_n\}$ we get

$$\gamma * e - h[f(x+e) - f(x)] = \tilde{\ell}. \tag{3.5}$$

A sequence $\{\mu_n\}$ is said to be a multiplier for the formula (ρ,σ) if 1) its Fourier transform $\hat{\mu}(\tau) := \Sigma \mu_n e^{ij\tau}$, τ real, satisfies $Re\hat{\mu}(\tau) > 0$ for all τ; 2) $\{\mu_n\}$ is ℓ_1-summable; and 3) the z-transform of $\{\mu_n\}$, $\sum_n \mu_n z^n$, is a rational function of z. Then we have input-output stability (see (3.1)) if, for some multiplier $\{\mu_n\}$, i) $Re\hat{\mu}(\tau)\hat{\gamma}(\tau) \geq 0$ for all τ and ii) the nonlinearity satisfies

$$\sum_{n=0}^{N} \langle \mu * e, f(x+e) - f(x)\rangle \leq 0. \tag{3.6}$$

4. Stability and Contractivity of Runge-Kutta Methods

A. Stability

Runge-Kutta (RK) methods for solving the system (1.11) are one-step methods defined by the relations

$$\begin{aligned} k_i &= f(t_n + c_i h,\, x_n + h\sum_{j=1}^{s} a_{ij}k_j),\ i = 1,2,\ldots,s \\ x_{n+1} &= x_n + h\sum_{i=1}^{s} b_i k_i. \end{aligned} \tag{4.1}$$

The RK-method depends on the parameters $A = (a_{ij})$, $\underset{\sim}{b} := (b_1,\ldots,b_s)^\dagger$, $\underset{\sim}{c} := (c_1,c_2,\ldots,c_s)^\dagger$, and s, the number of stages. For any given s there exist s-stage RK-methods of (maximal) order of accuracy $p = 2s$ [17]. When applied to the TE (1.2), the RK-method generates the difference equation $x_{n+1} = R(q)x_n$ with the amplification factor

$$R(q) = 1 + q\underset{\sim}{b}^{\dagger}(I - qA)^{-1}\underset{\sim}{e}; \tag{4.2}$$

here $q = h\lambda$ and $\underset{\sim}{e} := (1,1,...,1)^{\dagger}$, a vector of s components. The method (4.1) is A-stable iff $|R(q)| \leq 1$ for all q, $Re\, q \leq 0$. For every s the maximal order ($p = 2s$) s-stage RK-method, whose amplification factor $R(q)$ is the (s,s) diagonal entry of the Padé table, is A-stable [18]. More generally, an RK-method whose amplification factor is the (n,d) entry of the Padé table is A-stable iff $n \geq d \geq n - 2$ [19]. For $n - 1 \geq d \geq n - 2$ these methods are said to be L-stable [19], i.e. they have the property that $\lim_{q \to \infty} R(q) = 0$ (thereby mimicking the behavior of the amplification factor e^q of the true solution which satisfies $\lim_{h \to \infty} e^q = 0$ for $Re\lambda < 0$). Recently, considerable progress was made in discussing A-stability of RK-methods whose amplification factor $R(q)$ is not an element of the Padé table but some other rational approximation of e^q [20 – 22]. This progress came about particularly by an application of the "order-star" theory developed by Wanner et al. [22].

B. Contractivity

A RK-method is said to be BN-stable [23] if for any two solutions $\{x_n\}$ and $\{y_n\}$ of any nonlinear system satisfying

$$\langle x - y, f(t,x) - f(t,y) \rangle \leq 0 \tag{4.3}$$

(the special case $\mu = 0$ of the dissipativity condition (2.1)) we have

$$|x_n - y_n| \leq |x_{n-1} - y_{n-1}|, \quad n = 1,2,..., \tag{4.4}$$

where $|x|^2 = \langle x,x \rangle$. The following statement combines results of [23 – 26]: An irreducible [3] RK-method is BN-stable iff $b_i > 0$, $i = 1,...,s$, and $M = (m_{ij})$, with $m_{ij} = b_i a_{ij} + b_j a_{ji} - b_i b_j$, is positive semidefinite.

In [25] nonlinear contractivity results were given for RK-methods which are linearly contractive in circle-bounded regions.

C. Special Runge-Kutta methods

The RK-methods have the advantages that a) high orders of accuracy are compatible with A-stability; b) they are "self-starting" (i.e. they require only the one initial data provided by the initial condition for the first order system, in contrast to MS- and OL-methods which require $(k - 1)$ additional starting data); and c) they are insensitive to step size changes. However, the RK-methods are in general costly to implement requiring, for an mth order system, $O(m^3 s^3)$ operations per Newton step in solving the nonlinear equations for the k_i if the linearized problem is solved by elimination. It is possible to reduce this amount of work by using *special* RK-methods, noticeably the singly implicit methods [27,28], and the diagonally implicit methods [29,30] defined by $a_{ij} = 0,\ i<j$, or by using Rosenbrock methods which are linearly implicit variants of the RK-methods [31,32].

5. Extrapolation Methods

For certain second-order LM-methods and for nonlinearities $f \epsilon C^{2N+2}$ the global truncation error has an asymptotic expansion [3] in powers of h^2,

$$e(t,h) = \sum_{j=1}^{N} g_j(t) h^{2j} + 0(h^{2N+2}) \tag{5.1}$$

for all $h \epsilon [O,\hat{H}]$ and for some $\hat{H}>0$. Here $e(t,h) := \tilde{x}(t;h) - x(t)$ and $\tilde{x}(t;h)$ denotes the approximate solution at t calculated with a step h.

Examples of such formulas are 1) the Explicit Midpoint Rule [33]

$$\hat{x}(t_{n+1};h) - \hat{x}(t_{n-1};h) - 2hf(t_n, \hat{x}(t_n;h)) = 0 \tag{5.2}$$

followed by "smoothing":

$$\tilde{x}(t_n;h) := \frac{1}{4}[\hat{x}(t_{n+1};h) + 2\hat{x}(t_n;h) + \hat{x}(t_{n-1};h)]; \tag{5.3}$$

2) the Trapezoidal Rule (which may be followed by the same smoothing [1,34]; and 3) the semi-implicit midpoint rule discussed hereafter. Let $\{\tilde{x}_{i,1}\}$, $i = 1,2,...,$ be a sequence of second-order approximations of $x(H)$, where H is the "basic step", where the $\tilde{x}_{i,1} = \tilde{x}_{i,1}(H;h_i)$ are computed with steps $h_i := H/n_i$ using one of the formulas mentioned above, and where $\{n_i\}$ is some suitable sequence of integers. Then a tableau of higher order approximations $\{\tilde{x}_{ik}\}$ of $x(H)$ can be obtained by Richardson extrapolation, e.g. by using the Aitken-Neville algorithm

$$\tilde{x}_{ik} := \tilde{x}_{i,k-1} + (\tilde{x}_{i,k-1} - \tilde{x}_{i-1,k-1})\left[(n_i/n_{i-k+1})^2 - 1\right]^{-1}. \tag{5.4}$$

The error then satisfies

$$e_{ik} := \tilde{x}_{ik} - x(H) = o(H^{2k}), \tag{5.5}$$

i.e. the order of accuracy of the approximations in the kth column if $p = 2k$. A "locally optimal" order of approximation and a basic step size H can be determined simultaneously and cheaply by an algorithm due to Deuflhard [35] which minimizes some measure of the expected amount of work per unit time interval.

An integration method which lends itself to Richardson extrapolation for stiff problems is the Semi-implicit Midpoint Rule [36]. If we write $\dot{x} - Ax = \bar{f}(t,x) := f(t,x) - Ax$, where A is some constant approximation to the Jacobian matrix, then $\dot{x} - Ax$ is descretized by the expression $(1/2h)[(I - hA)\tilde{x}(t+h) - (I + hA)\tilde{x}(t-h)]$, where $I - hA \approx e^{-hA}$ and $I + hA \approx e^{hA}$. There exists an asymptotic h^2-expansion for $h \in [O, \hat{\bar{H}}]$, where $\hat{\bar{H}} = O(1/\bar{L})$ and where $\bar{L}$ is the "deflated" Lipschitz constant associated with $\bar{f}$ which, hopefully, is, smaller than the constant L associated with f. Recently, nonlinear contractivity results were given for some of the approximations obtained by this method and for nonlinear systems of the form

$$\dot{x} = Ax + \bar{f}(t,x) \tag{5.6}$$

where the linear part of (5.6) is dissipative,

$$<x,Ax> \leq \mu |x|^2, \tag{5.7}$$

and the nonlinearity is Lipschitz continuous,

$$|\bar{f}(t,x) - \bar{f}(t,y)| \leq \bar{L}|x-y|. \tag{5.8}$$

References

1. G. Dahlquist, "A special stability problem for linear multistep methods," BIT **3** (1963) 27-43.

2. G. Dahlquist, "Error analysis of a class of methods for stiff nonlinear initial value problems," *Numerical Analysis Dundee 1975,* Lecture Notes in Mathematics **506**, Springer Verlag, Berlin (1976), pp.60-72.

3. H. Stetter, *Analysis of Discretization Methods for Ordinary Differential Equations,* Springer Verlag, Berlin (1973).

4. O. Nevanlinna and W. Liniger, "Contractive methods for stiff ordinary differential equations. Part II," BIT **19** (1979) 53-72.

5. G. Dahlquist, W. Liniger and O. Nevanlinna, "Stability of two-step methods for variable integration steps," IBM Report RC 8494 (1980). To appear in SIAM J. Numer. Anal.

6. F. Odeh and W. Liniger, "On A-stability of second-order two-step methods for uniform and variable steps," Proc. IEEE Intl. Conf. Circuits and Computers 1980 (N. B. Rabbat, Ed.) Vol. 1, pp. 123-126.

7. R. Brayton and C. Cooley, "Some remarks on the stability and instability of the backward differentiation methods with non-uniform time steps," *Topics in Numerical Analysis,* Proc. Royal Irish Acad. Sci. (1972) pp.13-33.

8. O. Nevanlinna and W. Liniger, "Contractive methods for stiff ordinary differential equations. Part I," BIT **18** (1978) 457-474.

9. W. Liniger, "Zur Stabilität der numerischen Integrationsmethoden für Differentialgleichun-

gen," Doctoral Thesis, University of Lausanne, Switzerland (1957).

10. R. Brayton and C. Tong, "Stability of dynamical systems: A constructive approach," IEEE Trans. Circuits and Systems *CAS-26* (1979) 224-234.

11. W. Liniger, "A-contractivity of second-order multistep formulas with variable steps," IBM Report RC 9281 (1980). Submitted to SIAM J. Numer. Anal.

12. G. Dahlquist, "G-stability is equivalent to A-stability," BIT **18** (1978) 384-401.

13. F. Odeh and W. Liniger, "Nonlinear fixed-h stability of linear multistep formulae," J. Math. Anal. Appl. **61** (1977) 691-712.

14. G. Dahlquist and O. Nevanlinna, "ℓ_2-estimates of the error in the numerical integration of non-linear differential systems," Report TRITA-NA-7607, Royal Institute of Technology, Stockholm, Sweden (1976).

15. O. Nevanlinna, "On the numerical integration of nonlinear initial value problems by linear multipstep methods," BIT **17** (1977) 58-71.

16. O. Nevanlinna and F. Odeh, "Multiplier techniques for linear multistep methods," Numer. Funct. Anal. and Optimiz., **3** (1981) 377-423.

17. J. Butcher, "Implicit Runge-Kutta processes," Math. Comp. **18** (1964) 50-64.

18. G. Birkhoff and R. Varga, "Discretization errors for well-set Cauchy problems: 1., " J. Math. and Phys. **44** (1965) 1-23.

19. B. Ehle, "On Padé approximations to the exponential function and A-stable methods for the numerical solution of initial value problems," Report CSRR 2010, Univ. of Waterloo, Ontario, Canada (1969).

20. S. Nørsett, "C-polynomials for rational approximation to the exponential function," Numer. Math. **25** (1975) 39-56.

21. A. Iserles, "On the generalized Padé approximations to the exponential function," SIAM J. Numer. Anal. **16** (1979) 631-636.

22. G. Wanner, E. Hairer and S. Nørsett, "Order stars and stability theorems," BIT **18** (1978) 475-489.

23. K. Burrage and J. Butcher, "Stability criteria for implicit Runge-Kutta methods," SIAM J. Numer. Anal. **16** (1979) 30-45.

24. M. Crouzeix, "Sur la B-stabilité des méthodes de Runge-Kutta," Numer. Math. **32** (1979) 75-82.

25. G. Dahlquist and R. Jeltsch, "Generalized disks of contractivity for explicit and implicit Runge-Kutta methods," Report TRITA-NA-7907, Royal Institute of Technology, Stockholm, Sweden (1979).

26. W. Hundsdorfer and M. Spijker, "A note on B-stability of Runge-Kutta methods," Numer. Math. **36** (1981) 319-331.

27. J. Butcher, "On the implementation of implicit Runge-Kutta methods," BIT **16** (1976) 237-240.

28. S. Nørsett, "Runge-Kutta methods with a multiple real eigenvalue only," BIT **16** (1976) 388-393.

29. S. Nørsett, "Semi-explicit Runge-Kutta methods," Math. and Computation Report No. 6, University of Trondheim, Norway (1974).

30. R. Alexander, "Diagonally implicit Runge-Kutta methods for stiff O.D.E.s," SIAM J. Numer. Anal. **14** (1977) 1006-1021.

31. H. Rosenbrock, "Some general implicit processes for the numerical solution of differential equations," Computer J. **5** (1963) 329-330.

32. J. Cash, "Semi-implicit Runge-Kutta procedures with error estimates for the numerical integration of stiff systems of ordinary differential equations," J. ACM **23** (1976) 455-460.

33. W. Gragg, "On extrapolation algorithms for ordinary initial value problems," SIAM J. Numer. Anal. **2B** (1965) 384-403.

34. B. Lindberg, "On smoothing for the trapezoidal rule, an analytic study of some representative test examples," Report TRITA-NA-7131, Royal Institute of Technology, Stockholm, Sweden (1971).

35. P. Deuflhard, "Order and stepsize control in extrapolation methods," Report No. 93, Inst. for Appl. Math., Univ. of Heidelberg, Germany (1980).

36. G. Bader and P. Deuflhard, "A semi-implicit mid-point rule for stiff systems of ordinary differential equations," Report No. 114, Inst. for Appl. Math., Univ. of Heidelberg, Germany (1981).

NUMERICAL SOLUTION OF A SINGULAR PERTURBATION PROBLEM WITH TURNING POINTS

J. Lorenz
Fakultät für Mathematik der
Universität Konstanz
D775 Konstanz, F.R. of Germany

Abstract: Nonlinear singular perturbation problems with turning points are discretized on a uniform mesh. The scheme is second order on smooth solutions. Unique solubility of the difference equations is investigated.

1. Introduction. Consider a boundary value problem of the form

(1a) $\quad -\varepsilon u''+a(u)u'+b(x,u) = 0, \quad u(0) = \gamma_0, \quad u(1) = \gamma_1$

where $0<\varepsilon<<1$ and $a(\cdot)$ may have zeros. Equations of this type are supposed to model some of the mathematical difficulties of the equations for compressible flows (compare, e.g.[9]) where interior shock layers can occur. The differential equation gives rise to a nonlinear stiff system where the sign of the fast eigenvalue depends on the sign of $a(u)$, i.e. depends on the unknown solution [4]. In this paper we consider a difference scheme for (1a) on a uniform mesh with a mesh-size $h>>\varepsilon$. The difference equations are second order accurate in the smooth parts of the solution. We will not discuss asymptotic properties of the difference scheme here, but merely concentrate on the unique solubility of the nonlinear system of difference equations under the condition

(1b) $\quad b_u(x,u) \geq \mu > 0 \quad$ for all $\quad (x,u)\in[0,1]\times\mathbb{R}$.

(It is known that (1a,b) has a unique C^2-solution u_ε for all $\varepsilon>0$ and u_ε tends to a limit function in BV as $\varepsilon\to 0$, see[1,4].) For the discussion of the difference equations we use some results about M-functions. The difference scheme suggested is a modification of a scheme developed and investigated by Engquist and Osher [7,8]. The original E-O-scheme is only first order accurate on smooth solutions. We obtain second order accuracy by switching the zero order term $b(x,u)$ appropriately. Numerical results will be given.

2. The idea leading to the difference scheme. Let

$$f(u) = \int_0^u a(s)\,ds, \quad u \in \mathbb{R},$$

where $a(\cdot)$ is the coefficient function in (1a). Conservation laws

$$u_t(x,t) + f(u(x,t))_x = 0, \quad x \in \mathbb{R}, \ t \geq 0$$

have often been discretized in the form

$$\frac{1}{\Delta t}(u_j^{n+1} - u_j^n) + \frac{1}{\Delta x}\{g(u_{j+1}^n, u_j^n) - g(u_j^n, u_{j-1}^n)\} = 0$$

where $u_j^n \sim u(j\Delta x, n\Delta t)$ and g is a so-called numerical flux function. In [7,8] the choice

(2) $$g(u,v) = \int_0^u a_-(s)\,ds + \int_0^v a_+(s)\,ds$$

with $a_-(s)=\min(0,a(s))$, $a_+(s)=\max(0,a(s))$ was suggested leading to an upwind scheme in conservation form. The corresponding discretization for (1a) reads

(3a) $$\varepsilon h^{-2}(-u_{j-1}+2u_j-u_{j+1})$$

(3b) $$+h^{-1}\{g(u_{j+1},u_j)-g(u_j,u_{j-1})\} + b(jh,u_j) = 0$$

$(j=1,\dots,m)$,

$u_0=\gamma_0, \quad u_{m+1}=\gamma_1$

with $h=\Delta x=1/(m+1)$ (see [7,8]). Using (2), the term in{} becomes

(4) $$\{\dots\} = \int_{u_j}^{u_{j+1}} a_-(s)\,ds + \int_{u_{j-1}}^{u_j} a_+(s)\,ds .$$

Now let u_{j-1}, u_j, u_{j+1} be three values with

(5) $a(s) \geq 0$ between u_{j-1}, u_j, u_{j+1}.

Under this condition the part (3b) of the above discretization reads

$$h^{-1}\{f(u_j)-f(u_{j-1})\} + b(jh,u_j)$$

which is a first order substitution for

(6) $$a(u)u' + b(x,u) = f(u)' + b(x,u).$$

Similarly, if

(7) $a(s) \leq 0$ between u_{j-1}, u_j, u_{j+1},

then (3b) reduces to

$$h^{-1}\{f(u_{j+1})-f(u_j)\} + b(jh,u_j).$$

Our idea is to build a discretization where the corresponding terms discretizing (6) reduce to

(8) $$h^{-1}\{f(u_j)-f(u_{j-1})\} + \frac{1}{2}\{b((j-1)h,u_{j-1}) + b(jh,u_j)\}$$

or

(9) $$h^{-1}\{f(u_{j+1})-f(u_j)\} + \frac{1}{2}\{b(jh,u_j) + b((j+1)h,u_{j+1})\}$$

under the conditions (5) or (7) respectively.

For fixed $\varepsilon \geq 0$ and $h=1/(m+1)$ we set

(10) $$T_j u = \begin{cases} u_j, \ j=0, \ m+1 \\ \varepsilon h^{-2}(-u_{j-1}+2u_j-u_{j+1}) + h^{-1}\{g(u_{j+1},u_j) - g(u_j,u_{j-1})\} \\ +\beta_j^- b_{j-1} + \beta_j^0 b_j + \beta_j^+ b_{j+1}, \ j=1,\dots,m \end{cases}$$

where g is given by (2) and $b_i=b(ih,u_i)$. The coefficients $\beta_j^{-,o,+}$ are determined as follows: Let B be the smooth function

$$(11)\qquad B(r)=\begin{cases} 0 & , \quad r<0 \\ r^2 & , \quad 0\le r\le \frac{1}{2} \\ \frac{1}{2}-(1-r)^2 & , \quad \frac{1}{2}\le r\le 1 \\ \frac{1}{2} & , \quad 1<r \end{cases}$$

connecting the values 0 and $\frac{1}{2}$. With a parameter $p\ge 0$ we let

$$(12)\qquad \beta(\rho) = B(p\rho), \quad \rho\in\mathbb{R},$$

and set

$$\begin{aligned} \beta_j^- &= \beta(a(u_{j-1})/\sqrt{h}) \\ \beta_j^+ &= \beta(-a(u_{j+1})/\sqrt{h}) \\ \beta_j^o &= 1-\beta_{j+1}^- - \beta_{j-1}^+ \\ &= 1-\beta(a(u_j)/\sqrt{h})-\beta(-a(u_j)/\sqrt{h}). \end{aligned}$$

The value $\sqrt{h}$ is introduced into the arguments of β in order that p can be chosen independently of h later. The precise form of the function B will be unimportant, and other monotone C^1-functions connecting 0 and $\frac{1}{2}$ could be taken. The parameter p in (12) rules how fast β grows from 0 to $\frac{1}{2}$. The choice of p turns out to be crucial, both from a theoretical and a numerical point of view. Notice that the choice p=0 leads to

$$\beta_j^- = \beta_j^+ = 0, \quad \beta_j^o = 1,$$

and we get the original E-O-scheme. From a consistency point of view, one wants to choose p as large as possible in order to achieve the formulas (8) or (9) under the conditions (5) or (7) in most cases. But it turns out that the operator T defined in (10) is no longer an M-function, if p is too large. Numerically it is observed that Newton's method fails to converge very often for the system

$$(13)\qquad Tu = (\gamma_o,0,0,\dots,\gamma_1)^T ,$$

if p is too large. It might happen that (13) has several solutions for p large, but this is unproved.

3. Properties of the difference scheme. Let c_o, c_1 be real numbers with

$$(14a)\qquad b(x,c_o) \le 0 \le b(x,c_1), \quad x\in[0,1],$$

$$(14b)\qquad c_o \le \gamma_i \le c_1 \quad (i=0,1).$$

Such numbers can always be found if (1b) holds. Since the constant functions c_o and c_1 are lower and upper solutions of (1a), we have

$$c_o \le u_\varepsilon \le c_1 \quad \text{for all} \quad \varepsilon>0 .$$

We consider the operator T on the set

$$I_h = \{u\in\mathbb{R}^{m+2} : c_o\le u_i\le c_1,\ i=0,1,\dots,m+1\} .$$

For the next Lemma, condition (1b) is superfluous.

Lemma 1:

Assume $|b_u(x,u)| \le M_1$, $|a'(u)b(x,u)| \le M_2$ for all $0\le x\le 1$, $c_0\le u\le c_1$. If the parameter $p\ge 0$ in the definition of T is chosen such that

(15) $\quad p\sqrt{h}\ M_1\le 1,\quad 4p^2M_2\le 1,$

then $T'(u)$ has non-positive off-diagonal elements for all $u\in I_h$ and all $\varepsilon\ge 0$.

Proof:

For $j=1,\dots,m$ we have for $\varepsilon\ge 0$

$$\frac{\partial}{\partial u_{j-1}} T_j u \le -h^{-1}a_+(u_{j-1}) + B(pa(u_{j-1})/\sqrt{h})M_1 + B'(pa(u_{j-1})/\sqrt{h})M_2 p/\sqrt{h} =: R.$$

If $a(u_{j-1})\le 0$, then $R=0$. For $a(u_{j-1})>0$ multiply by $p\sqrt{h}$ and set $r=pa(u_{j-1})/\sqrt{h}$ to obtain $p\sqrt{h}R=-r+p\sqrt{h}B(r)M_1+p^2B'(r)M_2$. Using (15) we find

$$p\sqrt{h}R \le -r+B(r)+\tfrac{1}{4}B'(r).$$

An elementary discussion of the function B defined in (11) shows that

$$-r+B(r)+\tfrac{1}{4}B'(r) \le 0 \text{ for all } r\ge 0.$$

Similarly, it follows that

$$\frac{\partial}{\partial u_{j+1}} T_j u \le 0 \ . \qquad\qquad \text{q.e.d.}$$

Let us introduce some notations. For $u,v\in\mathbb{R}^k$, $u\le v$ will mean $u_j\le v_j$, $j=1,\dots,k$. For $y,z\in\mathbb{R}^k$ let

$$[y,z] = \{\, u\in\mathbb{R}^k : y\le u\le z\}.$$

An operator $T:[y,z]\to\mathbb{R}^k$ is called off-diagonally decreasing (or off-diagonally antitone), if $u\le v$, $u_j=v_j$ implies $T_jv\le T_ju$. If T is F-differentiable on $[y,z]$, then T is off-diagonally decreasing iff all off-diagonal elements of $T'(u)$ are non-positive for all $u\in[y,z]$. T is called inverse monotone (or inverse isotone) if $Tu\le Tv$ implies $u\le v$. An off-diagonally decreasing, inverse monotone operator is called an M-function [6]. δ denotes the vector with $\delta_j=1$ for all j. y (resp. z) is called a lower (upper) solution for the equation $Tu=r$, if $Ty\le r$ ($r\le Tz$). Different variants of the following theorem are well known (compare e.g. [6]).

Theorem 1

Let $y,z\in\mathbb{R}^k$, $y\le z$. Assume $T:[y,z]\to\mathbb{R}^k$ is continuous and off-diagonally decreasing; let $Ty\le r\le Tz$. Then the equation $Tu=r$ has a solution $u^*\in[y,z]$.

Proof:

Let $L=\{u\in[y,z]:Tu\le r\}$ be the set of all lower solutions. The vector u^* defined by $u_j^*=\sup\{\lambda\in\mathbb{R}:\text{there exists } u\in L \text{ with } u_j=\lambda\}$ can be shown to solve $Tu=r$. q.e.d.

Corollary 1

Under the conditions (14a,b) and (15) the system (13) has a solution

$u^* \in I_h = [c_0\delta, c_1\delta]$.

Proof:

Using (14a,b) it is clear that $c_0\delta$ and $c_1\delta$ are lower and upper solutions of the equation, because the coefficients $\beta_j^{-,o,+}$ are ≥ 0. Then the assertion follows by Lemma 1 and Theorem 1. q.e.d.

Assume now $b_u \geq \mu > 0$. Then the continuous solutions u_ε of (1a) are unique for all $\varepsilon > 0$ [4]. We show uniqueness of the solution u^* within I_h for all $\varepsilon \geq 0$ in the discrete case. Furthermore, a discrete stability inequality holds uniformly in ε and h. We use the notation

$$\|d\|_1 = h^{-1} \sum_{j=1}^{m} |d_j| \text{ for } d \in \mathbb{R}^{m+2} \text{ with } d_0 = d_{m+1} = 0.$$

Theorem 2

Assume (1b), (14a,b) and (15), where M_1, M_2 are defined in Lemma 1. Then the operator T given in (10) is an M-function on $I_h = [c_0\delta, c_1\delta]$ and the equation (13) has a solution $u^* \in I_h$ which is unique in I_h. For all $u, v \in I_h$ with $u_0 = v_0$, $u_{m+1} = v_{m+1}$ we have the stability inequality

$$(16) \qquad \|u-v\|_1 \leq \frac{1}{\mu} \|Tu-Tv\|_1 .$$

All statements hold true for all $\varepsilon \geq 0$ and all h .

Proof:

Existence of a solution has already been established in Corollary 1, and by Lemma 1 T is off-diagonally decreasing. Computing T'(u) we find in the rows for j=1,...,m:

$$\begin{aligned} &\varepsilon h^{-2}(-1,2,-1) \\ &+h^{-1}(-a_+(u_{j-1}),(a_+-a_-)(u_j),a_-(u_{j+1})) \\ &+ \quad (\beta_j^- b_{uj-1} \;,\; \beta_j^o b_{uj} \;,\; \beta_j^+ b_{uj+1}) \\ &+\frac{1}{\sqrt{h}}(\beta_{j-1}^{'+}(a'b)_{j-1} \;,\; (-\beta_j^{'+}+\beta_j^{'-})(a'b)_j \;,\; -\beta_{j+1}^{'-}(a'b)_{j+1}) \end{aligned}$$

with the abbreviations

$$b_{ui} = b_u(ih, u_i), \quad (a'b)_i = a'(u_i)b(ih, u_i),$$
$$\beta_i^{'+} = \beta'(a(u_i)/\sqrt{h}), \quad \beta_i^{'-} = \beta'(-a(u_i)/\sqrt{h}).$$

Since $\beta_{j-1}^+ + \beta_j^o + \beta_{j+1}^- = 1$, it follows that the j-th column (not row!)-sum of T'(u) equals

$$b_u(ju, u_j) \quad (\geq \mu > 0)$$

for j=2,...,m-1. For j=1 and j=m the column sums are even larger, because the missing entries in the sum are non-positive. Using the elements 1 in T'(u), which define the boundary conditions, we find a row vector

$$e = (k_0, 1, 1, \ldots, 1, k_1) \in \mathbb{R}^{m+2}, \quad k_i > 0,$$

with

(17) $eT'(u) \geq (1,\mu,\mu,\ldots,\mu,1)$

for all $u \in I_h$. Here k_o, k_1 may depend on h and ε, but can be taken independent of $u \in I_h$.

Now let $u, v \in I_h$ be fixed. With the matrix

$$A := \int_o^1 T'(u+s(v-u))ds$$

we can write

(18) $Tv-Tu = A(v-u)$.

All off-diagonal elements of A are non-positive, and furthermore by (17)

(19) $eA \geq (1,\mu,\mu,\ldots,\mu,1)$.

Thus A is an M-matrix; from $A^{-1} \geq O$ and (18) we find that $Tu \leq Tv$ implies $u \leq v$ which proves that T is inverse monotone.

To prove the stability inequaltiy (16), multiply the estimate (19) with the nonnegative matrix A^{-1} from the right. For $j=1,\ldots,m$ we find the estimates

$$(A^{-1})_{oj} + \mu \sum_{i=1}^{m} (A^{-1})_{ij} + (A^{-1})_{m+1,j} \leq 1,$$

especially

(20) $\sum_{i=1}^{m} (A^{-1})_{ij} \leq 1/\mu$.

Now multiply (18) by A^{-1} and get

(21) $|v_i-u_i| \leq \sum_{j=o}^{m+1} (A^{-1})_{ij} |T_j v - T_j u| \quad (i=O,\ldots,m+1)$.

We are reminded of $u_o=v_o$, $u_{m+1}=v_{m+1}$, $T_o u=T_o v$, $T_{m+1}u=T_{m+1}v$; sum (21) on i, use (20) and obtain the stability estimate (16). q.e.d.

Remark: To prove that T is an M-function, we also could have used Theorem 5.2 of [5] because $T'(u)$ is an M-matrix for all u. The proof of the stability estimate can analogously be given for the continuous equation [2,3].

4. Numerical example.

$$-\varepsilon u''-uu'+u = O, \quad u(O) = -1, \quad u(1) = .5.$$

Here the limit function of the continuous solutions u_ε is known to be (see [1], example (E3), or [4])

$$U(x) = \begin{cases} -1+x, & x < .75, \\ -.5+x, & .75 < x \end{cases} .$$

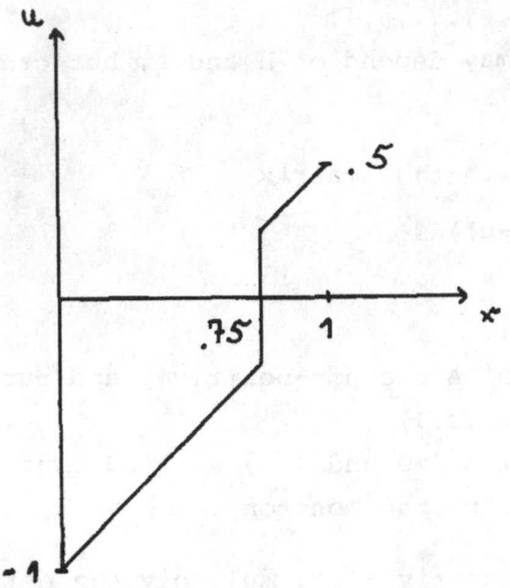

We can apply our results with

$$c_0 = -1, \quad c_1 = .5, \; \mu = M_1 = 1, \quad M_2 = 1.$$

Then condition (15) requires $0 \le p \le 1/2$, i.e. with any such p substituted in (12) we have unique solubility of the difference equations within $\{u \in \mathbb{R}^{m+2} : -1 \le u_j \le .5\}$ for any $\varepsilon \ge 0$. The following table shows the errors between U(x) and numerical solutions u* computed for h=.1, ε=0, and three values of p. p=1 is beyond our theory.

x	U-u*, p=0	U-u*, p=.5	U-u*, p=1
0	0	0	0
.1	.00499	0	0
.2	.01050	0	0
.3	.01664	0	0
.4	.02358	-.00025	0
.5	.03155	-.00418	0
.6	.04088	-.01206	0
.7	.01380	-.09264	-.06093
.8	.11796	.07943	.06093
.9	-.00990	.00777	0
1	0	0	0

h=.1, ε=0

The next table gives the errors U-u* for the same example at the same points x, but u* is computed with h=.05.

x	U-u*, p=0	U-u*, p=.5	U-u*, p=1
0	0	0	0
.1	.00256	0	0
.2	.00541	0	0
.3	.00861	0	0
.4	.01226	0	0
.5	.01651	0	0
.6	.02158	-.00054	0
.7	.02786	-.00486	0
.8	-.00013	.00486	0
.9	-.00524	.00054	0
1	0	0	0

$h=.05, \quad \varepsilon=0$

Remark: In this example, a formula (8) or (9) produces a consistency error zero for a continuous solution $U(x)=x+\text{const.}$. For that reason, the difference schemes with p>0 often give zero errors away from the interior layer. This will not be true in more complicated examples, but then again smaller errors away from layers are observed for p>0 as compared with p=0. This clearly is a hint that the higher order of consistency for p>0 also gives higher order of accuracy.

References

1. Howes, F.A.: Boundary-interior layer interactions in nonlinear singular perturbation theory. Memoirs AMS, No. 203 (1978).
2. Lorenz, J.: Zur Theorie und Numerik von Differenzenverfahren für singuläre Störungen. Habilitationsschrift, Universität Konstanz, 1980.
3. Lorenz, J.: Nonlinear singular perturbation problems and the Engquist-Osher difference scheme. University of Nijmegen, Report 8115, 1981.
4. Lorenz, J.: Nonlinear boundary value problems with turning points and properties of difference schemes. To appear in: Proc. Conf. Singuläre Störungstheorie mit Anwendungen, Oberwolfach 1981, Springer Lecture Notes.
5. Moré, J., Rheinboldt, W.: On P- and S-functions and related classes of n-dimensional nonlinear mappings. Lin. Alg. Appl. 6, 45-68 (1973).
6. Ortega, J.M., Rheinboldt, W.C.: Iterative solution of nonlinear equations in several variables. Academic Press, New York, London, 1970.
7. Osher, S.: Nonlinear singular perturbation problems and one sided difference schemes. SIAM J. Num. Anal. 18, 129-144 (1981).
8. Osher, S.: Numerical solution of singular perturbation problems and hyperbolic systems of conservation laws. In: Analytical and numerical approaches to asymptotic problems in analysis. S. Axelsson, L.S. Frank, A. van der Sluis (eds.), North-Holland, 1981, pp.179-204.
9. Stephens, A.B., Shubin, G.R.: Multiple solutions and bifurcation of finite difference approximations to some steady problems of fluid dynamics. SIAM J. Sci. Stat. Comput. 2, 404-415 (1981).

STABILITY IN HILBERT SPACES BY USING THE RICCATI EQUATION

J.-Cl. LOUIS
Department of Mathematics
Facultés Universitaires N.-D. de la Paix
B-5000 NAMUR BELGIUM

ABSTRACT

The regulator problem and the operatorial Riccati equation are discussed in Hilbert spaces. The results are then applied in the study of Liapunov stability for some non linear evolution equations.
In particular, we discuss a wave equation with friction coupled with an ordinary differential equation by using some energy functions, which are obtained via the Riccati equation.

1. INTRODUCTION

Let X and U be complex Hilbert spaces with inner product $<\cdot,\cdot>$ and norm $|\cdot|$. Let B be an element of the space $L(U, X)$ of bounded linear operators from U to X and A be the generator of a linear C_0-semigroup S on X with domain $D(A)$. Consider the Cauchy problem on $\mathbb{R}^+$

$$(1.1) \qquad \frac{dx}{dt}(\cdot) = A\,x(\cdot) + B\,u(\cdot) \quad , \quad x(0) = a \quad ,$$

where $a \in X$ and $u(\cdot) \in L^2(\mathbb{R}^+, U)$. The (mild) solution of (1.1) is

$$x(t) = S(t)a + \int_0^t S(t-s)\,Bu(s)\,ds \ , \quad t \geq 0 \ .$$

Let F be a continuous Hermitian form on $X \times U$,

$$F(x, u) = <F_1 x, x>_X + 2\mathrm{Re} < F_2 x, u >_U + < F_3 u, u >_U \ ,$$

where $F_1 \in L(X)$, $F_3 \in L(U)$ are selfadjoint and $F_2 \in L(X, U)$. Associated to F is the continuous Hermitian form J^+ defined on $\mathcal{H}^+ = L^2(\mathbb{R}^+, X) \times L^2(\mathbb{R}^+, U)$ by

$$J^+(y(\cdot), v(\cdot)) = \int_0^{+\infty} F(y(t), v(t))\,dt .$$

Finally let

$$M_a^+ \underset{\text{def}}{=} \{(x(\cdot), u(\cdot)) \in \mathcal{H}^+ : x(\cdot) \text{ is the solution of } (1.1)\}$$

We are first interested in the optimal control problem :
"for each $a \in X$, minimize the cost function J^+ on M_a^+" .

The weakest assumption for this problem to make sense is

L^2-controllability of (A, B) wich means that $M_a^+ \neq \emptyset$ for each $a \in X$

Alternative natural conditions for our problem to make sense are exponential stability of S (which means the existence of $\alpha < 0$ and $M \geq 1$ such that $|S(t)| \leq Me^{\alpha t}$, for all $t \geq 0$) or more generally, exponential stabilizability of (A, B). All the above conditions are special cases of the L^2-controllability of (A, B) which we assume in the sequel.

As well-known, in finite dimension, a fundamental contribution to the above problem has been given by R.E. Kalman and for the related stability problems by V.M. Popov.

In recent years, an increasing interest has been taken in the infinite dimensional setting under the assumption $F \geq 0$, see Curtain and Pritchard [1], Balakrisnan [2], Lions [3].

However little seems to be known in the infinite-dimensional case when the cost function is not sign definite. We would like to discuss this problem here. Our motivations lie in possible applications to systems requiring undbounded operators A (which are related to partial differential equations) and to some associated stability problems wich require the consideration of forms F without sign condition.

We extend to C_0-semigroups some results established earlier by Willems [6] and Molinari [5]. Our proofs, which will appear elsewhere, are based on some dynamic programming technics. We are also inspired by Yakubovich's approach with bounded Hilbert space operators [4].

2. STATEMENT OF MAIN RESULTS

2.1. The regulator problem

Theorem 1. Assume that : the pair (A, B) is L^2-controllable and there exists $\delta > 0$ such that

$$(2.1.1)\quad F(x, u) \geq \delta\,(|x|^2 + |u|^2), \quad \text{for all } (\omega, x, u) \in \mathbb{R} \times D(A) \times U \text{ with } i\,\omega x = Ax + Bu$$

Then :

(i) for each $a \in X$, there exists one and only one point $(x^+(\cdot, a), u^+(\cdot, a))$ minimizing J^+ on M_a^+ ;

(ii) the operator $a \to (x^+(\cdot, a), u^+(\cdot, a))$ from X to $\mathcal{H}^+$ is linear and continuous, so that the optimal cost

$$V^+(a) \underset{\text{def}}{=} J^+(x^+(\cdot, a), u^+(\cdot, a))$$

is a continuous Hermitian form on X ;

(iii) the selfadjoint $H^+ \in L(X)$ of the form V^+ (defined by $\langle H^+ a, a\rangle = V^+(a)$ for all $a \in X$) and $h^+ \underset{\text{def}}{=} -F_3^{-1}(B^* H^+ + F_2)$ satisfy :

$$(2.1.2.)\quad 2\mathcal{R}e \langle Ax + Bu, H^+ x\rangle_X + F(x, u) = |F_3^{1/2}(u - h^+ x)|_U^2, \ \forall\ (x, u) \in D(A) \times U$$

The C_0-semigroup S^+ generated by the operator $A^+ = A + B h^+$ is exponentially stable.

Moreover, $H \leqslant H^+$ for any selfadjoint solution $H \in L(X)$ of (2.1.2) .

Remarks. 1. Relation (2.1.2) is nothing else but the Algebraic (operator) Riccati equation (for short O.R.E.) written in terms of forms (see Willems [6] for the finite-dimensional case).

2. When S is exponentially stable, the above frequency domain Condition (2.1.1) may be written in its usual form : there exists $\hat{\delta} > 0$ such that for all $u \in U$ and $\omega \in \mathbb{R}$ we have

(2.1.3) $F((i\,\omega\, I_X - A)^{-1}\, Bu,\, u) \geqslant \hat{\delta}\, |u|^2$

where I_X is the identity on X .

3. An application of this results to stability may be found in [7] .

2.2. The controllable regulator

Additional results may be obtained when the pair (A, B) is exactly controllable on some interval [0, T] , T > 0, (with L^2-controls) which means that for each $(a, a') \in X^2$, there exists $u(\cdot) \in L^2([0, T], U)$ such that

$$S(T)\, a + \int_0^T s(T - s)\, B\, u(s)\, ds = a'$$

This condition is basic in finite-dimensional control theory but rather restrictive in infinite dimension, for it never applies for important classes of évolution equations, see Triggiani [8, 9] .
However it is wothwile to discuss its use in our setting too, since it holds for certain significant systems as, for instance, some controlled wave equations (see [1, chapter 2]),
So, as infinite dimension, under the controllability of (A, B) it is natural to consider also the minimization problem on $\mathbb{R}^-$. But first we have to introduce some definitions and notations.
Given an interval $I \subset \mathbb{R}$ and $u(\cdot) \in L^2(I, U)$ we say that a function $x(\cdot)$ from I to X is a (mild) solution of (1.1) on I if

$$x(t) = S(t - t_0)\, x(t_0) + \int_{t_0}^t S(t - s)\, Bu(s)\, ds, \ \forall\ t_0, t \in I, t_0 > t .$$

Consider the Hilbert space $\mathcal{H}^- \underset{def}{=} L^2(\mathbb{R}^-, X) \times L^2(\mathbb{R}^-, U)$ and the continuous Hermitian form J^- defined on $\mathcal{H}^-$ by

$$J^-(y(\cdot), v(\cdot)) = \int_{-\infty}^{0} F(y(t), v(t))dt$$

For each $a \in X$, we denote by M_a^- the set of couples $(x(\cdot), u(\cdot)) \in \mathcal{H}^-$ such that $x(\cdot)$ is the solution of (1.1) on $\mathbb{R}^-$ with control $u(\cdot)$ and terminal condition $x(0) = a$.

Clearly exact controllability of (A, B) implies that M_a^+ and M_a^- are not void. Indeed, if A generates a Co-group on X (which arises for some controlled wave equations) we may reverse the time in (1.1) and apply theorem 1 to obtain results which are much similar to those known in finite dimension. Let me mention the following :

Theorem 2. Assume that : A generates a Co-group on X ; the pair (A, B) is exactly controllable; and the coercive frequency condition (2.1.1) holds.
Then :

(i) for each $a \in X$, there exists one and only one point $(x^-(\cdot, a), u^-(\cdot, a))$ minimizing J^- on $\mathcal{M}_{\bar{a}}$;

(ii) the operator $a \to (x^-(\cdot, a), u^-(\cdot, a))$ from X to $\mathcal{H}^-$ is linear and continuous, so that the optimal cost

$$v^-(a) \underset{def}{=} -J^-(x^-(\cdot, a), u^-(\cdot, a))$$

is a continuous Hermitian form on X and its selfadjoint operator $H^- \in L(X)$ is the minimal solution of the ORE (2.1.2).

(iii) the selfadjoint operator $H^+ - H^-$ is coercive.

3. APPLICATION

We discuss the Liapunov type stability for the zero solution of the following system

$$(3.1) \qquad \begin{aligned} x_{tt} &= x_{\xi\xi} - \alpha\, x_t - a(\xi)\, \psi(\sigma) \qquad\qquad t > 0, \quad \xi \in \,]0, 1[\\ \sigma_t &= b(\xi)\, x_t - \rho\, \psi(\sigma) \end{aligned}$$

with

$$x(0, t) = x(1; t) = 0$$
$$x(\xi, 0) = x_o(\xi), \; x_t(\xi, 0) = y_o(\xi) \;, \sigma(\xi, 0) = \sigma_o(\xi) \;,$$

where $\psi \in C^1(\mathbb{R})$ has bounded derivative and $r\,\psi(r) \geq \beta\, r^2$ for some $\beta > 0$ and all $r \in \mathbb{R}$; $a(\cdot)$ and $b(\cdot)$ are continuous and $a(\xi) \geq a_o > 0$, $b(\xi) \geq 0$ on $[0, 1]$; $\alpha > 0$ and $\rho > 0$ are constants.

Consider the following abstract version of the above system

$$(3.2) \qquad \begin{cases} \dfrac{d}{dt}\binom{x}{y} = A\binom{x}{y} + B\,\psi(\sigma) \\[2ex] \dfrac{d\sigma}{dt} = b(\xi) y - \rho\, \psi(\sigma) \end{cases}$$

where $A = \begin{bmatrix} 0 & I \\ \dfrac{\partial^2}{\partial \xi^2} & -\alpha I \end{bmatrix}$ generates an exponentially stable C_o-group on $X \equiv H_o^1(0, 1) \times L^2(0, 1)$

$B = \begin{bmatrix} 0 \\ -a(\xi)\, I \end{bmatrix}$ from $U \equiv L^2(0, 1)$ to X

Since ψ is Lipschitz on $\mathbb{R}$, the Cauchy problem for (3.2) possesses a unique solution on $\mathbb{R}^+$ for each initial data.

It is easy to see that the pair (A, B) is exactly controllable.

Then, we look for a selfadjoint $H \in L(X)$ such that the function

$$W : X \times U \to \mathbb{R}, \; W(x, y, \sigma) = \left\langle -H\binom{x}{y}, \binom{x}{y} \right\rangle_X + \int_0^1 a(\xi)\, \Phi(\sigma(\xi))\, d\xi$$

where $\Phi(r) = \int_0^r \psi(s)\, ds$, is a Liapunov function of (3.2)

The derivative of W along the differentiable solutions of (3.2) is

$$\dot{W}(x, y, \sigma) = -(2 < A\binom{x}{y} + Bu, H\binom{x}{y} >_X + F(x, y, u)) ,$$

where $u = \psi \circ \sigma$ and F is a quadratic form defined on $X \times L^2$ by

$$F(x, y, u) = -\int_0^1 a(\xi)\, b(\xi)\, y(\xi)\, u(\xi)\, d\xi + \rho \int_0^1 a(\xi)\, u^2(\xi)\, d\xi \ .$$

Take now the complexifications of the above spaces and operators (We denote them by the superscript c) and look for an $\eta > 0$ and for a selfadjoint $H \in L(X^c)$ satisfying the following O.R.E. :

$$(3.3) \qquad 2\,\mathrm{Re} < A^c\binom{x}{y} + B^c u, H\binom{x}{y} >_{X^c} + F^c(x, y, u) - \eta(|\binom{x}{y}|^2_{X^c} + |u|^2_{U^c}) =$$

$$= |F_3^{1/2} (u - h\binom{x}{y})|^2_{U^c} \ , \quad \forall\, (x, y, u) \in D(A^c) \times U^c \ ,$$

where $F^c(x, y, u) = -\mathrm{Re} < aby, u >_{U^c} + \rho\, |a^{1/2} u|^2_{U^c}$ (the complexification of F).

To check the frequency condition, we note that system

$$i\,\omega\binom{x}{y} = A^c\binom{x}{y} + B^c u$$

is written here as

$$(3.4) \qquad \begin{cases} i\,\omega x = y \\ i\,\omega y = x_{\xi\xi} - \alpha y - a(\xi) u \end{cases}$$

By a simple calculation we see that : $-\mathrm{Re} < aby, u >_{U^c} = \alpha\omega^2 |b^{1/2} x|^2_{L^2}$, for all $(\omega, x, y, u) \in \mathbb{R} \times D(A^c) \times U^c$ satisfying (3.4) .

Moreover, since the C_0-semigroup generated by A is exponentially stable, (3.4) is equivalent to

$$\binom{x}{y} = (i\,\omega I^c - A^c)^{-1} B^c u \ .$$

It follows that for $\eta > 0$ sufficiently small the following coercive frequency condition holds :

$$F^c(x, y, u) - \eta(|{}^x_y|^2_X + |u|^2_X) \geq \delta(|\binom{x}{y}|^2_X + |u|^2_U) \ ,$$

for some $\delta > 0$ and all $(\omega, x, y, u) \in \mathbb{R} \times D(A^c) \times U^c$ satisfying (3.4).
It follows then by theorem 1 that the O.R.E. (3.3) possesses a selfadjoint solution

$H \in L(X^c)$. Moreover, since the Co-semigroup generated by A is exponentially stable, the maximal solution H^+ of this O.R.E. is $\leqslant 0$.

Now since (A, B) is exactly controllable, we may apply also theorem 2 to see that the above O.R.E. admits a minimal solution H^- and $\Delta = H^+ - H^-$ is coercive hence so is $-H^-$

So we have found an $\eta > 0$ and an operator $-H^- \in L(X^c)$ such that the derivative of the Liapunov function W satisfies

$$\dot{W}(x, y, \sigma) \leqslant -\eta(|\binom{x}{y}|^2_X + \int_0^1 \psi(\sigma(\xi))^2 \, d\xi) .$$

Now since $\psi(r)^2 \geqslant \beta^2 r^2$ for all $r \in \mathbb{R}$, we may find $\eta_1 > 0$ such that

$$\dot{w}(x, y, \sigma) \leqslant -\eta_1 |(x, y, \sigma)|^2_{X \times L^2} \tag{3.5}$$

The conditions on ψ imply also that for some $\beta_1, \beta_2 > 0$ we have

$$\beta_1 r^2 \leqslant \Phi(r) \leqslant \beta_2 r^2 \quad \text{for all } r \in \mathbb{R} ;$$

hence for some $\delta_1, \delta_2 > 0$ we have

$$\delta_1 |(x, y, \sigma)|^2_{X \times L^2} \leqslant W(x, y, \sigma) \leqslant \delta_2 |(x, y, \sigma)|^2_{X \times L^2}$$

$$\text{for all } (x, y, \sigma) \in X \times L^2$$

The above combined with (3.5) implies that W is a Liapunov function for the system (3.2) insuring uniform asymptotic stability in the large.

REFERENCES

[1] CURTAIN, R.F., PRITCHARD, A.J., "Infinite Dimensional Linear Systems Theory". Springer-Verlag, 1978.

[2] BALAKRISHNAN, A.V., "Applied Functional Analysis", Springer-Verlag, 1976.

[3] LIONS, J.L., "Optimal Control of Systems Governed by Partial Differential Equations, "Springer-Verlag, 1971.

[4] YAKUBOVICH, V.A., "A frequency theorem for the case in which the state and control spaces are Hilbert spaces with an application to some problems in the synthesis of optimal controls II", Siberian Math J. 16(1975). pp. 828-845.

[5] MOLINARI, B.P., "The time-invariant linear quadratic optimal control problem", Automatica 13(1977), pp. 347-357.

[6] WILLEMS, J.C., Least squares stationary optimal Control and the algebraic Riccati equation, IEEE Trans Autom. Control AC-16, n° 6, 1971, pp. 621-634.

[7] WEXLER, D., "On frequency domain stability for evolution equations in Hilbert spaces, via the algebraic Riccati equation", SIAM J. Math. Anal. 11 (1980), pp. 969-983.

[8] TRIGGIANI, R., On the lack of exact controllability for mild solution in Banach spaces, J. Math. Anal. Appl. 50 (1975), pp. 438-446.

[9] TRIGGIANI, R., A note on the lack of exact controllability for mild solutions in Banach spaces, SIAM J. Control and Optimization 15 (1977), pp. 407-411.

STABILITY ANALYSIS OF ABSTRACT HYPERBOLIC EQUATIONS USING FAMILIES OF LIAPUNOV FUNCTIONS

P. MARCATI*

Dipartimento di Matematica

Università di Trento

I-38050 POVO(TN) - ITALY

1. Introduction

In this paper we shall investigate the asymptotic behavior of the following nonlinear second order evolution equation

$$\frac{d}{dt}\left[A\left(\frac{du}{dt}\right)\right] + B\left(u, \frac{du}{dt}\right) + C(u) = 0 \tag{1.1}$$

with initial conditions

$$u(0) = \phi \qquad \frac{du}{dt}(0) = \psi \tag{1.2}$$

where A and C are nonlinear "potential" operators and B is "dissipative" in a suitable way. More precise hypotheses will be given later.
This kind of problems have been widely studied in the mathematical literature often using the Liapunov stability theory for infinite dimensional dynamical systems. The method introduced by La Salle [18] for autonomous systems of ordinary differential equations, has been extended in several directions and has found a wide variety of applications. Among the others, we mention the papers of Hale [13], Slemrod and Infante [29], Dafermos [6] [7], the book of Haraux [15] and its references. Related to this approach, when the resulting dynamical systems are constraction semigroups in Hilbert space, the methods of Dafermos-Slemrod [9] can be applied to an interesting class of examples like the vibrations of an elastic membrane with nonlinear damping terms or a viscous boundary support, linear thermoelasticity and linear viscoelasticity (see for instance Dafermos [8]).
Unfortunately these powerful techniques require the precompactness for the bounded

* This research was partially supported by CNR - GNAFA

orbits (which is not known "a priori" as in the finite dimensional case), giving possibly extra unnecessary conditions.
A different point of view was suggested for Lagrangian systems by Matrasov [22], provided the existence of an auxiliary Liapunov function acting on the set S where the Hamiltonian is constant along the orbits. The papers of Salvadori [27], [28] developed a constructive approach to the Matrasov theory using families of Liapunov functions. A successive modification of these results has been made by D'Onofrio [10] to study the asymptotic stability of the null solution and the almost periodic solutions of the dissipative wave equation.

2. Abstract stability results

Suppose that for the abstract equation (1.1) the following hypotheses are all valid.

(I) There exist three Banach spaces $P \subset D \subset K$ with dense continuous embeddings. (We shall denote by $|\ |_P$, $|\ |_D$, $|\ |_K$, the respective norms and by $<,>_P$, $<,>_D$, $<,>_K$, the duality forms between these spaces and their real conjugate spaces P^*, D^* and K^*)

(II) There exists a subset F of K such that $0 \in F$ and $A : F \to K^*$ is a continuous map which takes bounded sets of K into bounded sets of K^*. Moreover there exists a functional $f : K \to \mathbb{R}$ such that for all $w \in F$

$$Aw = \text{grad } f(w), \qquad A0 = 0$$

Let $a_0 \in C(\mathbb{R}_+, \mathbb{R}_+)$ be an increasing function such that $a_0(0) = 0$ and $a \in C([0,1] \times \mathbb{R}_+; \mathbb{R}_+)$ a continuous function such that for all $w \in F$, $\theta \in [0,1]$ $r \in \mathbb{R}_+$

$$f(w) \geq a_0(|w|_K) \tag{2.1}$$

$$< A(w) - A(\theta w), w > \geq a(\theta, |w|_K) , \quad \int_0^1 a(\theta,r)\,d\theta \geq a_0(r)$$

(III) Let $E \subset P$ and $0 \in E$, assume that $C : E \to P^*$ is a continuous operator and there exists a continuous functional $g : P \to \mathbb{R}$ having the following properties

$$\begin{aligned} &g(u) \geq c_0(|u|_E) \\ &C(u) = \text{grad } g(u) , \qquad C(0) = 0 \\ &< u, C(u) > \geq c_0(|u|_E) \end{aligned} \tag{2.2}$$

where $c_o : \mathbb{R}_+ \to \mathbb{R}_+$ is a continuous strictly increasing function such that $c_o(0) = 0$.

(IV) Let $E_1 \subset D \cap F$, $0 \in E_1$, and $B : E \times E_1 \to D^*$ a continuous function and there exist $\lambda \geq 2$, $b_o, b_1 \geq 0$ such that

$$\begin{aligned} &< v, B(u,v)> \geq b_o |v|_D^{\lambda} \\ &|B(u,v)|_{D^*} \leq b_1 \{|v|_D + |v|_D^{\lambda-1}\} \end{aligned} \tag{2.3}$$

provided that $(u,v) \in E \times E_1$.

(V) For any initial data $(\phi,\psi) \in E \times F$ and for any $T > 0$ there exists a unique solution

$$u \in C[0,T;\ P] \cap C^1[0,T;\ K] \cap H^{1,\lambda}[0,T;\ D] \tag{2.4}$$

and

$$\frac{d}{dt}[A(\frac{du}{dt})] \in L^{\lambda'}_{loc}(\mathbb{R}_+;\ P^*),\ \lambda' = \frac{\lambda-1}{\lambda} \tag{2.5}$$

such that the equation (1.1) is verified for almost all $t \in [0,T]$ and one has the continuous dependence upon the initial data in the above topologies.

<u>Proposition</u> (2.1) <u>Assume that I)...IV) hold. Then the null solution</u> $(u,u') \equiv (0,0)$ <u>is Liapunov stable in the norm of</u> $P \times K$. <u>If in addition</u> $b_o > 0$ <u>then it is asymptotically stable.</u>

To prove this proposition we shall make use of the following lemma in Liapunov stability theory due to Salvadori [27] [28] and D'Onofrio [10].

<u>Lemma (2.2)</u> <u>Let us denote by</u> (X,d) <u>a complete normed space. Given</u> $R > 0$ <u>if for all</u> $\mu \in (0,R)$ <u>there exists a function</u>

$$W_\mu : \mathbb{R}_+ \times X \to \mathbb{R} \tag{2.6}$$

<u>having the following properties</u>

(i) <u>there exists a map</u> $h : \mathbb{R}_+ \to \mathbb{R}_+$, $h(t) = O(t)$ <u>as</u> $t \to +\infty$ <u>such that</u>

$$W_\mu(t,x) + h(t) \geq 0 \qquad (t,x) \in \mathbb{R}_+ \times X \tag{2.7}$$

(ii) <u>For all</u> $x_o \in X$, $\mu \leq d(x_o,0) \leq R$, $t_o \in \mathbb{R}_+$

$$\hat{W}_\mu(t_o,x_o) \leq -c(\mu) < 0 \tag{2.8}$$

<u>Then if the null solution is stable it is asymptotically stable.</u>

Sketch of the proof of prop. (2.1)

To apply this lemma to the above proposition we consider the following Liapunov function

$$V(\phi,\psi) = < \psi, A\psi > - f(\psi) + g(\phi). \tag{2.9}$$

Since along the solution of (1.1) one has

$$\dot{V}(\phi,\psi) = - < \psi, B(\phi,\psi) > \leqq 0$$

we get easily the Liapunov stability.

Then we put

$$F_\varepsilon(t,\phi,\psi) = V(\phi,\psi) + \varepsilon[< \psi, A\psi > + \int_0^t < u(s), B(u(s),u'(s)) > ds]$$

where $(u(s),u'(s))$ is the solution having initial datum (ϕ,ψ).

Let $(\xi,\eta) \in E \times F$, the derivative along the trajectories of F_ε is given by

$$\begin{aligned} \dot{F}_\varepsilon(t,\xi,\eta) &= - < \eta, B(\xi,\eta) > + \varepsilon [< \eta, A\eta > - < \xi, C(\xi) >] \\ &\leqq - b_o|\xi|^\lambda + \varepsilon[l(R)|\eta| - c_o(|\xi|)] \end{aligned} \tag{2.10}$$

where $l(R) = \sup\{|A\eta| : |\eta| \leqq R\}$

If $\mu \leqq \sup(|\xi|,|\eta|) \leqq R$ one has after some calculations (see [21]) that there exists $\varepsilon(\mu) > 0$ such that

$$C(\mu) = \sup\{- bx^\lambda + \varepsilon(\mu)[l(R)y - c_0(x)] : \mu \leqq \sup(x,y) \leqq R\} < 0$$

Therefore if we set

$$W_\mu(t,\phi,\psi) = F_{\varepsilon(\mu)}(t,\phi,\psi)$$

the above lemma is fulfilled. #

The next proposition is concerned with exponential decay of the solutions to (1.1)

<u>Proposition</u> (2.3) <u>Let us assume the above hypotheses I) - V) and moreover suppose that the following requirements are fulfilled</u>

(VI) <u>there exists</u> $\gamma_0 > 0$ <u>such that for all</u> $(\phi,\psi) \in E \times F$ <u>one has</u>

$$< \phi, C(\phi) > \geqq g(\phi) \geqq \gamma_0|\phi|_P^2$$

(VII) <u>there exist</u> $\gamma_1, \gamma_2 > 0$ <u>such that</u>

$$< \psi, B(\phi,\psi) > \geq \gamma_1 < \psi, A(\psi) >$$

$$|B(\phi,\psi)|^2_{D^*} \leq \gamma_2 < \psi, A(\psi) >$$

(VIII) A <u>is a linear continuous operator from</u> K <u>to</u> K^* <u>and there exists</u> $\gamma_3 > 0$ <u>such that</u> $f(\psi) \geq \gamma_3 |\psi|^2_K$ <u>for all</u> $\psi \in K$.

<u>Therefore there exists</u> $\alpha > 0$ <u>such that we obtain</u>

$$V(u(t), u'(t)) = O(e^{-\alpha t}) \quad \underline{\text{as}} \quad t \to +\infty$$

For the proof we refer to [25]

3. EXAMPLES

(A) (*Semilinear Wave Equation*). This kind of problem is studied in Lions-Strauss [19], Ball [1], Ball-Slemrod [3],[4], Marcati [20] Haraux [14],[15] and many others. Let Ω a bounded open domain in $\mathbb{R}^N$ and $L[u]$ a uniformly strongly elliptic operator of order $2m$ in divergence form with $C_0^\infty(\Omega)$ coefficients. We shall consider the Cauchy problem associated to the following differential equation

$$y_{tt} + L[y] = F(x,y,y_t) \qquad x \in \Omega, \quad t \geq 0 \tag{3.1}$$

where we assume for all $(x,u) \in \Omega \times \mathbb{R}$ that $uF(x,u,0) \leq 0$ and if $N \geq 2m$, $|F(x,u,0| \leq$ $\leq d_0(|u| + |u|^{q-1})$ where $2 \leq q$ if $N = 2m$ and $2 \leq q \leq 2N/(N-2m)$ if $N > 2m$.

Moreover we impose the following dissipativness conditions

$$-b_1(|v|^\lambda + |v|^2) \leq v(F(x,u,v) - F(x,u,0)) \leq -b_0|v|^\lambda$$

for all $(x,u,v) \in \Omega \times \mathbb{R} \times \mathbb{R}$; $b_0, b_1 > 0$ and on λ we have assumptions similar to the above q .

(B) (*Strongly Dissipative Wave Equation*). For references on this type of equations see Nakao [23], [24], Webb [31], Ebihara [11] Caughey-Ellison [5] and Narazaki [25]. The equation is of the following form

$$y_{tt} + Ay + By_t + F_1(x,y\ldots D^\alpha y\ldots) + F_2(x,y_t\ldots D^\alpha y_t\ldots) = 0$$

where A,B are strongly uniformly elliptic operator in divergence form and F_1,F_2

are nonlinear operators of the form

$$F_i(x,u\ldots u_\alpha\ldots) = \sum_{|\alpha|\leq a_i} (-1)^{|\alpha|} D^\alpha C_\alpha(x,u_\alpha)$$

and smooth C_α's. For details see [21].

Remark (3.1) Some other different examples are possible in particular the quasilinear equation describing the transverse motion of an extensible beam (see Ball [2] and Fitzgibbon [12]). Some applications are possible also to quasilinear damped wave equation (see Nishida [26] and Yamada [32]) when smooth solutions exist globally in time. In this latter case the stability is given in terms of energy norm (see [21]). We think in the future to apply this method to equations of the type investigated by Larkin [17] and to almost periodic solutions of dissipative wave equation (see Haraux [16]).

REFERENCES

[1] BALL J. On the asymptotic behavior of generalized processes, with applications to nonlinear evolution equations. J. of Diff. Eq. 27 (1978), 224-265.

[2] BALL J. Stability theory for an Extensible Beam. J. of Diff. Eq. 14 (1973), 399-418.

[3] BALL J. - SLEMROD M. Nonharmonic Fourier Series. Comm. Pure and Appl. Math. 32 (1979), 555-587.

[4] BALL J. - SLEMROD M. Feedback stabilization of distributed semilinear control systems. Appl. Math. and Optimization 5 (1979), 169-179.

[5] CAUGHEY J.K. - ELLISON J. Existence uniqueness and stability of solutions of a class of nonlinear partial differential equations. J. Math. Anal. and Appl. 51 (1975), 1-32.

[6] DAFERMOS C.M. An invariance principle for compact process. J. of Diff. Eq. 9 (1971), 239-252.

[7] DAFERMOS C.M. Uniform processes and semicontinuous Liapunov functionals. J. of Diff. Eq. 11 (1973), 401-415.

[8] DAFERMOS C.M. Contraction semigroups and trend to equilibrium in continuum mechanics. Lecture Notes in Math. 503 Springer-Verlag, Berlin-Heidelberg-New York 1976.

[9] DAFERMOS C.M. - SLEMROD M. Asymptotic behavior of nonlinear contraction semigroups. J. Funct. Anal. 13 (1973), 97-106.

[10] D'ONOFRIO B.M. The stability problem for some nonlinear evolution equations. Boll. UMI 17-B (1980), 425-439.

[11] EBIHARA Y. On some nonlinear evolution equations with strong dissipation. J. of Diff. Eq. 30 (1978), 149-164.

[12] FITZGIBBON W.E. Strongly damped quasilinear evolution equations. J. Math. Anal. and Appl. 79 (1981), 536-550.

[13] HALE J.K. Dynamical systems and stability. J. Math. Anal. and Appl. 26 (1969), 39-50.

[14] HARAUX A. Comportment a l'infini pour certains systemes dissipatifs nonlineaires. Proc. Roy. Soc. Ed. 84 A (1979), 213-234.

[15] HARAUX A. Nonlinear Evolution Equations. Global Behavior of solutions. Lecture Notes in Math. 841 Springer-Verlag, Berlin-Heidelberg-New York 1981.

[16] HARAUX A. Dissipativity in the sense od Levinson for a class of second order, nonlinear evolution equation, Preprint 1982.

[17] LARKIN N.A. On some classes of nonlinear hyperbolic equations. This volume.

[18] LA SALLE J. On the stability of dynamical systems. SIAM Reg. Conf. Series in Appl. Math. 25, 1976.

[19] LIONS J.L. - STRAUSS W. Some nonlinear evolution equations. Bull. Soc. Math. France 93 (1965), 43-96.

[20] MARCATI P. Decay and stability for nonlinear hyperbolic equations. J. of Diff. Eq. 1983 (to appear).

[21] MARCATI P. Stability for second order abstract evolution equations (submitted).

[22] MATRASOV V.M. On the stability of motion PMM (J. Appl. Math. and Mech.) 26 (1962), 1337 - 1353.

[23] NAKAO M. Decay of solutions of some nonlinear wave equations in one space dimension. Funcialas Ekvacioj 29 (1977), 223-236.

[24] NAKAO M. Decay of solutions of some nonlinear evolution equations. J. Math. Anal. and Appl. 60 (1977), 542-549.

[25] NARAZAKI T. Existence and decay of classical solutions of some nonlinear evolution equations with strong dissipation. Proc. Tokai Univ. 25 (1979), 45-62.

[26] NISHIDA T. Publications Math. d'Orsay 1978.

[27] SALVADORI L. Famiglie ad un parametro di funzioni di Liapunov nello studio della stabilità. Symp. Math. IV Academic Press 1971.

[28] SALVADORI L. Sulla stabilità del movimento. Le Matematiche Catania 24 (1969), 218-239.

[29] SLEMROD M. - INFANTE E. An invariance principle for dynamical systems on Banach space. Instability of Continuous Systems (H. Leipholz, Ed.) pp 215-221. Springer-Verlag, Berlin-Heidelberg-New York 1971.

[30] WEBB G.F. A bifurcation problem for a nonlinear hyperbolic partial differential equation. SIAM J. Math. Anal. 10 (1979), 922-932.

[31] WEBB G.F. Estimates and asymptotic behavior for a strongly damped nonlinear wave equation. Canad. J. Math. 22 (1980), 631-643.

[32] YAMADA Y. Quasilinear wave equations and related nonlinear evolution equations. Nagoya Math. J. 84 (1981), 31-83.

ON CODIMENSION THREE BIFURCATIONS OF A FAMILY OF THREE-DIMENSIONAL VECTOR FIELDS

Milan Medveď

Mathematical Institute of the Slovak Academy of Sciences, 841 02 Bratislava, Czechoslovakia

Consider the vector field

$$\dot{x} = X_0(x) = Ax + G(x)\,, \tag{1}$$

where $x = (x_1, x_2, x_3)$, the matrix A is equivalent to the Jordan block with 1 above the diagonal and zeros elsewhere, G is smooth, $G(0) = 0$.

Using the method of Takens [10], it is possible to derive the following normal form

$$\begin{aligned} \dot{x}_1 &= f_1(x) = x_2 + a_{200}x_1^2 + a_{300}x_1^3 + o(\|x\|^3)\,, \\ \dot{x}_2 &= f_2(x) = x_3 + b_{200}x_1^2 + b_{110}x_1x_2 + b_{300}x_1^3 + o(\|x\|^3)\,, \\ \dot{x}_3 &= c_{200}x_1^2 + c_{020}x_2^2 + c_{110}x_1x_2 + c_{011}x_1x_3 + c_{300}x_1^3 + c_{030}x_2^3 + \\ &\quad + c_{210}x_1^2x_2 + c_{120}x_1x_2^2 + c_{201}x_1^2x_3 + c_{102}x_1x_3^2 + o(\|x\|^3)\,. \end{aligned} \tag{2}$$

We consider vector fields of the form (1) possessing symmetry under change of sign, $X_0(x) = -X_0(-x)$. These vector fields have the normal form (2), which does not contain even order terms. The unfoldings of the vector field (1) without any symmetry have been studied in [8], [9].

If $y_1 = x_1$, $y_2 = f_1(x)$, $y_3 = x_3$, then we obtain a vector field of the form (2), where the first equation does not contain nonlinear terms. Let (2) has already this property. Putting $y_1 = x_1$, $y_2 = x_2$, $y_3 = f_2(x)$, we obtain the vector field

$$\begin{aligned} &\dot{y}_1 = y_2,\ \dot{y}_2 = y_3, \\ &\dot{y}_3 = Ay_1^3 + a'y_2^3 + b'y_1^2y_2 + c'y_1y_2^2 + d'y_1^2y_3 + e'y_1y_3^2 + H(y)\,, \end{aligned} \tag{3}$$

where $H(y) = -H(-y)$, $H(y) = o(\|y\|^3)$. Using the Weierstrass preparation theorem (see [7]) similarly as in [1], it is possible to transform (3) into the same form, where $H(y_1, 0, 0) \equiv 0$. If $A \neq 0$, then introducing the change of coordinates $y \to (|A|)^{-1/2}y$ we obtain the vector field

$$(4) \qquad \begin{aligned} &\dot{y}_1 = y_2,\ \dot{y}_2 = y_3, \\ &\dot{y}_3 = \sigma y_1^3 + a y_2^3 + b y_1^2 y_2 + c y_1 y_2^2 + d y_1^2 y_3 + e y_1 y_3^2 + R(y), \end{aligned}$$

where $R(y) = -R(-y)$, $R(y) = o(\|y\|^3)$, $R(y_1, 0, 0) \equiv 0$, $\sigma = \operatorname{sign} A$.

Let us consider the following symmetric unfolding of the vector field (4):

$$v_\mu^\sigma : \quad \begin{aligned} &\dot{y}_1 = y_2,\ \dot{y}_2 = y_3, \\ &\dot{y}_3 = \sigma y_1(\mu_1 + y_1^2) + \mu_2 y_2 + \mu_3 y_3 + a y_2^3 + b y_1^2 y_2 + c y_1 y_2^2 + \\ &\qquad + d y_1^2 y_3 + e y_1 y_3^2 + R(y, \mu), \end{aligned}$$

where $R(y, 0) = o(\|y\|^3)$, $R(y_1, 0, 0, \mu) \equiv 0$.

The family v_μ^σ has $K = (0, 0, 0)$ as the unique critical point for $\mu_1 \geqq 0$ and if $\mu_1 < 0$, then it has three critical points $K_1 = (0,0,0)$, $K_2 = ((-\mu_1)^{1/2}, 0, 0)$, $K_3 = (-(-\mu_1)^{1/2}, 0, 0)$. The characteristic equations of the corresponding matrices $L(K)$, $L(K_1)$, $B = L(K_2) = L(K_3)$ of the linear parts of v_μ^σ are as follows:

(5) $\lambda^3 - \mu_3 \lambda^2 - \mu_2 \lambda - \sigma \mu_1 = 0$ for $L(K)$, $\mu_1 \geqq 0$,

(6) $\lambda^3 - \mu_3 \lambda^2 - \mu_2 \lambda - \sigma \mu_1 = 0$ for $L(K_1)$, $\mu_1 < 0$,

(7) $\lambda^3 - (\mu_3 - d\mu_1^2)\lambda^2 - (\mu_2 - b\mu_1)\lambda + 2\sigma \mu_1 = 0$ for B, $\mu_1 < 0$.

Denote $S_1 = \{\mu : \mu_1 = 0\}$, $Z_2^- = \{\mu : \mu_1 = \mu_2 = 0,\ \mu_3 < 0\}$, $Z_2^+ = \{\mu : \mu_1 = \mu_2 = 0,\ \mu_3 > 0\}$, $Z_2 = Z_2^- \cup Z_2^+$, $Z_{1c} = \{\mu : \mu_1 = \mu_3 = 0,\ \mu_2 < 0\}$.

The matrix $L(K)$ has

(1) zero as a simple eigenvalue if and only if $\mu \in S_1$, $\mu_2 \neq 0$

(2) zero as an eigenvalue of multiplicity 2 if and only if $\mu \in Z_2$

(3) zero as a simple eigenvalue and a couple of pure imaginary eigenvalues if and only if $\mu \in Z_{1c}$.

Denote by D_0, D_1, D_2 the discriminants of the equations (5), (6), (7) , respectively and let G_0, G_1, G_2 be the corresponding discriminant surfaces. Let D_j^+ (D_j^-) , j=0, 1, 2 be the set of all for which $D_j > 0$ ($D_j < 0$). Since $D_0 = p^3 + q^2$, where $p = -(1/3)(\mu_2 + (1/3)\mu_3^2)$, $q = -(1/2)(\mu_1 + (1/3)\mu_2\mu_3 + (2/27)\mu_3^3)$, we obtain that $G_0 = F_0^+ \cup F_0^-$, where $F_0^{\pm} = \{\mu : \mu_1 = F^{\pm}(\mu_2, \mu_3)$, $\mu_2 + (1/3)\mu_3^2 \geqq 0$, $\mu_1 \geqq 0\}$, $F^{\pm}(\mu_2, \mu_3) = -(1/3)(\mu_2\mu_3 + (2/9)\mu_3^2) \pm (2/\sqrt{27})(\mu_2 + (1/3)\mu_3^2)^{3/2}$ (similarly for G_1 and G_2). The sets G_j, j=0, 1, 2 are illustrated in Figure 1.

Let I_0 (I_1, I_2) be the set of all $\mu \in D_0^+$ (D_1^+, D_2^+) for which the equation (5) ((6) , (7)) has a couple of pure imaginary roots. The Cardan's formulas imply that $I_0 = \{\mu \in D_0^+ \; H_0(\mu_1, \mu_2, \mu_3) = 0$, $\mu_1 \geqq 0$, where $H_0(\mu_1, \mu_2, \mu_3) = 2\mu_3 - 3((-q + (D_0)^{1/2})^{1/3} + (-q - (D_0)^{1/2})^{1/3}$ (similarly for I_1, I_2). The function H_0 is of class C^1, $H_0(\mu) = 0$, $\frac{\partial H_0(\mu)}{\partial \mu_1} \neq 0$ for $\mu \in Z_2^- \cup Z_{1c}$. The Implicit function theorem implies that I_0 (I_1, I_2) is a two-dimensional C^1-manifold with the boundary $\partial I_0 = Z_2^- \cup Z_{1c} \cup \{0\}$ ($\partial I_1 = Z_2^- \cup Z_{1c} \cup \{0\}$, $\partial I_2 = Z_2^+ \cup Z_{1c} \cup \{0\}$) (see Figure 1).

For $\mu \in S_1$ there is the unique critical point K, for which the matrix L(K) has eigenvalues: $\lambda_1 = 0$, $\lambda_{2,3} = (1/2)(\mu_3 \pm D^{1/2})$, where $D = \mu_3^2 + 4\mu_2$. Therefore the bifurcation diagram in S_1 consists of the following components: $D_1 = \{\mu \in S_1 : \psi > 0, \; \mu_3 < 0\}$, $D_2 = \{\mu \in S_1 : \psi > 0, \; \mu_3 < 0, \; \mu_2 < 0\}$, $D_3 = \{\mu \in S_1 : \mu_2 > 0\}$, $D_4 = \{\mu \in S_1 : \psi > 0, \; \mu_3 > 0, \; \mu_2 < 0\}$, $D_5 = \{\mu \in S_1 : \psi < 0\}$, where $\psi(\mu_2, \mu_3) = \mu_3^2 + 4\mu_2$. From the signs of eigenvalues of the matrix L(K) in the regions D_1-D_5 and from the properties of roots of cubic equations, it is possible to determine the signs of all roots of the equations (5)-(7) in different regions of the bifurcation diagram illustrated in Figure 1.

Now, we analyse the bifurcations near the set Z_2. It suffices to consider the family with $\sigma < 0$. If $\sigma > 0$, then introducing the change of coordinates $y_1 \to y_1$, $y_2 \to -y_2$, $y_3 \to y_3$, $t \to -t$, $\mu_3 \to -\mu_3$, we obtain a family of the same form with $\sigma < 0$.

Let $\mu^0 = (0, 0, \mu_3^0) \in Z_2$. The linear change of coordinates $u = Cy$, where $C = (c_{ij})$, $c_{11} = -\gamma = -\mu_3^0$, $c_{12} = 1 - \gamma$, $c_{13} = c_{23} = 1$, $c_{22} = -\gamma$, $c_{21} = c_{13} = c_{23} = 0$, transforms the family v_μ^σ into the form

(8) $$\dot{u} = Nu + B_0(\mu)u + F(u,\mu),$$

where $N = (n_{ij})$, $n_{12} = 1$, $n_{33} = \gamma$ and the others elements of N are equal to zero, $B_0(\mu^0) = 0$, $F = (F_0, F_0, F_0)$, $F_0(u, \mu^0) = H_3(u) + o(\|u\|^3)$, $H_3(u)$ is a homogeneous polynomial of degree 3. The reduction to the central manifold has the form

(9) $$\dot{v} = Mv + B(\mu)v + Q(v,\mu),$$

where $v = (u_1, u_2)$, the matrix M has 1 above the diagonal and zeros elsewhere, $B(\mu^0) = 0$, $Q = (Q_1, Q_1)$, $Q_1(u_1, u_2, \mu^0) = A_{300}u_1^3 + A_{030}u_2^3 + A_{210}u_1^2u_2 + o(\|v\|^3)$, $A_{300} = \gamma^{-3}$, $A_{210} = 3\gamma^{-2}(\gamma^{-1} - \gamma^{-2}) - b\gamma^{-3}$. Let $z_1 = u_1$, $z_2 = u_2 + b_1(u_1, u_2, \mu) + Q_1(u_1, u_2, \mu)$, where b_1 is the first element of the vector $B(\mu)v$. Then

(10) $$\begin{aligned} \dot{z}_1 &= z_2, \\ \dot{z}_2 &= \alpha z_1^3 + \beta z_1^2 z_2 + A_{120} z_1 z_2^2 + A_{030} z_2^3 + o(\|z\|^3), \end{aligned}$$

where $\alpha = A_{300}$, $\beta = A_{210}$. Obviously, sign α = sign μ_3^0, sign β = sign $\gamma^4(3\gamma^{-2}(\gamma^{-1} - \gamma^{-2}) - b\gamma^{-3})$ and hence sign $\beta = -1$ if μ_3^0 is sufficiently small (we assume $\sigma < 0$). Therefore $\beta < 0$ and $\alpha > 0$ ($\alpha < 0$) if $\mu_3^0 > 0$ ($\mu_3^0 < 0$). Let $\mu_3^0 < 0$, i. e. $\mu^0 \in Z_2^-$, $\alpha < 0$, $\beta < 0$. Let P_0 be the plane parallel to the (μ_1, μ_2)-plane, crossing the set Z_2 transversally at μ^0. Let us restrict the set of parameters to a neighbourhood U_0 of μ^0 in P_0.

Let $\varepsilon = (\varepsilon_1, \varepsilon_2)$ be coordinates on U_0 and let $f(z, \varepsilon)$ be the unfolding corresponding to the vector field (10). Then $f(0, \varepsilon) \equiv 0$, $f(z, \varepsilon) = - f(-z, \varepsilon)$ and hence the hypothesis (H1)-(H3) from [2, Section 4. 2] are satisfied (see also [11]). The corresponding bifurcation diagram and bifurcations on the central manifold are the same as in [2, Figures 4, 6]. For $\mu^0 \in Z_2^+$ there is $\beta < 0$, $\alpha > 0$ and hence the corresponding bifurcations on the central manifold are the same as in [2, Figure 5].

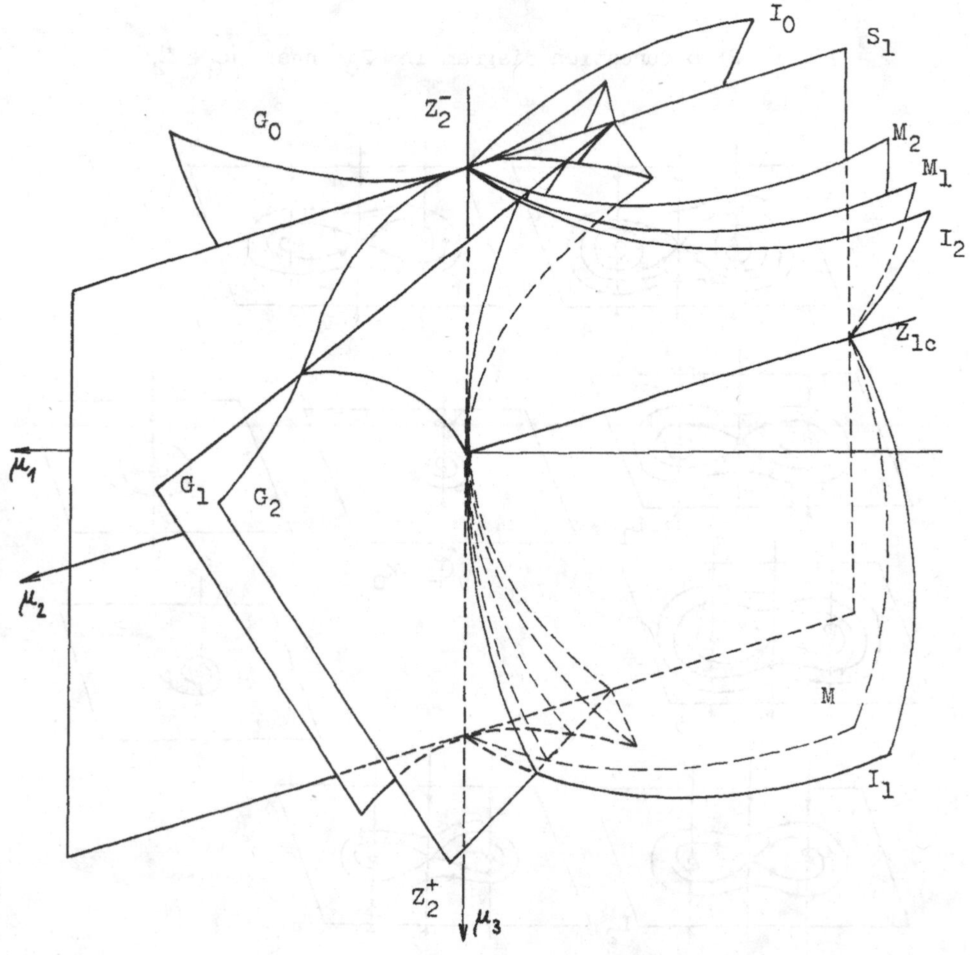

Figure 1

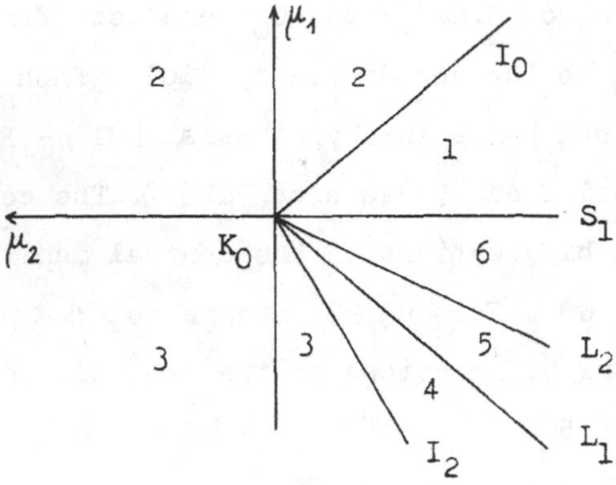

Figure 2: Bifurcation diagram in P_0 near $\mu_0 \in Z_2^-$

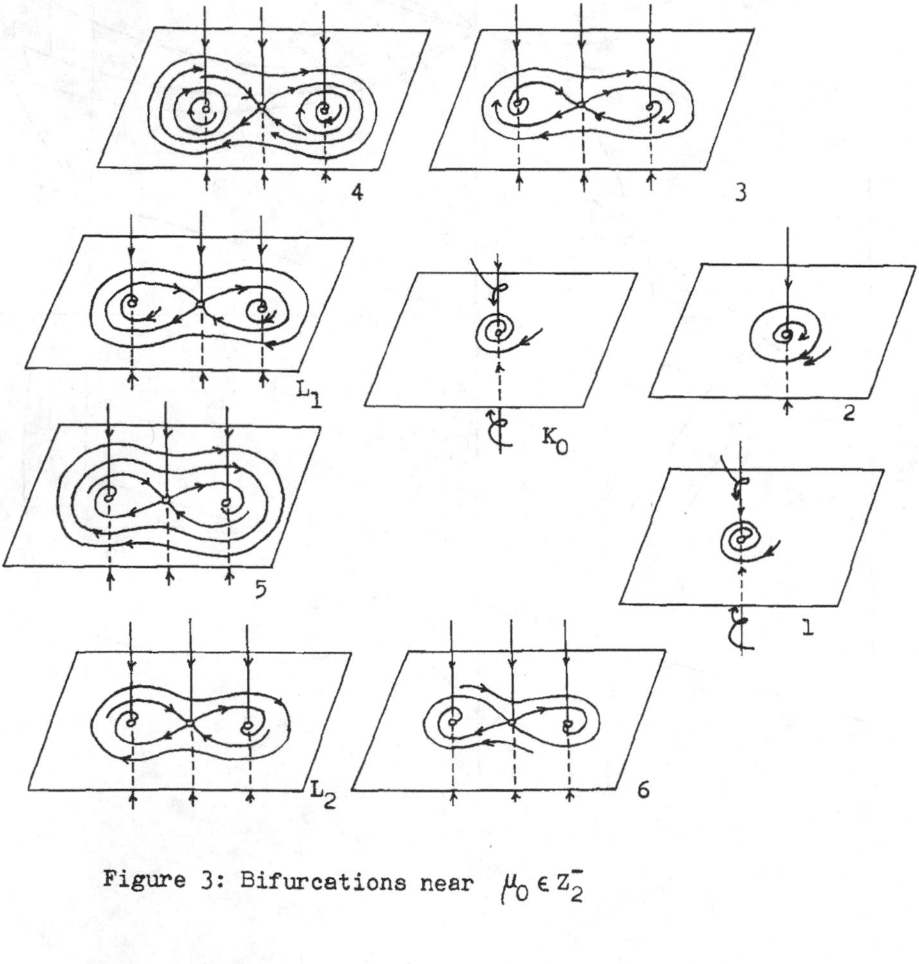

Figure 3: Bifurcations near $\mu_0 \in Z_2^-$

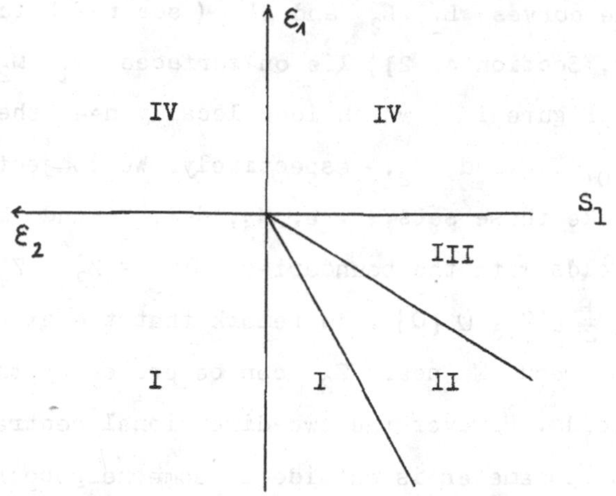

Figure 3: Bifurcation diagram in P_0 near $\mu_0 \in Z_2^+$

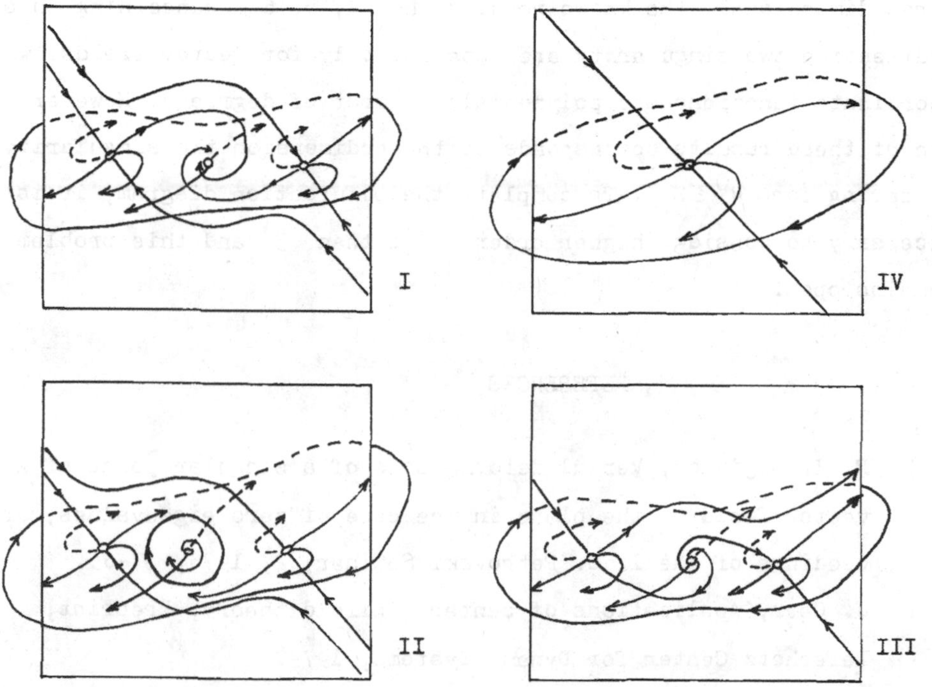

Figure 4: Bifurcations near $\mu_0 \in Z_2^+$

Remark. The curves L_1, L_2 and L (see the bifurcation diagrams from [2, Section 4. 2]) lie on surfaces M_1, M_2 and M, respectively (see Figure 1) , which look locally near the set Z_2 like the surfaces I_0, I_1 and I_2, respectively. We conjecture that they look globally like these sets, i. e. M_j, j=1, 2 and M are two-dimensional manifolds with the boundaries $\partial M_j = Z_2^- \cup Z_{1c} \cup \{0\}$, j= = 1, 2, $\partial M = Z_2^+ \cup Z_{1c} \cup \{0\}$. We remark that the existence of the surfaces M_1, M_2 and M near Z_2 can be proven by the methods of plane vector fields. However the two-dimensional central manifold is destroyed if the parameter is outside of some neighbourhood of the set Z_2. For $\mu^0 \in Z_{1c}$ the linear part of the vector field at the unique critical point has one zero eigenvalue and a couple of pure imaginary eigenvalues and hence the corresponding central manifold is three-dimensional. The known results [3, 4, 5, 6] concerning this codimension two singularity are obtained only for vector fields, whose coordinate functions are polynomials at most of degree 3. However no one of these results corresponds to the codimension two singularity appearing for $\mu^0 \in Z_{1c}$. To complete the bifurcation diagram, it is necessary to consider higher order terms than 3 and this problem remains open.

REFERENCES

[1] R. I. Bogdanov, Versal deformations of a singular point of a vector field on the plane in the case of zero eigenvalues, Proceedings of the I. G. Petrovski Seminar, 2, 1976, 37-65.

[2] J. Carr, Applications of center manifold theory, Preprint, Lefschetz Center for Dynam. Systems, 1979.

[3] P. Holmes, Unfolding a degenerate nonlinear oscillator: a codimension two bifurcation, Annals of the New York Acad. of Sci., Vol. 357, 1980, 475-488.

[4] P. Holmes, A Strange Family of Three-Dimensional Vector Fields Near a Degenerate Singularity, J. Differential Equations 37, 1980, 382-403.

[5] N. K. Gavrilov, On bifurcations of equilibrium points with one zero eigenvalue and a couple of pure imaginary eigenvalues, Proceedings, Gorkij 1978, 33-42.

[6] J. Guckenheimer, On a Codimension Two Bifurcation, Dynamical Systems and Turbulence, Warwick 1980, Proceedings, Springer-Verlag 1981, 99-142.

[7] B. Malgrange, Ideals of differentiable functions, Oxford Univ. Press, 1966.

[8] M. Medveď, Vector fields with a singularity of codimension 3 and their unfoldings, Proceedings of the IX. International Conference on Nonlinear Oscillations, Kijev 1981.

[9] M. Medveď, On a Codimension Three Bifurcation, submitted to Mathematica Slovaca.

[10] F. Takens, Singularities of vector fields, Publ. IHES 43, 1973, 47-100.

[11] F. Takens, Forced oscillations and bifurcations, Communications of Math. Inst., Rijksuniversiteit, Ultrecht, 3, 1974, 1-59.

THE COMPLETE INTEGRABILITY OF HAMILTONIAN SYSTEMS

Pierre van Moerbeke*

The discovery some fifteen years ago, that the Korteweg-de Vries equation could be integrated via spectral methods has generated an enormous number of new ideas in the area of completely integrable Hamiltonian systems; this had been a dormant subject for more than half a century. This work led to beautiful and fascinating connections between on the one hand infinite and finite dimensional Hamiltonian systems, and on the other hand Lie algebra theory, spectral theory and algebraic geometry; it has led to methods to generate and to linearize (integrate) completely integrable Hamiltonian systems. But the hard question remains : given a Hamiltonian system, how does one decide whether the system is completely integrable ? How does it fit into one of the schemes, from which complete integrability follows. This can easily be guessed in some examples, while in others it remains difficult. This is the question which I like to discuss here.

A Hamiltonian system

$$\dot{z} = J \frac{\partial H}{\partial z} \qquad J = J(z) \text{ antisymmetric, possibly depending on } z \in \mathbb{R}^{2n}$$

will be called *completely integrable*, if it has sufficiently many constants of the motion in involution; to be precise n constants of the motion $H_1,\ldots,H_n$ such that the Poisson brackets $\{H_i,H_j\}$ all vanish. Then a *theorem of Liouville* tells you that the compact and connected invariant manifolds are *tori* and there is a transformation to so-called action-angle variables, which maps the flow into a *straight line motion* on that torus.

Deciding whether a system is completely integrable in this C^∞-sense is quite hopeless at this stage, unless one imposes some extra structure; these extra features, which are quite typical, will be motivated by the following example : the Euler rigid body motion.

---

* University of Louvain, Louvain-la-Neuve, 1348 Belgium and Brandeis University, Waltham, Mass. 02254, USA; supported in part by NSF contract MCS-81-05576.

1. <u>The Euler rigid body motion</u>.

It expresses the free motion of a rigid body around a fixed point. Let $M = (M_1,M_2,M_3)$ be the angular momentum and $\lambda_1^{-1},\lambda_2^{-1},\lambda_3^{-1}$ the principal moments of inertia about the principal axes of inertia. Then the motion of the body is governed by the equations

$$\dot{M} = M \times \nabla H = (M_1,M_2,M_3) \times (\lambda_1 M_1,\lambda_2 M_2,\lambda_3 M_3) \tag{1}$$

where the Hamiltonian H = energy, has the form

$$H = \frac{1}{2}(\lambda_1 M_1^2 + \lambda_2 M_2^2 + \lambda_3 M_3^2).$$

Actually, equation (1) is Hamiltonian on the two-dimensional sphere obtained by putting the angular momentum $M_1^2+M_2^2+M_3^2$ equal to a constant, and for appropriate values of the constants, the invariant energy surface (ellipsoid) intersects the sphere according to circles, as expected from Liouville's theorem. But there is more to it : the problem can be integrated in terms of elliptic functions, as Euler discovered using his then newly invented theory of elliptic integrals. Indeed, from systems (1) or what is the same

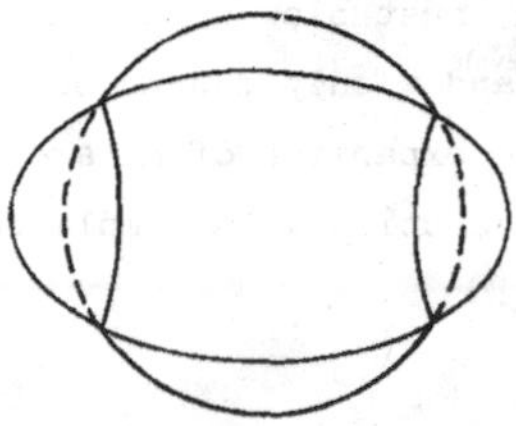

figure 1.

$$\begin{aligned} \dot{M}_1 &= (\lambda_3-\lambda_2)M_2M_3 \\ \dot{M}_2 &= (\lambda_1-\lambda_3)M_3M_1 \\ \dot{M}_3 &= (\lambda_2-\lambda_1)M_1M_2 \end{aligned} \tag{1'}$$

we have that

$$\frac{dM_1}{M_2M_3} = (\lambda_3-\lambda_2)dt \tag{2}$$

where M_1,M_2 and M_3 are related by

$$M_1^2 + M_2^2 + M_3^2 = c_1$$

$$\lambda_1 M_1^2 + \lambda_2 M_2^2 + \lambda_3 M_3^2 = c_2.$$

Therefore, if $\lambda_2 \neq \lambda_3$, M_2^2 and M_3^2 can expressed linearly in terms of M_1^2 and substituting these expressions into (2), we find after integration that the system (1) amounts to

$$(3) \qquad \int_{M_1(o)}^{M_1(t)} \frac{dM}{\sqrt{(\alpha^2-M^2)(\beta^2-M^2)}} = \gamma t$$

for appropriate constants α, β and γ, which depend only on c_1 and c_2, besides the fixed data λ_1, λ_2 and λ_3. We now think of *time* t as being a *complex* rather than real parameter; then $M_1(t)$ for $t \in \mathbb{C}$ is a doubly periodic meromorphic function of t with periods τ_1 and τ_2, which must have at least one pole t_∞, since otherwise $M_1(t)$ would be bounded and therefore constant.

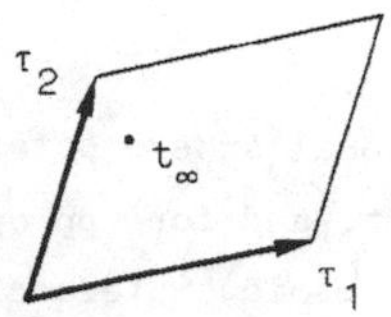

figure 2.

This is to say that, after a finite (complex) time t_∞, M_1 must blow up and $M_1(t)$ must have a Laurent expansion around t_∞. But since α, β and γ only depend on c_1 and c_2, these two free parameters must enter somewhere in the expansion of M_1 around the blow up point. Indeed, by looking for a Laurent series solution with simple pole of the system (1') you would find

$$M_i(t) = \frac{M_i^{(o)}}{t} + M_i^{(1)} + M_i^{(2)} t + \ldots;$$

the leading coefficient satisfies the equation

$$M^{(o)} + (M^{(o)} \times \lambda . M^{(o)}) = 0,$$

where $\lambda . M^{(o)}$ denotes the vector $(\lambda_1 M_1^{(o)}, \lambda_2 M_2^{(o)}, \lambda_3 M_3^{(o)})$. It has the following solution

$$(4) \qquad M_i^{(o)} = - \frac{1}{\sqrt{(\lambda_i - \lambda_{i+2})(\lambda_{i+1} - \lambda_i)}} \qquad 1 \leqslant i \leqslant 3 \text{ mod } 3.$$

The subsequent coefficients $M^{(k)}$ with $k \geqslant 1$ satisfy the linear equation

$$(M^{(k)} \times \lambda . M^{(o)}) + (M^{(o)} \times \lambda . M^{(k)}) + (1-k) M^{(k)} = \text{terms containing } M^{(j)} \text{ for } 1 \leqslant j < k$$

which amounts to

$$\begin{pmatrix} 1-k & (\lambda_3-\lambda_2)M_3^{(o)} & (\lambda_3-\lambda_2)M_2^{(o)} \\ (\lambda_1-\lambda_3)M_3^{(o)} & 1-k & (\lambda_1-\lambda_3)M_1^{o} \\ (\lambda_2-\lambda_1)M_2^{(o)} & (\lambda_2-\lambda_1)M_2^{(o)} & 1-k \end{pmatrix} \begin{pmatrix} M_1^{(k)} \\ M_2^{(k)} \\ M_3^{(k)} \end{pmatrix} = \text{terms containing } M^{(j)} \text{ for } 1 \leqslant j < k$$

with $M_i^{(o)}$ given by (4). This problem is uniquely solvable when the matrix on the left-hand side is invertible; an easy computation shows this is always so unless k=2 and then the rank equals 1. This shows that the coefficient $M_i^{(2)}$ (and only that one) in the expansion contains two free parameters, which account for c_1 and c_2.

To conclude, this example has much more structure : namely the circle of figure 1 extends to the complex torus $\mathbb{C}$/{Lattice defined by figure 2}, the flow is mapped by the integral (3) into a straight line motion on that torus and the coordinates $M_i(t)$ are meromorphic in t. The torus is also defined by the intersection of the two quadrics

$$\sum_1^3 M_i^2 = c_1 \quad , \quad \sum_1^3 \lambda_i M_i^2 = c_2$$

(since these expressions are preserved by the flow), except that one misses the points of blow up, which are captured automatically by replacing these equations by the homogeneous equations in $(M_o, M_1, M_2, M_3) \in \mathbb{P}^3$:

$$\sum_1^3 M_i^2 = c_1 M_o^2 \quad , \quad \sum_1^3 \lambda_i M_i^2 = c_2 M_o^2 .$$

Such a torus, also called elliptic curve, is known to have an algebraic addition law, i.e. the coordinates of the sum of two points are the zeros of a polynomial with coefficients, polynomial in the coordinates of the two points. Although this is quite a strong notion of complete integrability, almost all examples, classically known as integrable by quadrature,enjoy these properties. This state of affairs is summarized by the notion of *algebraic complete integrability*, which I now explain in general.

2. Algebraic complete integrability.

Consider the Hamiltonian system

$$\dot{z} = J\,\frac{\partial H}{\partial z}, \; z \in \mathbb{R}^{2n} \qquad (5)$$

with polynomial right hand side in z. Then the system will be called *algebraically completely integrable* when

1) it possesses n polynomial invariants $H_1,\ldots,H_n$ in involution ($\{H_i,H_j\} = 0$) and for most values of $c_i \in \mathbb{R}$, the invariant manifolds $\bigcap_{i=1}^{n} \{H_i = c_i\} \cap \mathbb{R}^{2n}$ are compact, connected and therefore real tori by Liouville's theorem.

2) Moreover, the real tori above are part of complex algebraic* tori $T \equiv \mathbb{C}^n/\text{Lattice}$; in the natural coordinates $(t_1,\ldots,t_n)$ (inherited from $\mathbb{C}^n$) of these tori, the flows defined by the Hamiltonian vector fields corresponding to the various constants of the motion are straight line motions; the addition law is algebraic and the coordinates $z_i = z_i(t_1,\ldots,t_n)$ are meromorphic in $t_1\ldots t_n$.

I now like to interpret this definition. Since the flow preserves the n constants of the motion $H_1,\ldots,H_n$ and since the tori are n-dimensional,

$$\bigcap_{i=1}^{n} \{H_i(z) = c_i,\ z \in \mathbb{C}^{2n}\}$$

defines at least that part of the torus T, where the coordinates z_i do not blow up; that means all but a codimension one subvariety D in T, since meromorphic functions have poles only along codimension one subvarieties, as schematically represented by figure 3. By the definition of algebraic complete integrability, the flow (5) is a straight line motion in T; it must therefore hit the manifold D in at least one place and the coordinates z_i must have a Laurent expansion around that point of intersection. Moreover through every point of D, there is a straight line motion and therefore a Laurent expansion in the time parameter t of that motion, which, of course, will depend on the point chosen on the n-1-dimensional submanifold of D; this is to say that the Laurent expansion must depend on n-1 parameters in addition to the n parameters c_i which define the torus (as in the example of the rigid body motion) : the total count is therefore 2n-1 ≡ dim (phase space) -1 parameters; this summarizes the proof of theorem 1.

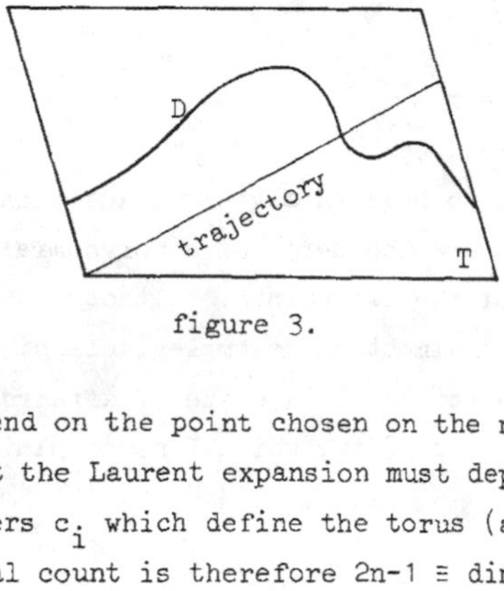

figure 3.

Theorem 1. (see [4]).

If the Hamiltonian flow (5) *is algebraically completely integrable, then the*

* i.e. defined globally by several homogeneous polynomial equations.

system of differential equations (5) *must admit solutions which are Laurent expansions in* t *such that*

1) *each* z_i *blows up for some value of* t.

2) *the Laurent expansions of* z_i *around the place where* z_i *blows up admit* dim (phase space) −1 *parameters.*

This necessary condition for algebraic complete integrability is implicit in a beautiful investigation of S. Kowalewski [8, 9] in 1889 for which she was awarded the Bordin prize by the French academy; there she finds all the completely integrable rigid body motions in the presence of gravity; to wit, the Euler rigid body motion, the Lagrange top and her famous "Kowalewski top".

Theorem 1 is only a necessary condition; I now show for a class of Hamiltonian systems how this condition is also sufficient and how it can be used to single out algebraically completely integrable systems.

3. The geodesic flow on SO(4) : necessary conditions for complete integrability.

We consider the geodesic flow on SO(4) for some left-invariant, diagonal, non-degenerate* metric $\sum_1^6 \lambda_i x_i y_i$. This example generalizes the 3-dimensional rigid body motion, which is nothing else but a geodesic motion on the group SO(3). In the first example, we have shown that all the geodesic flows on SO(3) for left-invariant metrics** are algebraically completely integrable. This is not the case anymore for SO(4), even if the metric is diagonal. The geodesic flow on SO(4) for a left-invariant metric can by translation be reduced to a motion on the tangent space at the origin, this is to say, on the Lie algebra so(4). With Arnold [5], the geodesic equations can be written as

$$\dot{X} = [X, \nabla H] = [X, \lambda . X], \tag{7}$$

where

---

* all λ_i's are different.

** a metric in SO(3) can always be diagonalized, such as to keep the form of the equations unchanged; this is not the case for the higher dimensional rotation groups.

$$X = (X_{ij})_{1\leqslant i,j\leqslant 4} = \begin{pmatrix} 0 & -x_3 & x_2 & -x_4 \\ x_3 & 0 & -x_1 & -x_5 \\ -x_2 & x_1 & 0 & -x_6 \\ x_4 & x_5 & x_6 & 0 \end{pmatrix} \in so(4)$$

and

$$\lambda . X = \begin{pmatrix} 0 & -\lambda_3 x_3 & \lambda_2 x_2 & -\lambda_4 x_4 \\ \lambda_3 x_3 & 0 & -\lambda_1 x_1 & -\lambda_5 x_5 \\ -\lambda_2 x_2 & +\lambda_1 x_1 & 0 & -\lambda_6 x_6 \\ \lambda_4 x_4 & \lambda_5 x_5 & \lambda_6 x_6 & 0 \end{pmatrix} \in so(4).$$

H is the kinetic energy and the phase space $so(4) \cong \mathbb{R}^6$ is 6-dimensional; the flow is Hamiltonian on the 4-dimensional space obtained by reducing so(4) by the trivial invariants

$$\sqrt{\det X} = x_1x_4 + x_2x_5 + x_3x_6 = c_1,$$

$$\frac{1}{2} \operatorname{Tr} X^2 \equiv \sum_1^6 x_i^2 = c_2.$$

On that 4-dimensional reduced phase space, if the problem is to be completely integrable, it must have on other constant of the motion, besides the kinetic energy $H = \frac{1}{2} \Sigma \lambda_i x_i^2$. We now pose the following question : for which metrics is this the case ? Theorem 1 provides a necessary condition for algebraic complete integrability; so, find metrics λ_i for which the system of differential equations (7), i.e.

$$\begin{aligned}
\dot{x}_1 &= x_2x_3(\lambda_3-\lambda_2) + x_5x_6(\lambda_6-\lambda_5)\\
\dot{x}_2 &= x_1x_3(\lambda_1-\lambda_3) + x_4x_6(\lambda_4-\lambda_6)\\
\dot{x}_3 &= x_1x_2(\lambda_2-\lambda_1) + x_4x_5(\lambda_5-\lambda_4)\\
\dot{x}_4 &= x_3x_5(\lambda_3-\lambda_5) + x_2x_6(\lambda_6-\lambda_2)\\
\dot{x}_5 &= x_3x_4(\lambda_4-\lambda_3) + x_1x_6(\lambda_1-\lambda_6)\\
\dot{x}_6 &= x_2x_4(\lambda_2-\lambda_4) + x_1x_5(\lambda_5-\lambda_1)
\end{aligned}$$

has Laurent expansion solutions

$$X = t^{-k}(X^{(0)} + X^{(1)}t + X^{(2)}t^2 + \ldots) \qquad k \geqslant 1$$

depending on dim (phase space) $-1 = 5$ free parameters. For this to happen $k = 1$

must hold as it turns out from a somewhat lengthy computation. The coefficients in this expansion are given as follows : in the differential equation $\dot{X} = [X,\lambda.X]$: at the 0th step, the coefficient of t^{-2} yields a non-linear equation

(8) $$X^{(o)} + [X^{(o)},\lambda.X^{(o)}] = 0$$

and at the jth step ($j \geqslant 1$), the coefficient of t^{j-2} leads to a system of linear equations in $X^{(j)}$:

(9) $$(L-jI)X^{(j)} = \begin{cases} 0 \text{ for } j = 1 \\ \text{quadratic polynomial in } X^{(j)},\dots,X^{(j-1)} \text{ for } j > 1 \end{cases}$$

where L denotes the linear map

(10) $$L(Y) = [Y,\lambda.X^{(o)}] + [X^{(o)},\lambda.Y] + Y \equiv \text{Jacobian map of (8).}$$

Observe that L depends on the solution $X^{(o)}$ of (8). By solving the systems (8) and (9) inductively, one observes that free parameters can only come

1) either from the non-linear equation; then its Jacobian map (10) is singular and L has a 0-eigenvalue.

2) or from the linear equations (9); then the operator $L-jI$ is singular for $j \geqslant 1$ and j is an eigenvalue for L.

To summarize, for $X^{(o)}$ satisfying $X^{(o)} + [X^{(o)},\lambda.X^{(o)}] = 0$, the operator L must have at least 5 integer ($\geqslant 0$) eigenvalues corresponding to 5 independent eigenvalues.

Now, it is easily checked that

$$(L+I)X^{(o)} = 0,$$

that

$$(L-2I)Y = 0 \text{ for } Y = (x_1^{(o)},-x_2^{(o)},0,x_4^{(o)},-x_5^{(o)},0) \text{ and } Y = (0,x_2^{(o)},-x_3^{(o)},0,x_5^{(o)},-x_6^{(o)}),$$

and it follows from a more involved argument that 1 is also an eigenvalue. In fact, one shows that

(11) $$\det(L-jI) = (j-2)^2(j^2-1)\,[(j-1)^2 - 4x_1^{(o)}x_2^{(o)}x_3^{(o)}K(\lambda)-1]$$

where $K(\lambda)$ is a cubic polynomial in the metric :

$$K(\lambda) \equiv \lambda_1\lambda_4(\lambda_2+\lambda_5-\lambda_3-\lambda_6) + \lambda_2\lambda_5(\lambda_3+\lambda_6-\lambda_1-\lambda_4) + \lambda_3\lambda_6(\lambda_1+\lambda_4-\lambda_2-\lambda_5).$$

Since Tr $L = 6$, by (10) and since L is to have 5 ($\geqslant 0$) integer eigenvalues, the spectrum of L must be

$$\text{(12)} \quad \begin{array}{ll} \text{either} & -1, 0, 1, 2, 2, 2, \\ \text{or} & -1, 1, 1, 1, 2, 2. \end{array}$$

In the first case, 0 is an eigenvalue; therefore at every point of $X^{(o)} + [X^{(o)},\lambda.X^{(o)}] = 0$, the Jacobian of this equation vanishes, so that $X^{(o)} + [X^{(o)},\lambda.X^{(o)}] = 0$ defines a curve, rather than isolated points. Along this curve, equation (11) evaluated at j=0 must vanish identically in the coordinate $X^{(o)} = (x_1^{(o)},\dots,x_6^{(o)})$ of the curve; hence

$$\det(L) = 16\, x_1^{(o)}x_2^{(o)}x_3^{(o)}K(\lambda) = 0$$

and, since for most points $X^{(o)}$ on that curve, $\prod_1^3 x_i^{(o)} \neq 0$, the metric must satisfy $K(\lambda) = 0$.

In the second case (12), the system $X^{(o)} + [X^{(o)},\lambda.X^{(o)}] = 0$ cuts out a discrete set of points and 1 is a triple eigenvalue which leads to a 3-dimensional eigenspace only if $\lambda_1-\lambda_4 = \lambda_2-\lambda_5 = \lambda_3-\lambda_6$. This is a degenerate metric of interest, which leads to a decoupling of the equations (7) into two Euler rigid body motions :

$$\begin{aligned} (x_1\pm x_4)^{\cdot} &= (\lambda_3-\lambda_2)(x_2\pm x_5)(x_3\pm x_6) \\ (x_2\pm x_5)^{\cdot} &= (\lambda_1-\lambda_3)(x_3\pm x_6)(x_1\pm x_4) \\ (x_3\pm x_5)^{\cdot} &= (\lambda_2-\lambda_1)(x_1\pm x_4)(x_2\pm x_5). \end{aligned}$$

4. Integrability of the geodesic flow and the existence of asymptotic expansions.

We now indicate how to establish conversely that if the metric satisfies $K(\lambda) = 0$, then the geodesic flow is algebraically completely integrable. To begin with, the meaning of $K(\lambda) = 0$ is easily seen to be the following : under that condition, there exists a vector $(\mu_1,\dots,\mu_6)$, independent from $(1,1,1,1,1,1)$ and $(\lambda_1,\dots,\lambda_6)$ such that

$$\text{(13)} \quad \det\begin{pmatrix} 1 & 1 & 1 \\ \lambda_i & \lambda_j & \lambda_k \\ \mu_i & \mu_j & \mu_k \end{pmatrix} = 0 \text{ for } (i,j,k) = (1,2,3),\ (1,5,6),\ (2,4,6) \text{ and } (3,4,5);$$

these three determinental relations imply that $(\sum_1^k \mu_i x_i^2)^{\cdot} = 0$, so that the flow (7) has, besides the invariants

$$Q_1 \equiv x_1x_4 + x_2x_5 + x_3x_6 = c_1$$

$$Q_2 \equiv \sum_1^6 x_i^2 = c_2$$

(14) $$Q_3 \equiv \sum_1^6 \lambda_i x_i^2 = c_3$$

a fourth independent invariant

$$Q_4 \equiv \sum_1^6 \mu_i x_i^2 = c_4,$$

with the μ_i related to the λ_i by (13). We need to show that

(15) $\bigcap_1^4 \{Q_i = c_i \text{ in } \mathbb{C}^6\}$ = two-dimensional complex algebraic torus $T \setminus$ smooth codimension one subvariety C.

Step 1.

This will be done by completing the commuting Hamiltonian vector fields

$$\frac{dX}{dt_1} = [X,\lambda.X] \text{ and } \frac{dX}{dt_2} = [X,\mu.X]$$

generated by Q_3 and Q_4, into two commuting (never vanishing) vector fields on a smooth compact complex manifold T such that

$$T \setminus \text{codimension one subvariety} = \bigcap_1^4 \{Q_i = c_i \text{ in } \mathbb{C}^6\}.$$

Then T must be a torus; indeed if one denotes the phase flows by $g^{t_1}(p)$ and $g^{t_2}(p)$, (i.e. the position on T after flowing with the completed vector fields, starting from p), then there is a 1-1 analytic map

$$\frac{\{(t_1,t_2) \in \mathbb{C}^2\}}{\text{discrete lattice } \{(t_1,t_2) \mid g^{t_2} \circ g^{t_1}(p) = p\}} \longrightarrow T;$$

since the left hand side must be compact in the same way as T, the lattice must be 4-dimensional over $\mathbb{Z}$ in $\mathbb{C}^2$; T is therefore a complex torus, which can also be shown to be algebraic.

So the main question is how to complete the manifold $\bigcap_1^4 \{Q_i = c_i\}$ and the differential equations on it ? A naive guess would be to take the natural completion in $\mathbb{P}^6$.

$$\bigcap_1^4 \{Q_i = c_i x_o^2 \text{ in } \mathbb{P}^6\};$$

unfortunately this manifold is singular along the locus $x_o = 0$, which turns out to be a whole curve. Finding the completion of the manifold and the vector fields is equivalent to proving the existence of the asymptotic expansions

$$X = t^{-1}(X^{(o)} + X^{(1)}t + X^{(2)}t^2 + \ldots)$$

for the differential equations (7) with 5 free parameters. So far we have found necessary conditions for the existence of the asymptotic expansions, we show next that these conditions are also sufficient.

Step 2.

To begin with, formal series solutions of differential equations with quadratic right hand side are automatically convergent as a consequence of the majorant method (see Hille [7]). So, it suffices to show that the inductive set of equations (8) and (9) has solutions with the required degrees of freedom. Whenever $L-jI$ is non singular, the solution exists and is unique; so the only problem may arise from those values of j for which $L-jI$ is singular, actually only for j=2, since the right hand side of (9) vanishes for j=1. This was carried out by L. Haine [6], by choosing a privileged vector field, generated by the Hamiltonian

$$H \equiv \sum_{i=2}^{4} \alpha_i Q_i$$

where $(\alpha_2, \alpha_3, \alpha_4)$ is required to satisfy

$$(\alpha_2, \alpha_3, \alpha_4) \begin{pmatrix} 1 & 1 & 1 \\ \lambda_1 & \lambda_2 & \lambda_3 \\ \mu_1 & \mu_2 & \mu_3 \end{pmatrix} = 0,$$

which is possible, in view of (13). Then, for appropriate constants ν_4, ν_5 and ν_6,

$$H = \nu_4 x_4^2 + \nu_5 x_5^2 + \nu_6 x_6^2$$

and the vector field takes the simple form

$$(16)\qquad \begin{aligned} \dot{x}_1 &= (\nu_6-\nu_5)x_5x_6 & \dot{x}_4 &= \nu_6x_2x_6 - \nu_5x_3x_5 \\ \dot{x}_2 &= (\nu_4-\nu_6)x_4x_6 & \dot{x}_5 &= \nu_4x_3x_4 - \nu_6x_1x_6 \\ \dot{x}_3 &= (\nu_5-\nu_4)x_4x_5 & \dot{x}_6 &= \nu_5x_1x_5 - \nu_4x_2x_4 . \end{aligned}$$

In these new variables, the non-linear equation (8) at the 0th step is easily shown to define an elliptic curve ($\equiv$ complex 1-dimensional torus), at the first step ($j=1$), L has a one-dimensional eigenvector Y_2 and now at the second step ($j=2$), a computation shows equation (9) reduces to

$$(L-2I)X^{(2)} = 0$$

with a vanishing right hand side, implying readily the existence of a 3-dimensional eigenspace spanned by vectors Y_3, Y_4 and Y_5. So, this establishes the existence of a Laurent series solution of (16) of the type

$$(17)\qquad X = t^{-1}(X^{(0)}(\alpha_1) + (X^{(1)}+\alpha_2Y_2)t + (X^{(2)}+\alpha_3Y_3+\alpha_4Y_4+\alpha_5Y_5)t^2 + O(t^3))$$

where $X^{(0)}$ depends non-linearly on a free parameter α_1 running over an elliptic curve, while the four remaining parameters $\alpha_2,\dots,\alpha_5$ appear linearly in the expansion. Expressing the fact that $X \in \bigcap_1^4 \{Q_i = c_i \text{ in } \mathbb{C}^6\}$ implies further non-linear relations between $\alpha_1,\alpha_2,\dots,\alpha_5$, which themselves define a curve C, which will turn out to be the smooth curve C suggested in (15).

Step 3.

Based on the expansion (17), we can now choose a new time parameter τ

$$\tau = x_1^{-1} = (x_1^{(0)})^{-1}t + \dots$$

and new space variables $u_2,\dots,u_6$, rational in $x_1,\dots,x_6$, such that

$$u_i(\tau) = \alpha_i + O(\tau)$$

has a nice Taylor expansion in τ (for τ small) with leading coefficient $\alpha_2,\dots,\alpha_5$ which themselves run over the smooth curve C. The manifold $\bigcap_1^4 \{Q_i = c_i \text{ in } \mathbb{C}^6\}$ is now completed by gluing to it the curve C, from which the flow sticks out at most points in a nice smooth transversal way; that turns it into a smooth manifold with a nice vector field defined all over. Any other Hamiltonian vector field generated by some linear combination of Q_i commutes, of course, with the distinguished one and it can be extended in a smooth transversal way to the completed manifold T which according to step 1 must be a complex torus.

Theorem 2. (see [4] and [6]).

The geodesic flow on SO(4) *for a left-invariant metric* $\sum_1^6 \lambda_i x_i^2$, *with* λ_i*'s all distinct, is algebraically completely integrable if and only if*

$$(18) \qquad K(\lambda) = \lambda_1\lambda_4(\lambda_2+\lambda_5-\lambda_3-\lambda_6) + \lambda_2\lambda_5(\lambda_3+\lambda_6-\lambda_1-\lambda_4) + \lambda_3\lambda_6(\lambda_1+\lambda_4-\lambda_2-\lambda_5) = 0$$

which amounts to the following parametrization for λ_i *in terms of new parameters* α_i *and* β_i

$$(18') \qquad \lambda_1 = \frac{\beta_2-\beta_3}{\alpha_2-\alpha_3} \qquad \lambda_4 = \frac{\beta_1-\beta_4}{\alpha_1-\alpha_4}$$

$$\lambda_2 = \frac{\beta_1-\beta_3}{\alpha_1-\alpha_3} \qquad \lambda_5 = \frac{\beta_2-\beta_4}{\alpha_2-\alpha_4}$$

$$\lambda_3 = \frac{\beta_1-\beta_2}{\alpha_1-\alpha_2} \qquad \lambda_6 = \frac{\beta_3-\beta_4}{\alpha_3-\alpha_4}.$$

This parametrization implies that the equations of the geodesic motion can be written in the form

$$(19) \qquad (X+\alpha h)^{\cdot} = [X+\alpha h, \lambda.X+\beta h]$$

where $\alpha = \mathrm{diag}(\alpha_1,\ldots,\alpha_4)$, $\beta = \mathrm{diag}(\beta_1,\ldots,\beta_4)$ *and* h *is a formal indeterminate. Under the condition* (18), *the motion has, besides the invariants* (14), *a fourth independent invariant*

$$Q_4 = \sum_1^6 \mu_i x_i^2$$

where the μ_i *can be parametrized in a similar fashion :*

$$\mu_1 = \frac{\gamma_2-\gamma_3}{\alpha_2-\alpha_3} \qquad \mu_4 = \frac{\gamma_1-\gamma_4}{\alpha_1-\alpha_4}$$

$$\mu_2 = \frac{\gamma_1-\gamma_3}{\alpha_1-\alpha_3} \qquad \mu_5 = \frac{\gamma_2-\gamma_4}{\alpha_2-\alpha_4}$$

$$\mu_3 = \frac{\gamma_1-\gamma_2}{\alpha_1-\alpha_2} \qquad \mu_6 = \frac{\gamma_3-\gamma_4}{\alpha_3-\alpha_4}.$$

The geodesic flow, expressed as (19) ties up with the theory of Kac-Moody Lie algebras : it can be identified as a Hamiltonian flow on a coadjoint orbit defined in a Kac-Moody Lie algebra; the latter is defined as the set of formal Laurent series with coefficients in one of the finite dimensional semi-simple Lie algebras; such algebras are themselves infinite dimensional. Hamiltonian flows on orbits in Kac-Moody Lie algebras are well understood from the point of view of complete integrability. Although these ideas have not been used in this paper, they provide an alternative way to tackle the problem of complete integrability; see [1] and [2].

Remark.

A free four-dimensional rigid body motion is a special case of geodesic motion on SO(4) for the metric (18'); it is obtained by putting $\alpha_i = \beta_i^2$, $X_{ij} \equiv M_{ij}$ the angular momentum and $(\lambda.X)_{ij} = \Omega_{ij}$ the angular velocity with $M_{ij} = (\beta_i+\beta_j)\Omega_{ij}$.

References.

1. Adler, M., van Moerbeke, P. : Completely integrable systems, Euclidean lie algebras, and curves. Adv. in Math. 38, 267-317 (1980)
2. Adler, M., van Moerbeke, P. : Linearization of Hamiltonian systems, Jacobi varieties, and representation theory. Adv. in Math. 38, 318-379 (1980)
3. Adler, M., van Moerbeke, P. : Kowalewski's asymptotic method, Kac-Moody Lie algebras and regularization. Comm. Math. Physics 83, 83-106 (1982)
4. Adler, M., van Moerbeke, P. : The algebraic integrability of geodesic flow on SO(4). Invent. Math. 67, 297-326 (1982)
   Mumford, D. : An appendix to the preceding article. Invent. Math. 67, 327-331 (1982)
5. Arnold, V.I. : Mathematical methods of classical mechanics. Springer-Verlag, New York (1978)
6. Haine, L. : Abelian surfaces and geodesic flow on SO(4). (to appear).
7. Hille, E. : Ordinary differential equations in the complex domain. NY : Wiley-Interscience 1976
8. Kowalewski, S. : Sur le problème de la rotation d'un corps solide autour d'un point fixe. Acta Mathematica 12, 177-232 (1889)
9. Kowalewski, S. : Sur une propriété du système d'équations différentielles qui définit la rotation d'un corps solide autour d'un point fixe. Acta Mathematica 14, 81-83 (1889)
10. van Moerbeke, P. : Complete integrability, Kac-Moody Lie algebras and algebraic geometry. Invited lecture, International Congress of Mathematicians 1983 (to appear).

LINEARIZED DYNAMICS OF SHEARING DEFORMATION PERTURBING REST IN VISCOELASTIC MATERIALS

A. Narain and D. D. Joseph
Dept. of Aerospace Engineering and Mechanics
University of Minnesota
Minneapolis, MN 55455

This paper extends our earlier work [6, 7] on the propagation of jumps in velocity and displacement for shearing deformations imposed impulsively at the boundary of viscoelastic fluids and solids obeying constitutive equations in integral form with arbitrary kernels of fading memory type. The earlier work is briefly reviewed in §1 and we give new results. In §2 we relate old results to experiments. The limiting velocity distribution for start-up of Couette flow between parallel plates is a linear shear. It is common practice to assume that the real motion is close to linear shear long before the stress approaches its asymptotic steady state value. When the simplified kinematics are assumed, the evolution of the wall shear stress is determined by material functions, independent of deformation. These material functions are then determined by experimental measurements. We argue that in some cases only very special features of the material functions can be determined by this method because (in all cases) the early time behavior of the motion is incorrectly given by the kinematic assumption. The assumption that the early part of the stress response can be ignored is at best an approximation when the dynamics shows the presence of a delta function singularity in the wall shear stress at time t=0 and at subsequent discrete times of reflection off bounding walls. This delta function contribution cannot be ignored even if the steady state is achieved rapidly. In fact the early time behavior of the material functions can be obtained from experiments only by using a correct theory based on dynamics rather than kinematical assumptions. When this is done it is possible to interpret data showing stress jumps with linear theories based on commonly used constitutive equations and to interpret early oscillations in the observed values of material functions in terms of repeated reflections off bounding walls. The foregoing remarks apply equally to the interpretation of stress relaxation experiments and other experiments involving impulsive changes in velocity and displacement. In §3 we derive formulas for the amplitude of jumps and reflections for fluids sheared between concentric cylinders. In §4 we develop integral methods of solution analogous to Duhamel integrals for inverting start up problems with arbitrary data perturbing rest. In §5 we apply our analysis to start up for viscoelastic solids and show how creep depends on the kernel of the integral equation.

§1. A Summary of Previous Work on Step Jumps of Velocity and Displacement.

In our earlier work [6], we treated the problems of step increase in velocity and displacement using a constitutive expression of the type:

$$\mathbf{T} = -p\mathbf{1} + \mu\mathbf{A}_1 + \int_0^\infty \tilde{\mu}(s)\mathbf{G}(s)\,ds \tag{1.1}$$

where, $\tilde{\mu}(s) \equiv \frac{dG}{ds}$ and $G:[0,\infty) \rightarrow \mathbf{R}^+ = \{x \in \mathbf{R} | x>0\}$

is assumed to be (i) strictly monotonically decreasing, (ii) continuous and piecewise continuously differentiable, (iii) of $O(e^{-\bar{\lambda}s})$ as $s\to\infty$ for some $\bar{\lambda}>0$ and, whenever needed, we may assume (iv) $G'(s) < 0$ is strictly monotonically increasing to $\lim_{s\to\infty} G'(s) = 0$.

Constitutive equations such as (1.1) may be justified in various ways (see Saut and Joseph [11] and Renardy [9]). We considered two singular problems in which the velocity is assumed to be in the form $\mathbf{V} = \hat{e}_y v(x,t)$ in the semi-infinite space above a flat plate and

$$\Omega = [x,y,z;\ 0<x<\infty,\ -\infty<y<\infty,\ -\infty<z<\infty].$$

At x=0 we imagine either a step-jump in velocity or displacement, satisfying

$$\mu\frac{\partial^2 v}{\partial x^2}(x,t) + \int_0^t G(s)\frac{\partial^2 v}{\partial x^2}(x,\ t-s)\,ds = \rho\frac{\partial v}{\partial t}(x,t). \tag{1.2}$$

$$v(x,\ 0) = 0,$$
$v(x,\ t)$ is bounded as $x,\ t\to\infty$.

And for step-increase in velocity at x=0

$$v(0,\ t) = H(t-0)\ . \tag{1.3}$$

For the step-increase in displacement of the bottom plate we have

$$v(0,t) = \delta(t). \tag{1.4}$$

§1.1 Linearized Simple Fluids of Maxwell Type (μ=0).

The solution of problem (1.2) and (1.3) is given in §4-6 of [6] as:

$$v(x,\ t) = f(x,\ t)\ H(t-\alpha x)$$

where

$$c = \frac{1}{\alpha} = \sqrt{G(0)/\rho}$$

and $f(x,t)$ is defined in (5.10) of [6]. Here it will suffice to note that (see [10], [6], and [2])

$$a(x) \overset{\text{def}}{=} f(x,\alpha x^{+}) = \exp(\alpha x G'(0)/2G(0)).$$

(1.5) $$\frac{\partial f}{\partial t}(x, \alpha x^{+}) = -\alpha x \exp\left(\frac{\alpha x G'(0)}{2G(0)}\right)\left[\frac{3}{8}\left(\frac{G'(0)}{G(0)}\right)^{2} - \frac{1}{2}\frac{G''(0)}{G(0)}\right].$$

$$\frac{\partial f}{\partial x}(x, \alpha x^{+}) = \alpha f(x, \alpha x^{+})\left[\frac{G'(0)}{2G(0)} + \alpha x\left\{\frac{3}{8}\left(\frac{G'(0)}{G(0)}\right)^{2} - \frac{1}{2}\frac{G''(0)}{G(0)}\right\}\right].$$

If $G(s) = ke^{-\mu s}$, then $\frac{3}{8}\left(\frac{G'(0)}{G(0)}\right)^{2} - \frac{1}{2}\frac{G''(0)}{G(0)} = -\frac{1}{2}\mu^{2} < 0$.

The solution of step-displacement problem (1.2) and (1.4) is given as (see in (10.7) of [6]).

(1.6) $$v(x, t) = \frac{\partial f}{\partial t}(x,t)\, H(t-\alpha x) + f(x, \alpha x^{+})\, \delta(t-\alpha x)$$

where $f(x, t)$ is the same as in (1.5).

§1.2 Special Kernels for Fluids of the Maxwell Type ($\mu=0$).

There are two special cases ($G'(0) = -\infty$, $G'(0)=0$):

(i) $G'(0) = -\infty$ and $0<G(0)<\infty$.

In this case the amplitude $a(x)$ of the shock (given in (1.5)) is zero. Thus the discontinuity of the data is removed but the support of the solution propagates with the speed $c = \frac{1}{\alpha}$.

In fact Renardy [8] has shown that for a kernel (used in certain molecular models)

$$G'(s) = -\sum_{n=1}^{\infty} \exp(-n^{\alpha}s), \quad \alpha>1,$$

$$G'(0) = -\infty,$$

$$G(0) = \sum_{n=1}^{\infty} \frac{1}{n^{\alpha}}$$

the solution is C^{∞} smooth at the support (see Fig. 1.1). It may be noted that the special kernel used by Renardy is such that all of its derivatives at $s=0$ are unbounded; that is the contact between the vertical axis and the curve $G(s)$ at $s=0$ is C^{∞} smooth. Some form of continuity of solution on kernels possessing nearly identical features globally might be expected. For example we may construct kernels with $G'(0)=-\infty$, and even with C^{∞} contact at the vertical axis whose graphs are indistinguishable from kernels for which $G'(0)$ is finite in all neighborhoods bounded away from $s=0$. This may lead to smooth, shock like solutions (see Fig. 1.1). Such problems are in some sense like the ones which are perturbed with a small viscosity μ. We shall remark in §1.3, that the small viscosity leads to a transition layer of size μ which collapses onto a shock as $\mu\to 0$. For small μ the solution is smooth, but shock like (see Figs. 1.1, 1.3). The heuristic argument for the equivalence of problems for kernels of type (i) with those perturbed by a small viscosity is as follows. We are given $G(s)$, $s>0$ such that $G(0)$ is finite, $G'(s) < 0$, $s > 0$, and $G'(0) = -\infty$. Now we implement the construction of a comparison kernel of Maxwell type. First choose a small time ε. Then, at $G(\varepsilon)$ draw the tangent $G'(\varepsilon)$. This tangent pierces $s=0$ at the value $G_M(0)$. Define $G_M(s)$

$$G_M(s) = \begin{cases} G'(\varepsilon)s + G_M(0), & s \le \varepsilon \\ G(s), & s > \varepsilon . \end{cases}$$

We may write

$$\int_0^t G(s)\frac{\partial^2 v}{\partial x^2}(x,t-s)\,ds = \int_0^t G_M(s)\frac{\partial^2 v}{\partial x^2}(x,t-s)\,ds$$

$$+ \int_0^{\varepsilon}(G(s) - G_M(s))\frac{\partial^2 v}{\partial x^2}(x,t-s)\,ds .$$

Using the mean value theorem the last integral may be written as

$$\varepsilon[G(\bar{s}) - G_M(\bar{s})]\ \frac{\partial^2 v}{\partial x^2}(x,\ t-\bar{s}),\ 0<\bar{s}<\varepsilon.$$

Then with $\varepsilon\to 0$ we get $\bar{s}(\varepsilon)\to 0$ and we approximate the perturbing term with

$$\varepsilon[G(0) - G_M(0)]\frac{\partial^2 v}{\partial x^2}(x,\ t).$$

The approximating problem is like one perturbed by a small viscosity $\mu=\varepsilon[C(0)-G_M(0)]$.

The reader may notice that the heuristic argument just given applies to any two kernels which coincide for $s>\varepsilon$. The implication is that an approximation to the solution corresponding to one kernel may be obtained by solving a problem with the other kernel, perturbed by a viscous term with a suitably selected viscosity coefficient.

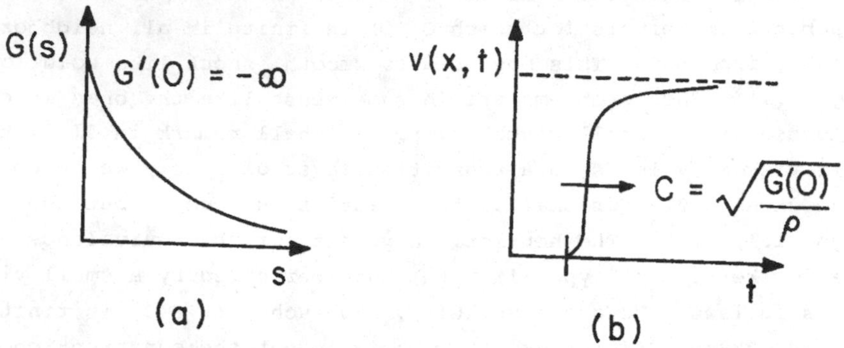

Fig. 1.1.: Propagating smooth solutions (b) occur when G(s), satisfying (i), is as sketched in (a).

To establish the above heuristic argument, we let $f(x,t) = f_1(x,t)$ in (5.10) of [6] be the solution for the kernel with $G'(0) = -\infty$ and let $f(x, t) = f_2(x, t)$ be the solution for the comparison kernel $\tilde{G}(s)$, $G(0)=\tilde{G}(0)$, $\tilde{G}'(s)$ is finite for $0\leq s<\varepsilon$ and $\tilde{G}(s) = G(s)$ for $s\geq\varepsilon$. Then by choosing small ε we reduce the value of $|\overline{G}(iy) - \overline{\tilde{G}}(iy)|$. Now invoking the continuity of (5.10) of [6] with respect to $r(y)$ and $p(v)$, we find that $|f_1(x,t) - f_2(x,t)|$ is small.

In the second special case we have

(ii) $G'(0) = 0.$

In this case, $a(x) = 1$, and

$$\frac{\partial f}{\partial t}(x, \alpha x^+) = \tfrac{1}{2}\alpha x\left[\frac{G''(0)}{G(0)}\right].$$

It is necessary that $G''(0) \geq 0$ if G is to be monotonically decreasing in $[0,\infty)$. For the case in which $G''(0) > 0$ there will be a velocity overshoot in the neighborhood of $t=\alpha x$ at all x.

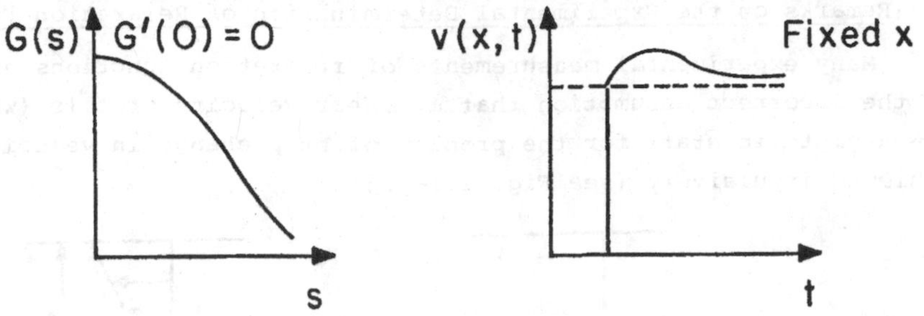

Fig. 1.2: Shock profile for the case G'(0) = 0.

§1.3 Viscosity and Transition Layers

Consider the problem of a step increase of velocity for Newtonian fluids ($\mu>0$, $\tilde{\mu}(s) \equiv 0$ in (1.1)). The classical solution of this problem ((1.2), (1.3)) is given by:

(1.7) $$v(x, t) = \operatorname{erfc}(x/\sqrt{4\nu t})$$

where $\nu=\frac{\mu}{\rho}$, and erfc is the complementary error function.

If $\mu>0$ is small and G has the assumed properties, it can be shown (see §18 of [6]) that there is a transition layer around the shock solution with $\mu=0$. This smooth transition layer exists in a bounded domain of $\{(x,t)\ x\geq 0$ and $t\geq 0\}$ and its thickness scales with μ. Thus:

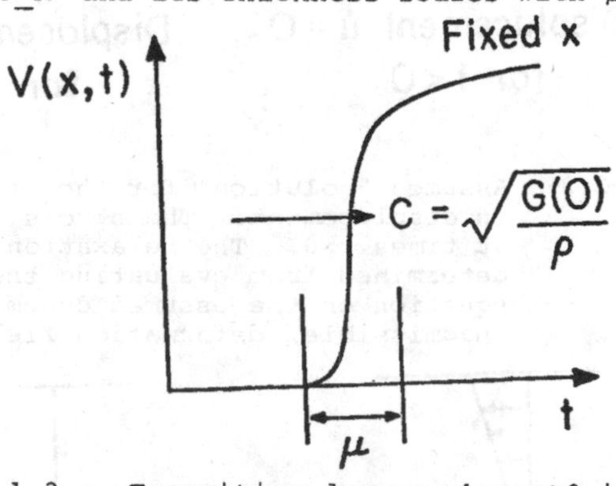

Fig. 1.3.: Transition layers when $\mu>0$ is small.

§2 Remarks on the Experimental Determination of Relaxation Functions.

Many experimental measurements of relaxation functions are based on the incorrect assumption that a linear velocity profile (which is the $t \to \infty$ asymptotic state for the problem of step change in velocity) can be achieved impulsively (see Fig. 2.1-2.3)

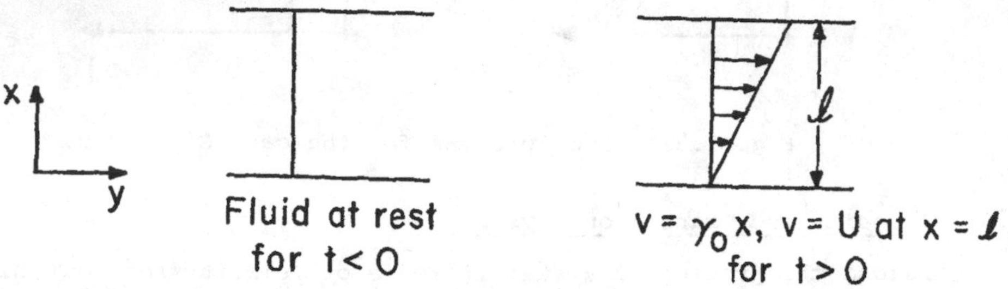

Fig. 2.1: Assumed "solution for the step increase in velocity. The stress is measured after times $t > 0$. The relaxation function is determined from the constitutive equation on the assumed, dynamically inadmissible, velocity field.

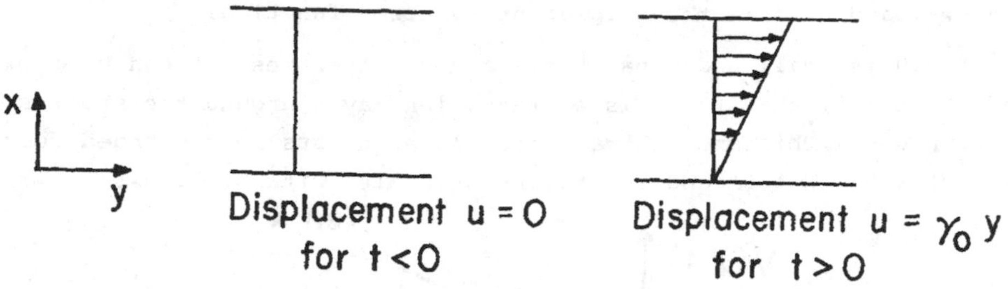

Fig. 2.2: Assumed "solution" for the step increase in displacement. The stress is measured at times $t > 0$. The relaxation function is determined from evaluating the constitutive equation on the assumed dynamically inadmissible, deformation field.

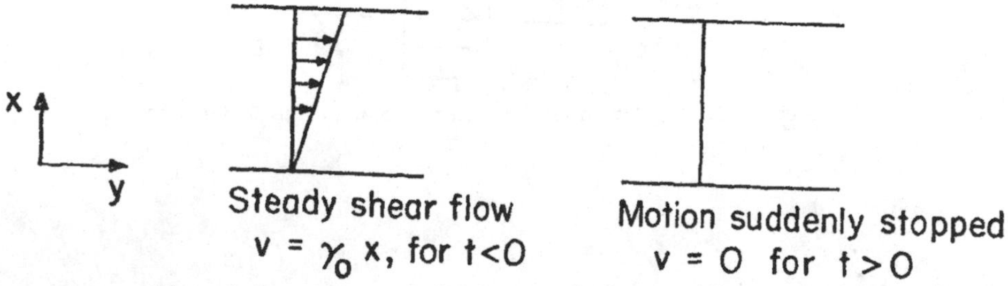

Fig. 2.3: Assumed "solution" for sudden cessation of motion. The stress is measured at times $t > 0$. The relaxation function is determined

from evaluating the constitutive equation on the assumed dynamically inadmissible deformation field.

However we have shown that the deformation assumed in Figs. 2.1 - 2.3 cannot be achieved at finite times on solutions of the initial-boundary value problem in the realm of linear viscoelasticity. The deformations assumed are in fact limiting cases for $t\to\infty$. It is therefore necessary to explain how and in what sense the customary methods of determining relaxation functions have validity. The following observations are important:

(1) The customary methods can always be used to measure "viscosity" ($\mu + \int_0^\infty G(s)ds$) by measurement at large times. But the test is inadequate to determine separately μ and $\int_0^\infty G(s)ds$.

(2) Suppose $\mu=0$, then the customary methods measure the stress on the stationary plate as a funciton of time. We are here concerned with the question whether the experimental measurement is going to be close to the relaxation function as indicated by the assumed kinematics of Fig. 1.4. In the context of linear viscoelasticity, we will show that this experimental measurement will never give the integral $\int_0^t G(s)ds$ for small time t near zero. However, this integral can be close to measured values for large times provided that the half life time of discontinuities is small. For simple Maxwell models with non-zero values of $|G'(0)|$, this time can be estimated as $-G(0)/G'(0)$.

To obtain expressions for the shear stress at the wall we consider the dynamics solution given in §8 of [6] for the step increase in velocity (see Fig. 2.1). In that solution the moving plate is at x=0 and the stationary plate is at $x=\ell$. For the case in which the moving plate is at $x=\ell$ we ultimately have simple shear $U(1-\frac{x}{\ell})$ as $t\to\infty$ with shear rate $\frac{\partial v}{\partial x} \overset{\text{def}}{=} -\dot{\gamma}_0 = -\frac{U}{\ell}$. The solution of this problem is:

$$(2.1)\qquad v(x,t) = U[f(x,t)H(t-\alpha x) + \{f(x+2\ell,t)H(t-\alpha(x+2\ell)) - f(2\ell-x,t)H(t-\alpha(2\ell-x))\} + \{\cdots\} + \cdots].$$

The stress at the wall x=0 and $x=\ell$ is given by:

$$(2.2)\qquad T^{\langle xy\rangle}(0,t) = \int_0^t G(s)\,\frac{\partial v}{\partial x}(0,t-s)\,ds$$

and

$$(2.3)\qquad T^{\langle xy\rangle}(\ell,t) = \int_0^t G(s)\,\frac{\partial v}{\partial x}(\ell,t-s)\,ds.$$

If we assume an instantaneous deformation as in Fig. 2.1, then (1.5) implies that

$$(2.4)\qquad T^{<xy>}(x,t) = -\frac{U}{\ell}\int_0^t G(s)\,ds, \quad x\in[0,\ell].$$

However (2.1) implies that

$$(2.5)\qquad \frac{\partial v}{\partial x}(0,t) = U[\{\frac{\partial f}{\partial x}(0,t)H(t-0) - \alpha f(0,t)\delta(t-0)\}$$

$$+2\{\frac{\partial f}{\partial x}(2\ell,t)H(t-(2\alpha\ell)) - \alpha f(2\ell,t)\delta(t-(2\alpha\ell))\}$$

$$+2\{\cdots\} + \cdots],$$

and

$$(2.6)\qquad \frac{\partial v}{\partial x}(\ell,t) = 2U[\frac{\partial f}{\partial x}(\ell,t)H(t-\alpha\ell) - \alpha f(\ell,t)\delta(t-\alpha\ell)$$

$$+\{\frac{\partial f}{\partial x}(3\ell,t)H(t-(3\alpha\ell)) - \alpha f(3\ell,t)\delta(t-(3\alpha\ell))\}$$

$$+\{\cdots\} + \cdots].$$

Combining (2.5) and (2.2), we find that in the time interval $0<t<\alpha(2\ell)$, the stress at the driving plate is

$$(2.7)\qquad T^{<xy>}(0,t) = U\int_0^t G(t-s)\frac{\partial f}{\partial x}(0,s)\,ds - U\alpha G(t)f(0,0^+),$$

but equation (1.5) implies that

$$Uf(0,0^+) = v(0,0^+) = U.$$

Hence,

$$(2.8)\qquad -T^{<xy>}(0,0^+) = U\sqrt{\rho G(0)}.$$

Combining (1.13) and (1.10) we get

$$(2.9)\qquad T^{<xy>}(\ell,t)=0 \text{ for } 0<t<\alpha\ell \text{ and}$$

$$(2.10)\qquad -T^{<xy>}(\ell,\alpha\ell^+) = 2U\sqrt{\rho G(0)}\,\exp(\frac{\alpha\ell G'(0)}{2G(0)}).$$

In general, for $t>(2n\alpha\ell)$; $n=1, 2, \ldots$ we find by combining

(2.5) and (2.6) with (2.2) and (2.3) that

$$(2.11)\qquad -T^{<xy>}(0,t) = [-\int_0^t G(t-s)\frac{\partial f}{\partial x}(0,s)\,ds + \alpha G(t)]$$
$$+2[-\int_{(2\alpha\ell)}^t G(t-s)\frac{\partial f}{\partial x}(2\ell,s)\,ds + G(t-(2\alpha\ell))\exp\left(\frac{G'(0)}{2G(0)}2\alpha\ell\right)]$$
$$+2[\cdots] + \cdots$$

and

$$(2.12)\qquad -T^{<xy>}(\ell,t) = 2U[\alpha G(t-\alpha\ell)f(\ell,\alpha\ell^{+}) - \int_{\alpha\ell}^t G(t-s)\frac{\partial f}{\partial x}(\ell,s)\,ds]$$
$$+2U[\alpha G(t-(3\alpha\ell))f(3\ell,(3\alpha\ell)^{+}) - \int_{(3\alpha\ell)}^t G(t-s)\frac{\partial f}{\partial x}(3\ell,s)\,ds]$$
$$+2[\cdots] + \cdots$$

In order to understand (2.11) and (2.12), we need to know some features of the function $\frac{\partial f}{\partial x}(2n\ell,t)$ for n=0, 1, 2,.... For a Maxwell fluid $G(s) = Ke^{-\mu s}$ and (see (7.3) of [6]):

$$(2.13)\qquad -\frac{\partial f}{\partial x}(x,t) = -U\sqrt{\frac{\rho\mu^2}{K}}\,\frac{\partial \hat{f}}{\partial \hat{x}}(\hat{x},\hat{t})\quad\text{where } x = \sqrt{\frac{K}{\rho\mu^2}}\,\hat{x}\ ,\ t = \frac{1}{\mu}\hat{t}$$

$$\frac{\partial \hat{f}}{\partial \hat{x}}(\hat{x},\hat{t}) = -\frac{1}{2}e^{-\frac{\hat{x}}{2}} - \frac{\hat{x}}{8}e^{-\frac{\hat{x}}{2}} + \frac{1}{2}\int_{\hat{x}}^{\hat{t}}\frac{e^{-\frac{\sigma}{2}}}{\sqrt{\sigma^2-\hat{x}^2}}\,I_1\left(\tfrac{1}{2}\sqrt{\sigma^2-\hat{x}^2}\right)d\sigma$$
$$+\frac{\hat{x}^2}{2}\int_{\hat{x}}^{\hat{t}}\frac{e^{-\frac{\sigma}{2}}}{(\sigma^2-\hat{x}^2)}\left\{\frac{I_1(\frac{1}{2}\sqrt{\sigma^2-\hat{x}^2})}{\sqrt{\sigma^2-\hat{x}^2}} - \frac{1}{2}I_1'\left(\tfrac{1}{2}\sqrt{\sigma^2-\hat{x}^2}\right)\right\}d\sigma$$

We also recall that when a steady state $v(x,\infty) = \frac{U(\ell-x)}{\ell}$ is approached we have

$$(2.14)\qquad \lim_{t\to\infty} T(x,t) = -\frac{U}{\ell}\int_0^\infty G(s)\,ds,\quad x\in[0,\ell].$$

There are two cases to consider: (i) $\sqrt{G(0)\rho} > \ell^{-1}\int_0^\infty G(s)ds$ and (ii) $\sqrt{G(0)\rho} < \ell^{-1}\int_0^\infty G(s)ds$. In the first case the initial value of the stress is larger than the final value (overshoot). A typical graph is sketched in Fig. 2.4(i). In the second case there is a jump of stress less than the steady state value. This case is sketched in Fig. 2.4(ii).

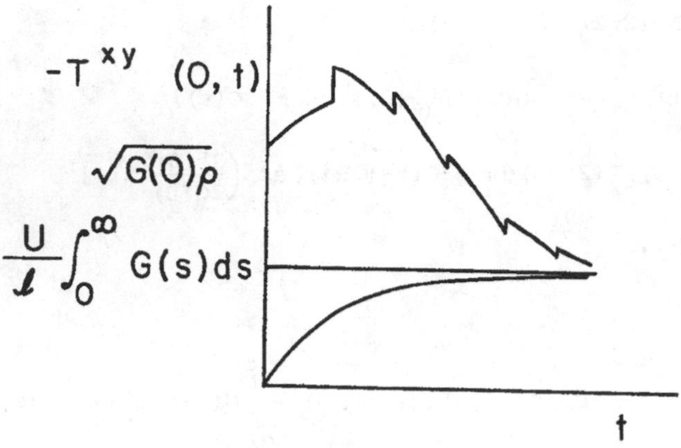

Fig. 2.4 (i)

Stress development at the lower wall of a channel filled with a viscoelastic fluid of Maxwell type under a step change of shear.

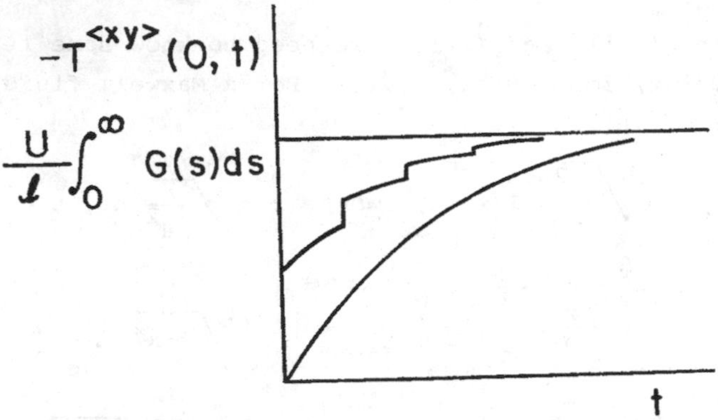

Fig. 2.4 (ii)

Stress development at the lower wall of a viscoelastic fluid of Maxwell type under a step change of shear.

Of course the amplitude of jumps in Fig. 2.4 (i), (ii) ultimately tend to steady state value. Moreover in the two special cases $G'(0) = -\infty$ or $\mu>0$ and small we will have essentially the same response as in Figs. 2.4 with smooth bumps replacing jumps. In any experiment the jumps (for $\mu=0$) would not be vertical because step changes at the boundary are discontinuous idealizations of smooth rapid changes and if $v(0,t)$ is a continuous function close to $UH(t-0)$, then $T^{<xy>}(0,0^+) = 0$ but $T^{<xy>}(0, \varepsilon_1) \cong U\sqrt{G(0)\rho}$ and $T^{<xy>}(\ell, \varepsilon_2) \cong 2U\sqrt{G(0)\rho}\, \exp(\frac{\alpha\ell G'(0)}{2G(0)})$ for some $\varepsilon_1, \varepsilon_2 > 0$ and small. This observation follows as a consequence of the continuous dependence of the solution on the data [6] and our solution for arbitrary initial data.

The aforementioned results may be applied to the interpretation of experiments by Meissner [5], Huppler et al [3], among others. They plot

$$\frac{T^{<xy>}(0,t)}{T^{<xy>}(0,\infty)} \underset{=}{\text{def}} \frac{\eta^+(t)}{\eta_0}$$

where

$$T^{<xy>}(0,t) \underset{=}{\text{def}} -\dot{\gamma}_0\,\eta^+(t)$$

$$= -\frac{U}{\ell}\eta^+(t),$$

$$\eta_0 = \int_0^\infty G(s)\,ds$$

Our analysis shows that at the driving plate

$$\frac{\eta^+(0^+)}{\eta_0} = \frac{\ell\sqrt{G(0)\rho}}{\int_0^\infty G(s)\,ds}$$

where $\dfrac{\eta^+(\infty)}{\eta_0} = 1.$

The stress response at the stationary wall is given by

$$\frac{T^{<xy>}(\ell,\alpha\ell^+)}{T^{<xy>}(\ell,\infty)} = \frac{\eta^+(\alpha\ell^+)}{\eta_0} = \frac{2\ell\sqrt{\rho G(0)}}{\int_0^\infty G(s)\,ds}\exp\left(\frac{\alpha\ell G'(0)}{2G(0)}\right)$$

where, $\dfrac{\eta^+(\infty)}{\eta_0} = 1$

Typical representations of experimental results of various authors are represented schematically in Fig. 2.5 (cf. Bird, Armstrong and Hassager, [1] Fig. A.4-9).

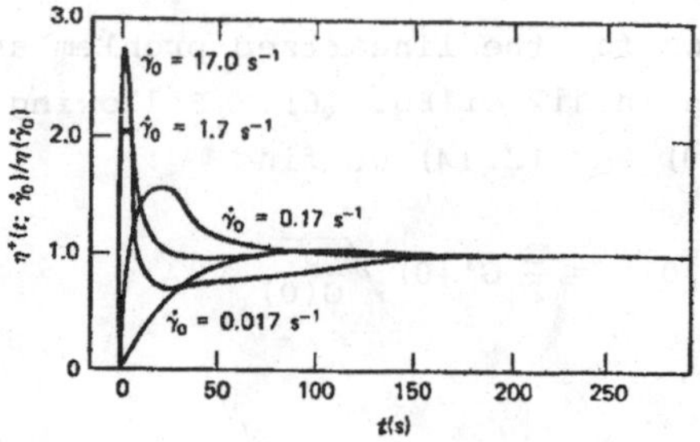

Fig. 2.5

Representations of stress development in a parallel plate channel under a step change of velocity. These representations are supposed to represent the results of experiments.

The experimental results represented in Fig. 2.5 do not exhibit the stress jumps, at small rates of shear, which are required by linearized dynamics. It is possible that the conditions of the experiments were such as to make the initial jumps in stress small relative to asymptotic ($t\to\infty$) levels of stress. However, stress overshoot could possibly occur even in the realm of linear theory. We cannot know whether or not overshoots do occur without reliable estimates of $G(0)$. The methods which are presently used to determine $G(0)$ are inadequate because they do not apply at small times. Some micro-molecular models like those of Kee and Carreau [4], have tried to explain this overshoot by allowing for such features in a "non-linear stress-strain history model" when evaluated at a kinematic assumption of Fig. 1.4. We believe it is now apparent that any such modeling on the above experimental data is meaningless if the dynamics are going to be neglected.

It is perhaps also possible to explain the oscillations at small times in the stress observed by Meissner [5] in terms of larger amplitudes of stress which are generated by reflections off bounding walls for fluids of the type which support shocks or near shocks (fluids with $G(0)<\infty$, $-G'(0)\leq\infty$ with or without a small viscosity.) Nonlinearity also participates in the results observed at high shears. For example, the narrowing of the width of peak region in the graphs shown in Fig. 2.5 may not be entirely explained by linear theory.

§2.1 The stress response for the step displacement problem

This problem is associated with Fig. 2.2. The kinematic assumption mentioned in the caption of that figure leads to a direct formula

$$\frac{U}{\ell}\, G(t) = -T^{\langle xy\rangle}(0,t)\ . \tag{2.15}$$

The dynamic solution for the linearized problem associated with that experiment is given in §12 of Eq. [6]. Following procedures used to obtain (2.8), (2.10) and (2.14) we find

$$-T^{\langle xy\rangle}(0,0^+) = \frac{U}{2}\, G'(0)\sqrt{\frac{\rho}{G(0)}} < 0 \tag{2.16}$$

and

$$T^{\langle xy\rangle}(0,\infty) = 0\,. \tag{2.17}$$

At the stationary plate we have

$$(2.18)\qquad -T^{<xy>}(\ell,\alpha\ell^+) = 2U[-G(0)\frac{\partial f}{\partial x}(\ell,\alpha\ell^+) + \alpha G'(0)\, f(\ell,\alpha\ell^+)]$$

and

$$T^{<xy>}(\ell,\infty) = 0 .$$

Eq. (2.15) may be a correct representation of linearized dynamics for large t but it is a false representation of linearized dynamics for small t.

§2.2 Summary

The asymptotic values of G(t) for large t can be obtained in the usual way using the kinematic assumptions exhibited in Figs. 1.5 and 1.6. The early time behavior of G(t) is not well represented by the asymptotic solution and at least should be correlated with the results of dynamic analysis. In the context of linearized dynamics which should be valid at least for small shears, we find that

$$-T^{<xy>}(0,0^+) = U\sqrt{G(0)\rho} ,$$

$$-T^{<xy>}(\ell,\alpha\ell^+) = 2U\sqrt{\rho G(0)}\ \exp\ (\alpha\ell\ G'(0)/2G(0))$$

for the step change in velocity. Here ℓ is the distance from the driving plate at x=0 to the stationary plate and $\alpha\ell$ is the time of first reflection. In the problem of the step change in displacement, we find that

$$-T^{<xy>}(0,0^+) = \frac{U}{2}\, G'(0)\sqrt{\frac{\rho}{G(0)}} < 0 ,$$

$$-T^{xy}(\ell,\alpha\ell^+) = 2U\,[-G(0)\ \frac{\partial f}{\partial x}(\ell,\alpha\ell^+) + \alpha G'(0)\ f(\ell,\alpha\ell^+)]$$

It may be useful to reinterpret existing experimental results in terms of the dynamic theory. For example, the constants κ_i and μ_i appearing in the Maxwell model with finitely many relaxation times

$$G(s) = \sum_{i=1}^{N}\kappa_i e^{-\mu_i s}$$

could, in principle, be determined by comparing experimental results with formulas which could be obtained from the analysis of the type of Kazakia and Rivlin [10].

It may be true that conclusions similar to the ones which we have considered here for experiments with viscoelastic fluids apply in the theory of viscoelastic solids [7].

§3 Cylindrical vortex sheets generated by sudden spin up of a cylinder in a fluid

The problem of spin up was considered in §14 of [6]. In this case the velocity of shearing motion is in circles

$$\underline{V}\ (\underline{x},t) = w(r,t)\underline{e}_\theta$$

and w(r,t) is defined in

$$\mathcal{D} = \{r \geq a,\ 0 \leq \theta \leq 2\pi,\ -\infty < z\}.$$

The boundary value problem for sudden spin up is given (see (14.8) of [6]) by

$$\rho\frac{\partial w}{\partial t}(r,t) = \int_0^t G(s)\left[\frac{\partial^2 w}{\partial r^2}(r,t-s) + \frac{1}{r}\frac{\partial w}{\partial r}(r,t-s) - \frac{w(r,t-s)}{r^2}\right] ds$$

(3.1)

$$w(a,t) = \begin{cases} a\,\Omega = 1 & \text{for } t > 0 \\ 0 & \text{for } t < 0, \end{cases}$$

$$w(r,0) = 0 \quad , \qquad r \geq a > 0,$$

w(r,t) is bounded as r, t→∞ .

We showed in [6] that the solution of (3.1) is given by

(3.2) $$w(r,t) = g(r,t)\ H(t-(r-a)\alpha).$$

where g(r,t) is defined in (14.16) of [6]. Here we derive a simpler form for $g(r,\ \alpha(r-a)^+)$ than the one given by (14.19) of [6]. This derivation follows along lines leading to the formulas (5.21), (5.23) in [6].

We know from (14.11) of [6] that

(3.3) $$w(r,t) = \frac{1}{2\pi i}\int_{\gamma-i\infty}^{\gamma+i\infty} \frac{e^{ut}}{u}\ \frac{K_1\left(r\sqrt{\frac{\rho u}{G(u)}}\right)}{K_1\left(a\sqrt{\frac{\rho u}{G(u)}}\right)}\ du; \qquad \text{Re } u > 0$$

where K_1 is a modified Bessel function whose asymptotic form is given by

$$(3.4)\qquad K_1(z) = \sqrt{\frac{\pi}{2z}}\ \exp(-z) + 0(\frac{1}{z}) .$$

The asymptotic expansion

$$(3.5)\qquad \bar{G}(u) = \frac{G(0)}{u} + \frac{G'(0)}{u^2} + 0(\frac{1}{u^3})$$

was established as (5.16) of [6].

It is easy to verify that:

$$(3.6)\qquad \sqrt{\frac{\rho u}{\bar{G}(u)}} = \sqrt{\frac{\rho}{G(0)}}\ u - \sqrt{\frac{\rho}{G(0)}}\ \frac{G'(0)}{2G(0)} + 0(\frac{1}{u})$$

$$= \alpha u - \frac{\alpha G'(0)}{2G(0)} + 0(\frac{1}{u}) .$$

Equation (3.4) and (3.6) imply that

$$(3.7)\qquad \frac{K_1\left(r\sqrt{\frac{\rho u}{\bar{G}(u)}}\right)}{K_1\left(a\sqrt{\frac{\rho u}{\bar{G}(u)}}\right)} = \sqrt{\frac{a}{r}}\ \exp\left[(r-a)\frac{\alpha G'(0)}{2G(0)}\right] \exp\left[(-\alpha u(r-a)\right] + 0(\frac{1}{u})$$

$$= \sqrt{\frac{a}{r}} \exp\left[(r-a)\frac{\alpha G'(0)}{2G(0)}\right] \exp\left[-\alpha u(r-a)\right] + 0(\frac{1}{u}) .$$

Substituting (3.7) into (3.3), we get:

$$(3.8)\qquad w(r,t) = \frac{1}{2\pi i} \exp\left[\frac{(r-a)\alpha G'(0)}{2G(0)}\right] \int_{\gamma-i\infty}^{\gamma+i\infty} \sqrt{\frac{a}{r}}\ \frac{e^{u\{t-\alpha(r-a)\}}}{u}\, du$$

$$+ \frac{1}{2\pi i}\int_{\gamma-i\infty}^{\gamma+i\infty} e^{ut}\ 0(\frac{1}{u^2})\ du$$

$$= \sqrt{\frac{a}{r}} \exp\left[\frac{(r-a)\ \alpha\ G'(0)}{2G(0)}\right] H(t-\alpha(r-a))$$

$$+ \frac{1}{2\pi i}\int_{\gamma-i\infty}^{\gamma+i\infty} e^{ut}\ 0(\frac{1}{u^2})\ du .$$

The last term in (3.8) is continuous $r \geq a$ and $t \geq 0$ because the integral is uniformly convergent for any fixed r,t. Comparing (3.8) with (3.2) while using the continuity of the second term in (3.8) we get

$$(3.9)\qquad g(r, \alpha(r-a)^+) = \sqrt{\frac{a}{r}} \exp\left[\frac{(r-a)\ \alpha\ G'(0)}{2G(0)}\right] .$$

The decay with r of cylindrical vortex sheets is more rapid than plane sheets which damp according to (1.5) without the factor $r^{-1/2}$.

We next consider the problem of reflections off the walls of concentric cylinders which bound a fluid occupying the region

$$\hat{\mathcal{D}} = \{a < r \leq b,\ 0 \leq \theta < 2\pi,\ -\infty < z < \infty\}.$$

The spin up problem may be stated as follows

$$(3.10)\quad \begin{aligned} \rho \frac{\partial w}{\partial t} &= \int_0^t G(s) \left[\frac{\partial^2 w}{\partial r^2}(r, t-s) + \frac{1}{r}\frac{\partial w}{\partial r}(r,t-s) - \frac{w(r,t-s)}{r^2} \right] ds, \\ w(a, t) &= \begin{cases} a\Omega = 1 & \text{for } t > 0 \\ 0 & \text{for } t \leq 0, \end{cases} \\ w(b, t) &= 0 \qquad \forall\, t \in \mathbb{R}, \\ w(r, 0) &= 0 \qquad \forall\, r \in [a, b], \\ &w(r, t) \text{ is bounded as } r, t \to \infty. \end{aligned}$$

We now utilize the method of Laplace transforms, following arguments given in §6 of [6] and find that

$$(3.11)\qquad w(r, t) = \frac{1}{2\pi i} \int_{\gamma-i\infty}^{\gamma+i\infty} e^{ut}\, \bar{w}(r, u)\, du$$

where

$$(3.12)\qquad \begin{aligned} \bar{w}(r, u) &= \frac{1}{u} \frac{I_1(b\eta(u))K_1(r\eta(u)) - K_1(b\eta(u))I_1(r\eta(u))}{K_1(a\eta(u))I_1(b\eta(u)) - K_1(b\eta(u))I_1(a\eta(u))}, \\ \eta(u) &= \sqrt{\frac{\rho u}{\bar{G}(u)}}. \end{aligned}$$

An asymptotic form for (3.11) follows from combining the asymptotic expressions for $|z| \to \infty$

$$(3.13)\qquad \begin{aligned} I_1(z) &= \frac{e^z}{\sqrt{2\pi z}} + 0(\tfrac{1}{z}), \\ K_1(z) &= \sqrt{\frac{\pi}{2z}}\, e^{-z} + 0(\tfrac{1}{z}) \end{aligned}$$

with (3.12). Thus

$$\bar{w}(r,\ u) = \sqrt{\frac{a}{r}}\ \frac{e^{(b-r)\ \eta(u)} - e^{-(b-r)\ \eta(u)}}{e^{(b-a)\ \eta(u)} - e^{-(b-a)\ \eta(u)}} + O(\frac{1}{u}).$$

Hence

$$(3.14) \qquad w(r,\ t) = \sqrt{\frac{a}{r}}\ \frac{1}{2\pi i} \int_{\gamma-i\infty}^{\gamma+i\infty} \frac{e^{ut}}{u}\ \frac{e^{(b-r)\eta(u)} - e^{-(b-r)\ \eta(u)}}{e^{(b-a)\eta(u)} - e^{-(b-a)\ \eta(u)}}\ du$$

$$+ \frac{1}{2\pi i} \int_{\gamma-i\infty}^{\gamma+i\infty} e^{ut}\ \ O(\frac{1}{u^2})\ du\ .$$

We next note that the first term in (3.14) is the same as in (8.3)-(8.7) of §8 of [6] if we set $r-a = x$, and $b-a = \ell$. The second term in (3.14), being uniformly convergent for any r and t, is a continuous function of r and t. Thus

$$(3.15) \qquad \begin{aligned} w(r,\ t) = \sqrt{\frac{a}{r}}\ \Big[& f(x,\ t)\ H(t-\alpha x) + \{f(x+2l,\ t) \\ & H(t-\alpha(x+2l)) - f(2l-x,\ t)\ H(t-\alpha(2l-x))\} \\ & + \cdots\cdots \Big] + \quad h(x,\ t)\ . \end{aligned}$$

The function f in (3.15) is the same f appearing in (1.5) while $h(x,t)$ is continuous for $x = r-a \in [0, \ell]$ and $t \geq 0$. It follows from (3.15) that discontinuities are reflected along the characteristic lines shown in Fig. 3.1.

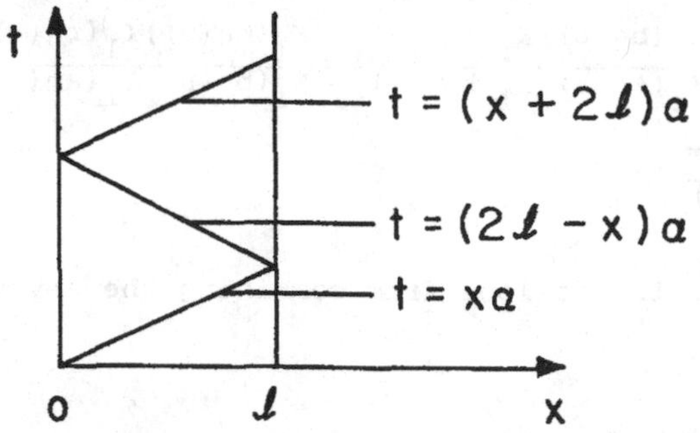

Fig. 3.1. Characteristic lines for reflection from the walls of concentric cylinders, $x = r-a$, $\ell = b-a$

The asymptotic steady state in case of flow governed by (1.20) is given by:

$$\lim_{t\to\infty} w(r,t) = \lim_{u\to 0} u\, \bar{w}(r,u) .$$

Using (1.22) and

$$\left.\begin{aligned} I_1(z) &\sim \frac{z}{4} \\ K_1(z) &\sim \frac{1}{z} \end{aligned}\right\} \quad \text{as } z \to 0 \text{ and } \operatorname{Re} z > 0$$

we get

$$(3.16) \qquad \lim_{t\to\infty} w(r,t) = \frac{\frac{b}{r} - \frac{r}{b}}{\frac{b}{a} - \frac{a}{b}} = \frac{a}{r}\,\frac{b^2-r^2}{b^2-a^2} \quad , \quad \text{for } a\Omega = 1.$$

§4 Solutions of start up problems with arbitrary boundary data by integrals of Duhamel's type

A shearing motion is initiated at x=0 by data of the form

$$V(0,t) = \begin{cases} g(t) & , t > 0 \\ 0 & , t < 0 \end{cases}$$

where g(t) is an arbitrary function (possessing a Laplace transform). The velocity V(x,t) then satisfies

$$(4.1) \qquad \int_0^t G(s)\frac{\partial^2 v}{\partial x^2}(x,t-s)\,ds = \rho\frac{\partial v}{\partial t}(x,t),$$

$$v(0,t) = g(t),\ \text{where } g(t) \equiv 0,\ t<0,$$

$$v(x,0) = 0,\quad x \geq 0,$$

$$v(x,t) \text{ is bounded}.$$

We shall solve (4.1) by superposition using the solution of the following singular problem:

$$\int_0^t G(s)\ \frac{\partial^2 u}{\partial x^2}(x,t-s)\,ds = \rho\ \frac{\partial u}{\partial t}(x,t),$$

$$(4.2) \qquad \begin{aligned} u(0,t) &= \delta(t-\tau),\ \tau\in(0,t), \\ u(x,0) &= 0,\quad x \geq 0, \\ u(x,t) &\ \text{is bounded for } x,\ t\to\infty. \end{aligned}$$

It is easy to see and not hard to prove that the solution of (4.2) is the time-derivative of the solution of (3.1) where

$$v(0,t) = \begin{cases} 1 \text{ for } t > \tau \\ 0 \text{ for } t \leq \tau . \end{cases}$$

It then follows that the solution of (4.2) is

$$(4.3) \qquad u(x,t) = \frac{\partial f}{\partial t}(x,\ t-T)\ H(t-\tau-\alpha x)$$

$$+\ f(x,\ \alpha x^{+}) \quad (t-\tau-\alpha x).$$

Of course (4.3) can be obtained directly as the inverse of the Laplace transform of (4.2). (The details of this type of calculation are given in §10 of [6]). We note that t in the upper limit of integration in the integral on the left of (4.2), may be replaced with $t+\delta$, $\delta>0$ because $u(x,-\delta) = 0$ for $\delta>0$. The interpretation of the δ function which this implies may be expressed as follows: for any h(s) such that $h(s) = 0$, $s<0$ we have

$$\int_{-\infty}^{\infty} h(s)\ \delta(s)\,ds = \int_{0}^{\infty} h(s)\ \delta(s)\,ds = h(0).$$

We now assert that the solution of problem (4.1) is a linear superposition (integration) of the function $g(\tau)\ u(x,t)$. This is true because

$$(4.4)\qquad v(0,t) = g(t) = \int_0^{\infty} g(t-\eta)\ \delta(\eta)\,d\eta = \int_0^{t} g(t-\eta)\ \delta(\eta)\,d\eta$$

$$= \int_0^{t} g(\tau)\ \delta(t-\tau)\,d\tau.$$

Using (4.3) and (4.4), we find that the solution of (4.1) is

$$(4.5)\qquad v(x,t) = \int_0^{t} g(\tau)\ u(x,t)\,d\tau$$

$$= \int_0^{t} g(\tau)\left[\frac{\partial f}{\partial t}(x,\ t-\tau)\ H(t-\tau-\alpha x) + f(x,\ \alpha x^{+})\,\delta(t-\tau-\alpha x)\right] d\tau.$$

It follows from (2.5) that if

$$(4.6)\qquad t-\alpha x<0 \text{ then } v(x,t) = 0.$$

This implies that the information of rest prior to start-up is always preserved. On the other hand, when $t-\alpha x > 0$, (4.5) gives

$$(4.7)\qquad v(x,t) = \int_0^{t-\alpha x} g(\tau)\ \frac{\partial f}{\partial t}(x,t-\tau)\ d\tau + f(x,\ \alpha x^{+}) \int_0^{t-\alpha x} g(t-\alpha x-\eta)\,\delta(\eta)\,d\eta$$

$$= \int_0^{t-\alpha x} g(\tau)\ \frac{\partial f}{\partial t}(x,t-\tau)\,d\tau + f(x,\ \alpha x^{+})\ g(t-\alpha x)$$

$$= \int_0^{t-\alpha x} g(\tau)\ \frac{\partial f}{\partial t}(x,t-\tau)\,d\tau + \exp\left(\frac{\alpha x G'(0)}{2G(0)}\right) g(t-\alpha x).$$

It is easy to verify that (4.7) reduces to (1.5) for $g(\tau) = H(\tau)$ and (1.6) for $g(\tau) = \delta(\tau)$. Eqs. (4.6) and (4.7) together constitute the solution of the problem posed in (4.1). Thus we conclude that discontinuities in the boundary values of g or its derivatives propagate into the interior with speed $C=1/\alpha$. Hence (4.7) also proves that any discontinuity in a start-up problem of linear viscoelasticity can come only through the boundary data. It is also clear from (4.7), that this propagating discontinuity is exponentially damped.

We turn next to the construction of the solution of start-up problems between parallel plates. The problem to be solved may be expressed as:

$$(4.8) \qquad \int_0^t G(s) \frac{\partial^2 v}{\partial x^2} (x, t-s)\, ds = \rho \frac{\partial v}{\partial t},$$

$$v(0,t) = g(t) \quad ; \quad \text{when } g(t) \equiv 0, \ t<0,$$

$$v(\ell,t) = 0,$$

$$v(x,0) = 0,$$

$$v(x,t) \text{ is bounded as } t \to \infty.$$

Proceeding as in the previous problem we first consider the case in which $g(t) = \delta(t-\tau)$. The $\hat{v}(x,t)$ for this singular problem is given by

$$\hat{v}(x,t) = [\psi(x,t-\tau) + \{\psi(x+2l, t-\tau) - \psi(2l-x,t-\tau)\}$$

$$+ \cdots\cdots]$$

where

$$(4.9) \qquad \psi(x,t) = \frac{\partial f}{\partial t}(x,t)\, H(t-\alpha x) + f(x,\alpha x^+)\, \delta(t-\alpha x).$$

The function $f(x, t)$ in (4.9) is defined by (1.5). We now use the principle of superposition to compose the solution of (4.8) in Duhamel form

$$(4.10) \qquad v(x,t) = \int_0^t g(\tau)\, \hat{v}(x,t)\, d\tau$$

$$= \int_0^t g(\tau) \left[\frac{\partial f}{\partial t}(x,t-\tau) + \left\{ \frac{\partial f}{\partial t}(x+2l, t-\tau) - \frac{\partial f}{\partial t}(2l-x,t-\tau) \right\} \right.$$

$$\left. + \cdots \right] H(t-\tau-\alpha x)$$

$$+ [f(x,\alpha x^+)\ g(t-\alpha x) + \{f(2l+x,\ \alpha(2l+x)^+)\ g(t-\alpha x)$$

$$- f(2l-x,\ \alpha(2l-x))\ g(t-\alpha x)\} + \cdots]\ d\tau$$

The dots in (4.10) represent similar terms arising out of repeated reflection between the walls at $x=0$ and $x=\ell$ of the original characteristic $t-\alpha x$ = const. It also follows from (4.10) that $v(x,t)=0$ when $t-\alpha x < 0$ and, when $t-\alpha x > 0$ we find that:

$$(4.11) \qquad v(x,t) = \int_0^{t-\alpha x} g(\tau) \left[\frac{\partial f}{\partial t}(x,t-\tau) + \left\{ \frac{\partial f}{\partial t}(x+2l,t-\tau) - \frac{\partial f}{\partial t}(2l-x,t-\tau) \right. \right.$$

$$\left. + \cdots \right] d\tau + \left[f(x,\alpha x^+) g(t-\alpha x) + \{f(2l+x,\ \alpha(2l+x)^+) \right.$$

$$\left. g(t-\alpha x) - f(2l-x,\alpha(2l-x)^+)\ g(t-\alpha x)\} + \cdots \right].$$

We may use (4.11) to study the interactions of multiple shocks generated by multiple discontinuities in the boundary data $g(t)$. For example, consider

$$(4.12) \qquad g(t) = \begin{cases} 0 & \text{for } t < 0, \\ 1 & \text{for } 0 \le t \le 1, \\ 0 & \text{for } t > 1. \end{cases}$$

It follows from (4.11) that the discontinuities of $g(t)$ propagate along the characteristic lines $t-\alpha x=0$ and $t-\alpha x=1$ and their repeated reflections, as in Fig. 4.1

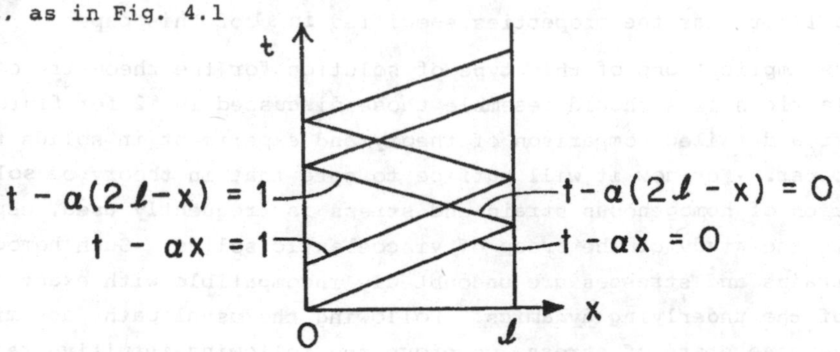

Fig. 4.1. Propagation of the singular data given by (4.12)

§5. Linear viscoelastic solids

In [7] we used Laplace transforms to study problems of singular boundary data in viscoelastic solids. If $G(0) > 0$ and $G'(0) < 0$ are finite, steps in displacement initiated at the boundary will propagate into the interior of the solid. We interpreted this to mean that the materia fails in shear. Stress relaxation experiments in solids are in some sense modeled by steps in displacement. As in fluids, it is necessary to understand the underlying dynamics of such problems. To pursue such an understanding we study the following initial-boundary value problem with smooth, but otherwise arbitrary boundary data $g(t)$, $t \geq 0$:

$$\rho \frac{\partial^2 v}{\partial t^2} = (\mu+G(0)) \frac{\partial^2 v}{\partial x^2} + \int_0^t \frac{dG}{ds}(s) \frac{\partial^2 v}{\partial x^2}(x,t-s)\,ds,$$

$$v(0,t) = g(t) \; ; \quad g(t) = 0 \quad t < 0, \tag{5.1}$$

$$v(x,0) = \frac{\partial v}{\partial t}(x,0) = 0,$$

$$v(x,t) \text{ is bounded as } x,\ t \to \infty.$$

We can solve (5.1) using the methods which led to the Duhamel type of integrals displayed in equations (4.1-4.7). Thus

$$v(x,t) = \int_0^t g(\tau) \frac{\partial \hat{f}}{\partial t}(x,t-\tau)\,d\tau + \hat{f}(x, \alpha x^+)\, g(t-\alpha x) \tag{5.2}$$

where

$$\alpha = \sqrt{\frac{\rho}{\mu+G(0)}}$$

and $\hat{f}(x,t) = f(x,t)$ where $f(x,t)$ is defined by equation (3.10,11) of [7] and $f(x,t)$ has the properties specified in §1 of this paper.

The implications of this type of solution for the rheometry of viscoelastic solids should resemble those discussed in §2 for fluids. We defer a detailed comparison of theory and experiment in solids to a later paper. For now it will suffice to note that in theory of solids the notion of homogeneous strain and stress is frequently used, especially in the study of the creep of viscoelastic solids. Such homogeneous strains and stresses are undoubtedly incompatible with exact analysis of the underlying dynamics. Following the usual path, assuming a homogeneous state of stress, we prove the following intuitive result: If the homogeneous stress in a linear viscoelastic solid relaxes monotonically in step-strain tests, then the longitudinal strain in the same solid increases monotonically in creep tests (see Fig. 5.1). To prove this we note that the stress T in a linear viscoelastic solid undergoing uni-axial strain $\varepsilon(x,t) = \frac{\partial u}{\partial x}(x,t)$ is given by $T = (\mu + G(0))\,\varepsilon(t) + \int_0^\infty \frac{dG}{ds}\,\varepsilon(t-s)\,ds$. A montonically decreasing stress relaxation for a homogeneous step-strain implies that G satisfies assumptions (i)-(ii) listed under (1.1). We have assumed either that $G'(0) \neq 0$ or $G''(0) \neq 0$. The strain ε defining creep is governed by

$$(5.3)\qquad T = (\mu + G(0))\,\varepsilon(t) + \int_0^t \frac{dG}{ds}(s)\,\varepsilon(t-s)\,ds = \begin{cases} 1 & \text{for } t > 0 \\ 0 & \text{for } t < 0 \,. \end{cases}$$

By taking various limits of (5.3) and its derivative we can show that

$$\left.\begin{aligned} \varepsilon(0^+) &= \frac{1}{\mu+G(0)}\,, \\ \varepsilon'(0^+) &= \frac{-G'(0)}{\{\mu+G(0)\}^2} > 0\,, \text{ if } G'(0) \neq 0 \\ \varepsilon''(0^+) &= -G''(0)/\{\mu+G(0)\}^2 \text{ if } G'(0) = 0 \text{ and } G''(0) \neq 0\,, \\ \lim_{t\to\infty} \varepsilon(t) = \varepsilon^* &= \frac{1}{\mu} \end{aligned}\right\}.$$

It is easy to verify, using (5.3) that $\varepsilon(t)$ is continuous and $\varepsilon'(t)$ exists for any $t > 0$. We want to prove that

$$(5.5)\qquad \varepsilon'(t) > 0\,, \quad \forall\ t > 0\,.$$

If (5.5) is not true, then (using (5.4)) there exists a $\bar{t} > 0$ such that

$$(5.6)\qquad \varepsilon'(\bar{t}) = 0 \quad \text{and} \quad \varepsilon'(t) > 0, \forall t \in [0,\bar{t}]\,.$$

By differentiating (5.3) once with respect to t, we find that

$$(5.7)\qquad (\mu + G(0))\,\varepsilon'(t) + G'(t)\,\varepsilon(0) + \int_0^t G'(s)\,\varepsilon'(t-s)\,ds = 0, \forall\ t > 0\,.$$

After evaluating (5.7) at $t = \bar{t}$, using (5.6), we get:

$$(5.8) \qquad G'(t)\, \varepsilon(0) \; + \int_0^{\bar{t}} G'(s)\, \varepsilon'(\bar{t}-s)\, ds = 0 .$$

But (5.8) then leads to a contradiction because the assumptions about $G(s)$ make the left side of (5.8) strictly negative. It follows that $\varepsilon'(t) > 0$ and not ≤ 0. It is not hard to demonstrate that $\varepsilon'(t) > 0$ when

$$(5.9) \qquad G(s) = a\, \delta(s) + h(s)$$

where $a > 0$, $h(s)$ satisfies the assumptions under (1.1) and $\delta(s)$ is a Dirac measure at the origin.

Graphical represntations of the monotonicity result are exhibited in Fig. 5.1 below:

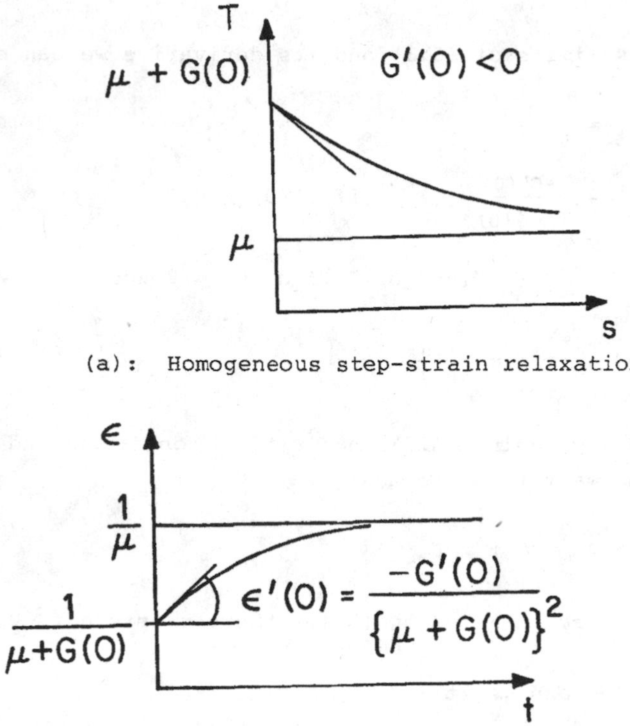

(a): Homogeneous step-strain relaxation

(b) Creep response to a homogeneous step in stress

Fig. 5.1: Relation between stress relaxation and creep.

When $a > 0$ in (5.9), the response to a step increase in stress is monotonic as in Fig. 5.1 (b), but it passes through the origin as in Fig. 5.2.

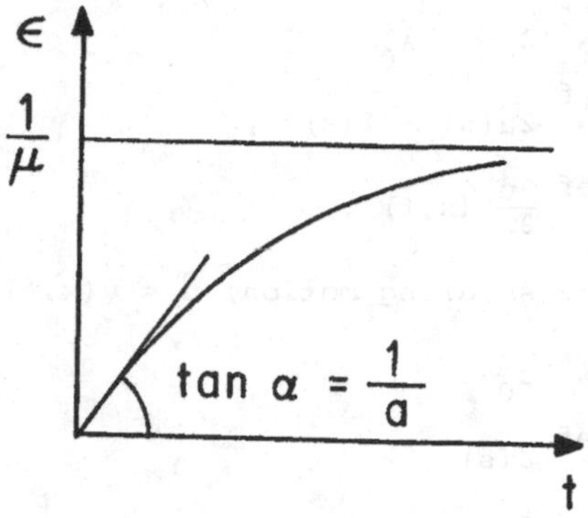

Fig. 5.2: Creep response for kernels of the type (5.9)

We close by reminding the reader that the type of response which we have described above depends tacitly on the unfounded and actually incorrect assumption that homogeneous step-strain (relaxation) and step-stress (creep) tests are admissible deformations compatible with dynamics.

Acknowledgement: The results of this paper are taken from the Ph.D. thesis of A. Narain. This work was supported by the U.S. Army.

Note added in proof: The following works which are relevant to the analysis of this paper have been brought to our attention:
P. W. Bucher and F. Mainaridi, Asymptotic expansions for transient viscoelastic waves, Journal de Mécanique, Vol. 14, No. 4, 1975 and R. M. Christensen, Theory of Viscoelasticity: An introduction, Academic Press, 1971. These works give expansions of solutions near the shock and develop full solutions for special choices of the kernel G(s).

We can extend the results of our work in [7] and in §5 of this paper to one dimensional longitudinal motion in the following way. If the symmetric part of the displacement gradient is denoted by $\hat{\nabla}\underset{\sim}{u}$, then the Cauchy stress $\underset{\sim}{T}$ for an isotropic linear viscoelastic solid is given by:

$$\underset{\sim}{T} = 2\mu_0\,\hat{\nabla}\underset{\sim}{u} + \lambda_0(\mathrm{tr}(\hat{\nabla}\underset{\sim}{u}))\,\underset{\sim}{1} + 2\int_0^\infty \frac{d\mu}{ds}(s)\,\{\hat{\nabla}\underset{\sim}{u}(x,t-s)-\hat{\nabla}\underset{\sim}{u}(x,t)\}ds$$

$$+\left\{\int_0^\infty \frac{d\lambda}{ds}(s)\,\{\mathrm{tr}(\hat{\nabla}\underset{\sim}{u}(x,\ t-s)) - \mathrm{tr}(\hat{\nabla}\underset{\sim}{u}(x,t))\}\,ds\right\}\underset{\sim}{1}$$

On the one hand a longitudinal displacement, $\underset{\sim}{u} = u(x,t)\hat{i}$ is governed by (5.3) with

$$\mu \overset{\text{def}}{=} 2\mu_0 + \lambda_0 \quad ,$$

$$G(s) \overset{\text{def}}{=} 2\mu(s) + \lambda(s) \quad ,$$

$$\varepsilon(t) \overset{\text{def}}{=} \frac{\partial u}{\partial x}(x,t) \quad .$$

On the other hand shearing motion, $\underset{\sim}{u} = v(x,t)\ \hat{j}$ satisfy (5.3) with

$$\mu \overset{\text{def}}{=} \mu_0 \ ,$$

$$G(s) \overset{\text{def}}{=} \mu(s) \ ,$$

$$\varepsilon(t) \overset{\text{def}}{=} \frac{\partial v}{\partial x}(x,t).$$

References

1. Bird, R.B., R.C. Armstrong, O. Hassager: Dynamics of Polymeric Liquids, Vol. I, John Wiley, New York, 1977.

2. Coleman, B.D. and M.E. Gurtin: Waves in Materials with Memory II. On the Growth and Decay of one dimensional Acceleration Waves, Arch. Rational Mech. Anal. 19, 239-265 (1965).

3. Huppler, J.D., I.F. MacDonald, E. Ashare, T.W. Spriggs, R.B. Bird and L.A. Holmes: Rheological Properties of three solutions. Part II. Relaxation and growth of shear and normal stresses. Trans. Soc. of Rheology, 11, 181-204 (1967).

4. Kee, D.D. and P.J. Carreau: A constitutive equation derived from Lodge's Network Theory. J. of Non-Newtonian Fluid Mechanics, 6, 127-143 (1979).

5. Meissner, J: Modifications of the Weissenberg Rheogonimeter for Measurement of Transient Rheological Properties of Molten Polyethylene under shear. Comparison with tensile data. J. of Appl. Polym. Sci., Vol. 16, pp. 2877-2899 (1972).

6. Narain, A. and D.D. Joseph: Linearized dynamics for step jumps of velocity and displacement of shearing flows of a simple fluid. Rheologica Acta 21, 228-250 (1982).

7. Narain, A. and D.D. Joseph: Classification of linear viscoelastic solids based on a failure criterion. Accepted and to appear in Journal of Elasticity (1982).

8. Renardy, M.: Some remarks on the propagation and non-propagation of discontinuities in linearly viscoelastic fluids. Rheol. Acta. 21, 251-254 (1982).

9. Renardy, M.: On the domain space for constitutive laws in linear viscoelasticity, (to appear).

10. Kazakia, J.Y. and Rivlin, R.S.: Run-up and spin-up in a viscoelasti fluid I. Rheol. Acta 20, 111-127 (1981).

11. Saut, J.C. and D.D. Joseph: Fading Memory, to appear in Arch. Rational Mech. and Anal. (1982).

CORRIGENDUM I

Linearized Dynamics for Step Jumps of Velocity and Displacement of Shearing Flows of A Simple Fluid, by A. Narain and D. D. Joseph (Rheol. Acta 21, 228-250 (1982)).

1) The quantities $G(s)$, $C_t(t)$ and $A_1(t)$ are tensors and should be in boldface.

2) The equations under (3.4) should read

$$\lambda^t(x,s) = 0 \quad \text{for} \quad t \leq 0$$

and

$$\frac{d\lambda^t}{ds}(x,s) = -\frac{\partial v}{\partial x}(x,t-s).$$

3) The first sentence under Fig. 5.1 should read "Now for $t-\alpha x<0$ ..."

4) Eqn (6.8) should be replaced by:

$$Mv_n \to Mv \equiv \int_{\alpha x}^{t} G(t-s) f_{xx}(x,s)ds - 2\alpha G(t-\alpha x) f_x(x,\alpha x^+)$$
$$+\alpha^2 G'(t-\alpha x) f(x,\alpha x^+) - \alpha^2 G(t-\alpha x) \frac{\partial f}{\partial t}(x,\alpha x^+)$$
$$- \rho\frac{\partial f}{\partial t}(x,t).$$

5) Eqn. (6.11) should be replaced by:

$$2\frac{\partial f}{\partial t}(x, \alpha x^+) + \frac{2}{\alpha}\frac{\partial f}{\partial x}(x,\alpha x^+) = \frac{G'(0)}{G(0)} f(x,\alpha x^+)$$

6) The left side of (14.11) should be replaced by

$$\frac{\omega(r,t)}{a\Omega}$$

7) Eqn. $(14.12)_{(ii)}$ is:

$$K_1(Z) \sim \sqrt{\frac{\pi}{2z}} \exp(-z) \quad \text{as} \quad |z|\to\infty.$$

8) Eqn. (16.5) can be ignored.

9) The sentence under Eqn. (4.5) should read "Eqs. (4.3, 4.5) imply ... ... half-plane Re $u>-\lambda$."

10) The equations between (14.3) and (14.6) should be numbered (14.4) and (14.5).

11) The left side of the equation above (14.6) should read $\mathcal{y}\langle r\theta\rangle(t)$ in place of $\mathcal{y}\langle \pi\theta\rangle(t)$.

12) The left side of Eqn (6.7) should read:

$$\frac{\partial^2 v_n}{\partial x^2}$$

13) The definition of $\eta(u)$ underneath (10.4) is

$$\eta(u) = \sqrt{\frac{\rho u}{\bar{G}(u)}}$$

14) The eqn. (5.17) should read

$$\left[1 + \frac{G'(0)}{G(0)u} + \frac{G''(0)}{G(0)u^2} + 0(\frac{1}{u^3})\right]^{-\frac{1}{2}}$$

$$= 1 - \frac{\bar{\gamma}'}{2u} + \frac{3}{8}\frac{\bar{\gamma}''}{u^2} + \frac{\bar{\gamma}''}{u^2} + 0(\frac{1}{u^3}).$$

where
$$\bar{\gamma}' \overset{\text{def}}{=} \frac{G'(0)}{G(0)}$$

$$\bar{\gamma}'' \overset{\text{def}}{=} -\frac{1}{2}\,\frac{G''(0)}{G(0)}$$

15) The right side of (12.5) should read:

$$v(x,t) = U[g(x,t) + \{g(x+2l, t) - g(2\ell-x,t\} + \{\cdot\cdot\} + \cdots\cdots]$$

CORRIGENDUM II

For the paper: Linearized Dynamics of Shearing Deformation Perturbing Rest in Viscoelastic Materials" by A. Narain and D.D. Joseph.

1) In §1.2, case (ii) of $G'(0) = 0$ we have $G''(0) \leq 0$. For the case of $G''(0) < 0$, Fig. 1.2 should show an undershoot as opposed to overshoot shown in the Figure.

2) Equation $(2.13)_2$ should read:

$$\frac{\partial\hat{f}}{\partial\hat{x}}(\hat{x},\hat{t}) = -\frac{1}{2}\exp(-\frac{\hat{x}}{2}) - \frac{\hat{x}}{2}\exp(-\frac{\hat{x}}{2}) + \frac{1}{2}\int_{\hat{x}}^{\hat{t}}\frac{\exp(-\frac{\sigma}{2})}{\sqrt{\sigma^2-\hat{x}^2}}\,I_1(\tfrac{1}{2}\sqrt{\sigma^2-\hat{x}^2})\,d\sigma$$

$$+ \frac{\hat{x}^2}{2}\int_{\hat{x}}^{\hat{t}}\frac{e^{-\frac{\sigma}{2}}}{(\sigma^2-\hat{x}^2)}\left\{\frac{I_1(\frac{1}{2}\sqrt{\sigma^2-\hat{x}^2})}{(\sigma^2-\hat{x}^2)^{-\frac{3}{2}}} - \frac{1}{2}\,I_1^1(\tfrac{1}{2}\sqrt{\sigma^2-\hat{x}^2})\right\}d\sigma$$

3) Equation (2.16) should read

$$-T^{<xy>}(0,0^+) = -\frac{U}{2}G'(0)\sqrt{\frac{\rho}{G(0)}} > 0.$$

4) **Equation (2.18) should read**

$$-T^{<xy>}(\ell,\alpha\ell^{+}) = -2U\alpha G(0)\exp\left(\frac{\alpha\ell G'(0)}{2G(0)}\right)\left[\alpha\ell\left\{\frac{3}{8}\left(\frac{G'(0)}{G(0)}\right)^{2} - \frac{G''(0)}{2G(0)}\right\} + \frac{G'(0)}{2G(0)}\right]$$

This correction should also be noted in the summary of §2.2

5) Equation (4.10) should read

$$v(x,t) = \int_0^t g(\tau)\,[\frac{\partial f}{\partial t}(x,t-\tau)H(t-\tau-\alpha x) + \{\frac{\partial f}{\partial t}(x+2\ell,t-\tau)H(t-\tau-\alpha(x+2\ell))$$

$$\frac{\partial f}{\partial t}(2\ell-x,t)H(t-\tau-\alpha(2\ell-x))\} + \cdots]d\tau$$

$$+[f(x,\alpha x^{+})\;g\,(t-\alpha x) + \{f(2\ell+x,\alpha(2\ell+x)^{+}g\,(t-\alpha(2\ell+x))$$

$$-f(2\ell-x,\alpha(2\ell-x)^{+})\,g(t-\alpha(2\ell-x))\} + \{\cdots\} + \cdots]$$

6) Equation (4.11) should read

$$v(x,t) = [\int_0^{t-\alpha x} g(\tau)\;\frac{\partial f}{\partial t}(x,t-\tau)d\tau + \{\int_0^{t-\alpha(x+2\ell)} g(\tau)\;\frac{\partial f}{\partial t}(2\ell+x,t-\tau)d\tau$$

$$-\int_0^{t-\alpha(2\ell-x)} g(\tau)\;\frac{\partial f}{\partial t}(2\ell-x,t-\tau)d\tau\} + \{\cdots\} + \cdots]$$

$$+[f(x,\alpha x^{+})g(t-\alpha x) + \{f(2\ell+x,\alpha(2\ell+x)^{+})$$

$$g(t-\alpha(2\ell+x) - f(2\ell-x,\alpha(2\ell-x)^{+})g(t-\alpha(2\ell-x))\}$$

$$+ \{\cdots\} + \cdots].$$

SEMIGROUPS OF OPERATORS IN BANACH SPACES

A. Pazy
Department of Mathematics
Hebrew University of Jerusalem
Jerusalem, Israel

0. PREFACE.

The purpose of this paper is to give a short and somewhat informal survey of the theory of semigroups of operators in Banach spaces. Due to lack of time and space the main stress will be put on the fundamental problem of the generation of such semigroups while other parts of the theory will be mentioned only very briefly and incompletely.

The paper will be divided into four parts:

1. Introduction
2. The linear theory
3. The nonlinear theory in Hilbert space
4. The nonlinear theory in Banach spaces.

1. INTRODUCTION.

Let X be a Banach space, $D \subset X$ be a subset of X.

<u>Definition</u>. A one parameter family of operators $\{S(t)\}_{t\geq 0}$, $S(t): D \to D$ is called a continuous (differentiable) semigroup of operators on D if

(i) $S(0) = I_{|D}$, $S(t+s) \equiv S(t)\circ S(s) = S(s)\circ S(t)$ (the semigroup property).

(ii) For every $x \in D$, $t \to S(t)x$ is continuous (differentiable) for $t \geq 0$.

The mapping $t \to S(t)$ is an algebraic homomorphism of the additive semigroup of nonnegative real numbers into the set of self-mappings of D and hence the name semigroup of operators.

The curve $t \to S(t)x$ is called the trajectory of the point x under $S(t)$ and the condition (ii) implies that all trajectories are continuous (differentiable) curves. We note that if $S(t)$ is a linear operator for every $t \geq 0$ then the continuity condition (ii) is referred to as strong continuity and $\{S(t)\}_{t\geq 0}$ is called a strongly continuous semigroup of operators.

One of the main sources of continuous semigroups are "well-posed" initial value problems. We pause now to explain this statement: Let $A: D(A) \subset X \to X$ be an operator in X and consider the initial value problem

$$\frac{du}{dt} = Au \quad \text{for} \quad t \geq 0, \qquad u(0) = x, \tag{1.1}$$

where $x \in X$. Assume that for every $x \in D(A)$ this initial value problem has a unique global solution $u(t;x)$ and define the solution operator $S(t)$, $t \geq 0$ by

$$(1.2) \qquad S(t)x = u(t;x) \quad \text{for} \quad x \in D(A).$$

Since $u(t;x)$ is a solution of (1.1) it is clear that $u(t;x) \in D(A)$ for $t \geq 0$ and therefore $S(t): D(A) \to D(A)$. Moreover, the assumed uniqueness of solutions of (1.1) implies readily that $S(t)$ has the semigroup property and since $t \to S(t)x$ is differentiable, $\{S(t)\}_{t\geq 0}$ is a differentiable semigroup on $D(A)$.

So far we have associated with an operator A, for which (1.1) is uniquely solvable for every $x \in D(A)$, a differentiable semigroup $\{S(t)\}_{t\geq 0}$ on $D(A)$.

In most cases we are not just interested in the existence and uniqueness of the solutions of (1.1) but also in their continuous dependence on the initial data $x \in X$. We will therefore restrict ourselves here to a very special class of initial value problems for which the solution operator $S(t)$ is Lipschitz continuous with constant $L(t)$, i.e.

$$(1.3) \qquad \|S(t)x - S(t)y\| \leq L(t)\|x-y\| \quad x, y \in D(A), t \geq 0$$

and $L(t)$ is bounded on bounded intervals. For such initial value problems the semigroup $\{S(t)\}_{t\geq 0}$ defined by (1.2) can be extended by continuity to a unique continuous semigroup $\{S(t)\}_{t\geq 0}$ on $\overline{D(A)}$. In the present lecture we will use the following somewhat nonstandard terminology.

<u>Definition</u>. The initial value problem (1.1) is "well-posed" on $\overline{D(A)}$ if for every $x \in D(A)$ it has a unique global solution $u(t;x)$ and the solution operator defined by (1.2) satisfies (1.3). In this case the extended continuous semigroup $\{S(t)\}_{t\geq 0}$ on $\overline{D(A)}$ is said to be infinitesimally generated by A and A is the infinitesimal generator of $\{S(t)\}_{t\geq 0}$.

We can now state two of the most fundamental problems of the theory of semigroups namely;

(I) Give sufficient and necessary conditions for an operator A to be the infinitesimal generator of a continuous semigroup $\{S(t)\}_{t\geq 0}$ on $\overline{D(A)}$ or, equivalently, for the initial value problem (1.1) to be well posed on $\overline{D(A)}$.

(II) Is every continuous semigroup $S(t)$ on a closed subset D of X infinitesimally generated by some operator A?

We note that once A satisfies sufficient conditions for (1.1) to be well-posed, the semigroup $\{S(t)\}_{t\geq 0}$ can be obtained from A by first solving (1.1) for $x \in D(A)$ and then extending $S(t)$ by continuity, to $\overline{D(A)}$. One very simple sufficient (but not necessary) condition for (1.1) to be well posed on X is that

A is Lipschitz continuous with some constant L on X. Indeed, if A is Lipschitz the initial value problem (1.1) possesses a unique global solution for every $x \in X$ by the classical Picard method and (1.3) follows from

$$\|S(t)x-S(t)y\| = \|x-y + \int_0^t (AS(\tau)x-AS(\tau)y)d\tau\| \leq \|x-y\| + L\int_0^t \|S(\tau)x-S(\tau)y\|d\tau$$

which by Gronwall's inequality implies

$$\|S(t)x - S(t)y\| \leq e^{tL}\|x-y\| .$$

In this particular case $D(A) = X$ and $\{S(t)\}_{t\geq 0}$ is actually a differentiable semigroup.

On the other hand given a semigroup $\{S(t)\}_{t\geq 0}$ on $D \subset X$ and assuming that it has an infinitesimal generator A, A is given by

(1.4) $$D(A) = \{x\colon x \in D, \lim_{t\to 0} \frac{S(t)x-x}{t} \text{ exists}\}$$

and

(1.5) $$Ax = \lim_{t\to 0} \frac{S(t)x-x}{t} \quad \text{for} \quad x \in D(A).$$

However, it is by no means clear that a given semigroup $\{S(t)\}_{t\geq 0}$ on a closed subset $D \subset X$ has an infinitesimal generator and indeed, in general, it does not have such an infinitesimal generator. It is therefore quite surprising that, by slightly restricting the problem, the answer to (II) is nevertheless positive in many cases.

Once the existence problems (I) and (II) have been (partially) settled the main object of the theory is to study the relations between the semigroup $\{S(t)\}_{t\geq 0}$ and its infinitesimal generator. Typical problems with which the theory deals are for example:

For a given $x \in X$ how fast does $S(t)x$ approach x as $t \to 0$? How should a sequence of infinitesimal generators A_n tend to A in order to assure the convergence of the corresponding semigroups $S_n(t)$ to $S(t)$ (the semigroup infinitesimally generated by A)? What properties of A make the function $t \to S(t)x$ differentiable for every $x \in D(A)$, C^∞, real analytic or given $t \to S(t)x$ other regularity properties. What properties of A are needed for the trajectories $S(t)x$, $t \geq 0$ to be bounded, compact, periodic or converge to a limit as $t \to \infty$?

2. THE LINEAR THEORY.

2.1. Linearity. In this Chapter we restrict ourselves to semigroups of linear operators on X and obtain rather complete answers to the problems (I) and (II).

We start by noting that from (1.4) and (1.5) it is obvious that if $S(t)$ is a linear operator on X for every $t \geqslant 0$ then $D(A)$ is a linear subspace of X and A is a linear operator. Slightly less obvious is the fact that if A, the infinitesimal generator of $\{S(t)\}_{t\geqslant 0}$, is linear so is the semigroup $\{S(t)\}_{t\geqslant 0}$, i.e. for every $t \geqslant 0$, $S(t)$ is a linear operator. This is a consequence of the required uniqueness of the solutions of (1.1). Indeed by the linearity of $\frac{d}{dt}$ and A we have for all scalars λ, μ and $x, y \in D(A)$

$$\frac{d}{dt}(\lambda S(t)x + \mu S(t)y) = A(\lambda S(t)x + \mu S(t)y),\ t \geqslant 0$$

and therefore by the uniqueness of the solution of (1.1) with x replaced by $\lambda x + \mu y$ it follows that

$$S(t)x + \ S(t)y = S(t)(\ x + y)$$

and hence $S(t)$ is a linear operator.

2.2. Uniformly continuous semigroups of linear operators. In the linear case the problems (I) and (II) become particularly simple if we insist that the semigroups with which we deal are continuous in the uniform operator topology on X rather than just strongly continuous as required by (ii). Semigroups satisfying this continuity condition are called uniformly continuous semigroups.

Let X be a Banach space and let $A: X \to X$ be a bounded linear operator on X. The initial value problem (1.1) has a unique solution for every $x \in X$ and we denote this solution by $S(t)x = e^{tA}x$. The existence and uniqueness of the solution of (1.1) here is a special case of the result described above for Lipschitz continuous operators A, or else can be proved directly be defining e^{tA} as a convergent power series and proving that $e^{tA}x$ is the unique solution of (1.1). Since A is bounded it is easy to show that $t \to e^{tA}$ is continuous in the uniform operator topology, Indeed, a simple computation shows that

$$\|e^{tA} - e^{sA}\| \leqslant |t-s|\, \|A\|\, e^{T\|A\|}$$

where $T = \max(|t|,|s|)$. Therefore every bounded linear operator is the infinitesimal generator of a uniformly continuous group of operators $\{e^{tA}\}$. A little more surprising but not difficult to prove is that every such group is infinitesimally generated by some bounded linear operator A on X. Combining the previous remarks we obtain the following complete answer to the problems (I) and (II) in the present setup.

THEOREM ([69], [50], [51]).

(i) Every bounded linear operator is the infinitesimal generator of a uniformly continuous group of bounded linear operators.

(ii) Every uniformly continuous group of bounded linear operators is infinitesimally generated by some bounded linear operator.

We note that this theorem provides us with a one to one correspondence between all uniformly continuous groups of bounded linear operators on X and all bounded linear operators on X. In this correspondence the operator A corresponds to the group e^{tA}, $-\infty < t < \infty$.

2.3. <u>The first result for unbounded generators</u> (Stone 1932). M. Stone studied groups of unitary operators in Hilbert space and proved the following remarkable result [62].

<u>THEOREM (Stone 1932)</u>.

(i) If A is a self adjoint (usually unbounded) operator in a Hilbert space H then iA is the infinitesimal generator of a group $e^{iAt} = U(t)$, $-\infty < t < \infty$ of unitary operators on H.

(ii) Every group of unitary operators on H is infinitesimally generated by an operator iA where A is self adjoint.

Stone's theorem gives us a one to one correspondence between all self adjoint operators in H and all groups of unitary operators on H. In spite of the formal similarity to the theorem of the previous section Stone's theorem is much deeper and considerably more difficult. The reason for this is that unlike the case where A is bounded, the infinitesimal generator iA here is not everywhere defined and it is badly discontinuous. For such operators the initial value problem (1.1) does not have an obvious solution as in the case of a bounded A. Moreover, the power series of $U(t)x = e^{iAt}x$ makes no sense, unless $x \in \bigcap_{n=1}^{\infty} D(A^n)$ and even in this case the convergence of the power series is not guaranteed.

The passage to unbounded operators however, is not done just for the sake of generality. In order to include all unitary groups on H in part (ii) of the theorem one is forced to deal also with those groups which are infinitesimally generated by unbounded operators.

2.4. <u>THE HILLE-YOSIDA THEOREM</u>

The study of semigroups of linear operators started in the early thirties but it was only in 1948 that E. Hille [35] and K. Yosida [70] succeeded independently to characterize the infinitesimal generators of strongly continuous semigroups of contractions, i.e. semigroups satisfying (1.3) with $L(t) \equiv 1$, on a Banach space X.

THEOREM (HILLE-YOSIDA)

(i) A densely defined linear operator A is the infinitesimal generator of a strongly continuous semigroup of contractions on X if $(\lambda I-A)^{-1}$ exists for all $\lambda > 0$ and $\|(\lambda I-A)^{-1}\| \leq \lambda^{-1}$ for $\lambda > 0$.

(ii) Every strongly continuous semigroup of contractions is infinitesimally generated by such an operator A.

The Hille-Yosida theorem gives a complete answer to the problems (I) and (II) for the case of semigroups of contractions. It turns out that the general case of strongly continuous semigroups of bounded linear operators can be reduced to the case of semigroups of contractions. In 1952, W. Feller [32], I. Miyadera [48] and R. S. Phillips [59] proved independently the following generalization of the Hille-Yosida theorem.

THEOREM (FELLER-MIYADERA-PHILLIPS)

A densely defined linear operator A is the infinitesimal generator of a strongly continuous semigroup of bounded linear operator on X if and only if there are numbers $\omega \in \mathbb{R}$ and $M \geq 1$ such that $(\lambda I-A)^{-1}$ exists for every $\lambda > \omega$ and $\|(\lambda I-A)^{-n}\| \leq M(\lambda-\omega)^{-n}$ for $\lambda > \omega$ and $n \geq 1$.

Here again we have a one to one corrrespondence between all strongly continuous semigroups of bounded linear operators on X and a class of unbounded closed linear operators in X. Thus for semigroups of bounded linear operators the problems (I) and (II) have a complete solution.

2.5. Some comments on the linear theory. After the characterization of the infinitesimal generators of semigroups of linear operators had been achieved the theory started to develop rather rapidly and it is up to now a rather active field of research.

1) Perturbation theory; R. S. Phillips [60] perturbations of the generator by a bounded operator, H. F. Trotter [66], K. Gustafson [34], P. Chernoff [20] Perturbations of semigroups of contractions, T. Kato [39], G. DaPrato [28], and others, Perturbations of analytic semigroups.

2) Regularity properties of semigroups; Analyticity, K. Yosida [71], E. Hille, R. S. Phillips [36], T. Kato [43], M. Crandall, A. Pazy, L. Tartar [26]. Differentiability, A. Pazy [52].

3) Continuous dependence of semigroups of operators on their generators, H. F. Trotter [65], P. Chernoff [19] and others.

4) Representation and approximation, E. Hille [35], K. Yosida [70], H. F. Trotter [65], P. Chernoff [19] and others.

5) Applications: Butzer and Berens [18] to approximation theory, W. Feller [33], E. B. Dynkin [29] to probability theory, H. Tanabe [64], K. Yosida [72], A. Pazy [54] to partial differential equations, H. Tanabe [64] to control theory, P. D. Lax and R. S. Phillips [44] to scattering theory and many more.

3. THE NONLINEAR THEORY IN HILBERT SPACE.

3.1. The characterization of the infinitesimal generator. The first successful attempt to study nonlinear semigroups of operators in a framework similar to the Hille-Yosida theory was done by Y. Komura [37] in 1967. In this work he gave sufficient conditions for an operator A defined in a real Hilbert space H to be the generator (in some sense) of a semigroup of not necessarily linear contractions. His conditions were essentially that A is monotone and maximal (in general such an operator need not be single valued). More precisely we assume that H is a real Hilbert space and define,

Definition. An operator $A: D(A) \subset H \to 2^H$ is monotone if $(x_1-x_2, y_1-y_2) \geq 0$ for every $y_i \in Ax_i$, $i = 1, 2$. A is maximal monotone if it has no proper monotone extension.

Komura showed that if $-A$ is maximal monotone then the initial value problem

$$\frac{du}{dt} + Au \ni 0 \quad \text{for} \quad t \geq 0, \quad u(0) = x \tag{3.1}$$

can be solved in some weak sense and that the solutions of (3.1) constitute a semigroup of contractions on $\overline{D(A)}$. The monotonicity of the operator $-A$ is equivalent to the contraction property of the semigroup. This can be seen formally as follows: Let u_1 and u_2 be solutions of (3.1) with initial data x_1 and x_2 respectively. The monotonicity of A implies that

$$\frac{1}{2}\frac{d}{dt}\|u_1-u_2\|^2 = \left(\frac{du_1}{dt} - \frac{du_2}{dt}, u_1-u_2\right) = -(v_1-v_2, u_1-u_2) \leq 0$$

where $v_i \in Au_i$. Therefore $\|u_1(t)-u_2(t)\| \leq \|x_1-x_2\|$ and $S(t)$ is a semigroup of contractions. On the other hand if $S(t)$ is a semigroup of contraction and $S(t)x$ satisfies (3.1) then

$$\frac{1}{t}(x_1-S(t)x_1 - x_2 + S(t)x_2, x_1-x_2) = \frac{1}{t}\|x_1-x_2\|^2 - \frac{1}{t}(S(t)x_1-S(t)x_2, x_1-x_2) \geq$$

$$\geq \frac{1}{t}\|x_1-x_2\|^2 - \frac{1}{t}\|S(t)x_1-S(t)x_2\| \cdot \|x_1-x_2\| \geq 0$$

and passing to the limit as $t \to 0$ this yields $(Ax_1 - Ax_2, x_1 - x_2) \geq 0$, i.e., the monotonicity of the infinitesimal generator A.

Soon after Komura's work T. Kato [40] clarified Komura's results and proved, in the case where A is single valued, that the initial value problem (3.1) has a strong solution for every $x \in D(A)$.

Finally, in 1969, a full characterization of the infinitesimal generators of semigroups of contractions on closed convex subsets of a Hilbert space H was achieved through the works of Y. Komura [38], T. Kato [41], M. Crandall and A. Pazy [24], [25].

THEOREM (Komura-Kato-Crandall-Pazy).

A (multivalued) operator -A in H is the infinitesimal generator of semigroup of contractions on a closed convex subset C of H if and only if $\overline{D(A)} = C$ and A is maximal monotone.

This theorem has several ingredients and we pause for a while to explain them. The sufficiency part of the theorem shows that if A is maximal monotone in H then $\overline{D(A)}$ is convex and the initial value problem (3.1) has a strong solution u for every $x \in D(A)$. Denoting $S(t)x = u(t)$ we obtain a semigroup of contractions on D(A) and this semigroup is extended by continuity to $\overline{D(A)}$. Moreover, it turns out that for $x \in D(A)$, $u(t) - S(t)x$ actually satisfies the equation

$$\frac{d^+u}{dt} + A^0u = 0 \tag{3.2}$$

where $\frac{d^+}{dt}$ is the right derivative and A^0 is the "lower section" of A defined by; A^0x is the unique element of minimum norm in the set Ax for $x \in D(A)$.

The necessity part of the thoerem shows that if $S(t): C \to C$ is a semigroup of contractions on the closed convex subset C of H then it has a densely defined infinitesimal generator -A'. This is perhaps the most tricky part of the proof and it was proved by Y. Komura in [38]. Komura's proof was considerably simplified by T. Kato in [41]. The infinitesimal generator -A' thus obtained is by definition single valued and by the above remarks A' is monotone. The fact that A' determines a unique maximal monotone operator A for which $A' = A^0$ was proved by M. Crandall and A. Pazy in [25]. This proof was subsequently simplified in [11].

The characterization of the infinitesimal generator of semigroups of contractions described above contains as a very special case the Hilbert space version of the Hille-Yosida theorem (see [24]). Here again restricting the problem to a Hilbert space and to semigroups of contractions we obtained a complete solution of (I) and (II)

3.2. Some results of the theory of nonlinear semigroups. After the characterization of the infinitesimal generators of semigroups of contractions was given in 1969 the theory was developed quite extensively. General references to this theory are the books of H. Brezis [10], V. Barbu [3], K. Yosida [72] and the lecture notes [53], [57].

To give some idea on the affluence of this rather recent theory we indicate some selected results.

It was realized quite early by T. Rockafellar [61] that the subdifferentials of lower semicontinuous convex functions from H into $\mathbb{R}$ are maximal monotone operators. It was H. Brezis who discovered in 1971 [7], that if $A = \partial\phi$ is the subdifferential of a lower semi continuous convex function ϕ then the semigroup $S(t)$ generated by $-A$ has a reularizing effect, namely $S(t): \overline{D(A)} \to D(A)$ for every $t > 0$. A similar regularizing effect holds for semi-groups generated by an operator $-A$ whose domain $D(A)$ has nonempty interior [10]. The characterization of generators of semigroups of contractions having this regularizing effect is still a wide open problem (In the linear case such a characterization exists even in Banach space [52].)

The maximal monotonicity of an operator A in H is of independent interest since it implies the solvability of the problem $u + \lambda Au \ni f$ for every $\lambda > 0$ and $f \in H$. An extensive theory of perturbations of maximal monotone operators was therefore developed, see, e.g. [24], [11], [10], [3] and others.

Results on the continuous dependence of semigroups of contractions on their infinitesimal generators which are completely analogous to the results of Trotter and Kato in the linear case were obtained for nonlinear semigroups of contractions by Brezis and Pazy [11] and Berrilan [4], see also [49].

As in the classical theory of ordinary differential equations, the asymptotic behavior of semigroups of contractions as $t \to \infty$ is both interesting and important. A large amount of effort is currently devoted to this subject, see e.g. [27], [16], [55], [56], [57].

The theory of semigroups of nonlinear contractions has important applications mainly to the solution of nonlinear initial value problems for partial differential equations. Many such applications are described in the basic paper [8] of H. Brezis.

In conclusion we mention that all the results concerning semi-groups of contractions in Hilbert space can be extended to semigroups satisfying

$$\|S(t)x - S(t)y\| \leq e^{\omega t}\|x-y\|$$

(see [53]). However, unlike the linear theory, we know essentially nothing about the generation theory of semigroups of Lipschitz operators with constant $L > 1$, i.e. semigroups satisfying

$$\|S(t)x - S(t)y\| \leq L\|x-y\| \quad \text{for} \quad t \geq 0.$$

In particular, we do not know whether or not such semigroups possess densely defined infinitesimal generators.

4. THE NONLINEAR THEORY IN BANACH SPACES.

4.1. Accretive operators. We turn now to the generation problem of semigroups of contractions in general Banach spaces. To this end the notion of monotonicity must be extended to Banach spaces. The appropriate notion is accretiveness which is defined as follows.

Definition. An operator $A: D(A) \subset X \to 2^X$ is accretive if

$$\|x_1-x_2+\lambda(y_1-y_2)\| \geq \lambda\|x_1-x_2\|$$

for all $\lambda > 0$ and $y_i \in Ax_i$, $i = 1, 2$. A is m-accretive if moreover the range $R(I+\lambda A)$ of $I+\lambda A$ is all of X for every $\lambda > 0$.

It is an easy exercise to check that if X is a Hilbert space the operator A is accretive in X if and only if it is monotone. Furthermore, it is also easy to see that if A is m-accretive and X is a Hilbert space then A is maximal monotone. The fact that maximal monotonicity in Hilbert space implies m-accretiveness is also true but not trivial (see Minty [47]).

Next we note that if A is a densely defined linear operator then $-A$ is the infinitesimal generator of a strongly continuous semigroup of contractions on X if and only if A is m-accretive. This, in fact, is the Lurner-Phillips [45] characterization of the infinitesimal generator of a strongly continuous semigroup of linear contractions.

The previous remarks indicate that one might hope that m-accretive operators ought to be the infinitesimal generators of semigroups of contractions in general Banach spaces. To what extent this hope is justified will be seen below.

4.2. Results in special Banach spaces. The first results on the generation of semigroups of nonlinear contractions in Banach spaces were obtained for special spaces. Already in 1967, T. Kato [42] showed that m-accretive operators generate semigroups of contractions in a Banach space X provided that X^* is uniformly convex. Similar results were also obtained by F. Browder [15]. Later, Crandall and Liggett [22] proved, somewhat indirectly, that m-accretive operators are the infinitesimal generators of semigroups of contractions in any reflexive Banach space.

If further assumptions are made on the space X, e.g. both X and X^* are uniformly convex one obatins existence results which are essentially the same as the results in Hilbert space, Crandall-Pazy [24], Kato [42], Brezis [9].

For a long time nothing was known about necessary conditions for an operator to generate a contraction semigroup even in the special situation where X and X^* are uniformly convex. The main difficulty in proving such results was to show that a semigroup of contractions in such a space has a densely defined infinitesimal generator. Only recently did B. Baillon succeed in proving this fact and thus obtained the following result.

THEOREM (Baillon [2]).

Let X be a Banach space with a uniformly convex adjoint X^* and let $S(t): C \to C$ be a semigroup of contractions on the closed convex set C. Then there exists a unique accretive operator A with the following properties:

(i) $\overline{D(A)} = C$, $R(I+\lambda A) \supset C$ for $\lambda > 0$.

(ii) For every $y \in Ax$ there is a $\lambda > 0$ such that $x + \lambda y \in C$.

(iii) $-A$ is the infinitesimal generator of $S(t)$.

Since $-A$ is the infinitesimal generator of $S(t)$, $u(t) = S(t)x$ satisfies

$$\frac{du}{dt} + Au \ni 0 \quad u(0) = x \quad \text{a.e. on} \quad t \geq 0 \tag{4.1}$$

for $x \in D(A)$ and therefore, in particular $S(t)$ has a densely defined infinitesimal generator. We conclude this section with the open problem of whether or not any semigroup of contractions on a closed convex subset of a reflexive Banach space possesses such an infinitesimal generator.

4.3. The theory in general Banach spaces. In general Banach spaces the situation is much more complicated than in the results that we have described so far. Let us start with the folloiwng instructive example: there exists a semigroup of contractions $S(t): X \to X$ on a Banach space X for which $t \to S(t)x$ is not differentiable to any $x \in X$ and $t \geq 0$ [22].

This example eliminates immediately the possibility of relating the semigroup $S(t)$ and its generator A through an initial value problem as we have done in all the previous cases. It is therefore also not surprising that in general Banach spaces it was impossible to prove that the initial value problem (4.1) has a strong solution provided that A is m-accretive and $x \in D(A)$.

In spite of these discouraging facts there is a way to associate with an m-accretive operator A in a general Banach space X, a unique semigroup of contractions $S(t)$ on $\overline{D(A)}$. This is the content of the fundamental result of Crandall and Liggett [22].

THEOREM 1. (Crandall-Liggett 1971)

Let A be m-accretive. If $x \in D(A)$ then

(4.2) $$\lim_{n\to\infty} (I + \frac{t}{n} A)^{-n} x = S(t)x$$

exists for every $t \geq 0$, the limit is uniform on bounded intervals of t and $S(t)$ is a semigroup of contractions on $\overline{D(A)}$.

Before we continue let us note that for the Crandall-Liggett theorem to hold one does not need the m-accretiveness of A, it suffices that A is accretive and $R(I+\lambda A) \supset D(A)$ for all $\lambda > 0$.

The equation (4.2) shows that $S(t)x$ is the limit of solutions of the implicit difference scheme

(4.3) $$\frac{u_i - u_{i-1}}{t/n} + Au_i \ni 0, \quad u_0 = x$$

for the initial value problem (4.1). Usually however, $S(t)x$ is not a solution of the initial value problem (4.1), since in general $S(t)x$ is not differentiable and does not satisfy $S(t)x \in D(A)$ for any $t \geq 0$. However, if it happens that $S(t)x$ is differentiable then it is also a solution of the initial value problem (4.1). More precisely we have:

THEOREM 2 (Crandall-Liggett).

Let A be m-accretive in X and let $S(t)$ be given by (4.2) then:

(i) If the initial value problem (4.1) has a strong solution u then $u(t) - S(t)x$.

(ii) If for some $t_0 > 0$, $S(t_0)x$ is differentiable then $S(t_0)x \in D(A)$ and

(4.4) $$\frac{d}{dt} S(t_0)x + AS(t_0)x \ni 0.$$

If $x \in D(A)$ and either X is reflexive or else A is linear, then for $x \in D(A)$, $t \to S(t)x$ is differentiable, a.e. on $t \geq 0$ (see [22]) and hence by Theorem 2, the initial value problem (4.1) possesses a strong solution. In particular Theorems 1 and 2 include the sufficiency part of the Hille-Yosida theorem as well as the sufficiency parts of the generation theorems in Hilbert space and in special Banach spaces.

The fact that in non reflexive Banach spaces $t \to S(t)x$ is not, in general, differentiable and therefore (4.1) does not have a strong solution is more an advantage than a handicap of the theory. It enables a much wider scope of applications to partial differential equations.

The theory in general Banach spaces still lacks in necessary conditions for an operator A to generate (in the above sense) a semigroup of contractions. In particular we do not know whether or not every semigroup of contractions is generated (in the above sense) by some accretive operator A satisfying $R(I+\lambda A) \supset D(A)$ for $\lambda > 0$. Thus, in the case of a general Banach space, we are still far from a satisfactory solution of the problems (I) and (II).

In spite of the still incomplete generation theory, the theory of semigroups of contractions in general Banach spaces developed and continues to develop rather rapidly and it contains many results analogous to those described above in the linear and Hilbert space cases. Much of the early development including Theorems 1 and 2 can be found in Barbu's book [3] while the more recent theory is given in [6].

As in the linear and Hilbert space cases, the theory of semigroups of contractions in general Banach spaces turned out to be extremely useful in the study of nonlinear initial value problems for nonlinear partial differential equations. Many such applications are descirbed in [31]. We mention here only a few: Applications to nonlinear diffusion including the porous media equation [5], [30], [58], semilinear parabolic equations [46],[67], a single conservation law in $\mathbb{R}^n$ [21], Hamilton-Jacobi equations [1], [17], [63], delay and integro-differential equations [68], [23], [14], and many more.

REFERENCES

1. Aizawa, S., A semi-group treatment of the Hamilton-Jacobi equations in one space variables, Hisohima Math. J., 3(1978), 367-386.

2. Baillon, B., Générateurs et semi-groupes dans les éspaces de Banach uniforement lisses, J. Func. Anal., 29(1978), 199-213.

3. Barbu, V., Nonlinear Semi-groups and Differential Equations in Banach Spaces, Nordhoof Publ. Leyden, 1976.

4. Benilan, Ph., Une remarque sur la convergence des semi-groupes non lineaires, C. R. Acad. Sci, 272(1971), 1182-1184.

5. Benilan, Ph., Operateur accrétifs et semi-groups dans les espaces L^p, Functional Analysis and Numerical Analysis, France-Japan Seminar, H. Fujita Ed., Tokyo, 1978.

6. Benilan, Ph., Crandall, M. G., Pazy, A., Nonlinear Evolution Governed by Accretive Operators, to appear.

7. Brezis, H., Propriété régularisante de certain semi-groupes non lineaires, Israel J. Math., 9(1971), 513-534.

8. Brezis, H., Monotonicity methods in Hilbert spaces and some application to nonlinear partial differential equations, 101-156, Contribution to nonlinear Func. Anal., E. Zarantonello, ed., Acad. Press, 1971.

9. Brezis, H., On a problem of T. Kato, Comm. Pure Appl. Math., 24(1971), 1-6.

10. Brezis, H., Opérateur Maximaux Monotones et Semigroupes de Contractions dans les Espaces de Hilbert, North Holland Publ. Com. Amseterdam, 1973.

11. Brezis, H., Pazy, A., Semigroups of nonlinear contractions on convex sets, J. Func. Anal., 6(1970), 237-281.

12. Brezis, H., Pazy, A., Accretive sets and differential equations in Banach spaces, Israel J. Math., 8(1970), 367-383.

13. Brezis, H., Pazy, A., Convergence and approximation of semi-groups of nonlinear operators in Banach spaces, J. Func. Anal., 9(1972), 63-74.

14. Bressan, R. V., Dyson, J., Functional differential equations and nonlinear evolution operators, Proc. Royal Soc. Edinburgh, 20(1975), 223-234.

15. Browder, F., Nonlinear equations of evolution and nonlinear accretive operators in Banach spaces, Bull. Amer. Math. Soc., 73(1967), 867-874.

16. Bruck, R., Asymptotic convergence of nonlinear contraction semi-groups in Hilbert space, J. Func. Anal., 18(1975), 15-26.

17. Burch, C., A semigroup treatment of the Hamilton-Jacobi equation in several space variables, J. Diff. Eqs., 23 (1977), 107-124.

18. Butzer, P. L., Berens, H., Semi-groups of Operators and Approximation, Springer Verlag, New York, 1967.

19. Chernoff, P. R., Note on product formula for operator semi-groups, J. Func. Anal., 2(1968), 238-242.

20. Chernoff, P. R., Perturbations of dissipative operators with relative bound one, Proc. Amer. Math. Soc., 33(1972), 72-74.

21. Crandall, M. G., The semigroup approach to first order quasilinear equations in several space variables, Israel J. Math., 12(1972), 108-132.

22. Crandall, M. G., Liggett, T., Generation of semigroups of nonlinear transformations on general Banach spaces, Amer. J. Math., 93(1971), 108-132.

23. Crandall, M. G., Nohel, J., An abstract functional differential equation and a related nonlinear Volterra equation, Israel J. Math., 29(1978), 313-328.

24. Crandall, M. G., Pazy, A., Semi-groups of nonlinear contractions and dissipative sets, J. Func. Anal., 3(1969), 376-418.

25. Crandall, M. G., Pazy, A., On accretive sets in Banach spaces, J. Func. Anal., 5(1970), 204-217.

26. Crandall, M. G., Pazy, A., Tartar, L., Remarks on generators of analytic semigroups, Israel J. Math., 32(1979), 363-374.

27. Dafermos, C., Slemrod, M., Asymptotic behavior of nonlinear contraction semi-groups, J. Func. Anal., 13(1973), 97-106.

28. DaPrato, G., Somma di generatori infinitesimali di semigrouppi analitici, Rend. Sem. Math. Univ. Padova, 40(1968), 151-161.

29. Dynkin, E. B., Markov Processes, Vol. I, Springer Verlag, Berlin, 1965.

30. Evans, L. C., Differentiability of nonlinear semigroups in L^1, J. Math. Anal. and Appl., 60(1977), 703-715.

31. Evans, L. C., Applications of nonlinear semigroup theory to certain partial differential equations, 163-188, Nonlinear Evolution Equations, M. G. Crandall, Ed., Academic Press, New York, 1979.

32. Feller, W., On the generation of unbounded semigroups of bounded linear operators, Ann. of Math., 58(1953), 166-174.

33. Feller, W., An Introduction to Probability Theory and its Applications, Vol. II, John Wiley, New York, 1966.

34. Gustafson, K., A perturbation lemma, Bull. Amer. Math. Soc., 72(1966), 334-338.

35. Hille, E., Functional Analysis and Semigroups, Amer. Math. Soc. Colloq. Publ., Vol. 31, New York, 1948.

36. Hille, E., Phillips, R. S., Functional Analysis and Semigroups, Amer. Math. Soc. Colloq. Publ., 31, Providence, R.I., 1957.

37. Komura, Y., Nonlinear semi-groups in Hilbert space, J. Math. Soc. Japan 19(1967), 493-507.

38. Komura, Y., Differentiability of nonlinear semi-groups, J. Math. Soc. Japan, 21(1969), 375-402.

39. Kato, T., Perturbation Theory of Linear Operators, Springer Verlag, New York, 1966.

40. Kato, T., Nonlinear semigroups and evolution equations, J. Math. Soc. Japan, 19(1967), 508-520.

41. Kato, T., Differentiability of Nonlinear Semi-groups, Global Anal. Proc. Symp. Pure Math. Amer. Math. Soc., 1970.

42. Kato, T., Accretive operators and nonlinear evolution equation in Banach spaces, Nonlinear Func. Anal. Proc. Symp. Pure Math. 18 Amer. Math. Soc. (1970), 138-161.

43. Kato, T., A characterization of holomorphic semi-groups, Proc. Amer. Math. Soc. 25(1970), 495-498.

44. Lax, P. D., Phillips, R. S., Scattering Theory, Acad. Press, New York, 1967.

45. Lumer, G., Phillips, R. S., Dissipative operators in a Banach space, Pacific J. Math., 11(1961), 679-698.

46. Massey, F. J., III, Semilinear parabolic equations with L^1 initial data, Indiana U. Math. J., 26(1977), 399-411.

47. Minty, G., Monotone operators in Hilbert space, Duke Math J., 29(1962), 341-346.

48. Miyadera, I., Generation of strongly continuous semi-groups of operators, Tohoku Math. J., 4(1952), 109-114.

49. Miyadera, I., Oharu, S., Approximation of semigroups of nonlinear operators, Tohoku Math. J., 23(1971), 245-258.

50. Nagumo, M., Einige analytische Untersuchungen in linearen metrischen Ringen, Japan J. Math., 13(1936), 61-80.

51. Nathan, D. S., One parameter groups of transformations in abstract vector spaces, Duke Math. J., 1(1935), 518-526.

52. Pazy, A., On the differentiability and compactness of semi-groups of linear operators, J. Math. and Mech., 17(1968), 1131-1141.

53. Pazy, A., Semi-groups of nonlinear contractions in Hilbert space, Problems in Nonlinear Anal., G. Prodi (Ed.), C.I.M.E., Varenna Cremonese (1971).

54. Pazy, A., Semi-groups of Linear Operators and Applications to Partial Differential Equations, Lectures Notes no. 10, Univ. of Maryland, 1974.

55. Pazy, A., The asymptotic behaviour of semigroups of nonlinear contractions having large sets of fixed points, Proc. Royal Soc. Edinburgh, 80A(1978), 261-271.

56. Pazy, A., Strong convergence of semigroups of nonlinear contractions in Hilbert space, J. d'Analyse Math., 34(1978), 1-35.

57. Pazy, A., Semigroups of nonlinear contractions and their asymptotic behaviour, pp. 36-134, Nonlinear anal. and mech. Heriot-Watt Symp., Vol. III, R. J. Knops (Ed.), Research notes in Math 30, Pitman, 1979.

58. Pazy, A., The Lyapunov method for semigroups of nonlinear contractions in Banach spaces, J. d'Analyse Math., 40(1980), 239-262.

59. Phillips, R. S., On the generation of semi-groups of linear operators, Pacific J. Math., 2(1952), 393-415.

60. Phillips, R. S., Perturbation theory of semi-groups of linear operators, Trans. Amer. Math. Soc., 74(1953), 199-221.

61. Rockafellar, T., Characterization of the subdifferential of convex functions, Pacific J. Math., 17(1966), 497-510.

62. Stone, M. H., On one-parameter unitary groups in Hilbert space, Ann. of Math., 33(1932), 643-648.

63. Tamburro, M. B., The evolution operator solution of the Cauchy problem for the Hamilton-Jacobi equation, Israel J. Math., 26(1977), 232-264.

64. Tanabe, H., Equation of Evolution, Pitman, London 1979.

65. Trotter, H. F., Approximation of semi-groups of operators, Pacific J. Math., 8(1958), 887-919.

66. Trotter, H. F., On the product of semi-groups of operators, Proc. Amer. Math. Soc., 10(1959), 545-551.

67. Veron, L., Effects regularisante de semi-groupes nonlinéaires dans les espaces de Banach, Annals Fac. Sci. Toulouse, 1(1979), 171-200.

68. Webb, G., Autonomous nonlinear functional differential equations and nonlinear semigroups, J. Math. Anal. and Appl., 46(1974), 1-12.

69. Yosida, K., On the group embedded in the metrical complete ring, Japan J. Math., (1936), 7-26.

70. Yosida, K., On the differentiability and the representation of one-parameter semi-groupes of linear operators, J. Math. Soc. Japan, 1(1948), 15-21.

71. Yosida, K., On the differentiability of semi-groups of linear operators, Proc. Japan Acad., 34(1958), 337-340.

72. Yosida, Functional Analysis, 6th edition, Springer Verlag, New York, 1980.

A New Method for Constructing Solutions of the Sine-Gordon Equation

Christoph Pöppe
Universität Heidelberg, SFB 123
Im Neuenheimer Feld 293
D-6900 Heidelberg

. Introduction

The Sine-Gordon equation (SGE) [4]

$$u_{xt} = \sin u \tag{1.1}$$

belongs to the family of so-called "soliton equations" [12] or "nonlinear evolution equations solvable by the inverse soectral transform" [5] . These equations are exceptional in that they exhibit a wealth of structure one expects to find only in linear PDEs: N-soliton solutions [6,7] , inverse scattering/ inverse spectral transform [1,2,3], Bäcklund transformations and a nonlinear superposition formula [8,9] .

Indeed, at least for the SGE, these properties can be deduced from corresponding properties of a linear PDE, called "base equation". For the SGE, the base equation is

$$f_{xt} = \frac{1}{2} f \quad . \tag{1.2}$$

The transformation which maps a solution u of (1.2) on a solution f of (1.1) essentially consists of constructing a set of linear Fredholm integral operators using f and computing their Fredholm determinants. This method not only recovers and slightly generalizes previously known results but also allows the explicit calculation of a new class of solutions, called "multipole solutions".

This paper reports on the author's Ph.D. thesis [10] where the above-mentioned results are treated in detail. It will be published elsewhere and is up to now available as a preprint [11] . The supervisor of the thesis was Prof. Willi Jäger whose support is gratefully acknowledged. The plots have been produced at the Zentralinstitut für angewandte Mathematik of the Kernforschungsanlage Jülich. This work has been supported by the Deutsche Forschungsgemeinschaft.

2. Construction of Solutions

Let f be a solution of the base equation (1.2) such that f, f_x, f_t, f_{tt} decay faster than $\frac{1}{|x|}$ for $x \to -\infty$. (These conditions can be slightly weakened.) For every x and t, define the Fredholm integral operator F by

$$F_{(x,t)}\phi\,(s) := \int_{-\infty}^{x} f(s+\sigma,t)\,\phi(\sigma)\,d\sigma \tag{2.1}$$

Then, for every real λ, the Fredholm determinant [13]

$$p(\lambda;x,t) = \det\,(1 + i\,\lambda\,F_{(x,t)}) \tag{2.2}$$

is well-defined and solves the "Sym equation" [14]

$$p\,p_{xt} - p_x\,p_t = \frac{1}{4}\,(p^2 - p^{*2}) \tag{2.3}$$

The key idea for proving this is expanding p as a power series w.r.t. λ (cf. [13]) and equating coefficients of λ^k on both sides of (2.3). However, this leads to extremely lengthy calculations which can be avoided by a suitable reformulation of (2.3). Then, the proof can be done essentially by algebraic calculations with operators.

The Sym equation in turn is the so-called bilinear form [6,7] of the SGE. From a solution p of (2.3), a solution u of the SGE can easily be calculated as

$$u = 4\,\arctan\frac{\mathrm{Im}\,p}{\mathrm{Re}\,p} \mod 2\pi\ . \tag{2.4}$$

3. N-Soliton and multipole solutions

If f is chosen so that the operator F has finite rank for all x and t, then p is identical to the ordinary determinant of the corresponding finite-dimensional operator. In this case, explicit formulas exist for p and, hence, for u.

Up to now, two classes of solutions of that type are known. The first of them is described by

$$(3.1) \qquad f(x,t) = \sum_{j=1}^{N} a_j \exp\left(\frac{a_j}{2} x + \frac{1}{a_j} t + \eta_{0j} \right), \mathrm{Re}\, a_j > 0,$$

f real-valued

leading to

$$(3.2) \qquad p = \sum_{P \subset \{1,\ldots,N\}} i^{\#P} \prod_{j<l \in P} \left(\frac{a_l - a_j}{a_l + a_j} \right)^2 \prod_{j \in P} \exp\left(a_j x + \frac{1}{a_j} t + \eta_{0j}\right).$$

These solutions generate by means of (2.4) the N-soliton solutions [6] which play a crucial role in any application of the SGE. They consist of N "solitary waves" ("solitons") each of which behaves like a stable particle moving asymptotically free. Interaction of solitons obeys the laws of energy and momentum conservation and leaves the particles essentially unchanged. Speed and asymptotic phase of each soliton are determined by its parameters a_j and η_{0j}. Soliton pairs with complex conjugate parameters a_j correspond to "bound states" or "breathers", pairs of solitons oscillating around eachother (cf. e.g. [12]).

The second class of "finite-dimensional" solutions is new and obtains by differentiating (3.1) w.r.t. one or more of its parameters a_j. Again, explicit, although somewhat lengthy, expressions for solutions of the SGE can be given. They are called "degenerate soliton" or "multipole" solutions and correspond to borderline cases between a bound state and a pair or any number of free particles. Alternatively, they may be obtained from N-soliton solutions by applying a suitable limiting process.

Some examples of explicit solutions are plotted below. To make the pictures more instructive, the physical coordinates are used:

$$(3.3) \qquad \begin{aligned} X &= x + t \\ T &= x - t \end{aligned}$$

in which the SGE takes the form

$$(3.4) \qquad u_{XX} - u_{TT} = \sin u \ .$$

Moreover, u_X is plotted instead of u . The lines show u_X as a function of X for different values of T.

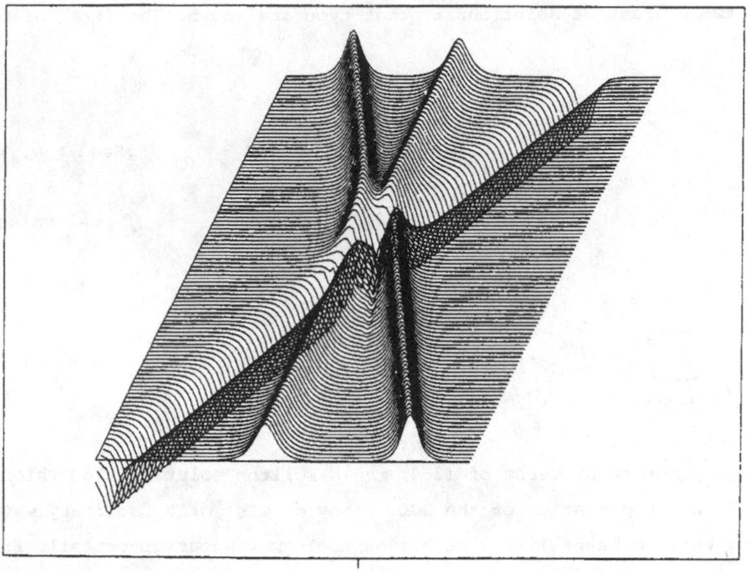

Fig. 1. Three-soliton solution.

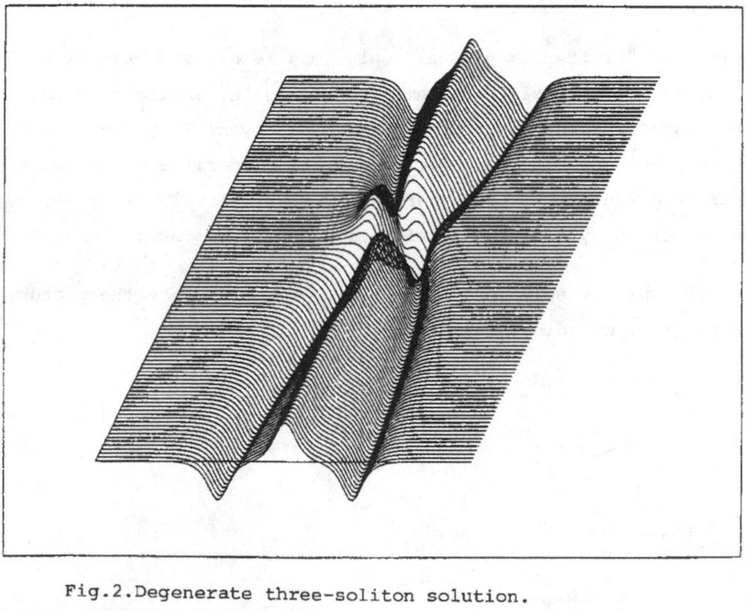

Fig.2.Degenerate three-soliton solution.

Note that in the degenerate case the particles even for large times don't move "freely", i.e. linearly in time, but with slowly decreasing speed.

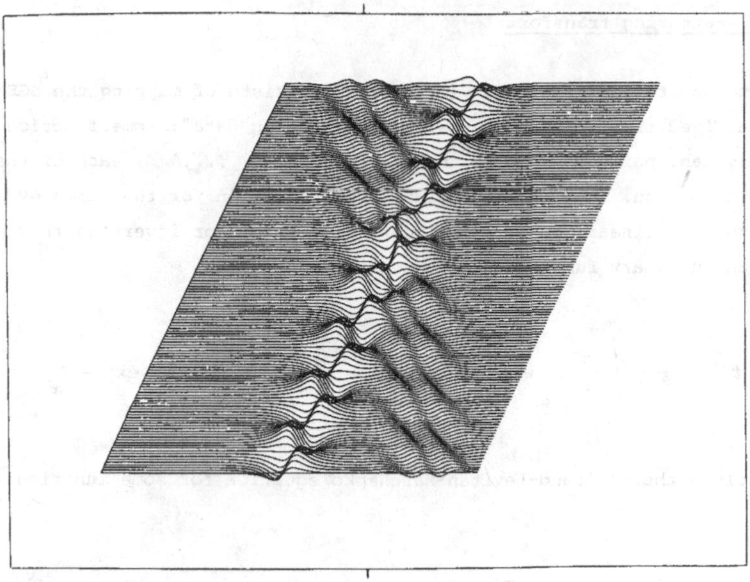

Fig.3. Moving breather meets stationary breather.

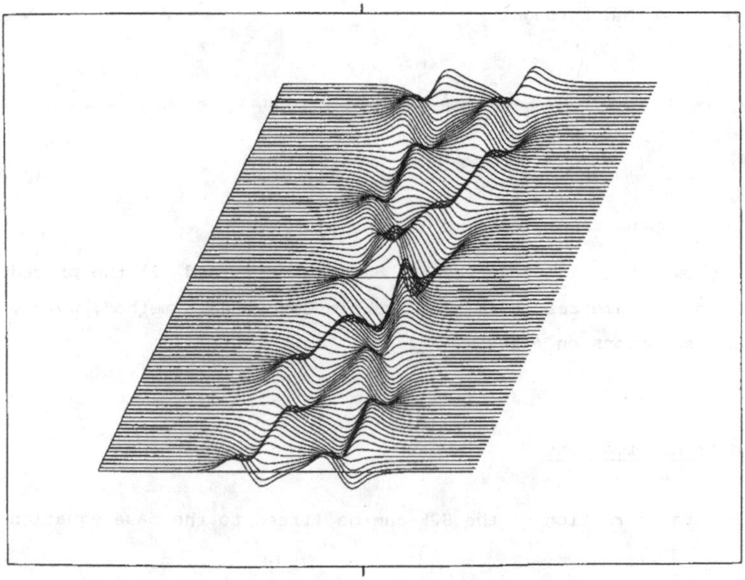

Fig.4. Degenerate two-breather solution.

No separation at all can be seen for the degenerate case.

4. Inverse scattering transform

The "inverse scattering transform" (IST) [2,3] consists of mapping the SGE solution $u(\cdot,t)$ at a fixed time t to the so-called "scattering data": some function $\rho_0(\xi)$, $\xi \in \mathbb{R}$ and finitely many pairs of complex numbers $(\rho_j, \zeta_j), j=1,\dots,N$, each of them corresponding to a soliton. The nonlinear Sine-Gordon equation for the time evolution of u transforms into linear ODEs for the scattering data. For inverting this map one first builds up an auxiliary function from the scattering data:

$$(4.1)\qquad f(x,t) := \frac{1}{2\pi}\int_{-\infty}^{\infty} \rho_0^*(\xi)\exp\left(-i\xi x - \frac{1}{2i\xi}t\right)d\xi - i\sum_{j=1}^{N}\rho_j^*\exp\left(-i\zeta_j x - \frac{1}{2i\zeta_j}t\right) ;$$

then one solves the Gel'fand-Levitan-Marchenko equation for some function $\tilde{k}$ of two variables

$$(4.2)\qquad \tilde{k}(x,y) - \tilde{f}(x+y) + \int_{-\infty}^{x}\int_{-\infty}^{x} \tilde{f}(y+s)\,\tilde{f}(s+\sigma)\,ds\,\tilde{k}(x,\sigma)\,d\sigma = 0 ;$$

finally u is determined through

$$(4.3)\qquad u_x = 4\,\tilde{k}(x,x)$$

($\tilde{f}$ and $\tilde{k}$ correspond resp. to G and $-L_1$ in [2] .)

It can be shown that 1) $\tilde{f}$ solves the base equation and 2) the procedure described above leads to the same result as the Fredholm determinant method, which, however, needs only weaker assumptions on f and u.

5. Bäcklund transformations

The Bäcklund transformation of the SGE can be lifted to the base equation in the following way.

If f and g fulfill the pair of ODEs

$$g_x - f_x = \frac{a_o}{2}\,(g + f)$$

(5.1)

$$g_t + f_t = \frac{1}{a_o}\,(g - f)$$

then both solve the base equation (1.2), as can be easily seen. Let now p and q be the solutions of the Sym equation (2.3) constructed resp. from f and g by the Fredholm determinant method. Then techniques similar to those of part 2 can be used to show that p and q solve

$$p\,q_x - p_x\,q = \frac{a_o}{2}\,(p\,q - p^*\,q^*)$$

(5.2)

$$p^*q_t - p_t^*\,q = \frac{1}{2a_o}\,(p^*q - p\,q^*)$$

These in turn generate via (2.4) the well-known Bäcklund-transformation equations for the SGE (v corresponds to q) :

$$\frac{v_x - u_x}{2} = a_o \sin\frac{v+u}{2}$$

(5.3)

$$\frac{v_t + u_t}{2} = \frac{1}{a_o} \sin\frac{v-u}{2}$$

Since eqs. (5.1) can be solved explicitly for g when f is known, results about the effect of a Bäcklund transformation can be easily obtained:

Generically, a Bäcklund transformation adds a soliton to a given solution while shifting the "old" components of the solution in space. In special cases, a soliton can be either destroyed or transformed to a degenerate double soliton.

The "superposition formula" for Bäcklund transforms [8] can as well be lifted to the Sym and base equation levels.

References

1. Ablowitz, M.J., Kaup, D.J., Newell, A.C., Segur, H.: Method for solving the Sine-Gordon equation. Phys.Rev.Lett. 30,1262(1973)

2. Ablowitz, M.J., Kaup, D.J., Newell, A.C., Segur, H.: The inverse scattering transform - Fourier analysis for nonlinear problems. Stud.Appl.Math. 53, 249(1974)

3. Ablowitz, M.J.: Lectures on the inverse scattering transform. Stud. Appl.Math. 58, 17 (1978)

4. Barone,A., Esposito, F., Magee, C.J., Scott, A.C.: Theory and applications of the Sine-Gordon equation. Riv. Nuov.Cim. 1, 227-267 (1971)

5. Calogero, F., Degasperis, A.: Nonlinear evolution equations solvable by the inverse spectral transform. I. Nuov.Cim. 32B,201 (1976); II. Nuov.Cim. 39B,1 (1977)

6. Hirota,R.: Exact solution of the Sine-Gordon equation for multiple collisions of solitons. J.Phys.Soc.Jap. 33,1459 (1972)

7. Hirota,R.: Direct method of finding exact solutions of nonlinear evolution equations. In [9] , p.40

8. Lamb, G.L.: Analytical description of ultrashort optical pulse propagation in a resonant medium. Rev.Mod.Phys. 43,99 (1971)

9. Miura,R. (ed.) : Bäcklund transformations, the inverse scattering method, solitons, and their applications. NSF Research Workshop on Contact Transformations. Springer Lecture Notes in Mathematics 515 (1976)

10. Pöppe, C.: Über die Konstruktion von Lösungen der Sine-Gordon-Gleichung mittels Fredholm-Determinanten. Dissertation, Universität Heidelberg 1981.

11. Pöppe, C.: Construction of solutions of the Sine-Gordon equation by means of Fredholm determinants. SFB 123 preprint no. 161, University of Heidelberg(1982)

12. Scott,A.C., Chu, F.Y.F., McLaughlin, D.W.: The soliton: a new concept in applied science. Proc. IEEE 61, 1443-1483 (1973)

13. Smithies, F.: Integral equations. Cambridge University Press 1965

14. Sym, A.: A simple derivation of some results connected with the Sine-Gordon equation. Phys. Lett. 58A,77 (1976)

PERIODIC SOLUTIONS OF A CLASS OF SEMILINEAR, STOCHASTIC DIFFERENTIAL EQUATIONS WITH RANDOM COEFFICIENTS

R. Riganti
Department of Mathematics
Politecnico di Torino
10129 Turin, Italy

1. Introduction

The analysis of the asymptotic behaviour of stochastic evolution equations with random coefficients is at present the interesting objective of many researches in the field of theoretical analysis and its applications to the study of physical systems. A systematic treatment of the problem may be found, for instance, in the book by Soong /1/ and in the works by Adomian /2,3/, Bellomo /4,5/ and related bibliography; in addition, qualitative results on the stability of suitable stochastic systems with random parameters have been obtained in /6,7/, and some statistical properties of the solution process have been analytically deduced in /8/ for a class of linear random differential equations and in /9/ for autonomuos systems with constant random parameters.

On the other hand, the periodic solutions of semilinear *deterministic* evolution equations have been also the objective of various Authors, see for instance the Chapter VII of the book by Sansone and Conti /10/ or the papers by Becker /11/, Ward /12/ which give sufficient conditions for the existence of periodic solutions of the above said class of differential equations.

More recently, the periodic solutions of *stochastic*, semilinear dynamical systems with random parameters have been studied in /13/ by applying the method of the stochastic Green's matrix proposed by Adomian /2/. On the ground of the same methodology, this paper deals with the study of a class of non-autonomous, stochastic differential equations characterized by the presence of a deterministic linear term and of a random nonlinear term, whose coefficients can be analytically described by suitable stationary random processes.

The mathematical description of the considered stochastic systems is given in section 2; its evolution equation is studied in section 3, where approximated analytical results concerning the first- and second-order statistics of the solution process are obtained. The properties of the periodic solution are deduced on the ground of the above quoted mathematical theory of deterministic, semilinear evolution equations, of the methods for studying stochastic differential equations, and by applying a regular perturbation technique. In the last sections, the developed theory is applied to the study of stochastic nonlinear oscillators with one degree of freedom, and quantitative results are obtained and discussed.

2. Mathematical description of the dynamical system

It is considered the class of semilinear equations of the form:

$$\dot{x} = [A]x + f(\nu t) + \varepsilon g(x; r(\omega,t)) \ , \quad \dot{x} = dx/dt \tag{1}$$

with

$$x = x(\omega,t) : \Omega\times T \rightarrow D_x \subset \mathbb{R}^n \ ; \ t \in T \ ; \qquad \omega \in (\Omega, F, \mu), \text{ a p.s.}$$

for which the following assumptions are made.

H.1 - *$[A]$ is a constant deterministic matrix;*

H.2 - *$f(\nu t) = \{f_i\}$, $i = 1,2,..n$ is a deterministic vector function of time with period $\tau = 2\pi/\nu$;*

H.3 - *$\varepsilon = o(1)$ is a small deterministic parameter;*

H.4 - *$g = \{g_i\}$, with $g_i \in C^m(D_x)$, is depending on a set $r(\omega,t)$ of stationary random processes, which can be analytically expressed in the form:*

$$r(\omega,t) = \{r_k(\omega,t)\} \ ; \quad r_k(\omega,t) = A_{ok} + \sum_{h=1}^{N} \{A_{hk} cos(h\nu t) + B_{hk} sin(h\nu t)\} \tag{2}$$

$h = 1,2,..N$; $k = 1,2,..M$, where A_{ok}, A_{hk}, B_{hk} are constant random variables defined in the complete probability space (Ω, F, μ) and are joined with known probability densities;

H.5 - *The deterministic differential equation corresponding to Eq.(1):*

$$\dot{x} = [A]x + f(\nu t) + \varepsilon g(x; r_o(t)) \tag{3}$$

with

$$r_o(t) = \{r_{ok}(t)\} \ ; \quad r_{ok}(t) = a_{ok} + \sum_{h=1}^{N} \{a_{hk} cos(h\nu t) + b_{hk} sin(h\nu t)\}$$

has a unique solution satisfying the initial conditions $x(0) = x_o$ for every $x_o \in \mathbb{R}^n$, $t \in T$ and for given values of $[A]$, a_{ok}, a_{hk}, b_{hk} and ε;

H.6 - *The linear, autonomous differential equation: $\dot{x} = [A]x$ has not any τ-periodic solutions, besides the null solution.*

The objective of this paper is to determine approximated τ-periodic solutions for such a class of semilinear stochastic equations, and to obtain quantitative results on the first- and second-order statistics of the related solution process. The theory which will be developed may be easily extended to the study of evolution equations where $f(\nu t)$ is a stochastic process of the type described by Eq.(2).

3. Analysis

The periodical behaviour of the stochastic system defined by Eq.(1) may be studied by considering (see ref./10/) that, as a consequence of the hypotheses H.5 and H.6, each τ-periodic solution of the deterministic equation (3) is also the solution of the integral equation

$$\underset{\sim}{x}(t) = \int_0^\tau [G(t,s)]\{\underset{\sim}{f}(\nu s) + \varepsilon \underset{\sim}{g}(\underset{\sim}{x}(s);\ \underset{\sim 0}{r}(s))\}ds \tag{4}$$

where $[G(t,s)]$ is the Green's matrix defined as:

$$\begin{aligned} [G(t,s)] &= [U(t)][U(s)]^{-1} - [U(t)][U(\tau) - I]^{-1}[U(\tau)][U(s)]^{-1}, & 0 \leq s \leq t \\ [G(t,s)] &= -[U(t)][U(\tau) - I]^{-1}[U(\tau)][U(s)]^{-1}, & t < s \leq \tau \end{aligned} \tag{5}$$

$[U(t)]$ being the principal matrix associated with the constant coefficient matrix $[A]$. On the ground of the methodology proposed by Adomian /2,3/ and recently applied in /13/, it follows that each solution of the *stochastic* equation (1) is also the solution of the stochastic integral equation:

$$\underset{\sim}{x}(\omega,t) = \int_0^\tau [G(t,s)]\{\underset{\sim}{f}(\nu s) + \varepsilon \underset{\sim}{g}(\underset{\sim}{x}(\omega,s);\ \underset{\sim}{r}(\omega,s))\}ds \tag{6}$$

with $[G(t,s)]$ defined by Eq.(5).

Let us now consider Eq.(6) and develop a regular perturbation technique in order to obtain approximated solutions of the above integral equation. By taking into account the hypothesis H.3, we search for this solution in the form:

$$\underset{\sim}{x}(\omega,t) \simeq \underset{\sim}{x}^*(\omega,t) = \sum_{j=0}^{m} \varepsilon^j \underset{\sim}{x}^{(j)}(\omega,t). \tag{7}$$

Owing to the hypothesis H.4, the nonlinear function g has the truncated power expansion

$$\underset{\sim}{g} \simeq \underset{\sim}{g}^* = \sum_{j=0}^{m-1} \varepsilon^j g^{(j)}\ ; \quad \|g - g^*\| \leq \delta = O(\varepsilon^m) \text{ for } \varepsilon \to 0 \tag{8}$$

where:

$$g^{(j)} = \frac{1}{j!}\left(\frac{d^j g}{d\varepsilon^j}\right)_{\varepsilon=0} = \{g_i^{(j)}\};\quad g_i^{(j)} = g_i^{(j)}(\underset{\sim}{x}^{(o)}, \underset{\sim}{x}^{(1)}, \ldots \underset{\sim}{x}^{(j)};\ \underset{\sim}{r}(\omega,t)) \tag{9}$$

and the norm is defined as

$$\|g - g^*\| = \max_{\substack{i=1..n \\ t \in T}} |g_i - g_i^*|. \tag{10}$$

Inserting Eqs.(7,8) into Eq.(6) it is obtained, by regular perturbation technique /16/:

$$\underset{\sim}{x}^{(o)}(t) = \int_0^\tau [G(t,s)]\, \underset{\sim}{f}(\nu s)\, ds \tag{11}$$

$$\underset{\sim}{x}^{(j)}(\omega,t) = \int_0^\tau [G(t,s)]\, \underset{\sim}{g}^{(j-1)}(\underset{\sim}{x}^{(o)}, \underset{\sim}{x}^{(1)}, \dots \underset{\sim}{x}^{(j-1)}; \underset{\sim}{r}(\omega,s))ds, \quad j = 1,..n \tag{12}$$

The solution of Eq.(11) gives the first term of the expansion (7), which is the solution of the corresponding *linear, deterministic* problem; whereas Eq.(12) supplies, by quadratures, the following terms of the approximated solution process.

The error bounds for the obtained periodic solution can be evaluated as a consequence of the Lemma proved in ref./13/ and by considering also Eq.(8); it results:

$$\|\underset{\sim}{x} - \underset{\sim}{x}^*\| = \|\varepsilon \int_0^\tau [G(t,s)]\, (\underset{\sim}{g} - \underset{\sim}{g}^*)ds\| \le \varepsilon\tau \|G(t,s)(\underset{\sim}{g} - \underset{\sim}{g}^*)\| \le$$

$$\le \varepsilon\tau \, \| G(t,s)\| \cdot \| \underset{\sim}{g} - \underset{\sim}{g}^* \| \le \varepsilon\tau\delta \|G(t,s)\| = O(\varepsilon^{m+1}). \tag{13}$$

The most important statistical properties of the solution process can be now deduced as follows. Define in the complete probability space the random vector:

$$\underset{\sim}{\alpha} = \{A_{ok}, A_{hk}, B_{hk}\} \in A \tag{14}$$

which, owing to the hypothesis H.4, is joined with a known probability density $P_\alpha(\underset{\sim}{\alpha})$ constant with time. The moments of each i-component of $\underset{\sim}{x}^*(\omega,t)$ are given by:

$$E\{x_i^\mu\} = \int_A \{x_i^*(\omega,t)\}^\mu P_\alpha(\underset{\sim}{\alpha})\, d\underset{\sim}{\alpha}. \tag{15}$$

In particular, the expected value of the i-th component is

$$\overline{x}_i(t) = E\{x_i^*\} = x_i^{(o)}(t) + \sum_{j=1}^m \varepsilon^j \int_A x_i^{(j)}(\omega,t) P_\alpha(\underset{\sim}{\alpha}) d\underset{\sim}{\alpha} \tag{16}$$

where $x_i^{(j)}$ is given by Eq.(12); and its variance can be calculated by

$$\mathrm{Var}\{x_i^*\} = \sum_{j=1}^m \varepsilon^{2j}\mathrm{Var}\{x_i^{(j)}\} + 2\sum_{\ell=1}^{m-1} \sum_{j=\ell+1}^m \varepsilon^{(j+\ell)} \mathrm{Cov}\{x_i^{(\ell)}, x_i^{(j)}\} \tag{17}$$

where $\mathrm{Cov}\{x_i^{(\ell)}, x_i^{(j)}\}$ is the covariance of each couple of random terms in the expansion (7).

4. Application

As an application of the developed theory to nonlinear vibrations, consider the class of stochastic oscillators whose motion is described by:

$$\ddot{x} + 2\gamma\dot{x} + \lambda^2 x + \varepsilon\{r_1(\omega,t)\, x^u + r_2(\omega,t)\, \dot{x}^v\} = f_o \sin(\nu t) \tag{18}$$

with u,v positive integers; $0 \le \gamma < \lambda$ and $r_k(\omega,t)$, $k = 1,2$, stochastic processes defined by Eq.(2). The Eq.(18) may be written in the form (1) with $\underset{\sim}{x} = \{x_1=x,\ x_2=\dot{x}\} \in D_x \subset R^2$, $f_1 = g_1 = 0$; $f_2 = f_0 \sin(\nu t)$; $g_2 = -r_1(\omega,t)x_1^u - r_2(\omega,t)x_2^v$; and:

$$[A] = \begin{bmatrix} 0 & 1 \\ -\lambda^2 & -2\gamma \end{bmatrix}. \tag{19}$$

Let us determine the ε-order approximated solution of Eq.(18) in the form:

$$\underset{\sim}{x}^*(\omega,t) = \underset{\sim}{x}^{(o)}(t) + \varepsilon \underset{\sim}{x}^{(1)}(\omega,t). \tag{20}$$

The terms of this solution are given by Eqs.(11,12), where the Green's matrix, which is defined by Eq.(5), has the following expression:

$$\begin{aligned} [G(t,s)] &= (1 - h\,e^{-2\gamma\tau})[U(t-s)] + h[U(t+\tau-s)] \quad , \quad 0 \le s \le t \\ [G(t,s)] &= -h\,e^{-2\gamma\tau}[U(t-s)] + h[U(t+\tau-s)] \quad , \quad t < s \le \tau \end{aligned} \tag{21}$$

being $[U(t)]$ the principal matrix associated to the constant matrix (19), and:

$$h = h(\gamma,\lambda,\nu) = \{1 - e^{-2\gamma\tau} - 2e^{-\gamma\tau}\cos(\sigma\tau)\}^{-1} \quad ; \ \sigma^2 = \lambda^2 - \gamma^2. \tag{22}$$

Inserting Eq.(21) into Eq.(11), the well-known solution of the deterministic, forced linear oscillator is obtained, whereas Eq.(12) supplies the following first-order perturbation term:

$$\underset{\sim}{x}^{(1)}(\omega,t) = \underset{\sim}{F}(t,t;\,\omega) \quad , \quad \underset{\sim}{F}(s,t;\,\omega) = \{F_1,\ F_2\} \tag{23}$$

$$F_1(s,t;\,\omega) = -\frac{1}{\sigma}\sum_{k=1}^{2}\sum_{h=0}^{N} \nu^{(k-1)}\{A_{hk}\,\Phi_{hk}(s,t) + B_{hk}\,\Psi_{hk}(s,t)\} \tag{24}$$

$$F_2(s,t;\,\omega) = -\gamma F_1(s,t;\,\omega) - \sum_{k=1}^{2}\sum_{h=0}^{N} \nu^{(k-1)}\{A_{hk}\,\Phi'_{hk}(s,t) + B_{hk}\,\Psi'_{hk}(s,t)\} \tag{25}$$

being Φ_{hk} the antiderivative, with respect to the variable s and with constant equal to zero, of the function:

$$\varphi_{hk}(s,t) = e^{-\gamma(t-s)}\sin\{\sigma(t-s)\}\cos(h\nu s)\{x_k^{(o)}(s)\}^{\eta} \tag{26}$$

where $\eta = u$ if $k = 1$ and $\eta = v$ if $k = 2$. The functions Ψ_{hk} are defined in the same way, upon substitution in Eq.(26) of $\cos(h\nu s)$ by $\sin(h\nu s)$, and the functions Φ'_{hk}, Ψ'_{hk} are obtained from the preceeding ones upon substitution of $\sin\{\sigma(t-s)\}$ by $\cos\{\sigma(t-s)\}$ respectively.

The moments of the two components of $\underset{\sim}{x}^*(\omega,t)$ are then obtained by inserting the above approximated solution into Eq.(15); in particular, the expected values are:

$$\overline{x}_i(t) = E\{x_i^*\} = x_i^{(0)}(t) + \varepsilon E\{x_i^{(1)}\} \ , \ i = 1, 2 \tag{27}$$

where, if A_{hk}, B_{hk} are *independent* random variables with mean values $<A_{hk}>$, $<B_{hk}>$:

$$E\{x_1^{(1)}\} = -\frac{1}{\sigma}\sum_{k=1}^{2}\sum_{h=0}^{N} \nu^{(k-1)}\{<A_{hk}>\Phi_{hk}(t,t) + <B_{hk}>\Psi_{hk}(t,t)\} \tag{28}$$

$$E\{x_2^{(1)}\} = -\sum_{k=1}^{2}\sum_{h=0}^{N} \nu^{(k-1)}\{<A_{hk}>(\gamma\Phi_{hk}(t,t)/\sigma - \Phi'_{hk}(t,t)) + <B_{hk}>(\gamma\Psi_{hk}(t,t)/\sigma - \Psi'_{hk}(t,t))\} \tag{29}$$

and moreover:

$$\mathrm{Var}\{x_i^*\} = \varepsilon^2 \mathrm{Var}\{x_i^{(1)}\} \ ; \quad \mathrm{Cov}\{x_1^*, x_2^*\} = \varepsilon^2 \mathrm{Cov}\{x_1^{(1)}, x_2^{(1)}\} \tag{30}$$

where $x_i^{(1)}$ are given by Eq.(23). Therefore, within the approximation introduced by the Eq.(20), the explicit determination of the first- and second-order moments of the solution process defined by Eq.(18) is reduced to the elementary quadratures of functions whose expression is defined by Eq.(26).

5. Example: stochastic Duffing equation

To visualize the results of the preceeding section, consider the differential equation describing the motion of an oscillator with small cubic nonlinearity of random nature:

$$\ddot{x} + 2\gamma\dot{x} + \lambda^2 x + \varepsilon r(\omega,t)\, x^3 = f_o \sin(\nu t) \tag{31}$$

and assume that the random coefficient $r(\omega,t)$ be a simplified version of the ones defined by Eq.(2), namely:

$$r(\omega,t) = C_o \cos(\nu t - \Theta) \tag{32}$$

with C_o, Θ known, independent random variables, being Θ uniformly distributed in $[-\pi/2, \pi/2]$. Application of Eqs.(28,29) with h=k=1; $A_{11} = C_o\cos\Theta$; $B_{11} = C_o\sin\Theta$ leads to the following expected values for the two components of the ε-order perturbation term $\underset{\sim}{x}^{(1)}(\omega,t)$:

$$E\{x_1^{(1)}\} = <C_o>(f_o\rho_1)^3 H_1(t)/(4\pi) \tag{33}$$

$$E\{x_2^{(1)}\} = <C_o>(f_o\rho_1)^3 \nu H_2(t)/(2\pi) \tag{34}$$

where $<C_o>$ is the mean value of C_o, and:

$$H_1(t) = \sin(2\nu t + \psi_1) + \sin(4\nu t + \psi_2) - 6\rho_1\gamma\nu/\lambda^2 \tag{35}$$

$$H_2(t) = \cos(2\nu t + \psi_1) + 2\cos(4\nu t + \psi_2) \quad (36)$$

$$\begin{cases} \sin\psi_1 = 8\rho_1\rho_2^2\gamma\nu\{\lambda^2 - \nu^2 + 4\rho_1^2\nu^2\gamma^2(\lambda^2 + 2\nu^2)\} & (37) \\ \cos\psi_1 = 2\rho_1\rho_2^2\{(\lambda^2 - \nu^2)(\lambda^2 - 4\nu^2)(1 + 8\gamma^2\nu^2\rho_1^2) + 64\,\gamma^4\nu^4\rho_1^2\} & (37') \end{cases}$$

$$\begin{cases} \sin\psi_2 = 2\rho_1\rho_4^2\gamma\nu\{4(\lambda^2 - \nu^2)(16\rho_1^2\gamma^2\nu^2 - 1) + (\lambda^2 - 16\nu^2)[4\rho_1^2(\lambda^2 - \nu^2)^2 - 1]\} & (38) \\ \cos\psi_2 = \rho_1\rho_4^2\{(\lambda^2 - 16\nu^2)(\lambda^2 - \nu^2)(16\rho_1^2\gamma^2\nu^2 - 1) - 16\gamma^2\nu^2[4\rho_1^2(\lambda^2 - \nu^2)^2 - 1]\} & (38') \end{cases}$$

$$\rho_q = \{(\lambda^2 - q^2\nu^2)^2 + 4q^2\gamma^2\nu^2\}^{-\frac{1}{2}}, \quad q \in \mathbb{N}. \quad (39)$$

The periodic behaviour of $E\{x_1^{(1)}\}$ given by Eq.(33) is plotted in the Fig.1 for $f_o = 6$, $\nu = 2$, $\lambda = 3$, $\gamma = .2$, $\varepsilon = .1$ and C_o uniformly distributed between 0 and 20. In order to visualize the random nature of the considered differential equation, it is compared with the term $\bar{x}_1^{(1)}(t, \langle C_o\rangle, \langle\Theta\rangle)$ of the solution of the corresponding deterministic problem, which should be obtained by assuming deterministic values of the random variables C_o, Θ equal to their mean values $\langle C_o\rangle$ and $\langle\Theta\rangle$.

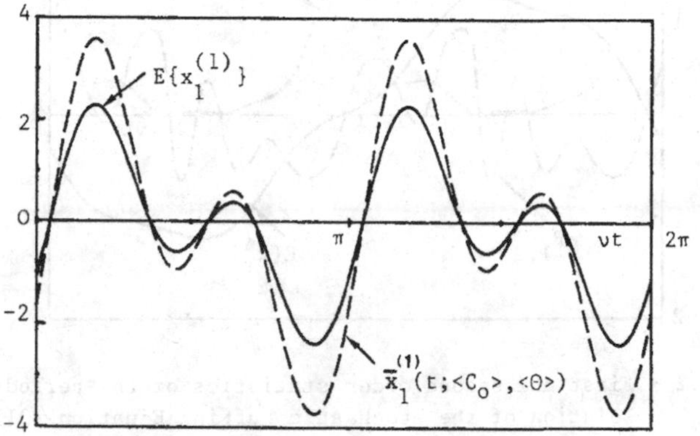

Fig.1 – ε-order perturbation term of the solution of Eq.(31) and comparison with the deterministic case.

The second-order moments of the solution process, which are supplied by Eq.(30), are:

$$\mathrm{Var}\{x_1^*\} = \varepsilon^2\{\langle C_o^2\rangle(f_o\rho_1/2)^6[H_1^2(t) + H_3^2(t)]/2 - E\{x_1^{(1)}\}\} \quad (40)$$

$$\mathrm{Var}\{x_2^*\} = \varepsilon^2\{2\nu^2\langle C_o^2\rangle(f_o\rho_1/2)^6[H_2^2(t) + H_4^2(t)] - E\{x_2^{(1)}\}\} \quad (41)$$

$$\mathrm{Cov}\{x_1, x_2\} = \varepsilon^2\{\nu\langle C_o^2\rangle(f_o\rho_1/2)^6[H_1(t)H_2(t) - H_3(t)H_4(t)] - E\{x_1^{(1)}\}E\{x_2^{(1)}\}\} \quad (42)$$

where $\langle C_o^2\rangle$ is the second-order moment of C_o, and:

$$H_3(t) = \cos(2\nu t - \psi_3) + \cos(4\nu t - \psi_4) + 3\rho_1(\lambda^2 - \nu^2)/\lambda^2 \tag{43}$$

$$H_4(t) = \sin(2\nu t - \psi_3) + 2\sin(4\nu t - \psi_4) \tag{44}$$

$$\sin\psi_3 = 4\rho_1\rho_2^2\gamma\nu\{4\rho_1^2(\lambda^2 - \nu^2)^3 - (\lambda^2 - 4\nu^2)[1 + 2\rho_1^2(\lambda^2 - \nu^2)^2]\} \tag{45}$$

$$\cos\psi_3 = -4\rho_1\rho_2^2\{\rho_1^2(\lambda^2 - \nu^2)^3(\lambda^2 - 4\nu^2) + 4\gamma^2\nu^2[1 + 2\rho_1^2(\lambda^2 - \nu^2)^2]\} \tag{46}$$

$$\sin\psi_4 = 2\rho_1\rho_4^2\gamma\nu\{4(\lambda^2 - \nu^2)(16\rho_1^2\gamma^2\nu^2 - 1) + (\lambda^2 - 16\nu^2)[4\rho_1^2(\lambda^2 - \nu^2)^2 - 1]\} \tag{47}$$

$$\cos\psi_4 = \rho_1\rho_4^2\{16\gamma^2\nu^2[4\rho_1^2(\lambda^2 - \nu^2)^2 - 1] - (\lambda^2 - \nu^2)(\lambda^2 - 16\nu^2)(16\rho_1^2\gamma^2\nu^2 - 1)\}. \tag{48}$$

The expectation and variance of the displacement $\underset{\sim}{x}_1^*(\omega,t)$, together with the correlation coefficient $R_{1,2} = \mathrm{Cov}\{x_1^*, x_2^*\}/[\mathrm{Var}\{x_1^*\}\mathrm{Var}\{x_2^*\}]^{\frac{1}{2}}$ of the two components of the solution process, are plotted in the Fig.2 as functions of νt and for the same values of the parameters of Fig.1.

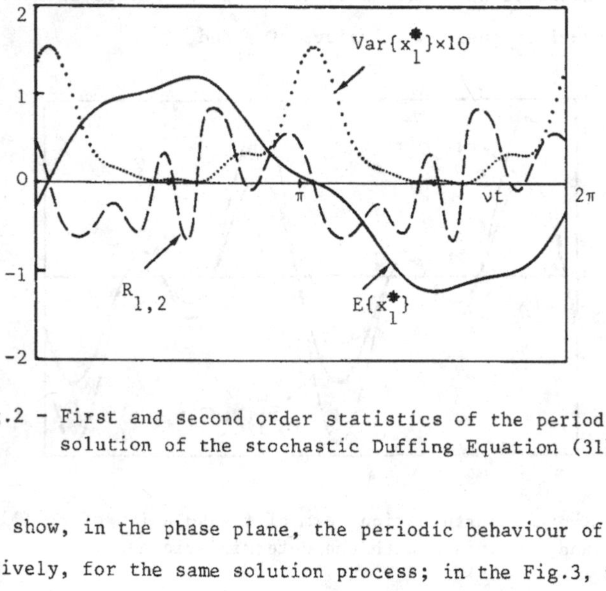

Fig.2 - First and second order statistics of the periodic solution of the stochastic Duffing Equation (31).

The Figs. 3 and 4 show, in the phase plane, the periodic behaviour of expectation and variance, respectively, for the same solution process; in the Fig.3, the probabilistic result is also compared with the well-known solution of the corresponding *deterministic* Duffing equation with $r(\omega,t) = \langle C_0 \rangle$, which is obtained, within the same approximation (see for instance /14/), by applying the perturbation method to solve deterministic nonlinear equations.

Concluding, let us remark that the proposed theory, which has been developed on the basis of known methods for studying stochastic evolution equations and semilinear deterministic differential equations, supply analytical approximated results concerning the periodical solutions of a class of stochastic differential equations with

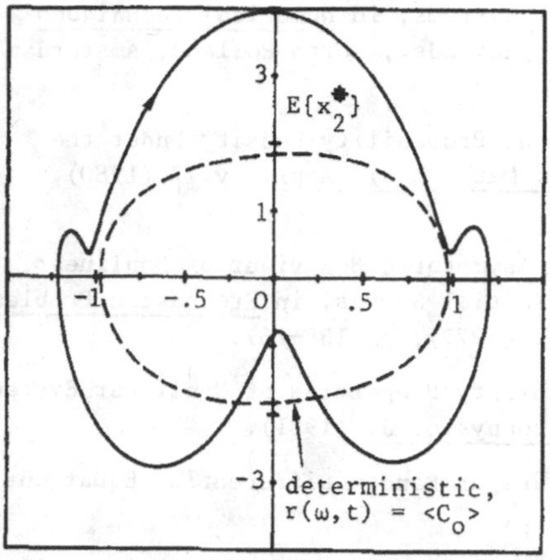

Fig.3- Expectations in the phase plane.

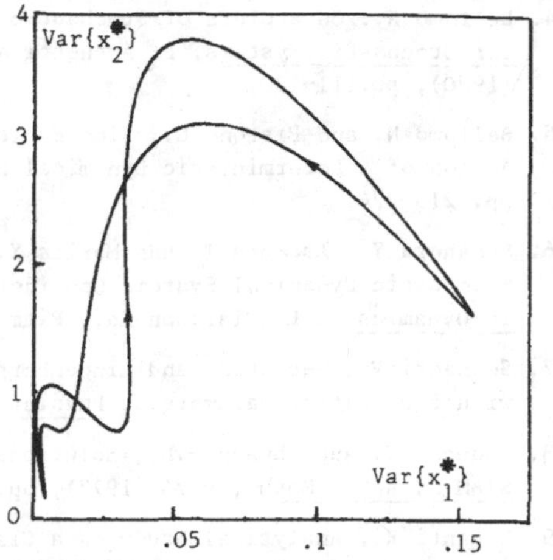

Fig.4- Variances in the phase plane.

small nonlinearities whose coefficients can be modeled by suitable stationary stochastic processes. The quantitative results obtained in the considered applications can be improved without any further conceptual difficulties, by determining higher-order terms of the approximated solution, and this can be made by calculating elementary integrals involving the product of exponential and trigonometric functions of the type defined by Eq.(26). The theory may be extended to the study of the the transient solution of the considered equation, with the further assumption of probabilistic initial conditions of motion and a more general analytical expression of the coefficient stochastic processes. The author followed this approach in /15/, where results concerning the moments and the evolution in time of the probability density of the solution process have been obtained in the transient regime of the class of stochastic oscillators which has been here considered in the applications.

Acknowledgments. This work has been realized within the activities of the Italian Council for the Research (C.N.R.), G.N.F.M.

References.

1. Soong T.T., Random Differential Equations in Science and Engineering, Academic Press, New York (1973).
2. Adomian G. and Sibul L., Stochastic Green's Formula and Application to Stochastic Differential Equations, J. Math. Anal. Appl., 3 (1977), pp.743-746.
3. Adomian G., Stochastic Systems, C.A.M. Report, Center of Applied Mathematics, University of Georgia, Athens, Georgia (July 1981).

4. Bellomo N., On a Class of Stochastic Dynamical Systems, in Numerical Techniques for Stochastic Systems, F. Archetti and M. Cugiani Eds., North Holland, Amsterdam (1980), pp.111-130.

5. Bellomo N. and Pistone G., Time-Evolution of the Probability Density Under the Action of a Deterministic Dynamical System, J. Math. Anal. Appl., v.77 (1980), pp. 215-224.

6. Sunahara Y., Asakura T. and Morita Y., On the Asymptotic Behaviour of Nonlinear Stochastic Dynamical Systems Considering the Initial States, in Stochastic Problems in Dynamics, B.L. Clarkson Ed., Pitman, London (1977), pp.138-167.

7. Seshadri V., West B.J. and Lindemberg K., Stability Properties of Nonlinear Systems with Fluctuating Parameters, Physica 338A, Europhysics J. (1981).

8. Soong T.T. and Chuang S.N., Solutions of a Class of Random Differential Equations, SIAM J. Appl. Math., v.24 (1973), pp. 449-459.

9. Riganti R., Analytical Study of a Class of Nonlinear Stochastic Autonomous Oscillators with One Degree of Freedom, Meccanica, v.14 (1979), pp.180-186.

10. Sansone G. and Conti R., Nonlinear Differential Equations, Pergamon Press, New York (1964).

11. Becker R.I., Periodic Solutions of Semilinear Equations of Evolution of Compact Type, J. Math. Anal. Appl., v.82 (1981), pp.33-48.

12. Ward J.R., Semilinear Boundary Value Problems in Banach Space, in Nonlinear Equations in Abstract Space, Academic Press, New York (1978), pp.469-477.

13. Adomian G., Bellomo N. and Riganti R., Semilinear Stochastic Systems: Analysis with the Method of Stochastic Green's Function and Application in Mechanics, J. Math. Anal. Appl., to be published.

14. Dinca F. and Teodosiu C., Nonlinear and Random Vibrations, Academic Press, New York (1973).

15. Riganti R., Transient Behaviour of Semilinear Stochastic Systems with Random Parameters, J. Math. Anal. Appl., to be published.

16. O'Malley R.E., Introduction to Singular Perturbations, Academic Press, New York (1974).

A DUALITY PRINCIPLE FOR NEUTRAL FUNCTIONAL DIFFERENTIAL EQUATIONS

Dietmar Salamon
Forschungsschwerpunkt Dynamische Systeme
Universität Bremen, 2800 Bremen 33
West Germany

1. INTRODUCTION

In this paper we present the basic ideas for a duality theory of linear neutral functional differential equations (NFDE) with general delays in the state- and input/output-variables.

We consider the controlled NFDE

$$(1) \qquad d/dt\Big(x(t) - Mx_t - \Gamma u_t\Big) = Lx_t + Bu_t$$

and the observed NFDE

$$(2) \qquad \dot{x}(t) = L^T x_t + M^T \dot{x}_t \,, \quad y(t) = B^T x_t + \Gamma^T \dot{x}_t \,,$$

which is obtained from (1) by transposition of matrices. In order to describe the duality relation between these two systems in a satisfactory way, we have to deal with two different state concepts.

The 'classical' way of introducing the state of a functional differential equation (FDE) is to specify an initial function of suitable length which describes the past history of the solution. An alternative (dual) state concept can be motivated as follows. The solution ($t \geq 0$) of a FDE can be derived from the initial function ($t \leq 0$) in two steps. First replace the initial function by an extra forcing term in the equation. Secondly determine the solution which corresponds to this forcing term. The dual state concept for the FDE is now obtained by defining the initial state to be such a forcing term of suitable length which determines the future behaviour of the solution (MILLER [9]). It has first been discovered by BURNS and HERDMAN [2] for Volterra integro-differential equations that these two state concepts are dual to each other via transposition of matrices. Corresponding results for retarded functional differential equations (RFDE) can be found e.g. in DIEKMANN [6], BERNIER-MANITIUS [1], MANITIUS [8], DELFOUR-MANITIUS [5], DELFOUR [4], VINTER-KWONG [11].

It is the purpose of this paper to apply a modified version of these ideas to NFDEs in the product space $\mathbb{R}^n \times L^p$ (BURNS-HERDMAN-STECH [3]) and in the Sobolev space $W^{1,p}$ (HENRY [7]).

NOTATION AND ASSUMPTIONS

We will always assume that $x(t)$, $x(t) \in \mathbb{R}^n$ and $u(t)$, $y(t) \in \mathbb{R}^m$ and we define $x_t(\tau) = x(t+\tau)$, $u_t(\tau) = u(t+\tau)$ for $-h \le \tau \le 0$ $(0 < h < \infty)$. Correspondingly L, M, B, Γ are bounded linear maps from $C = C([-h,0];\mathbb{R}^n)$ respectively $C([-h,0];\mathbb{R}^m)$ into $\mathbb{R}^n$. These can be represented by matrix functions $\eta(\tau)$, $\mu(\tau)$, $\beta(\tau)$, $\gamma(\tau)$ of bounded variation in the following way

$$L\varphi = \int_{-h}^{0} d\eta(\tau)\varphi(\tau) \quad , \quad M\varphi = \int_{-h}^{0} d\mu(\tau)\varphi(\tau) \quad , \quad \varphi \in C([-h,0];\mathbb{R}^n),$$

$$B\xi = \int_{-h}^{0} d\beta(\tau)\xi(\tau) \quad , \quad \Gamma\xi = \int_{-h}^{0} d\gamma(\tau)\xi(\tau) \quad , \quad \xi \in C([-h,0];\mathbb{R}^m).$$

The linear maps L^T, M^T, B^T, Γ^T from C into $\mathbb{R}^n$ respectively $\mathbb{R}^m$ are represented by the transposed matrix functions in an obvious manner. Without loss of generality we can assume that these matrix functions are normalized, i.e. vanish for $\tau \ge 0$, are constant for $\tau \le 0$ and left continuous for $-h < \tau < 0$. Moreover, we will always assume that

$$(3) \qquad -1 \notin \sigma\left(\lim_{\tau \uparrow 0} \mu(\tau)\right)$$

in order to guarantee existence and uniqueness of the solutions to (1) and (2).

Furthermore, we make use of the abbreviations $L^p = L^p([-h,0];\mathbb{R}^n)$ and $M^p = \mathbb{R}^n \times L^p$. The Sobolev space $W^{1,p} = W^{1,p}([-h,0];\mathbb{R}^n)$ will be identified with a (dense) subspace of M^p via the continuous embedding

$$\iota : W^{1,p} \to M^p \quad , \quad \iota\varphi = (\varphi(0) - M\varphi, \varphi) \quad .$$

Correspondingly, we define $\iota^T : W^{1,q} \to M^q$ by $\iota^T\psi = (\psi(0) - M^T\psi, \psi)$. The adjoint ι^{T*} of this mapping is a continuous embedding of $M^p = M^{q*}$ into the dual space $W^{-1,p}$ of $W^{1,q}$ $(1/p + 1/q = 1)$.

2. THE CLASSICAL STATE CONCEPT

It is well known that the observed system

$$\Omega^T \qquad \boxed{\dot{x}(t) = L^T x_t + M^T \dot{x}_t , \quad y(t) = B^T x_t + \Gamma^T \dot{x}_t ,}$$

admits a unique solution $x \in W^{1,q}_{loc}([-h,\infty);\mathbb{R}^n)$ for every initial condition

$$(4) \qquad x(\tau) = \psi(\tau) , \qquad -h \le \tau \le 0 ,$$

where $\psi \in W^{1,q}$ (HENRY [7]). Correspondingly,the state of system Ω^T at time $t \ge 0$ will be defined to be the function segment $x_t \in W^{1,q}$ which describes the past history of the solution at time t.

If the output $y(t)$ does not depend on the derivative of the solution, i.e. $\Gamma^T = 0$, then the above system can be considered in the product space M^q. For this sake it is useful to rewrite system Ω^T as

$$d/dt\left(x(t) - M^T x_t\right) = L^T x_t , \quad y(t) = B^T x_t ,$$

and to introduce the new variable $z(t) = x(t) - M^T x_t$. Then $z(t)$, $x(t)$ and $y(t)$ satisfy the following equations

$$\Sigma^T \qquad \boxed{\begin{array}{ll} \dot{z}(t) = L^T x_t , & x(t) = z(t) + M^T x_t , \\ y(t) = B^T x_t . & \end{array}}$$

It has recently been shown by BURNS, HERDMAN, and STECH [3] that this system admits a unique solution pair $z \in W^{1,q}_{loc}([0,\infty);\mathbb{R}^n)$ and $x \in L^q_{loc}([-h,\infty);\mathbb{R}^n)$ for every initial condition

$$(5) \qquad z(0) = \psi^0 , \quad x(\tau) = \psi^1(\tau) , \quad -h \le \tau \le 0 ,$$

where $\psi = (\psi^0,\psi^1) \in M^q$. Correspondingly, the state of system Σ^T at time $t \ge 0$ will be defined to be the pair $(z(t),x_t) \in M^q$.

Note that - in the case $\Gamma^T = 0$ - system Ω^T is nothing else than the restriction of system Σ^T to the dense subspace $W^{1,q}$ of M^q (recall the definition of the embedding $\iota^T : W^{1,q} \to M^q$).

3. THE DUAL STATE CONCEPT

We will define the state of the controlled NFDE (1) in a different way, namely through forcing terms. For this sake we rewrite (1) as a system of two equations in the following way

$$\Sigma \qquad \dot{w}(t) = Lx_t + Bu_t \ , \qquad x(t) = w(t) + Mx_t + \Gamma u_t \ .$$

Secondly, we represent the action of the initial functions of x and u on the right hand side of these equations through extra forcing terms. This leads to the following system

$$\tilde{\Sigma} \qquad \boxed{\begin{aligned} \dot{w}(t) &= \int_{-t}^{0} d\eta(\tau)x(t+\tau) + \int_{-t}^{0} d\beta(\tau)u(t+\tau) + f^1(-t) \ , \\ x(t) &= w(t) + \int_{-t}^{0} d\mu(\tau)x(t+\tau) + \int_{-t}^{0} d\gamma(\tau)u(t+\tau) + f^2(-t) \ , \\ w(0) &= f^0 \ , \end{aligned}}$$

where $f^1, f^2 \in L^p$ are given by

$$(6.1) \qquad f^1(-t) = \int_{-h}^{-t} d\eta(\tau)x(t+\tau) + \int_{-h}^{-t} d\beta(\tau)u(t+\tau) \ , \quad 0 \le t \le h \ ,$$

$$(6.2) \qquad f^2(-t) = \int_{-h}^{-t} d\mu(\tau)x(t+\tau) + \int_{-h}^{-t} d\gamma(\tau)u(t+\tau) \ , \quad 0 \le t \le h \ .$$

Now the initial state of $\tilde{\Sigma}$ is defined to be the bounded linear functional $\pi f \in W^{-1,p}$ on $W^{1,q}$ which is associated with the triple $f = (f^0, f^1, f^2) \in \mathbb{R}^n \times L^p \times L^p$ in the following way

$$\begin{aligned} &< \psi, \pi f >_{W^{1,q}, W^{-1,p}} \\ &= \psi^T(0)f^0 + \int_{-h}^{0} \psi^T(\tau)f^1(\tau)d\tau + \int_{-h}^{0} \dot{\psi}^T(\tau)f^2(\tau)d\tau \end{aligned}$$

(the lemma below shows that the solution $x(t)$ of $\tilde{\Sigma}$ with zero input vanishes for $t \ge 0$ if and only if $\pi f = 0$). Correspondingly the state of $\tilde{\Sigma}$ at time $t \ge 0$ is given by $\pi(w(t), w^t, x^t) \in W^{-1,p}$ where $w^t, x^t \in L^p$ denote the forcing terms of $\tilde{\Sigma}$ after a time shift. These are of the form

$$(7.1)\qquad w^t(\sigma) = \int_{\sigma-t}^{\sigma} d\eta(\tau)x(t+\tau+\sigma) + \int_{\sigma-t}^{\sigma} d\beta(\tau)u(t+\tau+\sigma) + f^1(\sigma-t) ,$$

$$(7.2)\qquad x^t(\sigma) = \int_{\sigma-t}^{\sigma} d\mu(\tau)x(t+\tau+\sigma) + \int_{\sigma-t}^{\sigma} d\gamma(\tau)u(t+\tau+\sigma) + f^2(\sigma-t) .$$

In the case $\Gamma = 0$ we may restrict system Σ to $W^{1,p}$-solutions. This can be expressed by rewriting Σ in the form

$$\Omega \qquad \dot{x}(t) = Lx_t + M\dot{x}_t + Bu_t .$$

Again we represent the action of the initial functions on the right hand side of Ω through an extra forcing term. Then equation Ω transforms into

$$\tilde{\Omega} \qquad \boxed{\begin{aligned} \dot{x}(t) &= \int_{-t}^{0} d\eta(\tau)x(t+\tau) + \int_{-t}^{0} d\mu(\tau)\dot{x}(t+\tau) \\ &+ \int_{-t}^{0} d\beta(\tau)u(t+\tau) + f^1(-t) , \qquad x(0) = f^0 . \end{aligned}}$$

where $f^1 \in L^p$ is given by

$$(8)\qquad f^1(-t) = \int_{-h}^{-t} d\eta(\tau)x(t+\tau) + \int_{-h}^{-t} d\mu(\tau)\dot{x}(t+\tau) + \int_{-h}^{-t} d\beta(\tau)u(t+\tau) .$$

The initial state of $\tilde{\Omega}$ is the pair $f = (f^0, f^1) \in M^p$ and the state $(x(t), x^t) \in M^p$ of $\tilde{\Omega}$ at time $t \geq 0$ is given by

$$(9)\qquad \begin{aligned} x^t(\sigma) &= \int_{\sigma-t}^{\sigma} d\eta(\tau)x(t+\tau+\sigma) + \int_{\sigma-t}^{\sigma} d\mu(\tau)\dot{x}(t+\tau+\sigma) \\ &+ \int_{\sigma-t}^{\sigma} d\beta(\tau)u(t+\tau+\sigma) + f^1(\sigma-t) , \qquad -h \leq \sigma \leq 0 . \end{aligned}$$

This expression can be obtained through a time shift of system $\tilde{\Omega}$.

The next lemma shows that - in the case $\Gamma = 0$ - system $\tilde{\Omega}$ is the restriction of system $\tilde{\Sigma}$ via the embedding $\iota^{T*} : M^p \to W^{-1,p}$.

<u>LEMMA</u> *Let* $\Gamma = 0$ *and let* $f \in M^p$, $\mathrm{f} \in \mathbb{R}^n \times L^p \times L^p$ *as well as* $u \in L^p_{loc}([0,\infty);\mathbb{R}^m)$ *be given. Moreover let* $x(t)$ *be the unique solution of* $\tilde{\Omega}$ *and* $w(t)$, $x(t)$ *the unique solution pair of* $\tilde{\Sigma}$. *Then* $x(t) = x(t)$ *for all* $t \geq 0$ *if and only if* $\pi\mathrm{f} = \iota^{T*} f$.

SKETCH OF THE PROOF

Step 1 Some straight forward computations show that $\pi f = \iota^{T^*} f$ if and only if

(10.1) $$f^0 + \int_{-h}^{0} f^1(\tau)d\tau = \left[I + \mu(-h)\right]f^0 + \int_{-h}^{0} f^1(\tau)d\tau \quad ,$$

(10.2) $$f^2(\sigma) + f^0 + \int_{\sigma}^{0} f^1(\tau)d\tau = \left[I + \mu(\sigma)\right]f^0 + \int_{\sigma}^{0} f^1(\tau)d\tau \quad .$$

Step 2 Introducing the functions

$$\tilde{x}(t) = x(t) - f^0, \quad \tilde{f}(t) = w(t) + f^2(-t) - f^0 - \mu(-t)f^0 \ ,$$

for $t \geq 0$, we obtain from the second equation in $\tilde{\Sigma}$ that

$$\tilde{x}(t) = \tilde{f}(t) + \int_{-t}^{0} d\mu(\tau)\tilde{x}(t+\tau) \quad , \qquad t \geq 0 \ .$$

Step 3 It follows from $\tilde{\Omega}$ and the first equation in $\tilde{\Sigma}$ that $x(t) = x(t)$ for all $t \geq 0$ if and only if $\tilde{x}(t)$ is absolutely continuous for $t \geq 0$ with $\tilde{x}(0) = 0$ and

$$\dot{\tilde{x}}(t) = \dot{w}(t) - f^1(-t) + f^1(-t) + \int_{-t}^{0} d\mu(\tau)\dot{\tilde{x}}(t+\tau) \quad , \qquad t \geq 0 \ .$$

Combining step 3 and step 2, we obtain that $x(t) = x(t)$ for all $t \geq 0$ if and only if

$$\tilde{f}(t) = \int_{0}^{t} \left(\dot{w}(s) - f^1(-s) + f^1(-s)\right)ds \quad , \qquad t \geq 0 \ .$$

This is equivalent to (10) and hence to $\pi f = \iota^{T^*} f$ (step 1).

Q.E.D.

4. THE MAIN RESULT

The following theorem describes the duality relation between the control system $\tilde{\Omega}$ (state space M^p) and the observed system Σ^T (state space M^q) respectively between the extended control system $\tilde{\Sigma}$

(state space $W^{-1,p}$) and the restricted observed system Ω^T (state space $W^{1,q}$).

THEOREM *Let* $u(.) \in L^p_{loc}([0,\infty);\mathbb{R}^m)$ *be given.*

(i) Let $f \in \mathbb{R}^n \times L^p \times L^p$ *and* $\psi \in W^{1,q}$. *Moreover, let* $\pi(w(t),w^t,x^t) \in W^{-1,p}$ *be the state of* $\tilde{\Sigma}$ *- defined by (7) - and let* $x(t)$ *be the unique solution of* Ω^T, *(4) with output* $y(t)$. *Then*

$$< \psi,\pi(w(t),w^t,x^t) > \;=\; < x_t,\pi f > + \int_0^t y^T(t-s)u(s)ds \quad , \qquad t \geq 0 \;.$$

(ii) Let $f \in M^p$ *and* $\psi \in M^q$. *Moreover, let* $(x(t),x^t) \in M^p$ *be the state of* $\tilde{\Omega}$ *- defined by (9) - and let* $z(t)$, $x(t)$ *be the unique solution of* Σ^T, *(5) with output* $y(t)$. *Then*

$$< \psi,(x(t),x^t) > \;=\; < (z(t),x_t),f > + \int_0^t y^T(t-s)u(s)ds \quad , \qquad t \geq 0 \;.$$

PROOF (i) Let $x(t) = 0$ and $u(t) = 0$ for $t < 0$. Then

$$\int_0^t \Big(x^T(t-s)Lx_s - [L^T x_{t-s}]^T x(s) \Big) ds$$

$$= \int_{-h}^0 \int_0^t x^T(t-s)d\eta(\tau)x(s+\tau)ds + \int_{-h}^0 \int_0^t x^T(t-s+\tau)d\eta(\tau)x(s)ds$$

$$= - \int_{-h}^0 \int_{t+\tau}^t x^T(t+\tau-s)d\eta(\tau)x(s)ds$$

$$= - \int_{-h}^0 \int_\tau^0 \psi^T(\tau-\sigma)d\eta(\tau)x(t+\sigma)d\sigma \;.$$

Analogous equations hold for M, B, and Γ. Moreover

$$\psi^T(0)w(t) - x^T(t)f^0 \;=\; \int_0^t \frac{d}{ds}\, x^T(t-s)w(s)ds$$

$$= \int_0^t x^T(t-s)\dot{w}(s)ds - \int_0^t \dot{x}^T(t-s)w(s)ds \;.$$

This implies

$$< \psi,\pi(w(t),w^t,x^t) >$$

$$= \int_{-h}^{0}\psi^T(\sigma)w^t(\sigma)d\sigma + \int_{-h}^{0}\dot{\psi}^T(\sigma)x^t(\sigma)d\sigma + \psi^T(0)w(t)$$

$$= \int_{-h}^{0}\int_{\tau}^{0}\psi^T(\tau-\sigma)d\eta(\tau)x(t+\sigma)d\sigma + \int_{-h}^{0}\int_{\tau}^{0}\psi^T(\tau-\sigma)d\beta(\tau)u(t+\sigma)d\sigma$$

$$+ \int_{-h}^{0}\int_{\tau}^{0}\dot{\psi}^T(\tau-\sigma)d\mu(\tau)x(t+\sigma)d\sigma + \int_{-h}^{0}\int_{\tau}^{0}\dot{\psi}^T(\tau-\sigma)d\gamma(\tau)u(t+\sigma)d\sigma$$

$$+ \int_{-h}^{0}\psi^T(\sigma)f^1(\sigma-t)d\sigma + \int_{-h}^{0}\dot{\psi}^T(\sigma)f^2(\sigma-t)d\sigma + x^T(t)f^0$$

$$+ \int_{0}^{t}x^T(t-s)\Big(Lx_s + Bu_s + f^1(-s)\Big)ds$$

$$- \int_{0}^{t}\dot{x}^T(t-s)\Big(x(s) - Mx_s - \Gamma u_s - f^2(-s)\Big)ds$$

$$= x^T(t)f^0 + \int_{-t}^{0}x^T(t+\tau)f^1(\tau)d\tau + \int_{-t}^{0}\dot{x}^T(t+\tau)f^2(\tau)d\tau$$

$$+ \int_{-h}^{-t}\psi^T(t+\tau)f^1(\tau)d\tau + \int_{-h}^{-t}\dot{\psi}^T(t+\tau)f^2(\tau)d\tau$$

$$+ \int_{0}^{t}[L^Tx_{t-s}]^Tx(s)ds + \int_{0}^{t}[M^T\dot{x}_{t-s}]^Tx(s)ds - \int_{0}^{t}\dot{x}^T(t-s)x(s)ds$$

$$+ \int_{0}^{t}[B^Tx_{t-s}]^Tu(s)ds + \int_{0}^{t}[\Gamma^T\dot{x}_{t-s}]^Tu(s)ds$$

$$= < x_t,\pi f > + \int_{0}^{t}y^T(t-s)u(s)ds .$$

(ii) If $\psi \in \operatorname{ran} \iota^T$, then statement (ii) is a direct consequence of statement (i) and the lemma above. In general, (ii) follows from a continuity argument.

Q.E.D.

Summarizing our results, we have to deal with the following four systems.

$$\begin{array}{cc} \tilde{\Sigma} & \Sigma^T \\ \tilde{\Omega} & \Omega^T \end{array}$$

The systems on the left hand side describe the controlled NFDE (1) in the state spaces $W^{-1,p}$ and M^p (dual state concept) and the systems on the right hand side describe the observed NFDE (2) in the state spaces M^q and $W^{1,q}$ (classical state concept). On each side the system below represents the restriction of the upper system to absolutely continuous solutions. The diagonal relation is described by the above duality theorem.

Results of this type have not been developed so far in the literature on NFDEs. They have several important consequences in the state space theory of neutral systems as well as for problems like completeness & small solutions, controllability & observability, feedback stabilization & dynamic observation (SALAMON [10]).

ACKNOWLEDGEMENT This work has been supported by the Forschungsschwerpunkt Dynamische Systeme.

REFERENCES

[1] C. BERNIER/A. MANITIUS
On semigroups in $\mathbb{R}^n \times L^p$ corresponding to differential equations with delays
Can. J. Math. 30(1978), 897-914

[2] J.A. BURNS/T.L. HERDMAN
Adjoint semigroup theory for a class functional differential equations
SIAM J. Math. Anal. 7(1976), 729-745

[3] J.A. BURNS/T.L. HERDMAN/H.W. STECH
Linear functional differential equations as semigroups in product spaces
Department of Mathematics, Virginia Polytechnic Institute and State University Blacksburg, Virginia 1981

[4] M.C. DELFOUR
Status of the state space theory of linear, hereditary differential systems with delays in state and control variables
in "Analysis and Optimization of Systems", A. Bensoussan, J.L. Lions, eds., pp. 83-96,
Springer-Verlag, New York, 1980

[5] M.C. DELFOUR/A. MANITIUS
The structural operator F and its role in the theory of retarded systems
Part 1: J. Math. Anal. Appl. 73(1980), 466-490
Part 2: J. Math. Anal. Appl. 74(1980), 359-381

[6] O. DIEKMANN
A duality principle for delay equations
Preprint, Mathematisch Centrum Report TN 100/81, Amsterdam 1981

[7] D. HENRY
Linear autonomous functional differential equations of neutral type in the Sobolev space $W_2^{(1)}$
Technical Report, Department of Mathematics
University of Kentucky, Lexington, Kentucky 1970

[8] A. MANITIUS
Completeness and F-completeness of eigenfunctions associated with retarded functional differential equations
J. Diff. Equations 35(1980), 1-29

[9] R.K. MILLER
Linear Volterra integro-differential equations as semigroups
Funkcial. Ekvac. 17(1974), 749-763

[10] D. SALAMON
On control and observation of neutral systems
Doctoral dissertation, Forschungsschwerpunkt Dynamische Systeme, Universität Bremen, Bremen 1982

[11] R.B. VINTER/R.H. KWONG
The finite time quadratic control problem for linear systems with state and control delays: an evolution equation approach
SIAM J. Control Opt. 19(1981), 139-153

CHARACTERIZATION OF PERIODIC SOLUTIONS OF SPECIAL DIFFERENTIAL DELAY EQUATIONS

Dietmar Saupe
Forschungsschwerpunkt "Dynamische Systeme"
Universität Bremen
2800 Bremen-33
West Germany

1. Introduction

The aim of this paper is to derive a simple finite dimensional characterization of periodic solutions of the differential delay equation

$$\dot{x}(t) = -\lambda f(x(t-1)), \quad \lambda > 0 \tag{1}$$

where $f : R \to R$ is an odd and piecewise constant function satisfying $xf(x) > 0$ for all $x \neq 0$. Here a solution x of (1) is a continuous piecewise linear function which solves the integrated version

$$x(t) = x(0) - \lambda \int_{-1}^{t-1} f(x(s))\, ds \tag{2}$$

of (1). From our results we obtain a suitable numerical procedure for the computation of periodic solutions of (1) and we report its performance on a test example.

Our task is motivated by the study of periodic solutions of (1) where f is a continuous nonlinearity, e. g. $f(x) = x/(1+x^8)$ (see [1, 3,6,7]). Recently H. Peters [4] has modelled this nonlinearity by a piecewise constant function which has two steps (for $x > 0$). He was able to completely and explicitly compute all periodic solutions. We will modify and extend his approach for nonlinearities f with n steps. Thus, we may use a better piecewise constant approximation to a continuous nonlinearity. However, we cannot expect to be able to explicitly compute nontrivial periodic solutions, because the computational complexity is too great even if n is of moderate size, say $n \geq 4$. Instead we have to rely on the computer which may numerically solve the problem.

Our approach has one promising aspect, namely that a solution to (1) is piecewise linear and therefore may be computed exactly up to round off errors. Hence, there are nodiscretization errors involved! We may say that the discretization has already occured in the choice of the piecewise constant nonlinearity f.

We now sketch one of the standard procedures of how the problem of computing periodic solutions may be cast into an operator equation. Let $\varphi \in C[-1,0]$ denote a continuous initial function for the initial value problem

$$(3) \qquad \begin{cases} \dot{x}(t) = -\lambda f(x(t-1)) & \text{for} \quad t \geq 0 \,, \\ x(t) = \varphi(t) & \text{for} \quad -1 \leq t \leq 0 \,. \end{cases}$$

There is a unique solution x_φ of (3) defined on the interval $[-1,\infty)$. We call a solution x of (1) or x_φ of (3) *slowly oscillating* if it has infinitely many zeroes and if the distance between any two zeroes is greater than 1. For the study of slowly oscillating, periodic solutions of (1) we may restrict to initial functions from the set

$$P = \{\varphi \in C[-1,0] \mid \varphi(-1) = 0 \,,\ \varphi \text{ is strictly monotonically increasing}\}\,.$$

Definition 1

For $\varphi \in P$ let x_φ denote the corresponding solution of (3). Then there is a zero $z_1 > 0$ of x_φ such that $x_\varphi(t) > 0$ for $-1 < t < z_1$. The *shift operator* $S_\lambda : P \to P$ is defined by $S_\lambda(\varphi) : t \to -x_\varphi(z_1+1+t)$.

Since f is odd we have that a fixed point of the shift operator S_λ or of one of its iterates S_λ^k, $k = 2,3,\ldots$ induces a slowly oscillating, periodic solution of (1), which we call an *S-solution* or *S^k-solution* respectively. Moreover, there exist *special S-solutions* x of (1) which have an additional symmetry : If z denotes a zero of x, then we have $x(z+t) = x(z+2-t)$ for all $t \in R$. Thus, these solutions are sinusoidal and they have the period 4.

Note that the nonlinearity f is only piecewise constant and thus, the shift operator S_λ is not continuous. However, in the next section we will see how the piecewise constant structure of f facilitates a modified, continuous, and even finite dimensional shift operator.

2. The shift operator for equivalence classes of initial functions

For $\varphi_1,\varphi_2 \in P$ let $\varphi_1 \sim \varphi_2$ if and only if $x_{\varphi_1}(t) = x_{\varphi_2}(t)$ for $t \geq 0$. "~" is an equivalence relation. For continuous nonlinearities f in (1) we have in general that the equivalence classes $[\varphi]$ trivially contain only one element, namely φ itself. But with piecewise constant nonlinearities $[\varphi]$ may have many elements. We show that the set of all equivalence classes $P/\sim$ is a finite dimensional set. Let us first define a suitable set of nonlinearities.

Definition 2

For $n \in \{1,2,...\}$ let F_n denote the set of real functions satisfying

(a) $f(-x) = -f(x)$ for $x \in R$.

(b) $xf(x) > 0$ for $x \neq 0$.

(c) There exist numbers $f_1,...,f_n > 0$ and a subdivision $0 = x_0 < x_1 < ... < x_n = \infty$ such that for $k = 1,...,n$ $f(x) = f_k$ holds if $x \in J_k = (x_{k-1},x_k]$.

(d) $f_k \neq f_{k+1}$ for $k = 1,...,n-1$.

We say that a function $f \in F_n$ has n steps. The following mapping I_f extracts the necessary information from an initial function $\varphi \in P$, which is relevant for the integration of (3) .

Definition 3

Let $I_f = (I_f^0,...,I_f^n) : P \to R \times R^n$ be defined by

(a) $I_f^0(\varphi) = \varphi(0)$.

(b) Let $x_0,...,x_n$ be as in Definition 2 and assume $\varphi(0) \in J_j = (x_{j-1},x_j]$.

(c) If $j = 1$, then set $I_f^1(\varphi) = 1$ and $I_f^k(\varphi) = 0$ for $k = 2,...,n$.

(d) If $j > 1$, then set

$$I_f^1(\varphi) = \varphi^{-1}(x_1) + 1 ,$$
$$I_f^k(\varphi) = \varphi^{-1}(x_k) + 1 - \sum_{i=1}^{k-1} I_f^i(\varphi) \quad \text{for } k = 2,...,j-1 ,$$
$$I_f^j(\varphi) = 1 - \sum_{i=1}^{j-1} I_f^i(\varphi) ,$$
$$I_f^k(\varphi) = 0 \quad \text{for } k = j+1,...,n .$$

Figure 1 illustrates the definition of I_f . Let $D = I_f(P)$. Obviously we have $D \subset R^+ \times \Delta_{n-1}$ where Δ_{n-1} denotes the standard $(n-1)$-simplex in R^n . The following lemma states that we can identify $P/\sim$ with the n-dimensional range of D of I_f .

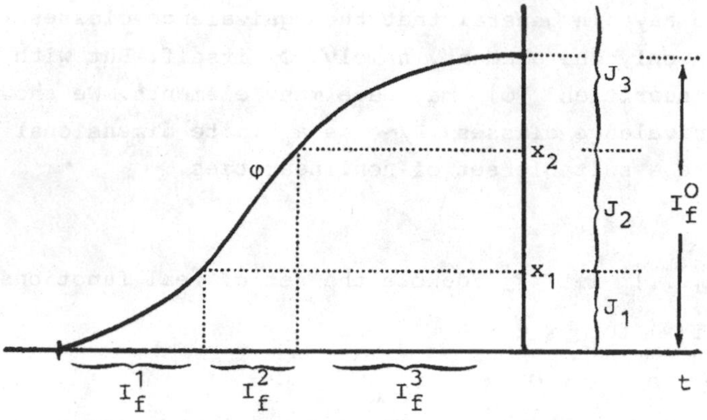

Figure 1 The definition of I_f

Lemma 4
Let $f \in F_n$, $\varphi_1, \varphi_2 \in P$. Then

$$\varphi_1 \sim \varphi_2 \quad \Leftrightarrow \quad I_f(\varphi_1) = I_f(\varphi_2) .$$

Proof: Assume $\varphi_1 \sim \varphi_2$ and that $k = \min \{k \mid I_f^k(\varphi_1) \neq I_f^k(\varphi_2)\}$ exists. If $k = 0$ then $\varphi_1(0) \neq \varphi_2(0)$ contradicting $x_{\varphi_1}(0) = x_{\varphi_2}(0)$. Otherwise we have for sufficiently small $\varepsilon > 0$ the contradiction $x_{\varphi_1}(t_0+\varepsilon) \neq x_{\varphi_2}(t_0+\varepsilon)$ where

$$t_0 = \sum_{i=1}^{k-1} I_f^i(\varphi_1) \; + \; \min \{I_f^k(\varphi_i) \mid i = 1,2\} .$$

If we assume $I_f(\varphi_1) = I_f(\varphi_2)$, then it follows from the construction of I_f that we have $\varphi_1 \sim \varphi_2$.

It is easy to see that $D = I_f(P)$ is a connected subset of $R^+ \times \Delta_{n-1}$. If $\varphi_0, \varphi_1 \in P$, then $\{\varphi_s = (1-s)\varphi_0 + s\varphi_1 \mid s \in [0,1]\} \subset P$ and I_f is continuous on this line segment of P. Therefore $[\varphi_0]$ and $[\varphi_1]$ can be connected by the path $s \to [\varphi_s]$.

We have that $\varphi_1 \sim \varphi_2$ implies $S_\lambda(\varphi_1) = S_\lambda(\varphi_2)$. Therefore we may define an induced shift operator $\overline{S}_\lambda$ on D via

$$\overline{S}_\lambda : P/\!\sim \; \to P/\!\sim$$
$$[\varphi] \to [S_\lambda(\varphi)] .$$

In contrast to the original shift operator $S_\lambda : P \to P$ we have that the induced map $\overline{S}_\lambda : P/\!\sim \; \to P/\!\sim$ is continuous. For the elementary but rather technical proof of this fact we refer to [6].

Concerning slowly oscillating, periodic solutions we now have :

<u>Corollary 5</u>

Let $k \in \{1,2,\ldots\}$.

(i) If $\varphi^* \in P$ such that $\varphi^* = S_\lambda^k(\varphi^*)$, then also $[\varphi^*] = \bar{S}_\lambda^k([\varphi^*])$.

(ii) If $\varphi \in P$ such that $[\varphi] = \bar{S}_\lambda^k([\varphi])$, then $\varphi^* = S_\lambda^k(\varphi^*)$ where $\varphi^* = S_\lambda(\varphi)$.

Therefore, S^k-solutions of (1) are in a one-to-one correspondence with the fixed points of the n-dimensional mappings $\bar{S}_\lambda^k$. Their numerical computation is a nonlinear fixed point problem, and in our case continuation methods seem to be the most appropriate tools for this task. We employ a predictor-corrector algorithm based on a piecewise linear approximation of the underlying mapping. It is called SCOUT and has been developed by H. Jürgens and the author (see [1,5,6]). Any other path following method may be used, provided that it is able to resolve singularities such as turning points and bifurcations. SCOUT is designed to compute the zeroes of a mapping $H : R^n \times R \to R^n$, and in order to facilitate its employment for our purpose we identify $R^+ \times \Delta_{n-1}$ with the n-dimensional nonnegative octant $[0,\infty)^n$ via $(u_0,\ldots,u_n) \to (u_0u_1,\ldots, u_0u_n)$ and extend $\bar{S}_\lambda^k$ to a continuous mapping $\tilde{S}_\lambda^k : R^n \to R^n$ such that the diagram

$$\begin{array}{ccccccc} D & \hookrightarrow & R^+ \times \Delta_{n-1} & \longrightarrow & [0,\infty)^n & \hookleftarrow & R^n \\ \downarrow \bar{S}_\lambda^k & & & & & & \downarrow \tilde{S}_\lambda^k \\ D & \hookrightarrow & R^+ \times \Delta_{n-1} & \longrightarrow & [0,\infty)^n & \hookleftarrow & R^n \end{array}$$

commutes and $\tilde{S}_\lambda^k(R^n) \subset D$. We then define

$$H : R^n \times R^+ \to R^n$$
$$(u,\lambda) \to u - \tilde{S}_\lambda^k(u) .$$

A first periodic solution as a start for the continuation method is trivially available : Let $f \in F_n$ and $x_1, f_1 > 0$ be as in Definition 2 . For $0 < \lambda < x_1/f_1$ we have that the initial function $\varphi(t) = \lambda f_1 (t+1)$ is a fixed point $\varphi = S_\lambda(\varphi)$ defining a special S-solution. Therefore $u = \bar{S}_\lambda(u)$ where $u = (\lambda f_1, 1, 0, \ldots, 0) \in D$.

If we approximate a given continuous nonlinearity by a piecewise constant function $f \in F_n$ then the equation $[\varphi] = S_\lambda^k([\varphi])$ is a *perturbed Galerkin equation* (in the sense of Krasnoselski´i et al [2]) for the computation of periodic solutions of the given delay equation.

3. Numerical results

All the computations were done with the SCOUT package using a triangulation with a mesh size of 0.001 . The predictor-corrector technique [5] was successfully employed such that the experiments could conveniently be done in an interactive time sharing session (on a Siemens 7.800).

In this section we are considering a special nonlinearity $f \in F_8$ designed to model the smooth function $g(x) = x/(1+x^8)$ (see Figure 2).

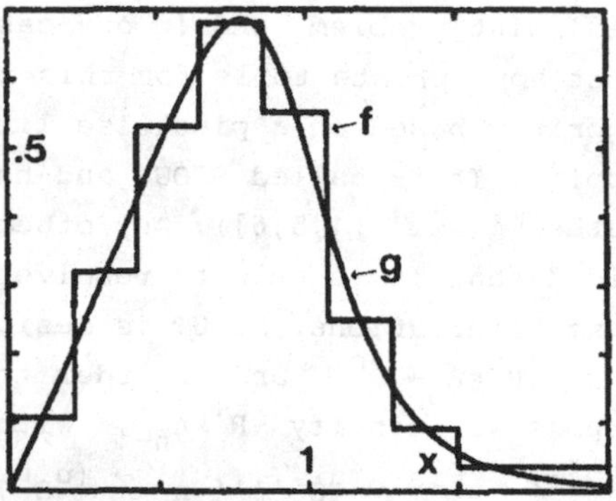

k	x_k	f_k
1	0.214	0.107
2	0.429	0.321
3	0.643	0.532
4	0.857	0.682
5	1.071	0.552
6	1.286	0.250
7	1.500	0.092
8	∞	0.035

Figure 2 The nonlinearity $f \in F_8$. We define $x_k = 3k/14$, $f_{k+1} = g(x_k+3/28)$ for $k = 0,\dots,7$ and $x_8 = \infty$.

Figure 3 shows a bifurcation diagram for S-solutions of (1) , where we have plotted the norm of the initial functions $\|\varphi\|$ versus λ . There is a continuum of special S-solutions emanating from the origin. The zig-

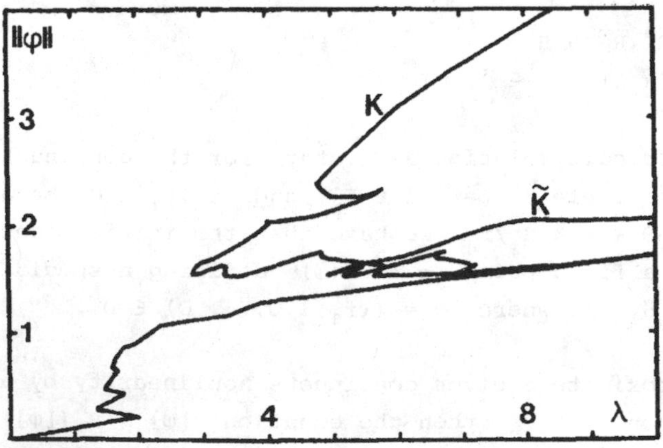

Figure 3 Bifurcation diagram of S-solutions of (1) .

zagging behaviour of this continuum for $1 < \lambda < 2$ is characteristic for equation (1) with piecewise constant nonlinearities such as f, and it has also been observed in [6,ch. 7]. At $\lambda = 12.6$ a bifurcation takes place : Two branches (K and $\tilde{K}$) of not special S-solutions bifurcate. K consists of S-solutions with periods greater than 4 whereas $\tilde{K}$ consists of S-solutions with periods less than 4. It can be seen as in [6,7] that K and $\tilde{K}$ are conjugate in the following sense. If (x,λ) denotes an S-solution with a period T, then we have that $(\tilde{x},\tilde{\lambda})$ is also an S-solution, where $\tilde{x}(t) = -x(-t(T-2)/2)$ and $\tilde{\lambda} = \lambda(T-2)/2$. With this notation we have that $(x,\lambda) \in K$ if and only if $(\tilde{x},\tilde{\lambda}) \in \tilde{K}$.

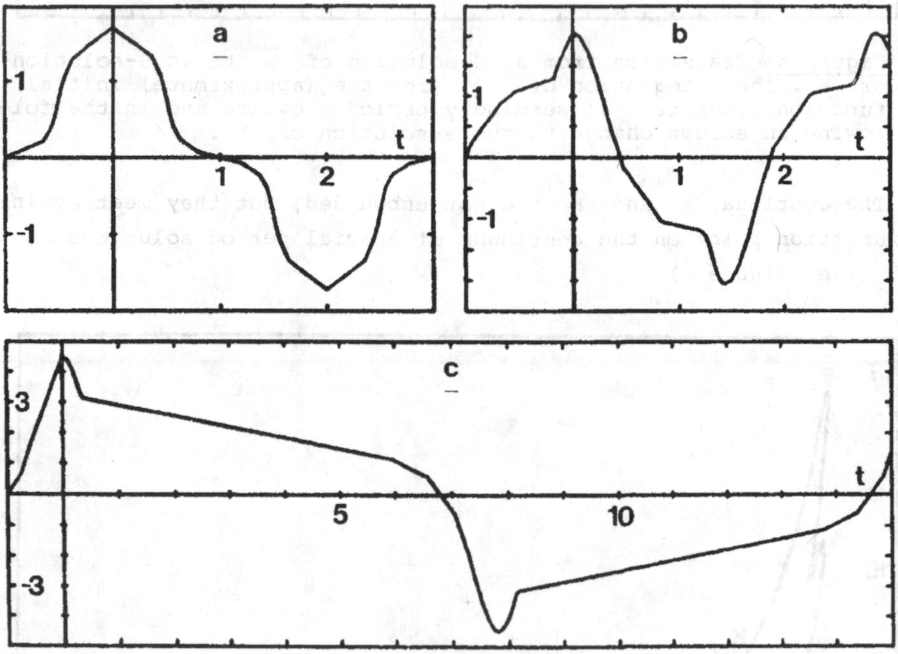

Figure 4 S-solutions of (1) for $\lambda = 10$. a) Special S-solution, b) S-solution of $\tilde{K}$ with a period $T = 2.875$, c) S-solution of K with a period $T = 15.622$. Note : For the shown initial functions φ we have $\varphi \sim S_\lambda(\varphi)$, not $\varphi = S_\lambda(\varphi)$.

Figure 4 shows three periodic solutions from the different branches at $\lambda = 10$. A stability test reveals that the S-solution of K is stable and attractive with respect to the integration of the delay equation, whereas the other two solutions are rather unstable (see Figure 5).

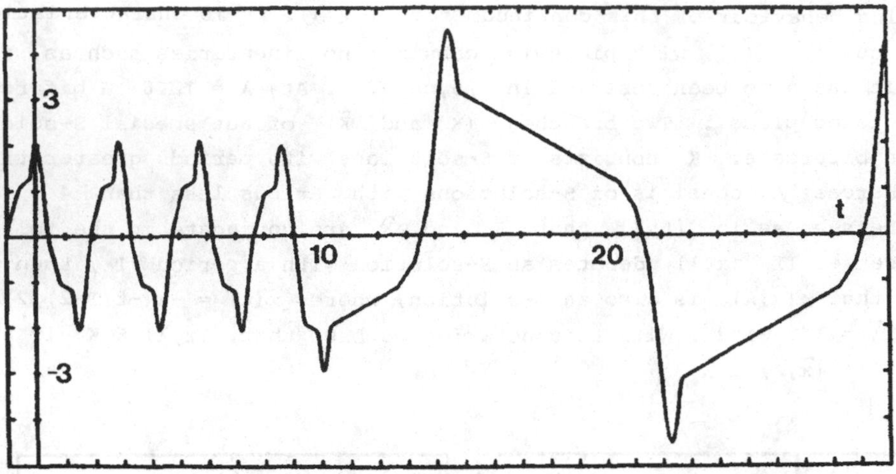

Figure 5 Transition from an S-solution of $\tilde{K}$ to an S-solution of K. The integration of (3) for the (approximate) initial function generates 3 seemingly periodic cycles and in the following an abrupt change to the S-solution of K.

The continua K and $\tilde{K}$ are not unbounded, but they meet again at a bifurcation point on the continuum of special period solutions at $\lambda = 84.95$ (see Figure 6).

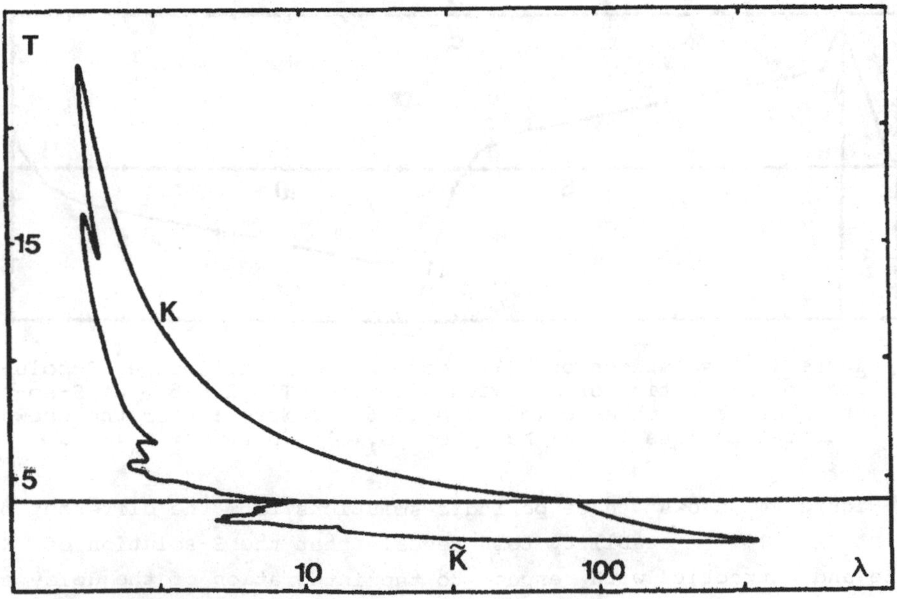

Figure 6 Bifurcation diagram of S-solutions of (1). Here the period T is graphed versus λ (in the log-scale).

In the last experiment we concentrate on S^2-solutions of (1). There are two such continua, the first one bifurcates from special S-solutions at $\lambda = 1.79$, forms a loop, and returns to the special S-solutions at $\lambda = 1.95$. The second continuum branches off at $\lambda = 2.35$ (see Figure 7 and 8).

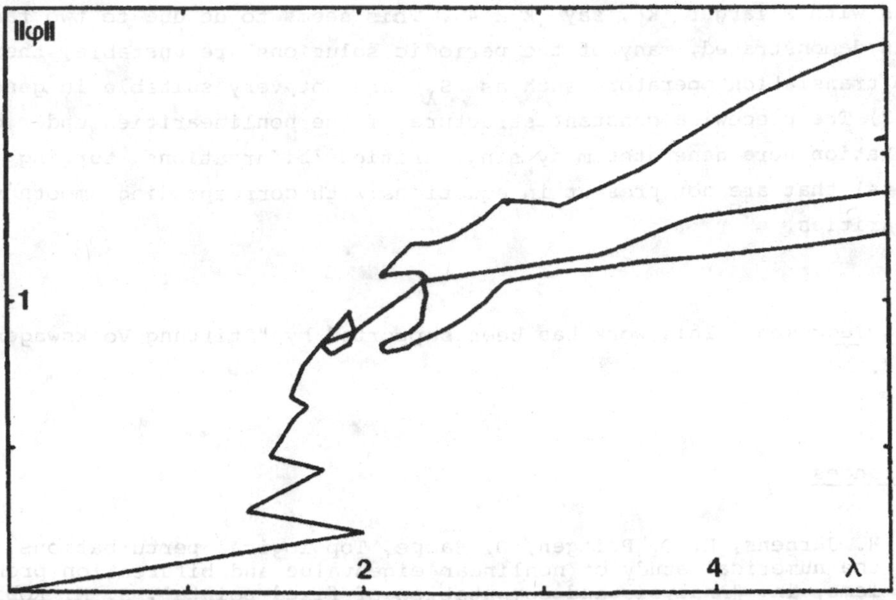

Figure 7 Bifurcation diagram of S^2-solutions of (1).

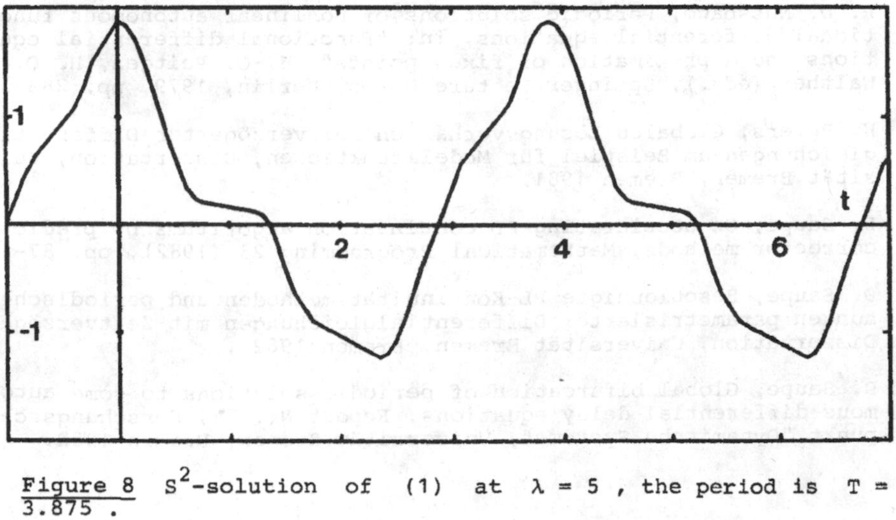

Figure 8 S^2-solution of (1) at $\lambda = 5$, the period is $T = 3.875$.

We remark that all of the above results agree with similar studies carried out in [1,6,7] for continuous nonlinearities. However, the approach of this paper as a numerical method for the computation of periodic solutions of such delay equations does not yield very satisfactory results. This is especially true for equations with nonlinearities $f \in F_n$ with a larger n, say $n \approx 20$ or for the computation of S^k-solutions with a larger k, say $k \geq 4$. This seems to de due to two facts. 1) As demonstrated, many of the periodic solutions are unstable, therefore translation operators such as S_λ are not very suitable in general. 2) The piecewise constant structure of the nonlinearities under consideration here generates many singularities (bifurcations, turning points) that are not present in equations with corresponding smooth nonlinearities.

Acknowledgement This work has been supported by "Stiftung Volkswagenwerk".

References

[1] H. Jürgens, H.-O. Peitgen, D. Saupe, Topological perturbations in the numerical study of nonlinear eigenvalue and bifurcation problems, in: "Analysis and computation of fixed points", S. M. Robinson (ed.), Academic Press, New York, 1980, pp. 139-182.

[2] M. A. Krasnosel'skii et al, Approximate solution of operator equations, Wolters-Noordhoff, Groningen, 1972.

[3] R. D. Nussbaum, Periodic solutions of nonlinear autonomous functional differential equations, in: "Functional differential equations and approximation of fixed points", H.-O. Peitgen, H. O. Walther (eds.), Springer Lecture Notes, Berlin, 1979, pp. 283-325.

[4] H. Peters, Globales Lösungsverhalten zeitverzögerter Differentialgleichungen am Beispiel für Modellfunktionen, Dissertation, Universität Bremen, Bremen 1981.

[5] D. Saupe, On accelerating PL continuation algorithms by predictor corrector methods, Mathematical Programming 23 (1982), pp. 87-110.

[6] D. Saupe, Beschleunigte PL-Kontinuitätsmethoden und periodische Lösungen parametrisierter Differentialgleichungen mit Zeitverzögerung, Dissertation, Universität Bremen, Bremen 1982 .

[7] D. Saupe, Global bifurcation of periodic solutions to some autonomous differential delay equations, Report Nr. 71, Forschungsschwerpunkt "Dynamische Systeme", Universität Bremen, Bremen 1982.

Uniform Stability of Almost Periodic Solutions of Delay-Differential Equations.

George Seifert
(Iowa State University, Ames, Iowa 50011)

1. Introduction. In a recent paper, conditions sufficient for the existence of almost periodic (a.p. for short) solutions for delay-differential equations with infinite delays were given [1]. More recently, conditions sufficient for the uniform stability of bounded solutions of such equations have been derived [2]. It is the purpose of this paper to show that the conditions for a.p. solutions are indeed also sufficient for their uniform stability. We also apply our basic result to an equation describing the population density of a species interacting with itself where infinite delays are involved.

2. Definitions and Notation. R^n denotes the set of real n-vectors, and $|x|$ any fixed norm for $x \in R^n$.

CB denotes the set of R^n valued functions continuous and bounded on $(-\infty,0]$; $\|\phi\| = \sup\{|\phi(s)| : s \leq 0\}$ for $\phi \in CB$.

$$CB_r = \{\phi \in CB : \|\phi\| \leq r\}.$$

If $x(s) : (-\infty,b) \to R^n$, for each $t < b$, we define $x_t(s) = x(t+s)$, $s \leq 0$.

If $f(t,\phi) : R \times CB \to R^n$, then $x(t) : (-\infty,b) \to R^n$ solves

$$(1) \qquad x'(t) = f(t,x_t)$$

on $[t_0,b)$ if for t in this interval:

(i) $x'(t)$ exists and is continuous;

(ii) $x_t \in CB$;

(iii) (1) holds.

If $f(t,0) = 0$ for all t, $x = 0$ is stable if given $t_0 \in R$, $\varepsilon > 0$, then exists $\delta = \delta(t_0,\varepsilon) > 0$ such that $\|\phi\| < \delta$ implies that any solution $x(t)$ of (1) with $x_{t_0} = \phi$ is defined for all $t \geq t_0$ and satisfies $|x(t)| < \varepsilon$ there; $x = 0$ is uniformly stable (on R) if it is stable and $\delta = \delta(\varepsilon)$.

If $\bar{x}(t)$ solves (1) on R, it is stable (uniformly stable) if $z = 0$ is stable (uniformly stable) for $z'(t) = f(t,\bar{x}_t + z_t) - f(t,\bar{x}_t)$.

We say $f(t,)$ is a.p. (Bohr) in t uniformly for $\phi \in S \in CB$ if (cf. [3])

(i) $|f(t,\phi)| \leq M$ for $(t,\phi) \in R \times S$;

(ii) f is a.p. in t for each $\phi \in S$, and for each $\varepsilon > 0$, the set of ε-translation numbers have a non-empty intersection over S which is relatively dense;

(iii) f is continuous in t uniformly for $(t,\phi) \in R \times S$.

3. The Basic Result.

Theorem 1. Let $F(t,x,\phi) : R \times R^n \times CB \to R^n$ be a.p. in t uniformly for (x,ϕ) in closed bounded sets in $R^n \times CB$. Suppose there exist $r > 0$, $M > 0$, $p > M/r$ such that $|F(t,0,\phi)| \leq M$ for $(t,\phi) \in R \times CB_r$, and for $h > 0$ and sufficiently small, suppose

(i) $|x - y + h(F(t,x,\phi) - F(t,y,\phi))| \leq (1 - ph)|x - y|$

for $|x| \leq r$, $|y| \leq r$, $\phi \in CB_r$, $t \in R$; and

(ii) $|x(t) - y(t) + h(F(t,x(t),x_t) - F(t,y(t),y_t))| \leq (1 - ph)\|x_t - y_t\|$

for $x(t)$, $y(t)$ such that $x_t \in CB_r$, $y_t \in CB_r$, $t \in R$. Also assume that if $\phi^k \in CB_r$ and $\phi^k(t) \to \phi(t)$ as $k \to \infty$ uniformly on compact sets in $(-\infty,0]$, then $F(t,x,\phi^k) \to F(t,x,\phi)$ as $k \to \infty$ uniformly for (t,x) in compact sets in $R \times R^n$.

Then

(2) $x'(t) = F(t,x(t),x_t)$

has a uniformly stable a.p. solution $\bar{x}(t)$ such that $|\bar{x}(t)| < r$ for all t.

Sketch of Proof. To show the existence of an a.p. solution $\bar{x}(t)$ of (2) such that $|x(t)| < r$ for all t, since $p > M/r$, there exists r_1, $0 < r_1 < r$, such that $p > M/r_1$. It is not difficult to show that the strict inequality $p > M/r_1$ allows us to relax the conditions $|x| \leq 2r_1$, $|y| \leq 2r_1$ to $|x| \leq r_1$, $|y| \leq r_1$ in Theorem 2 of [1], and consequently assert the existence of an a.p. solution $\bar{x}(t)$ of (1) such that $|\bar{x}(t)| \leq r_1 < r$ for all t. It is also not difficult to show that (H5) of this theorem in [1] is implied by our last condition together with (i); we omit the details.

To show that $\bar{x}(t)$ is uniformly stable, we use Theorem 1 in [2]. One of the hypotheses of this theorem is just (ii) in the hypotheses of the theorem we are proving. The other is essentially that $F(t,x(t),x_t)$ is continuous on R for any $x(t) : R \to R^n$ such that $x_t \in CB_r$ for $t \in R$. To show that this hypothesis is satisfied, we first note that it follows easily from (i) that there exists for fixed $h > 0$ with $ph < 1$ a constant L_0 such that

(3) $|F(t,x,\phi) - F(t,y,\phi)| < L_0|x-y|$

for $|x| \le r$, $|y| \le r$, $\phi \in CB_r$ and all $t \in R$; in fact, we may take $L_0 = (2-ph)/h$. Now let $x(t)$ be such that $x_t \in CB_r$ for all t and let $t_k \to t_0$ as $k \to \infty$. Then

(4) $$|F(t_k,x(t_k),x_{t_k}) - F(t_0,x(t_0)x_{t_0})| \le$$
$$|F(t_k,x(t_k),x_{t_k}) - F(t_0,x(t_k),x_{t_k})| +$$
$$|F(t_0,x(t_k),x_{t_k}) - F(t_0,x(t_0),x_{t_k})| +$$
$$|F(t_0,x(t_0),x_{t_k}) - F(t_0,x(t_0),x_{t_0})|$$

The first expression on the right in (4) becomes small as $k \to \infty$ because of condition (iii) in the definition of the a.p. property F has. The second also, because of the Lipschitz condition (3) on F. Finally the third gets small because of the last hypothesis on F of our theorem, since $x(t_k+s) \to x(t_0+s)$ uniformly for s in compact sets in $(-\infty,0]$. Hence we have the required continuity of $F(t,x(t),x_t)$ and conclude that the solution $\overline{x}(t)$ is uniformly stable. This completes our sketch of a proof of Theorem 1.

Remarks. The a.p. condition on F is implied by the following condition: (a) for each $S \subset R^n \times CB$, S closed and bounded, there exists $M(S)$ such that $|F(t,x,\phi)| \le M(S)$ for $(t,x,\phi) \in R \times S$, and (b) the set $\{F_\tau, \tau \in R\}$ is compact in the topology of uniform convergence on $R \times S$; here $F_\tau = F(t+\tau,x,\phi)$, cf. [3], Theorem 2.8. Also the last hypothesis in our theorem, required for the existence of the a.p. solution, is essentially that F is continuous in ϕ with respect to the compact open topology in CB, not the topology of the norm $\|\ \|$. It is a strong smoothness hypothesis on F but is usually satisfied if the delay in F is of fading memory type, cf. our last example.

Finally, if the Euclidean norm is used in R^n, condition (ii) in our theorem implies that if whenever for fixed t, we have $|x(t)-y(t)| \ge |x(s)-y(s)|$ for $s \le t$, then

$$(F(t,x(t),x_t) - F(t,y(t),y_t)) \cdot (x(t)-y(t)) \le 0$$

for $t \in R$; here $x_t \in CB_r$, $y_t \in CB_r$ for $t \in R$, and $x \cdot y = \sum_{j=1}^{n} x_j y_j$. By using a Liapunov-Razumikhin method (cf. [4] and Theorem 2 in [2]) uniform stability of the a.p. solution can then be established.

4. An application. Our theorem can be applied to population density equations in time fluctuating environments of the form:

(5) $$N'(t) = N(t)(a(t) - b(t)\, N(t) - r(t,N_t)),$$

where $N(t) \geq 0$ for $t \in R$, and $a(t)$, $b(t)$, and $r(t,\phi)$ are suitably a.p. in t. Note that $a(t)$, $b(t)$ and $r(t,\phi)$ can each be periodic in t, but unless the periods are rationally related, (5) will not have periodic, but a.p. time dependence.

The following result corrects and extends a theorem (Theorem 2) in [5].

Theorem 2. Let $a(t)$, $b(t)$ and $r(t,\phi)$ be real-valued functions, a.p. in t, the latter a.p. in t uniformly for ϕ in closed bounded sets in CB. Suppose there exist positive constants b_0, r, L such that

(a) $|r(t,\phi) - r(t,\psi)| \leq L\,\|\phi - \psi\|$ for $t \in R$, ϕ, ψ in $\widetilde{CB}_r$;

(b) $b(t) \geq b_0 > Le^{2r}$;

(c) $\sup\{|a(t) - b(t) - r(t,\phi)| \ : (t,\phi) \in R \times \widetilde{CB}_r\} < r(b_0 e^{-r} - Le^r)$;

(d) $r(t,\phi^k) \to r(t,\phi)$ as $k \to \infty$ uniformly for t in compact sets in R whenever $\phi^k(s) \to \phi(s)$ as $k \to \infty$ uniformly for s in compact sets in $(-\infty,0]$, and $\phi^k \in \widetilde{CB}_r$, $k = 1, 2, \ldots$; here $\widetilde{CB}_r = \{\phi \in CB \ ; \ e^{-r} \leq \phi(s) \leq e^t$ for $s \leq 0\}$.

Then (5) has a uniformly stable a.p. solution $\overline{N}(t)$ such that $e^{-r} \leq \overline{N}(t) \leq e^r$ for all t.

Proof. Under the change of variable $x = \log N$, (5) becomes

(6) $x'(t) = a(t) - b(t)\exp x(t) - r(t, \exp x_t)$; here

$\exp x_t = e^{x(t+s)}$, $s \leq 0$. We now apply Theorem 1 to (6) as follows: first choose $p = b_0 e^{-r} - Le^r$. Then from (c), $p > M/r$ where $M = M(r)$ is the right side of (c).

To check (ii) of Theorem 1 for (6) we have for $h > 0$ and sufficiently small

$$
\begin{aligned}
&|x(t) - hb(t)(e^{x(t)} - e^{y(t)}) - h(r(t,e^{x}t) - r(t,e^{y}t))| \leq \\
&\quad |x(t) - y(t)|(1 - hb_0 e^{\overline{x}(t)}) + hL\,\|e^{x_t} - e^{y_t}\| \leq \\
&\quad |x(t) - y(t)|(1 - hb_0 e^{-r}) + hLe^r\,\|x_t - y_t\| \leq \\
&\quad \|x_t - y_t\|(1 - hb_0 e^{-r} + hLe^r) = \\
&\quad \|x_t - y_t\|(1 - hp)
\end{aligned}
$$

for $t \in R$, $|x(t)| \leq r$, $|y(t)| \leq r$; here $x(t) \leq \overline{x}(t) \leq y(t)$.

To check (i) of Theorem 1, for $h > 0$ and sufficiently small, $|x - y - hb(t)(e^{x(t)} - e^{y(t)})| \leq$

$$|x - y|(1 - hb_0 e^{-r}) \leq |x - y|(1 - hp)$$

for $t, x(t), y(t)$ as above.

The rest of the proof is clear; we omit the details.

Remarks. From (b) of Theorem 2 it follows that if $r_0 = (\log(b_0/L))/2$ then $r < r_0$; thus the a.p. solution $\overline{N}(t)$ of (5) satisfies

$$e^{-r_0} < \overline{N}(t) < e^{r_0}, \text{ i.e., } (b_0/L)^{-1/2} < \overline{N}(t) < (b_0/L)^{1/2}$$

for all t.

If $\eta(t,x) : R^2 \to R$ is non decreasing in s for each $t \in R$, is a.p. in t uniformly for $s \in (-\infty, 0]$, and is such that the Stieltjes integral $\int_{-\infty}^{0} d_s\eta(t,x)$ converges uniformly for $t \in R$, it is not difficult to show that the function $r(t,\phi) = \int_{-\infty}^{0} \phi(s) d_s\eta(t,s)$ is a.p. in t uniformly for ϕ in closed bounded sets in CB. It also follows easily that condition (d) of Theorem 2 also is satisfied. Clearly (a) holds with

$$L = \sup\{\int_{-\infty}^{0} d_s\eta(t,s) : t \in R \text{ , and if } L < b_0, \text{ we may choose } r > 0$$

such that (b) holds. So for this special case of $r(t,\phi)$, our theorem applies if we can choose r such that (c) holds.

References

[1] G. Seifert, Almost periodic solutions for delay-differential equations with infite delays, J. Diff. Eqs. 41(3), (1981), 416-425.

[2] G. Seifert, Uniform stability for delay-differential equations with infinite delays, (submitted for publications).

[3] A. M. Fink, Almost Periodic Functions, Lecture Notes in Math. 377, Springer-Verlag, Berlin-Heidelberg-New York, 1974.

[4] R. D. Driver, Existence and stability of solutions of a delay-differential system, Arch. Rat. Mech. and Analysis, 10(5), (1962), 401-426.

[5] G. Seifert, Almost periodic solutions for single species population equations with infinite delays, Differential Equations and Applications in Ecology, Epidemics, and Population Problems, Busenberg & Cooke, edts. Academic Press, Inc. (1981).

VECTOR FIELDS IN THE VICINITY OF A COMPACT INVARIANT MANIFOLD*

George R. Sell

School of Mathematics
Institute for Mathematics and its Applications
University of Minnesota
Minneapolis, Minnesota 55455

I. Statement of Problem

Let us consider two vector fields

(1) $X' = F(X)$

(2) $Y' = G(Y)$

defined on a given Euclidean space E where F and G are of class C^{N+1}. Furthermore assume that there is a smooth compact manifold M smoothly imbedded in E and that M is invariant for both vector fields. Also that F and G agree on M, ie. $F|M = G|M$.

We wish to study the question of C^s - conjugacies between (1) and (2). Let s be a nonnegative integer. We shall say that F and G are C^s - conjugate near M if there are open neighborhoods V_1 and V_2 of M and a homeomorphism $H:V_1 \to V_2$ such that

A) H is a C^s-diffeomorphism for $s \geq 1$,

B) $H(X) = X$ for $x \in M$,

C) If $X(t)$ is a solution of (1) and $X(t) \in V_1$, for t in some interval I, then $Y(t) = H(X(t))$ is a solution of (2) for $t \in H$, and

D) Statement C) holds for $H^{-1}:V_2 \to V_1$.

---

* This research was supported in part by NSF Grants No. MCS 82-00765 and MCS 81-20789.

It is easy to see that for $s \geq 1$ the conditions C) and D) for a C^s - conjugacy can be restated as

$$DH(X)F(X) = G(H(X)) \quad , \quad X \in V_1$$

$$DH^{-1}(Y)G(Y) = F(H^{-1}(Y)) \quad , \quad Y \in V_2 .$$

More specifically we want to study the question of a C^s-conjugacy between (1) and (2) when G is the "linearized" vector field near M. We will define the linearized vector field shortly. Our general approach will be to find sufficient conditions in terms of $J(X) = DF(X)$, the Jacobian matrix of F, for $X \in M$ that guarantee that (1) and (2) are C^s - conjugate near M.

In the case that M is a fixed point or a periodic orbit, then the C^s- conjugacy question fits into the classical theory or ordinary differential equations and answers can be found in several sources including Belickii (1973, 1978), Grobman (1959, 1962), Hartman (1960, 1963, 1964), Nelson (1969), Palmer (1980) and Sternberg (1957, 1958).

Our interest here is primarily in the case that $\dim M \geq 2$. In this setting, rather little is known about any of these problems, however some contributions are especially relevant.

First there is the theorem of Pugh and Shub (cf. Hirsch et. al (1977) and Pugh-Shub (1970)) who give sufficient conditions for a C^0-conjugacy between (1) and (2). Specifically they show that if M is asymptotically stable and the flow near M is normally hyperbolic, then (1) and (2) are C^0-conjugate. (Incidentally, it is not difficult to show that their assumption of asymptotic stability can be dropped.)

Next there is the theorem of Robinson (1971) which can be applied to the question of a smooth conjugacy between (1) and (2). If, in addition to an assumption about the normal hyperbolicity, one assumes that F and G satisfy

$$D_1^P(F,G) = (0,0) \qquad (\text{at } X = 0)$$

for $0 \leq P \leq N$, where N is sufficiently large, then there is a C^s - conjugacy between (1) and (2). Robinson also describes a fairly complicated formula relating N , s and the spectral properties of (1) .

As noted above, we seek conditions in terms of the Jacobian matrix $J(X)$ alone which guarantee that there is a C^k - conjugacy between (1) and (2). This differs in an important way from Robinson's approach since he also made assumptions about the Taylor series expansion of the nonlinear part of F near M . Nevertheless one can take advantage of Robinson's Theorem.

It is convenient to simplify the discussion and assume that M is smoothly imbedded in E and that M has a trivial normal bundle. (The general problem can easily be reduced to this case.) It then follows that one can introduce curvilinear local coordinates so that in the vicinity of M the vector field (1) becomes

$$\begin{aligned} x' &= A(\theta)x + F(x,\theta) \\ \theta' &= g(\theta) + G(x,\theta) \end{aligned} \tag{3}$$

where θ represents local coordinates on M and $x \in R^k$ represents a normal vector to M . Furthermore F and G satisfy

$$(F,D_1F,G)(0,\theta) = (0,0,0)$$

where $D_1 = \partial/\partial x$. Also $A(\theta)$ is the linear part of F projected in the normal x-direction at the point $\theta \in M$. The equation $\theta' = g(\theta)$ describes the flow on the manifold M .

The linearized vector field near M is defined as the vector field

$$\begin{aligned} y' &= A(\phi)y \\ \phi' &= g(\phi) \end{aligned} \tag{4}$$

where $\phi \in M$ and $y \in R_k$. The linearized flow in the tangent bundle TM is given (in these coordinates) by

$$(5)\qquad \begin{aligned} v' &= B(\theta)v \\ \theta' &= g(\theta) \end{aligned}$$

where $B = D_2 g$, $D_2 = \partial/\partial\theta$ and $v \in R^p$ where $p = \dim M$.

The specific problem we are interested in here concerns the behavior of the flow in the vicinity of M . Specifically we seek sufficient conditions in terms of the matrices $A(\theta)$ and $B(\theta)$ in order that there exists a C^s-conjugacy H of the form

$$(6)\qquad y = x + u(x,\theta) \quad , \quad \phi = \theta + v(x,\theta)$$

which maps Eq. (3) to Eq. (4) in the vicinity of M . The restriction that H = identity on M means that $u(o,\theta) = 0$ and $v(0,\theta) = 0$.

II. The Spectra and Normal Hyperbolicity.

We shall use the spectral theory for flows developed in Sacker-Sell (1978, 1980). Let Σ_N denote the normal spectrum of M , that is Σ_N is the collection of all $\lambda \in R$ for which the linear skew-product flow

$$x' = (A(\theta) - \lambda I)x \quad , \quad \theta' = g(\theta)$$

fails to have an exponential dichotomy. Similarly let Σ_T denote the tangent spectrum of M , that is Σ_T is the collection of all $\lambda \in R$ for which

$$v' = (B(\theta) - \lambda I)v \quad , \quad \theta' = g(\theta)$$

fails to have an exponential dichotomy. Recall that if $\dim M \geq 1$, then $0 \in \Sigma_T$.

Next define $a \geq 0$ and $b \geq 0$ by

$$a = \inf\{\lambda \geq 0: \ \Sigma_T \subseteq [-\lambda,\lambda]\}$$
$$b = \sup\{\lambda \geq 0: \ \Sigma_N \subseteq (-\infty,-\lambda] \cup [\lambda,\infty)\}$$

The manifold M is said to be normally hyperbolic in the flow generated by (1) if $a < b$. M is normally hyperbolic of degree r, where r is positive integer, if $ra < b$.

Since the dimension of the normal bundle is k, it follows from the Spectral Theorem, Sacker-Sell (1978) that normal spectrum is the union of q nonoverlapping compact intervals, $I_1,\ldots,I_q$, where $1 \leq q \leq k$. Moreover associated with each spectral interval I_i there is an invariant spectral subbundle V_i of $R^k \times M$ with $\dim V_i(\theta) = n_i$. Furthermore n_i is independent of θ, $n_i \geq 1$ and $n_1 + \ldots + n_q = n$.

Next we wish to define the notion of an admissible k-type $(\lambda_1,\ldots,\lambda_k)$ from the spectrum Σ_N. What this means, in the case that A is a constant matrix with only real eigenvalues, is that the λ_i's are the eigenvalues of A repeated with their multiplicities. More generally we shall say that a given k-tuple of real numbers $(\lambda_1,\ldots,\lambda_k)$ is admissible provided

i) the mapping $j \to \lambda_j$ from $\{1,\ldots,k\}$ to R has its range in Σ_N, and

ii) $\mathrm{Card}\{j:\lambda_j \in I_i\} = n_i$, $\quad 1 \leq i \leq q$.

III. Statement of Main Result

In the statement of the smooth linearization theorem which we give below we shall use properties of the normal spectrum generated by $x' = A(\theta)x$. These properties are basically the generalization of the time-varying case of eigenvalue nonresonance conditions which arise in the study of linearization near a fixed point.

Theorem. Consider the equation (3)

$$x' = A(\theta)x + F(x,\theta)$$

$$\theta' = g(\theta) + G(x,\theta)$$

near M where the coefficients are of class C^{N+1} and M is normally hyperbolic of order r. Let a and b be defined as above. Assume that one has

1) $|\lambda - (m_1\lambda_1 + \dots + m_k\lambda_k)| > ra$

2) $|m_1\lambda_1 + \dots + m_k\lambda_k| > (r+1)a$

<u>for all</u> $\lambda \in \Sigma_N$, <u>and all admissible k-tuples</u> $(\lambda_1,\dots,\lambda_k)$ <u>and nonnegative integers</u> $m_1,\dots,m_k$ <u>that satisfy</u>

$$2 \leq (m_1 + \dots + m_k) \leq N$$

<u>If</u> $q = \min(r,N)$ <u>is sufficiently large then there is a</u> C^s – <u>conjugacy between</u> (3) <u>and</u> (4) .

The basic approach to this problem is to introduce a preliminary change of variables

(7) $$z = x + u(x,\theta) \quad , \quad \beta = \theta + v(x,\theta)$$

to reduce Eq. (3) to

$$\dot z = A(\beta)z + F(z,\beta)$$

$$\dot\beta = g(\beta) + g(z,\beta)$$

where $D^P(F,G) = (0,0)$ at $(0,\beta)$ for $0 \leq P \leq N$ and then to use Robinson's Theorem. The function u and v in Eq. (7) are chosen to be appropriate polynomials in the x-variable with coefficients that depend on θ . The smoothness of these coefficients is quaranteed by the following proposition concerning the solutions of inhomogeneous linear differential systems.

<u>Lemma</u>. <u>Let</u> M <u>be a smooth compact manifold with a flow</u> $\theta' = g(\theta)$ <u>given in local coordinates</u> θ , <u>and consider the linear inhomogeneous differential system over</u> M <u>given by</u>

$$x' = A(\theta)x + f(\theta) \quad , \quad x \in X$$

$$\theta' = g(\theta)$$

<u>where</u> X <u>is a finite dimensional Banach space, and</u> A, F <u>and</u> g <u>are of class</u> C^N <u>on</u> M . <u>Assume further that the manifold</u> M <u>in the vector</u>

field

$$x' = A(\theta)x \quad , \quad \theta' = g(\theta)$$

is normally hyperbolic of degree r. Then there is a unique continuous function $x : M \to X$ such that $x(\theta \cdot t)$ is a solution of $\dot{x} = A(\theta \cdot t)x + f(\theta \cdot t)$ and $\theta \cdot t$ is a solution of $\theta' = g(\theta)$, $\theta(0) = \theta$ on M. Moreover x is of class C^s on M where $s = \min(r,N)$.

BIBLIOGRAPHY

1. G.R. Belickii. (1973). Functional equations and the conjugacy of diffeomorphisms of finite smoothness class. Functional Anal. Appl. 7 268-277.
2. G.R. Belickii. (1978). Equivalence and normal forms of germs of smooth mappings. Russian Math. Surveys 33, 107-177.
3. D.M. Grobman. (1959). Homeomorphisms of systems of differential equation. Dokl. Akad. Nauk SSSR 128, 880-881.
4. D.M. Grobman. (1962). Topological classification of the neighborhood of a singular point in n-dimensional spacxe. Mat. Sb. (N.S.) 56 (98), 77-94.
5. P. Hartman. (1960). A lemma in the theory of structural stability of differential equations. Proc. Amer. Math. Soc. 11, 610-620.
6. P. Hartman. (1963). On the local linearlization of differential equations. Proc. Amer. Math. Soc. 14, 568-573.
7. P. Hartman. (1964). Ordinary Differential Equations. Wiley.
8. M.W. Hirsch, et. al. (1977). Invariant Manifolds. Springer-Verlag.
9. E. Nelson. (1969). Topics in Dynamics I. Flows. Princeton University Press.
10. K. Palmer. (1980). Qualitative behavior of a system of ODE near an equililbrium point. A generailization of the Hartman-Grobman Theorem. Technical Report, Institute fuer Angewandte Mathematik, University of Bonn.
11. C.C. Pugh and M. Shub. (1970). Linearization of normally hyperbolic diffeomorphisms and flows. Invent. Math. 10, 187-198.
12. C. Robinson. (1971). Differentiable conjugacy near compact invariant manifolds. Bol. Soc. Brasil. Mat. 2, 33-44.
13. R.J. Sacker and G.R. Sell. (1978). A spectral theory for linear differential systems, J. Diff. Eqns. 27, 320-358.
14. R.J. Sacker and G.R. Sell. (1980). The spectrum of an invariant submanifold J. Diff. Eqns. 38, 135-160.
15. S. Sternberg. (1957). Local contractions and a theorem of Poincare. Amer. J. Math. 79, 809-824.
16. S. Sternberg. (1958). On the structure of local homeomorphisms of Euclidean n-space. Amer. J. Math. 80, 623-631.

BIFURCATION FROM THE ESSENTIAL SPECTRUM

C.A. Stuart

1. INTRODUCTION

We consider the following non-linear eigenvalue problem:-

$$-\Delta u(x) \pm q(x)|u(x)|^{\sigma}u(x) = \lambda u(x) \quad \text{for} \quad x \in \mathbb{R}^N, \qquad (1\pm)$$

where

$$\text{(A1)} \begin{cases} N \geq 2 \\ \sigma \text{ is a positive constant} \\ q \in L^{\infty}_{loc}(\mathbb{R}^N) \text{ and } q > 0 \text{ a.e. on } \mathbb{R}^N. \end{cases}$$

A pair (λ,u) is called a (generalised) solution of $(1\pm)$ if and only if $\lambda \in \mathbb{R}$, $u \in H^1 \cap L^{\sigma+1}_{loc}$ and

$$\int \{\nabla u.\nabla v \pm q|u|^{\sigma}uv - \lambda uv\}dx = 0 \qquad \forall\, v \in C^{\infty}_{o}(\mathbb{R}^N).$$

Here and henceforth we use the usual notation for the spaces of real-valued functions, or equivalence classes of functions, on $\mathbb{R}^N$. Thus H^k denotes the Hilbert space $H^k(\mathbb{R}^N) = W^{k,2}(\mathbb{R}^N)$. When the domain of integration is not indicated, it is understood that the integration extends over all of $\mathbb{R}^N$.

A real number λ is an (L^2) bifurcation point for $(1\pm)$ if and only if $\exists$ a sequence $\{(\lambda_n,u_n)\}$ of solutions of $(1\pm)$ such that $u_n \not\equiv 0$ $\forall\, n \in \mathbb{N}$ and

$$\lambda_n \rightarrow \lambda, \ \|u_n\|_{H^1} \rightarrow 0 \quad \text{as} \quad n \rightarrow \infty.$$

We shall prove the following results concerning the existence and bifurcation of solutions of $(1\pm)$.

Theorem (1-)

Let (A1) hold and suppose that $q(x) \rightarrow 0$ as $|x| \rightarrow \infty$.

A (Existence) Suppose that $\sigma \in (0, \frac{4}{N-2})$.

For each fixed $\lambda < 0$, $\exists$ an infinite number of distinct solutions $\{(\lambda,u_n)\}_{n=1}^{\infty}$ of (1-).

B (Bifurcation) Suppose that $\sigma \in (0,\frac{4}{N})$ and that $\exists$ $A > 0$ and $t \in (0,2-\frac{N\sigma}{2})$ such that $q(x) \geq A(1+|x|)^{-t}$ a.e. on $\mathbb{R}^N$.

For each fixed $r > 0$, $\exists$ a solution (λ^r, u^r) of (1-) such that $\|u^r\|_{L^2} = r$ and $\lambda^r < 0$. As $r \to 0$, $\lambda^r \to 0-$ and $\|u^r\|_{H^1} \to 0$. Thus $\lambda = 0$ is a bifurcation point for (1-).

Theorem (1+)

Let (A1) hold and suppose that $\exists$ $A > 0$ and $t > \frac{N\sigma}{2}$ such that $q(x) \geq A(1 + |x|)^t$ a.e. on $\mathbb{R}^N$.

A (Existence) For each fixed $\lambda > 0$, $\exists$ an infinite number of distinct solutions $\{(\lambda, u_n)\}_{n=1}^{\infty}$ of (1+).

B (Bifurcation) For each fixed $r > 0$, $\exists$ an infinite number of distinct solutions $\{(\lambda_n^r, u_n^r)\}_{n=1}^{\infty}$ of (1+) such that $\forall\, n \in \mathbb{N}$, $\|u_n^r\|_{L^2} = r$ and $\lambda_n^r > 0$. As $r \to 0$, $\forall\, n \in \mathbb{N}$, $\lambda_n^r \to 0+$ and $\|u_n^r\|_{H^1} \to 0$. Thus $\lambda = 0$ is a bifurcation point for (1+).

We give a unified presentation of these theorems based upon a few fundamental results from critical point theory. Theorem (1-)A is related to earlier work on the case where q is constant by Berger, Strauss, Berestycki and Lions (P.L.) [7,8,9] whereas Theorem (1-)B is a special case of results due to the author [10,11]. Theorem (1+)A has been obtained by Benci and Fortunato [12,13] and Theorem (1-)B is due to Bongers, Heinz and Küpper [14]. Earlier work on related problems is contained in [15-25].

In section 2, we recall without proof some basic results about critical points. In sections 3 to 5, these results are applied to operator equations of the form:

$$Su \pm F(u) = \lambda u \quad , \qquad (2\pm)$$

where S is a self-adjoint linear operator and F is of higher order near $u = 0$. Finally, in sections 6 and 7, we apply the general results (obtained in sections 4 and 5) concerning (2±) to the special case (1±). By following this approach we hope to have clarified the rôles played by the various hypotheses of Theorems (1±) in establishing the conditions for the existence of critical points. The hypotheses (H1) to (H3) of section 3 apply to both (2-) and (2+). They make precise the variational structure which is assumed. To obtain the desired results on the existence and bifurcation of solutions, the equations (2-) and (2+) require different properties of F; namely (S1) to (S4) of section 4 for (2-) and (B1) to (B4) of section 5 for (2+). Nonetheless, these assumptions may be compared in the following way:

(S1)/(B1) : compactness,

(S2)/(B2) : comparison between $\langle F(u),u\rangle$ and $\phi(u)$,

(S3)/(B3) : F is higher order,

(S4)/(B4) : F is not too small.

The special form of nonlinearity in equation (1±) has been adopted to avoid unwanted technical discussion and yet retain the essential features of the problem. In particular, the homogeneity in u of the nonlinearity is not exploited and the approach can be used for the more general equations discussed in [11,14]. In fact, in sections 4 and 5, the nonlinearity is not required to be homogeneous. Finally we note that, for the existence results (Theorems (2±)A), F is required to be superlinear at infinity. This is ensured by the fact that $q > 2$ in (S2) and $\phi(u) > 0 \;\forall\, u \in H_T\setminus\{0\}$ in Theorem (2-)A, and by the requirement that $\lim_{t\to+\infty} g(t)/t = +\infty$ in (B4) for Theorem (2+)A.

Remarks

1. Using a slight generalisation (to admit the case where q is not constant) of Pohozaev's identity [30,4(I),32] and standard results on the non-existence of positive eigenvalues of Schrödinger's equation [31], it can be seen that the restrictions on σ and q in Theorem (1±) are more or less necessary [4,7-12,20-22]. This also shows that equation (1-) has no solution with $\lambda > 0$ (if $q \in L^\infty$) and that equation (1+) has no solution with $\lambda < 0$. Furthermore, in the context of Theorem (1-)B, $\lambda = 0$ is often the only bifurcation point of (1-) [20-22], whereas under the hypotheses of Theorem (1+)B, every $\lambda \geq 0$ is a bifurcation point for (1+) (and there are no others) [12].

2. When $q(x) \not\to 0$ as $|x| \to \infty$, Theorem (1-) remains true provided that q is bounded and radially symmetric. In particular, if q is a positive constant, $\lambda = 0$ is a bifurcation point for (1-) provided that $0 < \sigma < 4/N$. This sort of result is obtained by using the uniform decay of radially symmetric elements of H^1 [7-11]. At present, no results seem to cover cases where $q(x) \not\to 0$ as $|x| \to \infty$ and q is not radially symmetric.

3. Comparing Theorems (1±)B, one naturally asks how many solutions of (1-) there are on the sphere $\|u\|_{L^2} = r$. It seems that, at least in the case where q is radially symmetric, the method used by Berestycki-Lions [4(II)] might be adapted to verify the hypothesis (S4) of section 4 for $j > 1$, under appropriate restrictions on σ and q.

4. Let us stress that the conditions for bifurcation in Theorem

(1-)B pertain to the requirement $\|u_n\|_{H^1} \to 0$ (equivalently $\|u_n\|_{L^2} \to 0$). If L^2 is replaced by L^p, then the condition $t \in (0, 2-\frac{N\sigma}{2})$ should be modified, but this point has so far been resolved only for some special cases [27-29]. This situation is in contrast to the classical one (where $\mathbb{R}^N$ is replaced by a bounded domain on the boundary of which u vanishes) which reduces to finite dimensions (by the Lyapunov-Schmidt procedure) and all norms are equivalent.

5. So far, more precise results concerning the bifurcation of curves or continua of solutions are available only when special symmetries allow the asymptotic behaviour of solutions as $|x| \to \infty$ to be studied by O.D.E. methods [17,24,25,27-29].

6. The requirement that $u \in H^1$ for a solution of (1±) amounts to a boundary condition at infinity. For some semilinear equations on unbounded domains the study of solutions which are merely bounded has been undertaken by Kirchgässner and Scheurle [33-36].

2. CRITICAL POINT THEORY

We recall (without proof) some fundamental results concerning the existence of critical points. All the details and proofs may be found in [1-5]. Throughout this section, it is assumed that:-

(I) E is a real infinite dimensional Banach space,
(II) the functional $J \in C^1(E,\mathbb{R})$ is even and $J(0) = 0$.

2.A Unconstrained problems

For $b \in \mathbb{R}$, $K_b = \{u \in E : J(u) = b$ and $J'(u) = 0\}$ where $J'(u) \in E^*$ is the Fréchet derivative of J at u. If $J'(u) = 0$, then u is said to be a critical point of J. If $K_b \neq \phi$, then b is called a critical value of J.

Following Palais and Smale, J is said to satisfy the condition $(C)^-$ on E if every sequence $\{u_n\} \subset E$ which has the following two properties:

(1) $-\infty < \inf J(u_n) \leqslant \sup J(u_n) < 0$

(2) $\|J'(u_n)\|_{E^*} \to 0$ as $n \to \infty$,

has a subsequence converging in E.

If $-J$ satisfies $(C)^-$ on E, J is said to satisfy the condition $(C)^+$ on E.

Let $\Sigma = \{A \subset E\backslash\{0\} : A$ is closed in E and $A = -A\}$. The genus is the mapping $\gamma : \Sigma \to \mathbb{N} \cup \{0,+\infty\}$ defined by:

$\gamma(\phi) = 0$

$\gamma(A) = k$ if $\exists$ an odd mapping $h \in C(A,\mathbb{R}^k\setminus\{0\})$ and k is the smallest integer with this property

$\gamma(A) = +\infty$ if there is no integer k with the above property.

We note that:

(i) if $\gamma(A) \geqslant 2$, A contains an infinite number of elements

(ii) if there is an odd homeomorphism of A onto the unit sphere in $\mathbb{R}^k$, then $\gamma(A) = k$.

Setting $\Gamma_k = \{A \in \Sigma : \gamma(A) \geqslant k\}$, we have that $\Gamma_k \neq \phi \quad \forall\, k \in \mathbb{N}$, by (ii) since $\dim E = +\infty$.

Let $b_k = \inf\limits_{A\in\Gamma_k} \sup\limits_{u\in A} J(u)$.

Theorem 2.1

In addition to (I) and (II), suppose that J satisfies $(C)^-$ on E and that $-\infty < b_k < 0$. Then $K_{b_k} \neq \phi$ and, if $b_k = b_{k+1} = \ldots = b_{k+p-1}$, then $\gamma(K_{b_k}) \geqslant p$. In particular, if $p \geqslant 2$, K_{b_k} contains an infinite number of points.

Remarks

1. In this form the result is due to Clark [1,3].

2. If J is bounded below on E (i.e. $\inf\limits_{u\in E} J(u) > -\infty$), then $b_k > -\infty \quad \forall\, k \in \mathbb{N}$.

The following result due to Ambrosetti and Rabinowitz [1,2] is also based upon a "minimax" argument, but can be applied to functionals which are neither bounded above nor below.

Theorem 2.2

In addition to (I) and (II), suppose that J satisfies $(C)^+$ on E and the following two conditions:-

(i) $\exists\ \rho > 0$ and $\alpha > 0$ such that $\begin{cases} J(u) > 0 & \forall\ 0 < \|u\| < \rho \\ J(u) \geqslant \alpha & \forall\ \|u\| = \rho \end{cases}$

(ii) for every finite dimensional subspace Z of E, $Z \cap \{u \in E : J(u) \geqslant 0\}$ is bounded in E.

Then J has an infinite number of distinct critical values.

2.B Constrained problems

Throughout this part, we assume that (I) and (II) hold and also:

(III) E is reflexive and $E \hookrightarrow F$ where F is a real Hilbert space.

The notation $\hookrightarrow$ means that E is a dense subset of F and that $\exists\, C > 0$ such that $\|u\|_F \leq C\|u\|_E \quad \forall\, u \in E$.

For $r > 0$, let $M_r = \{u \in E : \|u\|_F = r\}$. In general, M_r is an unbounded subset of E. For $u \in E\backslash\{0\}$, $E_u \equiv \{v \in E : \langle u,v\rangle = 0\}$ is a closed subspace of E, and hence a Banach space with norm, $\|\cdot\|_E$. Its dual is denoted $(E_u)^*$. For $r > 0$ and $b \in \mathbb{R}$, set

$$K_b^r = \{u \in M_r : J(u) = b \quad \text{and} \quad (J|_{M_r})'(u) = 0\}$$

where $(J|_{M_r})'(u) \in (E_u)^*$ is the Fréchet derivative of the restriction of J to M_r. Following Palais and Smale, J is said to satisfy the condition (C) on M_r if every sequence $\{u_n\} \subset M_r$ which has the following two properties:

(1) $\{J(u_n)\}$ is bounded

(2) $\|(J|_{M_r})'(u_n)\|_{(E_{u_n})^*} \to 0$ as $n \to \infty$

has a subsequence converging in E.

Remarks

1. If J has the above property when (1) is replaced by

(1)$^-$ $-\infty < \inf J(u_n) \leq \sup J(u_n) < 0$,

then J is said to satisfy the condition $(C)^-$ on M_r.

2. For $u \in M_r$, let $PJ'(u) \in E^*$ denote the projection of $J'(u)$ onto the tangent plane E_u, defined by

$$PJ'(u)v = J'(u)v - \left\{\frac{J'(u)u}{r^2}\right\}\langle u,v\rangle_F \quad \forall\, v \in E.$$

Then, as is shown in [4(II)],

$$\|(J|_{M_r})'(u)\|_{(E_u)^*} \leq \|PJ'(u)\|_{E^*} \leq \left(1 + \frac{C\|u\|_E}{r}\right)\|(J|_{M_r})'(u)\|_{(E_u)^*} .$$

Thus, if (1)/(1)$^-$ implies that $\{u_n\}$ is bounded in E, then (2) is equivalent to

(2)' $\|PJ'(u_n)\|_{E^*} \to 0$ as $n \to \infty$.

For $r > 0$ and $k \in \mathbb{N}$, let

$$\Gamma_k^r = \{A \in \Gamma_k : A \subset M_r \quad \text{and} \quad A \text{ is compact in } E\}.$$

By (ii) of part 2.A, $\Gamma_k^r \neq \phi \quad \forall\, r > 0$ and $\forall\, k \in \mathbb{N}$.

Set $c_k^r = \inf_{A\in\Gamma_k^r} \max_{u\in A} J(u)$. Clearly $c_k^r < +\infty$.

Theorem 2.3

In addition to (I), (II) and (III), suppose that J satisfies the condition (C) on M_r and that $\inf_{u \in M_r} J(u) > -\infty$. Then $\forall\, k \in \mathbb{N}$, $K^r_{c^r_k} \neq \phi$ and, if $c^r_k = c^r_{k+1} = \ldots c^r_{k+p-1}$, then $\gamma\left(K^r_{c^r_k}\right) \geqslant p$. In particular, if $p \geqslant 2$, $K^r_{c^r_k}$ contains an infinite number of points. If (C) is replaced by $(C)^-$, the conclusion remains valid provided that $c^r_k < 0$.

Remark

In this form the result is due to Berestycki and Lions [4(II)]. Related work is contained in [1,5,6].

3. EQUATIONS $Su \pm F(u) = \lambda u$

Throughout this section, H denotes a real, infinite dimensional Hilbert space with scalar product $\langle\cdot,\cdot\rangle$ and norm $\|\cdot\|$.

(H1) $S : \mathcal{D}(S) \subset H \to H$ is a positive self-adjoint operator and $\inf \sigma_e(S) = 0$.

Here, $\sigma_e(S)$ denotes the essential spectrum of S. Since S is positive, $\inf \sigma(S) = 0$, too.

By (H1), $\exists$ a closed linear operator, $T : \mathcal{D}(T) \subset H \to H$ such that $\mathcal{D}(T)$ is dense in H and $T^*T = S$, where $T^* : \mathcal{D}(T^*) \subset H \to H$ is the adjoint of T in H. Let $(H_T, \|\cdot\|_T)$ be the Hilbert space obtained by equipping $\mathcal{D}(T)$ with the graph norm:

$$\|u\|_T = \{\|u\|^2 + \|Tu\|^2\}^{\frac{1}{2}} \quad \forall\, u \in \mathcal{D}(T).$$

Then $H_T \hookrightarrow H$ in the notation of 2.B, and identifying H with H^*, we can write $H_T \subset H = H^* \subset (H_T)^*$ and use $\langle\cdot,\cdot\rangle$ for the duality between $(H_T)^*$ and H_T. Since $T : H_T \to H$ is bounded, it has a conjugate, $T' : H^* = H \to (H_T)^*$ which is also bounded. Then $T'T : H_T \to (H_T)^*$ is a bounded linear operator such that $T'Tu = Su$ $\forall\, u \in \mathcal{D}(S)$ and

$$\mathcal{D}(S) = \{u \in H_T : T'Tu \in H\}.$$

All these points are discussed in more detail in [11].

(H2) $(Y, \|\cdot\|_Y)$ is a reflexive Banach space such that $Y \hookrightarrow H$ and $\phi \in C^1(Y, \mathbb{R})$ is an even functional with $\phi(0) = 0$, $\phi'(0) = 0$.

Remark

As above, we can write $Y \subset H = H^* \subset Y^*$ and use $\langle\cdot,\cdot\rangle$ for the duality between Y^* and Y. In particular,

$$\phi'(u)v = \langle F(u),v\rangle \quad \forall\ u,v \in Y$$

where $F : Y \to Y^*$ is continuous, $F(0) = 0$. Thus, by (H2), ϕ is a potential for F. The domain of F is related to that of S by the following hypothesis.

(H3) $H_T \cap Y$ is dense in H_T.

Let $(X,\|\cdot\|_X)$ be the Banach space defined by:

$$X = H_T \cap Y, \quad \|u\|_X = \{\|u\|_T^2 + \|u\|_Y^2\}^{\frac{1}{2}} \quad \forall\ u \in X.$$

From (H1) - (H3), it follows that X is reflexive and that $X \hookrightarrow H$. Once again we can write $X \subset H = H^* \subset X^*$ and use $\langle\cdot,\cdot\rangle$ for the duality between X^* and X.

A pair (λ,u) is called a generalised solution of $Su \pm F(u) = \lambda u$ if and only if $\lambda \in \mathbb{R}$, $u \in X$ and $T'Tu \pm F(u) = \lambda u$ in X^*. This makes sense since, with the identifications introduced above, $X \cup (H_T)^* \cup Y^* \subset X^*$. Observe also that if (λ,u) is a generalised solution of $Su \pm F(u) = \lambda u$ and $F(u) \in H$, then $u \in \mathfrak{D}(S)$ and $Su \pm F(u) = \lambda u$ in H.

Sections 4 and 5 deal with generalised solutions of $Su \pm F(u) = \lambda u$. In section 5, we shall use the following result which is similar to one proved in [14] under stronger assumptions.

Lemma 3.1

Let (H1) to (H3) hold. Then $\forall\ \varepsilon > 0$ and $\forall\ k \in \mathbb{N}$, $\exists$ a subspace Z_k of X such that $\dim Z_k = k$ and $\|Tu\| \leq \varepsilon\|u\|$ $\forall\ u \in Z_k$.

Proof

Fix $\varepsilon > 0$ and $k \in \mathbb{N}$. By (H1), $\exists\ \{e_1,\dots,e_k\} \subset H_T$ such that $\langle e_i,e_j\rangle = \delta_{ij}$ for $1 \leq i,j \leq k$ and

$$\|Tu\| \leq \varepsilon\|u\| \quad \forall\ u \in V_k \equiv \operatorname{span}\{e_1,\dots,e_k\}.$$

Let $\delta = \min\{\varepsilon,\frac{1}{2}\}$. By (H3), $\exists\ \{a_1,\dots,a_k\} \subset X$ such that

$$\left\{\sum_{i=1}^{k} \|a_i - e_i\|_T^2\right\}^{\frac{1}{2}} \leq \delta \quad \text{and} \quad \dim Z_k = k \quad \text{where} \quad Z_k = \operatorname{span}\{a_1,\dots,a_k\}.$$

Let $W : V_k \to Z_k$ be the isomorphism defined by

$$W(u) = \sum_{i=1}^{k} \langle u,e_i\rangle a_i.$$

Then $\|u - W(u)\|_T = \| \sum_{i=1}^{k} <u,e_i>(a_i - e_i)\|_T$

$$\leq \left\{\sum_{i=1}^{k} <u,e_i>^2\right\}^{\frac{1}{2}} \left\{\sum_{i=1}^{k} \|a_i - e_i\|_T^2\right\}^{\frac{1}{2}}$$

$$\leq \delta\|u\| \quad \forall\, u \in V_k \tag{3.1}$$

and

$$\|u\| \leq \|W(u)\| + \|u - W(u)\| \leq \|W(u)\| + \|u - W(u)\|_T$$

$$\leq \|W(u)\| + \delta\|u\| \quad \forall\, u \in V_k. \tag{3.2}$$

Since $\delta \leq \frac{1}{2}$, we have that $\|u\| \leq 2\|W(u)\| \quad \forall\, u \in V_k$. Hence for $v \in Z_k$, we have that

$$\|Tv\| \leq \|Tu\| + \|T(u - v)\| \quad \text{where} \quad u = W^{-1}(v) \in V_k$$

$$\leq \varepsilon\|u\| + \|u - W(u)\|_T \leq (\varepsilon + \delta)\|u\| \quad \text{by (3.1)}$$

$$\leq 2(\varepsilon + \delta)\|v\| \quad \text{by (3.2)}.$$

This proves the result.

4. EQUATION $T'Tu - F(u) = \lambda u$

In addition to (H1), (H2) and (H3) of section 3, we shall use the following hypotheses.

(S1) $H_T \hookrightarrow Y$ and, if $u_n \rightharpoonup u$ weakly in H_T, then

$$\phi(u_n) \to \phi(u) \quad \text{and} \quad \|F(u_n) - F(u)\|_{(H_T)^*} \to 0.$$

(This means that $\phi : H_T \to \mathbb{R}$ is weakly sequentially continuous and $F : H_T \to (H_T)^*$ is compact.)

(S2) $\exists\, q > 2$ such that $<F(u),u> \geq q\phi(u) \geq 0 \quad \forall\, u \in H_T$.

(This implies that $\phi(tu) \geq t^q\phi(u) \quad \forall\, u \in H_T \quad \forall\, t \geq 1$.)

(S3) $\exists\, K > 0$, $\alpha \geq 0$ and $\beta > 0$ with $\alpha + \beta > 2$ such that

$$<F(u),u> \leq K\|Tu\|^{\alpha}\|u\|^{\beta} \quad \forall\, u \in H_T.$$

(Together with (S2), this implies that $0 \leq \phi(u) \leq \frac{1}{2}K\|Tu\|^{\alpha}\|u\|^{\beta}$.)

(S4) $\exists\, j \in \mathbb{N}$ such that $c_j^r < 0 \quad \forall\, r > 0$ where c_j^r is as defined in section 2.B with

$$E = H_T, \quad F = H \quad \text{and} \quad J(u) = \tfrac{1}{2}\|Tu\|^2 - \phi(u).$$

(Note that (S4) is satisfied with $j = 1$ provided that

$$\inf\left\{\tfrac{1}{2}\|Tu\|^2 - \phi(u) : u \in H_T \text{ and } \|u\| = r\right\} < 0 \quad \forall\, r > 0.)$$

Theorem 4

Let (H1) to (H3) and (S1) to (S3) hold.

A (Existence) Suppose that $\phi(u) > 0 \quad \forall\, u \in H_T\backslash\{0\}$. For each fixed $\lambda < 0$, $\exists$ an infinite number of distinct solutions $\{(\lambda, u_n)\} \subset \mathbb{R}\times H_T$ of $T'Tu - F(u) = \lambda u$.

B (Bifurcation) Suppose that (S4) also holds and that $\alpha \in [0,2)$ in (S3). For each fixed $r > 0$, $\exists$ at least $2j$ distinct solutions $\left\{(\lambda_n^r, \pm u_n^r)\right\}_{n=1}^{j} \subset \mathbb{R}\times H_T$ of $T'Tu - F(u) = \lambda u$ such that $\forall\, n \in \{1,\ldots,j\}$

$$\|u_n^r\| = r \quad \text{and} \quad \lambda_n^r < 0\ ,$$

$$\|u_n^r\|_T \to 0 \quad \text{and} \quad \lambda_n^r \to 0- \quad \text{as} \quad t \to 0.$$

Proof of 4A

Fix $\lambda < 0$ and set $\|u\|_{T(\lambda)} = \{\|Tu\|^2 - \lambda\|u\|^2\}^{\frac{1}{2}} \quad \forall\, u \in H_T$. Then $\|\cdot\|_T$ and $\|\cdot\|_{T(\lambda)}$ are equivalent norms on H_T.

We shall apply Theorem 2.2 with $E = H_T$ and

$$J(u) = \tfrac{1}{2}\|Tu\|^2 - \phi(u) - \tfrac{\lambda}{2}\|u\|^2 = \tfrac{1}{2}\|u\|^2_{T(\lambda)} - \phi(u) \quad \text{for} \quad u \in H_T.$$

We prove that:

(a) J satisfies the condition $(C)^+$ on H_T.

(b) J satisfies the condition (i) of Theorem 2.2.

(c) J satisfies the condition (ii) of Theorem 2.2.

(a) Let $\{u_n\} \subset H_T$ be such that $J(u_n) \leq L \quad \forall\, n \in \mathbb{N}$ and $\|J'(u_n)\|_{(H_T)^*} \to 0$ as $n \to \infty$.

In particular, $\exists\ n_o \in \mathbb{N}$ such that $\|J'(u_n)\|_{(H_T)^*} \leq \min\{1,|\lambda|^{\frac{1}{2}}\}$ $\forall\, n \geq n_o$. Hence, $\forall\, n \geq n_o$,

$$\left|\|u_n\|^2_{T(\lambda)} - \langle F(u_n), u_n\rangle\right| = \left|\langle T'Tu_n - F(u_n) - \lambda u_n, u_n\rangle\right|$$

$$= |J'(u_n)u_n| \leq \min\{1,|\lambda|^{\frac{1}{2}}\}\|u_n\|_T \leq \|u_n\|_{T(\lambda)}\ ,$$

and so, $$\|u_n\|_{T(\lambda)} + \|u_n\|^2_{T(\lambda)} \geq \langle F(u_n), u_n\rangle \geq q\phi(u_n)$$

$$\geq q\left\{\tfrac{1}{2}\|u_n\|^2_{T(\lambda)} - L\right\}.$$

It follows that, $\forall\, n \geq n_o$,

$$\left(\frac{q}{2} - 1\right)\|u_n\|^2_{T(\lambda)} \leq qL + \|u_n\|_{T(\lambda)}$$

and hence we see that $\{u_n\}$ is bounded in H_T.

Therefore $\exists$ a subsequence $\{u_{n_i}\}$ of $\{u_n\}$ such that

$$u_{n_i} \rightharpoonup u_\infty \text{ weakly in } H_T \text{ as } i \to \infty.$$

By (S1), $\|F(u_{n_i}) - F(u_\infty)\|_{(H_T)^*} \to 0$ as $i \to \infty$.

Thus,

$$\|u_{n_i} - u_{n_j}\|^2_{T(\lambda)} = \langle T'T(u_{n_i} - u_{n_j}) - \lambda(u_{n_i} - u_{n_j}), u_{n_i} - u_{n_j}\rangle$$

$$= \langle J'(u_{n_i}) - J'(u_{n_j}) + F(u_{n_i}) - F(u_{n_j}), u_{n_i} - u_{n_j}\rangle$$

$$\leqslant \{\|J'(u_{n_i}) - J'(u_{n_j})\|_{(H_T)^*} + \|F(u_{n_i}) - F(u_{n_j})\|_{(H_T)^*}\}\|u_{n_i} - u_{n_j}\|_T$$

$$\to 0 \text{ as } i,j \to \infty.$$

This proves that $\{u_{n_i}\}$ converges in H_T and we have verified the condition $(C)^+$.

(b) For $u \in H_T$, $J(u) \geqslant \frac{1}{2}\|u\|^2_{T(\lambda)} - \frac{1}{2}K\|Tu\|^\alpha \|u\|^\beta$ by (S3)

$$\geqslant \frac{1}{2}\|u\|^2_{T(\lambda)}\left\{1 - \frac{K}{|\lambda|^{\beta/2}}\|u\|^{\alpha+\beta-2}_{T(\lambda)}\right\}.$$

Since $\alpha + \beta > 2$, the condition (i) of Theorem 2.2 is satisfied.

(c) Let Z be a finite dimensional subspace of H_T. Let us suppose that $\exists$ a sequence $\{u_n\} \subset Z$ such that

$$J(u_n) \geqslant 0 \quad \forall\, n \in \mathbb{N} \text{ and } \|u_n\|_{T(\lambda)} \to \infty.$$

Let $v_n = u_n/\|u_n\|_{T(\lambda)}$. Clearly $\exists$ a subsequence $\{v_{n_i}\}$ such that $v_{n_i} \to v_\infty$ strongly in Z as $i \to \infty$ and $\|v_\infty\|_{T(\lambda)} = 1$. But $\exists$ $n_0 \in \mathbb{N}$ such that $\|u_n\|_{T(\lambda)} \geqslant 1$ $\forall\, n \geqslant n_0$ and hence, by (S2),

$$\phi(u_n) = \phi(\|u_n\|_{T(\lambda)} v_n) \geqslant \|u_n\|^\alpha_{T(\lambda)}\phi(v_n) \quad \forall\, n \geqslant n_0.$$

Since

$$\frac{1}{2}\|u_n\|^2_{T(\lambda)} - \phi(u_n) = J(u_n) \geqslant 0 \quad \forall\, n \in \mathbb{N},$$

it follows that

$$0 \leqslant \phi(v_n) \leqslant \frac{1}{2}\|u_n\|^{2-\alpha}_{T(\lambda)} \quad \forall\, n \geqslant n_0.$$

Thus $\phi(v_n) \to 0$ as $n \to \infty$ and, in particular, $\phi(v_\infty) = 0$. This contradicts the fact that $\|v_\infty\|_{T(\lambda)} = 1$ and so we must conclude that $\{u \in Z : J(u) \geqslant 0\}$ is bounded.

The result now follows from Theorem 2.2.

<u>Proof of 4B</u>

Fix $r > 0$. We shall apply Theorem 2.3 with $E = H_T$, $F = H$ and $J(u) = \frac{1}{2}\|Tu\|^2 - \phi(u)$ for $u \in H_T$.

In view of (S4), we need only show that:

(a) $\inf_{u \in M_r} J(u) > -\infty$

(b) J satisfies the condition $(C)^-$ on M_r.

We begin by noting that, for $u \in M_r$ with $J(u) \leq 0$, we have

$\frac{1}{2}\|Tu\|^2 \leq \phi(u) \leq \frac{1}{2}K\|Tu\|^{\alpha}\|u\|^{\beta}$ and so

$$\|Tu\| \leq [Kr^{\beta}]^{\frac{1}{2-\alpha}} \quad \text{since } \alpha < 2. \tag{4.1}$$

(a) Suppose that $u \in M_r$ and that $J(u) \leq 0$. Then,

$$J(u) \geq -\phi(u) \geq -\frac{1}{2}K\|Tu\|^{\alpha}\|u\|^{\beta} \geq -\frac{1}{2}K^{\frac{2}{2-\alpha}} r^{\frac{2\beta}{2-\alpha}}, \tag{4.2}$$

by (4.1). Thus $\inf_{u \in M_r} J(u) > -\infty$.

(b) Consider a sequence $\{u_n\} \subset M_r$ such that $\sup J(u_n) < 0$ and $\|(J|_{M_r})'(u_n)\|_{(E_{u_n})^*} \to 0$ with $E = H_T$.

By (4.1), $\{\|Tu_n\|\}$ is bounded and hence we conclude that $\{u_n\}$ is bounded in $E = H_T$. Thus, in the notation of section 2.B, it follows that $\|PJ'(u_n)\|_{(H_T)^*} \to 0$ where

$$PJ'(u_n) = T'Tu_n - F(u_n) - \lambda_n u_n$$

with $\lambda_n = \dfrac{J'(u_n)u_n}{r^2} = \dfrac{\|Tu_n\|^2 - \langle F(u_n), u_n \rangle}{r^2}$.

Note that, by (S3),

$$\lambda_n \geq \frac{-K\|Tu\|^{\alpha}\|u\|^{\beta}}{r^2} \geq -K^{\frac{2}{2-\alpha}} r^{\frac{2(\alpha+\beta-2)}{2-\alpha}} \quad \text{by (4.1)} \tag{4.3}$$

and

$$\lambda_n \leq \frac{\|Tu_n\|^2 - 2\phi(u_n)}{r^2} \quad \text{by (S2)} \tag{4.4}$$

$$\leq \frac{2}{r^2} \sup J(u_n) < 0.$$

Thus we see that $\{\lambda_n\}$ is bounded and there exists a subsequence $\{u_{n_i}\}$ such that $u_{n_i} \rightharpoonup u_\infty$ weakly in H_T and

$$\lambda_{n_i} \to \lambda_\infty \leq \frac{2}{r^2} \sup J(u_n) < 0.$$

By (S1), $\|F(u_{n_i}) - F(u_\infty)\|_{(H_T)^*} \to 0$ as $i \to \infty$ and

$$T'Tu_\infty - F(u_\infty) = \lambda_\infty u_\infty \quad \text{in } (H_T)^*.$$

Thus, $\|T(u_{n_i} - u_\infty)\|^2 - \lambda_\infty\|u_{n_i} - u_\infty\|^2$

$$= \langle T'T(u_{n_i} - u_\infty) - \lambda_\infty(u_{n_i} - u_\infty),\ u_{n_i} - u_\infty\rangle$$

$$= PJ'(u_{n_i})(u_{n_i} - u_\infty) + \langle F(u_{n_i}) - F(u_\infty) + (\lambda_{n_i} - \lambda_\infty)u_{n_i},\ u_{n_i} - u_\infty\rangle$$

$$\leq \{\|PJ(u_{n_i})\|_{(H_T)^*} + \|F(u_{n_i}) - F(u_\infty)\|_{(H_T)^*}\}\|u_{n_i} - u_\infty\|_T + |\lambda_{n_i} - \lambda_\infty||\langle u_{n_i}, u_{n_i} - u_\infty\rangle|$$

$\rightarrow 0$ as $i \rightarrow \infty$, since $\{u_n\}$ is bounded in H_T.

Recalling that $\lambda_\infty < 0$, this proves that $\{u_{n_i}\}$ converges in H_T and so we conclude that J satisfies the condition $(C)^-$ on M_r.

According to (S4) and Theorem 2.3, for $n \in \{1,\ldots,j\}$, $K^r_{c^r_n} \neq \phi$ and $\exists\ u^r_n \in M_r$ such that

$$J(u^r_n) = c^r_n\ ,\quad (J|_{M_r})'(u^r_n) = 0 \text{ and } u^r_n \neq \pm u^r_m$$

for $n \neq m$. In particular, $PJ'(u^r_n) = 0$. That is,

$$T'Tu^r_n - F(u^r_n) - \lambda^r_n u^r_n = 0 \text{ where } \lambda^r_n = \frac{\|Tu^r_n\| - \langle F(u^r_n), u^r_n\rangle}{r^2}\ .$$

By (4.3) and (4.4),

$$-K^{\frac{2}{2-\alpha}}\ r^{\frac{2(\alpha+\beta-2)}{2-\alpha}} \leq \lambda^r_n \leq \frac{2c^r_n}{r^2} < 0.$$

Since $\alpha + \beta > 0$ and $\alpha < 2$, it follows that $\lambda^r_n \rightarrow 0$ as $r \rightarrow 0$.

By (4.1), $\|Tu^r_n\| \leq [Kr^\beta]^{\frac{1}{2-\alpha}}$ and so $\|u^r_n\|_T \rightarrow 0$ as $r \rightarrow 0$.

This completes the proof.

5. EQUATION $T'Tu + F(u) = \lambda u$

In addition to (H1), (H2) and (H3) of section 3, we shall use the following hypotheses.

(B1) X is compactly embedded in H.

(B2) $\exists\ Q > 0$ such that $\langle F(u),u\rangle \leq Q\phi(u)\quad \forall\ u \in X$.

(B3) $\phi(u)/\|u\|^2_X \rightarrow 0$ as $\|u\|_X \rightarrow 0$.

(B4) $\exists$ a continuous increasing function $g : (0,\infty) \rightarrow (0,\infty)$ such that $\langle F(u) - F(v), u - v\rangle \geq g(\|u - v\|_Y)\|u - v\|_Y\quad \forall\ u,v \in Y$.

(Together with (H2), this implies that $g(t) \rightarrow 0$ as $t \rightarrow 0$ and

$\langle F(u),u\rangle \geq g(\|u\|_Y)\|u\|_Y > 0\quad \forall\ u \in Y\backslash\{0\}$.)

Theorem 5

Let (H1) to (H3) and (B1) to (B4) hold.

A (Existence) Suppose that $g(t)/t \rightarrow +\infty$ as $t \rightarrow +\infty$. For each fixed $\lambda > 0$, $\exists$ an infinite number of distinct solutions

$\{(\lambda,u_n)\} \subset \mathbb{R}\times X$ of $T'Tu + F(u) = \lambda u$.

B (Bifurcation) For each fixed $r > 0$, $\exists$ an infinite number of distinct solutions $\{(\lambda_n^r,u_n^r)\}_{n=1}^{\infty} \subset \mathbb{R}\times X$ of $T'Tu + F(u) = \lambda u$ such that $\forall\, n \in \mathbb{N}$,

$$\|u_n^r\| = r \quad\text{and}\quad \lambda_n^r > 0\,,$$

$$\|u_n^r\|_T \to 0 \quad\text{and}\quad \lambda_n^r \to 0+ \quad\text{as}\quad r \to 0.$$

Proof of 5A

Fix $\lambda > 0$. We shall apply Theorem 2.1 with $E = X$ and $J(u) = \frac{1}{2}\|Tu\|^2 + \phi(u) - \frac{\lambda}{2}\|u\|^2$ for $u \in X$. We need only prove that:

(a) $\inf_{u\in X} J(u) > -\infty$,

(b) J satisfies the condition $(C)^-$ on X,

(c) $b_k < 0 \quad \forall\, k \in \mathbb{N}$.

(a) For $u \in X$,

$$\begin{aligned} J(u) &\geq \frac{1}{Q}\langle F(u),u\rangle - \frac{\lambda}{2}\|u\|^2 \quad \text{by (B2)}\\ &\geq \frac{1}{Q}g(\|u\|_Y)\|u\|_Y - \frac{\lambda C^2}{2}\|u\|_Y^2 \quad \text{by (H2) and (B2)}\\ &= \frac{\|u\|_Y^2}{Q}\left\{g(\|u\|_Y)/\|u\|_Y - \frac{\lambda C^2 Q}{2}\right\}. \end{aligned} \tag{5.1}$$

Since $g(t)/t \to +\infty$ as $t \to +\infty$, this proves that $\inf_{u\in X} J(u) > -\infty$.

(b) Consider a sequence $\{u_n\} \subset X$ such that $\sup J(u_n) < 0$ and $\|J'(u_n)\|_{X^*} \to 0$ as $n \to \infty$.

Then $0 > J(u_n) \geq \frac{1}{2}\|Tu_n\|^2 + \frac{\|u_n\|_Y^2}{Q}\left\{g(\|u_n\|_Y)/\|u_n\|_Y - \frac{\lambda C^2 Q}{2}\right\}$

by (5.1). It follows that $\{\|Tu_n\|\}$ and $\{\|u_n\|_Y\}$ are bounded sequences. Since $Y \hookrightarrow H$, this implies that $\{u_n\}$ is bounded in X. But X is reflexive, so $\exists$ a subsequence $\{u_{n_i}\}$ such that $u_{n_i} \rightharpoonup u_\infty$ weakly in X as $i \to \infty$.

By (B1), $\|u_{n_i} - u_\infty\| \to 0$ as $i \to \infty$.

Now,

$$\begin{aligned} &\|T(u_{n_i} - u_{n_j})\|^2 + g(\|u_{n_i} - u_{n_j}\|_Y)\|u_{n_i} - u_{n_j}\|_Y\\ &\leq \langle T'T(u_{n_i} - u_{n_j}),u_{n_i} - u_{n_j}\rangle + \langle F(u_{n_i}) - F(u_{n_j}),u_{n_i} - u_{n_j}\rangle\\ &= \langle J'(u_{n_i}) - J'(u_{n_j}),u_{n_i} - u_{n_j}\rangle + \lambda\|u_{n_i} - u_{n_j}\|^2\\ &\leq \|J'(u_{n_i}) - J'(u_{n_j})\|_{X^*}\|u_{n_i} - u_{n_j}\|_X + \lambda\|u_{n_i} - u_{n_j}\|^2 \to 0 \text{ as } i,j \to +\infty. \end{aligned}$$

This proves that $\{u_{n_i}\}$ converges in X and hence J satisfies the condition $(C)^-$ on X.[1]

(c) Set $\varepsilon = \lambda/4$ and fix $k \in \mathbb{N}$. By Lemma 3.1, $\exists$ a subspace Z_k of X such that $\dim Z_k = k$ and

$$\|Tu\|^2 \leq \varepsilon\|u\|^2 \quad \forall\, u \in Z_k.$$

In particular, $\exists$ $d(k) > 0$ such that $\|u\|_X \leq d(k)\|u\| \quad \forall\, u \in Z_k$. By (B3), $\exists$ $\delta(\varepsilon,k) > 0$ such that

$$|\phi(u)| \leq \frac{\varepsilon}{2d(k)^2}\|u\|_X^2 \quad \text{for } \|u\|_X \leq \delta(\varepsilon,k)$$

$$\leq \frac{\varepsilon}{2}\|u\|^2 \quad \text{for } \|u\| \leq \frac{\delta(\varepsilon,k)}{d(k)} \text{ and } u \in Z_k.$$

Let $A(\varepsilon,k) = \left\{u \in Z_k : \|u\| = \frac{\delta(\varepsilon,k)}{d(k)}\right\}$. Then, by section 2.A(ii), $A(\varepsilon,k) \in \Gamma_k$ and, for $u \in A(\varepsilon,k)$,

$$J(u) \leq \frac{\varepsilon}{2}\|u\|^2 + \frac{\varepsilon}{2}\|u\|^2 - \frac{\lambda}{2}\|u\|^2 = -\varepsilon\|u\|^2.$$

Thus we have that $b_k < 0$.

<u>Proof of 5B</u>

Fix $r > 0$. We shall apply Theorem 2.3 with $E = X$, $F = H$ and $J(u) = \frac{1}{2}\|Tu\|^2 + \phi(u)$ for $u \in X$.

We shall show that:

(a) $\inf\limits_{u\in M_r} J(u) > -\infty$

(b) J satisfies the condition (C) on M_r

(c) $\forall\, k \in \mathbb{N}$, $c_k^r/r^2 \to 0+$ as $r \to 0$.

First we note that, for $u \in X$,

$$J(u) \geq \frac{1}{2}\|Tu\|^2 + \frac{1}{\Omega}\langle F(u),u\rangle \quad \text{by (B2)}$$

$$\geq \frac{1}{2}\|Tu\|^2 + \frac{1}{\Omega}g(\|u\|_Y)\|u\|_Y \quad \text{by (B4)}. \tag{5.2}$$

(a) By (5.2), $J(u) \geq \frac{1}{\Omega}g\left(\frac{1}{C}\|u\|\right)\frac{1}{C}\|u\|$ since $Y \hookrightarrow H$.

Thus $\inf\limits_{u\in M_r} J(u) \geq \frac{r}{\Omega C}g(r/C) > 0$.

(b) Consider a sequence $\{u_n\} \subset M_r$ such that $J(u_n) \leq L \quad \forall\, n \in \mathbb{N}$ and

$$\|(J|_{M_r})'(u_n)\|_{(X_{u_n})^*} \to 0 \text{ as } n \to \infty.$$

By (5.2), $L \geq \frac{1}{2}\|Tu_n\|^2 + \frac{1}{\Omega}g(\|u_n\|_Y)\|u_n\|_Y$ from which it follows that $\{u_n\}$ is bounded in X. In the notation of section 2.B, we then have that $\|PJ'(u_n)\|_{X^*} \to 0$ where

$$PJ'(u_n) = T'Tu_n + F(u_n) - \lambda_n u_n$$

with $\lambda_n = \dfrac{J'(u_n)u_n}{r^2} = \dfrac{\|Tu_n\|^2 + \langle F(u_n), u_n\rangle}{r^2}$.

Noting that

$$0 < \lambda_n \leq \frac{\|Tu_n\|^2 + \Omega\phi(u_n)}{r^2} \leq \frac{(2+\Omega)J(u_n)}{r^2} \leq \frac{(2+\Omega)L}{r^2} , \tag{5.3}$$

we have that $\{\lambda_n\}$ is bounded. Since X is reflexive, $\exists$ a subsequence $\{u_{n_i}\}$ such that $u_{n_i} \rightharpoonup u_\infty$ weakly in X

$$\lambda_{n_i} \to \lambda_\infty \quad \text{as} \quad i \to \infty .$$

By (B1), $\|u_{n_i} - u_\infty\| \to 0$ as $i \to \infty$.

Now, $\|T(u_{n_i} - u_{n_j})\|^2 + g(\|u_{n_i} - u_{n_j}\|_Y)\|u_{n_i} - u_{n_j}\|_Y$

$$\leq \langle T'T(u_{n_i} - u_{n_j}) + F(u_{n_i}) - F(u_{n_j}), u_{n_i} - u_{n_j}\rangle$$

$$= \langle PJ'(u_{n_i}) - PJ'(u_{n_j}) + \lambda_{n_i}u_{n_i} - \lambda_{n_j}u_{n_j}, u_{n_i} - u_{n_j}\rangle$$

$$\leq \|PJ'(u_{n_i}) - PJ'(u_{n_j})\|_{X^*}\|u_{n_i} - u_{n_j}\|_X + |\lambda_{n_i} - \lambda_{n_j}||\langle u_{n_i}, u_{n_i} - u_{n_j}\rangle|$$

$$+ |\lambda_{n_j}|\|u_{n_i} - u_{n_j}\|^2 \to 0 \quad \text{as} \quad i,j \to \infty .$$

This proves that $\{u_{n_i}\}$ converges in X and so J satisfies the condition (C) on M_r.

(c) $\forall\, k \in \mathbb{N}$, $c_k^r \geq c_1^r = \inf\limits_{u\in M_r} J(u) > 0$ by (a).

Fix $\varepsilon > 0$ and $k \in \mathbb{N}$. As in the proof of part (c) for 5.A, $\exists$ a subspace Z_k of X such that $\dim Z_k = k$ and $\exists$ $\eta(\varepsilon,k) > 0$ such that $\|Tu\|^2 \leq \varepsilon\|u\|^2$ and $0 \leq \phi(u) \leq \frac{\varepsilon}{2}\|u\|^2$ for $u \in Z_k$ with $\|u\| \leq \eta(\varepsilon,k)$.

Let $A(r,k) = \{u \in Z_k : \|u\| = r\} = Z_k \cap M_r$.
Then, by section 2.A(ii), $A(r,k) \in \Gamma_k^r$ and, for $0 < r \leq \eta(\varepsilon,k)$,
$J(u) = \frac{1}{2}\|Tu\|^2 + \phi(u) \leq \varepsilon r^2$, $\forall\, u \in A(r,k)$.

Thus

$$0 < c_k^r \leq \varepsilon r^2 \quad \text{if} \quad 0 < r \leq \eta(\varepsilon,k).$$

This proves that $c_k^r/r^2 \to 0$ as $r \to 0$.

According to Theorem 2.3, $\forall\, n \in \mathbb{N}$, $K_{c_n^r} \neq \phi$ and $\exists$ $u_n^r \in M_r$ such that

$$J(u_n^r) = c_n^r\ ,\quad (J|_{M_r})'(u_n^r) = 0 \quad\text{and}\quad u_n^r \neq u_m^r \quad\text{for}\quad n \neq m.$$

In particular, $PJ'(u_n^r) = 0$ and so,

$$T'Tu_n^r + F(u_n^r) = \lambda_n^r u_n^r \quad\text{where}\quad \lambda_n^r = \frac{\|Tu_n^r\|^2 + \langle F(u_n^r),u_n^r\rangle}{r^2}.$$

By (5.3), $0 < \lambda_n^r \leqslant \dfrac{(2+Q)c_n^r}{r^2}$. Hence $\lambda_n^r \to 0+$ as $r \to \infty$. But,

$$\lambda_n^r \geqslant \frac{\|Tu_n^r\|^2 + g(\|u_n^r\|_Y)\|u_n^r\|_Y}{r^2}\ ,$$ from which it follows that

$\|u_n^r\|_X \to 0$ as $r \to \infty$.

6. PROOF OF THEOREM (1-)

In the setting of section 3, we put

$$H = L^2\ ,\quad \mathfrak{D}(S) = H^2 \quad\text{and}\quad Su = -\Delta u \quad\text{for}\quad u \in H^2.$$

Then $S: \mathfrak{D}(S) \subset H \to H$ satisfies (H1) and we take $T = S^{\frac{1}{2}}$, the positive self-adjoint square root of S. The Fourier transform shows that,

$$H_T = H^1 \quad\text{and}\quad \|Tu\| = \|\,|\nabla u|\,\|.$$

See [11] for details.

Let $Y = H_T$, $\phi(u) = \dfrac{1}{(\sigma+2)}\displaystyle\int q|u|^{\sigma+2}\,dx$

and $F(u) = q|u|^{\sigma}u.$

Lemma 6.1

Let (A1) hold and suppose that $\sigma \in (0,\frac{4}{N-2})$ and $q \in L^{\infty}(\mathbb{R}^N)$.

(a) Then $\phi \in C^1(H_T,\mathbb{R})$, $\phi(u) > 0 \quad \forall\, u \in H_T\backslash\{0\}$,

$F: H_T \to (H_T)^*$ is continuous,

$(\sigma+2)\phi(u) = \langle F(u),u\rangle$ and

$\phi'(u)v = \langle F(u),v\rangle \quad \forall\, u,v \in H_T$ and

$\exists\ K > 0$ such that $0 \leqslant \langle F(u),u\rangle \leqslant K\|Tu\|^{\alpha}\|u\|^{\beta} \quad \forall\, u \in H_T$ where $\alpha = \frac{N\sigma}{2}$ and $\alpha + \beta = \sigma + 2$.

(b) If, in addition, $q(x) \to 0$ as $|x| \to \infty$, then

$\phi: H_T \to \mathbb{R}$ is weakly sequentially continuous and

$F: H_T \to (H_T)^*$ is compact.

(c) If $\exists\ A > 0$ and $t \in (0,2-\frac{N\sigma}{2})$ such that

$q(x) \geqslant A(1+|x|)^{-t}$ a.e. on $\mathbb{R}^N$, then $\sigma \in (0,\frac{4}{N})$ and

$$c_1^r \equiv \inf\{\tfrac{1}{2}\|Tu\|^2 - \phi(u) : u \in H_T \text{ with } \|u\|_H = r\} < 0 \quad \forall\, r > 0.$$

Proof

(a) This follows from the Sobolev embeddings and the standard results about Nemytskii operators. The final inequality amounts to the multiplicative form of the Sobolev inequality due to Gagliardo. (See [10,11] for details.

(b) This follows from the decay of q and the compactness of the Sobolev embeddings on bounded domains. See [10,11] for details.

(c) This can be verified by using test functions of exponential type. See [10,11] for details.

Proof of Theorem (1-)

By Lemma 6.1, (H2) and (S1) to (S3) are satisfied. Since $Y = H_T = X$, we have that (H3) is satisfied.

Under the extra hypothesis of part (1-)B, it follows from Lemma 6.1(c) that (S4) is satisfied with $j = 1$ and that $\alpha < 2$ in (S3). Thus Theorem (1-) follows directly from Theorem 4.

7. PROOF OF THEOREM (1+)

Let H,S and T be as in section 6. Set

$$Y = \left\{u \in L^{\sigma+2} : \int q|u|^{\sigma+2}dx < \infty\right\} \text{ with } \|u\|_Y = \left\{\int q|u|^{\sigma+2}dx\right\}^{\frac{1}{\sigma+2}},$$

$$\phi(u) = \frac{1}{(\sigma+2)}\int q|u|^{\sigma+2}dx = \frac{\|u\|_Y^{\sigma+2}}{(\sigma+2)} \quad \text{and}$$

$$F(u) = q|u|^\sigma u.$$

Then $X = H_T \cap Y = \{u \in H^1 : \int q|u|^{\sigma+2} < \infty\}$ and

$$\|u\|_X = \left\{\|\,|\nabla u|\,\|^2 + \|u\|^2 + \|u\|_Y^2\right\}^{\frac{1}{2}}.$$

Lemma 7.1

Let (A1) hold and suppose that $q(x) \geqslant A > 0$ a.e. on $\mathbb{R}^N$.

(a) Then Y is a reflexive Banach space and $C_0^\infty(\mathbb{R}^N) \subset H_T \cap Y = X$.

(b) Furthermore, $\phi \in C^1(Y,\mathbb{R})$,

$F : Y \to Y^*$ is continuous,

$(\sigma+2)\phi(u) = \langle F(u),u\rangle$ and

$\phi'(u)v = \langle F(u),v\rangle \quad \forall\ u,v \in Y.$

(c) If, in addition, $\exists\ A > 0$ and $t > \frac{N\sigma}{2}$ such that $q(x) \geqslant A(1+|x|)^t$ a.e. on $\mathbb{R}^N$, then X is compactly embedded in H and $Y \hookrightarrow H$.

Proof

(a) Set $p = \sigma + 2$ and $\frac{1}{p} + \frac{1}{p'} = 1$. Then

$Y^* \cong \left\{u \in L^1_{loc} : \int q^{-p'/p}|u|^{p'}dx < \infty\right\}$. Hence Y is reflexive.

(b) This follows from the standard results about Nemystskii operators. For $u,v \in Y$, $\langle F(u) - F(v), u - v\rangle = \int q\left\{|u|^\sigma u - |v|^\sigma v\right\}(u-v)dx$

$$\geq \int q\, 2^{-\sigma}|u-v|^{\sigma+2}dx = 2^{-\sigma}\|u-v\|_Y^{\sigma+2}.$$

See [12-14] for details.

(c) For $u \in Y$ and $\omega \subset \mathbb{R}^N$,

$$\int_\omega u^2dx = \int_\omega q^{-\frac{1}{p}}q^{\frac{1}{p}}u^2dx \quad \text{where } p = \frac{\sigma+2}{2} \text{ and } \frac{1}{p} + \frac{1}{p'} = 1$$

$$\leq \left\{\int_\omega q^{-\frac{p'}{p}}dx\right\}^{\frac{1}{p'}}\left\{\int_\omega q|u|^{2p}dx\right\}^{\frac{1}{p}}$$

$$= \left\{\int_\omega q^{-\frac{2}{\sigma}}dx\right\}^{\frac{\sigma}{\sigma+2}}\|u\|_Y^2. \tag{7.1}$$

Since $q(x) \geq A(1+|x|)^t$ for some $t > \frac{N\sigma}{2}$, we see that

$$\int_{|x|>d} q^{-\frac{2}{\sigma}}dx \to 0 \quad \text{as} \quad d \to \infty.$$

From this and the compactness of the Sobolev embeddings on bounded domains, we conclude that X is compactly embedded in H. Furthermore, setting $\omega = \mathbb{R}^N$ in (7.1), we have that

$$\|u\| \leq C\|u\|_Y \quad \forall\, u \in Y \quad \text{where } C \equiv \left\{\int q^{-\frac{2}{\sigma}}dx\right\}^{\frac{\sigma}{2(\sigma+2)}} < \infty.$$

Thus $Y \hookrightarrow H$.

Proof of Theorem (1+)

By Lemma 7.1, we have that (H2) and (B1) - (B4) are satisfied with $g(t) = 2^{-\sigma}t^{\sigma+1}$ in (B4). Thus $g(t)/t \to +\infty$ as $t \to +\infty$. Since $C_o^\infty(\mathbb{R}^N)$ is dense in $H^1 = H_T$, we have that (H3) is also satisfied. Theorem (1+) now follows immediately from Theorem 5.

REFERENCES

[1] Rabinowitz, P.H.: Variational methods for nonlinear eigenvalue problems, C.I.M.E., Proc. editor G. Prodi, Edizioni Cremonese, Rome, 1974.

[2] Ambrosetti, A. and Rabinowitz, P.H.: Dual variational methods in critical point theory and applications, J. Functional Anal., 14 (1973), 349-381.

[3] Clark, D.C.: A variant of the Lusternik-Schnirelman theory, Ind. Univ. Math. J., 22 (1972), 65-74.

[4] Berestycki, H. and Lions, P.L.: Nonlinear scalar field equations I (Existence of a ground state) and II (Existence of infinitely many solutions). Univ. Paris VI preprints to appear in Arch. Rational Mech. Anal..

[5] Palais, R.S.: Lusternik-Schnirelman theory on Banach manifolds, Topology, 5 (1966), 115-132.

[6] Bongers, A.L.: Behandlung verallgemeinerter nichtlinearer Eigenwertprobleme mit Lusternik-Schnirelman Theorie, Doctoral dissertation, Mainz, 1979.

[7] Strauss, W.: Existence of solitaroy waves in higher dimensions, Comm. Math. Phys., 55 (1977), 149-162.

[8] Berger, M.S.: On the existence and structure of stationary states for a nonlinear Klein-Gordon equation, J. Functional Anal., 9 (1972), 249-261.

[9] Berestycki, H. and Lions, P.L.: Existence of stationary states in nonlinear scalar field equations, Bifurcation phenomena in mathematical physics and related topics, editors C. Bardos and D. Brézis, Reidel, 1980.

[10] Stuart, C.A.: Bifurcation for Dirichlet problems without eigenvalues, Proc. London Math. Soc., 45 (1982), 169-192.

[11] Stuart, C.A.: Bifurcation from the continuous spectrum in the L^2-theory of elliptic equations on $\mathbb{R}^n$, Recent Methods in Nonlinear Analysis and Applications, Proc. of SAFA. IV, Liguori, Naples 1981 (copies available from the author).

[12] Benci, V. and Fortunato, D.: Does bifurcation from the essential spectrum occur?, Comm. Partial Diff. Equat., 6 (1981), 249-272.

[13] Benci, V. and Fortunato, D.: Bifurcation from the essential spectrum for odd variational operators, Conf. sem. Bari, 178 (1981).

[14] Bongers, A., Heinz, H.P. and Küpper, T.: Existence and bifurcation theorems for nonlinear elliptic eigenvalue problems on unbounded domains, to appear in J. Diff. Equat.

[15] Nehari, Z.: On a nonlinear differential equation arising in nuclear physics, Proc. Royal Irish Acad., 62 (1963), 117-135.

[16] Ryder, G.H.: Boundary value problems for a class of nonlinear differential equations, Pacific J. Math., 22 (1967), 477-503.

17] Chiapinelli, R. and Stuart, C.A.: Bifurcation when the linearisation has no eigenvalues, J. Diff. Equat., 30 (1978), 296-307.

18] Stuart, C.A.: Des bifurcations sans valeurs propres, C.R. Acad. Sci. Paris, 284A (1977), 1373-1375.

19] Stuart, C.A.: Bifurcation pour des problèmes de Dirichlet et de Neunmann sans valeurs propres, C.R. Acad. Sci. Paris, 288A (1979), 761-764.

20] Stuart, C.A.: A variational method for bifurcation problems when the linearisation has no eigenvalues, Conf. Sem. Bari, Proc. SAFA III (1978), (162) 1979.

21] Stuart, C.A.: Bifurcation for variational problems when the linearisation has no eigenvalues, J. Functional Anal., 38 (1980), 169-187.

22] Stuart, C.A.: Bifurcation for Neumann problems without eigenvalues, J. Diff. Equat., 36 (1980), 391-407.

[23] Küpper, T.: The lowest point of the continuous spectrum as a bifurcation point, J. Diff. Equat., 34 (1979), 212-217.

[24] Küpper, T.: On minimal nonlinearities which permit bifurcation from the continuous spectrum, Math. Methods in Appl. Sci., 1 (1979), 572-580.

[25] Küpper, T. and Riemer, D.: Necessary and sufficient conditions for bifurcation from the continuous spectrum, Nonlinear Anal (TMA), 3 (1979), 555-561.

[26] Küpper, T. and Weyer, J.: Maximal monotonicity and bifurcation from the continuous spectrum, to appear in Nonlinear Anal..

[27] Toland, J.F.: Global bifurcation for Neumann problems without eigenvalues, J. Diff. Equat., 44 (1982), 82-110.

[28] Toland, J.F.: Positive solutions of nonlinear elliptic equations, existence and non-existence of solutions with radial symmetry in $L_p(\mathbb{R}^N)$, preprint.

[29] Amick, C.J. and Toland, J.F.: Nonlinear elliptic eigenvalue problems on an infinite strip, Global theory of bifurcation and asymptotic bifurcation, preprint.

[30] Pohozaev, S.I.: Eigenfunctions of the equation $\Delta u + \lambda f(u) = 0$, Soviet Math. Dokl. 5 (1965), 1408-1411.

[31] Reed, M. and Simon, B.: Methods of Modern Mathematical Physics Vol IV, Academic Press, New York, (1978).

[32] DeFigueiredo, D., Lions, P.L. and Nussbaum, R.D.: A priori estimates for positive solutions of semilinear elliptic equations, preprint to appear in J. Math. Pures et Appliquées.

[33] Kirchgässner, K. and Scheurle, J.: On the bounded solutions of a semilinear equation in a strip, J. Diff. Equat., 22 (1979), 119-148.

[34] Kirchgässner, K. and Scheurle, J.: Bifurcation from the continuous spectrum and singular solutions, Trends in Applications of Pure Math. to Mec. Vol. III, editor R.J. Knops, Pitman, London, 1980.

[35] Kirchgässner, K. and Scheurle, J.: Bifurcation of non-periodic solutions of some semilinear equations in unbounded domains, Editors H. Amann, N. Bazley and K. Kirchgässner, Pitman, London (1981).

SOME PROPERTIES OF NONLINEAR DIFFERENTIAL EQUATIONS WITH QUASIDERIVATIVES

M. Svec
Bratislava

In my communication I will speak about differential equations of the form

(E) $$L_n x + f(t)g(x) = 0,\ t \in J = (a,\infty)$$

where

(1) $$L_n x = (a_{n-1}(t)(a_{n-2}(t)(\dots(a_1(t)x')'\dots)')')'$$

and $f(t),\ a_i(t) \in C(J),\ a_i(t) > 0,\ i = 1, 2, \dots, n-1,\ f(t) > 0$

(2) $$\int_a^\infty a_i^{-1}(t)dt = \infty,\ i = 1, 2, \dots, n-1,$$

(3) $$g(x) \in C(-\infty,\infty),\ xg(x) > 0 \quad \text{for every} \quad x \neq 0.$$

The expressions

(4) $$L_0 x = x,\ L_1 x = (a_1(t)x', \dots, L_i x = a_i(t)L'_{i-1}x,\ i = 1, 2, \dots, n$$

with $a_n(t) = 1$, are said to be quasiderivatives with respect to the system of functions $a_i(t)$. Property (3) plays an important role in our considerations.

First I shall center my attention to the question: What kind of solutions can equation (E) have? Evidently it has zero solution on J. A solution $x(t) \in (E)$ with the property

$$\sup\{|x(t)|: t \geqq T\} > 0 \quad \text{for all} \quad T > a$$

will be called a proper solution of (E). The point ρ will be said to be a zero of multiplicity k of the solution $x(t) \in (E)$ iff $L_i x(\rho) = 0,\ i = 0, 1, \dots, k-1,\ L_k x(\rho) \neq 0$. A proper solution $x(t)$ (E) is called oscillatory if it has no last zero; otherwise, it is called nonoscillatory.

How may one classify nonzero solutions of (E)? Is it possible that it has a zero of multiplicity n? How many such zeros of multiplicity n can it have? P. Brunovsky and John Mallet-Paret [1] have recently proved that the equations

(5) $$x^{(4)} + |x|^\alpha \operatorname{sgn} x = 0,\ 0 < \alpha < 1$$

have solutions with zeros of multiplicity four (five). Such solutions have an infinity of zeros in every neighbourhood of the zero of multiplicity four. Using similar arguments as those used by Brunovsky and Mallet-Paret I have proved that the equation

$$x^{(4)} + f(t)|x|^{\alpha}\operatorname{sgn} x = 0, \ 0 < \alpha < 1$$

where $f(t) \in C(-\infty,\infty)$, $f(t) > 0$, $\int_{t_0}^{\infty} f(t)dt = \infty$, has also solutions with the zero of multiplicity four. On the other hand, it is easy to prove that the equation

$$x'' + g(x) = 0$$

has no nontrivial solutions with a double zero. In fact, multiplying this equation by $2x'$ and integrating between t_0 and t, where t_0 is a double zero of $x(t)$, we get

$$x'^2(t) + \int_0^{x(t)} g(z)dz = 0.$$

From this it follows that $x(t) = 0$ for all t.

We investigate the behavior of a proper solution $x(t)$ of (E) in the neighborhood of its zero ρ of multiplicity n. The following fundamental theorem appears in [2]:

Theorem 1. Let $x(t)$ be a proper solution of (E). Let ρ be its zero of multiplicity n. Then there is a right neighborhood $0^+ = (\rho, \rho + \varepsilon)$ $\varepsilon > 0$, such that either $x(t) \equiv 0$ on 0^+ or $x(t) \not\equiv 0$ on 0^+ and then it has there an infinity of zeros and ρ is their accumulation point. If n is even, then there exists also a left neighborhood $0^- = (\rho - \varepsilon_1, \rho)$, $\varepsilon_1 > 0$, such that either $x(t) \equiv 0$ on 0^- or $x(t) \not\equiv 0$ on 0^- and then it has there an infinity of zeros with ρ as accumulation point.

Remark. If n is odd it may happen that the behavior of the solution $x(t)$ of (E) with zero ρ of multiplicity n is different in the left neighborhood of ρ than Theorem 1 proposes. Let be $n = 3$ and $L_3x = (a(t)(a(t)x')')'$. Then we have the equation

$$L_3x + f(t)g(x) = 0. \tag{6}$$

Let $x(t)$ be a nontrivial solution of (6). Then

$$xL_3x = [xL_2x - \frac{1}{2} L_1^2x]' = -f(t)xg(x) < 0 \quad \text{for} \quad x \neq 0.$$

Thus, $F_3x(t) = x(t)L_2x(t) - \frac{1}{2} L_1^2x(t)$ is nonincreasing on J. Then evidently $F_3x(\rho) = 0$ and therefore $F_3x(t) \geqq 0$ for $a < t \leqq \rho$. It means that $x(t)$ is either identically zero for $a < t \leqq \rho$ or $x(t)\ L_1x(t).\ L_2x(t) \neq 0$ for $a < t < \rho$.

Another example gives us the equation $x' + \frac{3}{2} x^{1/3} = 0$. It has the solution $x(t) = (t_0 - t)^{3/2}$, $t \leqq t_0$. Evidently $x(t) \neq 0$ for $t < t_0$.

In general, there may occur the situations drawn in Figures 1-4 for n even, and figures 5-8 for n odd.

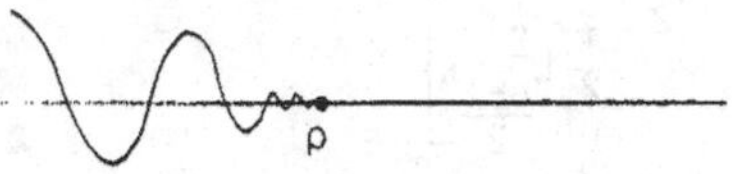

Figure 1

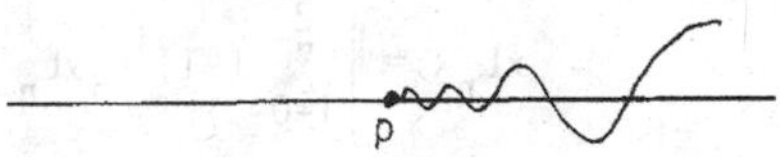

Figure 2

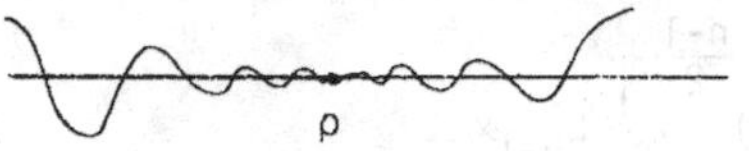

Figure 3

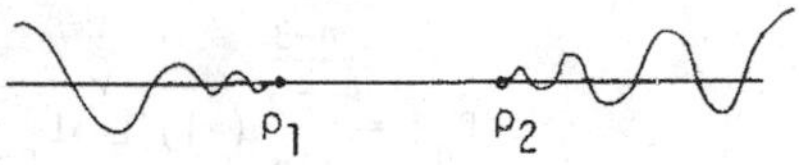

Figure 4

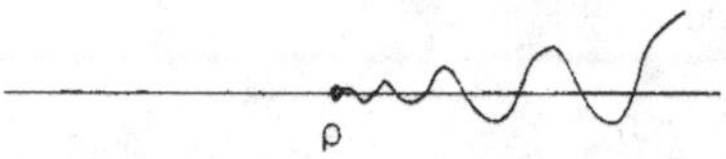

Figure 5

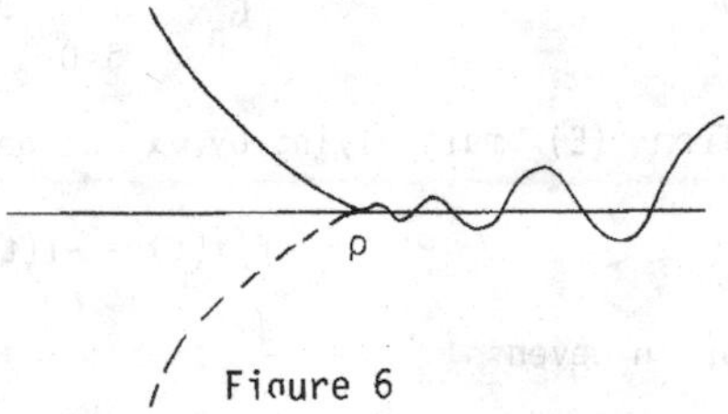

Figure 6

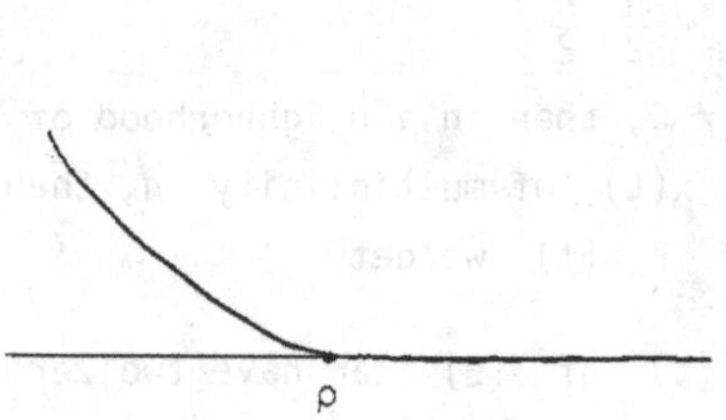

Figure 7

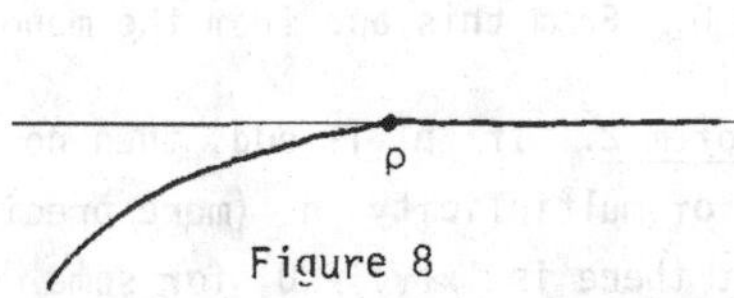

Figure 8

Now, I will discuss the following question: How many zeros of multiplicity n can a non-zero solution $x(t)$ of (E) have? We can get a partial answer:

Let the function $a_i(t)$ be such that

$$a_i(t) = a_{n-i}(t),\ i = 0, 1, \dots, n. \tag{7}$$

Then for n odd we have

$$xL_n x = \left[\sum_{i=0}^{\frac{n-3}{2}} (-1)^i L_i x L_{n-i-1} x + (-1)^{\frac{n-1}{2}} \frac{1}{2} L^2_{\frac{n-1}{2}} x\right]' . \tag{8}$$

and for n even

$$xL_n x = \left[\sum_{i=0}^{\frac{n}{2}-1} (-1)^i L_i x L_{n-i-1} x\right]' + (-1)^{\frac{n}{2}} L^2_{\frac{n}{2}} x \cdot a^{-1}_{\frac{n}{2}} . \tag{9}$$

Let's put

$$F_n x = \sum_{i=0}^{\frac{n-3}{2}} (-1)^i L_i x L_{n-i-1} x + (-1)^{\frac{n-1}{2}} \frac{1}{2} L^2_{\frac{n-1}{2}} x \tag{10}$$

$$G_n x = \sum_{i=0}^{\frac{n}{2}-1} (-1)^i L_i x L_{n-i-1} x. \tag{11}$$

Then from (E) multiplying by x we get for n odd

$$F'_n x(t) = -f(t)x(t)g(x(t)) \leqq 0 \tag{12}$$

and for n even

$$G'_n x(t) = -f(t)x(t)g(x(t)) - (-1)^{\frac{n}{2}} a^{-1}_{\frac{n}{2}} L^2_{\frac{n}{2}} x(t). \tag{13}$$

Thus, $F_n x(t)$ is nonincreasing on J, and if $x(\tau) \neq 0$, then in a neighborhood of τ it is strictly decreasing. If ρ is a zero of $x(t)$ of multiplicity n, then $F_n x(\rho) = 0$. From this and from the monotonicity of $F_n x(t)$ we get

<u>Theorem 2</u>. If n is odd, then no solution $x(t)$ of (E) can have two zeros $\rho_1 < \rho_2$, of multiplicity n (more precisely of multiplicity gerater than $(n-1)/2$) such that there is $x(\tau) \neq 0$ for some $\tau \in (\rho_1, \rho_2)$.

Similar reasoning gives us

<u>Theorem 3</u>. If $n = 4k$, $k \in \{1, 2, \dots\}$, then no solution $x(t)$ of (E) can

have two zeros, $\rho_1 < \rho_2$, of multiplicity n (more precisely of multiplicity greater $(1/n)-1$) such that there is $x(\tau) \neq 0$ for some $\tau \in (\rho_1, \rho_2)$.

Let's consider the fact that there may exist solutions of (E) such that they have the property: to $x(t) \in (E)$ there is a number $T_x > a$ such that $x(t)$ has no zero of multiplicity greater than $n-1$ on $[T_x,\infty)$. This suggests the definition.

We will say that the solution $x(t)$ of (E) is a regular solution if there exists a number $T_x > a$ such that $x(t)$ has no zero of multiplicity greater than $n-1$ on $[T_x,\infty)$. The set of all regular solutions of (E) we will denote by R.

Evidently it holds: If $x(t) \in R$, then for every point $\xi \in (T_x,\infty)$ there is a right neighborhood $(\xi, \xi+\varepsilon_1)$ and a left neighborhood $(\xi-\varepsilon_2, \xi)$ on which $L_i x(t)$, $i = 0, 1, \ldots$, has no zero.

Denote, following U. Elias [3], by $S(c_0, c_1, \ldots, c_n)$ the number of sign changes in the sequence $c_0, c_1, \ldots, c_n$) where $c_i \neq 0$, $i = 0, 1, \ldots, n$. Let $x(t) \in R$. Denote

$$S(x,\xi-) = \lim_{t-\xi-} S(L_0x(t), L_1x(t), \ldots, L_nx(t)) \tag{14}$$

$$S(x,\xi+) = \lim_{t-\xi+} S(L_0x(t), -L_1x(t), \ldots, (-1)^n L_nx(t)) \tag{15}$$

for $\xi \in (T_x,\infty)$. The existence of these two limits follow from the property introduced above. D. Sisolakova [2] proved the following theorem.

<u>Theorem 4</u>. For every regular solution $x(t)$ of (E) there exists the point t_1 such that $S(x,\xi+)$ and $S(x,\xi-)$ are constant for $t_1 < \xi < \infty$. If $S(x,\xi+) = k$ on (t_1,∞), then $S(x,\xi-) = n-k$ on (t_1,∞) and $0 \leqq k \leqq n$ is such that $(-1)^{n-k} f(t) < 0$. On (t_1,∞) the quasiderivatives $L_0x, L_1x, \ldots, L_{n-1}x$ may have only simple zeros.

This is a generalization of the theorem proved by U. Elias [3] for the linear case.

Using Theorem 4 we can define the classes of regular solutions of (E). Let

$$S_k = \{x(t) \quad R: S(x,t+) = k \text{ for all } t \text{ sufficiently large}\}, \tag{16}$$

$$k \in \{0, 1, \ldots, n\} \text{ such that } (-1)^{n-k} f(t) < 0.$$

In our case we have $(-1)^{n-k} < 0$, because $f(t)$ is supposed to be positive. Thus, $n-k$ is odd. Evidently the classes S_k are disjoint. We may examine the (asymptotic) properties of the solutions belonging to the class S_k and the structure of the set S_k, etc. We are going to do it for $n = 4$. Let us consider the equation

$$(17) \qquad (a_1(t)(a_2(t)(a_1(t)x')')')' + f(t)g(x) = 0.$$

Then

$$(18) \qquad G_4x(t) = x(t)L_3x(t) - L_1x(t)L_2x(t)$$

is nonincreasing for every solution of (17) and is strictly decreasing for all large t if $x(t)$ is regular. In the case $n = 4$, every proper solution of (17) is a regular solution and $R = S_1 \cup S_3$, $S_1 \cap S_3 = \phi$.

Theorem 5. Let $x(t) \in (17)$ be oscillatory. Then there is a number T_x such that on $(T_x,\infty)x(t)$, $L_1x(t)$, $L_2x(t)$, $L_3x(t)$ separate their zeros, i.e. between two consecutive zeros $(T_x) < \rho_1 < \rho_2$ of $L_ix(t)$ there is precisely one zero of $L_jx(t)$, $i, j = 0, 1, 2, 3$, $i \neq j$.

Proof. Let $x(t) \in (17)$ be oscillatory. Let T_x be such that $G_4x(t)$ has a constant sign on (T_x,∞) and such that $L_ix(t)$, $i = 0, 1, 2, 3$, has only simple zeros on (T_x,∞). We will use two following properties:

A) Between two consecutive zeros of the differentiable function $f(t)$ there is an odd number of zeros of $f'(t)$ if $f'(t)$ is continuous and $f'(t)$ has only isolated zeros.

B) If $f(t)$ is continuous and $f'(t)$ has isolated zeros and $f(t)$ is not identically zero on any subinterval of (α,β) and if $f(\alpha)f(\beta) > 0$, then the number of zeros of $f(t)$ between α and β is zero or even.

Let $x(t) \in (17)$ and let $(T_x <)\ t_1 < t_2$ be two consecutive zeros of $x(t)$. We are going to prove that in (t_1,t_2) there is precisely one zero of $L_1x(t)$. From Rolle's theorem it follows that in (t_1,t_2) there is at least one zero of $L_1x(t)$. Suppose that there is more than one zero of $L_1x(t)$. Then there must be at least three zeros, say $t_1 < \tau_1 < \tau_2 < \tau_3 < t_2$, of $L_1x(t)$. Rolle's theorem gives the existence of $\xi_1 \in (\tau_i,\tau_{i+1})$, $i = 1, 2$, such that $L_2x(\xi_i) = 0$, $i = 1, 2$. But $\operatorname{sgn} G_4x(\xi_1) = \operatorname{sgn} x(\xi_1)L_3x(\xi_1) = \operatorname{sgn} G_4x(\xi_2) = \operatorname{sgn} x(\xi_2)L_3x(\xi_2)$ and because $\operatorname{sgn} x(t) = \text{const.}$ on (t_1,t_2), we get that $L_3x(\xi_1)L_3x(\xi_2) > 0$. Then it follows from B) that $L_3x(t)$ has at least two zeros between ξ_1 and ξ_2. We note that the existence of one zero of $L_3x(t)$ between ξ_1 and ξ_2 follows from Rolle's theorem. Then from Rolle's theorem it follows also that between ξ_1 and ξ_2 lies at least one zero of $L_4x(t) = -f(t)g(x(t))$. But this is not possible because $L_4x(t)$ and $x(t)$ have the same zeros. This contradiction terminates the proof of our statement.

Now we are going to prove that between two consecutive zeros, say $(T_x<)t_1 < t_2$, of $L_1x(t)$ there is precisely one zero of $L_2x(t)$. Rolle's theorem gives us that

between t_1 and t_2 there is at least one zero of $L_2x(t)$. Suppose that there is more than one zero of $L_2x(t)$. Proceeding as in A), there are at least three zeros of $L_2x(t)$. Denote then $(t_1<)\tau_1 < \tau_2 < \tau_3(< t_2)$ and suppose that they are consecutive zeros. Then by Rolle's theorem there exist two zeros of $L_3x(t)$, denote them by ξ_1, ξ_2, such that $\tau_1 < \xi_1 < \tau_2 < \xi_2 < \tau_3$. Then we have $\operatorname{sgn} G_4x(\xi_1) = -\operatorname{sgn} L_1x(\xi_1)L_2x(\xi_1) = \operatorname{sgn} G_4x(\xi_2) = -\operatorname{sgn} L_1x(\xi_2)L_2x(\xi_2)$. Because $L_1x(t) \neq 0$ on (t_1,t_2), we get that $\operatorname{sgn} L_2x(\xi_1) = \operatorname{sgn} L_2x(\xi_2)$. But this is a contradiction, because $\operatorname{sgn} L_2x(t) \neq \operatorname{sgn} L_2x(t)$. This contradiction prove our statement.
$t\in(\tau_1,\tau_2) \quad t\in(\tau_1,\tau_3)$

The function $G_4x(t)$ can be written also in the form $G_4x(t) = a_1(t)[x(t)L_2'x(t) - x'(t)L_2x(t)]$. That means that $W(x,L_2x) = x(t)L_2'x(t) - x'(t)L_2x(t)$ has a constant sign on (T_x,∞). From this it follows that $x(t)$ and $L_2x(t)$ separate their zeros, i.e. between two consecutive zeros of $x(t)$ greater than T_x lies precisely one zero of $L_3x(t)$. Let $(T_x <)t_1 < t_2$ be two consecutive zeros of $L_2x(t)$. Then by Rolle's theorem between t_1 and t_2 lies at least one zero of $L_3x(t)$. Suppose that there is more than one zero of $L_3x(t)$. Following A) we know that the number of these zeros is odd. Let $(t_1 <)\ \tau_1 < \tau_2 < \tau_3\ (< t_2)$ be three consecutive zeros of $L_3x(t)$. Then, using Rolle's theorem we get firstly that there exist two zeros, ξ_1, ξ_2, of $L_4x(t)$ and $x(t)$ have the same zeros, using the above statement we get the existence of a zero of $L_2x(t)$ between ξ_1 and ξ_2 and therefore between t_1 and t_2. But this is a contradiction.

In the same way as above, using Rolle's theorem and the property A) it may be proved that also between two consecutive zeros of $L_3x(t)$ there is precisely one zero of $L_4x(t)$ and therefore precisely one zero of $x(t)$. We omit this proof.

In summary, we have proved that between two consecutive zeros of $L_ix(t)$ there is precisely one zero of $L_{i+1}x(t)$, $i = 0, 1, 2, 3$. This fact implies that also between two consecutive zeros of $L_{i+1}x(t)$ there is precisely one zero of $L_ix(t)$, $i = 0, 1, 2, 3$. In fact, let $(T_x <)\ \tau_1 < \tau_2$ be two consecutive zeros of $L_{i+1}x(t)$. Evidently, $L_ix(t)$ can not have two zeros between τ_1 and τ_2. If not, then $L_ix(t) \neq 0$ for $t \in (\tau_1,\tau_2)$ and because $x(t)$ is oscillatory, there are two consecutive zeros, say $t_1 < t_2$, of $L_ix(t)$ such that $t_1 < \tau_1 < \tau_2 < t_2$. But this is impossible, because between two consecutive zeros of $L_ix(t)$ there is precisely one zero of $L_{i+1}x(t)$.

From this last statement and from the fact that $x(t)$ and $L_4x(t)$ have the same zeros, and from the above statements also, we get that $L_ix(t)$ and $L_jx(t)$, $i, j = 0, 1, 2, 3$, $i \neq j$, separate their zeros.

<u>Theorem 6</u>. Let $x(t) \in (17)$ be an oscillatory solution. Let T_x be as in Theorem 5. Let there exist a zero $\rho > T_x$ of $x(t)$ such that

(19) $$\operatorname{sgn} L_1x(\rho) = \operatorname{sgn} L_3x(\rho) \neq \operatorname{sgn} L_2x(\rho).$$

Then for each zero ρ' of $x(t)$ such that $\rho' > \rho$ one has

(19') $$\operatorname{sgn} L_1x(\rho') = \operatorname{sgn} L_3x(\rho') \neq \operatorname{sgn} L_2x(\rho').$$

Proof. Let ρ_1 be the first zero of $x(t)$ following ρ. Assume (19) holds Then $x(t) \neq 0$ on (ρ,ρ_1) and $L_1x(t), L_2x(t), L_3x(t)$ have precisely one zero in (ρ,ρ_1) following Theorem 5. Therefore, $\operatorname{sgn} L_1x(\rho_1) = \operatorname{sgn} L_3x(\rho_1) \neq \operatorname{sgn} L_2x(\rho_1)$. The validity of (19') follows by induction.

Theorem 7. Let $x(t) \in (17)$ be an oscillatory solution. Let T_x be as in Theroem 5. Let there exist a zero ρ of $x(t)$, $\rho > T_x$ such that

(20) $$\operatorname{sgn} L_1x(\rho) = \operatorname{sgn} L_2x(\rho) = \operatorname{sgn} L_3x(\rho)$$

holds. Then for each zero ρ' of $x(t)$ such that $\rho' > \rho$

(20') $$\operatorname{sgn} L_1x(\rho') = \operatorname{sgn} L_2x(\rho') = \operatorname{sgn} L_3x(\rho')$$

holds.

Proof. Let ρ_1 be the first zero of $x(t)$ following ρ. Let (20) hold. Then $x(t) \neq 0$ on (ρ,ρ_1) and $L_1x(t), L_2x(t), L_3x(t)$ have precisely one zero in (ρ,ρ_1) following Theorem 5. Thus for $\rho' = \rho_1$, (20') holds. By induction we get the validity of our Theorem.

Definition. Denote by U_1 the set of all oscillatory solutions $x(t)$ of (17) such that there is a number A_x such that for every zero $\rho > A_x$ of $x(t)$, (19) holds. Denote by U_2 the set of all oscillatory solutions $x(t)$ of (17) such that there is a number B_x such that for every zero $\rho > B_x$ of $x(t)$, (20) holds.

Theorem 8. $U_1 \cap U_2 = \phi$ and $U_1 \cup U_2$ is the set of all oscillatory solutions of (17).

Proof. Let $x(t)$ be an oscillatory solution of (17) and let $x(t) \overline{\in} U_1$. Let T_x be such as in Theorem 5. Then for every $T > T_x$ there exists a zero ρ of $x(t)$ such that $\rho > T$ and (19) doesn't hold. Then there are three cases which can occur:

a) $\operatorname{sgn} L_1x(\rho) \neq \operatorname{sgn} L_3x(\rho) = \operatorname{sgn} L_2x(\rho)$

b) $\operatorname{sgn} L_1x(\rho) = \operatorname{sgn} L_2x(\rho) \neq \operatorname{sgn} L_3x(\rho)$

c) $\operatorname{sgn} L_1x(\rho) = \operatorname{sgn} L_2x(\rho) = \operatorname{sgn} L_3x(\rho)$.

The cases a) and b) cannot occur. In fact, let be $L_1x(\rho) > 0$, $L_3x(\rho) < 0$.

Let ρ_1 be the first zero of $x(t)$ following ρ. Then for $t \in (\rho,\rho_1)$ we have $x(t) \neq 0$ and $L_4x(t) < 0$. Thus, $L_3x(t)$ is decreasing and being negative in ρ it cannot have a zero in (ρ,ρ_1), which is a contradiction. The same contradiction will be reached if we suppose that $L_1x(\rho) < 0$, $L_3x(\rho) > 0$. Thus, case c) holds, i.e., (20) is proved. But then from Theorem 7 we have that in each case $\rho' > \rho$ of $x(t)$, (20) holds. Therefore $x(t) \in U_2$.

Theorem 9. An oscillatory solution $x(t)$ of (17) belongs to U_1 if and only if $G_4x(t) > 0$ for all $t \in J$.

Proof. Let $x(t)$ be an oscillatory solution of (17) and let $x(t) \in U_1$. Then for every zero $\rho > A_x$ of $x(t)$ we have $G_4x(\rho) = -L_1x(\rho)L_2x(\rho) > 0$. $G_4x(t)$ being os constant sign we have that $G_4x(t) > 0$ for all $t \in J$.

Let now $x(t)$ be an oscillatory solution of (17) and let $G_4x(t) > 0$ hold for all $t \in J$. Then in each zero ρ of $x(t)$ we have $G_4x(\rho) = -L_1x(\rho)L_2x(\rho) > 0$. Thus, $\operatorname{sgn} L_1x(\rho) \neq \operatorname{sgn} L_2x(\rho)$. Let $\rho > T_x$, where T_x is the number from Theorem 5 and let ρ_1 be the first zero of $x(t)$ following ρ. Then $x(t) \neq 0$ for $t \in (\rho,\rho_1)$ and $\operatorname{sgn} x(t) = \operatorname{sgn} L_1x(\rho) \neq \operatorname{sgn} L_4x(t)$. If $\operatorname{sgn} L_3x(\rho) \neq \operatorname{sgn} L_1x(\rho)$, then $L_3x(t)$ cannot have a zero in (ρ,ρ_1) which is a contradiction with Theroem 5. Thus, it follows that $\operatorname{sgn} L_1x(\rho) = \operatorname{sgn} L_3x(\rho)$. Applying Theorem 6 we get $x(t) \in U_1$.

As a complement to the Theorem 9 we get

Theorem 10. An oscillatory solution $x(t)$ of (17) belongs to U_2 if and only if $G_4x(t) < 0$ for all t great enough.

Let us not consider the relations between the sets U_1, U_2 and S_1, S_3. Let S_1^x be the set of all oscillatory solutions of (17) belonging to S_1 and S_3^x be the set of all oscillatory solutions of (17) belonging to S_3.

Theorem 11. $U_1 = S_1^x$, $U_2 = S_3^x$.

Proof. Let $x(t) \in S_1^x$. Then there exists a number t_0 such that $s(x,t+) = 1$ for all $t > t_0$ and $L_ix(t)$, $i = 0, 1, 2, 3$, have only simple zeros on (t_0,∞). Let be $x(\rho) = 0$, $\rho > t_0$. Suppose that $L_1x(\rho) > 0$. Then there exists $\varepsilon > 0$ such that for $t \in (\rho, \rho + \varepsilon)$ we have $L_ix(t) \neq 0$, $i = 0, 1, 2, 3$ and $x(t) > 0$, $L_1x(t) > 0$. Therefore, in the sequence $\{x(t), -L_1x(t), L_2x(t), -L_3x(t), L_4x(t)\}$ we have certainly one sign change on the first place. Because $x(t) \in S_1$ the sign changes cannot happen in other places. Thus, it must be $L_2x(t) < 0$, $L_3x(t) > 0$, $L_4x(t) < 0$ and in ρ we get the validity of (19). Similar reasoning gives us the validity of (19) also in the case if $L_1x(\rho) < 0$.

Let now $x(t)$ be an oscillatory solution of (17) and let T_0 be such that $S(x,t+) = \text{const.}$ for $t > T_0$. Suppose that in each zero $\rho > T_0$ of $x(t)$ (19) holds. Then there exists $\varepsilon_1 > 0$ such that for $t \in (\rho, \rho + \varepsilon_1)$ we have $x(t)L_1x(t)L_2x(t)L_3x(t) \neq 0$ and if $L_1x(\rho) > 0$ then $x(t) > 0$, $L_1x(t) > 0$, $L_2x(t) < 0$, $L_3x(t) > 0$, $L_4x(t) < 0$. It means that $S(x, \rho + 1) = 1$. Thus, $x(t) \in S_1^x$. If we suppose that $L_1x(\rho) < 0$, we will get $x(t) < 0$, $L_1x(t) < 0$, $L_2x(t) > 0$, $L_3x(t) < 0$, $L_4x(t) > 0$ for all $t \in (\rho, \rho + \varepsilon_1)$ and therefore $S(x, \rho+) = 1$ and $x(t) \in S_1^x$.

The equality $U_2 = S_3^x$ is then obvious.

<u>Theorem 12</u>. Suppose that in (17), $a_1(t) = a_2(t) = a(t)$. Then $U_1 = \{x(t) \in (17)$: $x(t)$ oscillatory, $|L_1x(t)|$ bounded$\} = \{x(t) \in (17)$: $x(t)$ oscillatory with $\lim_{t\to\infty} G_4x(t) = 0\}$.

<u>Proof</u>. Using the assumptions about the functions $a_i(t)$, $i = 1, 2$, the function $G_4x(t)$ can be written in the form $G_4x(t) = a(t)[x(t)L_2x(t) - L_1^2x(t)]'$. Suppose that $x(t) \in U_1$. Then $G_4x(t) > 0$ for all $t \in J$. Therefore, the function $x(t)L_2x(t) - L_1^2x(t)$ increases. Let $\rho_1 < \rho_2$ be two zeros of $L_2x(t)$. Then we get $-L_1^2x(\rho_1) < -L_x^2(\rho_2)$. It means that the sequence of maxima of $|L_1^2x(t)|$ decreases. Thus, $|L_1x(t)|$ is bounded.

Suppose now that $x(t) \in (17)$ is oscillatory and that $|L_1x(t)|$ is bounded and that $G_4x(t) < 0$ for all t great enough. Then there is $K > 0$ and $T_K > a$ such that

$$(21) \qquad a(t)[x(t)L_2x(t) - L_1^2x(t)]' < -K \quad \text{for all} \quad t > T_K.$$

Let $\rho_i > T_K$, $i = 1, 2, \ldots$ be the zeros of $\dot{x}(t)$ and $\lim \rho_i = \infty$ as $i \to \infty$. Then from (21) we get

$$-L_1^2x(\rho_i) + L_1^2x(\rho_1) > -K \int_{\rho_i}^{\rho_i} a^{-1}(t)dt \to -\infty \quad \text{for} \quad i \to \infty.$$

But this contradicts the assumption of boundedness of $L_1x(t)$.

Now suppose that $x(t) \in U_1$. Then $|L_1x(t)|$ is bounded. Suppose that $G_4x(t) > M > 0$ and that ρ_i, $i = 1, 2, \ldots$, is a sequence of zeros of $x(t)$ and $\lim_{i\to\infty} \rho_i = \infty$. Then integrating this last inequality we get the contradiction

$$-L_1^2x(\rho_i) + L_1^2x(\rho_1) > M \int_{\rho_i}^{\rho_i} a^{-1}(t)dt \to \infty \quad \text{as} \quad i \to \infty.$$

Thus, if $x(t) \in U_1$ then $\lim G_4x(t) = 0$ as $t \to \infty$.

Let now $x(t) \in (17)$ be oscillatory and assume $\lim_{t\to\infty} G_4x(t) = 0$. Then $G_4x(t)$ being nonincreasing must be positive. But then following Theorem 9, $x(t) \in U_1$ and therefore $|L_1x(t)|$ is bounded as it was proved above.

We have considered only the oscillatory solutions of (17). We now consider the nonoscillatory solutions of (17). In general we can divide the set of all nonoscillatory solutions of (E) into disjoint classes V_k, $k = 0, 1, \ldots, n-1$ (see [4]).

Definition. A nonoscillatory solution $x(t)$ of (E) belongs to the class V_k iff there exists a number T_x such that

a) $(-1)^{i+1}x(t)L_ix(t) > 0$, $i = k+1, k+2, \ldots, n-1$, for $t \geq T_x$,

b) $\lim L_kx(t)$ exists and is finite as $t \to \infty$,

c) $\lim L_ix(t) = 0$, $i = k+1, k+2, \ldots, n-1$, as $t \to \infty$,

d) $\lim L_ix(t) = \infty \operatorname{sgn} x(t)$ as $t \to \infty$, $i = 0, 1, \ldots, k-1$.

It is very easy to prove

Theorem 13. For $n = 4$: $V_0 \cup V_1 = S_1 - S_1^x$, $V_2 \cup V_3 = S_3 - S_3^x$.

Remark. Instead of the equation (E) we can consider the equation

$$\text{(E')} \qquad L_nx(t) + f(t)g(x(t), x'(t), \ldots, x^{(n-1)}(t)) = 0.$$

If we assume that $a_i(t) \in C^{(n-i)}(J)$, $a_i(t) > 0$, $i = 1, 2, \ldots, n-1$, $f(t) \in C(J)$ $f(t) > 0$, (2) holds and

$$\text{(3')} \quad g(x_0, x_1, \ldots, x_{n-1}) \in C(R^n),\ x_0g(x_0, x_1, \ldots, x_{n-1}) > 0 \text{ for all } x_0 \neq 0.$$

Then all that was said and proved for the equation (E) in this paper holds also for the equation (E').

REFERENCES

1. Brunovsky, P. and Mallet-Paret, J., Switching of optimal control and the equation $y^{(4)} + |y|^\alpha \operatorname{sgn} y = 0$, $0 < \alpha < 1$, to appear.

2. Sisolakova, D., Vlastnosti niektorych typov nelinearnych diferencialynch rovnic, Kandidatska praca, 1982, MFF-UK, Bratislava.

3. Elias, U., A classification of the solutions of a differential equation according to their asymptotic behavior, Proc. Royal Soc. Edinburgh, Sec. A.

4. Svec, M., Behavior of nonoscillatory solutions of some nonlinear differential equations, Acta Mathematica UC, XXXIX-1980, 115-130.

GLOBAL ASYMPTOTIC STABILITY IN EPIDEMIC MODELS

Horst R. Thieme
Universität Heidelberg, SFB 123
D-6900 Heidelberg, BR Deutschland

In [10] Hethcote, Yorke and Nold compare six prevention methods for gonorrhea using a multi-group epidemic model which only takes susceptible and infective individuals into account. Their study is based on the following global asymptotic stability result of Lajmanovich and Yorke in [12]: Either all solutions of the model vanish asymptotically as time tends to infinity or there exists a unique non-trivial equilibrium solution to which all non-trivial solutions of the model converge. A threshold condition determines which of these two possibilities actually occurs. This result allows, for the judgement of the effectivity of control methods, to confine oneself to the analysis of their impact on the equilibrium point.

The reduction of gonorrhea to a model only considering susceptible and infective individuals is justified in [10] by the argument that the periods of incubation and immunity are so short that they can be neglected. One aim of this paper consists in confirming this argument by showing that the introduction of short periods of incubation and immunity does not destroy the global stability of the model. Further we show that immigration into and emigration from the promiscuously active population (by individuals that start or cease to be promiscuously active) do not affect the global asymptotic stability of the model as long as immigration and emigration are balanced. Global asymptotic stability results concerning epidemic models for homogeneous populations (one-group models) have been derived in [9], [15], [8].

The other aim of this paper consists in presenting a method of transforming relatively complex epidemic models into a system of integral equations the right hand sides of which depend monotone increasing on the solution. This allows to derive renewal theorems (see [4], [7], [19]), to handle seasonal variations of the parameters of the model (see [1], [2], [14], [20]) and, if spatial spread is included, to prove the existence and uniqueness of travelling wave solutions (see [5], [7], [21]) and the existence of asymptotic speeds of spread (see [6], [16], [21]).

Our epidemic model can be represented schematically in the following way:

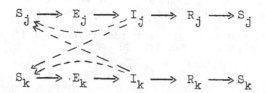

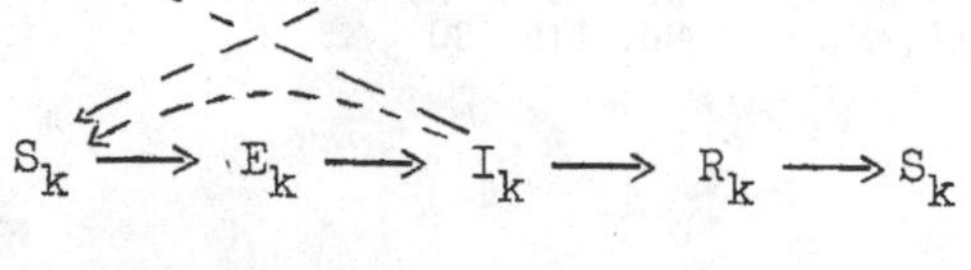

The population is assumed to be heterogeneous with respect to the disease , i.e. it consists of m subpopulations which react differently to the disease. In the gonorrhea case Hethcote et al. [10] subdivide the population according to sex, (promiscuous) sexual activity and symptomatic and asymptomatic infection. S_j, E_j, I_j, R_j denote the proportions of susceptible, exposed (incubating the disease), infected, and removed (immune) individuals in the jth subpopulation. The epidemic interaction between the subpopulations is due to the ability of infective individuals not only to infect susceptibles of their own, but also of other subpopulations. In the gonorrhea model [10], in which homosexual transmission is excluded, infectives cannot infect susceptibles of their own subpopulation. We confine our consideration to non-lethal diseases, but we include some population dynamics in so far as individuals may enter and leave the subpopulations (by birth and death or, in gonorrhea, by starting and ceasing to be promiscuously active). We assume, however, that the subpopulations remain constant in size.

The model equations have the following form:

$$S_j'(t) = -S_j(t)\,Q_j(t) - \mu_j\,S_j(t) + \mu_j + \rho_j\,R_j(t)$$

$$E_j'(t) = S_j(t)\,Q_j(t) - e^{-\mu_j \tau_j}\,(S_jQ_j)(t-\tau_j) - \mu_j E_j(t)$$

(1)

$$I_j'(t) = e^{-\mu_j \tau_j}\,(S_jQ_j)(t-\tau_j) - (\gamma_j + \mu_j)\,I_j(t)$$

$$R_j'(t) = \gamma_j I_j(t) - (\rho_j + \mu_j)\,R_j(t)\,.$$

(2) $$Q_j(t) = \sum_{j=1}^{m} \alpha_{jk}\,I_k(t)\,.$$

Hereby S_j, I_j are prescribed on $[-\tau_j, 0]$; further $E_j(0)$ and $R_j(0)$ are prescribed such that

(3) $$S_j(0) + E_j(0) + I_j(0) + R_j(0) = 1\,.$$

' denotes the derivative with respect to time. The number of infections S_jQ_j in group j is described by the usual law of mass action with Q_j indicating the infective impact on subpopulation j. α_{jk} gives the infective impact of infectives in subpopulation k on susceptibles in subpopulation j. The incubation period in subpopulation j is assumed to have a fixed length τ_j, whereas the durations of the periods of infectiousness and immunity are assumed to be exponentially distributed with mean durations $1/\gamma_j$ and $1/\varrho_j$. μ_j indicates the rate of immigration into and emigration from subpopulation j with all individuals that enter a subpopulation being susceptible.

Busenberg and Cooke [3] point out that systems like (1) do not describe epidemic models sufficiently. We add the integral condition

$$(4) \qquad E_j(0) = \int_{-\tau_j}^{0} S_j(s)\, Q_j(s)\, e^{\mu_j s}\, ds .$$

Adding the equations (1) and using (3) yields

$$(5) \qquad S_j(t) + E_j(t) + I_j(t) + R_j(t) = 1 .$$

By (4),

$$(6) \qquad E_j(t) = \int_{t-\tau_j}^{t} S_j(s)\, Q_j(s)\, e^{\mu_j(s-t)}\, ds \quad \text{for } t > 0,$$

because the right hand side of (6) also satisfies equation (1.2). In complete agreement with the epidemiological conception formula (6) states that those individuals incubate the disease at time t which have been infected between time $t-\tau_j$ and t and have not left the population. After these considerations the following result follows easily from Banach's fixed point theorem.

<u>THEOREM 1.</u> <u>Let</u> $\mu_j, \varrho_j, \gamma_j, \tau_j, \alpha_{jk}, E_j(0), R_j(0)$ <u>be non-negative</u>, $\mu_j + \gamma_j > 0$, <u>and let</u> $S_j(t), I_j(t)$ <u>be continuous non-negative functions of</u> $t \in [-\tau_j, 0]$ <u>such that</u> (3) <u>and</u> (4) <u>are satisfied</u> $(j,k = 1,\dots,m)$. <u>Then</u> S_j, E_j, I_j, R_j <u>can be uniquely extended to non-negative solutions of</u> (1), (2), (5), (6) <u>on</u> $(0,\infty)$.

In order to transform the system (1), (2), (5), (6) we fit (1.3) into (6) and integrate (1.4). Applying (5) and fitting the result into (1.3) yields

$$(7) \qquad I'(t) = [e^{-\mu\tau} - u(t) - I(t) - \int_0^t I(t-s)\, h(s)\, ds\,]\, Q(t) - (\gamma+\mu)\, I(t)$$

with

(8) $u(t) \longrightarrow 0$ for $t \longrightarrow \infty$, $u(t) \geq 0$ for $t \geq \tau$,

(9) $$h(s) = \gamma \begin{cases} e^{-\mu s} & ; \ 0 \leq s \leq \tau \\ e^{-\mu s + \varrho\tau - \varrho s} & ; \ s \geq \tau , \end{cases}$$

and the expression in square brackets being non-negative. Note that, for simplicity, we have omitted the index j. We write $I * a(t) = \int_0^t I(t-s)\, a(ds)$ with the measure $a(ds) = \delta_o(ds) + h(s)\, ds$. δ_o is the Dirac measure at zero. Then (7) can be formulated as

(10) $$I'(t) = f(t, I * a(t), Q(t)) - I * b(t)$$

with

(11) $$f(t,x,y) = \begin{cases} [e^{-\mu\tau} - u(t) - x]\, y + cx \ ; & \text{if } x \leq e^{-\mu\tau} - u(t), \ y \leq c \\ c[e^{-\mu\tau} - u(t)] \ ; & \text{otherwise} \end{cases}$$

for $x,y \geq 0$, and $b(ds) = c\, a(ds) + (\gamma + \mu)\, \delta_o(ds)$ with a constant $c \geq Q(t)$. We recall that $Q_j(t) \leq c_j := \sum_{k=1}^{m} \alpha_{jk}$. (10) can be transformed into the integral equation

(12) $$I(t) = I(0)\, r(t) + \int_0^t f(t-s, I * a(t-s), Q(t-s))\, r(s) ds$$

by using the resolvent kernel r(t) satisfying

(13) $r(0) = 1$, $r' = -r * b$.

The following lemma provides some important technical properties of f and r.

LEMMA 2. a) $f(t,x,y)$ *increases as* $x,y \geq 0$ *increase*.

b) *Let* $\int_0^\infty e^{\beta s}\, b(ds) \leq \beta < \infty$. *Then* $r(t) \geq e^{-\beta t}$ *and*

$$\int_0^\infty r(s) ds \cdot \int_0^\infty b(ds) = 1.$$

Proof: Part a) is obvious. For the proof of b) we define $r_\beta(t) = e^{\beta t}\, r(t)$, $\beta, t \geq 0$. Then $r_\beta(0) = 1$ and

$$(14) \qquad r_\beta'(t) = \beta\, r_\beta(t) - \int_0^t r_\beta(t-s)\, e^{\beta s}\, b(ds)\ .$$

If r_β is monotone increasing and β satisfies the assumption in b), then the right hand side of (14) is non-negative. Thus Banach's fixed point theorem implies that (14) has a monotone increasing solution r_β. r_β is the unique solution of (14). Thus $e^{\beta t}\, r(t) \geq 1$. The rest of the lemma follows by applying the Laplace transform to (13).

Before we formulate the main result of this paper we introduce the matrices

$$(15) \qquad A = (\alpha_{jk}), \quad B = (e^{-\mu_j \tau_j} (\gamma_j + \mu_j)^{-1}\, \alpha_{jk})\ .$$

Further we call the matrix A irreducible, if there exists some $n \in \mathbb{N}$ such that $\sum_1^n A^j$ is a strictly positive matrix. Epidemiologically this means that the disease affects all subpopulations regardless of the subpopulation in which it first breaks out.

THEOREM 3. Let the assumptions of theorem 1 be satisfied and let the spectral radius of B not exceed one. If it is one, let A be irreducible. Then $I_j(t) \longrightarrow 0$ for $t \longrightarrow \infty$.

THEOREM 4. Let the assumptions of theorem 1 be satisfied and let A be an irreducible matrix and the spectral radius of B exceed one.

a) Then there exists a unique strictly positive stationary solution of (1) and (2).

b) If $S_j(0) > 0$ for $j = 1,\dots,m$, $I_k(0) > 0$ for some $k \in \{1,\dots,m\}$ and if the ϱ_j are large and the τ_j small, then the solution of (1),...,(4) converges, for $t \longrightarrow \infty$, towards the strictly positive stationary solution.

In order not to consume too much space we only give a

Sketch of the proof of theorem 4b): It follows from (1) and (2) and the irreducibility of the matrix A that $I_j(t) > 0$ for $t \geq t_o$, with some $t_o > 0$. The theorem of Perron-Frobenius and the irreducibility of A imply the existence of $\lambda > 1$ and of a strictly positive vector $v \in \mathbb{R}^m$ such that $Bv = \lambda v$. See [13], I.6.2. We use this fact in order to show that $I_j(t) \geq \varepsilon$ for $t \geq t_o$, with some $t_o, \varepsilon > 0$. Then we define

$$\underset{\sim}{I} = \liminf_{t \to \infty} I(t), \quad \tilde{I} = \limsup_{t \to \infty} I(t)$$

and apply Fatou's lemma to (12). Take account of lemma 2. In this way we find a concave operator F on $[0, \infty)^m$ such that $\underset{\sim}{I} \geq F(\underset{\sim}{I})$, $\tilde{I} \leq F(\tilde{I})$ and $\check{I} = F(\check{I})$ with $\check{I}$ being the infective part of the stationary solution of (1), (2). Using the trick in [11], 6.2.2, we obtain $\underset{\sim}{I} = \check{I} = \tilde{I}$. An elaborate presentation of proving global asymptotic stability by monotone methods has been given in [17], [18], and [19] by the author.

Remark 5. The crucial points of this proof are the monotonicity of the functions f_j in (11) and the non-negativity of the kernels r_j in (13). Evaluating the assumption in lemma 2b) we can specify the assumptions in theorem 4 concerning ρ_j and τ_j:

(i) $$\rho_j - (c_j + \gamma_j) > 2\sqrt{c_j \gamma_j}$$

with $c_j = \sum_{k=1}^{m} \alpha_{jk}$, and

(ii) $$\tau_j \leq \sup\left\{ x^{-1} \ln(1 + \phi_j(x));\ c_j + \gamma_j < x < \rho_j \right\}$$

with

$$\phi(x) = \frac{x}{c\gamma\rho}\left([x - \rho][c + \gamma - x] - c\gamma\right).$$

(i) implies that

$$\phi((c + \rho + \gamma)/2) = \frac{c + \rho + \gamma}{8 c \gamma \rho}\left((\rho - [c + \gamma])^2 - 4c\gamma\right) > 0.$$

It is remarkable that the rates μ_j of immigration and emigration do not appear in these conditions. This means that balanced immigration and emigration into and from the subpopulations do not affect the global asymptotic stability of the model. In particular, if there is no incubation and no immunity ($\rho_j = \infty$, $\tau_j = 0$) the model is globally asymptotically stable.

ACKNOWLEDGEMENT: I thank S. Busenberg (Claremont) who posed the problem and helped to get this work started.

REFERENCES

[1] ARONSSON, G.; MELLANDER, I.: A deterministic model in biomathematics. Asymptotic behaviour and threshold conditions. Math. Biosciences 49 (1980), 207-222

[2] BUSENBERG, S.; COOKE, K.L.: Periodic solutions of a periodic nonlinear delay differential equation. SIAM J. Applied Math. 35 (1978), 704-721

[3] BUSENBERG, S.; COOKE, K.L.: The effect of integral conditions in certain equations modelling epidemics and population growth. J. Math. Biol. 10 (1980), 13-32

[4] DIEKMANN, O.: Limiting behaviour in an epidemic model. Nonlinear Anal., TMA, 1 (1977), 459-470

[5] DIEKMANN, O.: Thresholds and travelling waves for the geographical spread of infection. J. Math. Biol. 6 (1978), 109-130

[6] DIEKMANN, O.: Run for your life. A note on the asymptotic speed of propagation of an epidemic. J. Diff. Eq. 33 (1979), 58-73

[7] DIEKMANN, O.; KAPER, H.G.: On the bounded solutions of a nonlinear convolution equation. Nonlinear Anal., TMA, 2 (1978), 721-737

[8] GRIPENBERG, G.: On some epidemic models. Quart. Appl. Math. 39 (1981), 317-327

[9] HETHCOTE, H.W.: Qualitative analyses of communicable disease models. Math. Biosciences 28 (1976), 335-356

[10] HETHCOTE, H.W.; YORKE, J.A.; NOLD, A.: Gonorrhea modelling: A comparison of control methods. Math. Biosciences 58 (1982), 93-109

[11] KRASNOSEL'SKII, M.A.: Positive Solutions of Operator Equations. Groningen. Noordhoff 1964

[12] LAJMANOVICH, A.; YORKE, J.A.: A deterministic model for gonorrhea in a nonhomogeneous population. Math. Biosciences 28 (1976), 221-236

[13] SCHAEFER, H.H.: Banach Lattices and Positive Operators. Berlin - Heidelberg - New York: Springer 1975

[14] SMITH, H.L.: An abstract threshold theorem for one parameter families of positive noncompact operators. Preprint

[15] STECH, H.; WILLIAMS, M.: Stability in a class of cyclic epidemic models with delay. J. Math. Biol. 11 (1981), 95-103

[16] THIEME, H.R.: Asymptotic estimates of the solutions of nonlinear integral equations and asymptotic speeds for the spread of populations. J. reine angew. Math. 306 (1979), 94-121

[17] THIEME, H.R.: On a class of Hammerstein integral equations. Manuscr. math. 29 (1979), 49-84

[18] THIEME, H.R.: On the boundedness and the asymptotic behaviour of the non-negative solutions of Volterra-Hammerstein integral equations. Manuscr. math. 31 (1980), 379-412

[19] THIEME, H.R.: Renewal theorems for some mathematical models in epidemiology. Preprint

[20] VOLZ, R.: Global asymptotic stability of a periodic solution to an epidemic model. Preprint

[21] WEINBERGER, H.F.: Long-time behaviour of a class of biological models. Preprint

BIFURCATION AT MULTIPLE EIGENVALUES FOR EQUIVARIANT MAPPINGS

A. Vanderbauwhede
Instituut voor Theoretische Mechanica
Rijksuniversiteit Gent, Krijgslaan 281,
B-9000 Gent, Belgium

1. INTRODUCTION

In this contribution we give a number of bifurcation results at double eigenvalues for equivariant mappings. These results can in a sense be considered as generalizations of the Crandall-Rabinowitz theorem [1] on bifurcation from simple eigenvalues. The condition on the mixed derivative, appearing in the CR-theorem, will be replaced by a similar surjectivity condition. In the last section we will give a more geometrical interpretation of these technical conditions. Applications of our results include Hopf bifurcation and bifurcation of subharmonic solutions.

To set up the problem, let X and Z be two real Banach spaces, and G a compact group. Let $\Gamma : G \to L(X)$ and $\tilde{\Gamma} : G \to L(Z)$ be two representations of G, over X, respectively Z. (For more details, see e.g. [4,5]). We will consider nonlinear problems of the form

$$M(x,\lambda) = 0 \ , \qquad (1.1)$$

where $M : X \times \mathbb{R}^m \to Z$ is a smooth mapping satisfying the following :

(H1) (i) $M(0,\lambda) = 0 \quad , \quad \forall \lambda \in \mathbb{R}^m$;

(ii) M is *equivariant* with respect to $(G,\Gamma,\tilde{\Gamma})$, i.e.

$$M(\Gamma(g)x,\lambda) = \tilde{\Gamma}(g)M(x,\lambda) \quad , \quad \forall g \in G \ , \ \forall (x,\lambda) \ . \qquad (1.2)$$

The problem of bifurcation theory is to describe the nontrivial solutions of (1.1) near the branch of trivial solutions $\{(0,\lambda) \mid \lambda \in \mathbb{R}^m\}$. Because of (H1)(ii) such solutions will appear in *orbits* : if (x,λ) is a solution, then so is $(\Gamma(g)x,\lambda)$, for each $g \in G$.
In many applications the mapping M has the property that its linearization at the trivial solution, i.e. $L(\lambda) = D_xM(0,\lambda)$, is a Fredholm operator with zero index. Then (1.1) can only have nontrivial solutions near

those trivial solutions $(0,\lambda_0)$ for which $L(\lambda_0)$ has a nontrivial kernel. Let us assume that the origin is such a point, i.e. $L_0 = L(0)$ is a Fredholm operator, with $\dim N(L_0) = \operatorname{codim} R(L_0) = n > 0$. Since L_0 is equivariant one can find equivariant projections $P_0 \in L(X)$ and $Q_0 \in L(Z)$ such that :

$$R(P_0) = N(L_0) \quad , \quad N(Q_0) = R(L_0) \ . \tag{1.3}$$

It is easily seen that the action Γ leaves $N(L_0)$ invariant, and consequently Γ induces on $N(L_0)$ an n-dimensional representation of G. Similarly, $\tilde{\Gamma}$ induces on $R(Q_0)$ another n-dimensional representation of G; moreover, this representation is independent of Q_0 in the sense that for a different choice of Q_0 one obtains an equivalent representation. This allows us to formulate our second hypothesis, as follows :

(H2) (i) $L_0 = D_xM(0,0)$ is a Fredholm operator, with $\dim N(L_0) = \operatorname{codim} R(L_0) = n$;
(ii) the representation Γ is irreducible on $N(L_0)$;
(iii) if $Q_0 \in L(Z)$ is an equivariant projection with $N(Q_0) = R(L_0)$, then the representation $\tilde{\Gamma}$ on $R(Q_0)$ is equivalent to the representation Γ on $N(L_0)$.

To formulate our last hypothesis, consider the space $L_e(N(L_0),R(Q_0))$ of equivariant linear operators $A \in L(N(L_0),R(Q_0))$. This is a finite-dimensional space; in fact, one can show that $\dim L_e(N(L_0),R(Q_0)) = 1$ if n is odd, and = 1,2 or 4 if n is even. Now we assume :

(H3) (i) $m = \dim L_e(N(L_0),R(Q_0))$;
(ii) the map $Q_0D_\lambda D_xM(0,0)|_{N(L_0)} \in L(\mathbb{R}^m, L_e(N(L_0),R(Q_0)))$ is surjective.

This hypothesis ressembles very much the condition on the mixed derivative appearing in the CR-theorem; actually, by taking for G the trivial group our hypotheses (H1)-(H3) reduce exactly to the hypotheses of the CR-theorem. Remark also that if (H2)(iii) is not satisfied, then $\dim L_e(N(L_0),R(Q_0)) = 0$; so (H2)(iii) must be satisfied as soon as we can show that this dimension is strictly positive. A sufficient condition for this is the existence of $u_0 \in N(L_0)$ and $\tilde{\lambda} \in \mathbb{R}^m$ such that $D_\lambda D_xM(0,0).(u_0,\tilde{\lambda}) \notin R(L_0)$.

For a further discussion of our hypotheses we refer to section 5. In the next section we apply the Liapunov-Schmidt reduction to (1.1) and study the consequences of our hypotheses for the resulting bifurcation equations.

2. THE LIAPUNOV-SCHMIDT REDUCTION

Assume (H1)-(H3), and let P_0 and Q_0 be as in (1.3). By the standard argument of the Liapunov-Schmidt method the solution set of the equation $(I-Q_0)M(x,\lambda) = 0$ takes near $(0,0)$ the form of a finite-dimensional smooth manifold of the form $\{(u+v^*(u,\lambda),\lambda) \mid (u,\lambda) \in N(L_0) \times \mathbb{R}^m\}$, where $v^* : N(L_0) \times \mathbb{R}^m \to N(P_0)$ is a smooth equivariant mapping satisfying $v^*(0,\lambda) = 0$, $\forall\lambda$, and $D_u v^*(0,0) = 0$. Consequently, all solutions of (1.1) near the origin must belong to this same manifold, on which the equation (1.1) itself reduces to the *bifurcation equation* :

$$\tilde{F}(u,\lambda) \equiv Q_0 M(u+v^*(u,\lambda),\lambda) = 0 . \tag{2.1}$$

The mapping $\tilde{F} : N(L_0) \times \mathbb{R}^m \to R(Q_0)$ is smooth, equivariant and satisfies $\tilde{F}(0,\lambda) = 0$, $\forall\lambda$, and $D_u\tilde{F}(0,0) = 0$.

Because of (H2) there are isomorphisms $\chi \in L(\mathbb{R}^n, N(L_0))$ and $\zeta \in L(\mathbb{R}^n, R(Q_0))$ such that for each $g \in G$ we have $\chi^{-1}\circ\Gamma(g)\circ\chi = \zeta^{-1}\circ\Gamma(g)\circ\zeta \in O(n)$, the group of orthogonal operators on $\mathbb{R}^n$. Let

$$G_0 = \{g \in G \mid \Gamma(g)u = u, \ \forall u \in N(L_0)\} \tag{2.2}$$

and

$$H = \{\chi^{-1}\circ\Gamma(g)\circ\chi \mid g \in G\} . \tag{2.3}$$

G_0 is a normal subgroup of G, and all solutions of (1.1) near (0,0) will satisfy $\Gamma(g)x = x$, $\forall g \in G_0$. Also, the quotient group G/G_0 is isomorphic to H, a closed subgroup of $O(n)$ under which $\mathbb{R}^n$ is irreducible. Finally, we have that $\dim L_e(N(L_0),R(Q_0))$ equals the dimension of the space $L_e(\mathbb{R}^n) = \{A \in L(\mathbb{R}^n) \mid AS = SA, \ \forall S \in H\}$.

If we define $F : \mathbb{R}^n \times \mathbb{R}^m \to \mathbb{R}^n$ by $F(u,\lambda) = \zeta^{-1}F(\chi(u),\lambda)$, then equation (2.1) becomes equivalent to

$$F(u,\lambda) = 0 . \tag{2.4}$$

Summarizing, the mapping F has the following properties :

(P1) $F(0,\lambda) = 0$, $\forall\lambda \in \mathbb{R}^m$, and $D_uF(0,0) = 0$;

(P2) $F(Su,\lambda) = SF(u,\lambda)$, $\forall S \in H$;

(P3) $m = \dim L_e(\mathbb{R}^n)$, and $D_\lambda D_u F(0,0) \in L(\mathbb{R}^m, L_e(\mathbb{R}^n))$ is surjective.

This last property follows from (H3). In the next sections we will discuss the solution set of (2.4) near (0,0) for the cases n = 1 and n = 2; going the way back in the foregoing reduction procedure gives then the corresponding solution set of (1.1).

3. THE CASE n = 1

If n = 1, it follows from (P1) that F has the form

$$F(u,\lambda) = uF_1(u,\lambda) \ , \tag{3.1}$$

with $F_1(0,0) = 0$. Nontrivial solutions of (2.4) must satisfy $F_1(u,\lambda) = 0$. As for (P2) there are only two possibilities for the subgroup H of O(1) : either $H = SO(1) = \{1\}$, the trivial group, or $H = O(1) = \{1,-1\}$. In the first case (P2) is void, while in the second case we have $F_1(-u,\lambda) = F_1(u,\lambda)$. In both cases $\dim L_e(\mathbb{R}^n) = 1$, and the surjectivity property (P3) reduces to m = 1 and $D_\lambda F_1(0,0) \neq 0$. A simple application of the implicit function theorem gives then the following result :

<u>Theorem 1</u>. Let F in (2.4) satisfy (P1), (P2) and (P3), with n = 1. Then also m = 1, and the nontrivial solutions of (2.4) near the origin are given by a smooth curve of the form $\{(u,\lambda^*(u)) \mid u \in \mathbb{R}\}$, with $\lambda^*(0) = 0$.

Moreover, if $H = O(1)$, then this curve is symmetric with respect to the λ-axis : $\lambda^*(-u) = \lambda^*(u)$.

If H is trivial, theorem 1 gives us the Crandall-Rabinowitz theorem, while for $H = O(1)$ we obtain the classical pitchfork-bifurcation.

4. THE CASE n = 2

In order to treat the case n = 2 we will identify $\mathbb{R}^2$ with the complex plane $\mathbb{C}$, considered as a two-dimensional *real* vectorspace. So we consider F in (2.4) as a mapping from $\mathbb{C} \times \mathbb{R}^m$ into $\mathbb{C}$, and we will write $u \in \mathbb{R}^2 \cong \mathbb{C}$ in the form $\rho e^{i\theta}$. The basis of our analysis will be the equivariance property (P2), which will allow us to write F in an appropriate "normal form" from which we can obtain the solution set.

Up to a rotation in $\mathbb{C}$ there are essentially four different possibilities for the subgroup H : (1) O(2), the group of rotations and reflections in the plane; (2) SO(2), the group of rotations alone; (3) Δ_k (for some $k \geq 3$), the dihedral group generated by δ_k and σ, which are given by :

$$\delta_k u = e^{i2\pi/k} u \quad , \quad \sigma u = \bar{u} \; ; \tag{4.1}$$

and finally (4) ROT_k, the rotation group generated by δ_k, for some $k \geqslant 3$. In the cases (3) and (4) the condition $k \geqslant 3$ comes from the fact that $\mathbb{C}$ must be irreducible under the group action.

The following lemma's give us for each of these four possibilities the corresponding normal form of the mapping F; proofs will be given elsewhere.

Lemma 1. Let $F : \mathbb{C} \times \mathbb{R}^m \to \mathbb{C}$ be smooth and equivariant with respect to SO(2). Then there exists a unique smooth mapping $h : \mathbb{R} \times \mathbb{R}^m \to \mathbb{C}$ such that

$$F(u,\lambda) = h(\rho,\lambda)u \quad , \quad u = \rho e^{i\theta} \; ; \tag{4.2}$$

moreover :

$$h(-\rho,\lambda) = h(\rho,\lambda) \; . \tag{4.3}$$

In case F is equivariant with respect to O(2), then h is real-valued.

Lemma 2. Let $F : \mathbb{C} \times \mathbb{R}^m \to \mathbb{C}$ be smooth and equivariant with respect to ROT_k (for some $k \geqslant 3$). Then there exist unique smooth mappings h_i : $\mathbb{C} \times \mathbb{R}^m \to \mathbb{C}$ ($i = 1,2$) such that :

$$\text{(i)} \quad F(u,\lambda) = h_1(u,\lambda)u + h_2(u,\lambda).\bar{u}^{k-1} \; ; \tag{4.4}$$

$$\text{(ii)} \quad h_i(\tau u,\lambda) = h_i(u,\lambda) \quad , \quad \forall \tau \in \Delta_k \; , \; i = 1,2 \; . \tag{4.5}$$

If F is equivariant with respect to Δ_k ($k \geqslant 3$), then the mappings h_i are real-valued.

In our complex notation a general $A \in L(\mathbb{R}^2)$ takes the form $Au = \alpha.u + \beta.\bar{u}$, for some $\alpha,\beta \in \mathbb{C}$. From this it is easily seen that, for $H = SO(2)$ or ROT_k, we have $A \in L_e(\mathbb{R}^2)$ if and only if $\beta = 0$, while for $H = O(2)$ or Δ_k the condition becomes $\beta = 0$ and $\alpha \in \mathbb{R}$. It then follows from (P3) that $m = 1$ or $m = 2$, depending on H.

If $H = SO(2)$ or $H = O(2)$, then we can write F in the form (4.2), and nontrivial solutions of (2.4) must satisfy the equation :

$$h(\rho,\lambda) = 0 \; . \tag{4.6}$$

It follows from (P1) that $h(0,0) = 0$. If $H = O(2)$, then h is real-valued and (4.6) is just one single scalar equation. In that case (P3) implies that $m = 1$ and $D_\lambda h(0,0) \neq 0$, so that (4.6) can be solved by the implicit function theorem.

Theorem 2. Let F in (2.4) satisfy (P1)-(P3), with $n = 2$ and $H = O(2)$. Then $m = 1$, and the nontrivial solutions of (2.4) near the origin are given by a smooth submanifold of the form $\{(\rho e^{i\theta}, \lambda^*(\rho)) \mid \rho \in \mathbb{R},\ \theta \in \mathbb{R}\}$, where $\lambda^*(0) = 0$ and $\lambda^*(-\rho) = \lambda^*(\rho)$.

If $H = SO(2)$, then h in (4.6) is complex-valued, i.e. $h = h_1 + ih_2$, and we have to split (4.6) into its real and imaginary parts. Then (P3) implies that $m = 2$ and that the mapping $\lambda \mapsto (h_1(0,\lambda), h_2(0,\lambda))$ has rank 2 at $\lambda = 0$. Again the implicit function theorem gives the desired result :

Theorem 3. Let F in (2.4) satisfy (P1)-(P3), with $n = 2$ and $H = SO(2)$. Then $m = 2$, and the nontrivial solutions of (2.4) near the origin are given by a smooth submanifold of the form $\{(\rho e^{i\theta}, \lambda^*(\rho)) \mid \rho \in \mathbb{R},\ \theta \in \mathbb{R}\}$, where $\lambda^* : \mathbb{R} \to \mathbb{R}^2$ is such that $\lambda^*(0) = 0$ and $\lambda^*(-\rho) = \lambda^*(\rho)$.

An application of theorem 3 (or better : of the corresponding result for (1.1)) is the Hopf bifurcation theorem; although in its usual formulation Hopf bifurcation is studied for one-parameter families of equations, one has to introduce a second (time-scaling) parameter to bring the problem in the form (1.1) (see e.g. [5]). Our hypotheses (H1)-(H3) reduce for this particular problem to the usual hypotheses of Hopf's theorem; in particular, (H3) becomes equivalent to the transversality condition of that theorem. One obtains an example of the situation of theorem 2 when studying periodic solutions of autonomous reversible systems; the one parameter is then also a time-scaling parameter, giving the period of the solutions.

Next we turn to the case $H = ROT_k$ $(k \geq 3)$. Writing F in the form (4.4), multiplying by $\bar{u}$ and dividing by ρ^2, we see that nontrivial solutions have to satisfy the equation :

$$H(\rho,\theta,\lambda) \equiv h_1(u,\lambda) + \rho^{k-2} e^{-ik\theta} h_2(u,\lambda) = 0 . \qquad (4.7)$$

The function H is complex-valued ($H = H_1 + iH_2$), $H(0,\theta,0) = 0$ and $H(\rho,\theta+2\pi/k,\lambda) = H(\rho,\theta,\lambda) = H(-\rho,\theta+\pi,\lambda)$. (P3) implies that $m = 2$, while the mapping $\lambda \mapsto (H_1(0,\theta,\lambda), H_2(0,\theta,\lambda))$ has rank 2 at $\lambda = 0$. From this we get the following result.

Theorem 4. Let F in (2.4) satisfy (P1)-(P3), with $n = 2$ and $H = ROT_k$ ($k \geq 3$). Then $m = 2$, and the nontrivial solutions of (2.4) near the origin are given by a smooth submanifold of the form $\{(\rho e^{i\theta}, \lambda^*(\rho,\theta)) \mid \rho \in \mathbb{R}, \theta \in \mathbb{R}\}$, where $\lambda^*(0,0) = 0$ and $\lambda^*(\rho,\theta + 2\pi/k) = \lambda^*(\rho,\theta) = \lambda^*(-\rho,\theta+\pi)$.

The problem of bifurcation of subharmonic solutions for periodic differential equations leads to bifurcation equations of the form (4.7) (see [2] and [5]), and so it should be possible to apply theorem 4. Usually, this problem is formulated with a single scalar parameter, so that our hypothesis (H3) is not satisfied. Even then it is possible to obtain the usual results on bifurcation of subharmonic solutions from an analysis of the equation (4.7).

Now consider the last case $H = \Delta_k$ ($k \geq 3$). Again nontrivial solutions of (2.4) have to satisfy an equation of the form (4.7), but now h_1 and h_2 are real-valued. To simplify the discussion, let us suppose that $h_2(0,0) \neq 0$; this is a condition which involves the nonlinear part of M, contrary to our hypotheses (H2)-(H3) which involve only the linearization $L(\lambda)$. Then, for $\rho \neq 0$ and sufficiently small, $(\rho e^{i\theta}, \lambda)$ can only be a solution of (4.7) if $\sin k\theta = 0$. For such θ (4.7) reduces to a single scalar equation, which can be solved for λ. A detailed analysis gives the following result.

Theorem 5. Let F in (2.4) satisfy (P1)-(P3), with $n = 2$ and $H = \Delta_k$ ($k \geq 3$) Then $m = 1$, and (2.4) has near the origin at least the following branches of nontrivial solutions :

(i) if k is odd, there are k branches of the form $\{(\rho e^{i2\pi\ell/k}, \lambda^*(\rho)) \mid \rho \in \mathbb{R}\}$, with $\lambda^*(0) = 0$, and for $\ell = 0,1,\ldots,k-1$;

(ii) if k is even, there are again k branches, but they split into two times k/2 branches, of the form $\{(\rho e^{i2\pi\ell/k}, \lambda_1^*(\rho)) \mid \rho \in \mathbb{R}\}$, respectively $\{(\rho e^{i\pi(2\ell+1)/k}, \lambda_2^*(\rho)) \mid \rho \in \mathbb{R}\}$, with $\ell = 0,1,\ldots,k/2 - 1$; the functions λ_i^* satisfy $\lambda_i^*(0) = 0$ and $\lambda_i^*(-\rho) = \lambda_i^*(\rho)$, $(i = 1,2)$.

If $h_2(0,0) \neq 0$ in (4.7), these are the only nontrivial solutions of (2.4) near the origin.

As a possible application of this result we mention some recent work of Loud [3], who studied a problem of bifurcation of subharmonic solutions for periodic equations showing some additional symmetry. He finds precisely the kind of branches that are given by case (ii) of theorem 5.

5. AN ALTERNATIVE FORMULATION OF THE HYPOTHESES

In this last section we describe briefly an approach to the bifurcation problem (1.1) which will allow us to give a more geometrical interpretation of our hypotheses (H2)-(H3). We have already remarked that these hypotheses only involve the linearization $L(\lambda)$ of M at the branch of trivial solutions. This same linearization also plays a crucial role in the approach which we develop now.

Denote by $\mathcal{L}_e(X,Z)$ the space of equivariant operators $L \in \mathcal{L}(X,Z)$; let $\mathcal{F}$ be the open subset of $\mathcal{L}_e(X,Z)$ formed by the Fredholm operators with zero index, and denote by $\mathcal{F}_1$ the subset of all $L \in \mathcal{F}$ with $\dim N(L) > 0$. Let $L_0 \in \mathcal{F}_1$, and let $Q_0 \in \mathcal{L}(Z)$ be an equivariant projection such that $N(Q_0) = R(L_0)$. Then there exists an open neighbourhood U of L_0 in $\mathcal{L}_e(X,Z)$ such that $\mathcal{F}_1 \cap U$ is a finite union of disjoint smooth submanifolds of $\mathcal{L}_e(X,Z)$ with finite codimension. L_0 itself belongs to the sheet with the highest codimension, which is equal to $\dim \mathcal{L}_e(N(L_0),R(Q_0))$. All other elements L of this sheet have the same structure as L_0, in the sense that there exist equivariant automorphisms $S \in \mathcal{L}(X)$ and $T \in \mathcal{L}(Z)$ such that $L = T \circ L_0 \circ S$. If L_0 satisfies the hypotheses (H2)(ii) and (iii), then $U \cap \mathcal{F}_1$ coincides with the sheet we just described, and the codimension of this sheet is strictly positive. If not, then either $\mathcal{F}_1$ fills up the whole neighbourhood U, or $U \cap \mathcal{F}_1$ contains some further sheets with strictly lower codimension; along these other sheets one has $\dim N(L) < \dim N(L_0)$, and L_0 belongs to the closure of these sheets in $\mathcal{L}_e(X,Z)$.

Now suppose that M in (1.1) satisfies (H1), and define $L : \mathbb{R}^m \to \mathcal{L}_e(X,Z)$ by $L(\lambda) = D_x M(0,\lambda)$. Suppose that L takes its values in the open set $\mathcal{F}$. From the point of view of bifurcation we are interested in those parameter-values λ for which $L(\lambda) \in \mathcal{F}_1$. Now it is easily seen that under the hypotheses (H2)-(H3) the linearized equation

$$L(\lambda).x = 0 \tag{5.1}$$

has only nontrivial solutions for $\lambda = 0$, i.e. $\lambda = 0$ is an isolated point of $L^{-1}(\mathcal{F}_1)$. Furthermore, for generic mappings M the associated mapping L will at $\lambda = 0$ be transversal to the set $\mathcal{F}_1$; this transversality is analytically expressed by the surjectivity condition in (H3). With these elements at hand one can show that our hypotheses (H2)-(H3) are equivalent to the following :

(H*) $\lambda = 0$ is an isolated point of $L^{-1}(\mathcal{F}_1)$, and at $\lambda = 0$ the mapping L is transversal to $\mathcal{F}_1$.

From this we see that our hypotheses combine a generic condition (the transversality) with the condition that the solution set of the linearized problem (5.1) is as simple as possible.

ACKNOWLEDGEMENT

We like to thank Professor R. Mertens for his support and continuous interest in our work.

REFERENCES

1. M.G. CRANDALL & P.H. RABINOWITZ. Bifurcation from simple eigenvalues. J. Funct. Anal. 8 (1971), 321-340.

2. G. IOOSS & D. JOSEPH. Elementary stability and bifurcation theory. Springer-Verlag, New York, 1980.

3. W.S. LOUD. Subharmonic solutions of second order equations arising near harmonic solutions. Preprint 1982.

4. D.H. SATTINGER. Group theoretic methods in bifurcation theory. Lecture Notes in Math., Vol. 762, Springer-Verlag, Berlin, 1979.

5. A. VANDERBAUWHEDE. Local bifurcation and symmetry. Research Notes in Math., Pitman, London. To appear.

LINEAR STABILITY OF BIFURCATING BRANCHES OF EQUILIBRIA

José M. Vegas
Departamento de Ecuaciones Funcionales
Facultad de Matemáticas
Universidad Complutense
Madrid-3 SPAIN

ABSTRACT

We analyze some cases in which the Bifurcation Function obtained by applying the Liapunov-Schmidt method to a nonlinear O.D.E. problem can be used to discuss the stability properties of the equilibrium points.

1. INTRODUCTION AND STATEMENT OF RESULTS

We consider the differential equation

$$\dot{x} = X(x,y,\varepsilon) \qquad (1)$$
$$\dot{y} = Y(x,y,\varepsilon)$$

where $x \in R^n$, $y \in R^m$, ε is a small parameter in a Banach space E, X and Y are C^1 functions in all variables, and

(2) $X(0,0,\varepsilon) = 0$, $Y(0,0,\varepsilon) = 0$ for all ε,

(3) $X_x(0,0,0) = 0$,

(4) $Y_y(0,0,0)$ is a stable matrix.

(Subscripts denote partial differentiations. By a "stable matrix" we mean a matrix all whose eigenvalues have strictly negative real parts.)

In order to study the possible bifurcation of (nontrivial) equilibrium solutions of (1) at $\varepsilon = 0$, we apply the Liapunov-Schmidt method: By using condition (4) and applying the Implicit Function Theorem, we obtain a unique function $y = y^*(x,\varepsilon)$ as the solution of the equation $Y(x,y,\varepsilon) = 0$, i.e.,

(5) $Y(x,y^*(x,\varepsilon),\varepsilon) = 0$ for $|x|,|\varepsilon|$ small.

By substituting this function into the first equation in (1), we define

(6) $$\tilde{X}(x,\varepsilon) \overset{\text{def}}{=} X(x,y^*(x,\varepsilon),\varepsilon)$$

which is called the Bifurcation Function since (x_0,y_0,ε_0) is an equilibrium solution of (1) if and only if $y_0 = y^*(x_0,\varepsilon_0)$ and $\tilde{X}(x_0,\varepsilon_0) = 0$. (We are always assuming that $|x_0|$, $|y_0|$, $|\varepsilon_0|$ are small enough.)

Therefore, the zeros of $\tilde{X} = 0$ give us exactly the equilibrium solutions of (1). The problem is to obtain, if possible, some information about the stability of the equilibrium points thus computed. In other words, following de Oliveira and Hale [1] we consider the differential equation

(7) $$\dot{x} = \tilde{X}(x,\varepsilon)$$

and ask ourselves the following question: Is there any relationship between the stability properties of the equilibrium (x_0,ε_0) of (7) and the stability properties of $(x_0,y^*(x_0,\varepsilon_0),\varepsilon_0)$ as an equilibrium point of (1)?

Liapunov showed that, if (7) is a scalar equation (that is, n=1) and $\tilde{X}(x,0) = ax^q + O(|x|^{q+1})$, then the stability properties of $x = 0$ with respect to equation (7) are the same as those of $x = 0$, $y = 0$ with respect to system (1) (assuming $\varepsilon = 0$ in both cases)(see Bibikov [1]). The optimal result for the case $n = 1$ is due to de Oliveira and Hale [1], who show that (7) and the equation on the center manifold associated to (1) (see conditions (3) and (4)) have the same zeros and the same signs between zeros, whereby the flow defined by (7) and the flow on the center manifold are completely equivalent.

When $n \geq 2$, some generalizations of the method of Liapunov, based on the construction of a suitable Liapunov function, are available (see, for instance, Bibikov [1]). On the other hand, Golubitsky and Schaeffer [1] obtain an interesting result on the relationships between equation (7) and the equation on the center manifold: if $y = h(x,\varepsilon)$ is a local center manifold for (1), then there exists a matrix-valued smooth function $\sigma_{x,\varepsilon}$, with $\sigma_{0,0} = I$, such that

(8) $$\tilde{X}(x,\varepsilon) = \sigma_{x,\varepsilon} X(x,h(x,\varepsilon),\varepsilon).$$

When $n = 1$, this implies the theorem of de Oliveira and Hale, but if $n \geq 2$, (8) does not give any information with respect to the stability problem in general (see the counterexample below), although it can be very useful in some special cases (see Golubitsky and Schaeffer [1]).

In general, however, <u>no generalization of the result of de Oliveira and Hale is possible for</u> $n \geq 2$, as the following counterexample shows:

$$(9)\qquad \begin{bmatrix} \dot{x}_1 \\ \dot{x}_2 \\ \dot{y} \end{bmatrix} = \begin{bmatrix} 0 & \varepsilon & \varepsilon^3 \\ -\varepsilon & 0 & \varepsilon \\ \varepsilon & \varepsilon^3 & -1 \end{bmatrix} \begin{bmatrix} x_1 \\ x_2 \\ y \end{bmatrix}$$

Here, $n = 2$, $m = 1$, the origin $(0,0,0)$ is stable for $\varepsilon > 0$ sufficiently small (the characteristic polynomial is

$$\lambda^3 + \lambda^2 + (\varepsilon^2 - 2\varepsilon^4)\lambda + \varepsilon^2 - \varepsilon^3 + \varepsilon^7),$$

but

$$(10)\qquad \tilde{X}(x_1,x_2,\varepsilon) = \begin{bmatrix} \varepsilon^4 & \varepsilon+\varepsilon^6 \\ -\varepsilon+\varepsilon^2 & \varepsilon^4 \end{bmatrix} \begin{bmatrix} x_1 \\ x_2 \end{bmatrix}$$

and equation (7) is unstable for $\varepsilon > 0$ small.

As the discussion below will show, the reason for this to happen is that the eigenvalues of the linear operator (10) have the form $\pm\varepsilon i + O(\varepsilon^2)$, "too close" to the imaginary axis. Would it be possible, if we had some control on the distance of the eigenvalues to the imaginary axis, to show that a situation like the one just discussed cannot occur? The answer is "yes":

<u>Notation</u>: For a square matrix A, $|A|$ will denote its operator norm, $\sigma(A)$ its spectrum, and $d(A)$ = distance between $\sigma(A)$ and the imaginary axis.

<u>Theorem 1</u>. <u>For every</u> $r > 0$ <u>there exists</u> $\mu_0 = \mu_0(r)$ <u>such that if</u> (x_0,ε_0) <u>is a hyperbolic equilibrium point of equation (7) with</u> $|x_0| < \mu_0(r)$, $|\varepsilon_0| < \mu_0(r)$ <u>and the matrix</u> $\tilde{X}_x(x_0,\varepsilon_0)$ <u>satisfies</u>:

$$(11)\qquad d(\tilde{X}_x(x_0,\varepsilon_0)) > r|\tilde{X}_x(x_0,\varepsilon_0)|$$

<u>then</u> $(x_0,y^*(x_0,\varepsilon_0),\varepsilon_0)$ <u>is a hyperbolic equilibrium point of (1), and the dimensions of the corresponding unstable manifolds coincide.</u>

<u>In particular</u>, (x_0,ε_0) <u>is stable for (7) if and only if</u> $(x_0,y^*(x_0,\varepsilon_0),\varepsilon_0)$ <u>is stable for (1)</u>.

A useful consequence of this theorem is the following:

Corollary 2. *If $\varepsilon \in R$ and $x(\varepsilon)$ is a smooth branch of equilibrium solutions of (1) such that the matrix $\frac{\partial}{\partial\varepsilon}\tilde{X}_x(x(\varepsilon),\varepsilon)|_{\varepsilon=0}$ has no eigenvalue with zero real part and has k eigenvalues with positive real parts, then, for $|\varepsilon|$ sufficiently small, $(x(\varepsilon),y^*(x(\varepsilon),\varepsilon),\varepsilon)$ is a hyperbolic equilibrium point with a k-dimensional unstable manifold.*

2. PROOF OF THEOREM 1

Let (x_0,ε) be an equilibrium point of (7), and denote $y^*(x_0,\varepsilon)$ by y_0. Then we have

$$\tilde{X}_x(x_0,\varepsilon) = X_x(x_0,y_0,\varepsilon) + X_y(x_0,y_0,\varepsilon)y^*_x(x_0,\varepsilon) \tag{12}$$

From (5) we obtain

$$y^*_x(x_0,\varepsilon) = -Y_y(x_0,y_0,\varepsilon)^{-1}Y_x(x_0,y_0,\varepsilon) \tag{13}$$

Thus, (12) becomes

$$\tilde{X}_x(x_0,\varepsilon) = X_x - X_yY_y^{-1}Y_x \quad \text{at} \quad (x_0,y_0,\varepsilon) \tag{14}$$

Let us call $C(\varepsilon) = X_x$, $D(\varepsilon) = X_y$, $E(\varepsilon) = Y_x$, $F(\varepsilon) = Y_y$ (all functions evaluated at the point (x_0,y_0,ε). Then,

$$\tilde{X}_x(x_0,\varepsilon) = C(\varepsilon) - D(\varepsilon)F(\varepsilon)^{-1}E(\varepsilon) . \tag{15}$$

On the other hand, the linearization of (1) about the equilibrium point (x_0,y_0,ε) is given by the system

$$\begin{bmatrix} \dot{x} \\ \dot{y} \end{bmatrix} = \begin{bmatrix} C(\varepsilon) & D(\varepsilon) \\ E(\varepsilon) & F(\varepsilon) \end{bmatrix} \begin{bmatrix} x \\ y \end{bmatrix} \tag{16}$$

Our task is, therefore, to compare the eigenvalues of the matrices in (15) and (16). This is done in the following lemma, which concludes the proof of Theorem 1:

Lemma 3. Let F_0 be an $n\times n$ stable matrix. Then, for every $r > 0$ there exists $\mu_0 = \mu_0(r) > 0$ such that, if C, D, E and F have dimensions $n\times n$, $n\times m$, $m\times n$ and $m\times m$, respectively, and satisfy $|C|$, $|D|$, $|E|$, $|F-F_0| < \mu_0(r)$, and, furthermore, $d(C) > r|C|$, then the matrices $C - DF^{-1}E$ and $\begin{bmatrix} C & D \\ E & F \end{bmatrix}$ have the same number of eigenvalues with positive real parts, each of them counted according to its multiplicity.

<u>Proof of Lemma 3</u>: For an arbitrary $m \times n$ matrix we have

$$(17)\quad \begin{bmatrix} I & 0 \\ -M & I \end{bmatrix}\begin{bmatrix} C & D \\ E & F \end{bmatrix}\begin{bmatrix} I & 0 \\ M & 0 \end{bmatrix} = \begin{bmatrix} C+DM & D \\ -MC-MDM+E+FM & -MD+F \end{bmatrix}$$

This matrix will be upper triangular if and only if

$$(18)\quad -MC - MDM + E + FM = 0$$

or, if F is nonsingular,

$$(19)\quad M = F^{-1}(-E + MC + MDM)$$

By the Implicit Function Theorem, this equation has a unique solution $M = M^*(C,D,E,F)$ which is C^∞ (and even analytic) in a neighborhood of $(0,0,0,F_0)$ and satisfies

$$(20)\quad M^*(C,D,E,F) = -F^{-1}E + O(|C|^2+|D|^2+|E|^2+|F-F_0|^2)$$

If $|C|$, $|D|$, $|E|$ and $|F-F_0|$ are sufficiently small, all the eigenvalues of $-M^*D+F$ have negative real parts, and we can restrict ourselves to comparing the matrices $C - DF^{-1}E$ and $C + DM^*(C,D,E,F) = C - DF^{-1}E + O(|C|^2+|D|^2+|E|^2+|F-F_0|^2)$.

In order to do this, we change variables, defining $C_1 = C - DF^{-1}E$ in a neighborhood of $(0,0,0,F_0)$. Then, $M = M^*(C_1,D,E,F)$ satisfies:

$$(21)\quad M + F^{-1}E - F^{-1}MC_1 - F^{-1}MDF^{-1}E - F^{-1}MDM = 0 .$$

By implicit differentiation, we find that

$$(22)\quad M^*(C_1,D,E,F) = -F^{-1}E + H(C_1,D,E,F)$$

where $|H(C_1,D,E,F)| = O(|C_1|^2+|D|^2+|E|^2+|F-F_0|^2)$. But, for $C_1 = 0$, uniqueness implies that $M^*(0,D,E,F) = -F^{-1}E$; hence

$$(23)\quad |H(C_1,D,E,F)| = |C_1| O(|C_1|+|D|+|E|+|F-F_0|) .$$

Therefore, $C + DM^*$ has the following form:

$$(24)\quad C + DM^* = C_1 + DH(C_1,D,E,F)$$

Let $r > 0$ be given. By the continuity property of the spectrum, for every $\eta > 0$ there exists $\mu_1 = \mu_1(r,\eta)$ such that, for any $n \times n$ matrix satisfying $|C| = 1$, $d(C) \geq r$, and for any $n \times n$ matrix A with norm $|A| < \mu_1(r,\eta)$ we have

$$(25)\quad \sigma(C+A) \subset \sigma(C) + \eta \quad .$$

Let now C_1 satisfy $d(C_1) \geq r|C_1|$. Define $\tilde{C}_1 = \frac{1}{|C_1|} C_1$; for

$\eta = \frac{r}{2}$, we obtain $\mu_2(r) = \mu_1(r,\frac{r}{2})$. Thus, if $\mu_0(r)$ is such that $|C_1|$, $|D|$, $|E|$, $|F-F_0| < \mu_0(r)$ implies $||C_1|^{-1}DH(C_1,D,E,F)| < \mu_2(r)$, then, under this restriction, we would have

(26) $$\sigma(\tilde{C}_1 + |C_1|^{-1}DH(C_1,D,E,F)) \subset \sigma(\tilde{C}_1) + \frac{r}{2} ,$$

which implies

(27) $$\sigma(C_1 + DH(C_1,D,E,F)) \subset \sigma(C_1) + \frac{r}{2}|C_1| \subset \sigma(C_1) + \frac{d}{2}$$

by our hypothesis in C_1. (27) and the continuity properties of the eigenvalues of finite-dimensional matrices imply the result.

3. A SPECIAL CASE: GRADIENT FLOWS

As a final remark, it seems interesting to point out that for the special class of gradient flows, the requirements on the behavior of the ratio $|\tilde{X}_x|/d(\tilde{X}_x)$ which are part of the hypotheses of Theorem 1 are no longer necessary. This is due to the fact that, if system (1) has the form (dropping ε for simplicity)

(28) $$\dot{x} = -\frac{\partial V}{\partial x}(x,y) \qquad \dot{y} = -\frac{\partial V}{\partial y}(x,y)$$

for a given smooth function V, then the bifurcation function $\tilde{X}$ satisfies $\tilde{X}(x) = -\frac{\partial \tilde{V}}{\partial x}(x)$, where $\tilde{V}(x) = V(x,y^*(x))$. Therefore, <u>if (x_0,y_0) is a stable (resp. asymptotically stable) equilibrium point of (28), then</u> V as a minimum (resp. strict minimum) at (x_0,y_0), which implies that V has a minimum (resp. a strict minimum) at $x = x_0$; hence x_0 <u>is a stable (resp. asymptotically stable) equilibrium point of (29)</u>:

(29) $$\dot{x} = \tilde{X}(x) = -\frac{\partial V}{\partial x}(x)$$

<u>and conversely</u>: if x_0 is a stable equilibrium point of (29), then $V(x_0,y^*(x_0)) = \tilde{V}(x_0) \leq \tilde{V}(x) = V(x,y^*(x)) \leq V(x,y)$ for (x,y) near $(x_0,y^*(x_0))$ since $y^*(x)$ is precisely the $y \in R^m$ which minimizes (locally) $V(x,.)$; hence $(x_0,y^*(x_0))$ is a stable equilibrium point of (29); a similar argument holds for the case of asymptotic stability.

Acknowledgments: I wish to thank Reiner Lauterbach of Würzburg, who pointed out to me the need of important corrections in the original proof (and statement!) of Theorem 1. My gratitude also goes to Prof. Jack K. Hale of Brown University, who informed me about the paper of Golubitsky and Schaeffer, along with many other interesting ideas of his own.

REFERENCES

Bibikov, Yu. N., [1] : Local Theory of Nonlinear Analytic Ordinary Differential Equations. Lecture Notes in Math. 702, Springer Verlag 1979.

Golubitsky, M., and Schaeffer, D.: "Bifurcation Analysis near a double eigenvalue of a model chemical reactor, Arch. Rat. Mech. Anal. 75, (1981) 315 - 348 .

De Oliveira, J. C. and Hale, Jack K., [1]: "Dynamic behavior from bifurcation equations", Tôhoku Math. J., 32 (1980), 189 - 199.

EFFICIENT COMPUTATION OF STABLE BIFURCATING BRANCHES OF NONLINEAR EIGENVALUE PROBLEMS

H. Weber

Rechenzentrum und Fachbereich Mathematik
Johannes Gutenberg-Universität in Mainz
Postfach 3980, D-6500 Mainz 1, F.R.G.

1. Introduction

In this note we consider a simple but efficient numerical method for computing stable bifurcating branches of solutions of nonlinear elliptic eigenvalue problems. The algorithm consists of a nested Picard iteration using a multi-grid method for the arising linear problems and a certain correction of the bifurcation parameter on the lower levels, i.e. on the coarser grids. We present numerical examples for a nonlinear elliptic eigenvalue problem on the unit square and for von Kármán's equations for a simply supported rectangular elastic plate.

Recently some papers concerned with different approaches to the fast solution of similar nonlinear problems - related to stability and bifurcation - have been published. We refer the reader to [10,11,13] and the references given therein.

2. The Nonlinear Eigenvalue Problem

We consider nonlinear elliptic eigenvalue problems of the form

$$(1) \qquad Lu = f(\lambda,u) \qquad \text{on } \Omega \ , \ u = 0 \ \text{ on } \partial\Omega$$

where L is a 2m-th order linear elliptic differential operator, $\Omega \subset \mathbb{R}^n$ is a bounded domain with smooth boundary and $f: \mathbb{R} \times \mathbb{R} \longrightarrow \mathbb{R}$ is a smooth function satisfying $f(\lambda,0) = 0$. λ is a real bifurcation parameter.

We assume that $\lambda_0 \in \mathbb{R}$ is a simple bifurcation point in the sense of Crandall and Rabinowitz [6]. This means: the set of solutions of (1) near $(0,\lambda_0)$ consists of the trivial solution $(0,\lambda)$ and a smooth curve $(u(\varepsilon),\lambda(\varepsilon))$, $|\varepsilon| \leq \varepsilon_0$, of nontrivial solutions, which intersect only at the bifurcation point $(0,\lambda_0)$, see Fig. 1. Moreover there is a neighborhood of $(0,\lambda_0)$, such that these solutions are the only solutions of (1) contained in this neighborhood.

We assume furthermore that there is an exchange of stability at the bifurcation point, i.e. the trivial solution loses stability to the supercritical branch or branches bifurcating at $(0,\lambda_0)$, if λ increases beyond λ_0, see Fig. 1, where a pitchfork bifurcation is shown. Stability is understood in the sense of linearized stability with respect to the solutions of the evolution equation

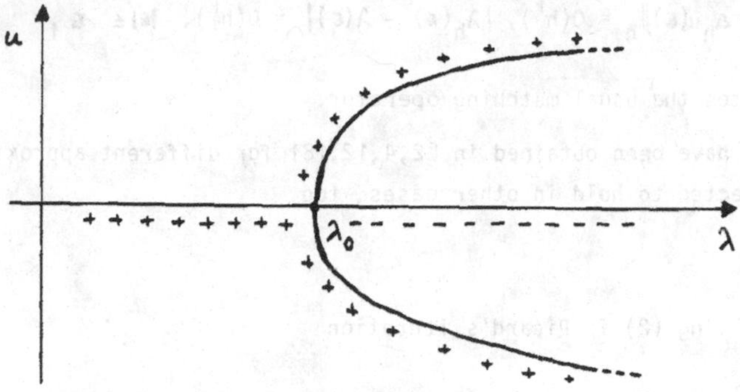

Fig. 1 Bifurcation and exchange of stability

$$\frac{du}{dt} + Lu - f(\lambda,u) = 0.$$

For the numerical solution of (1) a discretization is necessary, say by finite elements or by finite differences. For our purpose the actual type of discretization is not important. Of course it has to produce a large, sparse nonlinear system of equations. We write the discrete problem in the form

(2) $$L_h u_h = f_h(\lambda, u_h), \quad u_h \in \mathbb{R}^{N(h)}$$

where L_h is assumed to be a symmetric, positive definite, sparse matrix and $f: \mathbb{R} \times \mathbb{R}^{N(h)} \longrightarrow \mathbb{R}^{N(h)}$ is a smooth mapping, satisfying $f(\lambda,0) = 0$. h is a real discretization parameter, $N(h) \longrightarrow \infty$ for $h \longrightarrow 0$.

We make the hypothesis that (2) has similar bifurcation and stability properties as (1). This means: we assume that (2) exhibits simple bifurcation from the trivial solution at discrete bifurcation points λ_{0h} near λ_0 and that the discrete branches $(u_h(\varepsilon), \lambda_h(\varepsilon))$ approximate the branch $(u(\varepsilon), \lambda(\varepsilon))$ of the original problem:

Fig. 2 Diagram of bifurcating solutions of discrete and continuous problems, $h_1 > h_2 > h_3 > h_4$

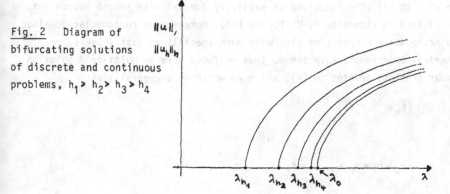

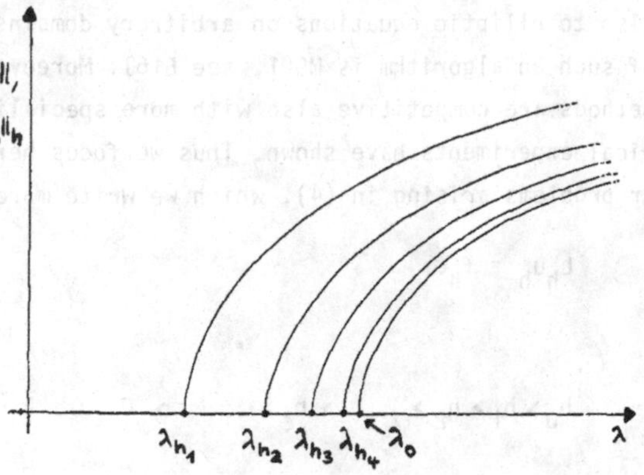

(3) $$\|u_h(\varepsilon)-\Delta_h u(\varepsilon)\|_h = O(h^k),\ |\lambda_h(\varepsilon)-\lambda(\varepsilon)| = O(h^k),\ |\varepsilon|\le \varepsilon_1$$

for some $k\in\mathbb{N}$. Δ_h denotes the usual matching operator.

Results of this form have been obtained in [2,4,12,18] for different approximation schemes and are expected to hold in other cases, too.

3. Picard's Iteration

A simple approach for solving (2) is Picard's iteration

(4) $$\lambda,\ u_h^{[0]} \text{ given,}\qquad \text{solve}\quad L_h u_h^{[i+1]} = f_h(\lambda, u_h^{[i]}),\quad i=0,1,\ldots$$

For the class of nonlinear elliptic eigenvalue problems considered here the stable solutions (in the sense discussed above) are also stable fixed points of the corresponding fixed point iteration

$$u^{[i+1]} = L^{-1} f(\lambda, u^{[i]}),\ i=0,1,\ldots$$

in the sense that the spectral radius of $L^{-1} f_u(\lambda,u)$ is less than one. Here we have to assume of course that the spectrum of L is real and positive. Under the condition that in the discrete case analogous results hold, the algorithm (4) has a desirable property: the 'selectivity'. This means that (4) converges for almost all starting values $u_h^{[0]}$ and λ near enough to λ_{oh} to the stable solution $u_h(\lambda)$, i.e. in the case of Fig. 1 to one of the discrete, supercritical, nontrivial solutions. For details we refer to Scheurle [15].

It is natural to choose a fast and efficient algorithm for solving the linear problem in each step of (4). There are various different choices possible. We think of fast elliptic solvers and multi-grid methods here. Whereas fast elliptic solvers, e.g. Buneman's algorithm (see [13]) are applicable directly only to elliptic problems on rectangular domains the multi-grid methods (cf. [3,8,9,10,16] are applicable also to elliptic equations on arbitrary domains with smooth boundaries. An example of such an algorithm is MG01, see [16]. Moreover on rectangular domains multi-grid methods are competitive also with more specialized fast solvers, as recent numerical experiments have shown. Thus we focus here on multi-grid solution of the linear problems arising in (4), which we write more generally as

$$L_h u_h = f_h.$$

Let

(5) $$h_0 > h_1 > h_2 > \ldots > h_\ell > \ldots > 0$$

be a sequence of discretization parameters, ℓ is the level number. For simplicity we choose $h_i = h_{i-1}/2$. The discrete problem with $h = h_\ell$ is then also denoted by

(6) $$L_\ell u_\ell = f_\ell$$

The solution u_ℓ of (6) as well as the right hand side f_ℓ belongs to a finite-dimensional normed vector space X_ℓ. The connection between grid functions on different levels is given by a prolongation $p_\ell : X_{\ell-1} \longrightarrow X_\ell$ and a restriction $r_\ell : X_\ell \longrightarrow X_{\ell-1}$. The characteristic feature of the multi-grid method is the combination of a smoothing step and a coarse-grid correction. During the smoothing step the defect is not necessarily decreased but smoothed. By the following correction step the discrete solution is improved by means of an auxiliary equation on a coarser grid. In fact this equation has to be of the same structure and sparsity pattern.It should be pointed out that the multi-grid method for certain elliptic operators is able to compute the approximate solution to truncation error on a grid of N points in O(N) arithmetic operations. The storage required is less than $2N/(1-2^n)$ units.

Since a detailed description of the multi-grid algorithm is available elsewhere ([3,9]) we give only a short explanation by means of a quasi-ALGOL program:

(7)
```
procedure multigrid(ℓ,u,f); integer ℓ; array u,f;
if ℓ=0 then  u:=L0^-1*f else
begin integer j;  array v,d;
for j:=1 step 1 until ν do u:=Gℓ(u,f);
d:= rℓ*(Lℓ*u - f); v:=0;
for j:=1 step 1 until γ do multigrid(ℓ-1,v,d);
u:=u - pℓ*v
end multigrid;
```

ℓ is the actual level number, f the actual right hand side, ν is the number of smoothing steps, γ is the number of multigrid iterations per level, G is the smoothing procedure. For an arbitrary input value $u = u_\ell^{[i]}$ the procedure multigrid computes $u = u_\ell^{[i+1]}$, the next multi-grid iterate.

In the following we shall assume that the convergence of the linear multi-grid algorithm, which is used, has been established, see [9].

The algorithm (4), together with multi-grid solution of the linear problems, works of course. However it is not very efficient, due to the large number of iteration steps which are necessary if $|\lambda - \lambda_{0\ell}|$ is small or if $|\lambda - \lambda_{0\ell}|$ is comparatively large. Even if one step is very cheap, an algorithm which needs, for example, 200 steps, may be too expensive, if compared with other developments.

4. A Nested Approach with λ-Correction

For getting an efficient algorithm one has to assure that only a very small number of calls of the linear multi-grid code are necessary on the finest grid, say on level ℓ_{max}. This implies the design of a nested approach. It has the form

$$(8)\qquad \begin{array}{l} \lambda \text{ fixed, } t\in\mathbb{N} \text{ small, } u_t^{[0]} \text{ given,} \\ \text{iteration (4) until convergence, giving } u_t^* \\ \ell = t+1(1)\ell_{max}: \begin{cases} \text{higher order interpolation } u_{\ell-1}^* \xmapsto{q_\ell} u_\ell^{[0]} \\ \text{iteration (4) on level } \ell \text{ until convergence, giving } u_\ell^* \end{cases} \end{array}$$

This algorithm worked satisfactorily but only relatively far from the bifurcation point. The reason becomes clear if we inspect Fig. 2: The parametrization of the different discrete branches by the same λ-scale is not adequate. This is true especially for discretizations of partial differential equations where h cannot be chosen very small.

Now let us recall some basic results from bifurcation theory, cf. e.g. [5,6]. The discrete solutions have the asymptotic expansion

$$(9)\qquad \begin{aligned} u_\ell(\varepsilon) &= \varepsilon(\phi_\ell + \varepsilon^p v_\ell) + O(\varepsilon^{p+2}) \\ \lambda_\ell(\varepsilon) &= \eta_{0\ell} + \varepsilon^p \eta_\ell + O(\varepsilon^{p+1}) \end{aligned}$$

where ϕ_ℓ is the discrete linearized eigenfunction corresponding to $\lambda_{0\ell}$. p+1 is the order of the first nonvanishing higher order term in the expansion of the nonlinearity and

$$\eta_\ell = -\frac{\langle \phi_\ell, Q_\ell(\lambda_{0\ell},\phi_\ell)\rangle_\ell}{\langle \phi_\ell, f_{\ell u_\ell \lambda}(\lambda_{0\ell},0)\phi_\ell\rangle_\ell}, \quad \langle\,,\rangle_\ell \text{ scalar product on } X_\ell,$$

$$Q_\ell(\lambda,u_\ell) = \frac{1}{(p+1)!}\frac{\partial^{p+1}}{\partial_u{}^{p+1}} f_\ell(\lambda,0)(u_\ell)^{p+1}$$

So, for given λ, we compute the asymptotic (real) amplitude of the discrete solution on level ℓ_{max} by

$$\varepsilon = \varepsilon_\lambda = \sqrt[p]{\frac{\lambda - \lambda_{0\ell_{max}}}{\eta_{\ell_{max}}}}$$

and set $\lambda^\ell = \lambda_{0\ell} + \varepsilon_\lambda{}^p \eta_\ell$. Instead of η_ℓ an approximation η will be sufficient. Our improved algorithm is now:

$$(10)\qquad \begin{array}{l} \lambda \text{ given, } t\in\mathbb{N} \text{ small, } u_t^{[0]} = \varepsilon_\lambda\phi_t, \\ \text{iteration (4) with } \lambda=\lambda^t \text{ until convergence, result } u_t^* \\ \ell = t+1(1)\ell_{max}: \begin{cases} \text{higher order interpolation } u_{\ell-1}^* \xmapsto{q_\ell} u_\ell^{[0]} \\ \text{iteration (4) with } \lambda=\lambda^\ell \text{ until convergence, result } u_\ell^* \end{cases} \end{array}$$

If $|\lambda-\lambda_0|$ increases, we suggest to use damping:

$$\lambda^{\ell}=\lambda_{0\ell}+\tau\varepsilon_{\lambda}^{p}\eta\,,\quad 0<\tau<1.$$

5. Numerical Examples

The first example to be presented here is

$$(11)\qquad -\Delta u=\lambda u-u^3 \text{ on } \Omega=(0,1)^2,\quad u=0 \text{ on } \partial\Omega,$$

which has been previously used as a test problem, too (cf. [12]). The linearized problem $-\Delta u=\lambda u$ on Ω, $u=0$ on $\partial\Omega$, has the eigenvalues $\pi^2(m^2+n^2)$ and eigenfunctions $\phi_{mn}(x,y)=\sin m\pi x\cdot\sin n\pi y$, $m,n\in\mathbb{N}$. (11) was discretized by the usual five-point difference star with uniform step width $h=1/N$, N an even integer. The first discrete eigenvalue is $\lambda_{11}^h=\frac{4}{h^2}(1-\cos\pi h)=2\pi^2+O(h^2)$. It is easily seen in this example that a supercritical stable bifurcation occurs at this point and that our algorithm (10) is applicable. The details of the linear multi-grid method used here are:

smoother G: pointwise Gauss-Seidel relaxation
prolongation p: linear interpolation
restriction r: injection

As higher order interpolation q quadratic interpolation was used. The program is based on Brandt's subroutines and uses an adaptive strategy, cf. [3]. The following table presents some typical results for $\ell_{max}=6$, $h=1/128$, $h_0=1/2$, $t=2$, $\lambda_{0h}=19.738217$, $\gamma_{0h}\approx 0.5625$ for small h. Note that the number of unknowns is 16129!

λ	τ	W_2	W_3	W_4	W_5	W_6	$W=\sum W_\ell$	u(0.5,0.5)
19.8	1	0.1	0.4	1.2	3.5	12.1	17.3	0.3310478
19.9	1	0.1	0.4	1.3	4.3	28.3	34.5	0.5353342
20.0	1	0.1	1.6	1.5	6.5	28.4	38.1	0.6808751
21.0	1	0.7	0.8	1.5	4.0	12.4	19.4	1.4887676
22.0	1	0.7	0.5	1.1	9.0	14.6	25.9	1.9842003
25.0	1	0.5	0.9	2.6	7.0	15.7	26.7	2.9860564
30.0	0.8	0.5	0.8	2.5	7.9	22.3	34.1	4.0855092

W_i is the accumulated relaxation work of the iterations on level i, where a sweep on the finest grid is taken as the work unit, cf. [3]. It is worthwhile to compare a typical value of the CPU-time (Honeywell-Bull HB 66/80, FORTRAN, single precision), say for $\lambda=25$: 23.35 sec, with the CPU-time required for solving Poisson's equation by the same multi-grid code. A typical value was 17.94 sec.

The results for this example are in good agreement with those given in [12].

As a less trivial example we have treated numerically von Kármán's equations for the buckling of a thin elastic simply supported rectangular plate, which is

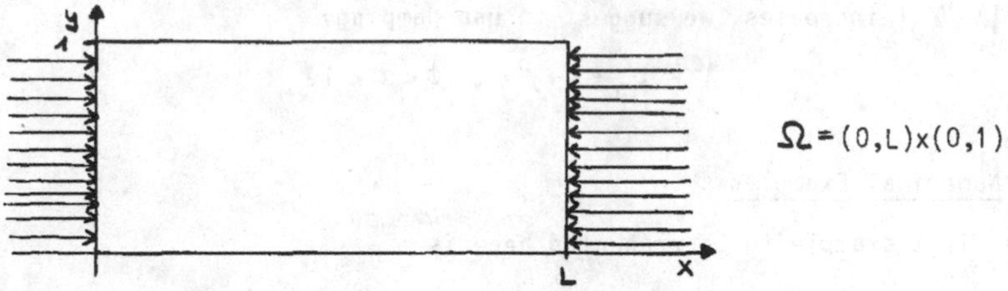

Fig. 3

subject to a compressive thrust applied along the short edges, see Fig. 3. In the dimensionless form von Kármán's equations for the deflection w(x,y) and the stress function f(x,y) are (cf. [1,5])

$$(12)\qquad \begin{aligned} \Delta^2 f &= -\tfrac{1}{2}[w,w] \quad \text{on } \Omega \quad, f = \Delta f = 0 \text{ on } \partial\Omega \\ \Delta^2 w + \lambda w_{xx} &= [f,w] \quad \text{on } \Omega \quad, w = \Delta w = 0 \text{ on } \partial\Omega \end{aligned}$$

where

$$[g,h] = g_{xx}h_{yy} + g_{yy}h_{xx} - 2g_{xy}h_{xy}$$

λ is proportional to the compressive force. (12) may be written in the form

$$(13)\qquad \Delta^2 w = -\lambda w_{xx} + C(w) \quad \text{on } \Omega\,, \quad w = \Delta w = 0 \quad \text{on } \partial\Omega$$

where C is a certain 'cubic' operator. Thus (13) is a generalization of equation (1) and could be treated analogously.

For numerical reasons, however, we introduce new variables

$$\Phi = \Delta f \qquad \text{and} \qquad \Psi = \Delta w$$

with Dirichlet boundary conditions. This leads to a mixed formulation of (12), consisting of four second order equations with zero boundary conditions. This problem may be solved iteratively by

$$(14)\qquad \begin{aligned} \Delta \Phi^{[i+1]} &= -\tfrac{1}{2}[w^{[i]},w^{[i]}] \text{ on } \Omega, & \Phi^{[i+1]} &= 0 \text{ on } \partial\Omega \\ \Delta f^{[i+1]} &= \Phi^{[i+1]} \quad \text{on } \Omega, & f^{[i+1]} &= 0 \text{ on } \partial\Omega \\ \Delta \Psi^{[i+1]} &= -w_{xx}^{[i]} + [f^{[i+1]},w^{[i]}] \text{ on } \Omega, & \Psi^{[i+1]} &= 0 \text{ on } \partial\Omega \\ \Delta w^{[i+1]} &= \Psi^{[i+1]} \quad \text{on } \Omega, & w^{[i+1]} &= 0 \text{ on } \partial\Omega \quad ,i=0,1,\ldots \end{aligned}$$

Of course a discrete version of (14) is actually used. We have again approximated the Laplace operator by the five-point difference star. The brackets [,] on the right hand sides were evaluated by central differences. The same holds for w_{xx}. For the solution of the linear problems the same multi-grid code as in the above example was used. The eigenvalues and eigenfunctions of the linearized problem are

$\lambda_{mn} = \frac{\pi^2}{L^2}\left[m + \frac{n^2L^2}{m}\right]^2$, $w_{mn}(x,y) = \sin\frac{m\pi x}{L}\sin n\pi y$ and $f_{mn} = 0$, $m,n \in \mathbb{N}$. For the square plate (L = 1) we have $\lambda_{11} = 4\pi^2$ and the corresponding discrete first eigenvalue is $\lambda_{11}^h = \frac{4}{h^2}(1 - \cos\pi h)^2/\sin^2(\pi h/2) = 4\pi^2 + O(h^2)$. An obvious generalization of algorithm (10) is applicable to the case of bifurcation from λ_{11}^h. The branch is stable and supercritical. η_h was determined experimentally to have a value near 1.6 for small h. The linear multi-grid code used here was the same as in the previous problem.

Deflection

Stress function

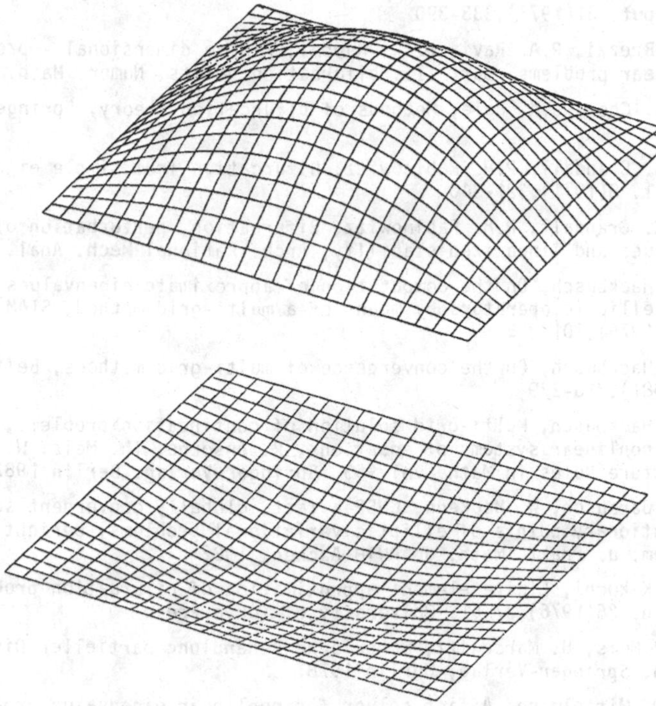

Fig. 4 Deflection and stress function of the square plate for $\lambda = 45$, h = 1/128

We present some typical results which where obtained for the square plate with $\ell_{max} = 6$, h = 1/128, $h_0 = 1/2$, t = 3, $\lambda_{0h} = 39.476561$ (64516 unknowns !):

λ	39.6	39.8	40	41	43
w(0.5,0.5)	0.27347502	0.44754198	0.58047975	1.0025525	1.5308431

45	50	60
1.9213904	2.6671492	3.7659858

The shape of the plate and the stress function for $\lambda = 45$ are shown in Fig. 4 .

A typical value of the CPU-time is 97.56 sec for $\lambda = 43$ (HB 66/80).
For more examples and details of the algorithm (12) we refer to [17].

References

[1] L. Bauer, E.L. Reiss, Nonlinear buckling of rectangular plates, J. SIAM 13 (1965),603-626

[2] W.J. Beyn, On discretization of bifurcation problems, Bifurcation problems and their numerical solution, H.D. Mittelmann, H. Weber (eds.), ISNM vol.54, pp.46-73, Birkhäuser-Verlag, Basel 1980

[3] A. Brandt, Multi-level adaptive solutions to boundary-value problems, Math. Comput. 31(1977),333-390

[4] F. Brezzi, P.A. Raviart, J. Rappaz, Finite dimensional approximation of nonlinear problems, part III: bifurcation points, Numer. Math. 38(1981),1-30

[5] S.N. Chow, J.K. Hale, Methods of bifurcation theory, Springer-Verlag, Berlin 1982

[6] M.G. Crandall, P.H. Rabinowitz, Bifurcation from simple eigenvalues, J. Funct. Anal. 8(1971),321-340

[7] M.G. Crandall, P.H. Rabinowitz, Bifurcation, perturbation of simple eigenvalues and linearized stability, Arch. Rational Mech. Anal. 52(1973),161-180

[8] W. Hackbusch, On the computation of approximate eigenvalues and eigenfunctions of elliptic operators by means of a multi-grid method, SIAM J. Numer. Anal. 16(1979),201-215

[9] W. Hackbusch, On the convergence of multi-grid methods, Beitr. Numer. Math. 9 (1981),213-239

[10] W. Hackbusch, Multi-grid solution of continuation problems, Iterative solution of nonlinear systems of equations, R. Ansorge, Th. Meis, W. Törnig (eds.), Lecture Notes in Math. vol.953, Springer-Verlag, Berlin 1982

[11] H. Jarausch, W. Mackens, CNSP a fast, globally convergent scheme to compute stationary points of elliptic variational problems, Bericht Nr.15, Inst. f. Geom. u. Prakt. Math. d. RWTH Aachen, 1982

[12] F. Kikuchi, Finite element approximation of bifurcation problems, Theor. Appl. Mech. 26(1976),37-51, University of Tokyo Press

[13] Th. Meis, U. Marcowitz, Numerische Behandlung partieller Differentialgleichungen, Springer-Verlag, Berlin 1978

[14] H.D. Mittelmann, A fast solver for nonlinear eigenvalue problems, Iterative solution of nonlinear systems of equations, R. Ansorge, Th. Meis, W. Törnig (eds.), Lecture Notes in Math. vol.953, Springer-Verlag, Berlin 1982

[15] J. Scheurle, Selective iteration and applications, J. Math. Anal. Appl. 59 (1977),596-616

[16] K. Stüben, MG01: a multi-grid program to solve $\Delta u - c(x,y)u = f(x,y)$ (on Ω), $u = g(x,y)$ (on $\partial\Omega$) on nonrectangular bounded domains Ω, Techn. Rep. IMA 82.02.02, GMD/IMA, Bonn 1982

[17] H. Weber, A multi-grid technique for the computation of stable bifurcation branches, Bericht No.1(1982), Rechenzentrum Univ. Mainz, submitted for publication

[18] R. Weiss, Bifurcation in difference approximations to two-point boundary value problems, Math. Comput. 29(1975),746-760

COVARIANCE ANALYSIS OF DISTRIBUTED SYSTEMS UNDER STOCHASTIC POINT FORCES

W. Wedig
University of Karlsruhe
Institute for Technical Mechanics
7500 Karlsruhe 1, BRD

1. Introduction of the problem

The present paper is dealing with elastic structures under stochastic point forces involved in identification problems as well as in ambient response monitoring systems. If the structures are sufficiently homogeneous, they can be described by piecewise holding partial differential equations and associated boundary and transition conditions. By means of the modal analysis such boundary value problems are reduced to systems of ordinary differential equations which then are investigated in the classical manner. In particular, the well established covariance analysis is applied in case of stochastic excitations. Although these investigation methods are often simply applicable and therefore extensively used, they have the significant disadvantage of a bad convergence so that only numerical evaluations are possible which don't clear up physical backgrounds.

Restricting our interest to a basic model of continuum mechanics, we are able to show that the double series of the modal covariance solutions have an important structure consisting in the fact that the diagonal elements of the double series can be represented by a piecewise analytical two-dimensional function. The same is approximately valid for the non-diagonal elements. To get a systematic approach, we finally set up the integral covariance equations associated to the given boundary value problem and solve them approximately by means of a Galerkin's method using a set of two-dimensional polynomials which are orthogonal and piecewise analytical in the given plane of the covariance distribution of the stationary string deflection processes.

2. A basic model of distributed systems

As a basic model of continuum mechanics we consider a uniformly distributed string with the mass μ per unit length preloaded by the axial force H_o and fixed at both ends of its length l.

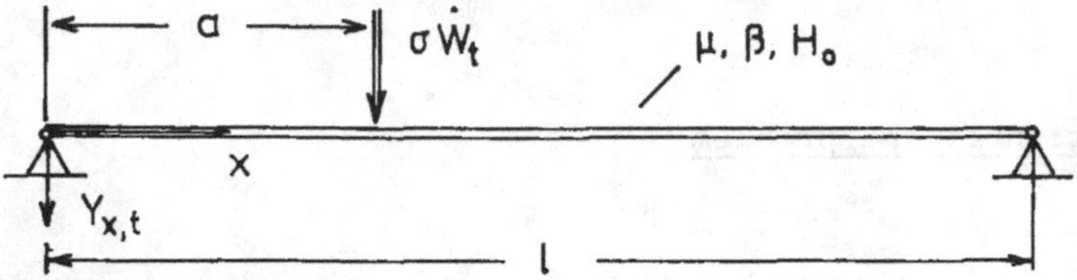

Fig. 1: Model of a preloaded string subjected to a random point force at x = a

As shown above, there is a stochastic point excitation by stationary white noise $\dot{W}_t$ with a vanishing mean value and a normed delta correlation function.

$$E(\dot{W}_t) = 0, \qquad E(\dot{W}_t \dot{W}_s) = \delta(t-s). \tag{1}$$

The parameter σ denotes its intensity. Thus the power spectrum of $\sigma \dot{W}_t$ is given by σ^2. Provided small deflections $Y_{x,t}$ of the string, the excited transverse vibrations are described by the following boundary value problem.

$$\begin{aligned} &\mu \ddot{Y}_{x,t} + \beta \dot{Y}_{x,t} - H_o Y''_{x,t} = Q_{x,t}, \qquad && Y_{o,t} = 0, \\ &Q_{x,t} = \sigma \dot{W}_t \,\delta(x-a), && Y_{l,t} = 0. \end{aligned} \tag{2}$$

Note that the independent variables x and t of the stochastic field processes $Y_{x,t}$ and $Q_{x,t}$ are written as indices. Dashes and dots denote partial derivations with respect to x or t, respectively. Hence, the first term in (2) represents the acceleration of the mass per unit length, the second is a viscous external damping β and the third term gives the restoring of the string. To avoid transition conditions, its excitation

$Q_{x,t}$ may be represented by the delta function $\delta(x-a)$ multiplied by the stochastic process $\sigma \dot{W}_t$. Finally, both boundary conditions of the stated problem are simply given by vanishing deflections at $x = 0$ and $x = 1$.

It is well known that the excitation and the response process can be expanded into the modal representation

$$Q_{x,t} = \sum_{n=1}^{\infty} Q_{n,t} V_n(x), \qquad Y_{x,t} = \sum_{n=1}^{\infty} Y_{n,t} V_n(x), \tag{3}$$

wherein $V_n(x)$ are deterministic eigenfunctions of the associated homogeneous problem.

$$V_n''(x) + \lambda_n^2 V_n(x) = 0, \qquad V_n(0) = V_n(1) = 0,$$
$$V_n(x) = \sin \lambda_n x, \qquad \lambda_n = \frac{\pi}{1} n, \qquad n = 1, 2, \ldots \tag{4}$$

Its solutions are simple sinusoidal functions satisfying the same vanishing boundary conditions. Furthermore, they satisfy the orthogonality condition.

$$\int_0^1 V_n(x) V_m(x)\, dx = \gamma_n 1\, \delta_{n,m}, \qquad \gamma_n = \frac{1}{2}, \qquad n = 1,2, \ldots \tag{5}$$

We make use of this important property in order to calculate the stochastic coefficient functions $Q_{n,t}$ in multiplying the expansion (3) by $V_m(x)$ and integrating it over the entire range $0 \leq x \leq 1$.

$$\int_0^1 Q_{x,t} V_m(x) dx = \sum_{n=1}^{\infty} Q_{n,t} \int_0^1 V_m(x) V_n(x) dx = \gamma_m 1\, Q_{m,t},$$
$$\mu \ddot{Y}_{n,t} + \beta \dot{Y}_{n,t} + H_0 \lambda_n^2 Y_{n,t} = \frac{1}{\gamma_n 1} \sigma \dot{W}_t V_n(a), \quad n=1,2\ldots \tag{6}$$

The same procedure is performable in the partial differential equation (2) leading to the uncoupled system (6) of ordinary differential equations. Finally, we introduce into (6) the state processes of the displacements $S_{n,t} = Y_{n,t}$ and of the velocities $T_{n,t} = \dot{Y}_{n,t}$ to rewrite (6) into the form of a first order system.

$$\dot{S}_{n,t} = T_{n,t}, \qquad n = 1,2, \ldots$$
$$\dot{T}_{n,t} = \frac{1}{\mu} [-\beta T_{n,t} - H_0 \lambda_n^2 S_{n,t} + \frac{\sigma}{\gamma_n 1} \dot{W}_t V_n(a)]. \tag{7}$$

The equations (6) or (7) are the starting point of the classical covariance analysis which first has been introduced by N. Wiener [1]. For some applications in structural mechanics see e.g. [2].

Instead of such modal expansions, we are more interested in integral methods which are started by the following integral representation of the given boundary value problem.

$$Y_{x,t} = \frac{1}{H_0} \int_0^1 G(x,\xi)(Q_{\xi,t} - \mu \ddot{Y}_{\xi,t} - \beta \dot{Y}_{\xi,t}) d\xi. \tag{8}$$

Herein, $G(x,z)$ is a Green's function given by

$$\begin{aligned} G(z,x) &= z\,(1-x/1), \qquad \text{for } 0 \leq z \leq x, \\ G(x,z) &= x\,(1-z/1), \qquad \text{for } z \leq x \leq 1. \end{aligned} \tag{9}$$

It is defined in the plane range $0 \leq x,z \leq 1$ and symmetric with respect to the diagonal line $x = z$. Note that $G(x,z)$ is the statical deflection at the position z of the string under a unit loading applied at x. It is therefore quite easy to calculate such Green's functions or to extend them to more general structures. To apply (9) to an integral covariance analysis, we finally introduce the integrated state processes

$$Z_{x,t} = \int_0^1 G(x,\xi)\, Y_{\xi,t} d\xi, \qquad X_{x,t} = \int_0^1 G(x,\xi)\, \dot{Y}_{\xi,t} d\xi \tag{10}$$

and go over to a first order system associated to (8).

$$\dot{Z}_{x,t} = X_{x,t}, \qquad \dot{X}_{x,t} = \frac{1}{\mu} [-\beta X_{x,t} - H_0 Y_{x,t} + \int_0^1 G(x,\xi) Q_{\xi,t} d\xi]. \tag{11}$$

In the case that $Q_{x,t}$ is a distributed white field process $\sigma W'_x \dot{W}_t$, the integral covariance analysis has been first applied in [3]. Some more general examples are given in [4] and [5] showing that this analysis is simply applicable and most effective. For concentrated stochastic excitations, it is effective as well but more complicated.

As already mentioned, the setting up of covariance equations is simply performable. Applying Itô's formula to the modal equations (7), we calculate the quadratic increments of the state processes $d(S_{n,t}T_{m,t})$ and then take the expectations in arriving at the four following moments equations [6].

$$\dot{E}(S_{n,t}S_{m,t}) = E(S_{n,t}T_{m,t}) + E(T_{n,t}S_{m,t}), \tag{12}$$

$$\dot{E}(S_{n,t}T_{m,t}) = E(T_{n,t}T_{m,t}) - \frac{\beta}{\mu} E(S_{n,t}T_{m,t}) - \frac{H_0}{\mu} \lambda_m^2 E(S_{n,t}S_{m,t}),$$

$$\dot{E}(T_{n,t}S_{m,t}) = E(T_{n,t}T_{m,t}) - \frac{\beta}{\mu} E(T_{n,t}S_{m,t}) - \frac{H_o}{\mu} \lambda_n^2 E(S_{n,t}S_{m,t}),$$

$$\dot{E}(T_{n,t}T_{m,t}) = - 2 \frac{\beta}{\mu} E(T_{n,t}T_{m,t}) - \frac{H_o}{\mu}[\lambda_n^2 E(S_{n,t}T_{m,t}) +$$

$$+ \lambda_m^2 E(T_{n,t}S_{m,t})] + (\frac{\sigma}{\mu})^2 \frac{1}{\gamma_n 1 \gamma_m 1} V_n(a)V_m(a), \quad n,m = 1,2,\ldots \quad (12)$$

Since we are only interested in stationary processes defined by $\dot{E}(.)=0$, the left-hand side of (12) is vanishing leading to algebrais equations with the solution as follows.

$$E(S_{n,t}S_{m,t}) = \frac{\sigma^2}{\beta H_o} \frac{1}{\gamma_n 1 \gamma_m 1} \frac{V_n(a)\ V_m(a)}{\lambda_n^2 + \lambda_m^2 + \varkappa(\lambda_n^2 - \lambda_m^2)^2}, \quad \varkappa = \frac{H_o \mu}{2\beta^2}. \quad (13)$$

Similar results are available for $E(S_{n,t}T_{m,t})$ and $E(T_{n,t}T_{m,t})$. Now, we can apply the modal representation (3) to the solution (13) which leads to the final result

$$E(Y_{x,t}Y_{z,t}) = \sum\sum_{n,m=1}^{\infty} E(S_{n,t}S_{m,t})\ V_n(x)V_m(z) = K(x,z),$$

$$K(x,z) = \frac{\sigma^2}{\beta H_o} \sum\sum_{n,m=1}^{\infty} \frac{V_n(a)V_m(a)V_n(x)V_m(z)}{\gamma_n 1 \gamma_m 1[\lambda_n^2 + \lambda_m^2 + \varkappa(\lambda_n^2 - \lambda_m^2)^2]}. \quad (14)$$

Obviously, the covariance K(x,z) of the stationary string deflections is a two-dimensional function. Similar as the Green's function, K(x,z) is symmetric because the space variables x and z are exchangeable.

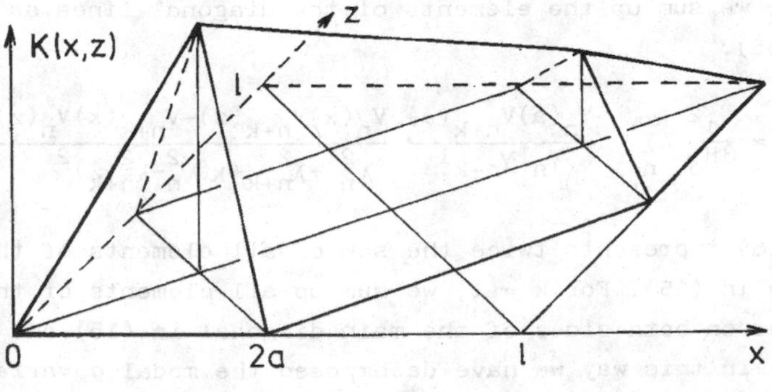

Fig. 2: Inhomogeneously distributed correlation of the stationary string deflection processes

For $x = z$, the space correlation function $K(x,z) = E(Y_{x,t}Y_{z,t})$ goes over to the square mean distribution $K(x,x) = E(Y^2_{x,t})$. Outside of the diagonal line $x = z$, we need therefore the measuring of deflections at two different positions in order to determine the non-homogeneous correlation function $K(x,z)$. Fig. 2 shows a simplified two-dimensional sketch of $K(x,z)$ in the plane range $0 \leq x,z \leq 1$. Note that $K(x,z)$ satisfies the vanishing boundary conditions at $x,z = 0,1$ in coincidence with the correspondent property of the Green's function. But in contrast with the fact that $G(x,z)$ is also symmetric in the diagonal line $x = z$, the mean square distribution $K(x,x)$ is strongly asymmetric in $x = z$. This is an important feature that complicates essentially the analytical analysis of $K(x,z)$.

3. Analytical structure of the modal covariances

To introduce an analytical analysis, we rearrange all terms of the series expansion (14) in the following matrix form.

$$\begin{array}{ccccc} (1,1) & (1,2) & (1,3) & (1,4) & \dots \\ (2,1) & (2,2) & (2,3) & (2,4) & k=2 \\ (3,1) & (3,2) & (3,3) & (3,4) & \\ k=2 & k=1 & k=0 & k=1 & \end{array} \qquad (15)$$

Subsequently, we sum up the elements of the diagonal lines as it is indicated in (15).

$$K_k(x,z) = \frac{\sigma^2}{\beta H_o} \sum_{n=1}^{\infty} \frac{V_n(a)V_{n+k}(a)}{\gamma_n l \gamma_{n+k} l} \; \frac{V_n(x)V_{n+k}(z)+V_{n+k}(x)V_n(z)}{\lambda_n^2 + \lambda_{n+k}^2 + \kappa(\lambda_n^2 - \lambda_{n+k}^2)^2} \quad . \qquad (16)$$

For $k = 0$, (16) represents twice the sum of all elements of the main diagonal line in (15). For $k = 1$, we sum up all elements of the two lines situated on both sides of the main diagonal in (15) and so on for $k = 2, 3, \dots$ In this way we have decomposed the modal covariance representation (14) in n different covariance parts $K_k(x,z)$ which have the following excellent properties.

Starting with $k = 0$, we obtain the main covariance part.

$$K_o(x,z) = \frac{\sigma^2}{\beta H_o} \sum_{n=1}^{\infty} \frac{V_n^2(a)}{(\gamma_n l)^2} \frac{1}{2\lambda_n^2} V_n(x)V_n(z). \tag{17}$$

Obviously, $K_o(x,z)$ is not influenced by the system parameter $\varkappa$ which is inversely proportional to the squared damping coefficient β. From this it follows that $K_o(x,z)$ is symmetric both with respect to the main diagonal line $x = z$ and the second diagonal line $z = l-x$ of its definition plane $0 \leq x,z \leq l$. Fig. 3 shows this important feature in a two-dimensional sketch.

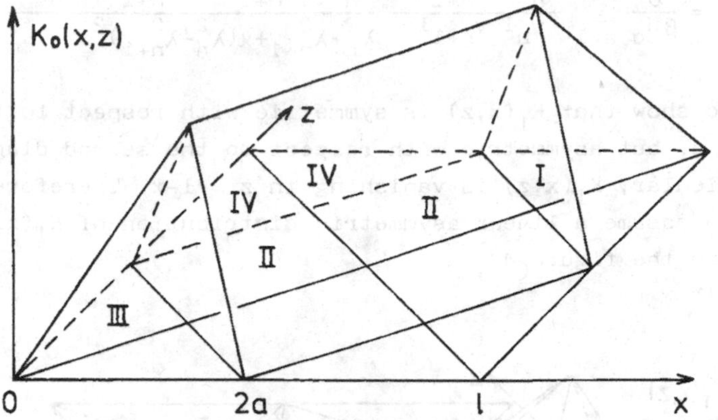

Fig. 3: Two-dimensional symmetric distribution of the diagonal modal covariance part $K_o(x,z)$

Furthermore, we recognize that the main diagonal covariance function $K_o(x,z)$ consists of eight plane surfaces. Therefore, $K_o(x,z)$ can be represented by the following linear functions

$$K_o^{I}(x,z) = \frac{\sigma^2}{2\beta H_o l}(l-z), \qquad K_o^{III}(x,z) = \frac{\sigma^2}{2\beta H_o l}\,x,$$

$$K_o^{II}(x,z) = \frac{\sigma^2}{4\beta H_o l}(x-z+2a), \qquad K_o^{IV}(x,z) = 0 \tag{18}$$

valid for $z \geq x$ and respectively for the subregions I, II, III and IV according to the figure 3. It is easy to expand this piecewise analytical functions into a series of the eigenfunctions $V_n(x)$

$$K_0(x,z) = \sum_{n,m=1}^{\infty}\sum a_{n,m} V_n(x) V_m(z),$$
$$a_{n,m} = \frac{1}{\gamma_n^1 \gamma_m^1} \iint_0^1 K_0(x,z) V_n(x) V_m(z)\, dx dz \tag{19}$$

in order to verify that the double sum in (19) is reduced to a single one which coincides with the result (17) previously derived. Hence, we have obtained two different representations of $K_0(x,z)$, the modal expansion (17) and the piecewise analytical functions (18).

We continue in studying the next covariance part $K_1(x,z)$ given by

$$K_1(x,z) = \frac{\sigma^2}{\beta H_0} \sum_{n=1}^{\infty} \frac{V_n(a) V_{n+1}(a)}{\gamma_n^1 \, \gamma_{n+1}^1} \frac{V_n(x) V_{n+1}(z) + V_{n+1}(x) V_n(z)}{\lambda_n^2 + \lambda_{n+1}^2 + \varkappa(\lambda_n^2 - \lambda_{n+1}^2)^2} . \tag{20}$$

It is easy to show that $K_1(x,z)$ is symmetric with respect to the diagonal line $x = z$, but asymmetric with respect to the second diagonal $z = 1-x$. In particular, $K_1(x,z)$ is vanishing in $z = 1-x$. Therefore, it is reasonable to assume a linear asymmetric distribution of $K_1(x,z)$ as it is sketched in the figure 4.

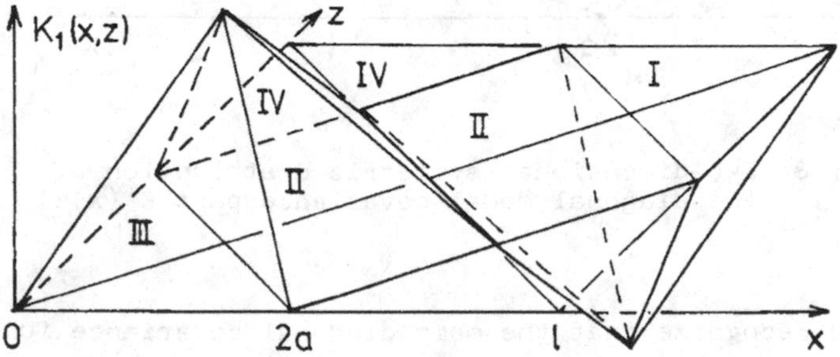

Fig. 4: Asymmetric linear covariance distribution of $K_1(x,z)$ - effect of the damping of the string

Accordingly, we have to set up four different functions as follows.

$$\bar{K}_1^{I}(x,z) = -B(\tfrac{1}{2} - a)(1 - z), \qquad \bar{K}_1^{III}(x,z) = B(\tfrac{1}{2} - a)\, x,$$
$$\bar{K}_1^{II}(x,z) = \tfrac{1}{4}B(2a - z + x)(1 - z - x), \qquad \bar{K}_1^{IV}(x,z) = 0. \tag{21}$$

Similar as in (18), they are piecewise analytical, continuous and respectively valid for one subregion I, II, III or IV. Particularly, they are vanishing if the external excitation is applied at a = 1/2 or a = 0. Because of the asymmetry, the modal expansion of $\bar{K}_1(x,z)$ yields a more complicated double series calculated to

$$\bar{K}_1(x,z) = 4B \sum_{n,m=1}^{n \neq m}\sum \frac{V_n(a)V_m(a)}{\gamma_n{}^1\gamma_m{}^1} \frac{1-(-1)^{n+m}}{(\lambda_n^2-\lambda_m^2)^2} V_n(x)V_m(z). \qquad (22)$$

The diagonal terms n = m of (22) are vanishing. In comparing the result (22) with the corresponding asymmetric covariance part (20), we find some similarities but no exact coincidence. This is confirmed by numerical evaluations of (20) shown in Fig. 5 for several excitation points a and for the diagonal distribution at x = z.

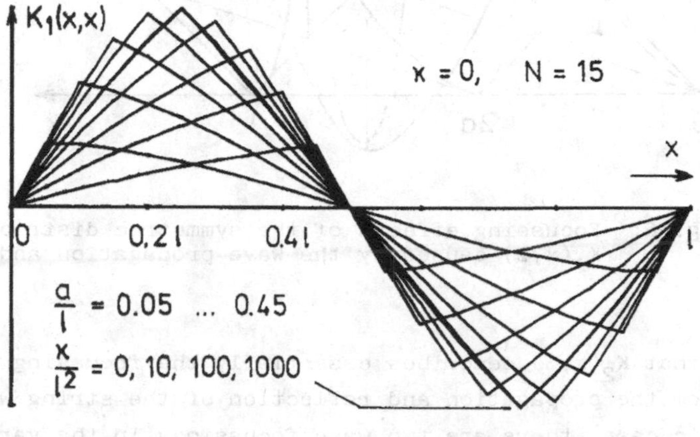

Fig. 5: Numerical evaluations of the modal representation of the asymmetric covariance part $K_1(x,z)$ at x = z

Clearly, the modal representation part (20) has some slight deviations from the straight lines expected from (21). However, the shaping of these deviations is nearly not influenced by the system parameter $\varkappa$, that is shown in Fig. 5 for an excitation point a = o.25l and the values $\varkappa$ = 0, 10, 100 and 1000. From this it follows that the modal result (20) $K_1(x,z)$ can be approximated by the piecewise analytical function (21) $\bar{K}_1(x,z)$ provided that the coefficient B in (21) is calculable by any additional appraoch in order to work out the intensity of the asymmetric damping effect in dependence of $\varkappa$.

Concluding such investigations, we still consider the third part $K_2(x,z)$ of the modal covariance distribution. By means of numerical evaluations, similarly as above, it can be shown that $K_2(x,z)$ is approximated by piecewise analytical functions with parabolic surfaces which are designed in Fig. 6 for the special excitation point a = o.25 l.

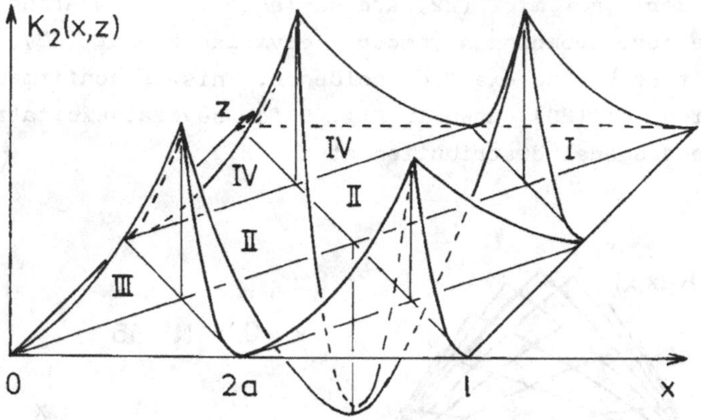

Fig. 6: Focussing effects of the symmetric distribution $K_2(x,z)$ caused by the wave propagation and reflection

We find out that $K_2(x,z)$ describes essentially the focussing effects following from the propagation and reflection of the string waves. In the stochastic case, there are two wave focussings in the variance distribution of the diagonal line x = z. The first is situated at the excitation point z = x = a. The second is its imaginary point at z = x=l-a. Besides, there are two further focussing points in the second diagonal line at x, z = $\pm$a + l/2 following from the corresponding effects in the covariance distribution. Particularly, this has the consequence that $K_2(x,z)$ is not vanishing in the subregion IV.

Finally, it is worth noting that $K_1(x,z)$ and $K_2(x,z)$ are only small corrections of the diagonal covariance $K_0(x,z)$. In the extreme case of a zero-valued parameter $\varkappa$, Fig. 7 shows a comparison of their magnitudes which all three parts possess at the diagonal line x = z. In practice, the involved elastic structures are slightly damped ($\beta \gg$) so that the system parameter $\varkappa = H_0\mu/2\beta^2$ is nearly not smaller than $\varkappa/l^2 = 1$. If $\varkappa$

tends to infinity, $K_1(x,z)$ and $K_2(x,z)$ are vanishing and the complete covariance function (14) goes over to the special diagonal form (17).

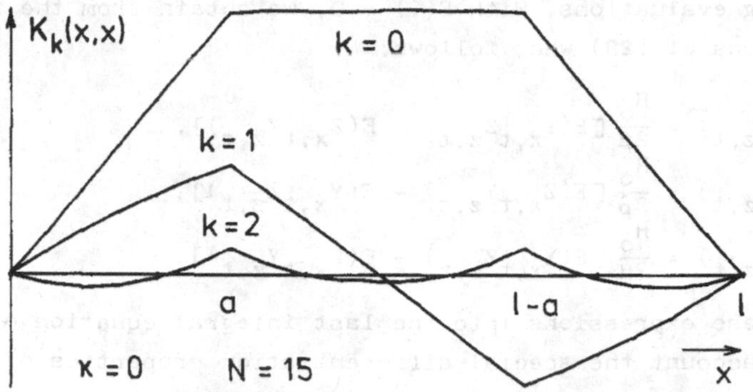

Fig. 7: Magnitude comparison of the modal variance parts $K_o(x,z)$, $K_1(x,z)$ and $K_2(x,z)$ for $\kappa = 0$

4. Integral and differential covariance equations

To calculate such piecewise analytical solutions in a systematic manner, we set up the integral covariance equations associated to the given stochastic boundary value problem (2). For this purpose, we make use of its integral representation (11) applying the Itô calculus in order to determine the quadratic increments of the state field processes $Z_{x,t}$ and $X_{x,t}$. Taking the expectations of these increments yields the four covariance equations as follows.

$$
\begin{aligned}
\dot{E}(Z_{x,t}Z_{z,t}) &= E(Z_{x,t}X_{z,t}) + E(X_{x,t}Z_{z,t}),\\
\dot{E}(Z_{x,t}X_{z,t}) &= E(X_{x,t}X_{z,t}) - \frac{\beta}{\mu}\, E(Z_{x,t}X_{z,t}) - \frac{H}{\mu}\circ\, E(Z_{x,t}Y_{z,t}),\\
\dot{E}(X_{x,t}Z_{z,t}) &= E(X_{x,t}X_{z,t}) - \frac{\beta}{\mu}\, E(X_{x,t}Z_{z,t}) - \frac{H}{\mu}\circ\, E(Y_{x,t}Z_{z,t}),\\
\dot{E}(X_{x,t}X_{z,t}) &= -\,2\,\frac{\beta}{\mu}\, E(X_{x,t}X_{z,t}) + \left(\frac{\sigma}{\mu}\right)^2 G(x,a)G(z,a)\, -\\
&\quad -\frac{H}{\mu}\circ\,[E(Y_{x,t}X_{z,t}) + E(X_{x,t}Y_{z,t})]. \qquad (23)
\end{aligned}
$$

Obviously, such a procedure can be performed in an exact correspondence to the well known covariance analysis of discrete dynamic systems.

Provided a finite non-vanishing damping coefficient β, the equations (23) possess stationary solutions to which we restrict our interest in the following evaluations. With $\dot{E}(.) = 0$, we obtain from the first three equations of (23) what follows.

$$\begin{aligned} E(Z_{x,t}X_{z,t}) &= \frac{H_o}{2\beta}[E(Y_{x,t}Z_{z,t}) - E(Z_{x,t}Y_{z,t})], \\ E(X_{x,t}Z_{z,t}) &= \frac{H_o}{2\beta}[E(Z_{x,t}Y_{z,t}) - E(Y_{x,t}Z_{z,t})], \\ E(X_{x,t}X_{z,t}) &= \frac{H_o}{2\mu}[E(Y_{x,t}Z_{z,t}) + E(Z_{x,t}Y_{z,t})]. \end{aligned} \tag{24}$$

Inserting these expressions into the last integral equation of (23) and taking into account the special differentiation properties of the Green 's function in $Z_{x,t}$ and $X_{x,t}$, yields finally

$$\begin{aligned} &[1 - \varkappa\frac{\partial^2}{\partial x^2}]\int_0^1 G(z,u)K(x,u)du - 2\varkappa K(x,z) + \\ &+ [1 - \varkappa\frac{\partial^2}{\partial z^2}]\int_0^1 G(x,u)K(z,u)du = \frac{\sigma^2}{\beta H_o} G(x,a)G(z,a). \end{aligned} \tag{25}$$

This is the basic equation of the integral covariance analysis for the determination of the complete covariance function $K(x,z) = E(Y_{x,t}Y_{z,t})$ of the stationary string deflection processes. The integral equation (25) is linear, inhomogeneous and symmetric in x and z.

First, we show that its solution coincides with the modal representation (14). For this purpose, we expand both, the unknown covariance distribution $K(x,z)$ as well as the given Green's function $G(x,z)$ into terms of the eigenfunctions $V_n(x)$ and $V_m(z)$.

$$\begin{aligned} K(x,z) &= \sum_{n,m}^{\infty}\sum c_{n,m}V_n(x)V_m(z), \qquad G(x,z) = \sum_n^{\infty} \frac{V_n(x)V_n(z)}{\gamma_n l\, \lambda_n^2}, \\ G(x,a)G(z,a) &= \sum_{n,m}^{\infty}\sum \frac{1}{\gamma_n l \gamma_m l}\frac{1}{\lambda_n^2\lambda_m^2} V_n(a)V_m(a)\ V_n(x)V_m(z). \end{aligned} \tag{26}$$

Subsequently, these expansions are inserted into the integral equation (25).

$$\begin{aligned} \sum_{n,m}\sum [\frac{1}{\lambda_m^2}(1 + \varkappa\lambda_n^2) + \frac{1}{\lambda_n^2}(1 + \varkappa\lambda_m^2) - 2\varkappa]\, c_{n,m}\ V_n(x)V_m(z) = \\ = \frac{\sigma^2}{\beta H_o}\sum_{n,m}\sum \frac{V_n(a)V_m(a)V_n(x)V_m(z)}{\gamma_n l\gamma_m l\ \lambda_n^2\lambda_m^2}\ . \end{aligned} \tag{27}$$

The comparison of all coefficients of $V_n(x)V_m(z)$ leads immediatly to $c_{n,m} = E(Y_{n,t}Y_{m,t})$ confirming that the integral equation (25) is simply solvable by means of the modal covariances derived in (14).

Next, we differentiate the basic integral equation (25) twice with respect to the coordinate x and then twice with respect to z. Using the special properties of the Green's function, this differentiation leads to a partial differential equation of the form

$$\varkappa\left[\frac{\partial^4}{\partial x^4} - 2\,\frac{\partial^4}{\partial x^2\partial z^2} + \frac{\partial^4}{\partial z^4}\right]K(x,z) - \left[\frac{\partial^2}{\partial x^2} + \frac{\partial^2}{\partial z^2}\right]K(x,z) = 0. \qquad (28)$$

It can easily be shown that (28) is also directly derivable by applying Itô's formula respectively the ordinary differential rules to the homogeneous partial differential equation of the initially stated boundary value problem (2). The obtained equation (28) is of fourth order, linear and homogeneous. The fourth order terms are hyperbolic; the second order terms are elliptic. It is therefore reasonable to introduce the new coordinates

$$u = \tfrac{1}{2}(z+x-1), \qquad z = u+v+\tfrac{1}{2},$$
$$v = \tfrac{1}{2}(z-x), \qquad x = u-v+\tfrac{1}{2}. \qquad (29)$$

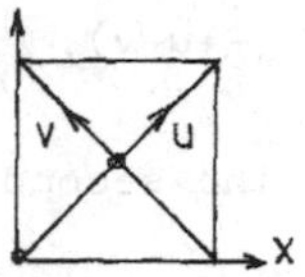

and to transform the differential equation (28) into the following normal form.

$$2\varkappa\frac{\partial^4}{\partial u^2\partial v^2}K(u,v) - \left[\frac{\partial^2}{\partial u^2} + \frac{\partial^2}{\partial v^2}\right]K(u,v) = 0. \qquad (30)$$

Obviously, the solution K(u,v) of the normal form (30) is now separable setting up

$$K(u,v) = P(u)Q(v), \qquad 2\varkappa\, P''\, \ddot{Q} = P''\, Q + P\, \ddot{Q} \qquad (31)$$

wherein dashes and dots denote derivations with respect to the space coordinates u or v, respectively. Inserting P(u)Q(v) into the partial equation (30), the separation is performable reducing (30) to an algebraic equation with two unknown eigenvalues α^2 and β^2.

$$2\varkappa = \frac{P(u)}{P''} + \frac{Q(v)}{\ddot{Q}}, \qquad 2\varkappa = \frac{1}{\alpha^2} + \frac{1}{\beta^2}. \qquad (32)$$

Consequently, the new coordinates u and v are most preferable for the following analytical approximation.

5. Application of a Galerkin's method

To get analytical and systematic approximations of a high rate of convergence, we make use of all structural properties, previously mentioned, setting up a series of two-dimensional covariance functions which are polynomials in u,v and piecewise valid in each subregion of 0=x,z=1.

$$K(u,v) = c_0 k_0(u,v) + c_1 k_1(u,v) + c_2 k_2(u,v) + \ldots \tag{33}$$

Herein, the coefficients c_i (i = 0,1,2, ...) are still undetermined and have to be calculated by means of the Galerkin's method. The polynomials $k_i(u,v)$ are constructed in such a way that they satisfy the boundary and transition conditions which we know from the structure of the modal covariances. Correpondingly, the first linear and symmetric set-up $k_0(u,v)$ has equivalent to (18) the form

$$\begin{aligned} k_0^I(u,v) &= (\tfrac{1}{2}-u-v), & k_0^{II}(u,v) &= (a-v), \\ k_0^{III}(u,v) &= (\tfrac{1}{2}+u-v), & k_0^{IV}(u,v) &= 0, \quad \text{for } v \geq 0. \end{aligned} \tag{34}$$

From (21) we get the second asymmetric set-up.

$$\begin{aligned} k_1^I(u,v) &= -(\tfrac{1}{2}-a)(\tfrac{1}{2}-u-v), & k_1^{II}(u,v) &= -u(a-v), \\ k_1^{III}(u,v) &= (\tfrac{1}{2}-a)(\tfrac{1}{2}+u-v), & k_1^{IV}(u,v) &= 0, \quad \text{for } v \geq 0. \end{aligned} \tag{35}$$

Note that both, the polynomials (34) and (35), include only linear terms in u and v representing bilinear forms. For the special case $a=\frac{1}{4}$, we may finally set up quadratic and symmetric polynomials as follows.

$$\begin{aligned} k_2^I(u,v) &= (\tfrac{1}{2}-u-v)(\tfrac{1}{2}-u+v), & k_2^{IV}(u,v) &= (\tfrac{1}{2}-v-u)(\tfrac{1}{2}-v+u), \\ k_2^{II}(u,v) &= 2(u^2+v^2)-(\tfrac{1}{4})^2-\gamma(\tfrac{4}{1})^2u^2v^2, & & \text{for } v \geq 0. \end{aligned} \tag{36}$$

The parameter γ in (36) is still undetermined and has to be used to satisfy a generalized orthogonality condition.

Following the Galerkin's method, we now insert the covariance series (33) into the basic integral equation (25). Then, we multiply (25) by the polynomials $k_j(u,v)$ and integrate it over the entire definition range $0 \leq x,z \leq 1$. Naturally, this procedure can be performed in the u, v - coordinates as well as in x and z.

$$\int_0^1\!\!\int_0^1 k_j(x,z)\Big[1-\varkappa\frac{\partial^2}{\partial x^2}\Big]\int_0^1 G(z,u)\sum_{i=0}^{\infty} c_i k_i(x,u)\,dudxdz \;+$$
$$+\int_0^1\!\!\int_0^1 k_j(x,z)\Big[1-\varkappa\frac{\partial^2}{\partial z^2}\Big]\int_0^1 G(x,u)\sum_{i=0}^{\infty} c_i k_i(z,u)\,dudxdz \;- \qquad (37)$$
$$-\,2\varkappa\int_0^1\!\!\int_0^1 k_j(x,z)\sum_{i=0} c_i k_i(x,z)\,dxdz = \frac{\sigma^2}{\beta H_o}\int_0^1\!\!\int_0^1 k_j(x,z)G(x,a)G(z,a)\,dxdz.$$

It is obvious that the two first polynomials $k_o(x,z)$ and $k_1(x,z)$ are orthogonal in the sense that on the left-hand side of (37) all terms involving a product of both are vanishing. By a suitable chosen parameter γ, this is also attainable for $k_2(x,z)$ and $k_o(x,z)$. Provided such a generalized orthogonality, the algebraic equations (37) are reducible to the following diagonal form.

$$[A_j - \varkappa(2B_j + C_j)]c_j = \frac{\sigma^2}{\beta H_o} D_j, \qquad \text{for } j = 0,1,2,\ \ldots \qquad (38)$$

The equation system (38) possesses the coefficients

$$A_j = \int_0^1\!\!\int_0^1\!\!\int_0^1 k_j(x,z)[G(z,u)k_j(x,u)+G(x,u)k_j(z,u)]\,dudxdz,$$
$$B_j = \int_0^1\!\!\int_0^1 k_j^2(x,z)\,dxdz, \qquad D_j = \int_0^1\!\!\int_0^1 k_j(x,z)G(z,a)G(x,a)\,dxdz, \qquad (39)$$
$$C_j = \int_0^1\!\!\int_0^1 k_j(x,z)\Big[\frac{\partial^2}{\partial x^2}\int_0^1 G(z,u)k_j(x,u)\,du + \frac{\partial^2}{\partial z^2}\int_0^1 G(x,u)k_j(z,u)\,du\Big]dxdz$$

which can be evaluated by a piecewise calculation.

To avoid such an extended calculation, we make use of the modal representations of $k_o(x,z)$ and $k_1(x,z)$. Inserting first the expansion (17) into the integrals (39), we obtain

$$A_o = 2l^2\sum_{n=1}^{\infty}\frac{1}{(\gamma_n l)^2\lambda_n^6}V_n^4(a), \qquad B_o = l^2\sum_{n=1}^{\infty}\frac{1}{(\gamma_n l)^2\lambda_n^4}V_n^4(a),$$
$$C_o = -2l^2\sum_{n=1}^{\infty}\frac{1}{(\gamma_n l)^2\lambda_n^4}V_n^4(a), \qquad D_o = l\sum_{n=1}^{\infty}\frac{1}{(\gamma_n l)^2\lambda_n^6}V_n^4(a). \qquad (40)$$

Consequently, the term $2B_o+C_o$ in (38) is vanishing and the coefficient

c_o is determined to

$$c_o = \frac{\sigma^2}{2\beta H_o l} . \tag{41}$$

Multipied by $k_o(x,z)$, this result coincides obviously with (18). Next, we apply the modal representation (22)

$$k_1(x,z) = \sum_{n,m}^{n\neq m}\sum \frac{c_{n,m}}{\gamma_n l \gamma_m l} V_n(x)V_m(z), \qquad c_{n,m} = 4\,\frac{1-(-1)^{n+m}}{(\lambda_n^2-\lambda_m^2)^2} V_n(a)V_m(a).$$

Inserting $k_1(x,z)$ into (38) immediately yields

$$c_1 = \frac{\sigma^2}{\beta H_o l^2}\,\frac{\alpha}{1+\gamma\varkappa/l^2}, \qquad \varkappa = \frac{\mu H_o}{2\beta^2} . \tag{42}$$

The parameter α and γ in (42) are evaluated by (39) to

$$\alpha = \frac{\sum_{n,m}^{n\neq m}\sum \frac{1}{\lambda_n^2\lambda_m^2}\, c_{n,m} V_n(a) V_m(a)}{\frac{1}{l^2}\sum_{n,m}^{n\neq m}\sum \left(\frac{1}{\lambda_n^2}+\frac{1}{\lambda_m^2}\right) c_{n,m}^2}, \qquad \gamma = \frac{\sum_{n,m}^{n\neq m}\sum \left(\frac{\lambda_n}{\lambda_m}-\frac{\lambda_m}{\lambda_n}\right)^2 c_{n,m}^2}{\frac{1}{l^2}\sum_{n,m}^{n\neq m}\sum \left(\frac{1}{\lambda_n^2}+\frac{1}{\lambda_m^2}\right) c_{n,m}^2} . \tag{43}$$

Therewith, we have calculated a first improvement of the diagonal solution (18) which takes into account the asymmetric property of the displacement covariance distribution. It has the explicit form

$$K_1^{I}(x,z) = \frac{\sigma^2}{\beta H_o l^2}\,\frac{\alpha}{1+\gamma\varkappa/l^2}\left(\tfrac{1}{2}-a\right)(z-1), \qquad K_1^{IV}(x,z) = 0,$$

$$K_1^{III}(x,z) = \frac{\sigma^2}{\beta H_o l^2}\,\frac{\alpha}{1+\gamma\varkappa/l^2}\left(\tfrac{1}{2}-a\right)x, \qquad \text{for } x \leq z, \tag{44}$$

$$K_1^{II}(x,z) = \frac{\sigma^2}{\beta H_o l^2}\,\frac{\alpha}{1+\gamma\varkappa/l^2}\,\tfrac{1}{4}(1-z-x)(2a-z+x).$$

Note that the coefficients α and γ in (44) are independent on the system parameter $\varkappa$. They only depend on the excitation point a.

Fig. 8 shows this dependence in the range $0 \leq a \leq 1$, above the parameter γ and below the parameter α . Both are numerically calculated from the series expansions (43) up to 400 terms of the double series in order to get sufficiently exact results. Obviously, there is only a weak influence of the excitation position a leading to a slight curvature in the graphs $\alpha(a)$ and $\gamma(a)$.

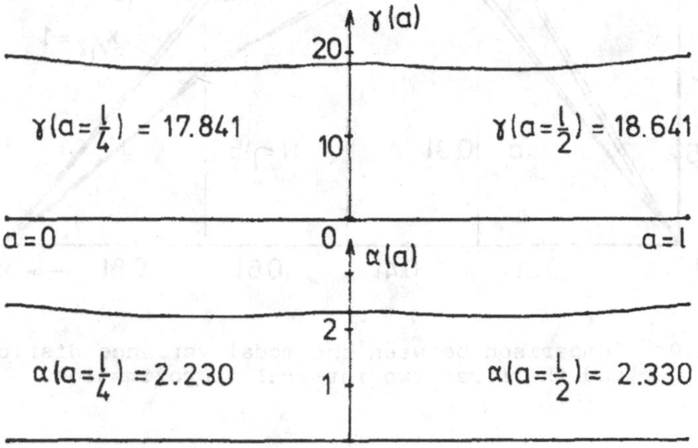

Fig. 8: Influence of the excitation point a on the parameter α and γ calculated with N = 20

It is therefore reasonable to fix both parameters e.g. with their values at a = 1/4 and to neglect their small variations for other excitation positions. With this laying down, the first two approximations of the covariance distribution K(x,z) can now easily be evaluated. Fig. 9 shows the associated square mean function related to $\sigma^2/\beta H_o$ for the sytem parameters $\varkappa/l^2$ = 100, 1 and o.1. A significant asymmetric influence is only observable for the parameter $\varkappa/l^2$ = o.1. For this value, we have also evaluated the modal covariance solution (14) with N = 15 eigenfunctions $V_n(x)$. The comparison with the integral approximation shows a good coincidence, particularly, in the excitation point x = a as well as in the imaginary point x = l-a in which we are most interested for practical applications. However, between both points, there are higher deviations following from the curvature of the modal covariances. To cover them, we clearly have to go further on to the next approximation of the applied Galerkin's approach including symmetric quadratic polynomials of the form (36). But, it is worth to note that such deviations are vanishing if the system parameter $\varkappa$ tends to infinity or inversely, if the damping β of the elastic structure tends to zero. In this limiting case, $K_o(x,z)$ of the first approximation represents the exact solution of the considered integral covariance equation.

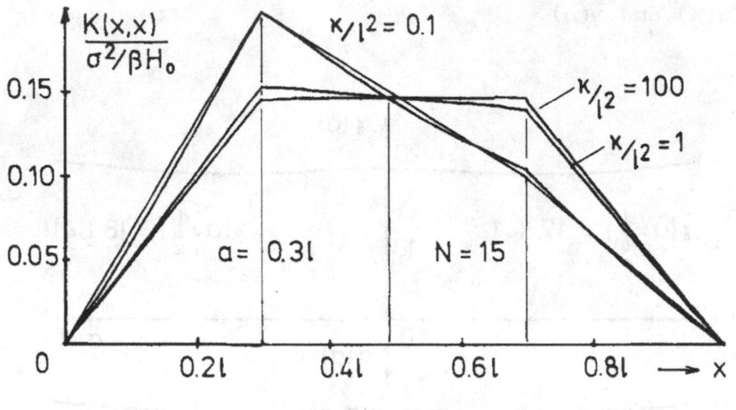

Fig.9: Comparison between the modal variance disribution and the first two integral approximations

7. Conclusions

A basi model of stochastic continuum mechanics is considered in order to analyse the wave processes of the elastic deflections in an one-dimensional finite structure. Inducing an asymmetric point force by white noise leads to a stationary wave propagation and reflection described by a two-dimensional, time-independent covariance distribution. Such statistical characteristics are available by measuring the deflections at two different positions of the structure and evaluating their mean squares or space correlations, respectively. The significant feature of the covariance distribution consists in the fact that there exist two focussing points in the mean square distribution whose intensities depend on the damping of the elastic structure. In slightly damped structures, we observe an almost symmetric square mean distribution. For a higher damping, the imaginary point is lower in comparison with the excitation point of the wave focussing. First numerical approximations and experimental verifications of such effects are given in [7].

Restricting our interest to analytical investigations, we show some new properties of the well established modal covariance analysis in decomposing the associated modal double series into single ones of the main diagonal terms, of the two neighbouring diagonals and so on. These modal covariance parts can be represented by two-dimensional polynomials which are piecewise valid in the definition range of the covariance distribution satisfying necessary boundary and transition conditions. To get a systematic approach for such approximations, we develop an integral covariance analysis. Starting from the well-known integral representation of the given boundary value problem, we set up new integral equations of the stationary covariance distribution including the geometric and dynamic boundary conditions in a compressed form. By means of the modal representation, we demonstrate the validity of the integral covariance equation. Differentiating it with respect to both space coordinates, we derive the associated partial differential equation the solution of which is separable by introducing a linear coordinate transformation. The new space coordinates are used in order to set up an orthogonal system of two-dimensional polynomials. Inserted into the integral equation and applying the Galerkin's method, we arrive finally at a fast approximation of the covariance distribution which coincides with the exact solution provided the external damping of the structure tends to zero.

References

1 Wiener, N., Generalized harmonic analysis, Acta Mathematica 55, 117 - 258 (1930)

2 Crandall, S.H., Yildiz, A., Random vibrations of beams, Journal Appl. Mech. 29, 267 - 275 (1962)

3 Wedig, W., Moments and probability densities of parametrically excited systems and continuous systems, ICNO VII (1975), Abhandlungen der Akademie der Wissenschaften der DDR, Akademie-Verlag, Berlin, Jg. 1977, Nr. 6N, 469 - 492 (1978)

4 Wedig, W., Stationäre Zufallsschwingungen von Balken - eine neue Methode zur Kovarianzanalyse, ZAMM 60, T89 - T91 (1980)

5 Wedig, W., Zufallsschwingungen von querangeströmten Saiten, Ing.-Archiv 48, 325 - 335 (1979)

6 Riemer, M., Wedig, W., Bauwerke unter Wellenlasten - Fokussierungseffekte bei Poissonerregungen, VDI-Berichte 419, VDI-Verlag, Düsseldorf, 201 - 207 (1981)

7 Crandall, S.H., Structured response patterns due to wide-band ran-

dom excitation, Stochastic Problems in Dynamics, Pitman, London, 366 - 389 (1977)

SUBHARMONIC OSCILLATIONS OF NONLINEAR SYSTEMS

Michel WILLEM
Institut Mathématique
2, Chemin du Cyclotron
B-1348 Louvain-la-Neuve (Belgium)

Introduction.

Using the dual action of CLARKE and EKELAND [4], we prove the existence of infinitely many subharmonics of hamiltonian systems and nonlinear string equations. Moreover we deduce from the variational characterizations some estimates on the amplitude and on the minimal period of the subharmonics. See [10] and [11] for a complete treatment.

1. Subharmonics of subquadratic hamiltonian systems.

Let us consider the hamiltonian system

$$J\dot{u} + \nabla H(t,u) = 0 \tag{1}$$

where

$$J = \begin{bmatrix} 0 & -I_n \\ I_n & 0 \end{bmatrix}$$

is the symplectic matrix and ∇ denotes the gradient with respect to u. Let $H : \mathbb{R} \times \mathbb{R}^{2n} \to \mathbb{R}$ be a continuous function, convex and differentiable with respect to u and such that ∇H is continuous on $\mathbb{R} \times \mathbb{R}^{2n}$.

THEOREM 1. *If* H *is* 2π*-periodic with respect to* t,

$$H(t,u) \to \infty, \quad |u| \to \infty \tag{2}$$

and

$$H(t,u)/|u|^2 \to 0, \quad |u| \to \infty \tag{3}$$

uniformly in t, *then, for every* $k \in \mathbb{N}^*$, *there exists a* $2k\pi$*-periodic solution* u_k *of* (1) *such that* $|u_k|_\infty \to \infty$, $k \to \infty$, *and the minimal period of* u_k *tends to* ∞.

Remarks. 1. Similar conclusions were obtained by RABINOWITZ [8] under the following stronger assumptions :

(i) H is strictly convex with respect to u.

(ii) There is $\nu \in]1,2[$ and $R > 0$ such that

$$|u| > R \Rightarrow 0 < (u, \nabla H(t,u)) \leqslant \nu H(t,u).$$

(iii) There is a, b > 0 and $s \in]1,\nu[$ such that

$$a|u|^s - b \leq H(t,u).$$

(iv) $|\nabla H(t,u)|/|u| \to 0$, $|u| \to \infty$.

2. The period 2π can be replaced by any other period.

3. If H is independent of t, it was proved by BREZIS and CORON [2] that, under assumptions (2) and (3), for every T sufficiently large, there is a solution of (1) with minimal period T. Our argument shows that the amplitude of these oscillations tends to infinity.

Notations. For $T > 0$, let us denote by $\mathcal{H}$ the Hilbert space $L^2(0,T;\mathbb{R}^{2n})$. Let L be defined by

$$D(L) = \{u : [0,T] \to \mathbb{R}^{2n} : u \text{ is absolutely continuous and } T\text{-periodic and } \dot{u} \in \mathcal{H}\},$$

$$Lu = J\dot{u}.$$

Then L is a self-adjoint operator, Ker L is the space of constant functions and $R(L) = \text{Ker } L^{\perp}$.

Let us write

$$K = L^{-1} : R(L) \to R(L) \cap D(L),$$

$$H^*(t,v) = \sup_{u\in\mathbb{R}^{2n}} [(v,u) - H(t,u)],\ v \in \mathbb{R}^{2n}, t \in \mathbb{R},$$

and

$$\phi(v) = \int_0^T [\tfrac{1}{2}(Kv,v) + H^*(t,v)]dt,\ v \in R(L).$$

The function H^* is the Legendre-Fenchel transform of H with respect to u and ϕ is the dual action.

LEMMA 1. *If there is* $\alpha, \beta > 0$ *and* $\gamma \in]0,2\pi/T[$ *such that*

$$\beta|u| - \alpha \leq H(t,u) \leq \gamma \frac{|u|^2}{2} + \alpha, \tag{4}$$

then there exist a solution of

$$Lu + \nabla H(t,u) = 0 \tag{5}$$

such that $-J\dot{u}$ *minimizes* ϕ *on* $R(L)$.

LEMMA 2. *If there is* $\alpha, \beta > 0$ *and* $\gamma \in]0,\pi/T[$ *such that* (4) *holds, then every solution of* (5) *is such that*

$$\int_0^T |\dot{u}|^2 \leq \frac{4\alpha\gamma\pi T}{\pi - \gamma T} \text{ and } \int_0^T |u| \leq \frac{1}{\beta}(2\alpha T + \int_0^T |\dot{u}|^2).$$

Proof of theorem 1. Assumption (2) and the convexity of H imply the existence of a,b > 0 such that

(6) $$b|u| - a \leqslant H(t,u).$$

It follows then from assumption (3) and lemma 1 applied to $T = 2k\pi$ that there is a $2k\pi$ solution u_k of (1) such that $-J\dot{u}_k$ minimizes

$$\phi_k(v) = \int_0^{2k\pi} [\tfrac{1}{2}(Kv,v) + H^*(t,v)]dt$$

on R(L). (Clearly K depends on k).

Let us estimate $c_k = \phi_k(-J\dot{u}_k)$ from above. Using (6) and the definition of H^*, we have

(7) $$|u| \leqslant b \Rightarrow H^*(t,v) \leqslant a.$$

Let $\rho \in \mathbb{R}^{2n}$ be such that $|\rho| = 1$. Since $h_k(t) = \frac{b}{2}(J\rho \cos \frac{t}{k} - \rho \sin \frac{t}{k}) \in R(L)$, it follows from (7) that

(8) $$c_k \leqslant \phi_k(h_k) \leqslant -\frac{b^2}{2}\pi k^2 + 2a\pi k.$$

If, for some subsequence (k_n), $|u_{k_n}|_\infty$ is bounded, it follows from (1) that $|\dot{u}_{k_n}|_\infty$ is also bounded. Thus there is $c' > 0$ such that $|u_{k_n}|_\infty |\dot{u}_{k_n}|_\infty \leqslant c'$. The definition of H^* implies that $H^*(t,v) \geqslant -c''$ where $c'' = \max_{t \in \mathbb{R}} H(t,0)$. We obtain $c_{k_n} = \phi_{k_n}(-J\dot{u}_{k_n}) \geqslant -2k_n(c'+c'')$, contrary to (8). Thus $|u_k|_\infty \to \infty$, $k \to \infty$.

Suppose now that, for a subsequence (k_n), the minimal period T_n of u_{k_n} is less than a constant $\tau > 0$. By assumption there is $\alpha, \beta > 0$ and $\gamma \in]0,\pi/\tau[$ such that (4) holds. It follows then from lemma 2 that

$$\int_0^{T_n} |\dot{u}_{k_n}|^2 \leqslant \frac{4\pi\alpha\gamma T_n}{\pi - \gamma T_n} \leqslant \frac{4\pi\alpha\gamma\tau}{\pi - \gamma\tau}$$

and

$$\int_0^{T_n} |u_{k_n}| \leqslant \frac{1}{\beta}[2\alpha T_n + \int_0^{T_n} |\dot{u}_{k_n}|^2].$$

It is easy to verify that those estimates and $T_n \leqslant \tau$ imply that $|u_{k_n}|_\infty$ is bounded. This is a contradiction. □

By the same argument one prove the following result.

THEOREM 2. *Let* $V : \mathbb{R} \times \mathbb{R}^n \to \mathbb{R} : (t,u) \to V(t,u)$ *be a continuous function, convex and differentiable with respect to* u *and such that* $\nabla V(t,u)$ *is continuous on* $\mathbb{R} \times \mathbb{R}^n$. *If* V *is* 2π *periodic with respect to* t,

$$V(t,u) \to \infty \text{ and } V(t,u)/|u|^2 \to 0, \ |u| \to \infty$$

uniformly in t, *then, for every* $k \in \mathbb{N}^*$, *there exists a* $2k\pi$*-periodic solution* u_k *of* $\ddot{u} + \nabla V(t,u) = 0$ *such that* $|u_k|_\infty \to \infty$, $k \to \infty$, *and the minimal period of* u_k *tends to* ∞.

__Remarks__. 1. The existence of infinitely many subharmonics was proved by CLARKE and EKELAND [5] under the following stronger assumptions :

(i) V is strictly convex with respect to u.

(ii) $\nabla V(t,0) = 0$.

(iii) There exist strictly positive constants d_k such that

$$d_1|u|^{1+d_2} - d_3 \leqslant V(t,u) \leqslant d_4(|u|^{2-d_5} + 1).$$

(iv) There is $d > 1$ and $\eta > 0$ such that

$$|u| \leqslant \eta \Rightarrow V(t,u) \geqslant d|u|^2/2.$$

2. Under the symmetry condition

$$\nabla V(-t,-u) = -\nabla V(t,u)$$

it is possible to treat non-convex potentials by an elementary argument (see [9]).

The case of a semilinear string equation with periodic-Dirichlet boundary conditions is similar but technically more complicated, since the kernel of the linear part is infinite dimensional. However there is no direct extension of lemma 2 in this context.

THEOREM 3. *Let* $g : \mathbb{R} \times [0,\pi] \times \mathbb{R} \to \mathbb{R} : (t,x,u) \to g(t,x,u)$ *be a continuous function, non decreasing with respect to* u. *If* g *is* 2π*-periodic with respect to* t, *if there is* $r, \delta > 0$ *such that, for every* t, x,

$$g(t,x,r) \geqslant \delta \text{ and } g(t,x,-r) \leqslant -\delta$$

and if

$$g(t,x,u)/|u| \to 0, \quad |u| \to \infty$$

uniformly in t, x, *then, for every* $k \in \mathbb{N}^*$, *there is a* $2k\pi$*-periodic* L^∞ *solution* u_k *of*

$$\text{(9)} \qquad \begin{cases} u_{tt} - u_{xx} + g(t,x,u) = 0 \\ u(t,0) = 0 = u(t,\pi) \end{cases}$$

such that $|u_k|_\infty \to \infty$, $k \to \infty$.

__Remarks__. 1. See [2] for the corresponding autonomous case.

2. The period 2π can be replaced by any rational multiple of π.

2. Subharmonics of a superquadratic wave equation.

Let us write $G(t,x,u) = \int_0^u g(t,x,s)ds$ and $\Omega_k =]0,2\pi k[\times]0,\pi[$.

THEOREM 4. *Let* $g(t,x,u)$ *be a continuous function,* 2π*-periodic with respect to* t. *If*

(A1) G *is strictly convex with respect to* u.

(A2) *There is* $q > 2$ *such that, for every* t,x,

$$q\,G(t,x,u) \leqslant g(t,x,u)u.$$

(A3) *There exist* $\alpha, \beta > 0$ *such that, for every* t,x,

$$\alpha|u|^q \leqslant G(t,x,u) \leqslant \beta|u|^q.$$

Then, for every $k \in \mathbb{N}^*$, *there exist a* $2k\pi$*-periodic non trivial solution* u_k *of* (9) *and* $|u_k|_{L^q(\Omega_k)} \to 0$, $k \to \infty$.

Remarks. 1. See [7] and [3] for the corresponding autonomous case. The existence of a non trivial 2π-periodic solution is proved under similar assumptions in [12] when $2 < q < 4$.

2. When $2 < q < 3$, $|u_k|_{L^\infty} \to 0$.

3. A similar argument applies to Hamiltonian systems (see [6] for the autonomous case).

4. Assumptions (A2) and (A3) are related, since (A2) implies the existence of $M,m > 0$ such that

$$|u| \leqslant 1 \Rightarrow G(t,x,u) \leqslant M|u|^q \text{ and } |u| \geqslant 1 \Rightarrow G(t,x,u) \geqslant m|u|^q.$$

Sketch of proof. The kernel N of the operator $Lu = u_{tt} - u_{xx}$ acting on $2k\pi$-periodic functions in L^q satisfying the Dirichlet conditions is given by

$$N = \{p(t+x) - p(t-x) : p \in L^q_{loc}(\mathbb{R}),\ p \text{ is } 2\pi\text{-periodic and } \int_0^{2\pi} p = 0\}.$$

Let us define

$$X = \{v \in L^p(\Omega_k) : \forall w \in N,\ \int_{\Omega_k} vw = 0\}$$

where $1/p + 1/q = 1$. For every $v \in X$ there is a unique function $u = Kv \in L^\infty$ such that $Lu = v$ and $\int_{\Omega_k} uw = 0$ for every $w \in N$. By the montain pass theorem [1] the functional

$$\phi_k(v) = \int_\Omega [\tfrac{1}{2}(Kv)v + G^*(t,x,v)]dt\,dx,\ v \in X,$$

has a non trivial critical point v_k, i.e. there is $w_k \in N$ such that $Kv_k + \nabla G^*(t,x,v_k) = w_k$. Then, by duality, $u_k = \nabla G^*(t,x,v_k)$ is a solution of (9).

By the minimax characterization of v_k we have

$$|v_k|_{L^p} \leq c\, k(\frac{2k+1}{k^2})^{\frac{1}{2-p}}. \quad \text{Thus } |u_k|_{L^q} \to 0. \qquad \Box$$

References.

[1] A. AMBROSETTI and P. RABINOWITZ, Dual variational methods in critical point theory and applications, *J. Funct. Anal.* 14 (1973) 349-381.

[2] H. BREZIS and J.M. CORON, Periodic solutions of nonlinear wave equations and hamiltonian systems, *American J. of Math.* 103 (1981) 559-570.

[3] H. BREZIS, J.M. CORON and L. NIRENBERG, Free vibrations for a nonlinear wave equation and a theorem of P. Rabinowitz, *Comm. Pure Appl. Math.* 33 (1980) 667-689.

[4] F.H. CLARKE and I. EKELAND, Hamiltonian trajectories having prescribed minimal period, *Comm. Pure Appl. Math.* 33 (1980) 103-116.

[5] F.H. CLARKE and I. EKELAND, Nonlinear oscillations and boundary-value problems for hamiltonian systems, preprint.

[6] I. EKELAND, Periodic solutions of hamiltonian equations and a theorem of P. Rabinowitz, *J. Diff. Equ.* 34 (1979) 523-534.

[7] P. RABINOWITZ, Free vibrations for a semilinear wave equation, *Comm. Pure Appl. Math.* 31 (1978) 31-68.

[8] P. RABINOWITZ, On subharmonic solutions of hamiltonian systems, Comm. Pure Appl. Math. 33 (1980) 609-633.

[9] M. WILLEM, Periodic oscillations of add second order hamiltonian systems, preprint.

[10] M. WILLEM, Subharmonic oscillations of convex hamiltonian systems, preprint.

[11] M. WILLEM, Subharmonic oscillations of a nonlinear wave equation, preprint.

[12] V. BENCI and D. FORTUNATO, The dual method in critical point theory. Multiplicity results for indefinite functionals, preprint.

Vol. 873: Constructive Mathematics, Proceedings, 1980. Edited by F. Richman. VII, 347 pages. 1981.

Vol. 874: Abelian Group Theory. Proceedings, 1981. Edited by R. Göbel and E. Walker. XXI, 447 pages. 1981.

Vol. 875: H. Zieschang, Finite Groups of Mapping Classes of Surfaces. VIII, 340 pages. 1981.

Vol. 876: J. P. Bickel, N. El Karoui and M. Yor. Ecole d'Eté de Probabilités de Saint-Flour IX – 1979. Edited by P. L. Hennequin. XI, 280 pages. 1981.

Vol. 877: J. Erven, B.-J. Falkowski, Low Order Cohomology and Applications. VI, 126 pages. 1981.

Vol. 878: Numerical Solution of Nonlinear Equations. Proceedings, 1980. Edited by E. L. Allgower, K. Glashoff, and H.-O. Peitgen. XIV, 440 pages. 1981.

Vol. 879: V. V. Sazonov, Normal Approximation – Some Recent Advances. VII, 105 pages. 1981.

Vol. 880: Non Commutative Harmonic Analysis and Lie Groups. Proceedings, 1980. Edited by J. Carmona and M. Vergne. IV, 553 pages. 1981.

Vol. 881: R. Lutz, M. Goze, Nonstandard Analysis. XIV, 261 pages. 1981.

Vol. 882: Integral Representations and Applications. Proceedings, 1980. Edited by K. Roggenkamp. XII, 479 pages. 1981.

Vol. 883: Cylindric Set Algebras. By L. Henkin, J. D. Monk, A. Tarski, H. Andréka, and I. Németi. VII, 323 pages. 1981.

Vol. 884: Combinatorial Mathematics VIII. Proceedings, 1980. Edited by K. L. McAvaney. XIII, 359 pages. 1981.

Vol. 885: Combinatorics and Graph Theory. Edited by S. B. Rao. Proceedings, 1980. VII, 500 pages. 1981.

Vol. 886: Fixed Point Theory. Proceedings, 1980. Edited by E. Fadell and G. Fournier. XII, 511 pages. 1981.

Vol. 887: F. van Oystaeyen, A. Verschoren, Non-commutative Algebraic Geometry, VI, 404 pages. 1981.

Vol. 888: Padé Approximation and its Applications. Proceedings, 1980. Edited by M. G. de Bruin and H. van Rossum. VI, 383 pages. 1981.

Vol. 889: J. Bourgain, New Classes of $\mathcal{L}^p$-Spaces. V, 143 pages. 1981.

Vol. 890: Model Theory and Arithmetic. Proceedings, 1979/80. Edited by C. Berline, K. McAloon, and J.-P. Ressayre. VI, 306 pages. 1981.

Vol. 891: Logic Symposia, Hakone, 1979, 1980. Proceedings, 1979, 1980. Edited by G. H. Müller, G. Takeuti, and T. Tugué. XI, 394 pages. 1981.

Vol. 892: H. Cajar, Billingsley Dimension in Probability Spaces. III, 106 pages. 1981.

Vol. 893: Geometries and Groups. Proceedings. Edited by M. Aigner and D. Jungnickel. X, 250 pages. 1981.

Vol. 894: Geometry Symposium. Utrecht 1980, Proceedings. Edited by E. Looijenga, D. Siersma, and F. Takens. V, 153 pages. 1981.

Vol. 895: J.A. Hillman, Alexander Ideals of Links. V, 178 pages. 1981.

Vol. 896: B. Angéniol, Familles de Cycles Algébriques – Schéma de Chow. VI, 140 pages. 1981.

Vol. 897: W. Buchholz, S. Feferman, W. Pohlers, W. Sieg, Iterated Inductive Definitions and Subsystems of Analysis: Recent Proof-Theoretical Studies. V, 383 pages. 1981.

Vol. 898: Dynamical Systems and Turbulence, Warwick, 1980. Proceedings. Edited by D. Rand and L.-S. Young. VI, 390 pages. 1981.

Vol. 899: Analytic Number Theory. Proceedings, 1980. Edited by M.I. Knopp. X, 478 pages. 1981.

Vol. 900: P. Deligne, J. S. Milne, A. Ogus, and K.-Y. Shih, Hodge Cycles, Motives, and Shimura Varieties. V, 414 pages. 1982.

Vol. 901: Séminaire Bourbaki vol. 1980/81 Exposés 561–578. III, 299 pages. 1981.

Vol. 902: F. Dumortier, P.R. Rodrigues, and R. Roussarie, Germs of Diffeomorphisms in the Plane. IV, 197 pages. 1981.

Vol. 903: Representations of Algebras. Proceedings, 1980. Edited by M. Auslander and E. Lluis. XV, 371 pages. 1981.

Vol. 904: K. Donner, Extension of Positive Operators and Korovkin Theorems. XII, 182 pages. 1982.

Vol. 905: Differential Geometric Methods in Mathematical Physics. Proceedings, 1980. Edited by H.-D. Doebner, S.J. Andersson, and H.R. Petry. VI, 309 pages. 1982.

Vol. 906: Séminaire de Théorie du Potentiel, Paris, No. 6. Proceedings. Edité par F. Hirsch et G. Mokobodzki. IV, 328 pages. 1982.

Vol. 907: P. Schenzel, Dualisierende Komplexe in der lokalen Algebra und Buchsbaum-Ringe. VII, 161 Seiten. 1982.

Vol. 908: Harmonic Analysis. Proceedings, 1981. Edited by F. Ricci and G. Weiss. V, 325 pages. 1982.

Vol. 909: Numerical Analysis. Proceedings, 1981. Edited by J.P. Hennart. VII, 247 pages. 1982.

Vol. 910: S.S. Abhyankar, Weighted Expansions for Canonical Desingularization. VII, 236 pages. 1982.

Vol. 911: O.G. Jørsboe, L. Mejlbro, The Carleson-Hunt Theorem on Fourier Series. IV, 123 pages. 1982.

Vol. 912: Numerical Analysis. Proceedings, 1981. Edited by G. A. Watson. XIII, 245 pages. 1982.

Vol. 913: O. Tammi, Extremum Problems for Bounded Univalent Functions II. VI, 168 pages. 1982.

Vol. 914: M. L. Warshauer, The Witt Group of Degree k Maps and Asymmetric Inner Product Spaces. IV, 269 pages. 1982.

Vol. 915: Categorical Aspects of Topology and Analysis. Proceedings, 1981. Edited by B. Banaschewski. XI, 385 pages. 1982.

Vol. 916: K.-U. Grusa, Zweidimensionale, interpolierende Lg-Splines und ihre Anwendungen. VIII, 238 Seiten. 1982.

Vol. 917: Brauer Groups in Ring Theory and Algebraic Geometry. Proceedings, 1981. Edited by F. van Oystaeyen and A. Verschoren. VIII, 300 pages. 1982.

Vol. 918: Z. Semadeni, Schauder Bases in Banach Spaces of Continuous Functions. V, 136 pages. 1982.

Vol. 919: Séminaire Pierre Lelong – Henri Skoda (Analyse) Années 1980/81 et Colloque de Wimereux, Mai 1981. Proceedings. Edité par P. Lelong et H. Skoda. VII, 383 pages. 1982.

Vol. 920: Séminaire de Probabilités XVI, 1980/81. Proceedings. Edité par J. Azéma et M. Yor. V, 622 pages. 1982.

Vol. 921: Séminaire de Probabilités XVI, 1980/81. Supplément: Géométrie Différentielle Stochastique. Proceedings. Edité par J. Azéma et M. Yor. III, 285 pages. 1982.

Vol. 922: B. Dacorogna, Weak Continuity and Weak Lower Semicontinuity of Non-Linear Functionals. V, 120 pages. 1982.

Vol. 923: Functional Analysis in Markov Processes. Proceedings, 1981. Edited by M. Fukushima. V, 307 pages. 1982.

Vol. 924: Séminaire d'Algèbre Paul Dubreil et Marie-Paule Malliavin. Proceedings, 1981. Edité par M.-P. Malliavin. V, 461 pages. 1982.

Vol. 925: The Riemann Problem, Complete Integrability and Arithmetic Applications. Proceedings, 1979-1980. Edited by D. Chudnovsky and G. Chudnovsky. VI, 373 pages. 1982.

Vol. 926: Geometric Techniques in Gauge Theories. Proceedings, 1981. Edited by R. Martini and E.M.de Jager. IX, 219 pages. 1982.

Vol. 927: Y. Z. Flicker, The Trace Formula and Base Change for GL (3). XII, 204 pages. 1982.

Vol. 928: Probability Measures on Groups. Proceedings 1981. Edited by H. Heyer. X, 477 pages. 1982.

Vol. 929: Ecole d'Eté de Probabilités de Saint-Flour X – 1980. Proceedings, 1980. Edited by P.L. Hennequin. X, 313 pages. 1982.

Vol. 930: P. Berthelot, L. Breen, et W. Messing, Théorie de Dieudonné Cristalline II. XI, 261 pages. 1982.

Vol. 931: D.M. Arnold, Finite Rank Torsion Free Abelian Groups and Rings. VII, 191 pages. 1982.

Vol. 932: Analytic Theory of Continued Fractions. Proceedings, 1981. Edited by W.B. Jones, W.J. Thron, and H. Waadeland. VI, 240 pages. 1982.

Vol. 933: Lie Algebras and Related Topics. Proceedings, 1981. Edited by D. Winter. VI, 236 pages. 1982.

Vol. 934: M. Sakai, Quadrature Domains. IV, 133 pages. 1982.

Vol. 935: R. Sot, Simple Morphisms in Algebraic Geometry. IV, 146 pages. 1982.

Vol. 936: S.M. Khaleelulla, Counterexamples in Topological Vector Spaces. XXI, 179 pages. 1982.

Vol. 937: E. Combet, Intégrales Exponentielles. VIII, 114 pages. 1982.

Vol. 938: Number Theory. Proceedings, 1981. Edited by K. Alladi. IX, 177 pages. 1982.

Vol. 939: Martingale Theory in Harmonic Analysis and Banach Spaces. Proceedings, 1981. Edited by J.-A. Chao and W.A. Woyczyński. VIII, 225 pages. 1982.

Vol. 940: S. Shelah, Proper Forcing. XXIX, 496 pages. 1982.

Vol. 941: A. Legrand, Homotopie des Espaces de Sections. VII, 132 pages. 1982.

Vol. 942: Theory and Applications of Singular Perturbations. Proceedings, 1981. Edited by W. Eckhaus and E.M. de Jager. V, 363 pages. 1982.

Vol. 943: V. Ancona, G. Tomassini, Modifications Analytiques. IV, 120 pages. 1982.

Vol. 944: Representations of Algebras. Workshop Proceedings, 1980. Edited by M. Auslander and E. Lluis. V, 258 pages. 1982.

Vol. 945: Measure Theory. Oberwolfach 1981, Proceedings. Edited by D. Kölzow and D. Maharam-Stone. XV, 431 pages. 1982.

Vol. 946: N. Spaltenstein, Classes Unipotentes et Sous-groupes de Borel. IX, 259 pages. 1982.

Vol. 947: Algebraic Threefolds. Proceedings, 1981. Edited by A. Conte. VII, 315 pages. 1982.

Vol. 948: Functional Analysis. Proceedings, 1981. Edited by D. Butković, H. Kraljević, and S. Kurepa. X, 239 pages. 1982.

Vol. 949: Harmonic Maps. Proceedings, 1980. Edited by R.J. Knill, M. Kalka and H.C.J. Sealey. V, 158 pages. 1982.

Vol. 950: Complex Analysis. Proceedings, 1980. Edited by J. Eells. IV, 428 pages. 1982.

Vol. 951: Advances in Non-Commutative Ring Theory. Proceedings, 1981. Edited by P.J. Fleury. V, 142 pages. 1982.

Vol. 952: Combinatorial Mathematics IX. Proceedings, 1981. Edited by E. Billington, S. Oates-Williams, and A.P. Street. XI, 443 pages. 1982.

Vol. 953: Iterative Solution of Nonlinear Systems of Equations. Proceedings, 1982. Edited by R. Ansorge, Th. Meis, and W. Törnig. VII, 202 pages. 1982.

Vol. 954: S.G. Pandit, S.G. Deo, Differential Systems Involving Impulses. VII, 102 pages. 1982.

Vol. 955: G. Gierz, Bundles of Topological Vector Spaces and Their Duality. IV, 296 pages. 1982.

Vol. 956: Group Actions and Vector Fields. Proceedings, 1981. Edited by J.B. Carrell. V, 144 pages. 1982.

Vol. 957: Differential Equations. Proceedings, 1981. Edited by D.G. de Figueiredo. VIII, 301 pages. 1982.

Vol. 958: F.R. Beyl, J. Tappe, Group Extensions, Representations, and the Schur Multiplicator. IV, 278 pages. 1982.

Vol. 959: Géométrie Algébrique Réelle et Formes Quadratiques, Proceedings, 1981. Edité par J.-L. Colliot-Thélène, M. Coste, L. Mahé, et M.-F. Roy. X, 458 pages. 1982.

Vol. 960: Multigrid Methods. Proceedings, 1981. Edited by W. Hackbusch and U. Trottenberg. VII, 652 pages. 1982.

Vol. 961: Algebraic Geometry. Proceedings, 1981. Edited by J.M. Aroca, R. Buchweitz, M. Giusti, and M. Merle. X, 500 pages. 1982.

Vol. 962: Category Theory. Proceedings, 1981. Edited by K.H. Kamps, D. Pumplün, and W. Tholen, XV, 322 pages. 1982.

Vol. 963: R. Nottrot, Optimal Processes on Manifolds. VI, 124 pages. 1982.

Vol. 964: Ordinary and Partial Differential Equations. Proceedings, 1982. Edited by W.N. Everitt and B.D. Sleeman. XVIII, 726 pages. 1982.

Vol. 965: Topics in Numerical Analysis. Proceedings, 1981. Edited by P.R. Turner. IX, 202 pages. 1982.

Vol. 966: Algebraic K-Theory. Proceedings, 1980, Part I. Edited by R.K. Dennis. VIII, 407 pages. 1982.

Vol. 967: Algebraic K-Theory. Proceedings, 1980. Part II. VIII, 409 pages. 1982.

Vol. 968: Numerical Integration of Differential Equations and Large Linear Systems. Proceedings, 1980. Edited by J. Hinze. VI, 412 pages. 1982.

Vol. 969: Combinatorial Theory. Proceedings, 1982. Edited by D. Jungnickel and K. Vedder. V, 326 pages. 1982.

Vol. 970: Twistor Geometry and Non-Linear Systems. Proceedings, 1980. Edited by H.-D. Doebner and T.D. Palev. V, 216 pages. 1982.

Vol. 971: Kleinian Groups and Related Topics. Proceedings, 1981. Edited by D.M. Gallo and R.M. Porter. V, 117 pages. 1983.

Vol. 972: Nonlinear Filtering and Stochastic Control. Proceedings, 1981. Edited by S.K. Mitter and A. Moro. VIII, 297 pages. 1983.

Vol. 973: Matrix Pencils. Proceedings, 1982. Edited by B. Kågström and A. Ruhe. XI, 293 pages. 1983.

Vol. 974: A. Draux, Polynômes Orthogonaux Formels – Applications. VI, 625 pages. 1983.

Vol. 975: Radical Banach Algebras and Automatic Continuity. Proceedings, 1981. Edited by J.M. Bachar, W.G. Bade, P.C. Curtis Jr., H.G. Dales and M.P. Thomas. VIII, 470 pages. 1983.

Vol. 976: X. Fernique, P.W. Millar, D.W. Stroock, M. Weber, Ecole d'Eté de Probabilités de Saint-Flour XI – 1981. Edited by P.L. Hennequin. XI, 465 pages. 1983.

Vol. 977: T. Parthasarathy, On Global Univalence Theorems. VIII, 106 pages. 1983.

Vol. 978: J. Ławrynowicz, J. Krzyż, Quasiconformal Mappings in the Plane. VI, 177 pages. 1983.

Vol. 979: Mathematical Theories of Optimization. Proceedings, 1981 Edited by J.P. Cecconi and T. Zolezzi. V, 268 pages. 1983.

Vol. 980: L. Breen. Fonctions thêta et théorème du cube. XII, 115 pages. 1983.

Vol. 981: Value Distribution Theory. Proceedings, 1981. Edited by I. Laine and S. Rickman. VIII, 245 pages. 1983.

Vol. 982: Stability Problems for Stochastic Models. Proceedings 1982. Edited by V. V. Kalashnikov and V. M. Zolotarev. XVII, 295 pages 1983.